HETEROGENEOUS REACTIONS: ANALYSIS, EXAMPLES, AND REACTOR DESIGN

Volume 1

HETEROGENEOUS REACTIONS:
Analysis, Examples, and Reactor Design

Volume 1: Gas–Solid and Solid–Solid Reactions

L. K. DORAISWAMY

Director
National Chemical Laboratory
Pune, INDIA

M. M. SHARMA

Professor of Chemical Engineering
Bombay University Department of Chemical Technology
Bombay, INDIA

A Wiley-Interscience Publication

JOHN WILEY & SONS

New York   Chichester   Brisbane   Toronto   Singapore

Library of Congress Cataloging in Publication Data:

Doraiswamy, L. K. (Laxmangudi Krishnamurthy)
    Heterogeneous reactions.

    Includes index.
    Contents: v. 1. Gas–solid and solid–solid reactions.
    1. Chemical reactors.   2. Chemical reactions.
I. Sharma, M. M. (Man Mohan)   II. Title.
TP157.D67 1983   660.2′99   82-19968
ISBN 0–471–05368–6 (v. 1)

Printed in the United States of America

10 9 8 7 6 5 4 3 2 1

Preface

The first major book on the design of certain classes of chemical reactors appeared in the late 1940s as Volume 3 of the now classical *Chemical Process Principles* by O. A. Hougen and K. M. Watson. This was followed by several books in the 1950s and 1960s: J. M. Smith's *Chemical Engineering Kinetics*, Octave Levenspiel's *Chemical Reaction Engineering*, G. Astarita's *Mass Transfer with Chemical Reaction*, P. V. Danckwerts's *Gas–Liquid Reactions*, H. Kramer's and K. R. Westerterp's *Elements of Chemical Reactor Design and Operation*, and K. Denbigh and J. C. R. Turner's *Chemical Reactor Theory*. Each book had as its main theme one or two classes of reactions, such as gas–solid (catalytic), gas–liquid, or homogeneous. The 1970s saw a series of books, again on practically the same systems, but with greater emphasis on analysis and mathematical modeling. Among these may be mentioned the books by J. J. Carberry; J. Szekely, J. W. Evans, and H. Y. Sohn; G. F. Froment and K. B. Bischoff; J. B. Butt; and Y. T. Shah. With the exception of *Chemical Reactor Analysis and Design* (by G. F. Froment and K. B. Bischoff), the strong point of almost all these books is largely analysis and/or design, with but a few examples interspersed to illustrate the theories. Particularly noteworthy books are the entirely theoretical *Chemical Reactor Theory*, edited by L. Lapidus and N. R. Amundson, the two practically oriented volumes of H. F. Rase, and the introductory text of C. G. Hill, Jr. . All these books have many commendable features.

Industrially important reactions are predominantly heterogeneous, and the notable absence of a comprehensive and connected discussion of such systems provided the basic motivation for knitting them into a coherent theme in a single presentation. Further, we were struck by the preponderance of hypothetical and arbitrary examples in many current books; since such examples lack the impact we intend for our readers, we decided at the outset to incorporate, as far as possible, real and industrially relevant examples, both qualitative and quantitative. Thus the two volumes were conceived and written. We believe that this attempt is unique and hope that it fulfills our objectives and meets the requirements of a wide cross section of readers. We also believe that this is one of the few sets of two volumes in which threads of analysis have been assiduously woven into a web of design and examples. Taken as a whole, they have no counterpart at present.

We have been acutely conscious that these volumes have taken an unusually long time to write, and have undergone revisions during which their contents have seen drastic changes, including realignment of chapters. These changes were a result of newer knowledge that became available as well as the increasing store of research, industrial, and consulting experience of the authors. Our keen desire to include information concerning the translation of laboratory data into commercial plants, albeit in a limited way, has also contributed significantly to the writing time. During this period both of us have had the pleasure of contributing research papers and state-of-art reviews in just about all the subjects that we have covered in the 35 chapters that make up the two volumes.

The main theme of these volumes is: how to dissect the often complex problems relating to the interaction between diffusion and chemical reaction into tractable parts through a systematic analysis and a rational process-design strategy. To buttress this approach we have given a surfeit of qualitative and quantitative examples.

More than ten heterogeneous systems have been considered in these volumes. Volume 1 deals essentially with systems in which a solid phase appears either as catalyst or as a reactant; the second reactant is a gas or a solid, but one case of a system in which the second reactant is a liquid (the immobilized enzyme system) has also been included, in view of the similarity of approaches. Volume 1 comprises 21 chapters, of which 17 pertain to gas–solid (catalytic) reactions. The eighteenth chapter includes some special reactors such as those for supported liquid-phase catalytic systems, polymer-bound catalytic systems, and immobilized enzyme systems. The next two chapters are devoted to gas–solid (noncatalytic) reactors and the one following is concerned with solid–solid reactions.

Volume 2 is concerned essentially with systems in which a liquid is involved as one of the reactants. The second reactant is either a gas or a liquid. In some systems the solid phase also appears, either as the second or third reactant or as a catalyst along with gas and liquid (slurry or trickle-bed reactors). The fourteen chapters comprising this volume cover fluid–fluid reactions; reactions with one gas and two liquid-phase reactants; simultaneous absorption and reaction of two gases; desorption with chemical reaction; simultaneous absorption and desorption with reaction; complex reactions; use of models in the simulation and design of reactors; solid–liquid reactions; reactions in fluid–fluid–solid systems; and solid-catalyzed fluid–fluid reactions. An extensive chapter is included to give pertinent details of a variety of contactors that are used for the reaction systems considered. An attempt has also been made to guide readers in selecting a contactor for a specified duty.

Although practically all industrially important heterogeneous systems are covered in the two volumes, three systems, namely, gas–solid (catalytic), gas–liquid and liquid–liquid, have received greater attention not only because they are truly ubiquitous but also because of the authors' greater personal involvement with these systems than with others.

The contents of the two volumes have been used for undergraduate and graduate courses in many countries. They have also been used for short intensive courses presented both in India and abroad. In addition, a wide cross section of chemical engineers and technologists have made valuable suggestions that have been incorporated.

These volumes are addressed not only to students of chemical engineering, chemical technology, and applied chemistry, but also to researchers, designers, and practicing engineers. We believe that all such readers of these volumes will find some directly useful material. We also believe that our coverage of the English-language literature up to 1981 is reasonably complete. We have made every effort to include literature from all over the world, but our lack of knowledge of Russian and Japanese prevented a fuller coverage of papers in these languages. We have covered some references that appeared in 1982 as well.

In some parts of these volumes, readers may discern an almost encyclopedic approach. While conceding such a disposition in selected areas, we would like to emphasize our main approach: we have cited in most parts of these books only those references which are necessary in building up a comprehensive structure for each system, but in doing so we have not denied the reader the advantage of references that might not be directly relevant but that can be usefully consulted if needed. The approach is not that of a standard textbook but rather that of an advanced treatise.

In a comprehensive effort of this kind there is a temptation to adopt a unified approach to all systems. We quickly discovered the severe limitations of such an effort and decided to adopt prevalent approaches to different systems. For example, in Volume 1 we use the concept of effectiveness factor, which denotes the effect of diffusion on reaction, while in Volume 2 we continue with the practice of using the enhancement factor, which denotes the effect of reaction on diffusion. It would perhaps have been possible to adopt a common approach, for example, one based on effectiveness factor, for all systems; but doing so would have made the volumes less useful and appealing to readers than we believe it is in the present form. In pursuit of this approach we have also adopted different systems of nomenclature for Volumes 1 and 2 consistent with the more commonly used notation, and these are clearly defined at the beginning of each volume; uncommon notation usually acts as an irritant and makes the reader wary. The choice of units presented another vexing problem, and after considerable thought we decided to adopt widely used units in preference to the SI units.

We have both drawn heavily from our students and associates in writing these volumes. Their attachment to this venture has been a source of great strength to us. It would not be possible for us to place on record our gratitude to all of them individually, but each of us would like to particularly thank a few co-workers who have toiled to help us complete this voluminous and ambitious document. Most of them have witnessed with mute sympathy the protracted evolution of this venture and share our relief at its completion.

LKD would like to thank B. D. Kulkarni for his invaluable assistance, always rendered cheerfully; without his continued help the author doubts if Volume 1 could ever have been completed. Kulkarni has assisted in many ways, such as literature search, careful editing, and stimulating discussions on various aspects of Volume 1. The author is also grateful to many of his other students and colleagues who have rendered willing assistance: R. V. Choudhary, V. R. Choudhary, R. K. Irani, V. K. Jayaraman, N. G. Karanth, S. D. Prasad, P. C. Prasannan, N. S. Raghavan, R. A. Rajadhyaksha,

P. A. Ramachandran, V. Ravi Kumar, A. Sadana, and S. S. Tamhankar. He would like in particular to record the assistance so cheerfully extended by Ravi Kumar and Jayaraman. He is thankful to Pratibha Khare for typing the manuscript from sheets of paper often covered with illegible writing and transforming them into readable material.

MMS would like to express his most sincere gratitude to a number of his students and colleagues: M. C. Chhabaria, G. C. Jagirdar, J. B. Joshi, V. V. Juvekar, V. V. Mahajani, R. A. Mashelkar, V. G. Pangarkar, P. A. Ramachandran, T. Vasudevan, and V. V. Wadekar. They have rendered invaluable assistance voluntarily.

We would like to express our gratitude to our wives Rajalakshmi and Sudha for having put up with our bursts of writing over a painfully long period of time and for mercifully refraining from expressing what they must undoubtedly have felt—that these volumes were never going to be completed! The satisfaction of not letting them down marks a particularly pleasing conclusion to an arduous undertaking.

Both of us profusely thank Professor Octave Levenspiel and Professor Peter Danckwerts for their sustained interest, constructive comments, and encouragement. Professor Danckwerts' advice, "Let the good not be the enemy of the best," made a deep impression on us, and it is in this spirit that we present these volumes to our readers.

L. K. DORAISWAMY

Pune, India

M. M. SHARMA

Bombay, India
January 1983

To my wife

Rajalakshmi

In whom simplicity, pure and elemental,
and courage, beautifully concealed,
combined to form the quintessence of life . . . .

Whose unspoken inspiration
sustained me through the
protracted evolution of the book—
the completion of which
I am not destined to share with her.

L.K.D.

Contents

ix

Notation

A = reactant species

A*l*, B*l*, . . . , *il* = adsorbed complexes of species A, B, . . . , *i*

A = general notation for a constant; or total area, cm^2; or group of variables

$\tilde{A}_{i-1,j}$ = parameter defined by Eq. 11.56

A' = group defined by Eq. 12.24 and Eq. 17.7

A'' = group defined in Eq. 11.64

A_a = area of the aperture in Fig. 4.3

$A_{bA}, A_{bB}, \ldots$ = surface areas of particles A, B, . . . in Chapter 21, cm^2

A_c = cross-sectional area of reactor, cm^2; or area of the cloud phase in Chapters 13 and 14; or area of contact in Chapter 21

A_f = frequency factor in Arrhenius equation, sec^{-1}

$A_{f,cb}$ = Arrhenius parameter for the carbon burning reaction

$A_{f,f}$ = Arrhenius frequency factor for the fouling reaction

A_{fp} = Arrhenius frequency factor, mol/cm^3 atm sec or mol/g atm sec

A_{fs} = Arrhenius frequency factor for sintering

A_h = heat transfer area per unit length of reactor, cm^2/cm

A'_1 = parameter defined by Eq. 11.81

$A_{i,j}$ = elements of the matrix defined by Eq. 7.3

$A'_{i,j}$ = matrix defined by Eq. 7.4

A_{max} = parameter defined in Fig. 12.18

A_p = surface area of a single pellet, cm^2

A_{po} = surface area of a single pellet at time zero, cm^2

A_r = surface area of reactor, cm^2

A_s = surface area per unit length, cm^2/cm

$A_{s,e}$ = effective contact area, cm^2

A = parameter defined by Eq. 13.37

$\mathscr{A}$ = agitation parameter, cm/sec

a = general symbol for a constant; or gas–solid interfacial area, cm^2/cm^3; or half the molecular size in Eq. 3.34; or initial concentration of solid in Chapter 21

a' = gas–solid interfacial area, cm^2/g

a_b = area of bubble per unit volume, cm^2/cm^3

a_e = area of emulsion per unit volume, cm^2/cm^3

a_h = heat transfer area per unit volume of reactor, cm^2/cm^3

a_{ij} = sum of the products of residuals of *i*th and *j*th responses in Chapter 2

General Note. Usually symbols occurring at least once are clearly defined in the text and are included in the Notation section. Such symbols are included only if they are considered sufficiently important. For each letter of the alphabet (for example, A) the following order is used: (a) roman capital, A; (b) italic capital, *A*; (c) italic lowercase *a*; (d) bold face, **A**; (e) script capital, $\mathscr{A}$. Under each of these the order of presentation is alphabetic with respect to the subscripts. Superscripts such as prime, hat, half moon, and so on, on any alphabet are not singled out and are included in the alphabetic order based on subscripts.

a_j	=	jet–emulsion contact area, cm^2/cm^3
a_p	=	area per unit volume of pellet, cm^2/cm^3
B	=	reactant or product species;
$(\text{Bi})_b$	=	Biot number for heat transfer based on radiant heat transfer coefficient
$(\text{Bi})_h$	=	Biot number for heat transfer, $h_{fp}R/k'_e$ or $hd_p/2k'_e$
$(\text{B}_i)_h^0$	=	static contribution to Biot number for heat transfer, $h_e^0 d_p/2k'_e$
$(\text{Bi})_h^*$	=	Biot number for heat transfer, defined as $h_w^* d_p/2k'_e$
$(\text{Bi})_m$	=	Biot number for mass transfer, $k_g R/D_e$ or $k_g d_p/2D_e$
$(\text{Bi})_w$	=	Biot number for wall heat transfer, $h_w R/k'_e$ or $h_w d_p/2k'_e$
B	=	general notation for a constant; or group of variables; or multicomponent diffusion parameter defined by Eq. 4.91
$\tilde{B}$	=	parameter defined by Eq. 11.56
B'	=	group defined by Eq. 12.24 and Eq. 17.14
B''	=	group defined in Eq. 11.64
$B(\mathbf{k}, s)$	=	Bayesian function in Chapter 2
b	=	general symbol for a constant; or dilution ratio in Chapter 9
$b_0, b_1, \ldots, b_6$	=	coefficients of Eq. 7.39
C	=	chemical species
CF	=	correction factor
CN	=	cyclone number
C	=	general term for concentration of any species, mol/cm^3
$\hat{C}$	=	dimensionless concentration in pellet, C/C_s
C''	=	group defined in Eq. 11.64
C^*	=	equilibrium concentration, mol/cm^3
C_0	=	inlet concentration to the reactor, mol/cm^3
C_1, C_2	=	intrinsic parameters for rate-model in Section 2.6.7
$\overline{\Delta C}$	=	confidence interval for concentration fluctuations (Eq. 13.38)

$C_A, C_B, \ldots$	=	concentration of A, B, $\ldots$, mol/cm^3
$\hat{C}_A, \hat{C}_B, \ldots$	=	dimensionless concentration of species A, B, $\ldots$ (C_A/C_{As}, C_B/C_{Bs}, $\ldots$)
$C_A^*, C_B^*, \ldots$	=	equilibrium concentration of A, B, $\ldots$, mol/cm^3
$C_{Ab}, C_{Bb}, \ldots$	=	concentration of A, B, $\ldots$ in the fluid bulk, mol/cm; or concentration of A, B, $\ldots$ in the bubble phase, mol/cm
$\hat{C}_{Ab}, \hat{C}_{Bb}$	=	dimensionless concentration of species A, B
C_{Ac}, C_{Bc}	=	concentration of A, B in the cloud phase, mol/cm^3
$C_{Ae}, C_{Be}, \ldots$	=	concentration of A, B, $\ldots$ in the exit stream, mol/cm^3, or concentration of A, B, $\ldots$ in the emulsion phase, mol/cm^3
$C_{Ai}, C_{Bi}, \ldots$	=	concentration of A, B, $\ldots$ at the interface, mol/cm^3
$\hat{C}_{Ai}$	=	dimensionless concentration of species A at the interface
$C_{Al}, C_{Bl}, \ldots$	=	concentration of adsorbed species A, B, $\ldots$ on the catalyst surface, mol/cm^3
$C_{A,ma},$ $C_{B,ma}, \ldots$	=	concentration of species A, B, $\ldots$ in the macropore, mol/cm^3
$\hat{C}_{A,ma}, \hat{C}_{B,ma}$	=	dimensionless concentration of species A, B in the macropore
$C_{A,mi}, C_{B,mi}$	=	concentration of species A, B in the micropore, mol/cm^3
$\hat{C}_{A,mi}, \hat{C}_{B,mi}$	=	dimensionless concentration of species A, B in the micropore
$C_{As}, C_{Bs}, \ldots$	=	concentration of A, B, $\ldots$ at the external surface of the catalyst, mol/cm^3
$C_{As}(l)$	=	local concentration of A on a catalyst plate at a distance l, mol/cm^3
$\hat{C}_{Asb}$	=	ratio of surface to bulk concentrations C_{As}/C_{Ab}
$C_{A0}, C_{B0}, \ldots$	=	inlet concentration of A, B, $\ldots$ to reactor, mol/cm^3
C_{A0}^0, C_{B0}^0	=	peak input pulse concentration of A, B, mol/cm^3

C_b = concentration in the fluid bulk, mol/cm^3; or concentration in the bubble phase, mol/cm^3

$C_{b,n}$ = concentration in bubble in the nth compartment, mol/cm^3

C_c = concentration in the cloud phase, mol/cm^3

C_{cp} = center-plane concentration of fresh catalyst, mol/cm^3

$\hat{C}_{cp}$ = dimensionless center-plane concentration for fresh catalyst in Eq. 8.13

$(C_{cp})_f$ = center-plane concentration in a deactivated catalyst, mol/cm^3

$(\hat{C}_{cp})_f$ = dimensionless center-plane concentration in a deactivated catalyst in Eq. 8.13

C_e = exit concentration, mol/cm^3; or concentration in the emulsion phase, mol/cm^3

$C_{e,n}$ = concentration in emulsion in the nth compartment, mol/cm^3

C_L = concentration at length L (corresponding to exit), mol/cm^3; or concentration of active centers in Chapter 2, mol/cm^3; or concentration of active centers at $t = 0$ in Chapter 8, mol/cm^3

C_{Lt} = concentration at the end of the dilute phase, mol/cm^3

C_l = concentration of vacant sites at time t, mol/cm^3

C_p = heat capacity of the fluid (gas), $cal/g°C$; or concentration of poison in Chapters 8 and 16, mol/cm^3

$\bar{C}_p$ = average heat capacity of the fluid, $cal/g°C$

C_{pe} = heat capacity of fluid at reaction equilibrium, $cal/g°C$

C_{pf} = heat capacity of fluid at frozen conditions, $cal/g°C$

C_{pg} = heat capacity of gas, used specifically when required to be distinguished from the heat capacity of solid, $cal/g°C$

C_{pm} = molar heat capacity of fluid, $cal/mol°C$

C_{ps} = heat capacity of solid, $cal/g°C$

C_{p0} = initial concentration of poison, mol/cm^3

C_S = concentration of substrate in Chapter 18, mol/cm^3

$\hat{C}_S$ = dimensionless concentration of substrate, C_S/C_{Sb}

C_{Sb} = concentration of substrate in the bulk, mol/cm^3 (Chapter 16)

C_{Ss} = concentration of substrate on the surface in Chapter 18, mol/cm^3

C_s = concentration at the external surface of the catalyst, mol/cm^3

C_t = total concentration of gases, mol/cm^3

C = cost factor

C_a = cost due to catalyst aging

C_r = fixed-interval replacement cost

C_T = total operating cost due to catalyst

$\bar{C}_T$ = average total cost per unit time

$\bar{C}_{Tm}$ = optimum total operating cost

c = dimensionless concentration in the reactor

$\mathbf{c}$ = vector of concentrations

D = chemical species

Da = Damköhler number, defined as kL/u or $k_S R/D_e$

Da' = local Damköhler number in Chapter 6; or any modified Damköhler number

Da_G = Damköhler number for reaction and diffusion in a grain

DF = dilution factor

D = general notation for diffusion coefficient, cm^2/sec; or group defined by Eq. 16.90

$\tilde{D}$ = interdiffusion coefficient in the case of solid–solid diffusion, cm^2/sec

$D(r_p)$ = diffusivity in a capillary of radius r_p, cm^2/sec

$D_A, D_B, \ldots, D_S$	=	diffusion coefficient of A, B, $\ldots$, S, cm^2/sec; or diffusivity of gas in solids A, B, $\ldots$, S, respectively, in Chapters 19 and 20, cm^2/sec
D_{AB}	=	diffusion coefficient of A in B, cm^2/sec
D_b	=	bulk diffusivity, cm^2/sec
D_{bA}, D_{bB}	=	bulk diffusivity of species A, B, cm^2/sec
D_c	=	combined diffusivity, cm^2/sec
D_{cA}, D_{cB}	=	combined diffusivity of species A, B, cm^2/sec
D_{cd}	=	configurational diffusivity, cm^2/sec
D_e	=	effective diffusivity, cm^2/sec
$D_{eA}, D_{eB}, \ldots, D_{eS}$	=	effective diffusivity of A, B, $\ldots$, S, cm^2/sec; or effective diffusivity of gas in solids A, B, $\ldots$ in Chapters 19 and 20, cm^2/sec
$D_{eA,ma}, D_{eB,ma}$	=	effective diffusivity of species A, B in macropore
$D_{eA,mi}, D_{eB,mi}$	=	effective diffusivity of species A, B in micropore
D_{eb}	=	effective bulk diffusivity, cm^2/sec
$D_{ebA}, D_{ebB}, \ldots$	=	effective bulk diffusion coefficient for species A, B, $\ldots$, cm^2/sec
D_{eG}	=	effective diffusivity of gas in a grain in the pellet, cm^2/sec
D_{eK}	=	effective Knudsen diffusion coefficient, cm^2/sec
$D_{eKA}, D_{eKB}, \ldots$	=	effective Knudsen diffusion coefficient for A, B, $\ldots$, cm^2/sec
D_{el}	=	effective axial diffusivity, cm^2/sec
$D_{e,ma}$	=	effective diffusivity in the macropore, cm^2/sec
$D_{e,mi}$	=	effective diffusivity in the micropore, cm^2/sec
D_{ep}	=	effective diffusivity of poison, cm^2/sec
D_{er}	=	effective radial diffusivity, cm^2/sec
D_{eS}	=	effective diffusivity for, or including, surface transport

$(D_e)_X, (D_e)_Y$	=	effective diffusivities in catalysts X and Y, respectively, cm^2/sec
D_g	=	gas-phase diffusivity in Chapters 13 and 14, cm^2/sec
D_{ij}	=	diffusivity of species j through stagnant i, cm^2/sec
D_K	=	Knudsen diffusion coefficient, cm^2/sec
D_{KA}, D_{KB}	=	Knudsen diffusion coefficient for species A, B, $\ldots$, cm^2/sec
$D_{K,ma}$	=	Knudsen diffusivity in macropore, cm^2/sec
$D_{K,mi}$	=	Knudsen diffusivity in micropore, cm^2/sec
D_L	=	liquid-phase diffusivity, cm^2/sec
D_l	=	axial diffusivity, cm^2/sec
D_M	=	molecular diffusivity, cm^2/sec
D_{ma}	=	bulk diffusivity in macropore, cm^2/sec
D_{mi}	=	bulk diffusivity in micropore, cm^2/sec
D_p	=	plate diameter in Chapter 13, cm
D_r	=	radial diffusivity, cm^2/sec; or diffusivity based on predominating pore radius
D_S	=	surface diffusion coefficient, cm^2/sec
$\mathbf{D}_{in}$	=	divergence between estimates of objective function for competing models
d	=	decay order; or exponent in Eq. 14.23
d_b	=	bubble diameter, cm
d_{bm}	=	maximum bubble diameter, cm
d_{b0}	=	initial bubble diameter, cm
d_{dl}	=	diameter of dip-leg, cm
d_e	=	effective diameter of packed column, cm; or equivalent bubble diameter in a reactor with internals, cm
d_o	=	outer diameter of horizontal immersed tube, cm
d_{or}	=	diameter of orifice, cm
d_{pa}	=	diameter of packing, cm
d_t	=	tube diameter
E	=	activation energy, kcal/mol; or a quantitative measure of the

stoichiometric presence of a second reactant in estimating the effectiveness factor, Section 4.5.1

$E(t)$ = residence time distribution function

E' = entrainment rate

E^* = elutriation constant

$\breve{E}$ = enzyme concentration, mol/cm^3

$\breve{E}_0$ = initial enzyme concentration, mol/cm^3

E_a = apparent (observed) activation energy, kcal/mol

E_b = activation energy for the reverse step

E_c = efficiency of fluidized-bed contact

E_d = activation energy for diffusion, kcal/mol

E_f = activation energy for the fouling step

$E_1^*(h)$ = $-\operatorname{Ei}(-h) = \int_h^t \dfrac{e^{-t}}{t}\,dt$ is the exponential integral

E_r = reactor efficiency

E_S = activation energy for surface diffusion, kcal/mol; or activation energy for sintering in Chapter 19

$E(t)$ = exit-stream age distribution

E_s^* = specific elutriation constant

FN = fouling number, defined by Eq. 8.40

Fr = Froude group, defined as $u^2/g\,d_b$; or u_{or}^2/gl_j when applied to jet

Fr$'$ = modified Froude group defined in Figure 13.5

F = molar feed rate, mol/sec; or represents a function

$F(r_p)$ = pore size distribution function

F_{Ms} = molal flow rate of solids per unit area of reactor, mol/cm^2 sec

F_R = recycle ratio

F_s = circulation rate of solids, g/sec; or weight rate of feed, g/sec

$F_{s,0}, F_{s,1}, F_{s,2}$ = quantity of material of a given size in the inlet, outlet, and elutriation streams, respectively

F_w = mass flow rate, g/sec

f = a function; or fugacity coefficient

$f(\mathbf{x}_m, \mathbf{k})$ = rate predicted by the model

f_a = free area of grid plate

f_B = void fraction of the packed bed

f_{bed} = used specifically for f_B in Chapter 20 to distinguish from the general nomenclature f_i, which stands for porosity of species i

f_c = void fraction of the catalyst pellet

f_e = fraction of total active area that constitutes the external surface

f_f = porosity of the fluidized bed

f_{fp} = voidage of the bed with unflooded packing

f_g = void fraction of catalyst particle occupied by gas in Chapter 18

f_i = fraction of the total area that constitutes the internal surface

f_j = fraction of total sites in the jth patch

f_{kj} = predicted rate in jth experiment using kth model

f_l = void fraction of catalyst pellet occupied by liquid in Chapter 18

f_{ma} = voidage due to macropores

f_{mi} = voidage due to micropores

f_{mf} = voidage of the fluidized bed at incipient fluidization

f_{min}, f_{max} = minimum and maximum bed voidage

f_p = void fraction of the particle

f_{ps} = ratio of volume of the particles moving with the bubble to the volume of the bubbles

f_{pa} = porosity of the packing, that is, voidage of an empty packed bed in a packed fluidized bed

f_s = volume fraction of the solid

f_w = volume fraction of wake

G = mass velocity, g/cm^2 sec; or parameter defined by Eq. 8.76

G' = group defined by Eq. 4.67

ΔG = change in free energy or chemical potential per mole of new phase, kcal/mol

$\Delta G'$ = free energy of activation per mole of nucleus growth, kcal/mol

$G(C_{As}, T_s)$ = a function of surface concentration and temperature as defined by Eq. 11.43

G_M = molar flow rate of gas, mol/cm^2 sec

G_{mf} = mass flow rate at minimum fluidization velocity, g/cm^2 sec

G'_{mf} = mass flow rate at minimum fluidization velocity in a packed fluidized bed, g/cm^2 sec

g = gravitational constant, cm^2/sec

g_c = conversion factor

g_I = constant defined by Eq. 11.81

H = group defined in Eq. 12.64; or Hamiltonian; or heat of reaction group defined by Eq. 12.2; or heat transfer group defined by Eq. 13.27; or the level differences in the calculation of the pressure drop in a fluid-bed reactor-regenerator system

H' = group defined by Eq. 4.67

ΔH = heat of reaction, kcal/mol

$\Delta H'$ = heat of reaction, kcal/g

$H(C_{As}, T_s)$ = function of surface concentration of A and temperature defined by Eq. 11.43

h = general term for heat transfer coefficient, cal/sec cm^2 °K or Planck's constant, J/sec

h_0 = a constant characteristic of the grid distributor, the height at which the bubble diameter is zero, cm

h_a = heat transfer coefficient due to convection, cal/sec cm^2 °K or kcal/m^2 hr °K

h_b = radiant heat transfer coefficient, cal/sec cm^2 °K or kcal/m^2 hr °K

h_C = heat transfer coefficient of control fluid, cal/sec cm^2 °K or kcal/hr m^2 °K

h_c = heat transfer coefficient due to conduction, cal/sec cm^2 °K or kcal/hr m^2 °K

h_{bs} = radiant heat transfer coefficient between solid particles, cal/sec cm^2 °K or kcal/hr cm^2 °K

h_{bv} = radiant heat transfer coefficient between voids, cal/sec cm^2 °K or kcal/m^2 hr °K

h_e = effective heat transfer coefficient for one-dimensional model, cal/sec cm^2 °K or kcal/hr m^2 °K

h_e^d = dynamic contribution to effective heat transfer coefficient, cal/sec cm^2 °K or kcal/hr m^2 °K

h_e^0 = static contribution to effective heat transfer coefficient, cal/sec cm^2 °K or kcal/hr m^2 °K

h_{fc} = heat transfer coefficient depicting the influence of fluid flow on conduction, cal/sec cm^2 °K or kcal/hr m^2 °K

h_{fp} = fluid–particle heat transfer coefficient, cal/sec cm^2 °K or kcal/hr m^2 °K

h_{max} = maximum heat transfer coefficient, cal/sec cm^2 °K or kcal/hr m^2 °K

h_{pc} = heat transfer coefficient for particulate motion as defined in Eq. 13.28, cal/sec cm^2 °K or kcal/hr m^2 °K

h_S = surface diffusion parameter defined by Eq. 4.86

h_w = heat transfer coefficient at the wall, cal/sec cm^2 °K or kcal/hr m^2 °K

h_w^d = dynamic contribution to heat transfer coefficient at the wall, cal/sec cm^2 °K or kcal/hr m^2 °K

h_w^* = heat transfer coefficient across a true fluid boundary layer, cal/sec cm^2 °K or kcal/m^2 hr °K

h_{wt}^d = total dynamic contribution to heat transfer coefficient at the wall, cal/sec cm^2 °K or kcal/hr m^2 °K

I = integral along the slug surface in Fig. 14.12;
or intensity of segregation

I_A, I_B = impulse functions in Eq. 12.52

I_0 = distance of separation between particles A and B within which they react

J = integral associated with the cloud (Eq. 14.31 and Fig. 14.3); or group defined by Eq. 11.60

J_p = jet penetration, cm

j_d = mass transfer factor defined by Eq. 6.2

j_h = heat transfer factor defined by Eq. 6.2

K = general constant;
or equilibrium constant;
or group defined by Eq. 8.29;
or a consolidated adsorption constant defined by Eq. 4.64c;
or multiple of half-particle diameter in the cell model

K' = overall constant defined as the sum of the resistances in Eq. 14.34 as $\left(\dfrac{1}{K_m} + \dfrac{1}{K_0(1-\delta)} \right)^{-1}$

K'' = effective rate constant, defined as $(k_v \rho_s W_e)$

K_1, K_2 = rate groups defined by Eq. 16.18 and 16.19

$K_A, K_B, \ldots, K_i$ = equilibrium constant for species A, B, $\ldots$, i, atm^{-1} or cm^3/mol

K_b = dimensionless reaction group defined in Eq. 14.27

K_{bc} = gas interchange coefficient between bubble and cloud, sec^{-1}

K_{be} = gas interchange coefficient between bubble and emulsion, sec^{-1}

$K_{be,n}$ = bubble–emulsion mass transfer coefficient in the nth compartment sec^{-1}

K_{ce} = gas interchange coefficient between cloud and emulsion, sec^{-1}

K_d = dimensionless reaction group for the dilute phase defined in Eq. 14.27

K_f = overall rate constant group, defined as $K_f' L_f / u_b$ for the fluidized bed (Eq. 14.13)

K_f' = overall rate constant defined in Eq. 14.12

K_f'' = group defined by Eq. 14.25

K_m = dimensionless mass transfer group defined in Eq. 14.27 as $k_{0b} a_b L_f / u$ or $K_{0b} L_f / u$; or Michaelis–Menten constant in Chapter 18

K_m' = apparent Michaelis–Menten constant

K_0 = dimensionless group defined as $K_v L_0 / u$ or $k_p PW / F$

K_0' = group defined by Eq. 14.26

K_{0b} = rate group defined as $k_{0b} a_b$, sec^{-1}

K_{0f} = dimensionless group defined as $k_v L_{mf} / u = k_v C_{A0} L_{mf} / u$

K_R' = overall constant for the successive contact model defined as $K' + K_b + K_d$ in Eq. 14.33

K_{we} = mass transfer coefficient between the wake and emulsion, sec^{-1}

k = general representation for rate constant;
or stage in a series;
or general representation of model parameters such as k, $k_A, \ldots$

k^p = pore inclination constant used in the Johnson–Stewart model

k^0 = general notation of constant at time zero, sec^{-1}

$\hat{k}$ = dimensionless rate group defined as $d_p A_f / u$ in Eqs. 11.57 and 11.58

$\bar{k}$ = parameter related to rate constant (e.g., in Eq. 20.29), sec

k_a = observed rate constant, cm/sec or sec^{-1}

k_b = bubble-side mass transfer coefficient in a fluidized bed, cm/sec or Boltzmann constant, cal/°K

k_{bc} = interchange coefficient between bubble and cloud phase, cm/sec

k_{be} = interchange coefficient between bubble and emulsion phases, cm/sec

k_{cb} = rate constant for carbon burning, cm^3/mol sec

k_{cbp} = rate constant for carbon burning, 1/sec atm

k_{ce} = interchange coefficient between cloud and emulsion phases, cm/sec

k_d = rate constant for deactivation, sec^{-1}

k_e = emulsion-side mass transfer coefficient in a fluidized bed, cm/sec

k_e' = effective thermal conductivity of the packed bed, cal/cm sec °K or kcal/m hr °K

$k_e'^0$ = static contribution to the effective thermal conductivity, ($k_{er}'^0$ and $k_e'^0$) cal/cm sec °K or kcal/m hr °K

$k_{ea}'^0, k_{eb}'^0, k_{ec}'^0$ = static contributions due to radiation, conduction, and convection, respectively, cal/cm sec °K or kcal/m hr °K

$k_{ea}'^d, k_{ec}'^d$ = effective thermal conductivity accounting for the dynamic contributions to convection and conduction, respectively, cal/cm sec °K or kcal/m hr °K

k_{eff} = effective rate constant, sec^{-1}

k_g' = fluid thermal conductivity, cal/cm sec °K or kcal/m hr °K

k_{el}' = effective axial thermal conductivity of the packed bed, cal/cm sec °K or kcal/m hr °K

$k_e'^0$ = static contribution to effective axial thermal conductivity, cal/cm sec °K or kcal/m hr °K

k_{er}' = effective radial thermal conductivity of the packed bed, cal/cm sec °K or kcal/m hr °K

$k_{er}'^d$ = dynamic contribution to effective radial thermal conductivity, cal/cm sec °K or kcal/m hr °K

$k_{er}'^0$ = static contribution to radial effective thermal conductivity, cal/cm sec °K or kcal/m hr °K

$(k_{er}')_{pf}$ = lateral thermal conductivity in a packed fluidized bed with horizontal flow, cal/cm sec °K or kcal/m hr °K

k_f = general notation for rate constant for the fouling reaction, sec^{-1}

k_g = phenomenological mass transfer coefficient, cm/sec

k_{gp} = mass transfer coefficient expressed in partial pressure units, mol/cm^3 atm sec

k_{gr} = mass transfer coefficient in the presence of chemical reaction, cm/sec

k_{gv} = volume-based mass transfer coefficient ($= k_g a$), cm/sec

k_j = equilibrium constant for species A on patch J

k_{je} = mass transfer coefficient from jet to emulsion, cm/sec

k_L = liquid–solid mass transfer coefficient, cm/sec

k_l = rate constant per active site

k_m = modified rate constant, defined as k_v/ω

k_{mw} = rate constant based on weight of reactant consumed, cm^3/g catalyst sec

k_{nf} = rate constant for nucleus formation

$k_{n,f}$ = rate constant for nth order decay in Eq. 8.74 (cm^3/mol)$^{n-1}$ sec^{-1}

$k_{n,f}''$ = rate constant defined by Eq. 8.72, sec^{-1}

k_{ob} = overall mass transfer coefficient, including k_b and k_e, in a fluidized bed, cm/sec

k_p = rate constant based on partial pressure, mol/g atm sec

k_p' = pellet conductivity, cal/sec cm °K

k_{pv} = rate constant based on partial pressure and volume of catalyst, mol/cm^3 atm sec

k_r = consolidated rate constant, defined as $k_v \left[(K+1)/K \right]$ for a reversible reaction, sec^{-1}

k_{rS} = consolidated rate constant based on surface $k_S[K/(K+1)]$, cm/sec

k_S = surface reaction rate constant, cm/sec

$k_{SA}, k_{SB}, \ldots$ = surface reaction rate constants for species A, B, . . . on catalyst, cm/sec

k_{sp} = surface rate constant for the poisoning reaction, cm/sec

k_s' = thermal conductivity of solids, cal/cm sec °K or kcal/hr m °K

k_V = rate constant based on reactor volume, sec^{-1}

k_v = rate constant based on catalyst volume, sec^{-1} (has the units of rate if the concentration is expressed in dimensionless form; thus in Chapter 17 the units are mol/cm^3 sec)

$k_{vA}, k_{vB}, \ldots$ = reaction rate constant for the species A, B, . . . , sec^{-1}

$(k_v)_A$ = rate constant for the geometry A, sec^{-1}

k_v^0 = rate constant at zero carbon content, sec^{-1}; or maximum value of rate constant corresponding to uniform activity, sec^{-1}

$\overline{k_v}$ = volume-averaged rate constant, sec^{-1}

$\hat{k}_v$ = dimensionless constant defined by Eq. 9.16

k_{ve} = rate constant for reaction in the emulsion phase, sec^{-1}

$k_{v,f}$ = rate constant for the fouling reaction, sec^{-1}

k_{vs} = rate constant at the temperature of the external surface, sec^{-1}

k_{v0} = rate constant at time zero

$k_{v1}, k_{v2}, \ldots$ = rate constant based on catalyst volume for reaction step 1, 2, . . . , sec^{-1}

k_w = rate constant based on weight of catalyst, cm^3/g catalyst sec

$\mathbf{k}$ = vector of model parameters in Chapter 2

$\hat{\mathbf{k}}$ = linearized value of vector $\mathbf{k}$

L = total length parameter (i.e., length of packed column), cm; or likelihood function in Chapter 2

$\hat{L}$ = dimensionless length parameter, l/L

L' = factor accounting for increased path flow (tortuous flow)

L_c = critical bed height in Eq. 14.70, cm

L_e = equilibrium line in the T–x plots

L_f = height of the fluidized bed, cm

L_{fb} = total length of freeboard region, cm

L_h = initial height of a fixed layer over the packing in a packed fluidized bed, cm

L_i = dimensionless distance of the reactant boundary from the center line in the case of pore-mouth poisoning

L_m = locus of the maximum rates in the T–x plots

L_{ma} = macropore length, cm

$\hat{L}_{ma}$ = dimensionless macropore length, L_{ma}/L

L_{mf} = height of bed at minimum fluidization, cm

L_{mi} = micropore length, cm

$\hat{L}_{mi}$ = dimensionless micropore length, L_{mi}/L

L_0 = height of fixed bed, cm

L_p = total length of pore, cm

L_{p0} = L_p at time zero, cm

L_s = length of slugging bed, cm

L_t = total height of fluidized bed, cm

L_w = horizontal distance traveled by particle (see design of cyclone), cm

l = length parameter, axial coordinate, cm; or depth of a plate, cm

l^* = dimensionless length defined by Eq. 20.9

l_b = height of the gas emanating from each perforation in a distributor, cm

l_j = length along the grid jet, cm

l_{fb} = length of freeboard region, cm

l_{ma}	=	macropore length coordinate, cm
l_{mi}	=	micropore length coordinate, cm
l_0	=	height of the fixed bed over the nonoperating orifice in a fluidized bed
l_0'	=	height of spouted bed over an operating orifice in a fluidized bed, cm
l_p	=	length of the pore, cm
$\hat{l}_R$	=	axial length normalized with respect to radius $(=l/R)$
l_s	=	length of slug, cm
l_t	=	distance between tube centers, cm
M	=	molecular weight; or dimensionless adiabatic temperature rise $\Delta T/T_0$; or momentum of the jet defined in Chapter 13, g cm/sec; or number of radial stages in the cell model
M_A, M_B	=	molecular weights of A, B
M_c	=	molecular weight of carbon
M_g	=	molecular weight of gas
M_0	=	initial molecular weight of the feed; or initial momentum, g cm/sec
M_p	=	molecular weight of poison
m	=	general symbol for constants or exponents; or solid content of the bubble phase; or temperature parameter defined in Eq. 7.35; or catalyst loading in a slurry reactor, g/cm^3 of slurry
m'	=	fraction of catalyst in mixed reactor in Chapter 12; or micropore modulus for a first-order reaction
m''	=	micropore modulus for a second-order reaction
m_1, m_2	=	roots of Eqs. 5.52 and 14.6
m_H	=	modulus used as a parameter in Fig. 14.3
m_i	=	index of reaction given by Eq. 21.18
m_n	=	parameter defined by Eq. 9.36

N	=	Avogadro's number; or flux of diffusing component; or product nM appearing in Eq. 8.77; or group defined by Eq. 11.60
N_1, N_2	=	groups defined by Eqs. 19.85 and 19.86
$\breve{N}_1', \breve{N}_2'$	=	groups defined by Eqs. 19.92 and 19.93
N_A, N_B	=	flux of diffusing component A, B due to volume diffusion relative to the fluid mixture at a given point, mol/cm^2 sec
N_A^0, N_B^0	=	combined diffusion and flow fluxes for species A, B
N_{Az}, N_{Bz}	=	flux of components A, B in the z direction
N_b	=	flux of diffusing component due to volume diffusion relative to stationary coordinates, mol/cm^2 sec
N_b'	=	flux of a diffusing component due to volume diffusion relative to the fluid mixture at a given point, mol/cm^2 sec
N_{bA}, N_{bB}	=	flux of diffusing component A, B due to volume diffusion with respect to stationary coordinates, mol/cm^2 sec
N_F	=	flux due to forced flow, mol/cm^2 sec
N_{KA}, N_{KB}	=	flux of diffusing component A, B due to Knudsen diffusion relative to stationary coordinates, mol/cm^2 sec
N_{or}	=	number of orifices per unit area
N_s	=	number of spirals traveled in a cyclone; or number of solid diffusion transfer units given by Eq. 16.48
N_t^0	=	total flux inclusive of Knudsen, bulk, and forced flow transport, mol/cm^2 sec
$N(q)$	=	distribution function characterizing the heats of adsorption
n	=	exponent appearing in various equations; or order of reaction, or number of sites involved in adsorption; or nth cell or compartment;

or number of fractions in a poly-dispersed phase

n_a	=	apparent (observed) reaction order
n_{or}	=	total number of orifices
n_p	=	number of cylindrical pores per unit external area
Pe_h	=	general definition of Peclet number for heat transfer $d_t \rho_g c_p u/k'$
$Pe_{hl}, Pe'_{hl}, Pe''_{hl}$	=	axial Peclet numbers for heat transfer
$Pe_{hr}, Pe'_{hr}, Pe''_{hr}$	=	radial Peclet numbers for heat transfer
Pe_m	=	general definition of Peclet number for mass transfer $d_t u/D$
Pe'_m, Pe'_h	=	Peclet number for mass and heat transfer based on the particle diameter $d_p u/D$ or $d_p \rho_g C_p u/k'$
Pe''_m	=	Peclet number based on interstitial gas velocity, $d_p u_i/D$ or $d_p u/f_B D$
$Pe_{ml}, Pe'_{ml}, Pe''_{ml}$	=	axial Peclet numbers for mass transfer
$Pe_{mr}, Pe'_{mr}, Pe''_{mr}$	=	radial Peclet numbers for mass transfer
$(Pe''_{mr})_M$	=	molecular Peclet number
Pr	=	Prandtl number, $C_p \mu/k'_g$
P	=	total pressure, atm; or group defined by Eq. 5.20; or multivariate probability density function in Chapter 2
P_1	=	dimensionless group defined as $k_{je} a_j A_c J_p/Q$
P_2	=	dimensionless group defined as $k_{be} a_b A_c J_p/Q$
ΔP_d	=	pressure drop across the distributor plate, atm
P_X, P_Y	=	groups defined by Eq. 15.25
$\mathbf{P}, \mathbf{P}_1, \mathbf{P}_2$	=	profit functions defined variously in Chapters 12 and 16
p	=	probability; or partial pressure, atm; or pressure differential across any two points, atm; or degree of poisoning
$\check{p}$	=	perimeter of the reactor tube

p^*	=	partial pressure of diffusing component at equilibrium, atm
$p_A, p_B, \ldots$	=	partial pressure of species A, B, $\ldots$, atm
$p_A^*, p_B^*, \ldots$	=	partial pressure of components A, B, $\ldots$ at equilibrium, atm
p_{As}	=	partial pressure at the surface, atm
p_{cr}	=	critical value of partial pressure, atm
p_{ij}	=	likelihood ratio for competing models i, j defined by Eq. 2.29
p_m	=	partial pressure at a hot spot or a point on the maxima curve, atm
p_{nm}	=	ratio of intrinsic rate of an nth-order reaction to that of an mth-order reaction
p'_0	=	lower limit of the partial pressure of reactant at the reactor inlet, atm
p_0^l, p_0^u	=	lower and upper limits of partial pressures of the reactant at the reactor inlet, atm
$p_0(k)$	=	relative probability density function in Chapter 2
p_s	=	partial pressure at the surface, atm
Q	=	volumetric flow rate, cm³/sec
Q_b	=	volumetric flow rate through the bubble phase, cm³/sec
Q_{be}	=	total gas exchange between bubble and emulsion phases, cm³/sec
Q_c	=	volumetric flow rate through the cloud phase, cm³/sec
Q_e	=	volumetric flow rate through the emulsion phase, cm³/sec
Q_h	=	total heat transferred, cal/sec
Q_{or}	=	total volume flowing through the nozzle, cm³/sec
q	=	heat of adsorption in Eq. 2.10, kcal/mol; or number of responses in Chapter 2; or volumetric flow rate between bubble and emulsion phases, cm³/sec

q_c	=	capacity of a single cyclone	$\mathbf{R}_{cb}$	=	dimensionless diffusion parameter in Eq. 8.83
R	=	reaction product			
Re	=	Reynolds number defined as $d_t u \rho_g / \mu$ or $d_t G / \mu$	$\mathbf{R}_{GA}$	=	total reaction in a grain of the pellet, mol/sec
Re$'$	=	Reynolds number defined as $d_p u \rho_g / \mu$ or $d_p G / \mu$	$\mathscr{R}$	=	global rate in unpoisoned pore, mol/sec
Re$''$	=	Reynolds number defined as $d_p u \rho_g / f_B \mu$ or $d_p G \rho_g / f_B \mu$	$\mathscr{R}_f$	=	global rate in poisoned pore, mol/sec
Re$'_{mf}$	=	Reynolds number at minimum fluidization velocity	$\mathscr{R}_{fb}$	=	value of $\mathscr{R}_f$ at the base temperature T_b, mol/sec
RF	=	reaction factor	r	=	radial coordinate, cm
R	=	radius of the pellet, cm;	r^*	=	neck radius defined in Chapter 21
$\hat{R}$	=	dimensionless radius, r/R			
R_B	=	bed expansion ratio	r_a	=	radius of aperture in Figure 4.3, cm
R_D	=	ratio of bulk to Knudsen diffusivities	r_b	=	radius of bubble, cm; or maximum micropore radius, cm
$(R_D)_e$	=	ratio of effective bulk to effective Knudsen diffusivities			
$(R'_D)_e$	=	ratio defined by Eq. 3.21	r_c	=	radius of the cloud, cm
R_g	=	universal gas constant, cal/mol	r_G	=	radial position in a grain of the pellet, cm
R'_g	=	universal gas constant, cm^3 atm/mol °K	$\hat{r}_G$	=	dimensionless radius in a grain of the pellet (r_G/r_{G0})
R_H	=	heat transfer factor defined by Eq. 11.23	r_{Gi}	=	position of moving interface in a grain of the pellet, cm
R_i	=	radius of the adsorption center, cm; or radius of moving interface	$\hat{r}_{Gi}$	=	dimensionless radius of moving interface in a grain of the pellet
$\hat{R}_i$	=	dimensionless radius of moving interface	r_{G0}	=	initial radius of a grain in the pellet, cm
R_M	=	mass transfer factor defined by Eq. 11.23; or smallest feed size for a growing particle or largest size for a shrinking particle in Chapter 20	r_{ma}	=	radial coordinate for macropore, cm
			r_{mi}	=	radial coordinate for micropore, cm
R_0	=	radius of pellet or grain before structural changes set in; or half the distance between two adsorbing centers on a catalyst surface, cm	r_0	=	radius of the packed bed, cm
			r_p	=	radius of the pore, cm
			$\bar{r}_p$	=	average radius of the pore, cm
			r^0	=	rate of main reaction at time zero
			$\hat{r}_A$	=	dimensionless rate of reaction, r_A/r_{As}
R_{ov}	=	overall resistance due to gas and liquid films, cm/sec	r_a	=	general notation for the observed rate
R_p	=	radius of unpoisoned pore, cm	r'_a	=	rate of increase in cost due to initial aging period
R_X, R_Y	=	radii of catalyst pellets X and Y, respectively, cm	r'_{cb}	=	rate of carbon burning, g/g catalyst sec
$\mathbf{R}$	=	total reaction rate per pellet, mol/sec;			
$\mathbf{R}_A, \mathbf{R}_B, \ldots$	=	total reaction rate per pellet for species A, B, $\ldots$, mol/sec	r_d	=	rate of mass transfer, mol/cm^2 sec

r_f^0	=	rate of fouling reaction at time zero	S_l	=	space velocity of liquid feed, $\sec^{-1}$
r_g	=	growth rate of nucleus			
r_S	=	rate of reaction based on surface area, $\mathrm{mol/cm^2\ sec}$	S_0	=	initial substrate concentration at the surface, $\mathrm{mol/cm^3}$
r_{Sa}	=	apparent (observed) rate of reaction based on surface area, $\mathrm{mol/cm^2\ sec}$	S_v	=	total surface area per unit volume, $\mathrm{cm^2/cm^3}$
$r_{SA}, r_{SB}, \ldots$	=	rate of reaction of A, B, $\ldots$ based on surface area, $\mathrm{mol/cm^2\ sec}$	S_{v0}	=	S_v at time zero, $\mathrm{cm^2/cm^3}$
			$\mathscr{S}$	=	selectivity function as variously defined in Chapter 5
r_V	=	rate of reaction, $\mathrm{mol/cm^3\ reactor\ sec}$	s	=	intrinsic selectivity ratio k_1/k_2; or stoichiometric number in Chapter 2; or any statistic (such as variance) in Chapter 3; or constant for a distributor plate in Chapters 13 and 14 (denotes spacing between openings in Chapter 13); or general class of nonlinear functions
r_v	=	rate of reaction, $\mathrm{mol/cm^3\ catalyst\ sec}$			
$r_{vA}, r_{vB}, \ldots$	=	rate of reaction of species A, B, $\ldots$, $\mathrm{mol/cm^3\ catalyst\ sec}$			
r_{vb}, r_{ve}	=	rate of reaction in bubble and emulsion phases, $\mathrm{mol/cm^3\ sec}$			
r_w	=	rate of reaction, $\mathrm{mol/g\ catalyst\ sec}$			
r_w'	=	rate of reaction ($= r_{cb}'$ when applied specifically to carbon burning), $\mathrm{g/g\ catalyst\ sec}$	s'	=	spacing between openings in the distributor, cm
$r_{wA}, r_{wB}, \ldots$	=	rate of reaction of species A, B, $\ldots$, $\mathrm{mol/g\ catalyst\ sec}$	s_{A_f}	=	selectivity based on Arrhenius preexponential factor (A_{f_1}/A_{f_2})
$r_{wA}', r_{wB}', \ldots$	=	rate of reaction of A, B, $\ldots$, $\mathrm{g/g\ catalyst\ sec}$	s_a	=	observed selectivity (i.e., selectivity in presence of diffusional effects)
r	=	vector of rates			
S	=	product species			
Sc	=	Schmidt number ($= \mu/\rho_g D$)	s_b	=	fraction of total solids in the bubble
SF	=	shape factor			
Sh	=	Sherwood number; or Biot number for mass, generally defined as $k_g R/D$	s_{bb}	=	volume fraction of solids in the bubbles with respect to bubble volume
St	=	Stanton number	s_{cb}	=	volume fraction of solids in the cloud with respect to bubble volume
St_w	=	Stanton number at the wall			
S	=	substrate; or group defined by Eq. 11.60; or shape constant defined by Eq. 4.1 (equal to 2 for sphere, 1 for cylinder, and 0 for plate); or least-squares objective function defined by Eq. 2.14	s_e	=	fraction of the external surface area
			s_{eb}	=	volume fraction of solids in the emulsion with respect to bubble volume
S'	=	constriction factor in the pores	s_{wb}	=	volume fraction of the wake solids with respect to bubble volume
ΔS	=	entropy change, $\mathrm{kcal/°K}$			
S_F	=	space velocity of feed, $\sec^{-1}$	T	=	temperature, °K
S_g	=	specific total surface area, $\mathrm{cm^2/g}$	T^*	=	equilibrium temperature, °K
S_i	=	internal surface associated with a pore, defined in Eq. 19.111	T	=	dimensionless temperature defined as T/T_s

$T(t)$	=	optimum temperature policy in Chapter 16
T_b	=	temperature in the bulk gas phase, °K; or base temperature, °K
$\hat{T}_b$	=	dimensionless temperature defined as T/T_b
T_C	=	control fluid temperature, °K
T_c	=	temperature of the cloud phase, °K
T_e	=	temperature of the emulsion phase, °K
T_F	=	dimensionless temperature defined by Eq. 16.81
T_{fu}	=	temperature defined in Table 11.4
T_i	=	temperature at the moving interface, °K
$\hat{T}_i$	=	dimensionless temperature defined as T_i/T_s
T_m	=	maximum temperature in the pellet, °K; or temperature corresponding to maximum partial pressure, °K
T_0	=	fluid inlet temperature, °K; or temperature at the end of a pore, °K
T_p	=	temperature in the pellet, °K
T_{pcr}	=	temperature in the pellet at the critical point, °K
T_{pI}	=	temperature in the pellet at the inflection point, °K
T_s	=	temperature at the external surface of the pellet, °K
T_u	=	highest temperature of operation, °K
T_w	=	temperature at the wall, °K
t	=	time, sec; or statistic in Chapter 2
t'	=	$(t - l/u_i)$ in Eq. 16.7
$\hat{t}$	=	general symbol for dimensionless time
$\hat{t}_{crp}$	=	dimensionless time for constant rate period
t^*	=	dimensionless time defined in Eqs. 19.60 and 20.10
$\bar{t}$	=	general definition of mean residence time, sec

t_A, t_B, t_p	=	dimensionless times defined in Eq. 8.39
$\hat{t}_A, \hat{t}_B$	=	dimensionless times defined in Table 8.4
t_C	=	dimensionless time based on t' defined by Eq. 16.4
t_c	=	constant activity loss period in Chapter 16
$\hat{t}_{cb}$	=	dimensionless time for carbon burning defined by Eq. 8.86
t_D	=	dimensionless time defined by Eq. 16.7
t_{DP}	=	time parameter in Chapter 19
t_E	=	dimensionless time defined by Eq. 16.52
t_e	=	retention time of the pulse at a time $1/e$ of pulse height
t_F	=	normalized time defined by Eq. 16.75
t_{Ft}	=	total regeneration time, sec
t_f	=	residence time in the fluid bed, sec; or initial aging period in Chapter 16
t_H	=	heat transit time defined by Eq. 16.93
t_M	=	time parameter in Chapter 19
t_m	=	retention time of the pulse maximum; or residence time in mixed reactor in Section 12.6
t_{0A}	=	duration of injection of the adsorbable gas in chromatography, sec
t_{p1}	=	time of run in fixed-bed reactor, sec
t_p	=	thickness of grid plate, cm; or production (on-stream) time, sec
t_{pc}	=	time for pore closure, sec
t_R	=	time parameter in Chapter 19
U	=	overall heat transfer coefficient, cal/cm² °K or kcal/m² hr °K; or group defined as $(1 - u_{mf}/u)$ in Davidson's model (Eq. 14.2)
u	=	superficial velocity, cm/sec
$\bar{u}$	=	mean molecular velocity
u_b	=	absolute rise velocity of bubbles, cm/sec

u_{bm} = maximum rise velocity of bubble, cm/sec

u_{bs} = rising velocity of a single bubble, cm/sec

u_c = velocity of the cloud phase, cm/sec

u_e = velocity of the emulsion phase, cm/sec

u_f = freeboard velocity, cm/sec

u_{fr} = upward velocity of gas with respect to emulsion, cm/sec

u_i = interstitial linear velocity of gas, cm/sec

u_m = critical minimum velocity defined in Section 13.8.4, cm/sec

u_{mb} = minimum bubbling velocity, cm/sec

u_{mf} = minimum fluidization velocity, cm/sec

u'_{mf} = minimum fluidization velocity in a packed fluidized bed, cm/sec

u_{ms} = minimum slugging velocity, cm/sec

u_{or} = velocity through the orifice or distributor opening, cm/sec

u_{opt} = optimum value of velocity at which heat transfer is maximum, cm/sec

u_p = fluidization velocity corrected for fixed packing voidage, cm/sec

u_s = velocity of the solids, cm/sec; or velocity of slugs, cm/sec

u_t = terminal velocity, cm/sec

u_{tr} = velocity at which a sharp decrease in pressure drop is recorded, denoting the onset of fast fluidization, cm/sec

V_b = volume of the bubble, cm^3

$V_{b,n}$ = volume of bubble phase in the nth compartment, cm^3

V_c = volume of the cloud phase, cm^3; or volume of catalyst, cm^3

V_e = volume of the emulsion phase, cm^3

$V_{e,n}$ = volume of emulsion phase in the nth compartment, cm^3

V_g = pore volume of catalyst, cm^3/g

V_{gn} = volume of growth nucleus, cm^3

V_m = product of rate constant and enzyme concentration in Chapter 18

V_p = volume of the pellet, cm^3

$(V_p)_p(V_p)_s$ = volume of pellet for plate and sphere, respectively, cm^3

V_R = chromatographic retention volume, cm^3

V_r = volume of reactor, cm^3

V_s = volume of slug, cm^3

V_t = total volume, cm^3

V'_v = volume of the voids, cm^3

v_b = volumetric fraction of the bubble; or volumetric fraction of gas

v_c = volume fraction of the continuous phase

v_e = volumetric flow rate through the emulsion phase, cm^3/sec

v_{pa} = volumetric fraction of the packing in the packed fluidized bed

v_w = volume of the wake phase, cm^3

W = weight of catalyst or solid, g

W_A, W_B = weight of particles of A, B, g

W_c = weight of coke on catalyst at any time, g/g catalyst

$\hat{W}_C$ = dimensionless coke concentration, W_c/W_{co}

W_{co} = coke concentration at saturation, g/g catalyst

W_e = width of pulse at time $1/e$ of the pulse height; or catalyst loading per unit volume of emulsion

$\hat{w}$ = weight ratio of particles, W_A/W_B

w_A = quantity of A adsorbed, g; or solids loading in the cyclone, g

w_B = quantity of B adsorbed, g

W_F = dimensionless wall heat transfer coefficient defined in Eq. 20.13

X = catalyst type

X = general distance parameter

$\hat{X}$ = dimensionless distance parameter, X/X_0

X_0	=	complete dimension of a given shape
X'_0	=	characteristic dimension of the pellet
$\mathbf{X}$	=	an $N \times P$ matrix defined in Section 2.6.6
x	=	fractional conversion; x_A, x_B, ... are sometimes used to denote specifically the conversion of A, B, ...; or concentration of particles of a given fraction; or distance parameter tangential to the surface in Figure 6.2
x^*	=	limiting concentration in a polydispersed system corresponding to y^*
$\bar{x}_A$	=	time-averaged conversion of A
x_{Ae}, x_{A0}	=	conversion at the exit and inlet
x_{av}	=	average concentration at the end of the operating time
x_{BA}	=	moles of B formed per mole of A in the feed
x_l	=	distance from the loading edge of a plate
x_m	=	conversion in the mixed reactor in Chapter 12
$\mathbf{x}$	=	vector of independent variables such as temperature or pressure in Chapter 2
Y	=	catalyst type
Y	=	group defined in Eq. 14.4
Y_a	=	actual point selectivity
Y_p	=	intrinsic point yield; or selectivity
Y_r	=	relative point yield; or selectivity
y	=	mole fraction; or concentration of particles of a given fraction at the outlet; or distance parameter normal to the surface in Figure 6.2. or observed response; or time-averaged mole fraction
y^*	=	limiting yield concentration
$y_A, y_B, \ldots$	=	mole fraction of A, B, ...
$y_{Ae}, y_{Be} \ldots$	=	mole fraction of A, B, ... at the outlet

$y_{AL}, y_{BL}, \ldots$	=	mole fraction of A, B, ... at $l = L$
$y_{A0}, y_{B0}, \ldots$	=	mole fraction of A, B, ... at $l = 0$
y_L	=	mole fraction at length $l = L$
y_m	=	observed response for the mth experiment in Chapter 2
$\mathbf{y}_m$	=	vector of y_m
$\mathbf{y}$	=	vector of y
Z	=	u_s/u; or term defined in Eq. 12.64; or normalized set of rate constants defined by Eq. 9.29
$\check{Z}$	=	ratio of interslug distance to bed parameter
Z_t	=	L_t/L_f
Z_v	=	volume-change parameter given by Eq. 19.105
z	=	dimensionless axial distance, l/L; or direction in the Johnson–Stewart model
z'	=	exponent on time in Eq. 21.18; or dimensionless length in the reactor, l/d_p
z_F	=	normalized length defined by Eq. 16.74.
z_f	=	dimensionless distance in a fluidized bed, l/L_f;
z_{ft}	=	dimensionless total distance in a fluidized bed including the dilute phase, L_t/L_f

GREEK NOTATION

α	=	Arrhenius parameter defined as $E/R_g T$; or u_{bs}/u_i in Chapters 13 and 14; or a general constant; or as defined in Eqs. 3.3 and 3.5: $1 + N_{bB}/N_{bA}$ and $1 - (M_A/M_B)^{1/2}$ respectively
α_b	=	$E/R_g T_b$
α_D	=	diffusivity ratio D_{eB}/D_{eA} for a reaction A → B
α_1	=	value of α at the influxion point
α_s	=	$E/R_g T_s$
α_{s1}, α_{s2}	=	value of α_s for reaction steps 1, 2, ...

$\alpha_{s,f}$ = $E_f/R_g T$

α_t = ratio of diffusion time in the macropore to that in the micropore region

α_w = dimensionless Arrhenius parameter based on wall temperature $(E/R_g T_w)$

α_1, α_2 = parameters defined by Eqs. 12.4 and 12.5

β = radial dispersion factor; or nondimensional temperature in Eq. 4.42; or enhancement due to reaction

β'' = ratio of the effective distance between particles to particle diameter

β_A = angle of repose of the solid, deg

β_H = Hatta number given in Eq. 14.31.

β_L = gas–liquid parameter defined by Eq. 4.105

β_m = thermicity factor defined by Eq. 4.44

β'_m = modified thermicity factor defined by Eq. 7.20

β_{mb} = thermicity factor based on bulk fluid conditions

β_{m1}, β_{m2} = thermicity factors for reaction steps 1 and 2

β'_o = dimensionless concentration defined as K'_m/C_{S0}

β_r = parameter defined in Eq. 14.31

β_S = nondimensional concentration defined as C_S/K'_m

Γ = parameter defined by Eq. 5.23

γ = heat generation parameter in Eq. 4.42; or dimensionless group $P_1 + P_2(H/J_p) - 1$ where P_1 and P_2 are as defined in Section 14.8.1

γ_p = function of pore parameter in Figure 3.1

Δ = non-key-component parameter

δ = experimental error; or stagnant fluid thickness; or volume fraction of bubbles in the bed; or ratio of Thiele modulus for

component B to that for A $(\phi_{s1})_B/(\phi_{s1})_A$

δ_e = derived parameter as defined in Eq. 9.46

δ_{fe} = defined in Eq. 14.31

δ_i = derived parameter as defined by Eq. 9.47

δ_{pe} = volume fraction of the particles in the emulsion phase

δ_{sl} = average liquid-layer thickness, cm

δ_0, δ_1 = contribution to second central moment in experimental chromatographic measurements, $\sec^{-1}$

ε = general notation for effectiveness factor

ε_c = integral combined effectiveness factors

ε'_c = local combined effectiveness factor for external and internal mass transfer

ε_f = effectiveness factor for fouling

$\bar{\varepsilon}_f$ = average effectiveness factor for fouling

ε_G = effectiveness factor for the gas phase in a slurry reactor

ε_{iso} = isothermal effectiveness factor used specifically to distinguish from the nonisothermal factor

ε_L = effectiveness factor for the liquid phase in a slurry reactor

ε_M = emissivity

ε_{ma} = effectiveness factor for the macropore

$(\varepsilon_{ma})_1, (\varepsilon_{ma})_2, \ldots$ = ε_{ma} for reaction steps 1, 2, ...

ε_{mi} = effectiveness factor for the micropore

$\varepsilon(r_p)$ = effectiveness factor for the pore

ε_v = effectiveness factor with volume change

$\varepsilon_1, \varepsilon_2$ = effectiveness factor for reactions 1, 2, ...

$\boldsymbol{\varepsilon}$ = $n \times 1$ vector of residuals

η = overall external effectiveness factor

η' = local external effectiveness factor

η_g = efficiency of the fluid phase

η_s = efficiency of the solid phase

θ = fraction of surface covered;
or dimensionless time $d_p t/L^2$ in Chapter 21;
or volume-change group defined by Eq. 4.57;
or as defined in Table 9.1;
or parameter defined by Eq. 11.71;
or catalyst distribution parameter in Chapter 14;
or dimensionless residence time

$\theta_A, \theta_B, \ldots$ = fractional surface coverage by A, B, . . .

θ_v = fraction of vacant sites

λ = mean free path, cm;
or parameter defined by Eq. 11.40;
or decay parameter defined in Eq. 17.13;
or activation-cum-growth parameter defined by Eq. 21.10

μ = viscosity, g/cm sec;
or parameter defined in Eq. 4.89;
or substrate modulus in Chapter 18

μ_g = viscosity of gas, g/cm sec

μ_n = nth central moment

μ'_n = nth absolute moment

μ_r = contact efficiency

μ'_1 = first absolute moment of the chromatographic peak or the adsorbable component, sec

μ_2 = second central moment of the chromatographic curve of the adsorbable components, sec^2

σ = Stefan–Boltzmann constant

σ' = ratio of the fluid concentrations defined in Eq. 19.44

σ^2 = experimental error variance

τ = tortuosity factor;
or space-time in Eq. 16.111, sec;
or dimensionless temperature T/T_0

τ_f = time parameter $WC_{A0}/F\rho_c$ defined in Eq. 9.26, sec

τ_w = dimensionless temperature at wall, T_w/T_0

ζ = ratio of heat capacities of reactant and control fluid streams given by Eq. 12.28

ρ = general notation for density, g/cm^3

ρ_A = density of solid A in Chapters 19 and 20, g/cm^3

ρ_B = bulk density, g/cm^3;
or density of solid B in Chapters 19 and 20

ρ_c = catalyst density, g/cm^3

ρ_F = density of any feed, g/cm^3

ρ_G = density of grain, g/cm^3

ρ_g = density of gas, g/cm^3

ρ_l = density of liquid, g/cm^3

ρ_{Mg} = molal density of gas, g/cm^3

ρ_s = density of any solid, g/cm^3

$\tau'_1, \tau'_2, \ldots, \tau'_m$ = reciprocal of absolute temperatures, $T_1, T_2, \ldots, T_m$, $^\circ K^{-1}$

ϕ = general term used for Thiele modulus;
or ratio of orifice area to the distributor area in Chapter 14

ϕ_b = Thiele modulus based on a base temperature T_b

$\phi_{i-1,j}$ = feed concentration in the ith, jth stage

ϕ_{ma}, ϕ_{mi} = general notation for Thiele modulus for the macropore and micropore, respectively

ϕ_n = shape-generalized Thiele modulus

ϕ_s = order-generalized Thiele modulus given by Eq. 4.21;
or Thiele modulus based on surface diffusivity

$\phi_{Sn,ma}$ = Thiele modulus for the macropore for an nth-order reaction in a pellet of general shape S

$\phi_{Sn,mi}$ = Thiele modulus for the micropore for an nth-order reaction in a pellet of general shape S

$\check{\phi}$ = size distribution function of solids

$\check{\phi}_0, \check{\phi}_1, \check{\phi}_2$ = size distribution function of solids in the inlet, outlet, and elutriation streams

$(\phi)_{A_f}$ = modified Thiele modulus based on preexponential factor

$(\phi)_m$ = modified Thiele modulus based on the modified rate constant k_m;
or defined by Eqs. 4.92 and 4.96; or any modified ϕ

$(\phi)_{mr}$ = modified Thiele modulus based on reaction rate per gross volume of catalyst

$(\phi)_r$ = Thiele modulus based on observed reaction rate instead of rate constant

$(\phi_S)_m$ = modified Thiele modulus based on surface diffusivity

(ϕ_{Sn}) = general term used for Thiele modulus; S represents the geometry and may be s for a sphere, c for a cylinder, or p for a flat plate; n represents the order of reaction; any modification to this definition appears outside the parentheses.

$(\phi_{Sn})_A, (\phi_{Sn})_B$ = Thiele modulus for geometry S and order n for species A, B

$(\phi_{Sn})_f$ = modified Thiele modulus for geometry S and nth-order reaction for fouling

$(\phi_{Sn})_m$ = modified Thiele modulus for shape S and order n

$(\phi_{Sn})_1, (\phi_{Sn})_2$ = Thiele modulus for geometry S and order n for reaction step 1, 2

$(\phi_s)_m$ = modified Thiele modulus defined in Eqs. 4.55 and 4.56 for a sphere

χ^2 = defined by Eq. 2.22 for the χ^2 test

ψ = parameter defined in Eqs. 4.89 and 4.90;
or dimensionless center-plane concentration (Eq. 8.12);
or functions defined in Eqs. 11.38 and 11.39;
or cumulative gas concentration as defined by Eq. 19.39;
or structural parameter defined by Eq. 19.114

ψ_I, ψ_{II} = structural parameters defined by Eqs. 19.126a, 19.126b

Ω = activity parameter

$\overline{\Omega}$ = average activity

Ω_0, Ω_f = initial and final activity

$\Omega(r)$ = rate constant density function as defined in Section 5.4.1

ω = adsorption parameter $K_A C_{AB}$;
or dimensionless radial distance r/R

ω' = dimensionless radial distance r/d_p

ω'_0 = dimensionless radial distance r_0/d_p

v = kinematic viscosity of fluid, cm^2/sec;
or ratio of the stoichiometric coefficients of the reactants, v_B/v_A

$v_A, v_B, \ldots$ = stoichiometric coefficient for species A, B, $\ldots$

HETEROGENEOUS REACTIONS:
ANALYSIS, EXAMPLES, AND REACTOR DESIGN
Volume 1

CHAPTER ONE

Introduction

This book is concerned with a variety of heterogeneous reaction systems. Volume 1 deals with heterogeneous reactions in which a solid phase is present either as a catalyst or as a reactant. The reactions considered may be broadly classified as fluid–solid catalytic or noncatalytic systems. The fluid involved is a gas except in the case of biochemical reactions, where it is a liquid.

Three principal classes of reactions are considered in this volume:

gas–solid catalytic;
gas–solid noncatalytic;
solid–solid.

The chief difference between a polyphase heterogeneous system such as those mentioned above and a homogeneous system is that transport of heat and/or mass across a surface is involved in the former. Thus any analysis of such a system requires simultaneous consideration of the reaction and transport rate processes.

A number of examples can be cited to illustrate the importance of the three classes of heterogeneous reactions considered in this volume. Gas–solid catalytic reactions occupy a prominent position in the chemical industry as a whole, including the refining, petrochemical, heavy organic, and inorganic industries. In a petrochemical complex in India with a total investment of over \$400 million, for example, more than 30% of the processes are based on the use of solid catalysts. These reactions produce chemicals that are used in the polymer, pesticide, drug, heavy organic, perfume, detergent, and other industries. The principal categories of reactions involved are oxidation, hydrogenation, isomerization, dehydrogenation, dehydration, cyclization, and cracking. Some industrially important gas–solid catalytic reactions are presented in Table 1.1. Many more reactions could be cited, but the purpose of the table is to show that this class of reactions covers a wide spectrum of the chemical industry. The conventional catalysts used in some major petrochemical processes are being (or will soon be) replaced by zeolite catalysts due to the shape selective behavior of the latter.

Gas–solid noncatalytic reactions are equally important. Reactions involved in metallurgical operations, pollution abatement, and gasification of coal belong in this category. The importance of these reactions in the context of the increasing emphasis on pollution control and of the need to develop fuels from coal and to design metallurgical units with high efficiency cannot be overemphasized. Gas–solid noncatalytic reactions involve either a solid reactant alone or solid and gaseous reactants. The product can be a gas or a solid, or a combination of the two. Thus here also an analysis that takes into account the simultaneous influence of reaction and heat and/or mass transport is necessary. A list of reactions belonging in this category is presented in Table 1.2. These reactions are classified according to the physical state (solid or gas) of the reactants and products.

Another class of reactions, to which chemical engineers have made surprisingly little contribution, is the solid–solid system. Design of reactors for reactions between two solid phases has always been on empirical lines, and very little information is available on the analysis of such reactions. Table 1.3 lists various industrial applications of solid–solid reactions.

An attempt is made in this book to analyze these three classes of reactions and to present methods for the design of reactors for carrying out these reactions on an industrial scale. We have resisted the temptation to be overly theoretical, and have attempted to present a balanced synthesis of theory, design, and examples. Considerable emphasis is placed on catalytic reactions, not only because of the intrinsic importance of such systems but also because of the vast volume of literature that has accumulated over the last 10–15 years. The principles of solid-catalyzed gas-phase reactions are outlined, with emphasis on the modeling of such reactions, and the role of mass and/or heat transport is discussed. The question of reactor design is then taken up in considerable detail. Different classes of reactors

TABLE 1.1 Some Examples of Industrial Catalytic Reactions

Reaction	Catalyst	Process conditions	Remarks
Desulfurization of petroleum feedstock Organic sulfur compound $+ H_2$ $\rightleftharpoons$ organic compound $+ H_2S$	Cobalt molybdate supported on alumina and zinc oxide	Temperature: 340–400°C Pressure: 20–100 atm	Exothermic reaction; heat produced depends partly on the sulfur type and partly on the actual compound; fixed-bed adiabatic reactor
Typical reaction: $C_4H_4S + 3H_2 \rightleftharpoons C_4H_8 + H_2S$ Catalytic cracking Hydrodewaxing Fischer–Tropsch synthesis $CO + H_2 \rightarrow$ hydrocarbons	Zeolites	Temperature: 450–550°C Pressure: 1 atm Temperature: 450–550°C Pressure: 20–50 atm Temperature: 300–400°C Pressure: 10–50 atm	Endothermic reaction; moving bed reactor Exothermic reaction; fixed-bed reactor Highly exothermic reaction; both fixed- and fluidized-bed reactors
Steam reforming of methane or naphtha $CH_4 + H_2O \rightleftharpoons CO + 3H_2$ $C_nH_{2n+2} + nH_2O \rightarrow nCO + (2n+1)H_2$ $CO + H_2O \rightleftharpoons CO_2 + H_2$	Nickel supported on alumina or magnesia with promoters (such as CaO, SiO_2, K_2O)	Temperature: up to 850°C for primary reformer and up to 1000°C for secondary reformer Pressure: up to 30 atm	Endothermic reaction; fixed-bed reactor
Water–gas shift (CO conversion) reaction $CO + H_2O \rightleftharpoons CO_2 + H_2$	Mixtures of Fe_3O_4 and Cr_2O_3 (high-temperature shift reaction); mixtures of CuO, ZnO, and Al_2O_3 (low-temperature shift reaction)	Temperature: 350–450°C (high-temperature shift catalyst); 200–250°C (low-temperature shift catalyst) Pressure: 1–30 atm	Exothermic reversible reaction; extent of maximum conversion limited by equilibrium governed by thermodynamic considerations; fixed-bed reactor with arrangement for cooling between reaction zones
Ammonia synthesis $N_2 + 3H_2 \rightleftharpoons 2NH_3$	Iron with promoters (such as K_2O, CaO, MgO, Al_2O_3, SiO_2, TiO_2, ZrO_2, V_2O_5)	Temperature: 300–450°C Pressure: 100–300 atm	Exothermic reaction; fixed-bed reactor with internal heat exchange by means of cooling tubes embedded in the catalyst
Methanol from synthesis gas $CO + 2H_2 \rightleftharpoons CH_3OH$	Zinc–chromium oxide with or without copper	Temperature: 320–370°C Pressure: 100 atm	Exothermic reversible reaction; fixed-bed reactor with arrangement for heat removal
Oxidation of sulfur dioxide $SO_2 + \frac{1}{2}O_2 \rightarrow SO_3$	Supported platinum or supported V_2O_5 with alkali metal promoters	Temperature: 400–450°C Pressure: 1 atm	Exothermic reaction; fixed-bed reactor with internal heat exchange by means of cooling tubes embedded in the catalyst or adiabatic reactors with external coolers between stages

Reaction	Catalyst	Conditions	Remarks
Oxidation of ammonia $4NH_3 + 5O_2 \rightarrow 4NO + 6H_2O$	Pt–Rh wire gauze	Temperature: 850–900°C Pressure: 1–8 atm	Exothermic reaction; reactor packed with layers of Pt–Rh wire gauze
Isomerization of xylenes	Pt–mordenite–Al$_2$O$_3$; ZSM-5; zeolite	Temperature: 350–550°C Pressure: 1–10 atm	Overall reaction slightly exothermic; adiabatic fixed-bed reactor
Isomerization of aliphatic hydrocarbons	Pt–SiO$_2$–Al$_2$O$_3$ (containing chlorine or fluorine)	Temperature: 450–550°C Pressure: 30–40 atm	Overall reaction endothermic; adiabatic fixed-bed reactor
Isobutene from n-butene $CH_3CH = CHCH_3 \rightleftharpoons (CH_3)_2C = CH_2$	Fluorinated Al$_2$O$_3$	Temperature: 400–450°C Pressure: 1 atm (in presence of N$_2$ or H$_2$)	Heat changes involved are negligible; fixed-bed adiabatic reactor
Catalytic reforming of naphtha	Pt–Re(Ir)–Al$_2$O$_3$ (containing chlorine); ZSM-5, zeolite	Temperature: 450–550°C Pressure: 1–10 atm	Overall reaction endothermic; adiabatic fixed-bed reactor
Reduction of nitrobenzene to aniline and nitrotoluenes to toluidines $C_6H_5NO_2 + 3H_2 \rightarrow C_6H_5NH_2 + 2H_2O$ $C_6H_4(CH_3)NO_2 + 3H_2 \rightarrow C_6H_4(CH_3)NH_2 + 2H_2O$	CuCrO$_3$	Temperature: 180–250°C Pressure: 1 atm	Exothermic reaction; fixed-bed multitubular reactor
Butadiene from n-butenes $C_4H_8 \rightleftharpoons C_4H_6 + H_2$	Ca–Ni phosphate stabilized with Cr$_2$O$_3$	Temperature: 620–700°C Steam/butene: 8–20/1 Pressure: 60–80 mm Hg Space velocity: 300–500	Endothermic reaction; adiabatic fixed-bed reactor
Butadiene from butane $C_4H_{10} \rightleftharpoons C_4H_6 + 2H_2$	Cr$_2$–Al$_2$O$_3$	Temperature: above 600°C Pressure: about 0.2 atm	Endothermic reaction; aidabatic fixed-bed reactor
Butadiene from ethanol 1. Two-step process: (i) $C_2H_5OH \rightarrow CH_3CHO + H_2$ (ii) $C_2H_5OH + CH_3CHO \rightarrow C_4H_6 + 2H_2O$	Supported Cu activated by Cr$_2$O$_3$ TaO on silica gel, Al$_2$O$_3$, etc.	Temperature: 250°C Pressure: 1 atm Temperature: 350°C Pressure: 1 atm	Endothermic reaction; adiabatic fixed-bed reactor
2. One-step process: $2C_2H_5OH \rightarrow C_4H_6 + H_2 + 2H_2O$	ZnO–Al$_2$O$_3$ (40:60)	Temperature: 425°C Pressure: 1 atm	Endothermic reaction; fluidized-bed reactor (adiabatic fixed-bed reactor can also be used)

3

(**continued**)

TABLE 1.1 (continued)

Reaction	Catalyst	Process conditions	Remarks
Isoprene from isopentane or 2-methyl butenes $CH_3CH(CH_3)\text{—}CH_2CH_3 \rightarrow$ $CH_2 = C(CH_3)\text{—}CH = CH_2 + 2H_2$ $CH_2 = C(CH_3)\text{—}CH_2\text{—}CH_3 \rightarrow CH_2$ $= C(CH_3)\text{—}CH = CH_2 + H_2$	$Cr_2O_3\text{–}Al_2O_3$	Temperature: above 600°C Pressure: 0.2–0.3 atm	Endothermic reaction; adiabatic fixed-bed reactor
Ethyl benzene by alkylation of benzene $C_6H_6 + C_2H_4 \rightarrow C_6H_5C_2H_5$	ZSM-5 zeolite	Temperature: 350–450°C Pressure: 10–30 atm	Highly exothermic reaction; fixed-bed reactor with quench
Dehydrogenation of ethyl benzene to styrene $C_6H_5CH_2CH_3 \rightarrow C_6H_5CHCH_2 + H_2$	Fe_2O_3 promoted with Al_2O_3 and potassium salts	Temperature: 580–610°C (isothermal reactor) 580–650°C (adiabatic reactor) in presence of steam Pressure: 1 atm	Endothermic reaction; adiabatic or isothermal (shell-and-tube heat exchanger type), reactors
Methyl ethyl ketone from sec-butyl alcohol (by dehydrogenation) $CH_3CH_2CH(OH)CH_3 \rightarrow CH_3CH_2COCH_3 + H_2$	ZnO or brass	Temperature: 400–500°C Pressure: 1 atm	Endothermic reaction; fixed-bed adiabatic reactor
MVP (methyl vinyl pyridine) from MEP (methyl ethyl pyridine) by dehydrogenation	Fe_2O_3 promoted with Cr_2O_3 and KOH	Temperature: 500–800°C in presence of N_2 and steam	Endothermic reaction; adiabatic or isothermal fixed-bed reactors
Acetone from isopropanol 1. By partial oxidation: $CH_3CH(OH)CH_3 + \frac{1}{2}O_2 \rightarrow CH_3COCH_3 + H_2O$	Ag or Cu	Temperature: 450–550°C Pressure: 1 atm	Exothermic reaction; fixed-bed multitubular reactor
2. By dehydrogenation: $CH_3CH(OH)CH_3 \rightarrow CH_3COCH_3 + H_2$	ZnO or Cu and its alloys	Temperature: 450–550°C Pressure: 1 atm	Endothermic reaction; fixed-bed multitubular reactor
Formaldehyde from methanol $CH_3OH + \frac{1}{2}O_2 \rightarrow HCHO + H_2O$	Ag or Cu gauze	Temperature: 450–650°C Pressure: 1 atm Up to 50 mol % methanol in the feed	Overall reaction exothermic; fixed-bed adiabatic reactor

Reaction	Catalyst	Conditions	Reactor
Hydrodealkylation of toluene $C_6H_5CH_3 + H_2 \rightarrow C_6H_6 + CH_4$	Acidic solids (e.g., fluorinated Al_2O_3 or SiO_2–Al_2O_3)	Temperature: above 500°C Pressure: 10–50 atm	Endothermic reaction; adiabatic fixed-bed reactor
Disproportionation of toluene $2C_6H_5CH_3 \rightarrow C_6H_4(CH_3)_2 + C_6H_6$	Pt–SiO_2–Al_2O_3; mordenite; ZSM-5 zeolite	Temperature: 300–500°C Pressure: 20–40 atm in presence of H_2	Single adiabatic fixed-bed reactor
Ethanol from ethylene $CH_2 = CH_2 + H_2O \rightarrow CH_3CH_2OH$	Phosphoric acid on an inert support	Temperature: 300°C Pressure: 66 atm Water/C_2H_4: 0.6/1.0	Exothermic reaction; fixed-bed reactor
Dehydration of ethanol $C_2H_5OH \rightarrow C_2H_4 + H_2O$	Al_2O_3	Temperature: 275–350°C Pressure: 1 atm	Endothermic reaction; adiabatic fixed-bed reactor
Dimethylaniline from aniline and methanol $C_6H_5NH_2 + 2CH_3OH \rightarrow C_6H_5N(CH_3)_2 + 2H_2O$	Al_2O_3 (acidic) or bauxite	Temperature: 250–300°C Pressure: 1 atm	Endothermic reaction; fixed-bed multitubular reactor
Monoethylaniline from aniline and ethanol $C_6H_5NH_2 + C_2H_5OH \rightarrow C_6H_5NH(C_2H_5) + H_2O$	Treated bauxite	Temperature: 270°C Pressure: 1 atm	Fluidized-bed reactor
Diphenylamine from aniline $2C_6H_5NH_2 \rightarrow C_6H_5NHC_6H_5 + NH_3$	Al_2O_3 or Al_2O_3 containing ammonium fluoborate	Temperature: 450–550°C Pressure: 6–8 atm	Endothermic reaction; fixed-bed reactor
Methylamines from methanol and ammonia $CH_3OH + NH_3 \rightarrow CH_3NH_2 + H_2O$	Al_2O_3 or Al silicate	Temperature: 450°C Pressure: 1 atm	Endothermic reaction; adiabatic fixed-bed reactor
Vinyl chloride from acetylene $C_2H_2 + HCl \rightarrow CH_2CHCl$	$HgCl_2$ on activated charcoal (with promoters, e.g., KCl or heavy metal salts)	Temperature: 80–250°C Pressure: 1 atm	Exothermic reaction; multitubular reactor
Chloromethanes from methane mainly $CCl_4 + CHCl_3$	Active carbon	Temperature: 250–350°C Pressure: 1 atm	Fluidized-bed reactor
Vinyl acetate from acetylene and acetic acid $C_2H_2 + CH_3COOH \rightarrow CH_3COOCH = CH_2$	Zn acetate or Cd salts supported on activated carbon or SiO_2 or Al_2O_3	Temperature: 180–210°C Pressure: 1 atm Acetylene/acetic acid: 4–5/1	Exothermic reaction; fixed-bed multitubular reactor
Methyl pyrole from furane and methylamine $C_4H_4O + CH_3NH_2 \rightarrow C_4H_4NCH_3 + H_2O$	Al_2O_3	Temperature: 400°C	

(continued)

5

TABLE 1.1 (continued)

Reaction	Catalyst	Process conditions	Remarks
Phosgene from carbon monoxide $CO + Cl_2 \rightarrow COCl_2$	Activated charcoal	Temperature: below 300°C Pressure: 1 atm	Exothermic reaction; fixed-bed multitubular reactor
Dimethyl sulfide from methanol and hydrogen sulfide $2CH_3OH + H_2S \rightarrow CH_3SCH_3 + 2H_2O$	γ-Al_2O_3 (containing K tungstate) or thorium on pumice	Temperature: above 300°C Pressure: 1 atm	Exothermic reaction
Dichloroethane by oxichlorination of ethylene $C_2H_4 + 2HCl + \frac{1}{2}O_2 \rightarrow ClCH_2CH_2Cl + H_2O$	Chloride of Cu, K, or La on a support		
Oxidation of ethylene to ethylene oxide $C_2H_4 + \frac{1}{2}O_2 \rightarrow C_2H_4O$	Supported Ag with promoters	Temperature: 250–300°C Pressure: 10–30 atm Contact time: 1–4 sec	Highly exothermic reaction; fixed-bed multitubular reactor
Oxidation of benzene to maleic anhydride $C_6H_6 + \frac{9}{2}O_2 \rightarrow C_2H_2(CO)_2O + 2CO_2 + 2H_2O$	Supported V molybdate with promoters	Temperature: 300–400°C Pressure: 1 atm Contact time: 0.01–5 sec	Highly exothermic reaction; fixed-bed multitubular reactor
Oxidation of toluene to benzaldehyde $C_6H_5CH_3 + O_2 \rightarrow C_6H_5CHO + H_2O$	Oxides of U and Mo supported on inert carrier	Temperature: 500°C Pressure: 1 atm	Exothermic reaction; fixed-bed multitubular reactor
Phthalic anhydride from o-xylene $C_6H_4(CH_3)_2 + 3O_2 \rightarrow C_6H_4(CO)_2O + 3H_2O$	V_2O_5 on inert support with promoters (such as K_2SO_4)	Temperature: 350–360°C Pressure: 1 atm Contät time: 4–5 sec	Highly exothermic reaction; fixed-bed multitubular reactor
Phthalic anhydride from naphthalene $C_{10}H_8 + \frac{9}{2}O_2 \rightarrow C_6H_4(CO)_2O + 2H_2O + 2CO_2$	Supported V_2O_5 with promoters	Temperature: 300–400°C Pressure: 1 atm	Highly exothermic reaction; fixed-bed multitubular or fluid-bed reactors
Furane from butadiene $C_4H_6 + O_2 \rightarrow C_4H_4O + H_2O$	Supported oxides of Mo, V, Sb, Bi (with P or Pb)	Temperature: 400–600°C Pressure: 1 atm	Exothermic reaction; fixed-bed multitubular reactor
Acrylonitrile by ammoxidation of propylene	Multicomponent molybdates (containing Ni, Co, Fe, Bi, P, K and Mo) supported on SiO_2	Temperature: 400–480°C Pressure: 1 atm	Fluidized-bed reactor

TABLE 1.2 Industrially Important Examples of Different Types of Noncatalytic Gas–Solid Reactions[a]

Type	General Reaction Scheme	Reaction
A	Solid + fluid → solid + fluid	Hydrofluorination of uranium oxide
		$UO_2(s) + 4HF(g) \rightarrow UF_4(s) + 2H_2O(g)$
		Oxidation of silicon carbide to obtain silica
		$SiC(s) + 2O_2(g) \rightarrow SiO_2(s) + CO_2(g)$
		Roasting of zinc ore
		$2ZnS(s) + 3O_2(g) \rightarrow 2ZnO(s) + 2SO_2(g)$
		Calcination of pyrites
		$4FeS_2(s) + 11O_2(g) \rightarrow 2Fe_2O_3(s) + 8SO_2(g)$
		Reduction of ferrous chloride by hydrogen
		$FeCl_2(s) + H_2(g) \rightarrow Fe(s) + 2HCl(g)$
		Reduction of iron oxide by carbon monoxide
		$FeO(s) + CO(g) \rightarrow Fe(s) + CO_2(g)$
		Production of uranium tetrachloride by chlorination of UO_2
		$UO_2(s) + CCl_4(g) \rightarrow UCl_4(s) + CO_2(g)$
		Reduction of uranium trioxide by methanol vapors
		$UO_3(s) + CH_3OH(g) \rightarrow UO_2(s) + CO(g) + H_2(g) + H_2O(g)$
		Purification of H_2S containing gas by calcined dolomite
		$(CaO + MgO)(s) + H_2S(g) \rightarrow (CaS + MgO)(s) + H_2O(g)$
		Selective chlorination of iron in ilmenite
		$FeTiO_3(s) + CO(g) + Cl_2(g) \rightarrow FeCl_2(g) + CO_2(g) + TiO_2(g)$
B	Solid + fluid → solid	Nitrogenation of calcium carbide to produce cyanamide
		$CaC_2(s) + N_2(g) \rightarrow CaCN_2(s) + C(s)$
		Oxidation of silver by atmospheric oxygen
		$4Ag(s) + O_2(g) \rightarrow 2Ag_2O(s)$
		Rusting reaction of iron
		$2Fe(s) + O_2(g) \rightarrow 2FeO(s)$
		Absorption of SO_2 by pyrolusite
		$MnO_2(s) + SO_2(g) \rightarrow MnSO_4(s)$
		Absorption of SO_2 by dry limestone injection
		$CaO(s) + SO_2(g) \rightarrow CaSO_3(s)$

7

(continued)

TABLE 1.2 **(continued)**

Type	General Reaction Scheme	Reaction
		Formation of calcium nitride $$2Ca(s) + N_2(g) \rightarrow 2CaN(s)$$ Absorption of CO_2 in solid K_2CO_3 $$K_2CO_3(s) + H_2O(g) + CO_2(g) \rightarrow 2KHCO_3(s)$$
C	Solid → fluid + solid	Calcination of limestone $$CaCO_3(s) \rightarrow CaO(s) + CO_2(g)$$ Calcination of siderite $$FeCO_3(s) \rightarrow FeO(s) + CO_2(g)$$ Decomposition of magnesium hydroxide $$Mg(OH)_2(s) \rightarrow MgO(s) + H_2O(g)$$ Decomposition of beryllium hydroxide $$Be(OH)_2(s) \rightarrow BeO(s) + H_2O(g)$$ Recovery of atmospheric oxygen by silver $$2Ag_2O(s) \rightarrow 4Ag(s) + O_2(g)$$ Decomposition of ferric sulfate $$2Fe_2(SO_4)_3(s) \rightarrow 2Fe_2O_3(s) + 6SO_2(g) + 3O_2(g)$$
D	Solid + fluid → fluid	Production of carbon disulfide $$C(s) + S_2(g) \rightarrow CS_2(g)$$ Chlorination of alumina by chlorine $$2Al_2O_3(s) + 6Cl_2(g) \rightarrow 4AlCl_3(g) + 3O_2(g)$$ Chlorination of alumina by hydrochloric acid gas $$Al_2O_3(s) + 6HCl(g) \rightarrow 2AlCl_3(g) + 3H_2O(g)$$ Chlorination of rutile to titanium tetrachloride $$TiO_2(s) + 2C(s) + 2Cl_2(g) \rightarrow TiCl_4(g) + 2CO(g)$$ Total chlorination of ilmenite $$FeTiO_3(s) + 3Cl_2(g) \rightarrow FeCl_2(g) + TiCl_4(g)$$ Oxidation of uranium tetrafluoride by fluorine $$UF_4(s) + F_2(g) \rightarrow UF_6(g)$$ Gasification of carbon $$C(s) + H_2O(g) \rightarrow CO(g) + H_2(g)$$

TABLE 1.2 (continued)

Type	General Reaction Scheme	Reaction
		Combustion of carbon
		$C(s) + O_2(g) \rightarrow CO_2(g)$ or $CO(g)$
E	Solid $\rightarrow$ fluid	Thermal decomposition of ammonium nitrate
		$NH_4NO_3(s) \rightarrow N_2O(g) + 2H_2O(g)$
		Thermal decomposition of oxalic acid
		$H_2C_2O_4(s) \rightarrow H_2O(g) + CO_2(g) + CO(g)$
		Decomposition of ammonium chloride
		$NH_4Cl(s) \rightarrow NH_3(g) + HCl(g)$
		Decomposition of ammonium sulfate
		$(NH_4)_2SO_4(s) \rightarrow 2NH_3(g) + SO_3(g) + H_2O(g)$
F	Fluid $\rightarrow$ solid + fluid	Mond process for nickel production
		$Ni(CO)_4(g) \rightarrow Ni(s) + 4CO(g)$
		Decomposition of iron carbonyl to iron powder
		$Fe(CO)_5(g) \rightarrow Fe(s) + 5CO(g)$
		Oxidation of silicon tetrachloride vapors
		$SiCl_4(g) + O_2(g) \rightarrow SiO_2(s) + 2Cl_2(g)$
		Burning of titanium tetrachloride to rutile
		$TiCl_4(g) + O_2(g) \rightarrow TiO_2(s) + 2Cl_2(g)$
		Decomposition of tantalum iodide to tantalum
		$2TaI_5(g) \rightarrow 2Ta(s) + 5I_2(g)$

[a]General reaction: $v_A A + v_B B = v_R R + v_S S$.

TABLE 1.3. Examples of Industrial Applications of Solid–Solid Reactions

Field of Application	Nature of Application and/or Reaction
Ceramics	Manufacture of ceramic and refractory materials; also, small quantities of certain compounds, added to improve the properties, react in the solid state.
Oxide semiconductors and ferrites	Manufacturing (and doping)
Cement production	In the dry manufacturing process the components react in the solid state. Typically, in the $MO-SiO_2$ and $MO-Al_2O_3$ systems ($M = Ca, Mg$, etc.) a number of product phases are formed and proper conditions are to be chosen to get desired compositions.
Catalyst preparation	Mixed oxide catalysts, which are known to be very selective, are sometimes prepared in the solid state.
Metallurgy	a. Preparation of solid solutions and alloys b. Reduction of ores, e.g., iron ore (hematite) by coke
Polymer chemistry	a. Polymerization in the solid state b. Preparation of large-polymer single crystals
Fertilizers	The compounds react in the solid state during processing as well as when added to the soil.
Drugs	If the adjuvants present in the drugs react with the active ingredients, the stability of the drug is affected, leading to storage problems.

are examined, principally the fixed-bed and fluidized-bed reactors, followed by a few other classes of less important reactors. A special chapter is included in which certain new types of catalysts are discussed, namely, supported liquid-phase catalysts, polymer-bound catalysts, and immobilized-enzyme catalysts.

Analogous procedures are followed for gas–solid noncatalytic and solid–solid reactions. In the case of gas–solid noncatalytic reactions, in particular, heavy emphasis is placed on reaction modeling. This has been motivated, as in the case of catalytic reactions, by the distinct possibility of designing industrial reactors for such systems from first principles. Perhaps it will be a while before this possibility becomes a reality, but the signs are propitious, and an attempt to compile an analysis (not exclusively mathematical) of heterogeneous reactions involving one or more solid phases, with equal stress on design and examples, seems clearly necessary. We emphasize that in all three of the cases considered the analysis is interspersed with examples, both qualitative and quantitative, to illustrate the principles that have been explained.

CHAPTER TWO

Catalysis, Kinetics, and Reaction Modeling

2.1. INTRODUCTION

Catalytic reactions are characterized by adsorption of reactants on the catalyst surface followed by chemical transformation on the surface and subsequent desorption of the reaction products. A theoretical analysis of chemical reactions is possible only for the simplest type of homogeneous reactions through the activated complex theory, collision theory, or the absolute reaction rate theory of Eyring, which makes possible the calculation of the kinetic constants from fundamental considerations. These theories are inadequate for reactions of even slightly greater complexity, and much more so for catalytic reactions.

In heterogeneous reactions the direct use of partial pressures would not be tenable, and a knowledge of the surface concentrations would be necessary. Also, whereas in homogeneous reactions one can conceive of a fixed reaction order, in the case of heterogeneous reactions the order may be sensitive to variables such as temperature. In a bimolecular reaction, for example, at some given temperature only one of the reactants may be strongly adsorbed, so that the kinetics would be first order. At some other temperature the adsorption of the second component might conceivably increase to such an extent that the reaction might then be second order.

Adsorption is known to be an essential precursor to surface catalysis. Halsey (1963) has listed the following adsorption conditions for reaction to occur: equilibrium adsorption, steady-state nonequilibrium adsorption, and unstable nonequilibrium adsorption. In the case of equilibrium and steady-state nonequilibrium adsorption there is no buildup of product on the surface, whereas in the case of unstable nonequilibrium adsorption there is a steady buildup of one of the reactants on the solid surface until ultimately the surface is completely covered with that reactant and the reaction rate becomes zero. In the present treatment of catalysis we shall largely be concerned with equilib-

rium and steady-state nonequilibrium adsorption.

We shall first consider homogeneous power law formulations, and then describe certain fundamental concepts in surface catalysis which are of direct relevance in engineering modeling and which will subsequently be invoked to formulate appropriate Langmuir–Hinshelwood (Hougen–Watson) models in later sections. Three other aspects of catalysis will also be briefly considered, namely, determination of catalyst properties under the actual conditions of catalysis, correlation of activity with catalyst properties, and catalyst development and engineering. Although the first two are not of direct relevance to design in their present state of development, they hold promise of a more fundamental approach to catalyst engineering in the future. The availability of such techniques as electron spectroscopy for chemical analysis (ESCA), low-energy electron diffraction (LEED), X-ray fluorescence, and scanning electron microscopy (SEM) has opened up possibilities of a clearer understanding of surface phenomena and therefore of the fundamental nature of catalysis; but a discussion of these is outside the scope of the present treatment.

2.2. RATE OF A CATALYTIC REACTION

2.2.1. Basic Definitions

One of the simplest ways of formulating a catalytic rate equation is to express the reaction rate in terms of unit catalyst surface (r_S) and then by appropriate transformations to get an expression for the volume rate (r_v) or weight rate (r_w) that is amenable to easy measurement. Thus for any first-order reaction A → R, relationships between the different rates (and rate constants) can be readily written; these are summarized in Table 2.1.

It is often necessary (and possible), in the interest of simplicity in design calculations (see Chapters 11 and

TABLE 2.1   Various Definitions of the Reaction Rate and Rate Constant

Reaction Rate[a]	Rate Constant[b]
$r_S, \dfrac{\text{mol}}{\text{cm}^2 \text{ catalyst sec}}$	$k_S, \dfrac{\text{cm}}{\text{sec}}$
$r_w = S_g r_S, \dfrac{\text{mol}}{\text{g catalyst sec}}$	$k_w = S_g k_S, \dfrac{\text{cm}^3}{\text{g catalyst sec}}$
$r_v = \rho_c r_w, \dfrac{\text{mol}}{\text{cm}^3 \text{ catalyst sec}}$	$k_v = \rho_c S_g k_S, \text{ sec}^{-1}$
$r_V = (1 - f_B)\rho_B r_w, \dfrac{\text{mol}}{\text{cm}^3 \text{ reactor sec}}$	$k_V = (1 - f_B)\rho_B S_g k_S, \text{ sec}^{-1}$
$r'_w = M S_g r_S, \dfrac{\text{g}}{\text{g catalyst sec}}$	$k_{mw} = M S_g k_S, \dfrac{\text{g cm}^3}{\text{mol g catalyst sec}}$

[a] Throughout, S_g is the surface area per gram of catalyst, M is the molecular weight, ρ_c and ρ_B are the catalyst and bulk densities, respectively, and f_B is the bed voidage.
[b] For a first-order reaction.

12), to treat many industrial reactions as first-order reactions. For this purpose a knowledge of the relationship between the parameters of the first-order equation and the true rate parameters would be necessary. By considering the reaction rate to be composed of parallel steps of different orders, it can be shown (deTar and Day, 1966) that the effect of neglecting the higher-order steps is often quite small.

Since the development of simple power law models is a straightforward procedure similar in all respects to the methods used in developing rate equations for homogeneous reactions, we shall make only a brief mention of this here. Rigorous procedures for obtaining the rate constants have been extensively documented (see, e.g., Kittrell et al., 1966a,b; Ball and Groenweghe, 1966; Mezaki and Butt, 1968; Himmelblau et al., 1967; Wright 1964; Wright and Ball, 1967; Kazragis, 1966; Draper et al., 1969; Akyurtlu and Stewart, 1978; Sica et al., 1978; Ng and Vermeulen, 1977; Yates and Best, 1976; Pirard and Kalitventzeff, 1978; Sundaram and Froment, 1978; Forni and Terzoni, 1977; Chang et al., 1977; Shah et al., 1977; Hsieh and Atwood, 1976; Bauerie et al., 1978a, b; Tsuruya et al., 1979; Soga et al., 1979). A simple and accurate method is to obtain the rates from differential or integral data (as described in Chapter 9) and to determine the kinetic parameters from nonlinear least-

squares analysis of the equation

$$r_{vA} = A_f \exp\left(-\frac{E}{R_g T}\right) C_A \tag{2.1}$$

in which the rate constant has been expressed in the usual Arrhenius form,

$$k_v = A_f \exp\left(-\frac{E}{R_g T}\right) \tag{2.2}$$

Note that data at different temperatures can be used simultaneously in this procedure.

2.2.2. Reduction of Complex Systems

It is often useful to have a compact method of representing the composition of a reaction mixture. For a simple reaction this can be written directly from its stoichiometry if the conversion is known, but for a complex reaction it is possible to invoke the methods of matrix mathematics to relate the concentrations of the various components involved to those of a minimum number of key components (see Aris, 1961, 1969; Lapidus and Amundson, 1977). The main point here is that the number of independent reactions is given by the rank of the reaction matrix.

The determination of the kinetic parameters of a complex reaction network is considerably more complicated than for a simple reaction. The most accurate method is that of Wei and Prater (1962); other methods (mostly multiresponse statistical) have also been described in the references cited in Section 2.2.1. The Wei–Prater method is restricted to a network of reversible first-order steps. Another method is to determine experimentally the rates of formation (or disappearance) of the individual components and then to calculate the rate of each step by material balance. A useful example is Reddy and Doraiswamy's (1969) analysis of the oxidation of toluene, in which this method was used to choose among parallel, consecutive, and triangular networks. Choice between competing networks can also be made by the kinetic tracer method of Neiman (Boudart, 1968; Neiman and Gal, 1970).

A strategy for network analysis (with particular reference to the catalytic oxidation of *o*-xylene) has been reported by Boag et al. (1975). It has been shown that product distribution alone is not adequate, since it is not a unique function of conversion. This is borne out by the fact that mutual inhibition effects are known to be important (Lyubarskii and Petoyan, 1967); since runs are usually carried out at different initial concentrations, these inhibitory effects can be significant.

2.2.3. The Rate-Determining Step

The controlling mechanism in a catalytic reaction is governed by the *slowest* in the sequence of consecutive steps involved—adsorption, reaction, and desorption. It is instructive to note, however, that in a system of parallel steps it is the *fastest* step that is significant.

Let us consider the reaction

$$A + B \rightarrow R + S$$

If *l* represents an active site, the different steps involved are outlined in Figure 2.1. If further the third step is assumed to be the slowest, this assumption merely implies that in each of the other four steps the magnitudes of the forward and reverse rates are so large that, while the difference between them continues to be equal to the net rate r, the ratio of the forward and reverse rates tends to be unity. Only for the third step is the ratio $r_{+3}/r_{-3} > 1$, so that this step is sufficiently removed from equilibrium to be rate determining. If now a rate model is developed for step 3, this would be the model for the overall reaction.

Certain complications can arise, however. Perona

and Thodos (1957) and Chakrabarty et al. (1979) have observed, for instance, that a transition in the controlling step occurs in the catalytic dehydrogenation of *sec*-butyl alcohol to methyl ethyl ketone over brass. Choudhary and Doraiswamy (1975) have observed a similar transition in the isomerization of 1-butene. In both these reactions, two steps may be simultaneously controlling in the transition region.

A useful method of finding the rate-determining step is by the transient-response technique, and because of its potential usefulness it is considered in Section 2.8. Another (less useful) method is the so-called rate-tracing method employing an isotope (Happel and Hnatow, 1973; Happel, 1975).

2.2.4. Stoichiometric Number

The stoichiometric number s is a characteristic number of any single elementary step, such as adsorption of a reactant, surface reaction, and desorption of a product, and defines the number of times the step occurs during the overall reaction. Thus, in the reaction

$$A + 2B \rightarrow 2R$$

the adsorption of A occurs once during the overall reaction ($s = 1$), while the adsorption of B, surface reaction, and desorption of R occur twice each during the same period ($s = 2$).

Another way of defining the stoichiometric number is through the free energy change: ΔG for the overall reaction will be proportional to ΔG_r for the rate-determining step, and this factor of proportionality is the stoichiometric number. Thus, if the stoichiometric number of the overall reaction is determined experimentally, it would correspond to the stoichiometric number of the rate-determining step.

It has been shown (Manes et al., 1954) that a plot of the net forward rate r (in the vicinity of equilibrium) vs. $-\Delta G/R_g T$ gives a slope of r_+/s (or r_-/s). Thus, the stoichiometric number can be readily determined provided the net rate near equilibrium and the forward or reverse rate are known. As the line passes through the origin, only one determination of r and $-\Delta G/R_g T$ is adequate in order to estimate r_+/s. A more elaborate method using this equation has been developed by Horiuti (1957, 1962) employing a radioactive tracer (see also Happel, 1967).

Purely kinetic methods can also be used in which activation energies are determined for both the forward and reverse steps by making measurements under conditions removed from equilibrium and calculating s

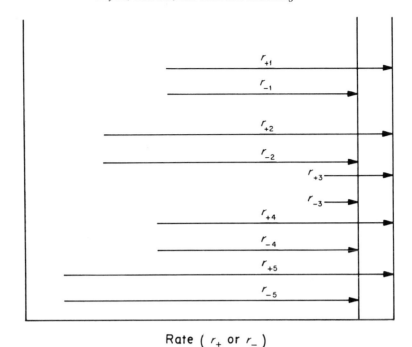

Steps

$$A + \mathit{l} \xrightarrow[r_{-1}]{r_{+1}} A\mathit{l}$$

$$B + \mathit{l} \xrightarrow[r_{-2}]{r_{+2}} B\mathit{l}$$

$$A\mathit{l} + B\mathit{l} \xrightarrow[r_{-3}]{r_{+3}} R\mathit{l} + S\mathit{l}$$

$$R\mathit{l} \xrightarrow[r_{-4}]{r_{+4}} R + \mathit{l}$$

$$S\mathit{l} \xrightarrow[r_{-5}]{r_{+5}} S + \mathit{l}$$

Figure 2.1. Schematic representation of the rate-determining step.

from $E_f - E_b = \Delta H / s$, where ΔH is the enthalpy change for the overall reaction (Boreskov, 1962).

Integral values of s have been reported for ammonia synthesis (Tanaka, 1966), sulfur dioxide oxidation (Kaneko and Odanko, 1965), the shift reaction (Kaneko and Oki, 1965), and dehydrogenation of butane (Happel et al., 1966).

2.3. SURFACE CONSIDERATION IN CATALYTIC KINETICS

2.3.1. Langmuir and Freundlich Isotherms

The simplest and most common method of analyzing a catalyst surface is to assume a homogeneous (ideal) surface with "active centers" of uniform activity. Recent evidence shows, however, that there might be a distribution of activity on the surface; this is referred to as *a priori* heterogeneity. Another assumption inherent in the ideal surface model is that the heat of adsorption is independent of the extent of surface coverage by adsorption. Based on these assumptions, Langmuir derived a relationship between the concentration of the reactant in the gas phase and its concentration on the catalyst surface. This isotherm, which has formed the basis of countless formulations in contact catalysis, is given by

$$C_{A\mathit{l}} = \frac{C_L K_A p_A}{1 + K_A p_A} \qquad (2.3)$$

where C_L is the concentration of active centers, C_{Al} is the concentration of adsorbed A, and K_A is the adsorption equilibrium constant $C_{Al}/p_A C_l$ (C_l being the concentration of vacant sites). It can be recast as

$$\frac{C_{Al}}{C_L} = \theta_A = \frac{K_A p_A}{1 + K_A p_A} \tag{2.4}$$

where θ_A is the fraction of active surface occupied by A. In terms of the quantity adsorbed (w), Eq. 2.4 becomes

$$w_A = \frac{w_{A0} K_A p_A}{1 + K_A p_A} \tag{2.5}$$

An equation similar to this can also be derived from statistical mechanical considerations (Fowler and Guggenheim, 1939).

These isotherms can be extended to any number of adsorbing components. Thus, if there are i components in an adsorbing system, the concentration of adsorbed A (one of the i components) is given by

$$\theta_A = \frac{C_{Al}}{C_L} = \frac{K_A p_A}{1 + \sum K_i p_i} \tag{2.6}$$

Equation 2.6 is strictly valid for an ideal surface only, and K_A can be expressed in the usual Arrhenius form.

Another way of expressing the fraction of occupied sites is through the Freundlich isotherm

$$\theta_A = a p_A^n \tag{2.7}$$

Many adsorption processes are known to follow this isotherm.

2.3.2. Nonideal Surfaces and Isotherms

The observed nonideal behavior of real surfaces can be explained on the basis of *a priori* heterogeneity of the surface, or interactions between the adsorbed species in a given neighborhood. Taylor and Lewis (1938), Taylor and Liang (1947), and Halsey and Taylor (1947) have demonstrated the existence of *a priori* heterogeneity.

The main consequences of nonideal behavior are that (a) the isosteric or differential heat of adsorption (at a given coverage) varies with the overall fractional coverage θ, and (b) the functional form of the isotherm relating θ to pressure and temperature changes. There is often a distribution of activity throughout the catalyst surface, and different processes can be rate controlling on different "patches" (Halsey, 1950). Each patch is characterized by a different but uniform activity.

Let us consider a surface consisting of j such independent patches. The total coverage on all the patches will be given by

$$\theta_A = \sum_j \frac{K_{jA} p_A}{1 + K_{jA} p_A} \tag{2.8}$$

Equation 2.8 is true only if the weight factors for all the patches are equal. Since this is not so in actual practice, we can include a weight factor f_j to obtain

$$\theta_A = \sum_j f_j \frac{K_{jA} p_A}{1 + K_{jA} p_A} \tag{2.9}$$

where f_j represents the fraction of total sites in the jth patch. Kolboe (1967, 1969) has used an equation analogous to 2.9 to identify five sets of active sites on a zinc oxide catalyst.

It is possible to define a continuous distribution function representing the abundance of sites with a heat of adsorption q. If $N(q)$ is the distribution function characterizing the heats, we have (Prasad and Doraiswamy, 1977)

$$\int_{q_1}^{q_2} N(q) \, dq = 1 \tag{2.10a}$$

$$\int_{q_1}^{q_2} \frac{N(q) dq}{1 + (e/p_A)^{\Delta S/R_g} e^{-q/R_g T}} = \theta \tag{2.10b}$$

where q_1 and q_2 represent the lower and upper limits of the heats. Also, the assumption is made that entropy change is independent of the heat of adsorption.

Often, however, an empirical trial isotherm of the type $\theta_A = f(p_A, T)$ is made to fit the experimental data. The corresponding distribution function is then found by solving the integral equation 2.10b. The problem is of the Fredholm type of the first kind for finite limits of the heat of adsorption, for which a solution exists only in special circumstances. Prasad and Doraiswamy (1977) have proposed criteria for the existence of solutions. The shapes of the distribution functions that have been reported, their analytical expressions, and the relations between q and coverage θ for various isotherms have been given by Prasad (1977).

Of the commonly used isotherms, the ideal Langmuir isotherm corresponds to a uniform surface with a sharp peak at q_0 (the initial heat of adsorption), whereas Temkin and Levich's isotherm corresponds to a constant distribution of sites in the range of heats $q_1 < q < q_2$. The Freundlich isotherm corresponds to an exponential distribution of active sites.

Perhaps the most reasonable model of the surface is that it consists of patches of distinctly different activities, with each patch characterized by a continuous distribution of activity. For this combination of the discrete and continuous types, a Stieltjes integral method may be used to get the resulting isotherm (Prasad and Doraiswamy, 1983). The following limiting isotherm valid for the high surface coverage region and which resembles the Fowler–Guggenheim isotherm has been derived:

$$\theta = \frac{1}{\left[1 + (e^{\Delta_s/R_g})(e^{-(A-B\theta)/R_g})/p_A\right]} \quad (2.11)$$

where A and B are constants. For small changes in pressure the variation of $B\theta$ is not significant, and Eq. 2.11 reduces to the familiar Langmuir isotherm. In other words, a heterogeneous surface can behave like a homogeneous one over narrow ranges of temperature and pressure. This accounts for the considerable success of the ideal Langmuir isotherm and the so-called paradox of heterogeneous kinetics.

There have also been a number of criticisms of the concept of *a priori* heterogeneity (Eucken, 1950). The same phenomenological results can be obtained on the basis of the interaction theory, as illustrated in the following references: Roberts (1935), Rideal and Herington (1944), Volkenstein (1948), and Miller (1949). Boudart (1952) has provided an alternative theory.

Prasad's (1977) table referred to earlier is important in that it gives physical explanation to otherwise empirical power law kinetics (e.g. the Freundlich isotherm) often encountered in practice. Although the Langmuir isotherm is the one most commonly used, based on which the so-called Langmuir–Hinshelwood models have been developed, power law models do not always lack mechanistic realism and can certainly be accepted wherever they are found to fit the data (see Weller, 1975). The statistical methods outlined in Section 2.6 can also be used to fit power law models for complex networks.

2.4. CATALYSIS UNDER REACTION CONDITIONS

In the Hougen–Watson models reaction rates are measured in appropriately planned kinetic experiments and the rate and adsorption constants are determined by statistical methods of parameter estimation (see Section 2.6). Rigorous procedures for obtaining the parameters have been extensively documented (Van Meerten and Coenen, 1977; Sica et al., 1978; Dumez et al., 1977; Dumez and Froment, 1976; Meier and Gut, 1978; Wisniak and Simon, 1979; Matson and Harriott, 1978; Farrell and Ziegler, 1979; Read et al., 1978). Kabel

and Johanson (1962) made some comparisons of the values of the adsorption parameters (in the dehydration of alcohol) determined by statistical methods with those obtained from adsorption studies (under nonreactive conditions), but no specific conclusions can be drawn from their results.

The results of Raghavan and Doraiswamy (1977) on the isomerization of *n*-butene (see Section 2.6.4) show that the values obtained from parameter estimation are essentially correct. Therefore, although a far better understanding of the elementary steps making up the overall reaction can be had by carrying out adsorption studies in the working state of the catalyst, modeling by parameter estimation can be used for reactor design purposes (provided the correct statistical procedures are used, as outlined in Section 2.6).

Tamaru (1964) has classified the experimental techniques used to study the catalyst properties in its working state into three categories: (1) volumetric methods, (2) gravimetric methods, and (3) gas chromatographic methods. The last method, which appears to be the most useful, is based on the relation

$$L/t_m u_i = 1/K_A \quad (2.11)$$

where L is the length of the packed column in centimeters, t_m is the retention time of the pulse maximum in seconds, u_i is the interstitial linear gas velocity in centimeters per second, and K_A is the adsorption equilibrium constant. Under special experimental conditions, such as high linear gas velocity and negligible dead volume in the system, the longitudinal diffusion and convectional effects can be neglected and the broadening of the pulse attributed entirely to adsorption effects. This equation can then be used for evaluating the adsorption equilibrium constant.

The gas chromatographic method is a very versatile one and has been used for determining—besides the equilibrium constant—many properties of the catalyst in its working state: surface area (Eberly and Spencer, 1961; Tamaru, 1959; Nakanishi and Tamaru, 1963); total number of active sites (Kokes et al., 1955); surface acidity and acid strength distribution (Misono et al., 1965); and transport properties (see Chapter 10).

The volumetric method has been the one most widely used so far. However, the chromatographic method, which is of more recent origin, is certainly the preferred one. Of the chromatographic techniques, that based on the analysis of moments (Schneider and Smith, 1968a, b) is particularly attractive and is considered in detail in Chapter 9 (see also Ramachandran and Smith, 1978).

2.5. LANGMUIR–HINSHELWOOD (HOUGEN–WATSON) MODELS AND THEIR APPRAISAL

Having outlined in the previous sections the important basic aspects of catalysis, we shall now consider the question of reaction modeling and the statistical methods of parameter estimation and model discrimination. These constitute the necessary first steps in any program of reactor design and modeling.

2.5.1. The Ideal Models

Let us consider the reaction

$$A \rightarrow B$$

The various consecutive steps involved in the catalytic reaction are

1. Adsorption of A: $A + l \underset{}{\overset{k_1}{\rightleftharpoons}} Al$

2. Surface reaction: $Al \overset{k_2}{\rightarrow} Bl$

3. Desorption of B: $Bl \overset{k_3}{\rightleftharpoons} B + l$

On the assumption that the surface reaction is controlling, the following typical Hougen–Watson (H–W) model can be derived:

$$r = \frac{k_2 K_A p_A}{1 + K_A p_A + K_B p_B} \qquad (2.12)$$

It can be seen that a large number of H–W models are possible, particularly for a reaction of the type $A + B \rightarrow R + S$, depending on the number of species adsorbed; and this number increases even further when provision has to be made for the dissociation of an adsorbed species, as in the case of hydrogen on nickel. In effect, the dissociation of any compound A leads to the term $\sqrt{K_A p_A}$ in the denominator instead of $K_A p_A$.

Another important situation arises when reaction occurs between the chemisorbed and van der Waals layers, since for all practical purposes the molecules in the van der Waals layer behave as though they are in the gas phase. In this model (usually referred to as the Rideal–Eley model) the term for the species that is not chemisorbed vanishes from the denominator of the rate equation.

A particularly interesting extension of the Rideal–Eley model is revealed when it is applied to a reaction in which two molecules of a given species are involved. In the dehydration of alcohol to ether, for instance, one of the theories suggests reaction between a chemisorbed molecule of alcohol and one in the van der Waals layer. Another example is the disproportionation of propylene (Begley and Wilson, 1967).

Yang and Hougen (1950) have summarized the principal postulations for the common reaction types by writing the rate model in the general form

$$\text{rate} = \frac{(\text{kinetic term})(\text{potential term})}{(\text{adsorption term})^n}$$

From the tables given by them, equations can be written for each of the terms. It may be noted that n represents the number of active centers participating in the reaction. Although this procedure is helpful in writing an H–W model for certain specific situations, it seems necessary to indicate a method by which all possible models for a reaction can be written. This is demonstrated below for a typical reaction.

2.5.2. Example: Development of Rate Equations for the Oxidation of Hydrocarbons

Consider the oxidation of benzene on a noble metal

$$B + O_2 \rightarrow P$$

where B is the hydrocarbon and P the product. It is known that in reactions of this type the product is usually not adsorbed, so that desorption of P need not be considered; this has the effect of reducing the number of possible mechanisms. Further simplification is introduced by neglecting the reverse reaction, which is fully justified for the present reaction. When oxidation occurs over a noble metal catalyst, the following adsorbed species may exist: benzene (B), dissociated hydrocarbon (D), molecular oxygen (O_2), and atomic oxygen (O). Conceivably, any possible combination of these adsorbed species can exist on the surface, and for each of these combinations several controlling mechanisms are possible. Barnard and Mitchell (1968) have suggested a systematic method of tabulating the various combinations.

First, a table is prepared in which the basic rate-controlling steps are written; thus in Table 2.2, 17 mechanisms are identified. In a second table, for each combination of adsorbed species, referred to as the adsorption system number, the various possible con-

TABLE 2.2. Possible Rate-Controlling Steps with Corresponding Rate Equations in Hydrocarbon Oxidation, $B + O_2 \rightarrow P$

Rate-Controlling Step[a]	Rate Equations[b]
Adsorption controlling:	
1. Adsorption of O_2	$r = k\theta_v p_{O_2}$
2. Immobile chemisorption of $O_2 \rightarrow 2O$	$r = k\left[\dfrac{Z}{Z - \theta_O}\right]\theta_v^2 p_{O_2}$
3. Dissociation of $O_2 \rightarrow 2O$	$r = k\theta_v^2 p_{O_2}$
4. Adsorption of hydrocarbon (B)	$r = k\theta_v p_B$
5. Adsorption of B (dual-site model)	$r = k\theta_v^2 p_B$
6. Dissociation of $B \rightarrow D^* + H^*$	$r = k\theta_v^2 p_B$
Surface reaction controlling:	
7. Reaction $B + O_2^*$	$r = k\theta_{O_2} p_B$
8. Reaction $B + O^*$	$r = k\theta_O p_B$
9. Reaction $B + 2O^*$	$r = k\theta_O^2 p_B$
10. Reaction $B^* + O_2$	$r = k\theta_B p_{O_2}$
11. Reaction $B^* + O_2^*$	$r = k\theta_B \theta_{O_2}$
12. Reaction $B^* + O^*$	$r = k\theta_B \theta_O$
13. Reaction $B^* + 2O^*$	$r = k\theta_B \theta_O^2$
14. Reaction $D^* + O_2^*$	$r = k\theta_D \theta_{O_2}$
15. Reaction $D^* + O^*$	$r = k\theta_D \theta_O$
16. Reaction $D^* + 2O^*$	$r = k\theta_D \theta_O^2$
17. Reaction $D^* + O_2$	$r = k\theta_D p_{O_2}$

[a]Adsorbed species: molecular oxygen, O_2^*; atomic oxygen, O^*; fuel, B^*; dissociated fuel, D^*.

[b]Coverages: θ_v, fraction of surface vacant; θ_{O_2}, fraction of surface covered by molecular oxygen; θ_O, fraction of surface covered by atomic oxygen; θ_B, fraction of surface covered by fuel; θ_D, fraction of surface covered by dissociated fuel; Z, number of nearest neighbors for a site. Note that the general notation r (without specification of units) has been used for the rate.

trolling mechanisms (as outlined in the first table) are listed; thus in Table 2.3 a total of 103 H–W models are postulated. Clearly, without a systematic tabulation in the manner just explained, some models can easily be missed. This can lead to a situation where no single model emerges as a satisfactory choice even after the various statistical procedures described later in this chapter are used.

2.5.3. Relaxation of Assumptions

It is possible to relax at a time any one (but not all) of the assumptions inherent in the ideal surface models and derive modified H–W models. Thus the following modified models may be envisaged:

1. More than one step is controlling (ideal: only one rate-controlling step).

2. More than one set of active centers is involved (ideal: only one set of active centers, i.e., all centers are equally active).

3. The enthalpy of adsorption decreases with surface coverage (ideal: the enthalpy of adsorption is independent of surface coverage).

Examples of reactions that have been modeled by relaxation of assumptions as outlined here are shown in Table 2.4.

2.5.4. An Appraisal of the Models

Though widely used, these models have been the subject of numerous criticisms. A critical appraisal of the models follows.

1. A maximum in the rate is sometimes observed at a certain temperature. Hougen–Watson models can explain such an observation, whereas the simple power law models cannot. It is also possible to develop criteria for the occurrence of a rate–temperature maximum (Maatman et al., 1968; Moffat and Clark, 1970). For example, in the case of disproportionation of propylene on a cobalt–molybdate–alumina catalyst the criterion is that the absolute value of the enthalpy of adsorption (of propylene) should be greater than half the activation energy of the reaction; and for the cracking of cumene on a silica–alumina catalyst it is that the enthalpy of desorption of the product should be greater than the activation energy of the reaction.

2. It is sometimes observed that the rate increases as the surface coverage θ_B (or p_B) increases for a bimolecular reaction $A + B \rightarrow$ products, reaches a maximum at $\theta_B = 0.5$, and then decreases with further increase in θ_B. This observation cannot be explained by simple power law models.

3. A serious objection to these models is that there are too many adjustable parameters (usually more than four) to permit the choice of a unique model. This situation was examined by Hutchinson et al. (1967) with respect to the simple *para-ortho* hydrogen shift reaction. They found that models based on individual control-

TABLE 2.3. Adsorbed Species and Controlling Mechanisms in Hydrocarbon Oxidation

Adsorption System Number	Adsorbed Species[a]				Possible Rate-Controlling Steps
	D	B	O_2	O	
0	NA	NA	NA	NA	
i	NA	AS	NA	NA	4, 10
ii	NA	NA	AS	NA	1, 7
iii	NA	NA	AS	AS	1, 2, 3, 7, 8, 9
iv	NA	AS	AS	NA	1, 4, 7, 10, 11
v	NA	AS	NA	AS	2, 4, 8, 9, 10, 12, 13
vi	NA	AS	AS	AS	1, 2, 3, 4, 7, 8, 9, 10, 11, 12, 13
vii	NA	NA	NA	AS	2, 8, 9,
viii	AS	NA	NA	NA	5, 17
ix	AS	AS	NA	NA	4, 5, 6, 10, 17
x	AS	NA	AS	NA	1, 5, 7, 14, 17
xi	AS	NA	AS	AS	1, 2, 3, 5, 7, 8, 9, 14, 15, 16, 17
xii	AS	AS	AS	NA	1, 4, 5, 6, 7, 10, 11, 14, 17
xiii	AS	AS	NA	AS	2, 4, 5, 6, 8, 9, 10, 12, 13, 15, 16, 17
xiv	AS	AS	AS	AS	1, 2, 3, 4, 5, 6, 7, 8, 10, 11, 12, 13, 14, 15, 16, 17
xv	AS	NA	NA	AS	2, 5, 8, 9, 15, 16, 17

[a]AS, Adsorbed species; NA, not adsorbed.

ling steps as well as on multistep control could be expressed in the same form, so that distinction was not possible. A more serious observation was that even this single model could not be upheld by the results. In contrast to this, Choudhary and Doraiswamy (1975) found that for the isomerization of *n*-butene to isobutylene, although single-step as well as dual-step control models gave identical rate forms, as in the case of the *ortho–para* shift reaction, experimental data could be satisfactorily represented by this model.

4. The choice of an H–W model is based on the hypothesis that all the adsorption constants must have a positive sign. When a gas A is adsorbed on a solid surface, a certain fraction of the surface is covered. If now a second gas B is introduced, the adsorption of A will be reduced. It is known, however, that in a few cases the presence of a second gas enhances the adsorption of the first (probably because of forces of attraction). Examples of this enhancing effect of adsorption are listed in Table 2.5. The chief consequence of this effect is that for a binary adsorbing system (e.g., A + B) the effect of the

second gas would apparently be represented by the term $(-K_B C_B)$ rather than $(+K_B C_B)$. However, the effect is very rare and can generally be ignored.

5. In spite of the advantages of H–W models presented in 1 and 2, the most serious practical objection is that simple power law models are often found to represent the data equally well— even for reactions for which H–W models have been fitted (Weller, 1956).

2.6. METHODS OF PARAMETER ESTIMATION AND MODEL DISCRIMINATION

The statistical methods of parameter estimation and model selection presented in this section apply equally to power law and H–W models.

Two methods of modeling are possible: (1) from integral data without extracting point rates, which involves estimating the parameters in a system of first-order differential equations that may be linear or nonlinear; and (2) from rate data obtained directly in a

TABLE 2.4 Examples of Relaxation of Assumptions in the Hougen–Watson Models

Relaxed Assumption	Reaction Type	Reaction Studied	Remarks on Models Derived	No. of Unknown Parameters Involved	Reference
Two-step control	$A \rightleftharpoons B$	Isomerization of *n*-butene to isobutene	Equations have been developed for the following combinations: adsorption of A and surface reaction; surface reaction and desorption of B; and adsorption of A and desorption of B. All can be expressed in the same form: $$r = \frac{\alpha(p_A - p_B/K)}{a + bp_A + cp_B}$$ where α, a, b, c are constants. Hence it is impossible to tell which combination of two steps is rate controlling. This expression is also identical with the H–W model for surface reaction controlling (single site), and thus this single-step model also cannot be distinguished from the two-step models.	5	Choudhary and Doraiswamy (1975)
Three-step control	$A \rightarrow B$	*Para–ortho* hydrogen shift reaction	An expression has been derived for the case in which all three of the single steps mentioned in the remarks above are simultaneously controlling. The rate equation obtained is also identical in form with the one above; hence no distinction can be made between two- and three-step control and surface reaction.	6	Hutchinson et al. (1967)
Two-step control	$A \rightarrow B + C$	Dehydrogenation of *sec*-butyl alcohol (studied by Thaller and Thodos, 1960)	A rather complicated equation has been obtained that cannot be used for integral reactor data. Extensive experimentation, covering wide ranges of variables, is necessary to consider the use of more than one rate-controlling step for this type of reaction. Statistical evaluation of rate parameters for this model has been discussed.	5	Bischoff and Froment (1962) Shah (1965)
No assumption of rate-controlling step (dual-site mechanism)	$A \rightarrow B + C$	Dehydrogenation of ethanol (studied by Franckaerts and Froment, 1964)	Improvement in the fit of the data with this highly complex model with 16 unknown parameters is negligible as compared to the fit with a simple H–W model.	16	Bradshaw and Davidson (1969)

Variable heat of A → B adsorption	*Para–ortho* hydrogen shift reaction	A complex model has been developed with seven unknown parameters as compared to only three in the H–W model and with an additional variable for surface coverage; but even this model does not fit the data satisfactorily.	7	Hutchinson et al. (1967)
Two different sets of active centers on the catalyst surface	A → B + C Dehydrogenation of *sec*-butyl alcohol (studied by Thaller and Thodos, 1960)	A model of the form $$r = \frac{k_1 p_A}{(1 + K_1 p_A)^n} + \frac{k_2 p_A}{(1 + K_2 p_A)^n}$$ has been developed where K_1 and K_2 are the adsorption constants for the two types of sites. A better fit of the data is claimed than in the original study (Thaller and Thodos), which involved a single species of active centers; however, it is difficult to discriminate among the three proposed models (i.e., with $n = 2, 3,$ and 4).	4	Kolboe (1967)
Five different sets of active centers on the catalyst surface	A → B + C Dehydrogenation of isopropyl alcohol over zinc oxide catalyst	A model of the form $$\sum_{i=1}^{5} \frac{k_i K_i p_A}{(1 + K_i p_A)^n}$$ where $n = 2, 3, 4,$ $\quad k_i = A_f \exp\left(\frac{-E_i}{R_g T}\right)$ and $$K_i = \exp\left(\frac{\Delta S_i}{R_g} - \frac{\Delta H_i}{R_g T}\right)$$ has been developed. The model has been fitted to the experimental rates using a set of rate constants and activation energies determined directly from adsorption studies for the same assumed number of active sites (five).	10	Kolboe (1969)

21

TABLE 2.5 Examples of Unusual Adsorption Behavior

Adsorbate	Catalyst	Observation	Reference
H_2	Iron	After rapid initial adsorption, the adsorption rate declined almost to zero within 2–3 hr, remained roughly constant for a further 50 hr, increased for a further 50 hr, then declined again.	Benton and White (1932)
H_2 and D_2	Copper	Adsorption rate was found to increase with amount adsorbed.	Beebe et al. (1935)
H_2	Nickel	When the catalyst was pretreated with H_2 the initial rate of adsorption of H_2 was found to be enhanced by the pretreatment, but there was decrease in the total adsorption and the ambient adsorption rate.	Eucken (1949)
H_2	Nickel[a]	With an increase in the amount of preadsorbed CO there was an increase in the rate of initial adsorption of H_2, but the total amount of adsorption and the ambient adsorption rate of H_2 decreased.	Iijima (1938)
H_2	Fischer–Tropsch catalyst[a]	Presence of CO on the surface enhanced the adsorption of H_2.	Sastri and Vishwanathan (1955)
H_2	Cobalt[a]	Total amount of H_2 adsorbed was less, but the rate of adsorption was found to be faster than that on clean catalyst (with no CO).	Agliardi and Marelli (1948)
H_2	Ru–Al$_2$O$_3$[a]	With increase in the amount of CO on the surface, total amount adsorbed (after 100 min) and rate of adsorption of H_2 were found to increase, while the initial adsorption rate of H_2 was retarded.	Low and Taylor (1959)
H_2	Ru–Al$_2$O$_3$[b]	Because of the presence of preadsorbed O_2, the adsorption of H_2 was found to be enhanced.	Low and Taylor (1959)
$H_2 + CO$ (25:1)	Cobalt	Adsorption of the mixture $(CO + H_2)$ was at first faster than that of pure H_2.	Agliardi and Marelli (1948)
$H_2 + CO$	Cobalt	Mutual enhancement in both the rates and the amounts of adsorption was observed.	Sastri et al. (1974)
Thiophene	Copper chromite	Gas chromatographic pulse experiments indicated an enhancement in the rate of irreversible adsorption and the amount of reversible adsorption with increase in surface coverage; a very complex but peculiar trend was observed.	Sansare et al. (1979)

[a]With preadsorbed CO.
[b]With preadsorbed O_2.

differential (or mixed) reactor or by differentiation of integral data (concentration vs. W/F), which involves estimating the parameters in a system of algebraic equations, linear or nonlinear.

Basically, all estimation methods define an objective function, which should be optimized with respect to model parameters. The various methods differ mainly in the nature of the objective function and the mathematical or structural features of the response relations.

2.6.1. Objective Functions

Two of the most widely used objective functions are based on the methods of least-squares analysis and maximum likelihood.

The Least-Squares Objective Function

For a general system of q responses the least-squares objective function may be defined as the determinant formed by the residual sum of squares of the individual responses and the sums of cross products of residuals taken two at a time:

$$S = \det \begin{bmatrix} a_{11} & \cdots & a_{1q} \\ \vdots & & \vdots \\ a_{q1} & \cdots & a_{qq} \end{bmatrix} \quad (2.13)$$

where a_{ij} is the sum of the products of the residuals of the ith and jth responses.

For a single-response system and n observations this reduces to

$$S = a_{i1} = \sum_{m=1}^{n} [y_m - f(\mathbf{x}_m, \mathbf{k})]^2 = \boldsymbol{\varepsilon}^T \boldsymbol{\varepsilon} \quad (2.14)$$

where y_m is the observed response (rate) for the mth experiment, $f(\mathbf{x}_m, \mathbf{k})$ is the rate predicted by the model, and $\boldsymbol{\varepsilon}$ is an $n \times 1$ vector of residuals. Since the controlled (or independent) variables $\mathbf{x}$ (such as temperature and pressure) are known, S is a function of model parameters $\mathbf{k}$ only. By use of a suitable optimization method (such as the Taylor series expansion), the functions S may be minimized with respect to $\mathbf{k}$ to yield the best set of parameter values (see e.g., Himmelblau, 1972).

Maximum-Likelihood Objective Function

The experimental error is randomly distributed and the probability density function is assumed to be a Gaussian function for the general case of q responses:

$$L = p[\mathbf{y}, f(\mathbf{x}, \mathbf{k}), s] \quad (2.15)$$

where p is a multivariate probability density function and s is a properly defined statistic—for example, variance. For a single response this reduces to

$$L = p[\mathbf{y} - f(\mathbf{x}, \mathbf{k}), s] \quad (2.16)$$

The function L, called the likelihood function, should be maximized with respect to the parameters to yield the maximum-likelihood estimates of the parameters. Sometimes it is computationally convenient to optimize the logarithm of this function, which is equivalent to optimizing the function itself.

When the error distribution is normal, both methods yield the same set of parameter values.

The Bayesian Function

The likelihood function does not take into account any available information concerning the values of the parameters. This *a priori* information can be expressed in the form of a relative probability density function $p_0(\mathbf{k})$, according to which the probability that $\mathbf{k} = \mathbf{k}_1$ is known in relation to the probability $\mathbf{k} = \mathbf{k}_2$.

A new function that is the product of the likelihood function and the prior distribution can now be defined:

$$\begin{aligned} B(\mathbf{k}, s) &= L(\mathbf{k}, s)p_0(\mathbf{k}) \\ &= p\{[\mathbf{y} - f(\mathbf{x}, \mathbf{k})], s\} p_0(\mathbf{k}) \quad (2.17) \end{aligned}$$

Here again for the sake of convenience the function log $B(\mathbf{k}, s)$ is maximized to obtain the best estimates of $\mathbf{k}$ for a given model.

Comments

In the Bayesian function the error distribution function as well as the initial probability should be known, whereas in the likelihood function a knowledge of the former is adequate. Since a normal error distribution is usually assumed and no prior information on parameter values is available, neither of these methods offers any advantage over the least-squares method for parameter estimation. The nonlinear least-squares method is therefore the recommended procedure for parameter estimation.

2.6.2. Response Relation

In the discussion above $f(\mathbf{x}, \mathbf{k})$ is referred to many times. This function is known as the response function and the model itself is called the response relation.

Differential Response Equations

Experimental data are usually available in the form of a product distribution for various values of W/F. We assume that the rate expression is of the form

$$\frac{d\mathbf{c}}{d(W/F)} = f(\mathbf{c}, \mathbf{k}) \quad (2.18)$$

where $\mathbf{c}$ is the vector of concentrations, $\mathbf{k}$ (as before) is the vector of rate constants, and $d\mathbf{c}/d(W/F)$ is the vector of rates. The functions $\mathbf{f}$ are a general class of nonlinear functions for which the analytical nature of the solution is not known. Hence the vector equation 2.18 is partially differentiated with respect to each member of $\mathbf{k}$. If there are q' independent reactions and p' parameters, this yields $q' \times p'$ additional differential equations, which may be readily expressed as a matrix equation:

$$\frac{d}{d(W/F)}\left(\frac{\partial \mathbf{c}}{\partial \mathbf{k}}\right) = \frac{\partial f}{\partial \mathbf{k}} + \frac{\partial f}{\partial \mathbf{c}}\frac{\partial \mathbf{c}}{\partial \mathbf{k}} \quad (2.19)$$

(Differentiation of a q'-dimensional vector with respect to a p'-dimensional vector produces a $q' \times p'$ matrix.) These $q' \times p'$ equations along with q' equations (2.18) are solved numerically for each W/F for which experimental data are available, on the assumption of an initial estimate of parameters. The sum of squares of errors is then minimized by means of a suitable optimization technique.

The procedure is repeated for all the candidate models. If unconstrained optimization is used, the following guidelines may be employed in selecting the models:

1. Adsorption and rate constants must all be positive.
2. Plots of observed **c** against **c** calculated by using the least-squares estimates must be straight lines with gradient close to unity.

Algebraic Response Equations

Here we have the experimental data in the form of rates for a set of product distributions,

$$\mathbf{r} = f(\mathbf{c}, \mathbf{k}) \tag{2.20}$$

where **r** is the vector of rates and f, **c**, **k** have the same significance as before. If the f are linear, least-squares parameter estimates are obtained without the need of initial estimates. If they are nonlinear, a sequential search for optimum values of parameters is undertaken. By using an initial estimate of parameters, the rates are calculated for all the sets of **c** for which experimental rates are known, and the sum of residual squares is minimized by a suitable optimization technique. The procedure is repeated for all the models. If unconstrained optimization is used, selection of models is based on the same guidelines as outlined for the differential response equations.

2.6.3. Preliminary Selection of Models

Several new design and analysis techniques for kinetic modeling with differential reactor data have been reported in the last 15–20 years. The procedures have been reviewed by Bard and Lapidus (1968), Kittrell (1970), Mezaki and Happel (1969), Himmelblau (1970), Reilly (1970), Froment (1975), and Prasad (1975); modifications and improvements continue to be introduced (Sica et al., 1978; Dumez and Froment, 1976; Meier and Gut, 1978; Atherton et al., 1975; Ng and Vermeulen, 1977; Yates and Best, 1976; Forni and

Terzoni, 1977; Chang et al., 1977; Sundaram and Froment, 1978).

Normally, a preliminary analysis of experimental data for discrimination of models is done without the aid of statistical methods (except in parameter estimation), and final discrimination is accomplished by statistical methods.

The preliminary choice of models from all the plausible models is based on the following general considerations: (1) All the model parameters should be significantly positive; and (2) the selected models should pass the F test for adequacy of fit (see Section 2.6.4). If more than one model is found to be acceptable by this procedure, they are subjected to the model discrimination methods described next.

From the Residual Sum of Squares

For a general system of q responses, the residual sum of squares (RSS) as defined by Eq. 2.13 is calculated for the models, and the model with the minimum RSS is taken as the most plausible one among the rival candidates. Equation 2.14 is used for a single-response system. Van Meerten and Coenen (1977), using a nonlinear least-squares computer fit program, selected one out of a possible three mechanisms for the hydrogenation of benzene.

The rate constant should always increase and the adsorption constant normally always decrease with temperature. In the case of activated adsorption, however, two situations (both of which are quite rare) are possible:

1. The adsorption constant increases with temperature (positive ΔH); a few reported instances of this behavior are listed in Table 2.6.
2. The adsorption constant passes through a maximum with increase in temperature, with transition from positive to negative ΔH, as observed by Choudhary and Doraiswamy (1975). Such a transition in the enthalpy of activated adsorption can be explained by considering the creation and destruction of active centers as the temperature is increased.

From Initial Rates

The complexity of the hyperbolic equations can be considerably reduced by writing them for the initial condition when all product terms get eliminated (Yang and Hougen, 1950; Kittrell, 1970). Then, by expressing the reactant partial pressure in terms of the mole fraction and the total pressure, the equations can be

TABLE 2.6. Typical Examples of Rate Models with Endothermic Adsorption

Reaction System	Rate Model	Temperature Coefficient (ΔH) of Adsorption Constant (cal/mol)	Reference
Hydrogenation of propylene over copper–magnesia catalyst	$r = \dfrac{kK_U p_H p_U}{(1 + K_U p_U + K_S p_S)^3}$	K_U: $\Delta H = 2{,}490$ K_S: $\Delta H = -1{,}870$	Sussman and Potter (1954)
	$r = \dfrac{kK_U p_H p_U}{(1 + K_U p_U + K_S p_S)^2}$ (U is propylene; H, hydrogen; S, propane)	K_U: $\Delta H = 5{,}040$ K_S: $\Delta H = 1{,}220$	Sussman and Potter (1954)
Hydrogenation of ethylene over copper–magnesia catalyst	$r = \dfrac{k p_H p_U}{(1 + K_U p_U)^3}$	K_U: $\Delta H = 1{,}107$	Sussman and Potter (1954)
	$r = \dfrac{k p_H p_U}{(1 + K_U p_U)^2}$ (U is ethylene)	K_U: $\Delta H = 4{,}110$	
Catalytic dehydrogenation of *sec*-butyl alcohol to methyl ethyl ketone over brass	$r = \dfrac{k(p_A - p_K p_H/K)}{p_K(1 + K_A p_A + K_{AK} p_A/p_K)}$ (A is alcohol; K, ketone)	K_A: $\Delta H = 12{,}230$	Perona and Thodos (1957)
Catalytic reduction of nitric oxide	$r = \dfrac{kK_{H_2} K_{NO} p_{H_2} p_{NO}}{(1 + K_{NO} p_{NO} + K_{H_2} p_{H_2})^2}$	K_{NO}: $\Delta H = 13{,}300$ K_{H_2}: $\Delta H = 15{,}400$	Ayen and Peters (1962)
Isomerization of *n*-butene to isobutene over fluorinated η-alumina	$r = \dfrac{k(p_A - p_B/K)}{1 + K_B p_B}$ (A is *n*-butene; B, isobutene)	K_B: ΔH is positive and negative	Choudhary and Doraiswamy (1975)

recast into forms that can lead to predictable trends in the behavior of the initial rate as a function of pressure or mole fraction.

Boudart (1967) has, however, demonstrated with examples the dangers of using initial rate data alone for the kinetic modeling of a catalytic reaction. Some of the reactions, such as the catalytic synthesis of ammonia from stoichiometric mixtures of nitrogen and hydrogen (Ozaki et al., 1960) and the oxidation of sulfur dioxide on vanadium pentoxide (Mars, 1958), are strongly inhibited by reaction products. For reactions of this type, obviously, initial rate data can give misleading results.

From Intrinsic Parameters

Parameters that are inherently present in the mathematical structure of a rate model are called intrinsic

parameters. These are much simpler functions of conversion and pressure than the parent models. The intrinsic parameters associated with a few typical models for the common reaction type $A + B \rightarrow R$ are summarized later in Table 2.10 together with methods of determining them and the criteria for model acceptance (e.g., in Section 2.6.7). Similar tables can be readily prepared for other reaction types.

A particularly powerful method of model selection using intrinsic parameters for a reaction in a slurry system has been proposed by Brahme and Doraiswamy. This is illustrated in Section 2.6.7 with respect to the hydrogenation of glucose.

From Nonintrinsic Parameters

A method involving the use of a parameter not inherently present in the rate model, the so-called nonintrinsic parameter, has been suggested by Mezaki and Kittrell (1966) for discriminating between two rival models in both the high and the low conversion regimes. However, this method is of little use (Choudhary and Doraiswamy, 1975) in model discrimination.

2.6.4. Example 1: Kinetic Model for the Isomerization of *n*-Butene to Isobutene

Raghavan and Doraiswamy (1977) studied the kinetics of isomerization of *n*-butene to isobutene over a fluorinated η-alumina catalyst. Their rate data for three temperatures can be represented by one of the 10 H–W models in Table 2.7. Which one adequately represents the data?

Solution

The Taylor series expansion method can be used in conjunction with the initial parameter values obtained from the linear least-squares analysis to evaluate the model parameters.

The t values may be calculated from the following equation:

$$t = \frac{b_i - 0}{\delta_{bi}}$$

where b_i is the value of the parameter and δ_{bi} is the standard deviation. If t_{calc} is greater than t_{table}, the parameter value can be considered to be significantly different from zero.

The models with positive parameter values are given in Table 2.8 for all three of the temperatures studied.

TABLE 2.7. Plausible Single Step Controlling Hougen–Watson Models (Isomerization of *n*-butene)

Model No.	Controlling Step[a]	Rate Model (Nonlinear Form)	Rate Model (Linear Form)
		Single-site Mechanism	
1	Adsorption of A $\underline{A + l \rightleftharpoons Al} \rightleftharpoons Bl \rightleftharpoons B + l$	$r = \dfrac{k(p_A - p_B/K)}{1 + K_B p_B}$	$\dfrac{p'}{r} = a + b p_B$
2	Surface reaction $A + l \rightleftharpoons \underline{Al \rightleftharpoons Bl} \rightleftharpoons B + l$	$r = \dfrac{k(p_A - p_B/K)}{1 + K_B p_B + K_A p_A}$	$\dfrac{p'}{r} = a + b p_B + c p_A$
3	Desorption of B $A + l \rightleftharpoons Al \rightleftharpoons \underline{Bl \rightleftharpoons B + l}$	$r = \dfrac{k(p_A - p_B/K)}{1 + K_A p_A}$	$\dfrac{p'}{r} = a + b p_A$
		Dual-Site Mechanism	
4	Adsorption of A $\underline{A + 2l \rightleftharpoons 2A_{1/2}l} \rightleftharpoons Bl + l$ $\overline{Bl \rightleftharpoons B + l}$	$r = \dfrac{k(p_A - p_B/K)}{(1 + K_B p_B)^2}$	$\sqrt{\dfrac{p'}{r}} = a + b p_B$
5	Surface reaction $A + 2l \rightleftharpoons \underline{2A_{1/2}l \rightleftharpoons Bl + l}$ $\overline{Bl \rightleftharpoons B + l}$	$r = \dfrac{k(p_A - p_B/K)}{(1 + \sqrt{K_A p_A} + K_B p_B)^2}$	$\sqrt{\dfrac{p'}{r}} = a + b p_A^{1/2} + c p_B$
6	Surface reaction $\underline{Al + l \rightleftharpoons Bl + l}$ $\overline{Al \rightleftharpoons A + l;\ Bl \rightleftharpoons B + l}$	$r = \dfrac{k(p_A - p_B/K)}{(1 + K_B p_B + K_A p_A)^2}$	$\sqrt{\dfrac{p'}{r}} = a + b p_B + c p_A$
7	Desorption of B $A + 2l \rightleftharpoons 2A_{1/2}l \rightleftharpoons \underline{2B_{1/2}l \rightleftharpoons B + 2l}$	$r = \dfrac{k(p_A - p_B/K)}{(1 + \sqrt{K_A p_A})^2}$	$\sqrt{\dfrac{p'}{r}} = a + b p_A^{1/2}$
		Half-site mechanism	
8	Adsorption of A $\underline{2A + l \rightleftharpoons A_2 l} \rightleftharpoons B_2 l \rightleftharpoons 2B + l$	$r = \dfrac{k(p_A^2 - p_B^2/K^2)}{1 + K'_B p_B^2}$	$\dfrac{p''}{r} = a + b p_B$

TABLE 2.7 (continued)

Model No.	Controlling Step[a]	Rate Model (Nonlinear Form)	Rate Model (Linear Form)
9	Surface reaction $$2A + l \rightleftharpoons A_2 l$$ $$A + l \rightleftharpoons \underline{Al \rightleftharpoons Bl} \rightleftharpoons B + l$$	$$r = \frac{k(p_A - p_B/K)}{1 + K_B p_B + K_A p_A + K'_A p_A^2}$$	$$\frac{p'}{r} = a + b p_B + c p_A + d p_A^2$$
10	Desorption of B $$2A + l \rightleftharpoons A_2 l$$ $$A + l \rightleftharpoons Al \rightleftharpoons \underline{Bl \rightleftharpoons B + l}$$	$$r = \frac{k(p_A - p_B/K)}{1 + K_A p_A + K'_A p_A^2}$$	$$\frac{p'}{r} = a + b p_A + c p_A^2$$

where

$$p' = p_A - \frac{p_B}{K}, \qquad p'' = p_A^2 - p_B^2/K^2$$

and K'_A, K'_B are the adsorption equilibrium constants for the half-site model.

[a] Controlling step is underlined

TABLE 2.8. Parameter Values of Hougen-Watson Models With All Positive Parameters (Isomerization of Butene)

Model No.	Rate Model	Model Parameters			t_{calc}	t_{table}
		Values of Parameters	Standard Error	95% Confidence Limits		
		Temperature: 605.5° K				
1	$r = \dfrac{k(p_A - p_B/K)}{1 + K_B p_B}$	$k = 0.185 \times 10^{-3}$ $K_B = 6.29$	0.38×10^{-5} 0.3779	$0.185 \times 10^{-3} \pm 0.105 \times 10^{-4}$ 6.29 ± 1.0505	47.6 16.55	2.78 2.78
6	$r = \dfrac{k(p_A - p_B/K)}{1 + K_B p_B + K_A p_A)^2}$	$k = 0.01248$ $K_B = 0.8022$ $K_A = 0.6376$	0.003761 0.6462 0.9456	0.01248 ± 0.01196 0.8022 ± 2.0549 0.6376 ± 3.0070	3.32 1.24 0.674	3.18 3.18 3.18
		Temperature: 621.0° K				
1	$r = \dfrac{k(p_A - p_B/K)}{1 + K_B p_B}$	$k = 0.34 \times 10^{-3}$ $K_B = 3.69$	0.64×10^{-5} 0.2162	$0.34 \times 10^{-3} \pm 0.1779 \times 10^{-4}$ 3.69 ± 0.6010	53.10 17.06	2.78 2.78
5	$r = \dfrac{k(p_A - p_B/K)}{(1 + \sqrt{K_A p_A} + K_B p_B)^2}$	$k = 0.02422$ $K_B = 0.8374$ $K_A = 0.2163$	0.01146 1.6129 0.8423	0.02422 ± 0.03644 0.8374 ± 5.1290 0.2163 ± 2.6785	2.11 0.52 0.26	3.18 3.18 3.18
		Temperature: 636.5° K				
1	$r = \dfrac{k(p_A - p_B/K)}{1 + K_B p_B}$	$k = 0.58 \times 10^{-3}$ $K_B = 1.29$	0.145×10^{-4} 0.1156	$0.58 \times 10^{-3} \pm 0.403 \times 10^{-4}$ 1.29 ± 0.3217	39.9 11.16	2.78 2.78
2	$r = \dfrac{k(p_A - p_B/K)}{1 + K_B p_B + K_A p_A}$	$k = 0.041$ $K_B = 4.3276$ $K_A = 0.3769$	0.1292 0.8411 1.4154	0.041 ± 0.4108 4.3276 ± 2.6747 0.3769 ± 4.5009	0.32 5.15 0.27	3.18 3.18 3.18
6	$r = \dfrac{k(p_A - p_B/K)}{(1 + K_B p_B + K_A p_A)^2}$	$k = 0.0327$ $K_B = 1.7752$ $K_A = 0.1467$	0.1071 0.3160 0.6616	0.0327 ± 0.3405 1.7752 ± 1.0049 0.1467 ± 2.1038	0.305 5.62 0.22	3.18 3.18 3.18

Calculation of F:

$$F = \frac{\text{lack-of-fit mean sum of squares}}{\text{pure-error mean sum of squares}}$$

The pure-error mean sum of squares can be calculated in the following way:

Run No.	$y \times 10^5$ (y = observed rate)
1	13.2961
2	12.5903
3	13.2993
Total 3	39.1857

Average: $\bar{y} = 13.0619 \times 10^{-5}$

Residual:

Sum of squares $\sum (y_i - \bar{y})^2 = 0.3736 \times 10^{-10}$

Degrees of freedom $n - 1 = 2$

Mean sum of squares (or variance) $\dfrac{\sum (y_i - \bar{y})^2}{n-1} = 0.1868 \times 10^{-10}$

Values of F_{calc} and F_{table} are given in Table 2.9 for all the models considered for selection.

At 605.5°K, as can be seen from Table 2.8, two models (1 and 6) have positive values for the parameters. In the case of model 6 the calculated t (Table 2.8) is less than the table t for the adsorption parameters and this indicates that the values are not significantly different from zero. However, this model finds a place in the final selection along with model 1, since the calculated F is less than its table value.

At 621.0°K, models 1 and 5 have positive values for the parameters. However, as can be seen from Table 2.8, the calculated t values for all three of the parameters of model 5 are less than the corresponding table values, which indicates that the values, though positive, are not significantly different from zero. Also, from Table 2.9 it can be seen that the calculated F value is higher than the table value for this model. These observations suggest that model 5 may be safely discarded, which leaves only model 1 as the choice at this temperature.

Three models 1, 2, and 6 are competing candidates for selection at 636.5°K. In the case of models 2 and 6, the values of the specific rate constant (k) and the adsorption equilibrium constant of n-butene (K_A) are not significantly different from zero, as indicated by their t values in Table 2.8. The adsorption equilibrium constant for isobutene (K_B) has a positive value for both these models. However, the calculated values of F for

TABLE 2.9. Analysis of Variance of Hougen–Watson Models with All Positive Parameters (Isomerization of Butene)

Model No.	Residual SS ($\times 10^{10}$)	DF	MSS ($\times 10^{10}$)	Lack of fit SS ($\times 10^{10}$)	DF	MSS ($\times 10^{10}$)	F_{calcd}	F_{table}
\multicolumn{9}{c}{*Temperature: 605.5° K*}
1	0.0102	4	0.00255	0.6420	2	0.3210	1.72	19.0
6	0.0320	3	0.0107	3.2246	1	3.2246	17.26	18.51
9	0.1238	3	0.0412	3.6292	1	3.6292	19.43	18.51
\multicolumn{9}{c}{*Temperature: 621.0° K*}
1	0.0373	4	0.0093	0.5180	2	0.2590	1.38	19.0
5	0.0631	3	0.0210	3.5228	1	3.5228	18.85	18.51
\multicolumn{9}{c}{*Temperature: 636.5° K*}
1	0.0941	4	0.0235	0.1920	2	0.096	0.514	19.0
2	0.2152	3	0.0717	4.0042	1	4.0042	21.43	18.51
6	0.2498	3	0.0833	3.7492	1	3.7492	20.07	18.51

[a]Pure error: SS = 0.3737×10^{-10}, DF = 2, MSS = 0.1868×10^{-10}, where SS is the sum of squares, DF the degrees of freedom, and MSS the mean sum of squares.
[b]At 95% confidence level.

the two models are higher than the corresponding table values (Table 2.9) and hence they can be discarded. Again we have model 1 as the most adequate one to represent the kinetic data at this temperature.

At 605.5°K, we have two models, models 1 and 6, to choose from. Since both models satisfy the F test, it is not possible to identify the better of the two. However, this problem can be overcome by evaluating the specific rate constant from the Arrhenius plot of $\ln k$ vs. $1/T$ with the other two temperatures, 621.0 and 636.5°K.

Interpolation of the Arrhenius plot drawn with values of k obtained for model 1 at 621.0 and 636.5°K to 605.5°K gives a value of 1.9×10^{-4} for model 1. This value is very close to the actual value of the rate constant for model 1, which is 1.85×10^{-4}, as compared to the value of the rate constant for the other model (model 6), which is 0.0125. This shows that model 1 fits the data more accurately than model 6.

Model 6 can also be rejected on the basis that its residual sum of squares is much higher than that of model 1, as seen from Table 2.9. Thus we have the model

$$r = \frac{k(p_A - p_B/K)}{1 + K_B p_B} \tag{2.21}$$

as the one adequately representing the kinetic data at all three temperatures.

2.6.5. Example 2: Model Discrimination in the Dehydrogenation of 1-Butene by a Simple Statistical Method

A method has been proposed by Hosten and Froment (1976) based on the concept that the minimum sum of squares of residuals divided by the appropriate number of degrees of freedom is an unbiased estimate of the experimental error variance for the correct model only. For all other models bias is introduced as a result of lack of fit. A test of the homogeneity of the estimates of the experimental error variance for each model should therefore provide an adequate discriminatory criterion. The χ^2 test can be used for the purpose:

$$\chi^2 = \frac{(\ln s^{-2}) \sum\limits_{i=1}^{m} (DF)_i - \sum\limits_{i=1}^{m} (DF)_i \ln s_i^2}{1 + \frac{1}{3(m-1)} \left[\sum\limits_{i=1}^{m} \frac{1}{(DF)_i} - \frac{1}{\sum\limits_{i=1}^{m} (DF)_i} \right]} \tag{2.22}$$

where $(DF)_i$ is the degrees of freedom associated with

the ith estimate of error variance plus lack of fit s_i^2; s^{-2} is the pooled estimate of variance plus lack of fit; and m is the number of models.

The calculation is quite straightforward. The calculated value of χ^2 with all the m models included is compared with the table value. If higher, the model corresponding to the largest estimate of error variance is discarded and reestimated. If again higher than the table value, another model is discarded and the procedure is repeated until one or more models emerge as the most probable candidates.

Dumez and Froment (1976) and Dumez et al. (1977) applied this method to discriminate between rival models in the dehydrogenation of 1-butene to butadiene. Twenty-four possible mechanisms were considered, including atomic and molecular dehydrogenation with hydrogen recombination on the surface or in the gas phase. Since the maximum number of parameters in the models is six, at least seven experiments had to be carried out to estimate the parameters and start the discrimination procedure. A number of models were eliminated even at this stage. The simple discriminatory criterion given by Eq. 2.36 was then used, together with the adequacy test given by Eq. 2.22. After each experiment the calculations were started from scratch (to avoid bias) and a total of seven designed experiments were carried out. Five models still remained (dual-site models), and the final choice had therefore to be made from other considerations. This strong possibility of several models passing the adequacy test is a major weakness of this simple method.

2.6.6. Example 3: Parameter Optimization for a Known model: Oxidation of o-Xylene

Often it may be possible to single out a particular model as the most acceptable one from a number of rival candidates. For example, in the oxidation of o-xylene, the redox mechanism or the steady-state adsorption model, both leading to the same kinetic form, has been widely used. The problem now is to determine the precise values of the parameters of the model. Such a situation can also arise when experiments are carried out by the transient-response technique outlined in Section 2.7; the search is often reduced to a few models (from which the final selection can be made from several considerations) or to a single model.

The redox model for the oxidation of o-xylene may be expressed as

$$r_w = \frac{k_a k_r C_a C_r}{k_a C_a + n k_r C_r} \tag{2.23a}$$

where k_r and k_a are the respective rate constants for the oxidation and reduction steps; n is the number of oxygen molecules used per mole of hydrocarbon oxidized; and C_r and C_a represent the concentrations of hydrocarbon and oxygen, respectively (see Mars and van Krevelen, 1954; Shelstad et al., 1960; Wainwright and Hoffman, 1977). Equation 2.23a may be expressed in the general form

$$r = f(\mathbf{c}, \hat{\mathbf{k}}) \qquad (2.23b)$$

Approximating this response function by Taylor series expansion about current estimates of the parameter and considering only the first-order terms, we can write

$$f(\mathbf{c}, \mathbf{k}) \approx f(\mathbf{c}, \hat{\mathbf{k}}) + \mathbf{X}(\mathbf{k} - \hat{\mathbf{k}}) \qquad (2.24)$$

where $\mathbf{X}$ is an $n \times p$ matrix whose nth row consists of p elements x_{ni} given by

$$x_{ni} = \left(\frac{\partial f(\mathbf{c}, \mathbf{k})}{\partial \mathbf{k}_i} \right)_{\mathbf{k} = \hat{\mathbf{k}}} \qquad (2.25)$$

Box and Lucas (1959) have suggested a design criterion for precise estimation of parameters in a nonlinear model. This criterion is based on choosing the settings of the operating variable $\mathbf{c}$ so as to minimize the determinant $|(\mathbf{X}^T\mathbf{X})^{-1}|$ or equivalently to maximize the determinant $|\mathbf{X}^T\mathbf{X}|$. Selecting experimental runs in this way effectively minimizes the variances for k.

In the case of o-xylene oxidation certain constraints exist on the oxygen and xylene concentrations. A maximum allowable conversion of 7% was arbitrarily chosen. The upper limit of temperature was set at 305°C. By using the sequential design technique outlined above, it was possible to estimate (Juusola et al., 1972) optimum values of the parameters k_a and k_r. The basic advantage of this model is that it reduces the correlation between the parameter estimates and provides a reliable prediction of the values of the parameters for carrying out additional runs.

It is interesting to note that Mars and van Krevelen (1954) proposed a similar redox model for the oxidation of benzene on vanadia catalysts. Whereas they found the rate of reoxidation of the catalyst to be of the same magnitude as that for the oxidation of naphthalene, Jaswal et al. (1969) found that the rate of reoxidation of the catalyst was four times lower for benzene oxidation than for naphthalene or toluene. Thus, even though the conversion data are well cor-

related by the rate equation of Juusola et al., (1970), there exists considerable doubt about the adequacy of the redox or steady-state adsorption model for benzene oxidation.

The rates of benzene oxidation are considerably lower than for o-xylene or naphthalene oxidation, and the rate constant for benzene oxidation is approximately 1000 times less than that for naphthalene oxidation and 200 times less than for o-xylene oxidation. The relative stability of the benzene molecule to oxidation on commercial vanadia catalysts is probably responsible for the high selectivities that are obtained in o-xylene and naphthalene oxidation.

2.6.7. Example 4: Model Discrimination Based on the Effect of Initial Reactant Concentration in a Slurry Reactor: Hydrogenation of Glucose

Bramhe and Doraiswamy (1976) studied the hydrogenation of glucose on Raney nickel in a slurry reactor. After preliminary screening of all probable models, five models were found to merit further consideration (Table 2.10). Discrimination between these rival models was then accomplished by a new method that is of particular value in slurry reactions.

Any Hougen–Watson model can be written in terms of an intrinsic parameter (C_1) that does not include the conversion and another (C_2) that is a multiplier for conversion. The equations for C_2 for the five rival models are presented in Table 2.10. It is possible to discriminate among these models on the basis of the behavior of the intrinsic parameter C_2 with respect to the initial glucose concentration C_{G0}.

Data were obtained at five initial glucose concentrations and the rate was determined at various values of C_G, including r_0, which corresponds to C_{G0}. From the values of r_0 and r, the C_2 for models 3, 4, 8, and 11 were calculated from the corresponding equations listed in Table 2.10. For model 10, a knowledge of C_1 is also required (see the table), and this was obtained by writing the rate equation in terms of C_1 and C_2:

$$r = \frac{P(1-x)}{C_1 + C_2 P} \qquad (2.26)$$

At initial condition this becomes

$$r_0 = \frac{P}{C_1 + C_2 P} \qquad (2.27)$$

or

$$\frac{P}{r_0} = C_2 P + C_1 \qquad (2.28)$$

TABLE 2.10. Probable Models and Equations for the Intrinsic Parameter C_2 (Hydrogenation of Glucose)

Model No.	Model Equation[a]	Equation for C_2	Equation for Calculating C_2 from Experimental Data
3	$r = \dfrac{C_G C_H K_G K_H k}{1 + C_H K_H + C_G K_G}$	$C_2 = \dfrac{1}{\alpha K_H k}$	$C_2 = \dfrac{P}{r_0 x} - \dfrac{P(1-x)}{rx}$
4	$r = \dfrac{C_G C_H K_G K_H k}{(1 + C_H K_H + C_G K_G + C_S K_S)^2}$	$C_2 = \dfrac{(K_G - K_S)C_{G0}^{1/2}}{(\alpha K_G K_H k)^{1/2}}$	$C_2 = \left(\dfrac{P}{r_0}\right)^{1/2}\dfrac{1}{x} - \left(\dfrac{P(1-x)}{r}\right)^{1/2}\dfrac{1}{x}$
8	$r = \dfrac{C_G C_H K_G K_H k}{[1 + (C_H K_H)^{1/2} + C_G K_G + C_S K_S]^3}$	$C_2 = \dfrac{(K_G - K_S)C_{G0}^{2/3}}{(\alpha K_G K_H k)^{1/3}}$	$C_2 = \left(\dfrac{P}{r_0}\right)^{1/3}\dfrac{1}{x} - \left(\dfrac{P(1-x)}{r}\right)^{1/3}\dfrac{1}{x}$
10	$r = \dfrac{C_G C_H K_H k}{1 + C_H K_H}$	$C_2 = \dfrac{1}{C_{G0} k}$	$C_2 = \dfrac{1-x}{r} - \dfrac{C_1}{P}$[b]
11	$r = \dfrac{C_G C_H K_G K_H k}{1 + C_H K_H + C_S K_S}$	$C_2 = \dfrac{K_S}{\alpha K_H k}$	$C_2 = \dfrac{P(1-x)}{rx} - \dfrac{P}{r_0 x}$

[a]Note that G stands for glucose, H for hydrogen, S for sorbitol; glucose + $H_2 \rightarrow$ sorbitol.
C_1 is obtained from the intercept of the plot of P/r_0 vs. P.

which shows a linear relationship between P/r_0 and P; C_1 is obtained from the intercept of this plot.

Models 3 and 11 require that C_2 be independent of C_{G0}. Since this was not found to be true, these models were rejected. For the other three models the equations listed in Table 2.10 require that the following plots be linear:

model 4 C_2 vs. $C_{G0}^{1/2}$

model 8 C_2 vs. $C_{G0}^{2/3}$

model 10 C_2 vs. $1/C_{G0}$

Figure 2.2 shows that model 10 alone passes the test. Further, for this model the line should pass through the origin and have a slope of $1/k$ where k is the rate constant for desorption of sorbitol. The values of k as obtained from nonlinear least-squares analysis and from Figure 2.2 are 3.2 and 3.4 hr^{-1}, respectively. The agreement is clearly satisfactory. Hence model 10, corresponding to desorption of sorbitol controlling, appears to be the most plausible one.

2.6.8. Statistical Discrimination among Models

The rival models are reduced to a manageable number by preliminary selection as discussed in the last section. Further discrimination is accomplished by statistical methods. We shall consider two of the most widely used discrimination methods, both of which can be readily adapted to modern machine computation: likelihood and Bayesian discrimination methods.

Likelihood Discrimination

The likelihood function was defined in Section 2.6.1. From the point of view of model discrimination the likelihood may be considered the "odds in favor" of the model's correctness. For choosing between competing models, however, a likelihood ratio is used. If $L_i(\mathbf{k})$ and $L_j(\mathbf{k})$ are the likelihood functions for models i and j, respectively, then

$$p_{ij} = \frac{L_i(\mathbf{k})}{L_j(\mathbf{k})} \tag{2.29}$$

which is called the likelihood ratio, is a measure of the preferability of model i over model j. It is obvious that only two models can be examined at a time, and that if m models are to be considered, then the likelihoods of all the models are computed and the ratio of the largest likelihood to the next largest likelihood is considered. Thus, if

$$p_{ij} \geq 100 \tag{2.30}$$

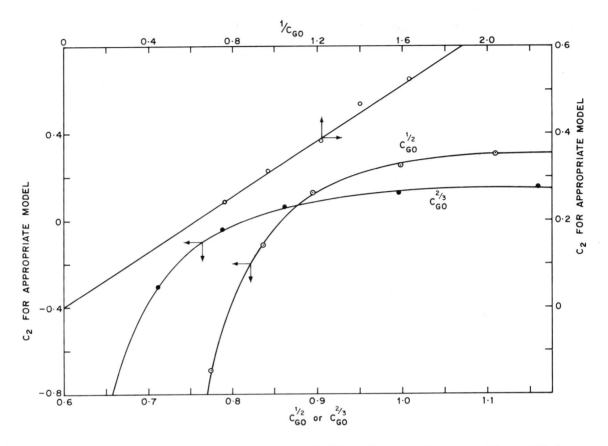

Figure 2.2. Hydrogenation of glucose: test of models 4, 8, and 10 (Table 2.10) from the dependence of C_2 on initial glucose concentration C_{G0} (Brahme and Doraiswamy, 1976).

then the ith model is 100 times more acceptable than the jth model.

Bayesian Discrimination

From the point of view of reaction modeling, the most important technique is the Bayesian method. Bayes's theorem, as applied to model discrimination, states that the probability of model j's being true, after all the current data **y** have been taken into account, is proportional to the product of the probability of the jth model before data **y** were acquired and the likelihood of model j after data **y** have been taken into account. This may be expressed mathematically as

$$p_j(\mathbf{y}) = \frac{p_j L_j(\mathbf{y})}{\sum\limits_{j=1}^{m} p_j L_j(\mathbf{y})} \qquad (2.31)$$

where $p_j(\mathbf{y})$ is the probability that model j represents data **y** after n experiments and is referred to as the posterior probability, p_j is the prior or initial probability (i.e., probability at the end of $n-1$ experiments), and $L_j(\mathbf{y})$ is the likelihood of model j after n experiments. The denominator is a proportionality constant, requiring that all the model probabilities add up to unity.

For constant standard deviation and Gaussian distribution the following expression for the likelihood in a single-response system may be written:

$$L_j(\mathbf{y}) = \frac{1}{(2\pi\sigma^2)^{n/2}} \exp\left[-\frac{1}{2\sigma^2} \sum_{i=1}^{n} (y_i - f_{ji})^2 \right] \qquad (2.32)$$

where σ^2 is the experimental error variance, y_i is the ith observation, and f_{ji} is the response predicted by the jth model in the ith experiment. For a multiple-response system the following expression for likelihood, suggested by Box and Hill (1967), may be used:

$$L_j(\mathbf{y}) = \frac{|\mathbf{N}|^{-1/2}}{(2\pi)^{q/2}} \exp\left[-\tfrac{1}{2}(\mathbf{y} - \mathbf{f}_j)^T \mathbf{N}_j^{-1}(\mathbf{y} - \mathbf{f}_j) \right] \qquad (2.33)$$

where, as before, q is the number of responses, $\mathbf{y}$ is a $q \times 1$ vector of experimental observations, and $\mathbf{f}_j$ is a $q \times 1$ vector of responses predicted by the jth model. $\mathbf{N}$ is the transformed variance–covariance matrix $(q \times q)$ and is defined by

$$\mathbf{N} = \mathbf{Z}\mathbf{M}^{-1}\mathbf{Z}^T + \mathbf{V} \qquad (2.34)$$

where $\mathbf{V}$ is the $q \times q$ variance–covariance matrix of responses, $\mathbf{Z}$ is the $q \times p$ matrix of partial derivatives, and $\mathbf{M}$ is the $p \times p$ variance–covariance matrix of parameters defined by

$$\mathbf{M}_j^{-1} = \left[\sum_{u=1}^{q} \sum_{v=1}^{q} \sigma^{uv}(\mathbf{X}_j^u)^T \mathbf{X}_j^v \right]^{-1} \qquad (2.35)$$

where σ^{uv} is the uvth element of $\mathbf{V}^{-1}$, $\mathbf{X}_j^u$ is the $q \times p$ matrix of partial derivatives for the jth model and uth response, and p is the number of parameters.

Using Eq. 2.31 it is possible to sequentially upgrade the probability of a model by acquiring more and more data until one of the models emerges as distinctly superior to the rest. Thus, at any stage, if the probability of a model approaches unity, it may be considered that this is the most acceptable model.

2.6.9. Sequential Discrimination

In both these methods the best model may not be overwhelmingly superior to the next best model. For example, at the end of Bayesian discrimination two models may have probabilities of 0.45 and 0.40, or at the end of a likelihood discrimination their likelihood ratio may be close to unity. Obviously in such cases the discrimination has to continue and more experimental data should be obtained. But these experiments must be designed and conducted under conditions (i.e., combination of independent variables) that would give maximum divergence between the models. This would eventually lead to one model's being overwhelmingly superior to all the rest.

A large experimental range is important for discrimination among the rival models. To fully appreciate this point, let us consider the two model responses plotted in Figure 2.3. It is impossible to discriminate among these models if the experiments are conducted only within the rectangle *abcd*, even with a large number of data points, but discrimination is possible with fewer experiments in the area *efgh*.

In order to design the "best" experiment it is necessary to postulate a criterion "which will provide a quantitative measure of the ability of each point in the

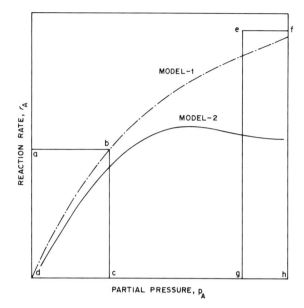

Figure 2.3. Typical response curves for model discrimination.

experimental network to accomplish discrimination." The chief objective of all the criteria is to give the maximum separation or divergence between the models. Of the two major criteria reported, those of Roth (1966) and Box and Hill (1967), the latter is distinctly superior and is outlined below. A multiple design criterion has also been proposed (Wentzheimer, 1969), in which two or more runs can be designed at a time, but this is not generally recommended.

A simple discriminatory criterion is the divergence between the estimates of the objective function (usually the reaction rate) for each of the two competing models 1 and 2:

$$D_{in} = f_{i1} - f_{i2}$$

where i represents the grid point on the x axis of Figure 2.3, and the nth experiment is designed so that the divergence D_{in} is maximal. When m models are present the divergence can be calculated from

$$D_{in} = \sum_{k=1}^{m} \sum_{l=k+1}^{m} (f_{ik} - f_{il})^2 \qquad (2.36)$$

where k and l represent the models. The double summation shows that each model is taken consecutively as a reference.

Figure 2.4 shows the response curves for two arbitrary candidate models together with the associated confidence regions. If the Roth criterion is to be used for the design of discriminatory experiments, position *a*

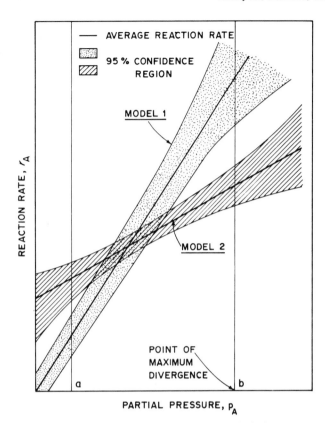

Figure 2.4. The Box–Hill criterion: a hypothetical case.

would be chosen, since this is where the separation between the response curves is maximal. This choice, however, presupposes an absolute precision in the prediction of reaction rates that is never achieved in practice, since there is a region of confidence always associated with every prediction. It would therefore be more appropriate to experiment at position b, where the distance between the confidence regions is maximal.

Based on the information theory, Box and Hill (1967) have postulated a criterion that maximizes the divergence between the confidence regions. Their criterion for m rival models is

$$u(\mathbf{x}_j) = \frac{1}{2} \sum_{k=1}^{m} \sum_{l=k+1}^{m} p_k p_l \left[\frac{(\sigma_{kj}^2 - \sigma_{lj}^2)^2}{(\sigma^2 + \sigma_{kj}^2)(\sigma^2 + \sigma_{lj}^2)} + (f_{kj} - f_{lj})^2 \left(\frac{1}{\sigma^2 + \sigma_{kj}^2} + \frac{1}{\sigma^2 + \sigma_{lj}^2} \right) \right] \quad (2.37)$$

where f_{kj} is the predicted rate in the jth experiment using the kth model, σ^2 is the error variance, and σ_{kj}^2 is the error variance for model k in the jth experiment and is calculated from the variance–covariance matrix for the parameters of the model from

$$\sigma_{kj}^2 = \mathbf{Z}_j (\mathbf{X}^T \mathbf{X})^{-1} (\mathbf{Z}_j)^T \sigma^2 \quad (2.38)$$

$\mathbf{Z}_j$ is a $1 \times p$ row vector whose ith element is given by

$$z_{ji} = \frac{\partial f_j}{\partial k_i}$$

where f_j represents the response evaluated at jth experimental conditions. Equation 2.37 is maximized over all the sets of possible experimental conditions to determine the best discriminatory experiment to be performed next. The corresponding multiresponse equation is

$$u(x_j) = \frac{1}{2} \sum_{k=1}^{m} \sum_{l=k+1}^{m} p_k p_l \left[\text{trace} (\mathbf{N}_k \mathbf{N}_l^{-1} + \mathbf{N}_l \mathbf{N}_k^{-1} - 2I_r) + (\mathbf{f}_k - \mathbf{f}_l)^T (\mathbf{N}_k^{-1} + \mathbf{N}_l^{-1})(\mathbf{f}_k - \mathbf{f}_l) \right] \quad (2.39)$$

where m is the total number of models; $\mathbf{N}_k$ is the transformed variance–covariance matrix of order equal to the number of responses q; $\mathbf{f}_k$ is a $q \times 1$ vector of predicted responses, computed using the kth model; and I_r is the indentity matrix of order q.

Box and Hill's method appears to be most useful in chemical engineering practice. Meeter et al., (1970) and Froment and Mezaki (1970) have confirmed its utility. But it has some important disadvantages: it handles the models only as truncated Taylor's series, and it is difficult in practical situations to assess the adverse effects of this linearization; further, it requires that the distribution be normal, with the variance precisely known in advance. Typical examples of the use of sequential discrimination are 1-butene hydrogenation (Dumez et al., 1977; Dumez and Froment, 1976), ethanol dehydrogenation (Franckaerts and Froment, 1964), pentene isomerization, and reaction of sulfur dioxide with methane over a bauxite catalyst (Helstrom and Atwood, 1978).

Note, too, that when using likelihood discrimination, the present likelihoods instead of probabilities may be used as weight factors in the above two criteria to determine the sequential design.

2.6.10. An Algorithm for Sequential Discrimination

To summarize, we give here the sequential discrimination algorithm using Bayes's theorem and starting from basic laboratory data.

STEP 1: Choose the initial sets of controlled variables and perform the experiments.

STEP 2: Interpolate and differentiate the data if it is desired to use algebraic response equations. (This step may be omitted either if data are obtained in a differential reactor, or if it is desired to use differential response equations.)

STEP 3: Make a preliminary analysis of data to reduce the number of competing models.

STEP 4: Assign prior probabilities to each model. Normally these are taken to be equal unless there is some evidence favoring one or more of the competing models. If m is the number of models after preliminary selection, take $1/m$ as the probability.

STEP 5: Estimate the components of the variance–covariance matrix if replicate data are available. Otherwise use a conservative (large) value. (It may be necessary to assume that measurements are not biased, that variances of all measurements are equal, and that errors are normally distributed.)

STEP 6: Make an initial guess of the parameters. It is usually adequate to use the linear least-squares estimates for this purpose. Estimate the parameters for all the models by means of one of the iterative (nonlinear or least squares) techniques until convergence. If no convergence is obtained, modify the initial guesses and repeat.

STEP 7: Using the current best parameters and a suitable design criterion (say, the Box–Hill criterion), find the next best experiment and perform that experiment.

STEP 8: Update the parameter values for all the models, including these new data.

STEP 9: Find the likelihood function for all the models.

STEP 10: Update the probabilities of all the models by using Bayes's theorem.

STEP 11: Conclude discrimination if the probability of any model exceeds a predetermined value, say 0.95. Otherwise go back to step 7.

2.6.11. Lumping of Reactions

The statistical procedures outlined in the forgoing sections can be used for a single reaction or for any reaction network. Some of the reactions involved in the petroleum and petrochemical industries—for example, thermal cracking of naphtha, and catalytic cracking and reforming—give rise to over 50 reaction products in different (often very small) quantities. A complete reaction network for such a system would be too complicated to build, and although it is theoretically possible to establish such a network, it is often not recommended. Thus it would be desirable to cut down the number of steps in the network in order to make the

analysis more tractable. The first attempt in this direction was reported as long ago as 1947 by Hougen and Watson, who considered the cracking of naphtha as a single-step reaction involving the rate of disappearance of the reactant mixture as a whole. This extreme form represents an oversimplification of a highly complicated network into a one-lump system. Obviously it is possible to make the analysis more rigorous by increasing the number of lumps.

Lumping procedures have been used for a variety of reactions. For example, the oxidation of hydrocarbons is normally represented by the triangular network (see Vaidyanathan and Doraiswamy, 1968)

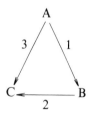

in which the desired product, B (phthalic anhydride, in the case of *o*-xylene oxidation), forms one corner of the triangle, and the reactant (A) and the final products of oxidation (C) form the other two corners. Intermediate compounds can be lumped either with B or with C, or with both.

The catalytic cracking of naphtha has also been represented as a triangular network in which the reactant A (gas oil) is converted to B (gasoline) in the first step, while the other two steps of the triangle lead to C_4, dry gas, and coke (represented by C). After a more extensive study, the group at Mobil Oil Corporation arrived at a 10-lump model for the reaction.

Similarly, for catalytic reforming Kmak (1971) proposed this scheme, based on five classes of pseudo-components for each carbon atom:

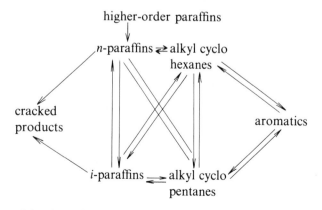

This kind of network was set up for carbon numbers C_5, C_6, and C_7, while for carbon numbers greater than

C_7 all the compounds of a group (e.g., *n*-paraffins) were lumped into one pseudo-component. A similar reforming network has been reported by Graziani and Ramage (1978) and Ramage *et al.* (1980).

2.7. ANALYSIS BY TRANSIENT-RESPONSE METHODS

The methods of kinetic analysis outlined above are based on the assumption of steady state. Since under steady-state conditions all the individual steps involved in a reaction occur at the same rate, a kinetic expression for the overall process is obtained that usually gives the kinetics of the rate-limiting step. No information can be extracted on the kinetics of the individual steps. In most cases, however, more than one model fits the experimental results reasonably well, and hence the ability to discriminate between rival models depends largely on the precision of the experimental results. Under transient conditions the individual steps occur at different rates and the kinetics of each of the steps can be determined.

In general there are two ways of carrying out the transient analysis of a catalytic reaction. The first method consists in perturbing the system from its steady state by varying any of the parameters (concentration of a reactant, temperature, pressure, flow rate, etc.) that define the kinetic state and studying the transient response of the system as a function of time until the steady state is restored. The other method is to study the behavior of the system as a function of time for a given input of the reactant (or adsorbate) concentration; here the system is always under transient conditions.

Consider now the catalytic reaction A → B. If A is strongly adsorbed, and surface reaction of adsorbed A is controlling (when the surface coverage of A is close to unity), different functional forms for the adsorption isotherm of A will give the same overall rate. Thus it would be necessary to carry out a separate transient analysis for determining the correct adsorption isotherm without the complicating presence of a reaction. The perturbation method is highly useful in determining catalytic reaction kinetics, while the second method is more advantageous for determining adsorption kinetics and adsorption isotherms.

2.7.1. Perturbation and Transient-Response Methods

The method outlined here consists in perturbing the system from its steady state by varying any of the variables that define the kinetic state (concentration of a reactant, temperature, pressure, flow rate, etc.) and studying the response of the system as a function of time. Although in principle any type of perturbation can be introduced, for the sake of mathematical simplicity the forcing function should be confined to one of the standard forms such as the step-rectangular pulse or Dirac delta function. The rectangular step is perhaps the simplest to analyze and has hence been used in most of the reported transient-response studies (Kobayashi and Kobayashi, 1974; Lee and Agnew, 1977; Burghardt and Smith, 1979; Stewart et al., 1978). Also, the concentration of the reactant in the feed has been the variable of choice because a step change in concentration can be easily accomplished.

The technique of square-wave cycling of the reactant ratio has been employed in a study of CO oxidation on a vanadium catalyst (Kareem et al., 1979). A maximum in the time-averaged reaction rate observed at a period of 20 minutes suggested the possibility of a resonance-type phenomenon. A dynamic pulse method has been used to measure the hydrogen uptake by supported metal catalysts (Wanke et al., 1979).

The transient method has been employed successfully to determine the mechanism of the oxidation of carbon monoxide (Kobayashi and Kobayashi, 1972a, b, c) and of the decomposition of nitrous oxide (Cutlip et al., 1972; Kobayashi and Kobayashi, 1973). Kobayashi and Futuya (1979) applied the method to the oxidation of propylene over bismuth molybdate at temperatures ranging from 190 to 350°C. They found that a stable intermediate exists on the surface during the reaction. Matsumoto and Bennett (1978) applied this technique to the reduction of CO and H_2 to methane on a commercial iron catalyst. The results indicate that the surface of the active catalyst is covered mostly by a carbon intermediate whose hydrogenation represents the rate-determining step. Transient-response studies on NO reduction by hydrogen on a rhodium–silicon dioxide catalyst were carried out recently (Bell and Savatsky, 1981). Evidence for dissociative adsorption of NO and facilitation of NO adsorption by coadsorption of hydrogen was obtained. The mechanistic details of the decomposition of ammonia on a W surface and of the Fischer–Tropsch synthesis over a Ru catalyst have been worked out.

These methods greatly help in elucidating the mechanism of heterogeneous catalytic reactions (Tamaru, 1981). A surprising fact emerging from transient-response studies is that under actual conditions of catalysis the properties of catalyst surfaces are remarkably different from those of the clean surfaces.

Since in the transient method one more dimension is added to the analysis, in order to retain the simplicity of the mathematical description, differential (plug-flow) or stirred (fully mixed) reactors (See Chapter 9) are preferred.

2.7.2. Measurement of Adsorption Transients

We outline here a method to measure the adsorption parameters important in a catalytic study (Prasad and Doraiswamy, 1981). This method includes a self-consistent experimental strategy for the measurement of all quantities and parameters of interest in an adsorption study. This is done by carrying out a transient analysis of the stirred gas–solid reactor under conditions typical of activated adsorption. A significant feature of the method is its mathematical simplicity. The vexing problem of solving partial differential equations in order to analyze the breakthrough curves of an adsorbate at the exit of a packed bed of adsorbent is thereby avoided. The surface concentrations are readily obtained by the integration of a few ordinary differential equations.

The gas phase is assumed to be ideally mixed, and preliminary experiments are carried out in a stirred batch reactor. When external mass transfer is not limiting, a nonzero final gas-phase concentration points to the operation of reversible linear exchange kinetics. Discrimination between irreversible linear and Elovich models can be effected by noting the drastic difference in the rates of adsorption.

The kinetic model is then validated by a positive F test. Since the surface-phase concentration is small, the chance for desorption is small, and we obtain the adsorption kinetic rate law (except in the case of linear exchange kinetics). Similarly, negative F tests unambiguously give a desorption kinetic model. Combination of these two gives a theoretical isotherm (i.e., equating the rate of adsorption to the rate of desorption).

The stirred reactor is then operated under conditions of adsorption equilibrium. The adsorption equilibrium parameters are measured by the positive F test. These are reaffirmed by negative F tests. The adsorption isotherm originally postulated is thus confirmed. Hence, in principle, it is possible to solve problems involving the measurement of adsorption parameters and discrimination between adsorption kinetic models.

When external mass transfer and intraparticle diffusion resistances are present, the moments of the exit-stream concentration–time curve can in principle be used to estimate the rate parameters (Tobias et al., 1979).

REFERENCES

Agliardi, N., and Marelli, S. (1948), *Gazz. Chim. Ital.*, **78**, 707.

Akyurtlu, J. F., and Stewart, W. E. (1978), *J. Catal.*, **51**, 101.

Aris, R. (1961), *The Optimal Design of Chemical Reactors*, Academic Press, New York, Chapter 3.

Aris, R. (1969), *Elementary Chemical Reactor Analysis*, Prentice-Hall, Englewood Cliffs, New Jersey, p. 11.

Atherton, R. W., Schainker, R. B., and Ducot, E. R. (1975), *AIChE J.*, **21**, 441.

Ayen, R. J., and Peters, M. S. (1962), *Ind. Eng. Chem. Process Design Dev.*, **1**, 204.

Ball, W. E., and Groenweghe, L. C. D. (1966), *Ind. Eng. Chem. Fundam.*, **5**, 181.

Bard, Y., and Lapidus, L. (1968), *Catal. Rev.*, **2**, 67.

Barnard, J. A., and Mitchell, D. S. (1968), *J. Catal.*, **12**, 376.

Bauerie, G. L., Wu, S. C., and Nobe, K. (1978a), *Ind. Eng. Chem. Process Design Dev.*, **17**, 117.

Bauerie, G. L., Wu, S. C., and Nobe, K. (1978b), *Ind. Eng. Chem. Process Design Dev.*, **17**, 123.

Beebe, R. A., Low, G. W., Jr., Wildner, E. L., and Goldwasser, S. (1935), *J. Am. Chem. Soc.*, **57**, 2527.

Begley, J. W., and Wilson, R. T. (1967), *J. Catal.*, **9**, 375.

Bell, A. T., and Savatsky, B. J. (1981), *Stud. Surf. Sci. Catal.*, **7**, 1486.

Benton, A. F., and White, T. A. (1932), *J. Am. Chem. Soc.*, **54**, 1820.

Bischoff, K. B., and Froment, G. F. (1962), *Ind. Eng. Chem. Fundam.*, **1**, 195.

Boag, I. F., Bacon, D. W., and Downie, J. (1975), *J. Catal.*, **38**, 375.

Boreskov, G. K. (1962), *Kinet. Katal.*, **3**, 3556.

Boudart, M. (1952), *J. Am. Chem. Soc.*, **74**, 3556.

Boudart, M. (1967), *Chem. Eng. Sci.*, **22**, 1387.

Boudart, M. (1968), *Kinetics of Chemical Processes*, Prentice-Hall, Englewood Cliffs, New Jersey.

Box, G. E. P., and Hill, W. J. (1967), *Technometrics*, **4**, 30.

Box, G. E. P., and Lucas, H. L. (1959), *Biometrica*, **46**, 77.

Bradshaw, R. W., and Davidson, B. (1969), *Chem. Eng. Sci.*, **24**, 1519.

Bramhe, P. H., and Doraiswamy, L. K. (1976), *Ind. Eng. Chem. Process Design Dev.*, **15**, 130.

Burghardt, A., and Smith, J. M. (1979), *Chem. Eng. Sci.*, **34**, 267.

Chakrabarty, T., Silveston, P. L., and Hudgins, R. R. (1979), *Can. J. Chem. Eng.*, **57**, 651.

Chang, F. W., Fitzgerald, T. F., and Park, J. Y. (1977), *Ind. Eng. Chem. Process Design Dev.*, **16**, 59.

Choudhary, V. R., and Doraiswamy, L. K. (1975), *Ind. Eng. Chem. Process Design Dev.*, **14**, 227.

Cutlip, M. B., Yang, C. C., and Bennett, C. O. (1972), *AIChE J.*, **18**, 1073.

de Tar, D. F., and Day, V. M. (1966), *J. Phys. Chem.*, **70**, 495.

Draper, N. R., Kanemasu, H., and Mezaki, R. (1969), *Ind. Eng. Chem. Fundam.*, **8**, 423.

Dumez, F. J., and Froment, G. F. (1976), *Ind. Eng. Chem. Process Design Dev.*, **15**, 291.

Dumez, F. J., Hosten, L. H., and Froment, G. F. (1977), *Ind. Eng. Chem. Fundam.*, **16**, 298.

Eberly, P. E., Jr., and Spencer, E. H. (1961), *Trans. Faraday Soc.*, **57**, 289.

Eucken, A. (1949), *Z. Electrochem.*, **53**, 702.

Eucken, A. (1950), *Disc. Faraday Soc.*, **8**, 128.

Farrell, R. J., and Ziegler, E. N. (1979), *AIChE J.*, **25**, 447.

Forni, L., and Terzoni, G. (1977), *Ind. Eng. Chem. Process Design Dev.*, **16**, 288.

Fowler, R. H., and Guggenheim, E. A. (1939), *Statistical Thermodynamics*, Cambridge Univ. Press, New York and London.

Franckaerts, J., and Froment, G. F. (1964), *Chem. Eng. Sci.*, **19**, 807.

Froment, G. F. (1975), *AIChE J.*, **21**, 1041.

Froment, G. F., and Mezaki, R. (1970), *Chem. Eng. Sci.*, **25**, 293.

Graziani, K. R., and Ramage, M. P. (1978), *ACS Symp. Ser.*, **65**, 282.

Halsey, G. D., Jr. (1950), *Disc. Faraday Soc.*, **8**, 54.

Halsey, G. D., Jr. (1963), *J. Chem. Phys.*, **67**, 2038.

Halsey, G. D., Jr., and Taylor, H. S. (1947), *J. Chem. Phys.*, **15**, 624.

Happel, J. (1975), *AIChE J.*, **21**, 602.

Happel, J. (1967), *Chem. Eng. Symp. Ser.* (No. 72), **63**, 31.

Happel, J., and Hnatow, M. (1973), *Ann. N. Y. Acad. Sci.*, **213**, 206.

Happel, J., Blanck, H., and Hamill, T. D. (1966), *Ind. Eng. Chem. Fundam.*, **5**, 289.

Helstrom, J. J., and Atwood, G. A. (1978), *Ind. Eng. Chem. Process Design Dev.*, **17**, 114.

Himmelblau, D. M. (1970), *Process Analysis by Statistical Methods*, Wiley, New York.

Himmelblau, D. M., Jones, C. R., and Bischoff, K. B. (1967), *Ind. Eng. Chem. Fundam.*, **6**, 539.

Himmelblau, D. M. (1972), *Applied Nonlinear Programming*, McGraw-Hill, New York.

Horiuti, J. S. (1957), *Adv. Catal.*, **9**, 239.

Horiuti, J. S. (1962), *Adv. Catal.*, **1**, 199.

Hosten, L. H., and Froment, G. F. (1976), *Proc. 2nd Int. Symp. Chem. React. Eng.*, Hiedelberg, Dechema.

Hsieh, Y., and Atwood, G. A. (1976), *Ind. Eng. Chem. Process Design Dev.*, **15**, 359.

Hutchinson, H. L., Barrick, P. L., and Brown, L. F. (1967), *Chem. Eng. Prog.*, **63**, 18.

Hougen, O. A., and Watson, K.M. (1947), *Chemical Process Principles*, Wiley, New York.

Iijima, S. I. (1938), *Rev. Phys. Chem. Japan*, **12**, 1.

Jaswal, I. S., Mann, R. F., Juusola, J. A., and Downie, J. (1969), *Can. J. Chem. Eng.*, **47**, 284.

Juusola, J. A., Bacon, D. W., and Downie, J. (1972), *Can. J. Chem. Eng.*, **50**, 796.

Juusola, J. A., Mann, R. F., and Downie, J. (1970), *J. Catal.*, **17**, 106.

Kabel, R. L., and Johanson, L. N. (1962), *AIChE J.*, **8**, 621.

Kaneko, Y., and Odanko, H. (1965), *J. Res. Inst. Catal. (Hokkaido Univ.)*, **12**, 29.

Kaneko, Y., and Oki. (1965), *J. Sci. Inst. Catal. (Hokkaido Univ.)*, **13**, 55.

Kareem, H. K. A., Silveston, P. L., and Hudgins, R. R. (1979), *Can. Symp. Catal.*, **6**, 236.

Kazragis, A. P. (1966), *J. Phys. Chem. (USSR)*, **40**, 1220.

Kittrell, J. R. (1970), Mathematical methods in chemical reactions, *Adv. Chem. Eng.*, **8**, 98.

Kittrell, J. R., and Erjavec, J. (1968), *Ind. Eng. Chem. Process Design Dev.*, **7**, 321.

Kittrell, J. R., Hunter, W. G., and Watson, C. C. (1966a), *AIChE J.*, **12**, 5.

Kittrell, J. R., Hunter, W. G., and Mezaki, R. (1966b), *AIChE J.*, **12**, 1014.

Kmak, W. S. (1971), AIChE national meeting, Houston, Texas.

Kobayashi, M., and Futuya, R. (1979), *J. Catal.*, **56**, 73.

Kobayashi, M., and Kobayashi, H. (1972a), *J. Catal.*, **27**, 100.

Kobayashi, M., and Kobayashi, H. (1972b), *J. Catal.*, **27**, 108.

Kobayashi, M., and Kobayashi, H. (1972c), *J. Catal.*, **27**, 114.

Kobayashi, M., and Kobayashi, H. (1973), *J. Chem. Eng. Japan*, **6**, 438.

Kobayashi, M., and Kobayashi, H. (1974), *Shokubai (Tokyo)*, **16**, 8.

Kokes, R. J., Tobin, H., and Emmett, P. H. (1955), *J. Am. Chem. Soc.*, **77**, 5860.

Kolboe, S. (1967), *Ind. Eng. Chem. Fundam.*, **6**, 169.

Kolboe, S. (1969), *J. Catal.*, **13**, 208.

Lapidus, L., and Amundson, N. R., eds. (1977), *Chemical Reactor Theory—A Review*, Prentice-Hall, Englewood Cliffs, New Jersey.

Lee, R. S. H., and Agnew, J. B. (1977), *Ind. Eng. Chem. Process Design Dev.*, **16**, 495.

Low, M. J. D., and Taylor, H. A. (1959), *J. Electrochem.*, **106**, 138.

Lyubarskii, A. G., and Petoyan, V. P. (1967), *Kinet. Katal.*, **8**, 235.

Maatman, R. W., Blankespoor, R., Lightenberg, K., and Verghage, H. (1968), *J. Catal.*, **12**, 398.

Manes, F., Hoffer, L. J. E., and Weller, S. (1954), *J. Chem. Phys.*, **22**, 1612.

Mars, P. (1958), Doctoral dissertation, Delft; *Chem. Abstr.*, **54**, 23673d (1960).

Mars, P., and van Krevelen, D. W. (1954), *Chem. Eng. Sci. Spec. Suppl.*, **3**, 41.

Mars, P., and Maessen, J. G. H. (1965), *Proc. 3d Int. Cong. Catal.*, **1**, 266.

Matson, S. L., and Hariott, P. (1978), *Ind. Eng. Chem. Process Design Dev.*, **17**, 322.

Matsumoto, H., and Bennett, C. O. (1978), *J. Catal.*, **53**, 331.

Meeter, D., Pirie, W., and Biot, W. (1970), *Technometrics*, **12**, 457.

Meier, H., and Gut, G. (1978), *Chem. Eng. Sci.*, **33**, 123.

Mezaki, R., and Butt, J. B. (1968), *Ind. Eng. Chem. Fundam.*, **7**, 120.

Mezaki, R., and Happel, J. (1969), *Catal. Rev.*, **3**, 241.

Mezaki, R., and Watson, C. C. (1966), *Ind. Eng. Chem. Process Design Dev.*, **5**, 62.

Mezaki, R., and Kittrell, J. R. (1966), *Can. J. Chem. Eng.*, **44**, 285.

Miller, A. R. (1949), *The Adsorption of Gases on Solids*, Cambridge Univ. Press, New York and London.

Misono, M., Saito, Y., and Yoneda, Y. (1965), *Proc. 3d Int. Cong. Catal.*, **1**, 408.

Moffat, A. J., and Clark, A. (1970), *J. Catal.*, **17**, 264.

Nakanishi, J., and Tamaru, K. (1963), *Trans. Faraday Soc.*, **59**, 1470.

Neiman, M. B., and Gal, D. (1970), *Application of Radioactive Isotopes in Chemical kinetics (Kinetic Isotopic Methods)*, Nauka, Moscow.

Ng, T. H., and Vermeulen, T. (1977), *Ind. Eng. Chem. Fundam.*, **16**, 125.

Ozaki, A., Taylor, H., and Boudart, M. (1960), *Proc. Roy. Soc. (London)*, **A258**, 47.

Perona, J. J., and Thodos, G. (1957), *AIChE J.*, **3**, 230.

Pirard, J. P., and Kalitventzeff, B. (1978), *Ind. Eng. Chem. Fundam.*, **17**, 11.

Prasad, K. B. S. (1975), NCL report.

Prasad, R., and Kar, A. K. (1976), *Ind. Eng. Chem. Process Design Dev.*, **15**, 170.

Prasad, S. D. (1977), Ph.D. thesis, University of Pune.

Prasad, S. D., and Doraiswamy, L. K. (1977), *Phys. Lett.* **A60**, 11.

Prasad, S. D., and Doraiswamy, L. K. (1983), *Phys. Lett.* **A94**, 219.

Prasad, S. D., Doraiswamy, L. K. (1981), *J. Catal.*, **67**, 21.

Raghavan, N. S., and Doraiswamy, L. K. (1977), *J. Catal.*, **48**, 21.

Ramachandran, P. A., and Smith, J. M. (1978), *Ind. Eng. Chem. Fundam.*, **17**, 17.

Ramage, M. P., Graziani, K. R., and Krambeck, F. J. (1980), *Chem. Eng. Sci.*, **35**, 41.

Read, J. F., Chan, Y. T., and Conrad, R. E. (1978), *J. Catal.*, **55**, 166.

Reddy, K. A., and Doraiswamy, L. K. (1969), *Chem. Eng. Sci.*, **24**, 1415.

Reilly, P. M. (1970), *Can. J. Chem. Eng.*, **21**, 655.

Rideal, E. K., and Herington, E. F. G. (1944), *Trans. Faraday Soc.*, **40**, 505.

Roberts, J. K. (1935), *Proc. Roy. Soc. (London)*, **152**, 445.

Roth, P. M. (1966), Ph.D. thesis, Princeton University, New Jersey.

Sansare, S. D., Choudhary, V. R., and Doraiswamy, L. K. (1979), *J. Catal.*, **60**, 21.

Sastri, M. V. C., and Vishwanathan, T. S. (1955), *J. Am. Chem. Soc.*, **77**, 3967.

Sastri, M. V. C., Gupta, R. B., and Vishwanathan, B. (1974), *J. Catal.*, **32**, 325.

Schneider, P., and Smith, J. M. (1968a), *AIChE J.*, **14**, 762.

Schneider, P., and Smith, J. M. (1968b), *AIChE J.*, **14**, 886.

Shah, M. J. (1965), *Ind. Eng. Chem.*, **57**, 18.

Shah, Y. T., Huling, G. P., Paraskos, J. A., and McKinney, J. D. (1977), *Ind. Eng. Chem. Process Des. Dev.*, **16**, 89.

Shelstad, K. A., Downie, J., and Graydon, W. F. (1960), *Can. J. Chem. Eng.*, **38**, 102.

Sica, A. M., Valles, E. M., and Gigola, C. E. (1978), *J. Catal.*, **51**, 115.

Soga, K., Imamura, H., and Ikeda, S. (1979), *J. Catal.*, **56**, 119.

Stewart, W. E., Sorensen, J. P., and Teeter, B. C. (1978), *Ind. Eng. Chem. Fundam.*, **17**, 221.

Sundaram, K. M., and Froment, G. F. (1978), *Ind. Eng. Chem. Fundam.*, **17**, 174.

Susmann, M. V., and Potter, C. (1954), *Ind. Eng. Chem.*, **46**, 457.

Tamaru, K. (1959), *Nature (London)*, **183**, 319.

Tamaru, K. (1964), *Adv. Catal.*, **15**, 65.

Tamaru, K. (1981), *Stud. Surf. Sci. Catal.*, **7**, 566.

Tanaka, K. (1966), *J. Res. Inst. Catal.*, *(Hokkaido Univ.)*, **13**, 119.

Taylor, H. S., and Lewis, J. R. (1938), *J. Am. Chem. Soc.*, **60**, 877.

Taylor, H. S., and Liang, S. C. (1947), *J. Am. Chem. Soc.*, **69**, 1306.

Thaller, L. H., and Thodos, G. (1960), *AIChE J.*, **6**, 369.

Tobias, G. G., Ocana, J. J., and Rosales, M. A. (1979), *Can. Symp. Catal.*, **6**, 228.

Tsuruya, S., Okamoto, Y., and Kuwada, T. (1979), *J. Catal.*, **56**, 52.

Vaidyanathan, K., and Doraiswamy, L. K. (1968), *Chem. Eng. Sci.*, **23**, 537.

Van Meerten, R. Z. C., and Coenen, J. W. E. (1977), *J. Catal.*, **46**, 13.

Volkenstein, F. E. F. (1948), *Lenin Uspekhi Khim.*, **17**, 174.

Wainwright, M. S., and Hoffman, T. W. (1977), *Can. J. Chem. Eng.*, **55**, 557.

Wanke, S. E., Lotochinsky, B. K., and Sidewell, H. C. (1979), *Can. Symp. Catal.*, **6**, 78.

Wei, J., and Prater, C. D. (1962), *Adv. Catal.*, **13**, 203.

Weller, S. (1956), *AIChE J.*, **2**, 59.

Weller, S. W. (1975), *Adv. Chem. Ser.*, **148**, 26.

Wentzheimer, W. W. (1969), Ph.D. thesis, Univ. of Pennsylvania.

Wisniak, J., and Simon, R. (1979), *Ind. Eng. Chem. Process Design Dev.*, **18**, 50.

Wright, B. S. (1964), *Brit. Chem. Eng.*, **9**, 758.

Wright, J. D., and Ball, W. E. (1967), *Ind. Eng. Chem. Fundam.*, **6**, 475.

Yates, Y. G., and Best, R. J. (1976), *Ind. Eng. Chem. Prod. Res. Dev.*, **15**, 239.

Yang, K. H., and Hougen, O. A. (1950), *Chem. Eng. Prog.*, **46**, 146.

CHAPTER THREE

Diffusion in Solid Catalysts

3.1. INTRODUCTION

Diffusion of reactants from the surface of a catalyst to the interior of its pores constitutes one of the resistances in a reaction system catalyzed by a solid surface. Since chemical reaction takes place simultaneously, concentration and temperature gradients tend to be established within the pores.

From the design point of view what is important is the extent of accessibility of the entire internal surface of the catalyst at pore-mouth conditions. This is usually expressed in terms of an effectiveness, or utilization, factor ε, which is defined as the ratio of the actual reaction rate in the presence of diffusional effects to the rate under conditions of surface concentration and temperature (i.e., no mass and heat transport resistance within the pellet). The diffusion of gas in a solid matrix is a complex phenomenon and is not amenable to accurate estimation. The best that we can do is to postulate various diffusion models and obtain an estimate of the diffusion coefficient.

The existing models that describe the transport of reactants into the porous catalyst fall into two main classes. In the first type the porous medium is modeled as a network of interconnected capillaries. The flux relations based on models of this first type can be deduced by sound physical arguments in the limiting case of Knudsen and bulk diffusion regimes (see Section 3.2.2). The lack of a solution to the problem of transport in a capillary whose diameter is comparable to the mean free path of the diffusing molecules (the transition region), however, hampers the use of these models in the transition region.

Models of the second type, based on the dusty-gas concept, circumvent this problem. The n species present in a diffusing gas mixture are supplemented by an $(n + 1)$th dummy species called the dust. The interaction of the gas molecule with the dust molecules simulates the interaction of the immobile solid matrix of a porous medium with the gas species. The model has, however, certain drawbacks in that some features of the real medium, such as the void fraction or pore distributions, have no concrete analogues in the model. The proportion of the dust particles dispersed in the gas can of course be varied so that we can cover the entire range from bulk diffusion, where collisions between pairs of molecules predominate, to Knudsen diffusion, where collisions between molecules and wall predominate.

Most of the reported literature until the early seventies was concerned primarily with the diffusion of a single species. In general, the developments were based on models of the first type. More recently, however, the question of multicomponent diffusion in real systems has been receiving increasing attention. Both types of models have been found to be useful in the analysis of multicomponent systems. In view of the absence of proven methods of using the equations for multicomponent diffusion, our attention will be primarily restricted to single-component diffusion. At the end of the chapter, however, a brief reference will be made to multicomponent diffusion.

3.2. MODES OF DIFFUSION IN CATALYST PORES

3.2.1. General Considerations

We shall primarily be concerned with the bulk properties of pellets prepared by subjecting catalyst particles to compression. Although in general the physical effects of compression alone (such as change in porosity, pore size distribution, and density) are important, the possibility of chemical changes occurring within a pellet as a result of compression must also be kept in mind. The results of Ogino and co-workers (1965, 1966a, b) have shown, for instance, that the surface acidity of solids varies on compression, resulting in a change in their catalytic activity. It has also been shown (Ogino and Nakajima, 1967) that compression of $ZnO-Cr_2O_3$ catalysts results in changes in the catalytic activity of the solids during the decomposition of methanol, there being a maximum activity at a compaction pressure of 2500 kg/cm².

The transport of a gas inside a porous pellet may take place by one or more of the following mechanisms: bulk diffusion; Knudsen diffusion; flow due to a pressure gradient; and surface diffusion. In the following treatment of diffusion, mainly the bulk and Knudsen modes of transport (which are known to be the more important ones) are considered. In view of the increasing realization of the role of surface diffusion and flow due to a pressure gradient, a brief discussion of these modes is also presented.

3.2.2. Bulk Diffusion

Bulk diffusion in catalyst pores constitutes diffusion according to the laws of gaseous diffusion, and the walls of the pores are not involved in this process. The local molar diffusion flux of a component A in a system consisting of components A and B is defined by Fick's law:

$$N_{bA} = -D_{bA}\frac{dC_A}{dl} \qquad (3.1)$$

where D_{bA} is the bulk diffusion coefficient (cm²/sec) of the component A. In the case of flow in chemical reactors or similar equipment, the design of the equipment is facilitated if this flow is expressed in relation to the equipment concerned, that is, with respect to stationary coordinates. Thus the conservation equation represented by (3.1) may be modified to

$$N_{bA} = -\frac{D_{bA}}{1-\alpha y_A}\frac{dC_A}{dl} \qquad (3.2)$$

where

$$\alpha = \left(1 + \frac{N_{bB}}{N_{bA}}\right) \qquad (3.3)$$

and y_A is the mole fraction of A in the mixture. It has been shown by several investigators (principally Hoogschagen, 1955; McCarty and Marson, 1960; and Dullien and Scott, 1962) that equimolal counterdiffusion does not exist (i.e., $N_{bB}/N_{bA} \neq 1$). Hoogschagen further showed that the flux ratio N_{bB}/N_{bA} can be related to molecular weights by the equation

$$\frac{N_{bB}}{N_{bA}} = -\sqrt{\frac{M_A}{M_B}} \qquad (3.4)$$

or

$$\alpha = \left(1 - \sqrt{\frac{M_A}{M_B}}\right) \qquad (3.5)$$

Using the boundary conditions $y_A = y_{A0}$ at $l = 0$, and $y_A = y_{AL}$ at $l = L$, we can readily derive the following integral form of the equation:

$$N_{bA} = D_{bA}\frac{P}{R_g TL}\left(\ln\frac{1-\alpha y_{AL}}{1-\alpha y_{A0}}\right) \qquad (3.6)$$

In general, bulk diffusion will be promoted (for simple molecules) in pores whose diameter is greater than 1000 Å (at atmospheric pressure). Bulk diffusion is independent of pore diameter.

The following theoretical relationship for the bulk diffusion coefficient, first proposed by Stefan and Maxwell, can also be derived (see also Present, 1958; Loeb, 1961).

$$D_{bA} = \left(\frac{Nk_b}{M^{1/2}}\right)^{1.5}\left(\frac{T}{P}\right)^{3/2} \qquad (3.7)$$

In Eq. 3.7, N represents Avogadro's number (6.023×10^{23}) and k_b the Boltzmann constant $[1.38 \times 10^{-16}\ \text{g cm}^2/(\text{g mol})^3\ {}^\circ\text{K}]$.

It is possible to estimate D_{bA} from generalized correlations. Quite often, more than two components are involved, and in such situations the diffusion coefficient of any component A can be estimated from a knowledge of the diffusivities in all the binary systems containing A.

3.2.3. Knudsen Diffusion

For gaseous molecules diffusing into pores whose radius is considerably less than the mean free path, the probability of collisions between the molecules and the pore walls increases in relation to the frequency of collisions between the molecules themselves, and the transport caused by this gas–solid collisional process is termed Knudsen diffusion.

From the kinetic theory of gases Knudsen derived the following equation for the diffusion of a gas inside a single infinitely long pore of radius r_p (neglecting end effects):

$$N_{KA} = -D_{KA}\frac{dC_A}{dl} = -D_{KA}\left(\frac{P}{R_g T}\right)\frac{dy_A}{dl} \qquad (3.8)$$

where

$$D_{KA} = \frac{2}{3}r_p\bar{u} = 9700r_p\left(\frac{T}{M}\right)^{1/2} \qquad (3.9)$$

Here $\bar{u}$ represents the mean molecular velocity and D_{KA} the diffusion coefficient of A for Knudsen diffusion.

TABLE 3.1. Effect of Operating Variables on Bulk and Knudsen Diffusivities and Fluxes

	Temperature	Pressure[a]	Pore Radius[a]
Diffusivity			
Bulk	$T^{3/2}$	$1/P$	ind
Knudsen	$T^{1/2}$	ind	r_p
Flux			
Bulk	$T^{1/2}$	ind	ind
Knudsen	$T^{-1/2}$	P	r_p

[a]ind = independent.

3.2.4. Effect of Temperature, Pressure, and Pore Radius on Bulk and Knudsen Diffusion

The effect of temperature and pressure on the bulk diffusion coefficient can be predicted from Eq. 3.7, and on the Knudsen diffusion coefficient from Eq. 3.9. But the behaviors of the two fluxes follow a different pattern. The conclusions are summarized in Table 3.1. An interesting feature of the effect of temperature on the flux in a capillary in which both the mechanisms are operative is that it would be independent of temperature at some well-defined pressure.

3.2.5. Diffusion in the Transition Region

Approximate criteria for the bulk and Knudsen modes of transport are as follows:

Knudsen:	$KN < 10$	(3.10a)
Transition:	$KN = 0.1–10$	(3.10b)
Bulk:	$KN > 0.1$	(3.10c)

where KN is the Knudsen number, defined as $KN = \lambda/2r_p$. In view of the uncertainty in the estimation of the mean pore radius, these criteria are at best crude. An extensive treatment of effective diffusivity in the transition region has been presented by Rothfeld (1960), Henry et al. (1967), and Cunningham and Geankoplis (1968).

The diffusion coefficient in the transition region is generally computed from the law of additive resistances (Bosanquet, 1944):

$$\frac{1}{D_{cA}} = \frac{1}{D_{bA}} + \frac{1}{D_{KA}} \qquad (3.11)$$

which is strictly valid for self-diffusion only (Pollard and Present, 1948; see also Wheeler, 1955). The following equation for bulk diffusion can now be formulated for the transition region for equimolal diffusion (see Pollard and Present, 1948; Scott and Dullien, 1962; Spiegler, 1966):

$$N_A = -\frac{D_{cA}P}{R_g T}\frac{dy_A}{dl} \qquad (3.12)$$

which on integration leads to

$$N_A = \frac{D_{bA}P}{\alpha R_g TL}\ln\left(\frac{1 - \alpha y_{AL} + R_D}{1 - \alpha y_{A0} + R_D}\right) \qquad (3.13)$$

where

$$R_D = \frac{D_{bA}}{D_{KA}} \qquad (3.14)$$

This is an important equation, for it represents completely the diffusion process within a capillary, containing all the information on the Stefan–Maxwell as well as the Knudsen equations. It has two interesting features: (1) it reduces to the Knudsen equation for small radii and to the Stefan–Maxwell equation for large values; and (2) it gives the Bosanquet interpolation formula for the special case of self-diffusion.

3.3. MODELS FOR EFFECTIVE DIFFUSION IN A POROUS CATALYST

It is known that diffusion from one side of a pellet to the opposite side follows a devious path. When this is combined with the fact that pellets are usually characterized by a distribution of pore radii, the uncertainty involved in determining the diffusion of a gas in a porous matrix becomes apparent. Thus an effective diffusion coefficient (for determining which a knowledge of the diffusion length is obviously necessary) would have to be defined that would be a function of pore characteristics. In addition to tortuous flow, three other factors influence gaseous diffusion in a porous matrix: fraction of dead-end pores, absence of diffusion through the solid portion of the porous pellet (namely, porosity), and the shape factor of the pellet.

Two approaches are available for defining an effective diffusion coefficient: (1) define a "labyrinth," or "tortuosity," factor, which would account for the lack of linear passage through the pellet; or (2) evaluate the effective diffusivity from computational models based on a distribution of pore lengths and radii.

3.3.1. Models Based on Tortuosity Factor

The Parallel-Pore Model

The parallel-pore model assumes straight parallel pores of equal radius running through the particle at an angle of 45° to the direction of flow. Based on this model, the tortuosity factor may be shown to be 2.

An approximate value of the pore radius can be obtained from data on pore volume and total pore area. On the assumption that the pores are cylindrical, Wheeler (1955) recommends the following equation for the average radius $\bar{r}_p$:

$$\bar{r}_p = 2\left(\frac{V_g}{S_g}\right) \qquad (3.15)$$

While Eq. 3.15 can be used to obtain an estimate of the mean pore radius for the simple parallel-pore model, for more complex systems a larger number of parameters would have to be defined. For any system, however, the mean pore radius can be defined by

$$\bar{r}_p = \frac{1}{\gamma_p}\left(\frac{2V_g}{S_g}\right) \qquad (3.16)$$

where γ_p is a function of the parameters of the pore. For nonintersecting straight cylindrical pores, γ_p equals

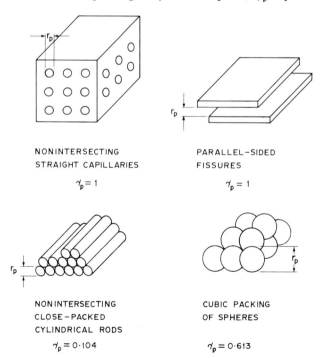

NONINTERSECTING
STRAIGHT CAPILLARIES

$\gamma_p = 1$

PARALLEL-SIDED
FISSURES

$\gamma_p = 1$

NONINTERSECTING
CLOSE-PACKED
CYLINDRICAL RODS

$\gamma_p = 0.104$

CUBIC PACKING
OF SPHERES

$\gamma_p = 0.613$

Figure 3.1. Some single-parameter pore models with their γ_p values.

unity. The values of γ_p for four single-parameter models, as computed by Everett (1958), are given in Figure 3.1, which also shows the pore structures considered. Two other pore structures that have received attention recently, the so-called spherical and cylindrical ink-bottle pores, are described in greater detail in Chapter 4. Another, more general, method of estimating the average pore radius is from a knowledge of the pore volume distribution, which can only be determined experimentally. Thus

$$\bar{r}_p = \int_0^{V_g} \frac{r_p \, dV}{V_g}. \qquad (3.17)$$

Lippert and Schneider (1973) have tested the parallel-pore model by using experimental diffusivity data for hydrogen in three catalysts. Although this model was found to give reasonably good agreement, the distributed-pore models (to be described later) were found to be slightly superior.

Effective Diffusivities in the Bulk and Knudsen Regimes

Recognizing that a tortuosity factor τ has to be incorporated in the analysis and that the actual diffusion flux would be N_{bA}/f_c (and not N_{bA}), we can define the following "effective bulk diffusion coefficient":

$$D_{ebA} = \frac{D_{bA} f_c}{\tau} \qquad (3.18)$$

The corresponding equation for Knudsen diffusion is

$$\log D_{eKA} = \log\left(\frac{19,400}{\tau}\right) + \log\left[\frac{f_c^2}{S_g \rho_c}\left(\frac{T}{M}\right)^{1/2}\right] \qquad (3.19)$$

where D_{eKA} is the effective diffusivity for Knudsen diffusion. This equation suggests that the effective diffusion coefficient for Knudsen diffusion is proportional to f_c^2. Weisz and Schwartz (1962) have proposed a model that postulates a system of overlapping spherical cells, each partially open to adjoining cell space. According to this model D_{eKA} is proportional to f_c^3.

It can be shown by considering the ratio of the circumference to the diameter of a body that the tortuosity factor should lie between 1 and 2. Higher values have been reported (Satterfield, 1970), sometimes as high as 20 (Scott and Dullien, 1962; de la Rue and

Tobias, 1955; see also Petersen, 1958), but no acceptable theoretical interpretation is available for such large values.

In a more rational approach to tortuosity, Dullien (1975a, b) expresses τ as the product $L'S'$ where L' accounts for increased path length (tortuous flow) and S' accounts for constrictions in the pores. If the porous matrix is assumed to consist of a network of capillary tubes, then one third of the tubes act as conducting or diffusing tubes in every orientation. The reciprocal of this factor is L'. Based on a network permeability model he has derived an expression for the constriction factor S', so that τ can be calculated from $\tau = L'S'$. Since, however, considerable experimental data on pore structure are necessary to compute S', this is not a readily usable method.

Transition Regime

When the equations for combined diffusivity developed earlier for the transition region are extended to diffusion in a catalyst pellet, one has to use appropriate tortuosity factors and voidage fractions to obtain a meaningful effective diffusion coefficient. If tortuosity correction is to be employed for both bulk and Knudsen diffusion, then R_D as defined earlier should be replaced by $(R_D)_e$:

$$(R_D)_e = \frac{D_{ebA}}{D_{eKA}} \tag{3.20}$$

On the other hand, if a tortuosity factor is defined for bulk diffusion only, based on the assumption that diffusion is restricted to macropores, which act as distribution channels for reaction to occur in micropores, then $(D_{eKA}) = D_{KA}$ and R_D becomes

$$(R'_D)_e = \frac{D_{ebA}}{D_{KA}} \tag{3.21}$$

Thus four equations can be written for D_e, and these are presented in Table 3.2. In order to use these equations the value of D_{ebA}, as well as that of D_{eKA}, for the first two equations in the table must be known. As the pressure is raised the diffusion coefficient reaches an asymptotic value which is equal to D_{bA} (or D_{ebA} in the case of a pellet), and $(R_D)_e$ as well as $(R'_D)_e$ become so small as to be negligible. Thus Eq. 3.13 can be used together with one experimental measurement of the flux at a high pressure and the corresponding values of y_{AL} and y_{A0}. Equations 3 and 4 of Table 3.2, which assume a tortuosity factor for bulk diffusion only, can

then be used to obtain D_{eA}. If Eqs. 1 and 2 of the table are to be used, then the value of D_{eKA} should also be known. This can be obtained from

$$N_A = \frac{D_{eKA} P}{R_g T L}(y_{A0} - y_{AL}) \tag{3.22}$$

by carrying out an experiment at a sufficiently low pressure where Knudsen diffusion would prevail.

Effect of Dead-End Pores

So far we have assumed that the entire pore volume of a pellet is available for diffusion; but such an assumption is likely to be in error, since there may be pores that are open at one end but closed at the other—cul-de-sac pores. Pores that are open at both ends are called flow pores, whereas those that have only one passage open to the exterior (i.e., one end is closed) are called dead-end pores.

In general, dead-end pores that are short relative to the pellet size (say $< 5\%$) would not contribute to diffusion, and in such cases it would be necessary to use flow porosity in determining D_e. Hedley et al. (1966) have proposed an elegant equation for estimating the length of dead-end pores on the assumption that in a system of straight intersecting pores (the Wheeler model) the average length of dead-end pores would be of the same order of magnitude as the average distance between pore intersections, given by $(3 \times 10^4 \pi \bar{r}_p / 4f_c)$. Goodknight et al. (1960) and Goodknight and Fatt (1961, 1963) have shown, however, from theoretical and experimental studies that the cul-de-sac pores affect only the transient-response transport characteristics of the porous material, so that their influence on steady-state operation may be neglected. On the other hand, the results of Wakao and Nardse (1974) suggest a significant influence of dead-end pores in the context of catalysis.

3.3.2. Rigorous Computational Models

A unique relationship cannot exist between the average radius (and such other gross properties as surface area and porosity) and pore size distribution, and therefore the use of an average radius in determining D_{KA} for estimating D_{eA} leads to erroneous results. The effect of pore size distribution on diffusion in a porous matrix is therefore of fundamental importance, and has been studied by Johnson and Stewart (1965) through a simple mathematical procedure. Wakao and Smith (1962) have proposed a random-pore model for a

TABLE 3.2. Mass Flux and Effective Diffusivity Equations in the Transition Region

Tortuosity Factor	Mass Flux	Effective Diffussivity	Remarks
When tortuosity is defined for both bulk and Knudsen diffusion	$N_A = \dfrac{(D_{ebA})P}{\alpha R_g TL}\ln\left[\dfrac{1-\alpha y_{AL}+(R_D)_e}{1-\alpha y_{A0}+(R_D)_e}\right]$	$D_{eA} = D_{ebA}\cdot\dfrac{\ln\left[\dfrac{1-\alpha y_{AL}+(R_D)_e}{1-\alpha y_{A0}+(R_D)_e}\right]}{\ln\left(\dfrac{1-\alpha y_{AL}}{1-\alpha y_{A0}}\right)}$	Integrated form (1)
		$D_{eA} = D_{ebA}\dfrac{\ln\left[\dfrac{1-\alpha y_{AL}+(R_D)_e}{1-\alpha y_{A0}+(R_D)_e}\right]}{\alpha(y_{A0}-y_{AL})}$	Integrated form (2) based on Fick's law
When tortuosity is defined for bulk diffusion only	$N_A = \dfrac{D_{ebA}P}{\alpha R_g TL}\ln\left[\dfrac{1-\alpha y_{AL}+(R'_D)_e}{1-\alpha y_{A0}+(R'_D)_e}\right]$	$D_{eA} = D_{ebA}\dfrac{\ln\left[\dfrac{1-\alpha y_{AL}+(R'_D)_e}{1-\alpha y_{A0}+(R'_D)_e}\right]}{\ln\left(\dfrac{1-\alpha y_{AL}}{1-\alpha y_{A0}}\right)}$	Integrated form (3)
		$D_{eA} = D_{ebA}\dfrac{\ln\left[\dfrac{1-\alpha y_{AL}+(R'_D)_e}{1-\alpha y_{A0}+(R'_D)_e}\right]}{\alpha(y_{A0}-y_{AL})}$	Integrated form (4) based on Fick's law

bidispersed pore system that has been extended to multidispersed systems by Cunningham and Geankoplis (1968).

A Simple Mathematical Procedure for Including the Effect of Pore Size Distribution

The basis of the simple analysis with respect to the pellet as a whole is illustrated in Figure 3.2. Diffusion occurs in the z direction, but within the pellet it occurs in the direction l in any given pore (note that there is no restriction on the diameter or length of the pore). In

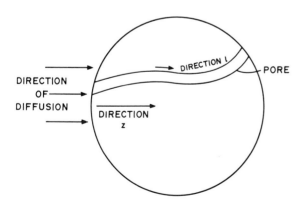

Figure 3.2. Johnson and Stewart's model (1965) for taking pore size distribution into account in estimating D_e.

order to include the effect of the pore matrix, two factors must be considered: pore inclination and pore size distribution. On the assumption of a constant value for pore inclination, Johnson and Stewart (1965) have derived the following equation for the flux in the direction z:

$$N_A = N_{Az} = -\frac{k^p P}{\alpha R_g TL}$$
$$\int_0^\infty \ln\left(\frac{1-\alpha y_{A0}+R_D}{1-\alpha y_{AL}+R_D}\right)f'(r_p)\,dr_p \tag{3.23}$$

where k^p is the pore inclination constant and $f'(r_p)dr_p$ the volume fraction of pores in the range from r_p to $r_p + dr_p$. Equation 3.23 can be used to calculate N_{Az} and then D_{eA} can be estimated from the integrated form of Fick's equation (see also Satterfield and Cadle, 1968).

In computing N_{Az} from Eq. 3.23, data on pore size distribution should be known, so that integration can be performed for several increments. The procedure is illustrated in Section 3.3.3.

This model is in essence an extension of the Wheeler model. It postulates straight cylindrical pores that do not intersect with one another, and assumes further that one third of the total number of pores are aligned in each of the three dimensions, that is, $k^p = \frac{1}{3}$. This leads to a tortuosity factor of 3, as in the Dullien model referred to earlier (with the constriction factor $S' = 1$). But for accurate work k^p should be experimentally determined from a diffusion measurement. Petty (1973) has suggested a simplification of the method, but in view of the constraints involved it is of limited use.

The Random-Pore Model

The random-pore model, first postulated by Wakao and Smith (1962) and extended subsequently by Cunningham and Geankoplis (1968), is based on the micro-macro structure of the pellet. The area involved in the diffusive process as well as the operative modes of diffusion are determined by the probability of a macropore lining up with another macropore or with a micropore. Mathematically the analysis, although leading to cumbersome equations, is quite straightforward. Its chief assumption is that the area voidage of a pellet is the same as its volume voidage.

Equations can be readily written for the three different pore alignments that are possible: (1) macropore–macropore; (2) micropore–micropore (i.e., through the particles); and (3) macropore–micropore. The areas involved (normal to the direction of diffusion) in each of these additive (parallel) modes of transport can be easily derived if it is remembered that

$f_{ma} + f_{mi} + f_s = 1$ (where f_s is the volume fraction occupied by the solid). The final expression is given later (in Table 3.4), and its use is demonstrated in Section 3.3.3. The foregoing treatment can be extended to several combinations of micro- and macropore regions, but this is not usually warranted.

Convergent–Divergent Pore Model

Houpeurt (1959) pointed out that a pore can be regarded as a series of convergent–divergent sections (see also Batra et al., 1970). Foster and Butt (1966) and Foster et al. (1967) used the convergent–divergent model to incorporate the effect of pore size distribution by assuming that the entire pore volume of a pellet is made up of two major conical arrays of pores: (1) a centrally convergent array, which is narrowest at the center, and (2) a centrally divergent array, which is widest at the center; each being the mirror image of the other. The pore size distribution is accounted for by assigning the volume corresponding to a given radius to segments of the two arrays of that radius.

3.3.3. Example: Effective diffusivity of Isobutene in Alumina: Comparison of the Wakao–Smith and Stewart–Johnson Models

Figure 3.3 shows the pore size distribution of an alumina catalyst. The micropore and macropore vo-

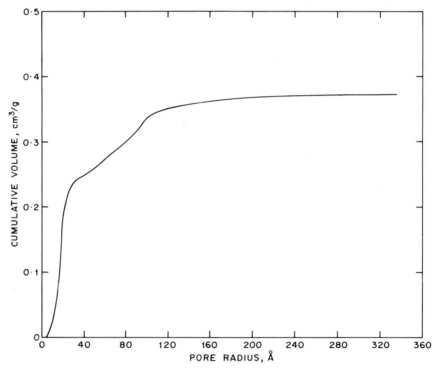

Figure 3.3. Pore size distribution of the alumina catalyst under consideration in the example.

lumes of the pellet are 0.25 and 0.13 cm^3/g, respectively. Calculate the effective diffusivity of isobutene in the pellet at 638°K and 1 atm pressure by using Wakao and Smith's model and Johnson and Stewart's model. Assume that pure isobutene and pure hydrogen are present on either side of the pellet, the bulk density of the catalyst is 1.04, and the molecular diffusivity is 1.13 cm^2/sec.

Solution

Wakao and Smith's Model

1. Calculation of macropore and micropore void fraction:

Macropore void fraction is given by

$$f_{ma} = \frac{\text{volume of macropores/g}}{\text{volume of the pellet/g}} = \frac{0.13}{1/1.04} = 0.135$$

Similarly, $f_{mi} = 0.25 \times 1.04 = 0.260$.

2. Estimation of mean micropore and macropore radius:

Let dv be the micropore volume, consisting of micropores whose radius ranges from r_{mi} to $r_{mi} + dr_{mi}$; then

$$dv = F(r_p)\, dr_p \qquad (3.24)$$

where $F(r_p)$ is the pore size distribution function. The values of $F(r_p)$ at various values of r_p can be obtained by differentiating the cumulative pore size distribution given in Figure 3.3. The function $F(r_p)$ so obtained is plotted in Figure 3.4. Hence

$$\bar{r}_{mi} = \frac{\int_0^{r_b} r_p F(r_p)\, dr_p}{\int_0^{r_b} F(r_p)\, dr_p} \qquad (3.25)$$

where r_b is the maximum micropore radius.

In Figure 3.4 a minimum is observed at $r_p = 42$ Å, and hence this is taken as the maximum j micropore radius r_b. The integrals in Eq. 3.25 can be evaluated numerically, and the value of the mean micropore radius so obtained is 21 Å.

Similarly, the mean macropore radius is given by

$$\bar{r}_{ma} = \frac{\int_{r_b}^{\infty} r_p F(r_p)\, dr_p}{\int_{r_b}^{\infty} F(r_p)\, dr_p} \qquad (3.26)$$

Numerical integration gives $\bar{r}_{ma} = 108.5$ Å.

3. Evaluation of micropore and macropore Knudsen diffusivities:

The Knudsen diffusivity is given by

$$D_{KA} = (9.7 \times 10^{-5})\bar{r}_p \left(\frac{T}{M_A}\right)^{1/2} \qquad (3.27)$$

where $\bar{r}_p$ is in Angstroms.
Hence

$$D_{Kma} = 3.56 \times 10^{-2}\,\text{cm}^2/\text{sec}$$
$$D_{Kmi} = 6.9 \ \times 10^{-3} \ \ \text{cm}^2/\text{sec}$$

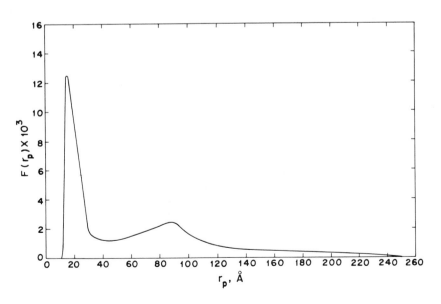

Figure 3.4. Plot of the function $F(r_p)$ vs. the pore radius of the catalyst in the example.

4. Estimation of the effective diffusion coefficient:

$$\frac{D_{eA}}{D_{bA}} = \left[\ln\left(1 - \alpha y_{AL}\right)/\left(1 - \alpha y_{A0}\right)\right]^{-1}\left\{ f_{ma}^2 \ln\left[\frac{1 - \alpha y_{AL} + \left(D_{bA}/D_{Kma}\right)}{1 - \alpha y_{A0} + \left(D_{bA}/D_{Kma}\right)}\right]\right.$$

$$+ f_{mi}^2\left[\frac{\alpha\left(y_{A0} - y_{AL}\right)}{1 - \alpha\left(\dfrac{y_{A0} + y_{AL}}{2}\right) + \dfrac{D_{bA}}{D_{Kmi}}}\right] + \left[\frac{4f_{ma}\left(1 - f_{ma}\right)}{1 + \dfrac{\left(1 - f_{ma}\right)^2}{f_{mi}^2}}\right]$$

$$\left. \times \frac{\alpha\left(y_{A0} - y_{AL}\right)}{1 - \alpha\left(\dfrac{y_{A0} + y_{AL}}{2}\right) + \dfrac{D_{bA}/D_{Kmi}}{1 + f_{mi}^2/\left(1 - f_{ma}\right)^2}}\right]\right\} \qquad (3.28)$$

Here $\alpha = 1 - \sqrt{56/2} = -4.3$

Substituting the numerical values in Eq. 3.28, we get $D_{eA}/D_{bA} = 3.15 \times 10^{-3}$ or $D_{eA} = \left(3.15 \times 10^{-3}\right)\, 1.13 = 3.56 \times 10^{-3}\ \text{cm}^2/\text{sec}.$

Johnson and Stewart's Model

The effective diffusivity in this case is given by

$$\frac{D_{eA}}{D_{bA}} = \frac{k^p P}{\left(M_A/M_B\right)^{1/2} - 1}\int_0^\infty \ln\left\{\frac{1 + y_{A0}\left[\left(M_A/M_B\right)^{1/2} - 1\right] + D_{bA}/D_{eA}}{1 + y_{AL}\left[\left(M_A/M_B\right)^{1/2} - 1\right] + D_{bA}/D_{eA}}f'\left(r_p\right)\,dr_p\right\} \quad (3.29)$$

where $f'(r_p)\,dr_p$ is the fraction of the pore volume consisting of pores with radii from r_p to $r_p + dr_p$.

Now $F(r_p)\,dr_p$ is the volume (in cubic centimeters per gram) consisting of pores with radii from r_p to $r_p + dr_p$. Hence

$$f'(r_p)\,dr_p = \frac{F(r_p)\,dr_p}{\text{total volume per gram}} \qquad (3.30)$$

That is, $f'(r_p) = \rho_B F(r_p) = 1.04\, F(r_p)$. The value of k^p for an isotropic medium is $\frac{1}{3}$; $y_{A0} = 1$; $y_{AL} = 0$.

The average pore radius is given by

$$\bar{r}_p = \frac{\displaystyle\int_0^\infty r_p f'(r_p)\,dr_p}{\displaystyle\int_0^\infty f'(r_p)\,dr_p} \qquad (3.31)$$

By numerical integration, $\bar{r}_p = 69.9\ \text{Å}$. Then

$$D_{KA} = \left(9.7 \times 10^{-5}\right)\left(\frac{T}{M}\right)^{1/2}\bar{r}_p, \qquad (\bar{r}_p \text{ in Angstroms}) \qquad (3.32)$$

$$= 2.29 \times 10^{-2}\ \text{cm}^2/\text{sec}$$

Substituting all these values in Eq. 3.24 and integrating numerically, we get $D_{eA}/D_{bA} = 2.57 \times 10^{-2}$ or $D_{eA} = 2.91 \times 10^{-2}\ \text{cm}^2/\text{sec}$.

It may be noted that there is a difference of almost one order of magnitude between the predictions of the two models. This emphasizes the need for choosing the right model for the purpose of estimation.

3.3.4. Role of Chemical Reaction

According to the experimental results of Wakao (1968) and Sterrett and Brown (1968), the effective diffusivities calculated from kinetic data (i.e., in the presence of a reaction) differ considerably from those determined under nonreactive conditions (say, from any of the models described above). A model has therefore been developed by Wakao (1968) that postulates an average pore radius for the so-called reaction void of a catalyst pellet as against the usual average pore radius calculated from the Wheeler model (Eq. 3.15).

The combined diffusivity (in a single uniform capillary) as defined in Section 3.2.5 can be calculated from Eq. 3.11 for both reactive and nonreactive situations in a pellet, with the difference that Hoogschagen's inverse square root molecular weight relation does not hold for

TABLE 3.3. Effective diffusivity and Equivalent Pore Radius Under Reactive and Nonreactive Conditions (Wakao Model)[a]

Condition	Effective Diffusivity	Equivalent Pore Radius
Reactive conditions	$D_{cA}^R(r_p) = \left[\dfrac{3}{2\bar{u}r_p} + \dfrac{y_B - (N_{bB}/N_{bA})}{D_{bA}} \right]^{-1}$	By simultaneous solution of $$\int_0^{V_t} \frac{dV}{r_p} = \left(\frac{f_c^R}{\bar{r}_p^R} \right) \frac{1}{\rho_c}$$ $$\int_0^{V_t} \left[\frac{D_{cA}^R(r_p)}{r_p} \right]^{1/2} dV = \left[\frac{D_{cA}^R(\bar{r}_p R)}{\bar{r}_p R} \right]^{1/2} \frac{f_c^R}{\rho_c}$$
Nonreactive conditions	$D_{cA}^D(r_p) = \left[\dfrac{3}{2\bar{u}r_p} + \dfrac{1 - (1 - \sqrt{M_A/M_B})y_A}{D_{bA}} \right]^{-1}$ For the known pore distribution $$D_{cA}^D = \rho_c \int_0^{V_t} D_{cA}^D(r_p)\, dV$$	$D_{cA}^D = f_c^D D_{cA} \bar{r}_p$

[a]Superscripts R and D denote reaction and diffusion (nonreaction) conditions, respectively.

the reactive case. The corresponding equations for the diffusion coefficient are given in Table 3.3, together with those for the pore radius based on Wakao's model.

It has been shown by Wakao (1968) that the diffusivities and pore parameters calculated by means of the equations in Table 3.3 for reactive and nonreactive pellets are quite different from each other.

3.3.5. Miscellaneous Models, Empirical Equations, and Experimental Observations

In addition to the more important models outlined above, a few other models for the pore structure of a pellet have also been proposed (Barrer and Gabor, 1959; Hewitt, 1966; Schlosser, 1966; Nicholson and Petropoulos, 1968; Strazsko and Schneider, 1974). These and several other aspects of the structure of porous materials have been discussed by Dullien and Batra (1970) and Jackson (1977).

A number of experimental studies have been reported that attempt to verify some of the theoretical models presented earlier. A few studies also attempt to provide semiempirical expressions that can be used for predicting the effective diffusivity. Thus, based on the results obtained on a Cr–Fe oxide catalyst in the temperature range 360–400°K, Plygunov et al. (1971)

suggest the following semiempirical equation:

$$\log D_{eA} = \log\left(\frac{D_{bA} f_c}{\tau} \right) \tanh\left(\frac{\bar{r}_p}{\lambda} \right) + 0.941 \log A_f + \frac{0.94E - E_a}{2.303 R_g T - 8.52} \tag{3.33}$$

where E and E_a represent the true and apparent activation energies, respectively. Grachev et al. (1971) compared various methods of measuring effective diffusivity and concluded that the "chemical method" is superior to the others. For the eight samples of Fe–Mo and vanadium catalysts tested they recommend the equation

$$D_e = (0.1-0.2)D_r$$

where D_r is the diffusivity calculated on the basis of the predominating pore radius.

3.3.6. Configurational Diffusion

Thus far we have considered the development of equations for the cases of bulk, Knudsen, and a combined type of diffusion. Bulk diffusion prevails when the dimensions of the pores are much larger than the diffusing molecule. On the other hand, Knudsen diffusion prevails when the pore size continues to be larger than the diffusing molecule but is much smaller than the mean free path. Extension of this condition brings us to the case where the dimensions of the pore and those of the diffusing molecule are about the same. This type of diffusion has been termed configurational diffusion (Weisz, 1973), and a rigorous analysis of it requires a consideration of complexities such as the details of the force field and wall-to-molecule interactions. No comprehensive theory has yet been developed for these problems; however, because of the practical importance of configurational diffusion, arising out of the use of a newer class of catalysts such s zeolites, it has been extensively investigated in the last few years (Brown et al., 1971; Riekert, 1970; Barrer, 1971; Bean, 1972).

Anderson and Quinn (1974), in analogy with the motion of spherical particles in a liquid flowing through a capillary, have suggested the following equation for configurational diffusion:

$$\frac{D_{cd}\tau}{f_c D_b} = \left(1 - \frac{a}{r_p}\right)^{2+n}, \qquad a < r_p \qquad (3.34a)$$

where a is half the molecular size and $n = 2$ for a sphere on the center line and 4 for a sphere off the center.

Satterfield et al. (1973) and Colton et al. (1975) suggest the following empirical equation:

$$\log_{10}\left(\frac{D_{cd}\tau}{f_c D_b}\right) = -2\left(\frac{a}{r_p}\right) \qquad (3.34b)$$

based on extensive experimental data.

3.4. ROLE OF FORCED FLOW AND SURFACE DIFFUSION

As already pointed out in Section 3.2.1, the influence of forced flow and surface diffusion is generally neglected in the various models proposed for the effective diffusion coefficient. In the following paragraphs we shall briefly examine the role of these two modes of transport.

3.4.1. Transport Due to Forced Flow

In the presence of a pressure gradient across a capillary, transport occurs as a result of both diffusion and flow, and Eq. 3.12 cannot be integrated, as was done for the case of constant pressure to give Eq. 3.13, without a knowledge of the variation of P with l. Account must also be taken of the fact that the bulk diffusion coefficient is itself a function of pressure and that the inverse square root molecular weight relationship (Eq. 3.5) no longer holds. A review of the procedures employed for including the effect of forced flow has been presented by Youngquist (1970).

To analyze the role of forced flow the following basic equation may be written for the total flux N_t°:

$$N_t^\circ = N_t + N_F = N_A + N_B + N_F = N_A^\circ + N_B^\circ \qquad (3.35)$$

where N_A° and N_B° are the combined diffusion and flow fluxes for species A and B, respectively, and N_F is the forced flow term. From the developments of Dullien and Scott (1962) and Wakao et al. (1965), the following final equation can be derived:

$$N_A^\circ = \frac{-D_{bA}\, dC_A/dl}{1 + R_D - \alpha y_A} + \frac{N_F y_A}{1 + R_D}$$

$$(1 - \alpha)N_A^\circ + N_B^\circ = -\frac{Y}{R_g T}\frac{dp}{dl} \qquad (3.36)$$

where

$$Y = \frac{(1-\alpha)D_{KA}}{[1+(1-\alpha)y_A]R_D} + \left[\frac{r_p^2 P}{8\mu} + \frac{D_{KA}P}{4[y_A + y_B/(1-\alpha)]}\right]$$
$$\times \left(\frac{1-\alpha y_A}{1+R_D} + \frac{1-y_A}{1+R_D}\right)$$

Equation 3.35 may now be integrated simultaneously by a trial-and-error procedure to give the total flux through the capillary (i.e., for bulk, Knudsen, and forced flow). Arithmetic averages for the composition and pressure may also be used without significant loss of accuracy to estimate the composition- and temperature-dependent terms in the expression for Y.

Equations 3.35 and 3.36 are for flow through a capillary, but can also be used for a catalyst merely by replacing D_{KA} by D_{eKA} and R_d by $(R_d)_e$. The numerical values of D_{KA} and D_{bA} can be determined by carrying out experiments at low and high pressures as explained in Section 3.3.1.

Haring and Greenkorn (1970) have proposed a model for flow in a porous structure based entirely on flow due to a pressure gradient by assuming a two-parameter probability distribution function (in the form of a β function) for both the radius and the length of the pore. In previous models based on a pressure gradient the porous solid was assumed to be homogeneous and isotropic.

For multicomponent diffusion the necessary condition that determines whether or not there will be a pressure gradient is given by $\Sigma v_i \sqrt{M_i} = 0$. For binary diffusion this simply reduces to $\Sigma v_i = 0$, leading to the common association of pressure gradients with change in number of moles (volume) during the reaction.

The lower and upper bounds for the pressure variations in simple cases have been obtained by Hite and Jackson (1977) for Knudsen as well as bulk diffusion control systems undergoing a first-order reaction (A → v_BB). For Knudsen diffusion the pressure rises from a value p_s at the surface to a value $\sqrt{v_B}\,p_s$ deep within the pellet. v_B represents the stoichiometric coefficient and for a value equal to 2 the maximum pressure variation could be as high as 40%. For bulk diffusion control the pressure within the pellet always lies between the bounds

$$
p_s \quad \text{and} \quad \left[p_s + \left(\ln \sqrt{\frac{M_A}{M_B}} \right) \frac{8}{3} \frac{\rho_B \bar{u} \lambda \tau D_{ebA}}{f_c r_p^2} \right]
$$

The effective diffusivity D_{eA} in the transition region is given by

$$
\frac{1}{D_{eA}} = \frac{1}{D_{ebA} p_s} + \frac{1}{D_{eKA}} \tag{3.37}
$$

where the ratio $D_{ebA} p_s / D_{eKA}$ has a value of approximately 0.01 and 1.0 for the bulk and transition regimes, respectively, and is greater than 1.0 for Knudsen diffusion.

3.4.2. Surface Diffusion

Surface diffusion is believed to occur as a result of partial desorption, which may be regarded as a "loosening" of the bond that is formed on adsorption to such an extent that some energetic molecules move to another active site on which the bonding will be stronger. In the absence of an active site nearby, the molecule will be desorbed into the gas phase. Surface diffusion is significant only in the case of pellets with very small pores, that is, in the region where Knudsen diffusion prevails. It has yet to be established fully whether surface transport is a momentum transfer or a diffusion phenomenon (Field et al., 1963; Barrer, 1963). In any case, for purposes of engineering convenience (particularly to enable a composite treatment of diffusion including volume and surface transport), it is expedient to retain the concept of a surface diffusion coefficient. This diffusion is to be differentiated from surface self-diffusion, in which an atom of the solid surface diffuses. Such self-diffusion rates can be enhanced by the presence of adsorbates. We are not concerned here with this aspect of surface diffusion, which has been reviewed by Bonzel (1973) and Rhead (1975).

Consider a single capillary in which surface migration occurs. The entire process of adsorption and desorption is so rapid that the two steps can be assumed to be in equilibrium. With this assumption we can write $C_{As} = K_A C_{Ab}$, where K_A is an adsorption equilibrium constant. It is thus possible to express surface concentration in terms of the gas-phase concentration. The following equation (which also accounts for surface diffusion along with bulk and Knudsen diffusion in defining a combined diffusion coefficient) can then be developed:

$$
\frac{1}{D_{cA}} = \frac{1 - \alpha y_A}{D_{bA}} + \frac{1}{D_{KA}} + \frac{r_p}{2 K_A D_S} \tag{3.38}
$$

Since bulk diffusion is usually negligible in the region where surface and Knudsen diffusion are operative, the first term on the right-hand side of Eq. 3.38 can be neglected.

There are several studies and reviews on surface diffusion, including those of Barrer (1963), Dacey (1965), Field et al. (1963), Bienert and Gelbin (1967), Kammermeyer and Rutz (1959), Sladek (1967), Satterfield (1970), Roberts and McKee (1978), and King (1980). Kammermeyer and Rutz (1959) propose a particularly interesting correlation in which the effect of surface diffusion is presented as a quantity in excess of that for Knudsen flow.

The following guidelines can be laid down for the absence of surface diffusion: (1) the partial pressure of the reacting species should be quite low; (2) the temperatures involved should be sufficiently high; and (3) the total pressure should be kept as low as possible. Whenever these prescriptions are not strictly followed surface diffusion cannot be ignored *a priori* and tests must be conducted to ensure its absence. Thus flux measurements with helium as the diffusing gas can be carried out and the applicability of Hoogschagen's relationship examined. While in general the applicability of this inverse square root law with helium as the diffusing gas can be regarded as a conclusive test for the absence of surface diffusion, some doubts have also been expressed regarding the correctness of this test (Barrer and Strachan, 1955).

Surface diffusion values usually range from 10^{-3} to 10^{-4} cm^2/sec at ambient temperature. Some typical values for commercial catalysts are given by Barrer and Gabor (1959, 1960), Smith and Metzner (1964), and Gelbin (1966).

3.4.3. Multicomponent Diffusion

Multicomponent diffusion of gases in porous solids is a fairly common situation encountered in many practical systems, and a number of mathematical models for these systems have been reported in the literature. For gaseous systems, the model of Mason and Evans (1969) and Gunn and King (1969) is simple and fairly accurate. These and other models (e.g., Feng and Stewart's model, 1973), however, do not aim at providing an explicit relationship for the effective diffusivity but are concerned with total gas transport.

The simple model of Mason et al. (1967) considers the flow and diffusion of n species in the l direction along a single pore within the solid. The total flux equation is written by adding viscous flow, gaseous diffusion, and surface diffusion terms:

$$N_t^\circ = \underset{\substack{\text{viscous} \\ \text{flow}}}{N_F} + \underset{\substack{\text{gaseous} \\ \text{diffusion}}}{N_t} + \underset{\substack{\text{surface} \\ \text{diffusion}}}{N_S} \qquad (3.39)$$

The terms for viscous flow and gaseous and surface diffusion require a knowledge of the gradients dp/dl and dC_i/dl. By properly accounting for these gradients and combining the result with Eq. 3.39 the flux equation for a single pore can be obtained.

This model has been extended by Feng and Stewart (1973) to a thoroughly cross-linked main pore network with singly attached dead-end pores. Among other models, mention may be made of those that take into account the pore size distribution. For this purpose, integration with a constant tortuosity factor for all pore sizes or a two-point quadrature to allow for a pore-size-dependent tortuosity factor can be used. Alternatively, pore size data can be used directly.

Although many diffusion models for multicomponent systems are available, they are largely untested because of lack of experimental data. From the data of Feng et al. (1974) on the system $He-N_2-CH_4$ on pelleted γ-alumina, it would appear that even the simple model using a single pore size with surface diffusion neglected gives surprisingly good predictions. Such a model therefore seems adequate for all practical purposes.

The models described thus far belong to the first type (see Section 3.1) in that they are based on the laws of transport in a single capillary. In the second type (the dusty-gas model), the ratio of the dust to gas-phase concentration is used to characterize the system. By varying the concentration of dust, the transport regime can be moved from Knudsen to bulk diffusion (see Jackson, 1977). For more recent developments in this area reference may be made to the papers by Haynes (1978), Ovenston and Walls (1982), and Fott and Schneider (1982).

Use of a Mean Effective Binary Diffusivity

For an ideal gas the kinetic theory leads to the Stefan–Maxwell equations:

$$N_j = -\sum_{i=1}^{N-1} C_t D_{ji} \nabla y_i + y_j \sum_{i=1}^{N} N_{bi} \qquad (3.40)$$

which are too complex for many engineering calculations. It seems desirable for most engineering purposes to define a mean effective binary diffusivity in order to simplify the computations.

Equation 3.40 can be written as

$$N_j = -C_t \overline{D}_j \nabla y_j + y_j \sum_{i=1}^{N} N_{bi} \qquad (3.41)$$

where $\overline{D}_j$ represents the mean binary diffusivity and can be obtained from the following considerations. For ideal gases the first terms in Eq. 3.41 can be written as

$$-C_t \nabla y_j = \sum_{\substack{j=1 \\ i \neq j}}^{N} \frac{1}{D_{ji}} (y_i N_j - y_j N_{bi}) \qquad (3.42)$$

Comparison of Eq. 3.41 and 3.42 gives

$$\frac{1}{\overline{D}_j} = \frac{\displaystyle\sum_{i=1}^{N} \frac{1}{D_{ji}} \left(y_i - y_j \frac{N_{bi}}{N_j} \right)}{1 - y_j \displaystyle\sum_{i=1}^{N} \frac{N_{bi}}{N_j}} \qquad (3.43)$$

Equation 3.43 for the diffusion of species 1 through stagnant species 2, 3, etc. leads to the well-known Fairbanks and Wilke (1950) equation:

$$\frac{1}{\overline{D}_j} = \frac{1}{1 - y_i} \sum_{i=2,3}^{N} \frac{y_i}{D_{1i}} \qquad (3.44)$$

This equation, however, cannot be used for reactive mixtures. Also, all other species are not necessarily stagnant. In such cases the steady-state flux ratios are determined by reaction stoichiometry and Eq. 3.43 becomes

$$\frac{1}{\overline{D}_j} = \frac{\displaystyle\sum_{i \neq j}^{N} \frac{1}{D_{ji}} \left(y_i - y_j \frac{v_i}{v_j} \right)}{1 - y_j \displaystyle\sum_{i=1}^{N} \frac{v_i}{v_j}} \qquad (3.45)$$

TABLE 3.4. Equations for Estimating Diffusivities in Various Regimes

Serial Number	Flow Regime	Parameters to be Estimated	Equation	Input Data
1.	Bulk diffusion	Bulk diffusivity	$D_{AB} = \dfrac{0.0018583\,T^{3/2}}{P\sigma^2\Omega}\left(\dfrac{1}{M_A} + \dfrac{1}{M_B}\right)^{1/2}$	Molal weights, temperature, and Leonard–Jones constants σ and Ω (Satterfield, 1970)
2.	Bulk diffusion in a porous pellet	Effective bulk diffusivity D_{ebA}	$D_{ebA} = \dfrac{f_c}{\tau}\,D_{bA}$	all the forgoing plus the tortuosity factor
3.	Knudsen diffusion in a straight uniform capillary	a. Knudsen diffusivity D_{KA}	$D_{KA} = 9700\,r_p\left(\dfrac{T}{M}\right)^{1/2}$	Capillary radius, temperature, and molecular weight
		b. Pore radius	$r_p = \bar{r}_p = \dfrac{2V_g}{S_g};\quad \bar{r}_p = \dfrac{\int_0^{V_g} r_p\,dV}{V_g}$	Surface area and pore volume per gram, pore volume distribution
4.	Knudsen diffusion in a porous pellet	Effective Knudsen diffusivity D_{eKA}	$D_{eKA} = 9700\,\bar{r}_p\left(\dfrac{T}{M}\right)^{1/2} = \dfrac{19400\,f_c^2}{S_g\rho_c\tau}$	$(T/M)^{1/2}$, mean radius (from 3b), temperature, molecular weight.
			$D_{eKA} = \dfrac{\alpha_A D_{KA}(y_{A0} - y_{AL})}{\ln\left(\dfrac{1 - \alpha_A y_{AL}}{1 - \alpha_A y_{A0}}\right)}$	Knudsen diffusivity from 3a and α_A from experiment or the Hoogschagen relation $\dfrac{N_{bA}}{N_{bB}} = -\left(\dfrac{M_B}{M_A}\right)^{1/2} = \alpha - 1$
5.	Transition regime a. Macropore–micropore model	Effective diffusivity	(see equation below)	Bulk diffusivity, molecular weights, macro- and micro porosities, mean macro- and micropore radii
	b. Rigorous equation accounting for pore size distribution	Effective diffusivity	$D_e = \dfrac{k^p D_{bA}}{\alpha(y_{A0} - y_{AL})}\displaystyle\int_0^\alpha \ln\left(\dfrac{1 - \alpha y_{A0} + R_D}{1 - \alpha y_{AL} + R_D}\right) f'(r_p)\,dr_p$	

Equation for row 5a:

$$\frac{D_e}{D_{bA}} = \frac{1}{\ln(1 - \alpha y_{AL})/(1 - \alpha y_{A0})}\; f_{ma}^2\ln\left[\frac{1 - \alpha y_{AL} + \dfrac{D_{bA}}{D_{Kma}}}{1 - \alpha y_{A0} + \dfrac{D_{bA}}{D_{Kma}}}\right]$$

$$+ \left\{ f_{mi}^2\,\frac{\dfrac{\alpha(y_{A0} - y_{AL})}{2}}{1 - \alpha\dfrac{(y_{A0} + y_{AL})}{2} + \dfrac{D_{bA}}{D_{Kmi}}}\right\}$$

$$+ \left\{ \frac{\dfrac{4f_{ma}(1 - f_{ma})}{1 + \dfrac{(1 - f_{ma})^2}{f_{mi}^2}}\;\alpha(y_{A0} - y_{AL})}{1 - \alpha\dfrac{(y_{A0} + y_{AL})}{2} + \dfrac{D_{bA}/D_{Kmi}}{1 + f_{mi}^2/(1 - f_{ma})^2}}\right\}$$

3.4.4. Example: Calculation of Mean Effective Diffusivity of Ammonia

Estimate the mean effective diffusivity of ammonia at an operating pressure of 226 atm and temperature of 385°C with the following data (the range of industrial operation is outside the region where Knudsen diffusion is important).

Subscripts A, B, and C denote ammonia, nitrogen, and hydrogen, respectively; $y_A = 0.0276$; $y_B = 0.2219$; $y_C = 0.6703$; inerts = 0.0802; $D_{AB} = 0.161$ cm^2/sec at 0°C and 1 atm; $D_{AC} = 0.629$ cm^2/sec at 0°C and 1 atm; and $f_c = 0.52$.

The bulk diffusion coefficient of ammonia can be obtained by Wilke's equation

$$\frac{1}{\overline{D}_A} = \frac{1}{1 - y_A} \sum_{i = B, C}^{N} \frac{y_i}{D_{Ai}}$$

$$\overline{D}_A = 0.3979 \text{ cm}^2/\text{sec}$$

Temperature and pressure correction are applied by using

$$\left(\frac{1}{\overline{D}_A}\right)_{T, P} = \left(\frac{1}{\overline{D}_A}\right)_{NTP} \left(\frac{T}{273}\right)^{1.75} \frac{1}{P}$$

$$(\overline{D}_A)_{T, P} = 8.2048 \times 10^{-3} \text{ cm}^2/\text{sec}$$

The mean effective diffusivity can then be calculated by using Wheeler's equation:

$$D_{eA} = \tfrac{1}{2}\overline{D}_A f_c$$
$$= 2.1342 \times 10^{-3} \text{ cm}^2/\text{sec}$$

3.5 RECOMMENDED EQUATIONS

The recommended equations are summarized in Table 3.4. When using Eq. 2 of Table 3.4 for bulk diffusion, a knowledge of the tortuosity factor is essential. Several investigators have reported values of τ ranging from 1.7 to 7.5 for various catalyst systems. Any value of τ between 2 and 4 would be a reasonable assumption. For unconsolidated particles τ is about 1.2 (Currie, 1960; see also Hoogschagen, 1955).

In the case of Knudsen diffusion, in addition to tortuosity, the average pore radius should be known. As for bulk diffusion, however, the choice of a suitable value for the tortuosity, is necessary. The more important values have been summarized by Satterfield (1970), from which it would appear that τ varies from 1.3 to 6.

As in the case of bulk diffusion, any value between 2 and 4 would be a reasonable assumption in engineering calculations.

The greatest difficulty arises in estimating the effective diffusivity in the transition regime. The simplest way to estimate this value is to use the Bosanquet equation referred to earlier (Eq. 3.11) but written for the effective values of bulk and Knudsen diffusivities. An average value of τ between 2 and 4 may be used.

The Johnson–Stewart and Wakao–Smith models have been shown to be quite accurate (see Satterfield and Cadle, 1968). However, Brown et al. (1969) have found that for extrapolation of data from one pressure to another the Johnson–Stewart model is superior. Raghavan and Doraiswamy (1977) have determined the effective diffusivity of isobutene in a fluorinated η-alumina catalyst by the gas chromatographic method and concluded that the Johnson–Stewart model is superior to the Wakao–Smith model and predicts D_e to within about 5%. In applying this model a theoretical orientation factor of $\tfrac{1}{3}$ corresponding to $\tau = 3$ may be used.

A simple rule of thumb for estimating D_e is

$$\frac{D_{eA}}{D_{bA}} = 0.05\text{--}0.10 \tag{3.46}$$

Thus the following recommendations are considered appropriate:

1. Equation 3.46 for a quick and approximate estimate of D_e.
2. Equation 5a from Table 3.4 (Wakao and Smith's) if the pore size distribution is distinctly bimodal.
3. Equation 5b from Table 3.4 (Johnson and Stewart's) if a single pore size distribution is adequate, by assuming a tortuosity factor of 3 (i.e., an orientation factor of $\tfrac{1}{3}$).

Note that, in using Eqs. 5a and 5b, since no experiments are to be conducted, precise values of y_{A0} and y_{AL} will not be known. However, little error is introduced by assuming $y_{A0} = 1$ and $y_{AL} = 0$ in solving these equations.

REFERENCES

Anderson, J. L., and Quinn, J. A. (1974), *Biophys. J.*, **14**, 130.

Barrer, R. M. (1971), *Adv. Chem.*, **102**, 1.

Barrer, R. M. (1963), *Appl. Mater. Res.*, **2**, 129.

Barrer, R. M., and Gabor, T. (1959), *Proc. Roy, Soc. (London)*, **A251**, 353.

Barrer, R. M., and Gabor, T. (1960), *Proc. Roy. Soc. (London)*, **A256**, 267.

Barrer, R. M., and Strachan, E. (1955), *Proc. Roy. Soc. (London)*, **A231**, 52.

Batra, V. K., Fulford, G. D., and Dullien, F. A. L. (1970), *Can. J. Chem. Eng.*, **48**, 622.

Bean, C. P. (1972), *The Physics of Porous Membranes*, I: *Membranes— A Series of Advances.*, ed. Eiseman G., Vol. I, Marcel Dekker, New York.

Bienert, R., and Gelbin, D. (1967), *Chem. Tech. (Berlin)*, **19**, 207.

Bonzel, H. P. (1973), *Structure and Properties of Metal Surfaces* (Honda Memorial Series on Material Science, No. 1, Maruzen, Tokyo.

Bosanquet, C. H. (1944), Brit. TA report BR-507.

Brown, L. F., Haynes, H. W., and Manogue, W. H. (1969), *J. Catal.*, **14**, 220.

Brown, L. M., Sherry, H. S., and Krambeck, F. J. (1971), *J. Phys. Chem.*, **75**, 3846.

Colton, C. K., Satterfield, C. N., and Lai, C. J. (1975), *AIChE J.*, **21**, 289.

Cunningham, R. S., and Geankoplis, J. (1968), *Ind. Eng. Chem. Fundam.*, **7**, 535.

Currie, J. A. (1960), *Brit. J. Appl. Phys.*, **11**, 318.

Dacey, J. R. (1965), *Ind. Eng. Chem.*, **57**(6), 27.

de la Rue, R. E., and Tobias, C. W. (1955), *Conductivities and Random Dispersions* (Proc. 107th Meeting Electrochem. Soc., Cincinnati).

Dullien, F. A. L. (1975a), *AIChE J.*, **21**, 299.

Dullien, F. A. L. (1975b), *AIChE J.*, **21**, 820.

Dullien, F. A. L., and Batra, V. K. (1970), *Ind. Eng. Chem.*, **62**(10), 25.

Dullien, F. A. L., and Scott, D. S. (1962), *Chem. Eng. Sci.*, **17**, 771.

Everett, D. H. (1958), *Proc. Symp. Colston Research Soc.*, **10**, 95.

Fairbanks, D. F., and Wilke, C. R. (1950), *Ind. Eng. Chem.*, **42**, 471.

Feng, C. F., and Stewart, W. E. (1973), *Ind. Eng. Chem. Fundam.*, **12**, 143.

Feng, C. F., Kostrov, V. V., and Stewart, W. E. (1974), *Ind. Eng. Chem. Fundam.*, **13**, 5.

Field, G. J., Watts, H., and Weller, K. R. (1963), *Rev. Pure Appl. Chem.*, **13**, 2.

Foster, R. N., and Butt, J. B. (1966), *AIChE J.*, **12**, 180.

Foster, R. N., Butt, J. B., and Bliss, H. (1967), *J. Catal.*, **7**, 191.

Fott, P., and Schneider, P. (1982), *Ind. Chem. Eng.*, (in press).

Gelbin, D. (1966), *Chem. Tech. (Berlin)*, **18**, 200.

Goodknight, R. C., and Fatt, I. (1961), *J. Phys. Chem.*, **65**, 1709.

Goodknight, R. C., and Fatt, I. (1963), *J. Phys. Chem.*, **67**, 949.

Goodknight, R. C., Klikoff, W. A., Jr., and Fatt, I. (1960), *J. Phys. Chem.*, **64**, 1162.

Grachev, G. A., Beskov, V. S., Ione, K. G., Malinovskaya, O. A., and Slinko, M. G. (1971), *Kinet. Katal.*, **12**, 1301.

Gunn, R. D., and King, C. J. (1969), *AIChE J.*, **15**, 507.

Hedley, W. H., Lavacot, F. J., Wank, S. L., and Armstrong, W. P. (1966), *AIChE J.*, **12**, 321.

Haring, R. E., and Greenkorn, R. A. (1970), *AIChE J.*, **16**, 477.

Haynes, Jr. H. W. (1978), *Can. J. Chem. Eng.*, **56**, 582.

Henry, J. P., Cunningham, R. S., and Geankopolis, C. J. (1967), *Chem. Eng. Sci.*, **22**, 11.

Hewitt, G. F. (1966), *Chem. Phys. Carbon*, **1**, 73.

Hite, R. H., and Jackson, R. (1977), *Chem. Eng. Sci.*, **32**, 703.

Hoogschagen, J. (1955), *Ind. Eng. Chem.*, **47**, 906.

Houpeurt, A. (1959), *Rev. Inst. Franc Petrole Ann. Combust. Liquides*, **14**, 1468.

Jackson, R. (1977), *Transport in Porous Catalysts*, Elsevier, Amsterdam.

Johnson, M. F. L., and Stewart, W. E. (1965), *J. Catal.*, **4**, 248.

Kammermeyer, K. A., and Rutz, L. O. (1959), *Chem. Eng. Progr. Symp. Ser.*, **55** (24), 163.

King, D. A. (1980), *J. Vac. Sci. Technol.*, **17**, 241.

Lippert, E., and Schneider, P. (1973), *Proc. Int. Symp. RILEM– IUPAC*, Part 2, E39–E55.

Loeb, L. B. (1961), *The Kinetic Theory of Gases*, 3rd ed., Dover, New York.

Mason, E. A., and Evans, R. B., III, (1969), *J. Chem. Educ.*, **46**, 358.

Mason, E. A., Malinauskas, A. P., and Evans, R. B., III, (1967), *J. Chem. Phys.*, **46**, 3199.

McCarty, K. P., and Mason, E. A. (1960), *Phys. Fluids*, **3**, 908.

Nicholson, D., and Petropoulos, J. H. (1968), *Brit. J. Appl. Phys.*, **1**, 1379.

Ogino, Y., and Kawakami, T. (1965), *Bull. Chem. Soc. Japan*, **38**, 972.

Ogino, Y., and Nakajima, S. (1967), *J. Catal.*, **9**, 251.

Ogino, Y., Kawakami, T., and Tsurumi, K. (1966a), *Bull. Chem. Soc. Japan*, **39**, 639.

Ogino, Y., Kawakami, T., and Matsuoka, T. (1966b), *Bull. Chem. Soc. Japan*, **39**, 859.

Ovenston, A., and Walls, J. R. (1982), *Chem. Eng. Sci.*, **37**, 1.

Petersen, E. E. (1958), *AIChE J.*, **4**, 343.

Petty, C. A. (1973), *Chem. Eng. Sci.*, **28**, 119.

Plygunov, A. S., Denisov, A. A., and Zhidkov, D. A. (1971), *Katal. Katal.*, **8**, 41.

Pollard, W. G., and Present, R. D. (1948), *Phys. Rev.*, **73**, 762.

Present, R. D. (1958), *Kinetic Theory of Gases*, McGraw-Hill, New York.

Raghavan, N. S., and Doraiswamy, L. K. (1977), *Ind. Eng. Chem. Process Design Dev.*, **16**, 519.

Rhead, G. (1975), *Surf. Sci.*, **47**, 207.

Riekert, L. (1970), *Adv. Catal.*, **21**, 281.

Roberts, M. W., and McKee, C. S. (1978), *Chemistry of the Metal–Gas Interface*, Oxford (Clarendon Press), New York and London.

Rothfeld, L. B. (1960), Ph.D. thesis, Univ. of Wisconsin, Madison.

Satterfield, C. N. (1970), *Mass Transfer in Heterogeneous Catalysis*, MIT Press, Cambridge, Massachusetts.

Satterfield, C. N., and Cadle, P. J. (1968), *Ind. Eng. Chem. Fundam.*, **7**, 202.

Satterfield, C. N., Colton, C. K., and Pitcher, W. H., Jr. (1973), *AIChE J.*, **19**, 628.

Schlosser, J. (1966), *Nucl. Sci. Eng.*, **24**, 123.

Scott, D. S., and Dullien, F. A. L. (1962) *AIChE J.*, **8**, 113.

Sladek, K. J. (1967), D. Sc. thesis, Massachusetts Institute of Technology.

Smith, R. K., and Metzner, A. B. (1964), *J. Phys. Chem.*, **68**, 2741.

Spiegler, K. S. (1966), *Ind. Eng. Chem. Fundam.*, **5**, 529.

Sterrett, J. S., and Brown, L. F. (1968), *AIChE J.*, **14**, 696.

Strazsko, J., and Schneider, P. (1974), *Chem. Prum.*, **24**, (7–8), 381.

Thomas, J. M., and Thomas, W. J. (1967), *Introduction to the Principles of Heterogeneous Catalysis*, Academic Press, New York.

Wakao, N. (1968), *Bull. Fac. Eng. Yokohama Natl. Univ.*, **16**, 101.

Wakao, N., and Nardse, Y. (1974), *Chem. Eng. Sci.*, **29**, 1304.

Wakao, N., and Smith, J. M. (1962), *Chem. Eng. Sci.*, **17**, 825.

Wakao, N., Otani, S., and Smith, J. M. (1965), *AIChE J.*, **11**, 435.

Weisz, P. B. (1973), *Chem. Tech. (Berlin).*, 498.

Weisz, P. B., and Schwartz, A. B. (1962), *J. Catal.*, **1**, 399.

Wheeler, A. (1955), *Catalysis*, **2**, 105.

Youngquist, G. R. (1970), *Ind. Eng. Chem.*, **62**(8), 52.

CHAPTER FOUR

Role of Pore Diffusion in Simple Reactions

The concept of the effectiveness factor ε has already been explained in Chapter 3. The use of this concept in reactor design computations is now well recognized. In this chapter methods are described for estimating catalyst effectiveness factors under different conditions for simple reactions, and worked-out examples illustrating the application of the important procedures described to typical situations encountered in reacting systems are presented. The treatment is extended to complex reactions in Chapter 5.

4.1. GENERALIZED CONSERVATION EQUATIONS

For a reaction A → products, the continuity equation for the diffusion of A inside a catalyst pellet accompanied by a chemical reaction is given by

$$\frac{d^2 C_A}{dX^2} + \frac{S}{X}\frac{dC_A}{dX} = \frac{r_{vA}}{D_{eA}} \quad (4.1)$$

Note that the general distance variable X (i.e., $X = r$ or l) has been used, and S is a shape constant with the following values: 0 for an infinite slab, 1 for a cylinder, and 2 for a sphere. The corresponding equation for energy conservation is

$$\frac{d^2 T}{dX^2} + \frac{S}{X}\frac{dT}{dX} = -\frac{r_{vA}(-\Delta H)}{k'_e} \quad (4.2)$$

Equations 4.1 and 4.2 can be rendered dimensionless through the following normalized parameters:

$$\hat{C}_A = \frac{C_A}{C_{As}}, \qquad \hat{T} = \frac{T}{T_s}, \qquad \hat{X} = \frac{X}{X_0} \quad (4.3)$$

where the subscript s denotes surface conditions and

X_0 is either the radius R or thickness L. The resulting expressions are

and

$$\frac{d^2\hat{C}_A}{d\hat{X}^2} + \frac{S}{\hat{X}}\frac{d\hat{C}_A}{d\hat{X}} = \frac{X_0^2 r_{vAs}}{D_{eA}C_{As}}\hat{r}_{vA} \quad (4.4)$$

$$\frac{d^2\hat{T}}{d\hat{X}^2} + \frac{S}{\hat{X}}\frac{d\hat{T}}{d\hat{X}} = \frac{(-\Delta H)X_0^2 r_{vAs}}{k'_e T_s}\hat{r}_{vA} \quad (4.5)$$

4.2. ISOTHERMAL EFFECTIVENESS FACTORS

4.2.1. First-Order Reaction in a Sphere

Consider the reversible reaction $A \rightleftharpoons B$. The reaction rate per unit volume of catalyst pellet is given by

$$r_{vA} = -k_v\left(C_A - \frac{C_B}{K}\right) \quad (4.6)$$

This equation can be readily recast in terms of a single component A to give

$$r_{vA} = -\frac{k_v(K+1)}{K}(C_A - C_A^*) = -k_r(C_A - C_A^*) \quad (4.7)$$

where C_A^* represents the equilibrium concentration and k_r is a consolidated constant that replaces the term $k_v(K+1)/K$ but has the same units as k_v ($\sec^{-1}$). For a spherical pellet the conservation equation takes the form

$$\frac{d^2 C_A}{dr^2} + \frac{2}{r}\frac{dC_A}{dr} = \frac{k_r}{D_{eA}}(C_A - C_A^*) \quad (4.8)$$

If the reaction is assumed to be homogeneous, the

boundary conditions are

$$C_A = C_{As} \quad \text{at} \quad r = R; \qquad \frac{dC_A}{dr} = 0 \quad \text{at} \quad r = 0$$
$$\text{(4.9)}$$

At this stage a dimensionless quantity (the Thiele modulus) may be formulated (Thiele, 1939; Zeldovich, 1939; Wheeler, 1951; Weisz, 1956, 1957):

$$\phi_{sn} = R\left[\frac{k_r(C_{As} - C_A^*)^{n-1}}{D_{eA}}\right]^{1/2}$$
$$= R\left[\frac{k_p(p_{As} - p_A^*)^{n-1} R_g T}{D_{eA}}\right]^{1/2} \quad \text{(4.10)}$$

where ϕ_{sn} represents the modulus for an nth-order reaction in a sphere, k_p is the rate constant expressed in partial pressure units, namely, $\text{mol/cm}^3 \text{ sec (atm)}^n$, and R_g is the gas constant in units of $\text{cm}^3 \text{ atm/mol} \,^\circ\text{K}$. If the reaction is assumed to be irreversible and first order, Eq. 4.10 reduces to

$$\phi_{s1} = R\left(\frac{k_v}{D_{eA}}\right)^{1/2} = R\left(\frac{k_p R_g T}{D_{eA}}\right)^{1/2} \quad \text{(4.11)}$$

Equation 4.8 can now be recast in dimensionless form as (for a reaction of order n)

$$\frac{d^2\hat{C}_A}{d\hat{X}^2} + \frac{S}{\hat{X}} \frac{d\hat{C}_A}{d\hat{X}} = \phi_{sn}^2 \hat{C}_A^n \quad \text{(4.12)}$$

where

$$\hat{C}_A = \begin{cases} \dfrac{C_A - C_A^*}{C_{As} - C_A^*} & \text{(for a reversible reaction)} \\[2ex] \dfrac{C_A}{C_{As}} & \text{(for an irreversible reaction)} \end{cases}$$
$$\text{(4.13)}$$

For the sphere, $S = 2$ and the boundary conditions expressed in nondimensional form are $\hat{C}_A = 1$ at $\hat{R} = 1$, and $d\hat{C}_A/d\hat{R} = 0$ at $\hat{R} = 0$. If the reaction is assumed to be first order, the solution is easily found as

$$\hat{C}_A = \frac{\sinh(\phi_{s1}\hat{R})}{\hat{R} \sinh \phi_{s1}} \quad \text{(4.14)}$$

This equation expresses the dimensionless concentration $\hat{C}_A$ as a function of dimensionless distance $\hat{R}$ inside the catalyst. The effectiveness factor can be

written as

$$\varepsilon = \frac{\text{actual reaction rate in the pellet}}{\text{rate in the pellet at pore-mouth concentration}}$$
$$\text{(4.15)}$$

where

$$\text{actual rate} = 4\pi R^2 D_{eA}\left[-\left(\frac{dC_A}{dr}\right)_{r=R}\right] \quad \text{(4.16)}$$

We can obtain $-(dC_A/dr)_{r=R}$ by differentiating Eq. 4.14 and setting $r = R$. Thus ε can be written as

$$\varepsilon = \frac{4\pi R^2 D_{eA}(-dC_A/dr)_{r=R}}{\frac{4}{3}\pi R^3 k_r(C_{As} - C_A^*)} \quad \text{(4.17)}$$

which, on simplification, leads to the following expression for the isothermal effectiveness factor for a first-order reaction in a catalyst pellet of spherical geometry:

$$\varepsilon = \frac{3}{\phi_{s1}}\left(\frac{1}{\tanh \phi_{s1}} - \frac{1}{\phi_{s1}}\right)$$
$$= \frac{3}{\phi_{s1}^2}(\phi_{s1} \coth \phi_{s1} - 1) \quad \text{(4.18)}$$

Clearly, an increase in the diffusion length (R) and reaction velocity constant or a decrease in effective diffusivity tends to decrease the effectiveness factor.

A further examination shows two asymptotic regions

$$\varepsilon \simeq 1, \qquad \phi_{s1} < 0.5$$
$$\varepsilon \simeq \frac{3}{\phi_{s1}}, \qquad \phi_{s1} > 5.0 \quad \text{(4.19)}$$

Thus we can easily estimate the effectiveness factor provided that $\phi_{s1} > 5.0$, whereas for values of $\phi_{s1} < 0.5$ pore diffusion can be neglected. The intermediate region (between the asymptotes) is not amenable to such generalization.

4.2.2. Effect of Reversibility

The Thiele modulus for a first-order reaction as given by Eq. 4.11 can be rewritten in the form

$$\phi_{s1} = R\left\{\left[\frac{k_v(K+1)}{K}\right]\frac{1}{D_{eA}}\right\}^{1/2}$$

For a reversible reaction, the term $(K+1)/K$ cannot be

neglected and the Thiele modulus will have a higher value than for an irreversible reaction.

Solutions similar to the forgoing have been presented by Wicke and Brotz (1949), Boreskov (1947), Boreskov and Slinko (1952), and Carberry (1962a) for a reaction in a catalyst slab, and by Wagner (1943) and Smith and Amundson (1951) for a reaction in a spherical catalyst particle. Other significant general treatments are those of Weisz and Prater (1954), Weisz and Swegler (1955), Petersen (1965), Drott and Aris (1969), Rester and Aris (1969), Rester et al. (1969), Satterfield (1970), Aris (1975), and Carberry (1976).

4.2.3. Effect of Reaction Order

For reactions of order higher than one, analytical solutions to the differential equations become difficult, since they involve combination of the linear diffusion process with a nonlinear chemical reaction, and numerical solutions become necessary. For the asymptotic portion of an nth order reaction corresponding to the falling ε region, the effectiveness factor becomes

$$\varepsilon = \frac{a(S+1)}{\phi_{sn}} \qquad (4.20)$$

The values of the constant a for various orders are zeroth, $\sqrt{2}$; first, 1; and second, $\sqrt{\frac{2}{3}}$. Petersen (1965) has shown that a generalized modulus can be postulated. Thus, for a sphere

$$\phi_s = \sqrt{A}\,\phi_{sn}, \qquad \text{where} \quad A = \frac{n+1}{2} \qquad (4.21)$$

If ϕ_{sn} in Eq. 4.20 is replaced by ϕ_s, this equation can be generalized and the asymptotic portions of the curves for different orders will coincide with one another (the value of a being unity). Aris (1957) and Bischoff (1965) have also formulated similar generalized moduli.

So far we have restricted the treatment to spherical geometry. Analogous derivations can be made for other geometric shapes: infinite slab and infinite cylinder. The solutions to the continuity equations for the three shapes, together with the equations for the respective moduli, are given in Table 4.1.

Generalization of reaction order is possible for other shapes also. Thus, for an infinite slab (plate), the curves for reactions of zero, first, and second order can be brought together (to within about 6%) by the following empirical equation:

$$\phi_p = \left(\frac{n+2.5}{3.5}\right)\phi_{pn} \qquad (4.22)$$

4.2.4. Effect of Catalyst Shape

The effect of catalyst shape is more pronounced than that of reaction order and has been the subject of several studies. Aris (1957) has shown that of all catalyst shapes of fixed volume for isothermal first-order reactions the spherical pellet has the lowest effectiveness factor; and Luss and Amundson (1967) have presented a mathematical proof of this hypothesis.

A practical consequence of this hypothesis is that for an exothermic reaction occurring inside a catalyst pellet, the intraphase temperature gradient (from the center to the surface of the shape) would be largest for a spherical pellet. This increased temperature gradient leads to an enhancement of the reaction rate, so that for a fixed volume the spherical shape would appear to be the best. By similar reasoning it can be seen that for simple endothermal reactions, the spherical shape would be the least desirable. For complex reactions, on the other hand, no such generalizations are possible, and each reaction would have to be analyzed separately. For example, in the reaction

$$A \xrightarrow{\text{exothermic}} B \xrightarrow{\text{endothermic}} C$$

the second step would be accentuated in the pellet interior, leading to a faster disappearance of B; consequently, if B is the desired product, the spherical shape would be the least desirable.

An examination of the modulus equations for the different shapes for a first-order reaction presented in Table 4.1 shows that all of these equations can be put in the general form

$$\phi_1 = X_0'\left(\frac{k_v}{D_e}\right)^{1/2} \qquad (4.23)$$

in which X_0' is a characteristic dimension dependent on the pellet shape (see Section 4.1). Aris (1957) has shown that for a first-order reaction at low values of ε the curves for all the shapes can be brought together if X_0' is defined as

$$X_0' = \frac{\text{volume of shape } V_p}{\text{surface area } A_p} \qquad (4.24)$$

TABLE 4.1. Intraphase Diffusion Parameters and Equations for a First-Order Reaction in Different Catalyst Shapes

Shape and Definition[a]	Thiele Modulus	Normalized Thiele Modulus	Equation for Concentration Profile	Equation for ε
Sphere $\hat{R} = \dfrac{r}{R}$	$\phi_{s1} = R\left(\dfrac{k_v}{D_e}\right)^{1/2}$	$\dfrac{R}{3}\left(\dfrac{k_v}{D_e}\right)^{1/2}$	$\hat{C}_A = \dfrac{\sinh(\phi_{s1}\hat{R})}{\hat{R}\sinh\phi_{s1}}$	$\varepsilon = \dfrac{3}{\phi_{s1}^2}\left(\phi_{s1}\coth\phi_{s1} - 1\right)$
Infinite slab a. One end open b. Both ends open $\hat{L} = \dfrac{l}{L}$	$\phi_{p1} = L\left(\dfrac{k_v}{D_e}\right)^{1/2}$ $\phi_{p1} = \dfrac{L}{2}\left(\dfrac{k_v}{D_e}\right)^{1/2}$	$L\left(\dfrac{k_v}{D_e}\right)^{1/2}$ $\dfrac{L}{2}\left(\dfrac{k_v}{D_e}\right)^{1/2}$	$\hat{C}_A = \dfrac{\cosh\phi_{p1}(1 - \hat{L})}{\cosh\phi_{p1}}$	$\varepsilon = \dfrac{\tanh\phi_{p1}}{\phi_{p1}}$
Infinite cylindrical rod (sealed ends) $\hat{R} = \dfrac{r}{R}$	$\phi_{c1} = R\left(\dfrac{k_v}{D_e}\right)^{1/2}$	$\dfrac{R}{2}\left(\dfrac{k_v}{D_e}\right)^{1/2}$	$\hat{C}_A = I_0(\hat{R}\phi_{c1})/I_0(\phi_{c1})$, or $\hat{C}_A = \exp[\hat{R}(\phi_{c1})]\exp(-\phi_{c1})$ for $\phi_{c1} > 4$	$\varepsilon = \dfrac{I_1(\phi_{c1})}{\phi_{c1}I_0(\phi_{c1})}$ where I_0 and I_1 are modified Bessel functions of the first kind of order 0 and 1, respectively.
Single pore (open ends) $\hat{L} = \dfrac{l}{L_p}$	$\phi_{p1} = L_p\left(\dfrac{k_v}{D_e}\right)^{1/2}$ $\left[= L_p\left(\dfrac{2k_s}{r_pD_e}\right)^{1/2}\right]$		Same as for infinite slab with $\hat{R}$ replaced by $\hat{L}\left(=\dfrac{l}{L_p}\right)$.	

Hollow cylinder (finite)

$$\alpha = \frac{R_1}{R_0}; \quad \beta = \frac{1}{R_0}; \quad Z = \frac{h\phi_{hc1}^2}{R_0}$$

γ, β_n, λ represent eigenvalues:

$$\gamma = \sqrt{1 + \left(\frac{n\pi R_0}{H\phi_{hc1}^2}\right)}; \quad \lambda = \sqrt{1 + \beta_n^2}$$

$$Y_0\left(\frac{\beta_n R_1 \phi_{hc1}^{1/2}}{R_0}\right) J_0\left(\beta_n \phi_{hc1}^{1/2}\right)$$

$$-Y_0\left(\beta_n \phi_{hc1}^{1/2}\right) J_0\left(\frac{\beta_n R_1 \phi_{hc1}^{1/2}}{R_0}\right) = 0$$

J_0 is the Bessel function of the first kind of order 0; Y_0 is the Bessel function of the second kind of order 0.

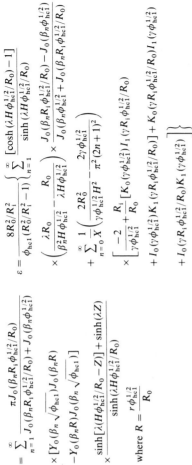

$$\phi_{hc1} = R_0\left(\frac{k_v}{D_e}\right)^{1/2} \qquad \frac{(1-\alpha)\beta\phi_{hc1}^{1/2}}{2(1-\alpha+\beta)}$$

$$\hat{C}_A = \sum_{n=1}^{\infty} \frac{\pi J_0(\beta_n R_1 \phi_{hc1}^{1/2}/R_0)}{J_0(\beta_n R_1 \phi_{hc1}^{1/2}/R_0) + J_0(\beta_n \phi_{hc1}^{1/2})}$$

$$\times \left[Y_0(\beta_n\sqrt{\phi_{hc1}}) J_0(\beta_n R) - Y_0(\beta_n R) J_0(\beta_n\sqrt{\phi_{hc1}})\right]$$

$$\times \frac{\sinh[\lambda(H\phi_{hc1}^{1/2}/R_0 - Z)] + \sinh(\lambda Z)}{\sinh(\lambda H\phi_{hc1}^{1/2}/R_0)}$$

$$\text{where } R = \frac{r\phi_{hc1}^{1/2}}{R_0}$$

$$\varepsilon = \frac{8R_0^2/R_i^2}{\phi_{hc1}(R_0^2/R_i^2 - 1)}\left\{\sum_{n=1}^{\infty} \frac{[\cosh(\lambda H\phi_{hc1}^{1/2}/R_0) - 1]}{\sinh(\lambda H\phi_{hc1}^{1/2}/R_0)}\right.$$

$$\times \left(\frac{\lambda R_0}{\beta_n^2 H\phi_{hc1}^{1/2}} - \frac{R_0}{\lambda H\phi_{hc1}^{1/2}}\right) \times \frac{J_0(\beta_n R_1\phi_{hc1}^{1/2}/R_0) - J_0(\beta_n\phi_{hc1}^{1/2}/R_0)}{J_0(\beta_n\phi_{hc1}^{1/2} + J_0(\beta_n R_1\phi_{hc1}^{1/2}/R_0)}$$

$$+ \sum_{n=0}^{\infty} \frac{1}{X}\left(\frac{2R_0^2}{\gamma\phi_{hc1}^{1/2}H^2} - \frac{2\gamma\phi_{hc1}^{1/2}}{\pi^2(2n+1)^2}\right)$$

$$\times \left[\frac{-2}{\gamma\phi_{hc1}^{1/2}} + \frac{R_1}{R_0}\left[K_0(\gamma\phi_{hc1}^{1/2})I_1(\gamma R_1\phi_{hc1}^{1/2}/R_0)\right] + K_0(\gamma R_1\phi_{hc1}^{1/2}/R_0)I_1(\gamma\phi_{hc1}^{1/2})\right.$$

$$\left.\left. + I_0(\gamma R_1\phi_{hc1}^{1/2}/R_0)K_1(\gamma\phi_{hc1}^{1/2})\right]\right\}$$

Solid cylinder (finite)

$$\gamma = \left(\frac{2R}{h}\right)^{1/3}; \quad \alpha = \frac{R}{h};$$

$$J_0(\sqrt{\lambda_n}) = 0$$

$$\phi_{sc1} = R\left(\frac{k_v}{D_e}\right)^{1/2} \qquad \left(\frac{2}{3}\right)^{1/3}\frac{\gamma\phi_{sc1}}{\gamma^3 + 2}$$

—

$$\varepsilon = 1 - \frac{32\phi_{sc1}^2}{\pi^2}\sum_{n=1}^{\infty}\sum_{m=1}^{\infty}\left\{(2m+1)^2\lambda_n[\lambda_n + \alpha^2(2m+1)^2\pi^2 + \phi_{sc1}^2]\right\}^{-1}$$

Paralleleopiped

a. With square ends

$$a = b, \quad \delta = \left(\frac{a}{c}\right)^{1/3}$$

b. Ends with constant[b] area

$$bc = a^2; \quad \delta = \frac{b}{c} = \frac{a}{c}; \quad \Delta = \frac{1}{\delta}\frac{a}{b}$$

$$\phi_{ps1} = R_e\left(\frac{k_v}{D_e}\right)^{1/2} \qquad \left(\frac{\pi}{6}\right)^{1/3}\frac{\delta\phi_{ps1}}{\delta^3 + 2}$$

$$\phi_{pc1} = R_e\left(\frac{k_v}{D_e}\right)^{1/2} \qquad \left(\frac{\pi}{6}\right)^{1/3}\frac{\delta\phi_{ps1}}{\delta^2 + \delta + 1}$$

—

$$\varepsilon = 1 - \frac{512\phi_{ps1}^2}{\pi^8}\sum_{m=1}^{\infty}\sum_{n=1}^{\infty}\sum_{p=1}^{\infty}\left[(2m+1)^2(2n+1)^2(2p+1)^2\right]^{-1}$$

$$\times \left[(2m+1)^2 + \Delta^2(2n+1)^2 + \delta^2(2p+1)^2 + \frac{\phi_{ps1}^2}{\pi^2}\right]^{-1}$$

[a] This column gives the leading dimensions as well as the definitions of the parameters of proportionality used $(\alpha, \beta, \gamma; \delta, \Delta)$ and eigenvalues of solutions.

[b] One dimension is held equal to the radius of the equivalent sphere (R_0), and the other two allowed to vary at fixed volume.

where A_p and V_p are the external surface area and volume of the shape, respectively. Thus $X_0' = L$ for an infinite slab, $R/3$ for a sphere, and $R/2$ for an infinite cylinder.

Based on Eqs. 4.23 and 4.24, a generalized expression can now be derived for ε as a function of ϕ_1 which will be independent of shape. Thus Eq. 4.17 becomes (for a first-order reaction)

$$\varepsilon = -\frac{1}{\phi_1^2}\left(\frac{d\hat{C}_A}{d\hat{X}}\right)_{\hat{X}=1} \qquad (4.25)$$

A plot of the effectiveness factor for a first-order reaction is presented in Figure 4.1 for the three shapes under consideration as a function of the shape-generalized modulus defined by Eq. 4.23. It will be seen that the points coincide at low values of ε and also as $\varepsilon \to 1$. There is a slight spread in the intermediate region, but even here the error is only of the order of 6–8% and is not significant.

Effectiveness factors for complicated shapes such as finite hollow cylinders, considered by Gunn (1967), and solid cylinders and parallelepipeds, considered by Luss and Amundson (1967), are also included as points in Figure 4.1 (see Table 4.1). It will be noted that the generalized modulus adequately represents the data for these shapes also.

While analytical solutions have been given to many of the shapes mentioned above, such solutions become more and more approximate as the shape irregularity increases. In such cases resort to numerical solution is necessary in order to obtain the effectiveness factor for even a simple first-order reaction. Recently some complicated shapes have been suggested in the patent literature to improve the reactor efficiency (see Suzuki and Uchida, 1979). When the reaction is limited by pore diffusion, the apparent reaction rate can be enhanced by reducing the particle size. This, however, leads to a corresponding increase in pressure drop. Ring catalysts have been used to overcome this problem, but these catalysts suffer from reduced crushing strength and manufacturing difficulties. Kasaoka and Sakata (1966) have observed that the Aris approximation leads to considerable deviations in effectiveness factor predictions for these catalysts.

An alternative solution is to use solid noncylindrical extruded catalysts with different—for example, circular, L-shaped, dumbbell, trilobe, or quadrulobe—cross sections. Suzuki and Uchida (1979) have computed effectiveness factors for extruded catalysts with these cross-sectional shapes by obtaining numerical solutions to the corresponding diffusion-cum-reaction equations. Their computations show that cylindrical and noncylindrical extrudates have the same effectiveness factor at a given generalized Thiele modulus defined using Eq. 4.24. This demonstrates the utility of the shape-independent modulus suggested by Aris as applied to such shapes. The only exception to the use of the generalized Thiele modulus for first-order reactions appears to be the ring catalyst.

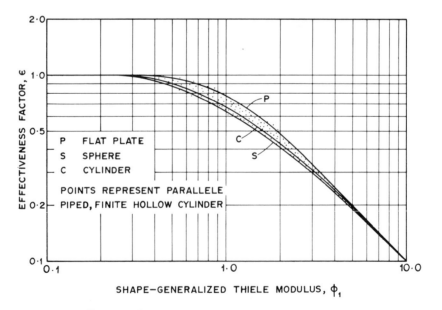

Figure 4.1. Effectiveness factor as a function of shape-generalized Thiele modulus (Rester and Aris, 1969).

A similar shape-generalized curve, based on the Aris definition, is not possible for reactions where the order $n \neq 1$. Sadana and Doraiswamy (1971) have suggested an empirical definition of the diffusion length (instead of Aris's ratio of volume to external surface area) which brings together the curves for all the shapes with reasonable accuracy.

4.2.5. Micropore and Macropore Effectiveness Factors

The effectiveness factor estimated by the methods outlined in Section 4.2.4 relate largely to the macropores. While this situation is essentially true for pellets prepared from silica gel, it may not be true for other pellets (e.g., of alumina). It would therefore be desirable to define and estimate effectiveness factors for both micropores and macropores. Henry et al. (1961), Mingle and Smith (1961), Carberry (1962), Tartarelli and Morelli (1967), and Tartarelli et al. (1970) have attempted to incorporate the effect of micropore diffusion. Two distinct methods of approach have been adopted: (1) based on separate definitions of the Thiele modulus and effectiveness factor for the macropores and micropores; and (2) from a generalized Thiele modulus.

Separate ϕ and ε for Macropores and Micropores

The overall effectiveness factor of a pellet is given by the product of the micro- and macroeffectiveness factors: $\varepsilon = \varepsilon_{mi} \cdot \varepsilon_{ma}$. Thus for a flat plate we have

$$\frac{\varepsilon}{\varepsilon_{mi}} = \frac{\tanh \phi_{p1,ma}}{\phi_{p1,ma}}$$

where

$$\phi_{p1,ma} = L_{ma}\left(\frac{k_v \varepsilon_{mi}}{D_{eA,ma}}\right)^{1/2}, \qquad (4.26)$$

$$\phi_{p1,mi} = L_{mi}\left(\frac{k_v}{D_{eA,mi}}\right)^{1/2}$$

In these equations the subscripts mi and ma represent micropores and macropores, respectively.

Far more data are required, however, in order to use Eq. 4.26 for evaluating $\varepsilon/\varepsilon_{mi}$; these data include pore length, diffusivities (separately) in the micropores and macropores, and a separate experimental program with particles (which constitute the pellet) to determine ε_{mi}.

Ors and Dogu (1979) have also obtained an analytical expression for the effectiveness factor for a bidispersed porous catalyst when a simple first-order reaction occurs in a spherical pellet with spherical microporous particles. The effectiveness factor is expressed in terms of the usual Thiele modulus and a parameter α_t that defines the ratio of diffusion time in the macropore to that in the micropore region. The value of this parameter for industrial systems can lie in the range of 1–1000, and therefore ε–ϕ_{s1} curves were generated for several values of α_t in this range. An interesting conclusion from this work is that ε varies greatly with the value of α_t. The conventional methods that take $\alpha_t \to \infty$ can therefore give substantially inaccurate predictions. The analysis has been extended by Kulkarni et al. (1981) to the case of a general-order reaction in an arbitrary shape. The final asymptotic expression derived by them is

$$\varepsilon = \sqrt{A}\,(1+S)^2\,\alpha_t^{-1/2}\,\phi^{-3/2} \qquad (4.27a)$$

where

$$\sqrt{A} = \frac{(2/n+1)^{1/2}}{n+\frac{3}{4}} \qquad (4.27b)$$

As in the case of macropore diffusion (Eq. 4.21), $\sqrt{A}$ generalizes the equation with respect to reaction order, with the difference that it has now been redefined to include the effect of micropore diffusion.

Generalized Thiele Moduli

It is possible to construct effectiveness factor plots on the basis of a generalized Thiele modulus (ϕ_{m-m}) for a bimodal pore structure. Table 4.2 summarizes the definitions of the Thiele modulus as well as the various nondimensional quantities (including M, m', and m'') used in these definitions. An analytical solution exists for the first-order case (given in the table), whereas for the second-order case resort to numerical methods is necessary.

Based on the numerical solutions given by Tartarelli et al. (1970), effectiveness factor plots are presented in Figure 4.2 for various values of the geometric modulus. In constructing these plots the appropriate definitions of the Thiele modulus for the first- and second-order reactions have been used and mean curves drawn. The maximum deviation of these curves from the precise first- and second-order plots is less than 6% and occurs in the intermediate range of the Thiele modulus.

As pointed out earlier for the case of separate moduli for the macropores and micropores, here also a great deal more information is required than for the single-pore model.

TABLE 4.2 Macro–Micro Effectiveness Factors: Equations for the Generalized Thiele Modulus for Different Reaction Types

Reaction	Generalized Thiele Modulus	Effectiveness Factor	Reference
A → products	$\phi_{m-m} = 2^{1/2}(m')^{3/2}M$	$\left[\tanh\left(\dfrac{\phi_{m-m}^2}{2M^2}\right)^{1/3}\right]^{1/2}\tanh\left\{(2\,\phi_{m-m}M^2)^{1/3}\right\}$ $\times\left[\tanh\left(\dfrac{\phi_{m-m}^2}{2M^2}\right)^{1/3}\right]^{1/2}\right\}\times\dfrac{1}{\phi_{m-m}}$	Tartarelli and Morelli (1967)
2A → products	$\phi_{m-m} = \left(\dfrac{75}{8}\right)^{1/4}(m'')^{3/2}M$	No analytical solution	Tartarelli et al (1970)
A + B → products	$\dfrac{\alpha'_D(m'')^{3/2}M(\hat{C}_{BA})_s}{\left\{\left[(\hat{C}_{BA})_s - \dfrac{\alpha'_D}{3}\right]^{3/2} - \dfrac{3}{5\alpha'_D}\left[(\hat{C}_{BA})_s\dfrac{\alpha'_D}{3}\right]^{5/2} + \dfrac{3[(\hat{C}_{BA})_s - \alpha'_D]^{5/2}}{5\alpha'_D}\right\}^{1/2}}$	No analytical solution	Tartarelli et al (1970)

$$m' = l_{mi}\left(\frac{2k_s}{r_{mi}D_{eA}}\right)^{1/2} \qquad (\hat{C}_{BA})_s = \frac{C_{Bs}}{C_{As}}$$

$$m'' = l_{mi}\left(\frac{2k_S C_{As}}{r_{mi}D_{eA}}\right)^{1/2} \qquad \alpha'_D = \frac{D_{eA}}{D_{eB}}$$

$$M = \frac{D_{eA\cdot mi}}{D_{eA\cdot ma}}\left(\frac{l_{ma}^2}{r_{ma}l_{mi}}\right)^{1/2}$$

(M is the geometric modulus for ma–mi pore structure.)

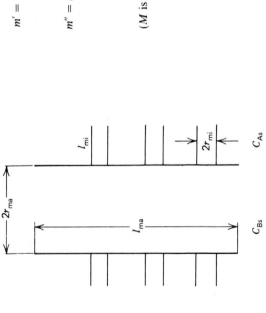

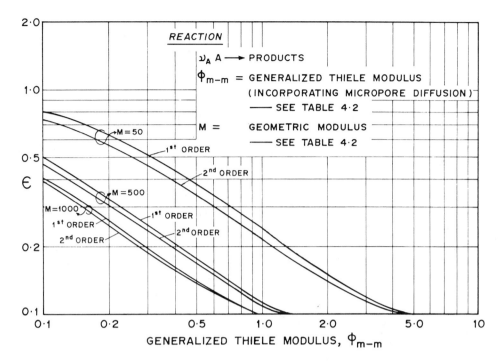

Figure 4.2. Effectiveness factor as a function of a generalized Thiele modulus for micropore–macropore diffusion (drawn from the data of Tartarelli et al., 1970).

4.2.6. Miscellaneous Effects

The effectiveness factor of a catalyst is influenced by such factors as transverse diffusion within a pore, multicomponent diffusion, pore size distribution, pore shape, inhomogeneity of the reaction, and flow regime. Although quantitatively these effects are usually not significant, an understanding of their role is necessary in order to appreciate the restraints and assumptions associated with the effectiveness factor plots developed earlier. Accordingly, a brief discussion of these effects is presented next.

Transverse Diffusion

Bischoff (1965) has shown that the effect of transverse diffusion is negligible except at low values of l/R (say, < 50) and may have to be considered only for very thin slabs, a situation not very common in industrial practice.

Pore Size Distribution

The developments presented so far have been based on the assumption that a single value of the effective diffusivity is adequate to characterize diffusion within a solid pellet. An alternative approach, developed by Schmalzer (1969), is to use experimentally determined distribution data directly in estimating the effectiveness factor, thus obviating the need to compute an effective diffusion coefficient. The only diffusion data required in such a procedure would be the bulk and Knudsen diffusivities, and the effectiveness factor so computed would obviously be a function of pore radius and pore length dispersion.

The effectiveness factor plots prepared by Schmalzer (1969) show that the effect of radial dispersion is small in the chemical control regime (low ϕ), whereas in the diffusion-influenced regime increased radial dispersion has the effect of lowering the effectiveness factor. The effect of pore length dispersion is similar but less marked. However, if an experimental value of D_e is available, this method has no advantage to offer.

Flow Regime

The development of effectiveness factors so far has been based on the concept of an effective diffusion coefficient that accounts for bulk and Knudsen modes of flow. It would be interesting, however, to examine the effect of flow regime on the effectiveness factor.

Scott (1962) and subsequently Otani et al. (1965) have developed the following equation for the effectiveness factor for a first-order reaction in a capillary (which is the same as for a flat plate):

$$\varepsilon = \frac{\tanh\left[\phi_{pl}(1 + D_{bA}/D_{KA})^{1/2}\right]}{\phi_{pl}(1 + D_{bA}/D_{KA})^{1/2}} \quad (4.28)$$

When the ratio of the bulk to Knudsen diffusivities $D_{bA}/D_{KA} \ll 1$, bulk diffusion dominates and Eq. 4.28 reduces to the equation for an infinite slab given in Table 4.1, which is the commonly used effectiveness factor equation. The effectiveness factor decreases considerably as the regime of flow changes from bulk to Knudsen.

Pore Shape (*Ink-Bottle Pores*)

It was pointed out in Chapter 3 that several types of pore structures are possible, and some of them (single-parameter pores) were illustrated in Figure 3.1. In all these structures the characterization of a pore through a single length parameter, usually the radius, was assumed. The possibility of an entirely different but more realistic two-dimensional pore structure, the so-called ink-bottle pore, was first suggested by Kraemer (1931), and elaborated subsequently by Broekhoff and de Boer (1968), who proposed two general types of ink-bottle pores: one essentially spheroidal with a circular aperture, and the other cylindrical with a constriction at the entrance. These two types of pores are schematically represented in Figure 4.3.

Obviously the effectiveness factor calculated for ink-bottle pores would be considerably different from that for cylindrical pores. Chu and Chon (1970) have presented effectiveness factor plots for a first-order irreversible isothermal reaction occurring in these two

TYPE I INK BOTTLE

(a) SPHERICAL
    INK BOTTLE

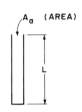

(b) AREA EQUIVALENT
    CYLINDRICAL PORE

TYPE II INK BOTTLE

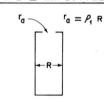

(a) CYLINDRICAL
    INK BOTTLE

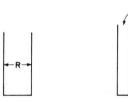

AREA EQUIVALENT CYLINDRICAL PORES

(b) $R' = R$ (c) $R' = r_a = \rho_1 R$

TYPE III INK BOTTLE

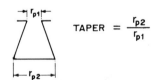

Figure 4.3. Parameters of ink-bottle pores. A_a, r_a, e_1, R', r_{p1} and r_{p2} represent the parameters of area equivalent shapes.

TABLE 4.3   Dimensions of Tapered Pores (Idealized Ink-Bottle Pores) in Silica–Alumina Catalysts[a]

Catalyst	Type	Mean Radius of the Pore Mouth r_{p1} (Å)	Mean Radius of the Pore Bottom r_{p2} (Å)	Mean Taper r_{p2}/r_{p1}	Total Porosity (cm³/g)
1	Commercial (13 % Al_2O_3)	974	1400	1.44	0.494
2	Same as catalyst 1 activated (1 % MgO)	1730	1990	1.15	0.504
3	Commercial (13 % Al_2O_3)	630	1155	1.85	0.505
4	Same as catalyst 2 activated (5 % ZnF_2)	866	1415	1.63	0.590

[a]From Ferraiolo et al. (1973).

types of pores on the assumption that bulk diffusion prevails, from which it would appear that the effectiveness factor for a Type I ink-bottle pore is much higher than that for a cylinder without a narrow neck (i.e., an ordinary straight cylindrical pore).

In another approach, the ink-bottle pores have been idealized by being represented as tapered pores (Type III ink bottle in Figure 4.3). Evidently the ratio of the pore radii at the two ends (r_{p2}/r_{p1}) is a measure of the effect of ink bottling. Ferraiolo et al. (1973) have shown that the activity of many silica–alumina catalysts decreases with increase in this ratio. A few typical values of the ratio are given in Table 4.3.

Inhomogeneity of Surface

It is of interest to note that the mass conservation equation can also be solved on the assumption that the reaction occurs on the surface. In this simple method of accounting for inhomogeneity, the boundary condition at $X = 0$ (i.e., $\hat{X} = 0$) is written as

$$-D_{eA}\frac{dC_A}{dX} = k_v C_A, \quad X = 0 \qquad (4.29)$$

instead of $dC_A/dX = 0$, which implies that there is no concentration gradient at the closed end. Based on the solutions worked out by Chu and Chon (1970) using this boundary condition in conjunction with the normal second condition $C_A = C_{As}$ at the surface, it may be concluded that this simple mathematical method of accounting for heterogeneity differs little from the pseudohomogeneous assumption and may be dis-

carded in favor of the more elaborate methods (based on H–W models) to be described in Section 4.6.

4.2.7. Practical Modification of the Effectiveness Factor Plot

In using the effectiveness factor plots derived and discussed in the earlier sections, a knowledge of the rate constant is necessary for calculating the Thiele modulus. Since a rate equation may not always be readily available, a modified Thiele modulus, originally suggested by Wagner (1943) and Weisz and Prater (1954), can be advantageously employed. This modulus $(\phi)_r$ is based on reaction rate rather than rate constant:

$$(\phi)_r^2 = X_0'^2 \left(\frac{r_{vA}}{D_{eA}C_{As}}\right) = \left(\frac{X_0'^2}{D_{eA}C_{As}}\right)\varepsilon k_v C_{As}^n \qquad (4.30)$$

where r_{vA} (or $r_{wA}\rho_B$) is the rate of reaction, expressed as moles reacting per unit time per unit gross volume of catalyst (mol/sec cm³ catalyst). Combining this with Eq. 4.23 (for a pellet of any shape) yields

$$(\phi)_r = \varepsilon^{1/2}\phi \qquad (4.31)$$

ε can now be plotted as a function of $(\phi)_r$ (Weisz and Prater, 1954)

4.2.8. Effectiveness Factors for a Bimolecular Reaction

If the concept of the effectiveness factor is extended to systems in which more than one reacting species is involved, the number of parameters and their combinations become too large for generalized graphical representation, and numerical solution for individual cases would be necessary. Solutions for some specific parameter values for the reaction

$$A + B \rightleftharpoons R + S$$

have been given by Maymo and Cunningham (1966).

Several industrial bimolecular reactions, such as the oxidation of hydrocarbons and hydrogenation of organic compounds, are of the type $A + v_B B \rightarrow$ products, and it would be useful to determine the extent to which the presence of B influences the reaction if it is assumed to be first order in A. Reactant A will be referred to as the key reactant and B as the non-key reactant. As a result of the reaction, either the key component or the non-key component can be exhausted at a certain depth of penetration. If the key component is exhausted, no special treatment is necessary. Gioia et al. (1970) established a criterion (for an isothermal pellet) to determine whether the non-key component is limiting. Karanth et al. (1974) have extended the treatment to a nonisothermal pellet with both inter- and intraphase gradients and have proposed a non-key-component parameter Δ whose value provides a measure of the influence of this component:

$$\Delta = 1 - \frac{C_{Bs}}{C_{As}} \frac{D_{eB}}{D_{eA}} \frac{1}{v_B} \quad (4.32)$$

As C_{Bs}/C_{As} approaches zero, Δ approaches unity. On the other hand, increasing values of C_{Bs}/C_{As} lead to a decrease in Δ, which now approaches zero as C_{Bs} approaches $C_{As} D_{eA} v_B / D_{eB}$. Throughout the region represented by

$$0 < \Delta < 1$$

the component B may be limiting, that is, an inert inner core of constant concentration and temperature may exist. Under these conditions Karanth et al. (1974) have shown that the effectiveness factor would be lower than that calculated on the assumption of Δ equal to unity. The condition

$$\frac{C_{Bs}}{C_{As}} \frac{D_{eB}}{D_{eA}} \frac{1}{v_B} > 1$$

leads to negative values of Δ, which means that the key component is now limiting, so that the first-order assumption would be valid.

It may be concluded that pseudo-first-order behavior is justifiable only when the key component is limiting. When the non-key component is limiting, Δ must be estimated from Eq. 4.32 and the influence of this component then determined as follows.

The parameter Δ represents in reality the reduced concentration of the key component C_A at the depth where the non-key component is exhausted. Karanth et al. (1974) have provided solutions that give the effectiveness factor as a function of both the Thiele modulus and the parameter Δ (in addition to the parameter α_s and γ, as discussed in Section 4.3). These can be used to obtain ε for given values of Δ and ϕ. In most practical cases, however, condition 4.32 would be valid and no correction for Δ would be necessary.

4.2.9. Effect of the External Surface

At very high values of the Thiele modulus (say, greater than 10^6) the resistance to diffusion is so large that the reactant penetrates only a short distance into the pellet before it is completely consumed. Since in this situation the total internal surface involved is small, the contribution of the external surface cannot be neglected. In contrast to this case, when the depth of penetration is larger, the internal surface associated with the reaction is also correspondingly larger; thus it is justifiable to ignore the external surface.

The chief practical consequence of this situation is that where costly metals such as platinum are used as catalysts, it would be desirable to coat this material on the external surface of a nonporous support. This will avoid the wasteful penetration of the metal into regions whose contribution to the total catalytic action would be insignificant.

In the analysis of catalyst effectiveness on porous supports, all of the catalytically active area is assumed to be internal. However, there exist examples, such as the partial oxidation of ethylene, in which it is preferable that the support material be nonporous or have a low internal area. In such cases the external area constitutes a small but finite fraction of the area for catalysis. The problem of finite external and internal areas for the purpose of catalysis has been considered by Goldstein and Carberry (1973) and subsequently by Varghese et al. (1978), who obtained the following asymptotic expressions for the effectiveness factor:

1. $\phi_{p1} \rightarrow 0$

$$\varepsilon = 1 - (1 - f_e) \left[\frac{1 - f_e}{3} \frac{1}{(Bi)_m} \right] \phi_{p1}^2 + (1 - f_e)$$

$$\left[\frac{1}{(Bi)_m^2} + \frac{(1 - f_e)(2 - f_e)}{3(Bi)_m} + \frac{2}{15}(1 - f)^2 \right] \phi_{p1}^4 \quad (4.33)$$

2. $\phi_{p1} \to \infty$

$$e = \frac{1}{\phi_{p1}}, \quad \text{for} \quad (Bi)_m \to \infty, \quad f_e = 0$$

$$\varepsilon = f_e + \frac{\sqrt{1-f_e}}{\phi_{p1}}, \quad \text{for} \quad (Bi)_m \to \infty, \quad f_e \neq 0 \quad (4.34)$$

$$\varepsilon = \frac{(Bi)_m}{\phi_{p1}^2}\left[1 - \frac{(Bi)_m}{\phi_{p1}} + \left(\frac{(Bi)_m}{\phi_{p1}}\right)^2\right.$$
$$\left. + 0\left(\frac{1}{\phi_{p1}^3}\right)\right], \quad \text{for} \quad (Bi)_m \text{ finite}, \quad (4.35)$$
$$f_e = 0$$

$$\varepsilon = \frac{(Bi)_m}{\phi_{p1}^2}\left[1 - \frac{(Bi)_m}{f_e \phi_{p1}^2} + o\left(\frac{1}{\phi_{p1}^3}\right)\right], \quad \text{for} \quad (Bi)_m \text{ finite},$$
$$f_e \neq 0 \quad (4.36)$$

where f_e is the fraction of the total active area that constitutes the external surface and $(Bi)_m$ is the Biot number for mass transfer $(k_g L/D_e)$.

4.3. NONISOTHERMAL EFFECTIVENESS FACTORS

The assumption of isothermicity is valid for most endothermal reactions (which are usually characterized by relatively low heats of reaction) occurring on solid catalysts of high thermal conductivity, but not for most exothermal reactions (which usually involve high heats of reaction) occurring on catalysts of low thermal conductivity (e.g., supported catalysts).

4.3.1. Examples of Intraphase Temperature Gradients

The temperature gradient inside a pellet is a strong function of the pellet thermal conductivity. Since the ratio of mass to heat Biot numbers is an indication of the relative thermal conductivities of the pellet and the fluid film surrounding it, Kehoe and Butt (1972) varied this ratio from 20 to 200 by using pellets whose thermal conductivities differed by a factor of 10. A substantial temperature gradient was found to exist in the pellet for $(Bi)_m/(Bi)_h \simeq 20$, whereas for $(Bi)_m/(Bi)_h \simeq 200$ the entire temperature rise was localized in the film and the pellet was essentially isothermal. Thus the question of pellet nonisothermicity is closely linked with system properties. Examples of pellet nonisothermicity are given by (among others) Irving and Butt (1967), Miller

and Deans (1967) and Koh and Hughes (1974); intraparticle temperature gradients of $10°$–$50°$ C have often been observed.

4.3.2. Formulation of Generalized Continuity and Conservation Equations

The following equation can be obtained for the concentration of reactant at any point in a pellet as a function of temperature at that point (Damköhler, 1943; Weisz and Prater, 1954):

$$(C_{As} - C_A) = \frac{k'_e}{D_{eA}(-\Delta H)}(T - T_s)$$

or

$$\hat{T} = \frac{T}{T_s} = 1 + \frac{(-\Delta H)D_{eA}C_{As}}{k'_e T_s}(1 - \hat{C}_A) \quad (4.37)$$

This equation is of general applicability in that it is independent of particle shape or size or kinetic expression. The temperature (T/T_s) is sometimes referred to as the Prater temperature.

We shall now consider the two approaches that have been employed for dealing with the differential equations to compute nonisothermal effectiveness factors. Essentially these approaches are based on eliminating all concentration or temperature terms (only the boundary value of the eliminated term is retained). Östergaard (1963) and Weisz and Hicks (1962) eliminated the temperature terms, while Tinkler and Metzner (1961) and Gunn (1966) eliminated the concentration terms (see Aris, 1975). In both these approaches it is necessary to express the rate constant at any point within the pellet in terms of the surface value. This can be done through the Arrhenius equation,

$$k_v = A_f \exp\left(-\frac{E}{R_g T}\right) \quad (4.38)$$

The following three forms can be obtained:

$$\frac{k_v}{k_{vs}} = \exp\left[\left(\frac{E}{R_g}\right)\frac{T_s - T}{T_s^2}\right] \quad (4.39a)$$

[Note that (T/T_s) has been replaced by T_s^2.]

$$\frac{k_v}{k_{vs}} = \exp\left[\alpha_s\left(1 - \frac{1}{\hat{T}}\right)\right] \quad (4.39b)$$

$$\frac{k_v}{k_{vs}} = \exp\left[\alpha_s\left(\frac{\Delta T}{T_s}\right)\left(1 + \frac{\Delta T}{T_s}\right)^{-1}\right] \quad (4.39c)$$

where

$$\alpha_s = \frac{E}{R_g T_s} \tag{4.40a}$$

$$\hat{T} = \frac{T}{T_s} = \frac{T_s + \Delta T}{T_s} \tag{4.40b}$$

Note that Eq. 4.39a involves an approximation, whereas the other two do not and are merely different forms of the Arrhenius equation expressed in terms of surface conditions.

The following generalized nondimensional continuity and conservation equations applicable to different shapes can now be developed by eliminating the concentration term and using Eq. 4.39a for expressing k_v/k_{vs}:

$$\frac{d^2\beta}{d\hat{X}^2} + \frac{S}{\hat{X}}\frac{d\beta}{d\hat{X}} + \exp(-\beta\gamma)(1 - n\phi_n^2\beta) = 0 \tag{4.41}$$

where

$$\beta = \frac{(T - T_s)k_e'}{(-\Delta H)R^2 k_{vs} C_{As}^n}$$

$$\gamma = \frac{E(-\Delta H)D_{eA}C_{As}}{R_g T_s^2 k_e'} \tag{4.42}$$

These represent the temperature and heat generation terms, respectively. As against this two-parameter (ϕ, γ) equation, the following three-parameter $(\phi, \beta_m, \alpha_s)$ equation can be obtained by eliminating the temperature term:

$$\frac{d\hat{C}_A}{d\hat{X}^2} + \frac{S}{\hat{X}}\frac{d\hat{C}_A}{d\hat{X}} + \phi_n^2\hat{C}_A^n \exp\left[\alpha_s\beta_m\frac{1 - \hat{C}_A}{1 + \beta_m(1 - \hat{C}_A)}\right] = 0 \tag{4.43}$$

where

$$\beta_m = \frac{(-\Delta H)C_{As}^n D_{eA}}{k_e' T_s} \tag{4.44}$$

These equations are based on the applicability of power law kinetics. If, however, the equations are to be useful for any kind of kinetics (including the H–W models described in Chapter 2), the rate term should be preserved as an independent function of concentration, say $f(\hat{C}_A)$. Thus the following general set of equations (based on three parameters) can be written for concentration and temperature:

$$\frac{d^2\hat{C}_A}{d\hat{X}^2} + \frac{S}{\hat{X}}\frac{d\hat{C}_A}{d\hat{X}} = \phi^2 f(\hat{C}_A) \exp\left[\alpha_s\beta_m\frac{1 - \hat{C}_A}{1 + \beta_m(1 - \hat{C}_A)}\right] \tag{4.45}$$

$$\frac{d^2\hat{T}}{d\hat{X}^2} + \frac{S}{\hat{X}}\frac{d\hat{T}}{d\hat{X}} = -\beta_m\phi^2 f(\hat{C}_A)$$

$$\times \exp\left[\alpha_s\beta_m\frac{1 - \hat{C}_A}{1 + \beta_m(1 - \hat{C}_A)}\right] \tag{4.46}$$

We shall have occasion to use these equations while considering the H–W models in Section 4.5.

4.3.3. Solutions

Numerical Solutions to the Three-Parameter Equations

Exact numerical solutions to Eq. 4.43 in terms of the three parameters β_m, ϕ_{sn}, and α_s provide an accurate estimate of the effectiveness factor under different conditions. Solutions for certain specific values of these parameters have been worked out by Weisz and Hicks (1962) and Weisz (1959) for a first-order reaction in a slab, while Carberry (1961) has considered both first- and second-order equations for slab geometry. Solutions have also been provided by Östergaard (1963) and Gunn (1966) for a first-order reaction, by Hlavacek and Marek (1967, 1968) for the zero-order case, and by Carberry and Wendel (1963), McGinnis (1965), Liu (1967), and Kesten (1969) for other situations. The subject has been fully reviewed by Aris (1975).

Multiple Solutions

Some of the solutions of Weisz and Hicks (1962) are presented in Figure 4.4. It will be noticed that the effectiveness factor passes through a maximum in the low Thiele modulus region and then falls monotonically as the modulus is further increased. The first region corresponds to chemical control, whereas in the second diffusional limitation is involved. It is further evident from the figure that for highly exothermic reactions three solutions are possible, and in the case of a zero-order reaction in a spherical pellet (Hlavacek and Marek, 1968) more than three solutions (e.g., five) are possible.

If to this intraparticle problem interphase resistance is also added, the nature of the solutions can be completely altered (see Section 6.7). Thus McGreavy and Cresswell (1969) have observed that where a single solution is obtained in the intraparticle problem, imposition of external film resistances can result in three solutions. Hatfield and Aris (1969) have further shown that more than two regions of multiple solutions can exist for the general case where both internal and external resistances are involved, resulting in three

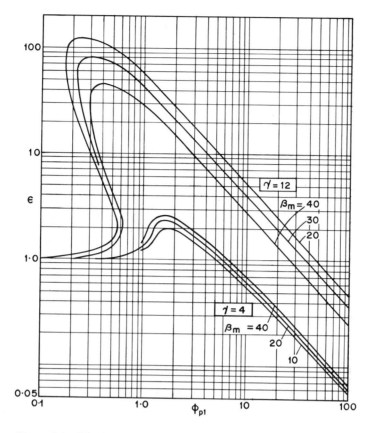

Figure 4.4. Effectiveness factor plots prepared from solutions of the three-parameter equation (redrawn from Aris, 1966).

steady-state values. When the regions overlap the number of steady states rises to five. The case of a first-order irreversible reaction occurring in a catalyst pellet with external transport limitations under assumed pellet isothermicity has been analyzed by several workers (Cresswell, 1970; Hlavacek and Kubicek, 1970; Lee et al., 1972; McGreavy and Thornton, 1970, and Pereira et al., 1979). The number of stable steady states must be distinguished from the number of solutions, for not all solutions represent steady-state values. Many attempts have been made (see, e.g., Hlavacek et al., 1968; Cresswell, 1970; Copelowitz and Aris, 1970; Michelsen and Villadsen, 1972; Aris, 1975, van den Bosch and Luss, 1977a, b; Luss, 1977; Tsotsis and Schmitz, 1979; Hlavacek and van Rompay, 1981) to determine the number of steady states and to derive criteria for establishing the uniqueness of the steady state and the bounds within which multiple solutions occur; these are discussed in Chapter 12 while dealing with the general problem of multiple solutions.

The numerical solutions above are based on the assumption that the parameter β_m can be calculated at

surface conditions. Although strictly β_m is a function of temperature, the temperature dependence of β_m being negligible, this assumption is quite justifiable.

Solutions to the Two-Parameter Equation

Computer solutions of Eq. 4.41 for reactions of different orders involving catalyst pellets of different shapes have been published by Tinkler and Metzner (1961); those for a first-order reaction in a sphere are shown in Figure 4.5. In this figure the Thiele modulus is based on V_p/A_p and should therefore be applicable to all shapes.

Many reactions that are diffusionally limited fall in the Thiele modulus range greater than about 2. Thus the straight decreasing portion of the effectiveness factor plots would be of greater value for design purposes, and it would be useful to develop equations that are asymptotically exact in terms of the two parameters ϕ and γ. Under asymptotic conditions reaction is restricted to the surface layer, so that all pellet shapes may be treated as slabs. The following

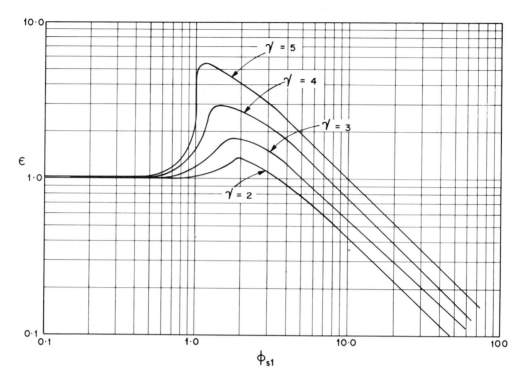

Figure 4.5. Effectiveness factor plots prepared from solutions of the two-parameter equation (Tinkler and Metzner, 1961).

asymptotic solutions can be derived (see Petersen, 1962, 1965; Patterson and Cresswell, 1971):

First-order reactions:

$$\varepsilon = \frac{3\sqrt{2}e^{\gamma/2}}{|\gamma|\phi_{s1}}[1 - e^{-\gamma}(1+\gamma)]^{1/2} \qquad (4.47)$$

Second-order reactions:

$$\varepsilon = \frac{3\sqrt{2}e^{\gamma/2}}{\phi_{s1}}\left\{-\frac{2}{\gamma^3}[e^{-\gamma}(\gamma-1)-1]-\frac{e^{-\gamma}}{\gamma}\right\}^{1/2} \qquad (4.48)$$

These equations are exact only at the asymptotic extreme, that is, at $\phi \to \infty$, but can be used for all values of $\phi > 1.0$ in reactor design computations.

Limits of Validity of the Two-Parameter Solutions

It will be recalled that the two-parameter solutions of Tinkler and Metzner are based on an approximate form of the Arrhenius relation given by Eq. 4.39a. This equation is exactly equivalent to the Arrhenius equation only when $T = T_s$, and for the values of $T/T_s \neq 1$

the equivalence is only approximate, the disparity being greater the greater the deviation of T/T_s from unity (see Aris, 1966).

Figure 4.4, which contains a few selected effectiveness factor plots of Weisz and Hicks (1962), shows that for a given value of the parameter γ, several solutions are possible, depending on the values of α_s. Evidently the approximation is seriously in error for large values of γ, while it seems to be quite adequate for low values of γ (say, < 4).

It has been shown by Luss (1968) that the Tinkler–Metzner approximation for a given value of γ is in fact the upper limit for the exact solutions based on specific values of α_s and β_m, the product $\alpha_s\beta_m$ being equal to γ. In an extension of this theorem, the lower bound is shown to be given by the approximation $\gamma = \alpha_s\beta_m/(1+\beta_m)$. The exact solution for any combination of α_s and β_m then lies between these two bounds. At values of $\gamma < 6$, the region of approximation has been shown to reduce to a narrow band which can be used for obtaining reasonably reliable estimates of the effectiveness factor. Where the heat effect is higher, either numerical solutions should be obtained, or the Weisz–Hicks plots should be used. In either case, three-parameter solutions must be resorted to.

4.3.4. Generalized Empirical Equations

Wedel and Luss (1980) have proposed a general semirigorous correlation for estimating ε. A simpler, though somewhat more approximate and restrictive, correlation is that of Churchill (1977), which is applicable to an isothermal irreversible reaction of general order in a pellet of any geometry.

$$\frac{1}{\varepsilon} - \frac{k_v C_{Ab}^{n-1} V_p}{k_g A_p} = \left[\left(\frac{C_{Ab}}{C_{As}} \right)_0^{2n-2} + \frac{(n+1) k_v C_{Ab}^{n-1} V_p^2}{2(CF) D_{eA} A_p^2} \left(\frac{C_{Ab}}{C_{As}} \right)^{n-1} \right]^{1/2}$$

(4.49)

Valdman and Hughes (1976) have also proposed a simple expression for calculating the effectiveness factor for the H–W type of kinetics. A single-point collocation has been used to obtain the results for the specific rate form $r_v = k_v C_A / (1 + K_A C_A)^2$.

Carberry (1961) and Liu (1970) have developed the following two-parameter approximate equations for estimating the effectiveness factor under nonisothermal conditions:

First-order reactions:

Exothermal: $\varepsilon = \dfrac{\exp(\gamma/5)}{\phi_1}, \qquad \begin{array}{l} \gamma < 6 \\ \phi_1 > 2.5 \end{array}$ (4.50)

Endothermal: $\varepsilon = \dfrac{\tanh \phi_1 (1-\gamma)^{0.3}}{\phi_1 (1-\gamma)^{0.3}}$ (4.51)

Second-order reactions:

Exothermal: $\varepsilon = \exp(0.133 \phi_2 \gamma) - 1.0$

$$+ \frac{\tanh(1.33 \phi_2)}{1.33 \phi_2}, \qquad \begin{array}{l} \gamma > 3.5 \\ \phi_2 < 1.15 \end{array}$$

(4.52)

$$\varepsilon = \frac{\exp(\gamma/5)}{1.33 \phi_2}, \qquad \text{other conditions}$$

Endothermal:

$$\varepsilon = \frac{\tanh(1.33 \phi_2)(1-\gamma)^{0.33}}{1.33 \phi_2 (1-\gamma)^{0.33}}$$

(4.53)

Rajadhyaksha and Vasudeva (1974) have suggested the following three-parameter empirical equation:

$$\varepsilon = \{ \exp[1.172 \beta_m (\alpha_s - 8)]^{1/2} - 1 \} [\exp(\varepsilon_{iso}) - 1] + \varepsilon_{iso}$$

(4.54)

It is only necessary to estimate the isothermal effectiveness factor by the methods described in Section 4.2 and compute the nonisothermal effectiveness factor from this relationship. The advantage of this method is that, although it holds strictly for the first-order case only, it can be extended to any order by plotting $\varepsilon - \varepsilon_{iso}$ vs. $\exp(\varepsilon_{iso}) - 1$. In fact, the procedure is also applicable to reactions following H–W kinetics. The method of Jouven and Aris (1972), though slightly more accurate, is far more tedious to use.

Equation 4.54 can be used for reactions involving nth-order or H–W kinetics by employing the following modified moduli (for a sphere) in place of the usual Thiele modulus:

nth-order kinetics:

$$(\phi_s)_m = \phi_{s1} \sqrt{\bar{n}}$$

(4.55)

H–W kinetics:

$$(\phi_s)_m = \phi_{s1} \left(\frac{1}{1 + K C_{As}} \right) \left(\frac{K C_{As}}{\ln(1 + K C_{As})} \right)^{1/2}$$

(4.56)

4.3.5. Effect of a Nonuniform Environment

Where steep temperature gradients are involved in a reactor, there might be considerable variation in the temperature within the short distance represented by the catalyst particle, particularly where pellets of reasonable size are concerned (Wicke et al., 1968; Liu and Amundson, 1963). It is therefore likely that under certain conditions the assumption of uniform temperature and concentration over the surface of a catalyst will not be strictly valid. This problem has been considered by Bischoff (1968) and Copelowitz and Aris (1970). The general finding appears to be that the effect of a nonuniform environment is not serious and that even under conditions where steep temperature gradients exist within a reactor the effectiveness factor as calculated by the assumption of a uniform environment differs by only 10–15% from that calculated on the basis of a nonuniform environment.

4.3.6. Example: Effectiveness Factor in the Oxidation of Benzene to Maleic Anhydride over V_2O_5–MoO_3

Maleic anhydride is produced by the vapor-phase oxidation of benzene with air. Vaidyanathan and Doraiswamy (1968) studied the kinetics of this reaction over V_2O_5–MoO_3 catalyst deposited on silica gel, and proposed the following kinetic expression at 350°C for the conversion of benzene (with B representing benzene):

$$r_{wB} = (1.537 \times 10^{-3})p_B \quad \frac{mol}{g \text{ catalyst hr}}$$

The activation energy in the range from 310°C to 350°C was found to be 19.1 kcal/mol. The other system constants are $(-\Delta H) = 6.6 \text{ kcal/mol}$, $D_{eB} = 1.57 \times 10^{-3} \text{ cm}^2/\text{sec}$, $k'_e = 0.44 \times 10^{-3} \text{ cal/sec}$ cm °C, $\rho_c = 1.25 \text{ g/cm}^3$, and pellet radius = 5 mm. Calculate the effectiveness factor assuming the concentration of benzene at the surface to be 1.86×10^{-2} atm.

Solution

$$C_{Bs} = \frac{p_{Bs}}{R_g T} = \frac{1.86 \times 10^{-2}}{82 \times 623} = 3.65 \times 10^{-7} \quad \text{mol/cm}^3$$

$$\begin{aligned}
\beta_m &= \frac{D_{eB}(-\Delta H)C_{Bs}}{k'_e T_s} \\
&= \frac{(1.57 \times 10^{-3})(6660)(3.65 \times 10^{-7})}{(0.44 \times 10^{-3})(623)} \\
&= 1.395 \times 10^{-5}
\end{aligned}$$

This value of β_m is extremely low and hence the effect of nonisothermicity can be ignored. This is because of the extremely low concentration of benzene (1.86%) in the fluid phase.

Evaluation of the Thiele modulus:

The rate constant k_p for the reaction is 1.537×10^{-3} mol/hr g catalyst atm or

$$\begin{aligned}
k_v &= \left(\frac{1.537 \times \rho_c \times R_g T}{3600}\right) \times 10^{-3} \\
&= 2.726 \times 10^{-2} \quad \text{sec}^{-1}
\end{aligned}$$

The Thiele modulus is given by $\phi_{s1} = R(k_v D_{eB})^{1/2}$, or

$$\phi_{s1} = 2.08$$

Evaluation of effectiveness factor:

$$\varepsilon = \frac{3}{2.08}\left(\frac{1}{\tanh 2.08} - \frac{1}{2.08}\right) = 0.795$$

Thus the effectiveness factor is close to 0.8. In other words, the reaction is taking place under conditions approaching kinetic control.

4.4. EFFECT OF VOLUME CHANGE ON EFFECTIVENESS FACTOR

Let us consider the general reaction $A \rightarrow v_B B$, in which a change of volume occurs. Analytical solutions have been obtained by Lin and Lih (1971, 1973) for zero-order reactions both for isothermal and nonisothermal cases, but we shall be concerned here mainly with the more important case of first-order reactions (which involve numerical solution).

4.4.1. Isothermal Case

In analyzing a reaction with volume change, the following new group has been introduced by Weekman and Gorring (1965):

$$\theta = (v_B - 1)x_{As} \qquad (4.57)$$

which vanishes to zero for a constant-volume reaction. They have set up and solved the appropriate differential equations and presented charts for $(\varepsilon)_v/\varepsilon$ as a function of the volume-change modulus θ and the Thiele modulus both for plate and sphere geometries [$(\varepsilon)_v$ is the effectiveness factor under conditions of volume change]. More recently Jayaraman et al. (1982a) have extended the analysis to include the effect of micropore-macropore structure. Besides θ and ϕ they defined a parameter

$$\alpha_t = (1 + S)(1 - f_c)\frac{D_{e,mi}}{D_{e,ma}}\frac{R^2}{R_{mi}^2}$$

that characterizes the bipore system. The general conclusions of the analysis indicate that the effectiveness factor $(\varepsilon)_v$ always falls with increase in the value of any one of the parameters α_1, ϕ, and θ. The effect of volume change is shown to be less significant under diffusion-controlled conditions and also somewhat insensitive to changes in α_t.

In the analysis presented thus far the nonisobaric nature of the pellet has been ignored. However, for a binary system involving a change in volume, it was shown in Chapter 3 that pressure gradients would exist within the pellet. The rigorous formulation of the problem should then include terms to account for mass transport by viscous flow (Kehoe and Aris, 1973). By neglecting the contribution of viscous flow Haynes (1978) has obtained the pressures drops within the pellet and effectiveness factor values in close agreement with those including viscous flow. It seems safe therefore to conclude that the viscous flow terms in the calculations of diffusion and reaction in porous media can be neglected for all practical purposes.

It is known that a change in moles due to reaction induces a pressure gradient within the pellet that retards the net influx into the catalyst particle, lowering its effectiveness. A similar but opposite effect can be considered when the diffusive flow is accompanied by a convective flow generated as a result of a pressure differential across the pellet. Nir and Pismen (1977) have considered such a case. For a simple first-order reaction in a plate, cylinder, or sphere they have shown that a considerable improvement in the value of the effectiveness factor results in the intermediate range of the Thiele modulus. For asymptotic values of the modulus no improvement is noticeable. The conclusions, however, cannot be generalized to more complex reactions, and each case will have to be treated separately.

4.4.2. Nonisothermal Case

The problem of diffusion-cum-reaction accompanied by volume change in a nonisothermal pellet is considerably more complicated than that in an isothermal pellet. Weekman (1966) has considered this problem and prepared plots of $(\varepsilon)_v$ vs. ϕ for various values of α_s, β_m and ϕ. On the basis of several such plots some interesting conclusions can be drawn. These provide useful fine points in the rational design of a reactor and are summarized here:

1. In the case of exothermic reactions volume expansion depresses ε and raises the value of β_m for the appearance of multiple solutions. Volume contraction has the opposite effect; in other words, even a mildly exothermic reaction can exhibit multiplicity.
2. In the case of endothermic reactions, the effect of volume change is negligible.

4.4.3. Influence of Temperature and Pressure Gradients: Thermal Transpiration

In general, pressure gradients become important when the reaction involves a change in the number of moles; but even with no change in volume, pressure gradients can exist as a result of the existence of temperature gradients. The phenomenon involved is known as thermal transpiration.

The conventional equation for the molar flux of the species i, $N_i = -D_{ei}\,dC_i/dl$, with constant effective diffusivity becomes inapplicable for the nonisothermal

nonisobaric pellet. Wong et al. (1976) have therefore used the dusty-gas model and incorporated the terms corresponding to thermal transpiration and thermal diffusion in order to produce effectiveness factor plots for a first-order reaction in a spherical pellet. Differences as high as 30 % in pellet effectiveness factors over those reported by Weisz and Hicks have been found for both highly exothermic and highly endothermic reactions, thus indicating the rather strong influence of pressure gradients.

4.5. EFFECTIVENESS FACTORS FOR HOUGEN–WATSON RATE MODELS

Consider a simple reaction $A \rightarrow B$, whose rate is given by

$$r_{vA} = \frac{k K_A C_A}{1 + K_A C_A + K_B C_B} \qquad (4.58)$$

in accordance with a typical H–W model (see Chapter 2). If reactant adsorption is negligible, then $K_A C_A$ vanishes. In the presence of a strong diffusional effect, the product concentration at the pore mouth would be lower than at the interior, so that even if the adsoption constant K_B is high, the term $K_B C_B$ may be neglected and Eq. 4.58 can be written as a simple first-order rate equation, $r_{vA} = (k K_A)C_A = k_v C_A$. Thus the reaction would exhibit first-order kinetics at the pore mouth, whereas in the interior H–W kinetics would prevail.

Obviously the most direct method of computing effectiveness factors for H–W kinetics would be the use of numerical techniques (Schilson and Amundson, 1961; Chu and Hougen, 1962; Rozovskii and Schekin, 1960; Akehata et al., 1961; Aris, 1975). In our present treatment we shall primarily be concerned with generalized methods that lead to graphical display of results in terms of easily measurable nondimensional parameters.

4.5.1. Isothermal Reactions

For the sake of clarity in presentation, the effect of adsorption of one reaction component will be considered first, and this will be followed by a treatment of simultaneous adsorption of all the reaction components.

Adsorption of Reactant Alone

Consider the reaction $A \rightarrow B$ again, with the product not adsorbed; the rate equation is given by Eq. 4.58 with the term $K_B C_B$ eliminated. If slab geometry is assumed, the concerned continuity equation in dimensionless form is

$$\frac{d^2 \hat{C}_A}{d\hat{L}^2} - \phi_{p1}^2 \frac{\hat{C}_A}{1 + (K_A C_{As})\hat{C}_A} = 0 \qquad (4.59)$$

where

$$\phi_{p1} = L \left(\frac{kK_A}{D_{eA}}\right)^{1/2} \qquad (4.60)$$

The effectiveness factor can then be computed from the usual relation, $\varepsilon = (1/\phi_{p1}^2)(d\hat{C}_A/d\hat{L})\hat{L} = 1$. Numerical solutions to Eq. 4.59 in which $K_A C_{As}$ appears as an additional parameter have been provided by Krasuk and Smith (1965).

Adsorption of More Than One Component

As the number of components increases, the number of nondimensional groups (i.e., $K_A C_{As}$, $K_B C_{As}$) that must be independently fixed in order that the results can be displayed graphically also increases correspondingly (see, e.g., the solution of Hutchings and Carberry, 1966). To avoid this cumbersome situation, a generalized treatment of the problem has been proposed (Roberts and Satterfield, 1965, 1966).

The following two common reaction types will be considered:

$$A \rightarrow v_1 B_1 + v_2 B_2 + \ldots + v_n B_n \qquad (I)$$

$$A + A_1 \rightarrow v_1 B_1 + v_2 B_2 + \ldots + v_n B_n \qquad (II)$$

The H–W rate equations for these types are as follows.

Type I:

$$r_{vA} = \frac{k_v C_A}{1 + K_A C_A + \sum_i K_i C_i} \qquad (4.61)$$

Type II:

$$r_{vA} = \frac{k_v C_A C_{A_1}}{1 + K_A C_A + \sum_i K_i C_i} \qquad (4.62)$$

where $\Sigma_i K_i C_i$ represents the summation of $K_i C_i$ for all products and reactants other than A. These two rate equations should be regarded purely as mathematical models, and no particular significance should be attached to the physical basis of these models.

Note that, relative to the model involving the adsorption of A alone, the complication in the use of these equations arises from the term $\sum_i K_i C_i$. It would therefore be desirable to define a new parameter K such that $K_A C_A + \sum_i K_i C_i$ can be put in the form KC_A, resulting in an equation with a denominator of the form $1 + KC_A$, which corresponds mathematically to the case where the reactant alone is adsorbed (described earlier in this section) and which is therefore more tractable. The final expression obtained is

$$r_{vA} = \frac{k_m C_A}{1 + KC_A} \qquad (4.63)$$

where

$$k_m = \frac{k_v}{\omega} \qquad (4.64a)$$

$$\omega = 1 + \sum_i K_i(C_{is} + C_{As} v_i D_{eA}/D_{ei}) \quad (4.64b)$$

$$K = \frac{K_A - D_{eA} \sum_i K_i v_i /D_{ei}}{\omega} \qquad (4.64c)$$

It is now expedient to define a modified Thiele modulus based on k_m rather than k_v:

$$(\phi)_m = L \left(\frac{k_m}{D_{eA}}\right)^{1/2} \qquad (4.65)$$

Further, a practically more useful modulus is that based on the reaction rate per gross volume of catalyst. Thus we have in analogy with Eq. 4.31.

$$(\phi)_{mr} = \varepsilon^{1/2} (\phi)_m \qquad (4.66)$$

Roberts and Satterfield (1965) have presented their results as plots of ε vs. $(\phi)_m$ and ε vs. $(\phi)_{mr}$ for various values of the parameter KC_{As}. The former are reproduced in Figure 4.6. The effect of adsorption, under conditions where the reactant is more strongly adsorbed than the products, is brought out through a series of curves bounded by the zero-order curve (at $KC_{As} \rightarrow \infty$) on one side and the first-order curve (at $KC_{As} \rightarrow 0$) on the other. When KC_{As} is negative (corresponding to product adsorption), the curves move away to the left of the zero- and first-order curves.

Though the Aris approximation for different shapes does not hold for reactions following H–W kinetics, values to within about 20% can be obtained by using this approximation.

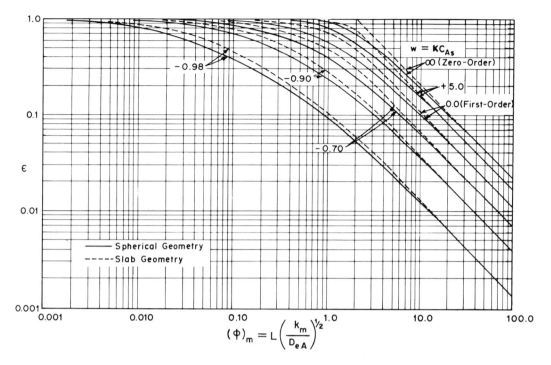

Figure 4.6. Effectiveness factor vs. Thiele modulus for various values of the parameter KC_{As}.

For type II reactions a parameter

$$E = \frac{-D_{eA_1} P_{A_1s}}{v_{A_1} D_{eA} P_{As}} - 1$$

has been defined to provide a quantitative measure of the stoichiometric presence of the second reactant A_1 and solutions obtained for $E = 0$, 1, and 10 (Roberts and Satterfield, 1966).

Reversible Reactions

Kao and Satterfield (1968) extended the method of Roberts and Satterfield (1965) and presented plots of ε vs. $(\phi)_m$ for different values of the following two parameters:

$$G' = \frac{C_A^*}{C_{As}}, \qquad H' = \frac{KC_{As}(1 - G')}{1 + KC_{As}G'} \qquad (4.67)$$

where K can be deduced from

$$K = \frac{K_A - K_B (D_{eA}/D_{eB})}{\omega} \qquad (4.68)$$

$$\omega = 1 + K_B\left[C_{Bs} + C_{As}\left(\frac{D_{eA}}{D_{eB}}\right)\right] + \sum_j K_j C_j \qquad (4.69)$$

Comparison of these parameters with those used for irreversible reactions reveals that H' now replaces KC_{As} and an additional "equilibrium parameter" G' is involved. The limits of G' are zero for irreversible reactions and unity for reversible reactions at equilibrium. For a given value of H', the effectiveness factor has been shown to decrease as G' increases from zero to unity, that is, as the reaction becomes increasingly reversible. (Reference may also be made to the method of Schneider and Mitschka, 1966a, b.)

4.5.2. Nonisothermal Reactions

Solutions can be obtained for Eqs. 4.45 and 4.46 presented earlier by incorporating the appropriate H–W model in place of $f(\hat{C}_A)$ in these equations, on the assumption that the adsorption equilibrium constants are essentially temperature independent. For slab geometry and a simple irreversible reaction A → B with both A and B adsorbed, the mass balance equation to be solved (with $S = 0$) is

$$\frac{d^2\hat{C}_A}{d\hat{L}^2} - \phi_{p1}^2 \exp\left[\alpha_s\beta_m \frac{1 - \hat{C}_A}{1 + \beta_m(1 - \hat{C}_A)}\right] f(\hat{C}_A) = 0$$
$$(4.70)$$

Rajadhyaksha et al. (1976) have considered Eq. 4.63 in solving 4.70 for different values of β_m, α_s, and the

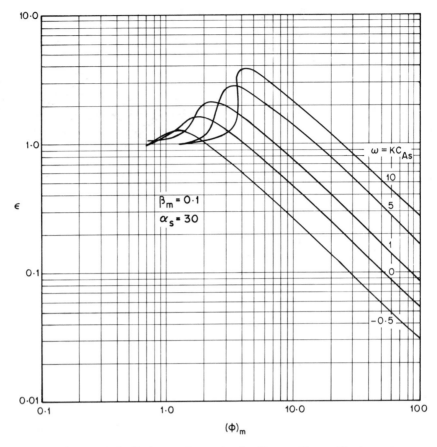

Figure 4.7. A typical effectiveness factor plot for Hougen–Watson kinetics under nonisothermal conditions (Rajadhyaksha et al., 1976).

adsorption parameter KC_{As}. Some typical results are presented in Figure 4.7 as ε–$(\phi)_m$ plots with KC_{As} as parameter. An interesting feature of the solutions is that the effect of adsorption on the effectiveness factor is seen to be different for different values of the Thiele modulus. Because of the opposite trends at low and high values of the modulus, intersection of the ε–$(\phi)_m$ curves is observed in the intermediate range.

The differential mass and heat balance equations given by Eqs. 4.45 and 4.46 have also been solved by Elnashaie and Mahfouz (1978) by incorporating rate equations involving activated adsorption. An interesting feature of activated adsorption in unimolecular reactions is that it gives multiplicity even for zero apparent activation energy of the reaction and is referred to as "adsorption multiplicity." No such multiplicity is possible for nonactivated adsorption. Surface coverage also has an important influence on the effectiveness factor and region of multiplicity. It increases the range of Thiele modulus over which multiplicity

occurs and enhances the value of the effectiveness factor considerably.

For bimolecular reactions, multiple steady states are possible even under isothermal conditions. However, the range of multiplicity and number of steady states depend strongly on surface coverage.

4.5.3. Example: Effectiveness Factor in the Isomerization of n-Butene to Isobutene (with H–W Kinetics)

Choudhary and Doraiswamy (1975) studied the kinetics of isomerization of n-butene to isobutene over alumina catalyst; the kinetic expression derived by them is

$$r_A = \frac{k(p_A - p_B/K)}{1 + K_B p_B}$$

where subscripts A and B represent n-butene and

isobutene, respectively. The values of the constants at $638°K$ are as follows: $k = 1.1935 \times 10^{-3}$ mol/hr g catalyst atm, $K_B = 0.611$ atm^{-1}, $K = 2.25$, $D_{eA} = 1.70$ cm^2/sec, $\rho_c = 1.2$ g/cm^3. Calculate the effectiveness factor at $638°K$ when the vapor-phase is pure n-butene at atmospheric pressure and the reaction is carried out over (a) a $\frac{1}{4}$-in.-thick slab of catalyst; and (b) a $\frac{1}{2}$-in.-diameter spherical pellet.

Solution

a. Since the gas phase consists of pure n-butene and the equilibrium constant is fairly high (2.25), the reversibility of the reaction can be ignored. The rate expression then becomes

$$r_A = \frac{k p_A}{1 + K_B p_B} \qquad (4.71)$$

Since $p_A + p_B = $ constant $= 1$ atm, $p_B = 1 - p_A$. Hence

$$r_A = \frac{k p_A}{1 + K_B - K_B p_A} = \frac{a p_A}{1 - K p_A}$$

where

$$a = \frac{k}{1 + K_B}, \qquad K = \frac{K_B}{1 + K_B}$$

In the present case

$$a = \frac{1.1935 \times 10^{-3}}{1 + 0.611}$$

$$= 0.74 \times 10^{-3} \text{ mol/hr g catalyst atm}$$

$$= \frac{(0.74 \times 10^{-3})(82)(638)(1.2)}{3600}$$

or 12.9×10^{-3} sec^{-1}

$$K = \frac{K_B}{1 + K_B} = \frac{0.611}{1 + 0.611} = 0.38 \text{ atm}^{-1}$$

Roberts and Satterfield (1966) have presented numerical solutions (Figure 4.6) for the effectiveness factor for reactions obeying the kinetic expression given by Eq. 4.71, when the reaction is carried out in a slab of catalyst. The parameters in the solution are (ϕ_m) (which in this case equals ϕ_{p1}) and $K p_{As}$. These parameters in the present case can be evaluated as follows:

$$(\phi)_m = L \left(\frac{k_v}{D_{eA}} \right)^{1/2} = 0.63 \left(\frac{12.9 \times 10^{-3}}{1.7 \times 10^{-3}} \right)^{1/2} = 1.74$$

$$K p_{As} = -0.38 \times 1 = -0.38$$

From Figure 4.6, by interpolation,

$$\varepsilon = 0.38 \text{ for } K p_{As} = -0.38, \quad (\phi)_m = 1.74$$

b. In the case of first-order reactions it has been shown that the solutions for flat geometry can be adapted to spheres by using $R/3$ as the characteristic dimension. Although this has not been shown to be true for the H–W type of kinetic expressions, it is expected that the curves for different geometries would be very close to each other at the low value of $(\phi)_m$ encountered in the present example. Hence the effectiveness factor is evaluated by using $R/3$ as the characteristic dimension.

$$(\phi)_m = \frac{0.63}{3} \left(\frac{12.9 \times 10^{-3}}{1.7 \times 10^{-3}} \right)^{1/2} = 0.58$$

$$K p_{As} = -0.38$$

By interpolation

$$\varepsilon = 0.76 \text{ for } (\phi)_m = 0.58, \quad K p_{As} = -0.38$$

Rajadhyaksha et al. (1976) obtained numerical solutions for spherical geometry and showed that the curves for different values of $K p_{As}$ coincide with the first-order curve if the Thiele modulus is redefined as

$$(\phi)_m = \phi_{s1} \frac{1}{1 + K p_{As}} \left[\frac{K p_{As}}{\ln(1 + K p_{As})} \right]^{1/2}$$

where

$$\phi_{s1} = R \left(\frac{k_v}{D_{eA}} \right)^{1/2}$$

In the present case $\phi_{s1} = 1.74$, and

$$(\phi)_m = 1.74 \frac{1}{1 + (-0.38)} \left[\frac{-0.38}{\ln(1 + -0.38)} \right]^{1/2} = 2.5$$

Hence the effectiveness factor will be the same as that for a first-order reaction with $\phi = 2.5$ carried out in a spherical pellet. The effectiveness factor will then be given by

$$\varepsilon = \frac{3}{(2.5)} \left[\frac{1}{\tanh(2.5)} - \frac{1}{2.5} \right] = 0.731$$

4.5.4. Reactions with Negative Order

Negative-order reactions, typified by the oxidation of CO over noble metals, differ from normal or positive-order reactions in that the observed rate is enhanced by both concentration and temperature gradients. The conservation equations for heat and mass for a spherical pellet have been solved by Smith et al. (1975) for the following H–W kinetic model:

$$r_{vA} = \frac{k_v C_A}{(1 + K_A C_A)^2} \qquad (4.72)$$

The results of the analysis presented as ε–ϕ_s curves indicate that the usual adverse effect of diffusion turns to advantage for negative-order kinetics. As will be expected, $\beta_m > 0$ proves useful in enhancing the effectiveness factor and the maximum now occurs at lower values of the Thiele modulus. In general, these reactions are much more thermally sensitive to different parameters than those described by a positive reaction order. This peculiarity indicates that such reactions should rarely be viewed as isothermal. Uppal and Ray (1977) have also presented a lumping procedure to obtain effectiveness factors under isothermal conditions for bimolecular H–W kinetics, which is frequently used to represent oxidation and hydrogenation reactions.

4.6. GENERALIZED ESTIMATION PROCEDURES

4.6.1. Collocation Method

An elegant method for the estimation of effectiveness factors is to use the principle of collocation. This method is based on the weighted residual technique. The solution is written as a trial function (whose dependence on position is intuitively chosen) that includes one or more adjustable parameters. The parameters are then evaluated by collocation, that is, by stipulating that the trial functions satisfy the differential equation at selected points. Usually the number of points selected equals the number of parameters in the model.

The trial function is substituted into the differential equation that describes the physical situation and the resultant is set to zero at the collocation points. The Laplacian and the first differentials are expressed in terms of the matrices B and A, which can be determined by a procedure described by Finlayson (1972). In general, a large number (N) of interior collocation points can be chosen to improve the accuracy of the method. In many instances, however, even a single-point collocation gives reasonably accurate results.

The method has found widespread use in different areas in chemical engineering. As far as the catalyst pellet is concerned, its application to different situations has been summarized by Finlayson (1974) and Villadsen and Michelsen (1978).

4.6.2. Conversion of Boundary Value Problem to Initial Value Problem (The Ibanez and Namjoshi–Kulkarni–Doraiswamy Transformations)

It seems feasible sometimes to convert an original two-point boundary value problem into a simpler, equivalent initial value problem by defining suitable invariant transformations. Thus, for the case of a general nth-order reaction in a spherical pellet, Ibanez (1979) defined the transformations presented in Table 4.4 and solved the resulting initial value problem to obtain information regarding the stability of the state. Jayaraman et al. (1982b) have employed these transformations to solve the case of diffusion cum reaction in a micropore–macropore system (Table 4.5). The saving in computational time is shown to be clearly about ten fold.

The transformations defined by Ibanez (1979) are valid only for the power law type of rate forms. Namjoshi et al. (1982) have defined new sets of transformations that can be employed for H–W rate forms. These are also included in Table 4.4, along with the rate forms. The use of these simple transformations allows one to scan the entire range of parameter values without any trial-and-error or numerical difficulties.

4.6.3. "Isothermalization" of Pellet[†]

The problem of simultaneous inter- and intraparticle transport limitations will be considered in greater detail in Chapter 6, but for the purpose of the present treatment we shall anticipate the conclusion that several systems of industrial significance exhibit high values of the ratio of Sherwood to Nusselt numbers. Thus the mathematical approach to the problem can be based on equating the heat and mass transport rates across the fluid film and stipulating that this rate also be equal to the rate of reaction (thus bringing the effectiveness factor into the equations). The central feature of this method is that although the pellet may not be truly isothermal, it is conceptually "isothermalized" by demanding that the heat effect be localized in the film (McGreavy and Thornton, 1970). Indeed, Pereira et al. (1979) have shown mathematically that this is so, thus

[†] See also Section 6.4.

TABLE 4.4. Sets of Transformations for Several Rate Forms

Transformations	Rate Form	Parameter Definitions
$u = \dfrac{R\,dz/dR}{1+z}$ $v = \phi^2 R^2 (1+z)^{n-1}$ $w = \ln \dfrac{1+z}{1+z_0}$	$\phi^2 (1+z)^n$	$\phi = R\left(\dfrac{k_v}{D_e}\right)^{1/2}$ $1+z = C_A$ $z_0 = $ value of z at the pellet center $K' = K_A C_{As}$
$u = \dfrac{R\,dz/dR}{1+K'(1+z)}$ $v = \dfrac{\phi^2 R^2 (1+z)}{[1+K'(1+z)]^2}$ $w = \ln \dfrac{1+K'(1+z)}{1+K'(1+z_0)}$	$\dfrac{\phi^2 (1+z)}{1+K'(1+z)}$	
$u = \dfrac{R\,dz/dR}{[1+K'(1+z)]^{z-n}}$ $v = \dfrac{\phi^2 R^2 (1+z)}{[1+K'(1+z)]^2}$ $w = \left[\dfrac{1+K'(1+z)}{1+K'(1+z_0)}\right]^{n-1}$	$\dfrac{\phi^2 (1+z)}{[1+K'(1+z)]^n}$	

TABLE 4.5. Mass Balance Equations for the General nth-Order Case in a Micropore–Macropore System

$$\nabla_x^2 C_{A,\,mi} = \phi_{sn}^2 C_{A,\,mi}^n \tag{1}$$

$$\nabla_y^2 C_{A,\,ma} = \alpha_t \left(\frac{dC_{A,\,mi}}{dx}\right)_{x=1} \tag{2}$$

with the boundary conditions

$$x=0: \frac{dC_{A,\,mi}}{dx} = 0; \qquad x=1: \ C_{A,\,mi} = C_{A,\,ma} \tag{3}$$

$$y=0: \frac{dC_{A,\,ma}}{dy} = 0; \qquad y=1: \ C_{A,\,ma} = 1 \tag{4}$$

Using the transformation corresponding to this case in Table 4.4, we obtain

$$\frac{du}{dw} = \frac{v}{u} - u - 1 \tag{5}$$

$$\frac{dv}{dw} = \frac{2v}{u} + (n+1)v \tag{6}$$

with initial conditions

$$u = v = 0 \quad \text{at} \quad w = 0 \tag{7}$$

This represents an initial value problem and can be solved to obtain the flux $(dC_{A,\,mi}/dx)_{x=1}$. Equation (2) can now be solved using the method of Weisz and Hicks (1962) to get the effectiveness factor for several values of ϕ_{sn} and α_t.

confirming the earlier observations of Kehoe and Butt (1972) and Butt et al. (1977).

Consider a reversible reaction of the type represented by A $\rightleftharpoons$ B occurring in a spherical pellet. The mass balance is given by Eq. 4.1 with r_{vA} defined by Eq. 4.6. For the heat balance, instead of using the usual expression, the following equation based on the localization of heat transfer in the film can be written:

$$\begin{aligned}
-h_{pf}(T_b - T_p) &= (-\Delta H)k_g(C_{Ab} - C_{As}) \\
&= (-\Delta H)k_g(C_{Bs} - C_{Bb}) \\
&= \frac{R}{3}(-\Delta H)r_{vA} \\
&= \frac{R}{3}(-\Delta H)\varepsilon(T_p)k_v \frac{K+1}{K}\left(C_{As} - \frac{C_{Bs}}{K}\right)
\end{aligned} \tag{4.73}$$

Thus the surface concentrations are given by

$$C_{As} = C_{Ab} + \frac{h_{pf}}{k_g(-\Delta H)}(T_b - T_p) \tag{4.74}$$

$$C_{Bs} = C_{Bb} + \frac{h_{pf}}{k_g(-\Delta H)}(T_b - T_p) \tag{4.75}$$

Then, combining Eqs. 4.74 and 4.75, we get

$$\varepsilon = \frac{3}{\phi_{s1}(T_p)}\left[\frac{1}{\tanh \phi_{s1}(T_p)} - \frac{1}{\phi(T_p)}\right] \quad (4.76)$$

Equation 4.73 for ε can be recast as follows for the temperature T_p:

$$\frac{R(-\Delta H)}{\phi_{s1}(T_p)}\left[\frac{1}{\tanh \phi_{s1}(T_p)} - \frac{1}{\phi_{s1}(T_p)}\right]\frac{k_v(T_p)[K(T_p)+1]}{K(T_p)}$$

$$\left\{C_{Ab} - \frac{C_{Bb}}{K(T_p)} + \left[1 + \frac{1}{K(T_p)}\right]\frac{h_{pf}}{k_g(-\Delta H)}(T_b - T_p)\right\}$$

$$+ h_{pf}(T_b - T_p) = 0 \quad (4.77)$$

This is an implicit equation in T_p that can be readily solved by iteration, and the effectiveness factor can then be directly obtained from Eq. 4.76. Note that in solving Eq. 4.77, the rate constant k_v and diffusivity D_{eA} should be calculated at T_p for obtaining $\phi_{s1}(T_p)$ and the equilibrium constant K also estimated at T_p.

This method can be easily extended to any reaction type and is simple to use and quite accurate. As will be pointed out in Chapter 11, it is particularly suited for fixed-bed reactor design calculations in which a single value of the effectiveness factor cannot be assumed and point-to-point estimation is necessary.

4.7. EFFECT OF SURFACE DIFFUSION

In Chapter 3 we saw how surface diffusion can influence the value of the effective diffusion coefficient.

It has often been shown that the divergence between experimental rate data and the rates calculated from models ignoring surface diffusion can largely be attributed to mass transfer due to surface migration. A few examples of this will be discussed at the end of this section.

Two methods of approach can be visualized for incorporating the effect of surface diffusion: (1) include a term for surface diffusion in the Thiele modulus; and (2) postulate a new surface diffusion group, which would appear as a parameter in the usual effectiveness factor plots. It is also possible that chemical reaction is influenced by surface diffusion alone and transport limitations due to bulk and Knudsen diffusion are not involved.

4.7.1. Including Surface Diffusion in the Thiele Modulus

A simple method of accounting for surface diffusion, as pointed out earlier, would be to include the surface diffusion coefficient in a modified definition of the Thiele modulus. Restricting our attention to an isothermal system, we can readily write the following equation:

$$\frac{d}{dl}(N_b + N_s) + r_{vA} = 0 \quad (4.78)$$

where

$$N_s = -D_s\frac{dC_{As}}{dl} = -D_s k_A\frac{dC_{Ab}}{dl} \quad (4.79)$$

and

$$D_s = \text{surface diffusivity}$$

Substituting for N_b, N_s (see Chapter 3), and r_{vA}, we obtain

$$D_{bA}\left(\frac{d^2C_{Ab}}{dl^2}\right) + D_s K_A\left(\frac{d^2C_{Ab}}{dl^2}\right) - k_v K_A C_{Ab} = 0 \quad (4.80)$$

Note that the reaction is assumed to be first order and the rate term $k_v C_{As}$ has been replaced by $k_v K_A C_{Ab}$. Equation 4.80 can now be rearranged as

$$D_{eS}\left(\frac{d^2C_{Ab}}{dl^2}\right) - k_v K_A C_{Ab} = 0 \quad (4.81)$$

where D_{eS} is an overall diffusivity (which includes surface transport) and is given by

$$D_{eS} = D_{bA} + D_s K_A \quad (4.82)$$

The solution (with the usual boundary conditions) is given by

$$\varepsilon = \frac{\tanh \phi_S}{\phi_S} \quad (4.83)$$

where ϕ_S is a new Thiele modulus given by

$$\phi_S = L\left(\frac{k_v K_A}{D_{eS}}\right)^{1/2} \quad (4.84)$$

Krasuk and Smith (1965) suggest that the modified modulus can be recast as

$$\phi_S = \phi_{p1}\left(\frac{1}{1+h_S}\right) \quad (4.85)$$

where

$$h_S = \frac{D_s K_A}{D_{bA}} \quad (4.86)$$

This "split" brings out distinctly the effect of surface migration, and yet a separate parameter is not involved.

4.7.2. Use of an Independent Surface Diffusion Group

Consider a differential segment of a single cylindrical capillary, in which the simple reaction A → B is allowed to occur. On the assumption that surface transport occurs in parallel with gas-phase (essentially Knudsen) transport, the following material balance can be written:

$$-\frac{d}{dl}(N_b r_p^2 + N_S r_p^2) - A_f \exp\left(\frac{-E}{R_g T}\right) K_A C_{Ab} = 0 \tag{4.87}$$

where C_{As} has been expressed as $K_A C_{Ab}$.

Foster and Butt (1966) have solved Eq. 4.87 on the basis of certain drastic assumptions; the solutions are not discussed here.

4.7.3. Surface Diffusion Effects in the Absence of Other Modes of Transport

A typical reforming catalyst consists of alumina of about 100 m²/g specific area impregnated with about 1% platinum. The diameter of the platinum center is of the order of 100 Å, and the ratio of the distance between the centers and the diameter is approximately 20. This shows that the centers are fairly widely separated. It may be postulated that reaction occurs by the adsorption of the reactant on a center and its diffusion along the surface to a point where it reacts either with another reactant in the gas phase or by itself. The geometric configuration envisaged is shown in Figure 4.8. The ratio R_0/R_i represents the ratio of the distance between centers to their diameter. Using this ratio as an important factor in the analysis of surface diffusion, Sohn et al. (1970) have presented effectiveness factor plots with a modified Thiele modulus, defined as

$$(\phi_S)_m = \left(\frac{n+1}{2}\right)^{1/2} R_i \left(\frac{k_v C_0^{n-1}}{D_S}\right)^{1/2} \tag{4.88}$$

where n is the reaction order. Surface diffusion by itself is of no significant importance; hence, this approach serves only to emphasize the similarity in treatments for all modes of transport inside a pellet.

An opposite view of the problem has been discussed by Aris (1971), who assumes that the adsorption occurs on the entire surface and the species then diffuses to the reaction site where the reaction A → B occurs. The conservation equation for this situation can be solved subject to appropriate boundary conditions to obtain an expression for the effectiveness factor:

$$\varepsilon = 1 - \frac{\mu(\psi)}{\psi K_1(\psi) + \mu K_0(\psi)} \tag{4.89}$$

where

$$\mu = \frac{k_S R_i}{D_S} \quad \text{and} \quad \psi^2 = \frac{k_d R_i^2}{D_S}(1 + K_A p_A)$$

As before, the parameter R_i represents the radius of the reaction site and k_S and k_d are the rate constants for reaction and desorption, respectively; K_0 and K_1 are the modified Bessel functions of the second kind of order 0 and 1, respectively.

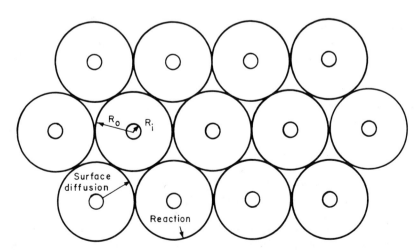

Figure 4.8. Catalyst surface model with uniformly distributed adsorption centers.

Equation 4.89 for the effectiveness factor can be generalized to a variety of situations, such as when the products are also adsorbed on the surface along with the reactants, or when a change in the number of moles is involved, provided that appropriate modifications are made in the definition of ψ. The parameter ψ for the general purpose can be defined as (Jayaraman and Kulkarni, 1981)

$$\psi^2 = \frac{k_d R_i^2}{D_S}\left(1 + \frac{K_A p_A}{1 + \sum_i K_i p_i}\right) \tag{4.90}$$

where the subscript i stands for the product.

4.7.4.   Examples of the Role of Surface Diffusion

In general, surface diffusion is assumed to be operative if D_e obtained from experimental rate measurement is considerably different from the value calculated from theory.

As pointed out in Chapter 3, another criterion for the influence of surface diffusion is that the experimental value of D_e should decrease with increase in temperature. It is also possible that D_e for a given pellet would exhibit an optimum value in a given temperature range in the presence of surface diffusion. Based essentially on these criteria, a few reactions are believed to be influenced by surface diffusion. Thus in the case of the hydration of isobutylene on Dowex-50, Gupta and Douglas (1967) reported a decrease in effective diffusivity with increase in temperature. Similar variations have been reported by Bienert and Gelbin (1967), Miller and Kirk (1962), and Krasuk and Smith (1965) in the dehydration of alcohol on alumina and silica–alumina.

A cooperative effect induced by surface diffusion has been noted by Sinfelt and Lucchesi (1963) and Bond (1962) in the hydrogenation of ethylene on Pt–Al$_2$O$_3$ polyfunctional catalyst. The experimental data on *ortho–para* hydrogen conversion on FeO also show the influence of surface diffusion (Sterrett and Brown, 1968). Surface-diffusion-induced "spillover" has been noted by several workers (see, e.g., Levy and Boudart, 1974, and the references quoted therein). Another influence of surface diffusion is in decreasing the number of unavailable sites for reaction with increase in temperature, for example, in the hydrogenation of propylene on Pt. In the gasification of carbon (Yang and Wong, 1981, and the references quoted therein) the rate data clearly indicate the role of surface diffusion in H–W kinetics.

4.8.   EFFECTIVENESS FACTORS IN MULTICOMPONENT DIFFUSION

The majority of catalytic processes involve multicomponent reactions. In the case of multicomponent systems a complex problem arises because the ratio of the fluxes of the individual components is fixed by the reaction stoichiometry. This leads to a gradient of pressure, so that the diffusion equation should account for both flow due to concentration and total pressure gradients (see Chapter 3). The simple Stefan–Maxwell equation valid for the isobaric pellet then becomes inapplicable. Schneider (1975) has proposed a useful analysis for such a situation.

In the analysis of multicomponent systems the use of H–W models to describe the rate is usually preferred to the power law kinetics. This is mainly due to the incapability of the power law equations to describe the rate when the products inhibit the reaction rate, besides the empiricism involved in such formulations. For a rate equation of the form

$$r = \frac{k(1+B)C}{1+BC} \tag{4.91}$$

where

$$B = \frac{1 + C_t \sum_i K_i y_{is}}{y_{is} C_t \sum_i K_i \delta_i} - 1, \qquad i = 1, \ldots, n$$

$$\delta_i = \frac{E}{(\Delta H_{ad})_i}$$

and y_{is} is the mole fraction of species i at the surface, the governing equations for mass and heat conservation have been solved by Schneider (1975) to obtain the values of the effectiveness factor shown in Figure 4.9. A modified Thiele modulus, which provides a more compact representation of the effect of internal diffusion, is used:

$$(\phi)_m = \left(\frac{\phi_1}{S+1}\right)[2(1+B)]^{-1/2}|B|$$

$$\times [B - \ln(1+B)]^{-1/2} \tag{4.92}$$

where $S = 0$, 1, and 2 for the slab, cylinder, and sphere, respectively.

The forgoing analysis was discussed to some extent in Section 4.5.1. The present treatment is more general in that the geometric factor is incorporated in the formulation; also, the flow term due to pressure

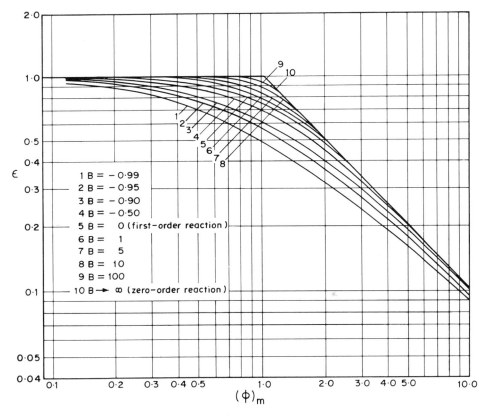

Figure 4.9. Effectiveness factor as a function of generalized modulus (Eq. 4.92) for an infinite slab of catalyst (Schneider, 1975). B is defined by Eq. 4.91.

gradients arising in a multicomponent system has been included.

The parameter B characterizes the effect of adsorption of A_1, and for extreme values (0 and ∞) Eq. 4.91 reduces, respectively, to zero- and first-order kinetics. The effectiveness factor curves in Figure 4.9 are therefore plotted for various values of B. For any positive value of B lying between 0 and ∞ the curves always lie within this zero- and first-order envelope. The parameter B can also take negative values when the strongly adsorbed products inhibit the reaction.

At high and low values of $(\phi)_m$ (representing diffusion control and chemical control, respectively) the curves for all values of B coincide, indicating no influence of B in this region. For intermediate values of $(\phi)_m$ (transition regime), however, there is a distinct resolution of the single curve, showing higher effectiveness factors for higher values of B.

For the reaction rate given by Eq. 4.91 under nonisothermal conditions the influence of various parameters such as β_m, γ, and B on the effectiveness factor has been analyzed by Schneider (1975). As

expected, for positive β_m the effectiveness factor exceeds unity and multiple solutions seem possible.

4.9. EXAMPLES OF PORE DIFFUSION CONTROL WITH TYPICAL EFFECTIVENESS FACTOR VALUES

Empirical equations are often found to be more useful than analytical or exact numerical solutions. Several empirical equations that can be advantageously used in the design of fixed-bed reactors (discussed in Chapters 11 and 12) are available. Most of the important equations have been summarized by Carberry (1961), Liu (1970), Jouven and Aris (1972), and Rajadhyaksha and Vasudeva (1974) (see Section 4.3.4).

Several reactions (some of which are discussed in greater detail in Chapter 7) are known to be controlled by pore diffusion. A list of reactions for which effectiveness factors have been reported is presented in Table 4.6. This table may be regarded as an extension of a similar table presented by Satterfield (1970).

TABLE 4.6 Effectiveness Factors for Some Typical Systems

Reaction	Catalyst	Pellet Size (cm)	Effectiveness Factor	Reference
Cracking of cumene at 420°C and 1 atm	Silica–alumina pellets	0.46 0.35 0.01	0.15 0.20 1.00	Weisz and Prater (1954)
Cracking of gas oil at 480°C and 1 atm	Silica–10% alumina pellets	0.44	0.55	Johnson et al. (1957)
Cracking of cumene at 510°C and 1 atm	Silica–alumina Thermofor catalytic cracking catalyst	0.045 0.33 0.43 0.53	0.72 0.16 0.12 0.09	Prater and Lago (1956)
Cracking of n-hexane at 320°C	Selective offretite catalysts	—	0.5	Miale et al. (1966)
Dehydration of alcohols (considered as a single reaction) at 260 and 370°C	a. Silica–alumina Thermofor catalytic cracking catalyst[a]	0.50 0.23 0.04	(260°C) (370°C) 0.15 0.10 0.24 0.16 0.98 0.90	Miller and Kirk (1962)
	b. Activated Indian bauxite (grains), 300–375°C[b]	0.0135–0.0345		Abhyankar and Doraiswamy (1979)
Synthesis of ammonia	a. Promoted iron catalyst (2.5–3.5% promoter) at			
	(i) 400–450°C, 1 atm	0.014–0.24	0.20	Bokhoven and van Raayen (1954) from data of Larsen and Tour (1922)
	(ii) 425–450°C, 100 atm	0.24–0.33	0.80	
	b. Promoted iron catalyst (2.9% Al_2O_3 and 1.1% K_2O) at			
	(i) 325°C, 1 atm	0.24–0.28 0.05–0.07	0.78 1.0	Bokhoven and van Raayen (1954)
	(ii) 400°C, 30 atm	0.24–0.28 0.05–0.07	0.95–1.0 1.0	
	(iii) 500°C, 30 atm	0.24–0.28 0.05–0.07	0.40 0.89	
	c. Promoted iron oxide (K_2O, CaO, Al_2O_3) at 450°C, 214 atm	0.57 0.50	0.21–0.75 (along the reactor) 0.5–0.98 (along the reactor)	Nielsen (1968)

Reaction	Catalyst	Size	Effectiveness factor	Reference
Synthesis of methanol at 330–410°C, 280 atm	Zinc oxide–chromium oxide pellets	0.5–1.6	0.52–0.95	Pasquon and Dente (1962)
Decomposition of ammonia at 387–467°C and 1 atm	Promoted iron–alumina pellets (10.2% Al_2O_3)	10–14 mesh to 35–40 mesh	1.0	Love and Emmett (1941)
Hydration of isobutylene to t-butanol at 70–115°C and 1 atm	Dowex 50 W × 8	100–115 mesh 48–65 mesh	0.96 (74°C) 0.57 (97°C) 0.87 (74°C) 0.35 (97°C)	Gupta and Douglas (1967)
Isomerization of 1-butene at 530–550°C and 1 atm	γ-Alumina (Alcoa F-100)	0.64 50–60 mesh	0.04 1.0	Forni et al. (1971)
Isomerization of n-butene at 335–400°C and 1 atm	Fluorinated η-alumina (containing 1% F by weight)	40–60 mesh	1.0	Choudhary and Doraiswamy (1975)
Ortho–para hydrogen conversion at 196°C and 1 atm	5% NiO on SiO_2 7% NiO on Al_2O_3	540 μm radius 2.54 cm in diameter 0.64 cm thick	0.89 0.25	Raja Rao and Smith (1963)
Ortho–para hydrogen conversion at 76°K and 32 psi	Ferric oxide gel	30–100 mesh	0.78–0.96	Sterrett and Brown (1968)
Ortho–para hydrogen conversion at 198°C	Aluminium oxide catalyst impregnated with nickel	0.32 cm cylindrical pellets	0.449 (100 psi) 0.429 (400 psi)	Wakao et al. (1962)
Oxidation of ethylene and propylene at 130–150°C	Silica-supported noble metals Pt, Pd, Ru, and Rh	20–30 mesh	1	Cant and Hall (1970)
Vapor-phase oxidation of methanol at 250–460°C and 1 atm	Manganese dioxide–molybdenum trioxide	0.0525 cm	1	Mann and Hahn (1969)
Vapor-phase oxidation of o-xylene	Potassium sulfate-promoted vanadium pentoxide on silica carrier	—	1	Juusola et al. (1970)
Oxidation of sulfur dioxide at 400–450°C	Platinum on alumina, and V_2O_5 formulations used by Mars and Maessen (1964); 400–450°C	1 cm 1 cm	0.14 at 0.2 ft of reactor length 0.40 at 0.2 ft of reactor length	Minhas and Carberry (1969)
Oxidation of toluene (to benzaldehyde) at 450°C	Molybdenum oxide (15%) and tungstic oxide (5%) on alumina	80–100 mesh	0.84 (for the overall reaction)	Reddy and Doraiswamy (1969)
Oxidation of benzene at 350°C	Vanadium pentoxide–molybdenum trioxide, 350°C	5–12 mesh 36–60 mesh	0.87 1.00	Vaidyanathan and Doraiswamy (1968)

87

(continued)

TABLE 4.6 (continued)

Reaction	Catalyst	Pellet Size (cm)	Effectiveness Factor	Reference
Hydrogenation of ethylene at 80–140°C	Nickel on aluminum oxide	0.2–0.5	0.5 (calculated 0.07–0.19)	Rozovskii and Shchekin (1960)
Hydrogenation of nitrobenzene at 80°C	Palladium–silver catalyst	—	1	Metcalfe and Rowden (1971)
Dehydrogenation of butane at 530°C	Chromia (12%)–alumina pellets	0.32	0.70	Blue et al. (1952)
Dehydrogenation of cyclohexane	a. Chromia–alumina pellets; 478°C, 1 atm	0.62 0.37	0.48 0.65	Weisz and Swegler (1955)
	b. Platinum on alumina, 340–490°C, 200 psig	0.3	0.38 (430°C)	Barnett et al. (1961)
Decomposition of NH_3 at 650–750°C and 1 atm	Copper, Hastelloy C, and Inconel surfaces (lined on stainless steel)		1	Nozik and Behnken (1965)

[a]The effectiveness factor decreases with increase in temperature, increases with increase in pressure, and is independent of the alcohol used.
[b]A correlation has been developed for the overall conversion which holds for all four of the alcohols used: ethyl, butyl, hexyl, and octyl. General observations are similar to those reported under a.

4.9.1. Example: Hypothetical Problem to Illustrate Calculation of Effectiveness Factor for a Complex Situation

The reaction $A \to B$ obeys the kinetic expression

$$r = \frac{k_v C_A}{1 + K_A C_A}$$

where

$$k_v = (1.6 \times 10^{10}) \exp\left(-\frac{19,800}{R_g T}\right) \quad \text{sec}^{-1}$$

and

$$K_A = (8.71 \times 10^4) \exp\left(\frac{20,000}{R_g T}\right) \quad \text{cm}^3/\text{mol}$$

The other system constants are as follows: $D_{eA} = 3.6 \times 10^{-2}$ cm^2/sec, $-\Delta H = 30$ kcal/mol, $k_e' = 2.64 \times 10^{-4}$ cal/cm sec°C, and $d_p = 5$ mm. Calculate the effectiveness factor when the gas phase consists of 50% A and the temperature at the surface of the pellet is 500°K using

1. the numerical solutions given by Kao and Satterfield (1968);
2. the correlation given by Eq. 4.54.

Solution

1. Calculation of the dimensionless parameters:

a. Thiele modulus:

$$k_v = (1.6 \times 10^{10}) \exp\left(-\frac{19,800}{1.98 \times 500}\right) = 32.4 \text{ sec}^{-1}$$

$$\phi_{s1} = R \left(\frac{k_v}{D_{eA}}\right)^{1/2} = 15$$

b. Calculation of β_m:

Since the gas phase contains 50% A, the concentration of A at the pellet surface will be given by

$$C_{As} = \frac{0.5}{R_g T} = \frac{0.5}{82 \times 500} = 1.22 \times 10^{-5} \text{ mol/cm}^3$$

Hence

$$\beta_m = \frac{3.6 \times 10^{-2} \times 30,000 \times 1.22 \times 10^{-5}}{2.64 \times 10^{-4} \times 500} = 0.10$$

c. Calculation of α_s:

$$\alpha_s = \frac{E}{R_g T_s} = \frac{19,800}{1.98 \times 500} = 20.0$$

d. Calculation of KC_{As}:

The adsorption constant K_A at 500°K is

$$K_A = (8.71 \times 10^4) \exp\left(\frac{20,000}{1.98 \times 500}\right)$$

$$= 6.55 \times 10^5 \text{ cm}^3/\text{mol}$$

The maximum temperature that can be reached in the pellet is

$$T_{max} = T_s(1 + \beta_m) = 550°\text{K}$$

The adsorption constant at 550°K is 5.47×10^5.

The corresponding value of $K_A C_{As}$ at 500°K is 8; at 550°K, $K_A C_{As}$ is 6.68. Hence the variation of $K_A C_{As}$ is sufficiently small to enable us to assume a constant average value, that is, $K_A C_{As} = 7.34$.

2. Calculation of the effectiveness factor from the numerical solution:

From the figure of Kao and Satterfield (1968), by interpolation $\varepsilon = 1.0$.

3. Calculation of the effectiveness factor from Eq. 4.54:

The modified Thiele modulus is estimated from

$$(\phi)_m = \frac{\phi}{1 + K_A C_{As}} \left[\frac{K_A C_{As}}{\ln (1 + K_A C_{As})}\right]^{1/2} = 3.35$$

Thus $(\phi)_m > 2$,

$$\beta_m \alpha_s = 20 \times 0.1 = 2 < 6$$

and

$$\alpha_s = 20 > 8$$

Hence Eq. 4.54 can be used for the evaluation of the effectiveness factor:

$$\varepsilon = \left\{\exp^{1.172 \beta_m \sqrt{(\alpha_s - 8)^{1/2}}} - 1\right\} \left[\exp^{(\varepsilon_{iso})} - 1\right] + \varepsilon_{iso}$$

$$\varepsilon_{iso} = \frac{3}{(\phi_s)_m} \left[\frac{1}{\tanh (\phi_s)_m} - \frac{1}{(\phi_s)_m}\right] = 0.629$$

$$\varepsilon = \left\{\exp^{[1.172 (0.1) \sqrt{(20 - 8)^{1/2}}]} - 1\right\}$$

$$\times \left[\exp^{(0.629)} - 1\right] + 0.629$$

$$= 1.066$$

Hence the effectiveness factors calculated by the two methods agree to within about 6%.

4.10. SOME SPECIAL CASES

The concept of the effectiveness factor as explained in the preceding sections of this chapter can be extended to a variety of diffusion-cum-reaction situations. A few important ones are described below.

4.10.1. Effectiveness Factors for Size-Dispersed Catalysts

By analogy with the definition of the effectiveness factor for particles and/or pellets of uniform size considered so far in this chapter, the effectiveness factor for a mixture of particle sizes can be defined. Thus Pratt and Wakeham (1975) have shown that the degree of size dispersion and the upper and lower bounds of particle size in a size-dispersed catalyst can have a profound influence on the overall effectiveness factor for the catalyst. The influence becomes more marked as the Thiele modulus corresponding to the geometric mean particle size crosses unity, or as the size dispersion of the catalyst becomes larger. A reaction-equivalent Thiele modulus has been defined in terms of the geometric mean Thiele modulus, standard deviation of the log normal particle size distribution, and the lower and upper particle size limits.

4.10.2. Effectiveness Factors for Catalysts with Nonuniform Distribution of Activity

In the developments presented so far our attention has been restricted to catalysts characterized by a uniform activity distribution. Sometimes uniform distribution of catalyst activity is undesirable, especially when the cost of the catalyst is the main consideration, as in the case of precious metal catalysts. Two kinds of activity distribution have commonly been used. In one the catalyst consists of a thin, impregnated, active outer shell with an inert core (such catalysts are also referred to as partially impregnated catalysts; Minhas and Carberry, 1969); in the other there is a continuous decrease of activity from the surface to the center (Kasaoka and Sakata, 1968; Yazdi and Petersen, 1972; Corbett and Luss, 1974; Villadsen, 1976; Becker and Wei, 1977; Coughlin and Verykios, 1977). Acting on the practical thought that catalysts with a continuously varying distribution of activity might not be so easy to prepare, Becker and Wei (1977) and Wang and Varma (1978, 1980) have investigated several types of discontinuous activity distribution: "egg yolk," "egg white," and "eggshell" modes of deposition. In all these cases an *a priori* activity distribution was assumed and

the performance of the catalyst pellet evaluated for a given situation. Comparison of the results therefore helps in identifying the best activity distribution. The same problem has been considered by Morbidelli and Varma (1982), who reformulated it as follows: For a given amount of catalytic material, what would be the optimum distribution for realizing the best utilization of the catalyst pellet? Their results, obtained for bimolecular H–W kinetics, show that the optimum activity distribution is simply a Dirac delta function, and an expression to obtain the optimum location in terms of the parameters of the system has been derived. Their results further indicate that about 5% deviation on either side of the optimum location has little influence on the performance of the catalyst. Experimental methods to exactly locate such an activity are not yet standardized, however. Analyses of specific systems based on sharp interphase activity distribution have been reported by Friedrichsen (1969) and Smith and Carberry (1975) for the phthalic anhydride reactor and by Horvath and Engasser (1973) for enzyme reactors.

In this section we shall evaluate the performance of a supported catalyst with a nonuniform distribution of activity by considering a simple reaction of first order as well as of general order. The effect of a nonuniform distribution of catalyst activity on selectivity in a complex reaction with and without deactivation will be considered in Chapters 5 and 8, respectively.

The distribution obtained by impregnation of a porous bed depends largely on (1) the nature of the catalyst, (2) the nature of the support, (3) the method of impregnation, and (4) the composition of the impregnating solution. A general form of activity distribution has been defined by Yazdi and Petersen (1972):

$$k_v = k_v^0 \hat{R}^q \qquad (4.93)$$

where q is some suitably chosen exponent, k_v is the local volumetric rate constant, and k_v^0 is the value of k_v at uniform activity.

Alternatively, the following activity profile, suggested by Nystrom (1978), may be used:

$$
\begin{aligned}
k_v &= k_v^0 \left[\left(\frac{r - R^*}{R - R^*} \right)^n (1 - s) + s \right] && \text{for} \quad R^* \leqslant r \leqslant R \\
&= 0 && \text{for} \quad 0 \leqslant r \leqslant R^*
\end{aligned}
$$
$$(4.94)$$

In this equation R^* represents that value of r at which the activity reaches zero, and n and s are constants characterizing the shape and relative activity, respectively, in the center of the pellet.

The mass balance equations for a simple irreversible reaction A → B can be easily written, say, for a spherical pellet. The boundary condition arising out of the symmetry of the pellet (namely, $dC/dr = 0$ at $r = 0$) can be used along with the surface concentration to solve these equations. The case of a first-order irreversible reaction has been solved by Yazdi and Peterson (1972). An approximate solution for an nth-order reaction can also be obtained and the result expressed as the ratio of the effectiveness factor for nonuniform distribution to that for uniform distribution:

$$\frac{\varepsilon_{\text{nonuniform}}}{\varepsilon_{\text{uniform}}} = \frac{3}{q+3} \qquad (4.95)$$

when the condition $n - q - 3 < 0$ is satisfied.

The results of both the analyses indicate that non-uniform distribution of the catalyst causes some decrease in the net rate of consumption of A and production of B (per unit volume of porous pellet). Thus concentrating the activity in the external shell lowers the effectiveness of the system. This is of course to be expected because the inner part of the pellet is not utilized completely. However, as will be shown in Chapter 5, the rate of production of B per unit mass of catalyst is improved significantly in the case of a consecutive reaction.

The case of a first-order reaction has also been solved by Nystrom (1978) using Eq. 4.94. The great variations in the ε–ϕ curves necessitate a modified definition of the Thiele modulus that brings the different curves together. The Thiele modulus for this purpose is defined as

$$(\phi)_{\text{m}} = R\left[\frac{k_{\text{v}}(\text{CF})}{D_{\text{e}}}\right]^{1/2} \qquad (4.96)$$

where CF is the correction factor for the shorter diffusion path in the shell catalysts and can be obtained from

$$\text{CF} = 4.85\frac{\delta}{R} \qquad (4.97)$$

where δ represents the depth that divides the pellet into two parts with equal activity. Equating the rate thus obtained with the average rate per pellet leads to

$$2\int_{R}^{R-\delta} k_{\text{v}}\, 4\pi r^2\, dr = \overline{k_{\text{v}}}\,\frac{4\pi R^3}{3} \qquad (4.98)$$

where $\overline{k_{\text{v}}}$ is the average value of k_{v}; ε can thus be estimated.

4.10.3. Effectiveness Factor for a Diluted Catalyst Pellet

The distribution of catalyst activity on a support in accordance with a predetermined pattern poses manufacturing difficulties. A simple way to overcome the problem is to use a diluted catalyst pellet, as proposed by Ruckenstein (1970), in which microspheres of active catalyst are embedded in an inert porous medium of such characteristics as would permit more efficient utilization of the active material. The chief advantage of the diluted pellet is that the degree of freedom represented by the choice of the inert porous medium may be exploited in order to design a catalyst of improved performance characteristics.

Varghese and Wolf (1980) have carried out an analysis of a diluted pellet and derived the following expression for the effectiveness factor:

$$\frac{\text{rate in diluted pellet}}{\text{rate in undiluted pellet}} = \left(\frac{f_2 D_{\text{e}}}{D_2}\right)^{1/2}$$
$$\times \frac{\tanh\left[\phi\,(f_2 D_2/D_{\text{e}})^{1/2}\right]}{\tanh \phi} \qquad (4.99)$$

where the effective diffusivity D_{e} is given by the following expression:

$$D_{\text{e}} = f_2 D_2 + (1 - f_2)D_1 \qquad (4.100)$$

In these equations subscripts 1 and 2 represent the inert porous medium and active porous particles, respectively, and f_2 represents the volume fraction of the active material present. Figure 4.10 shows a plot of the ratio of the reaction rate of the diluted pellet to that of the undiluted pellet as a function of the fraction of the active metal for various values of the ratio of the inert and active material diffusivities. It will be noted that beyond a certain value of f_2 this ratio exceeds unity and passes through a maximum. Clearly, dilution can be employed as a practical means of increasing the activity of the pellet as a whole for a diffusion-influenced reaction. This observation can best be exploited in the case of a deactivating pellet.

4.10.4. Effectiveness Factors for Reactions with Configurational Diffusion

Effectiveness factors for reactions involving configurational diffusion have not been estimated to any

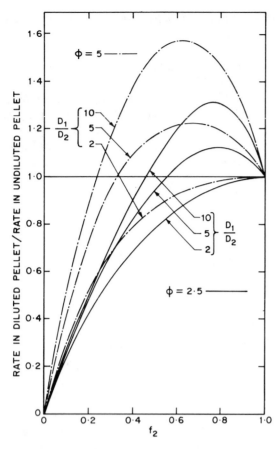

Figure 4.10. Reaction rate of the diluted catalyst relative to the reaction rate of the undiluted catalyst vs. the fraction of active metal (Varghese and Wolf, 1980).

significant extent. The conservation equation for the species undergoing this type of diffusion can be written in analogy with Eq. 4.1, and for a first-order reaction in a slab the following well-known expression for the effectiveness factor results:

$$\varepsilon = \frac{\tanh \phi}{\phi} \qquad (4.101)$$

where now ϕ is formally defined as

$$\phi = L\left\{ k_v \left[\frac{f_c D_b}{\tau}\left(1 - \frac{a}{r_p}\right)^4 \right]^{-1} \right\}^{1/2}, \qquad a < r_p \qquad (4.102)$$

However, where configurational diffusion is involved, it becomes necessary to account for pore size distribution. Employing the parallel cross-linked pore model, we can obtain the same expression for the effectiveness factor

with ϕ redefined as

$$\phi = L\left[\frac{\int (2/r_p)k_S(r_p)f(r_p)\,dr_p}{(D_{eA}/\tau)\int(1 - a/r_p)^4 f(r_p)\,dr_p} \right], \qquad a < r_p \qquad (4.103)$$

There seems to be growing interest in the area, particularly because of the development of the technology of hydrotreating processes, which require removal of sulfur and metals from residual oil, coal- and shell-derived fuels. In most cases the rate of demetallation is first order with respect to the metal concentration (Sato et al. 1971; Chang and Silvestri, 1974; Spry and Sawyer, 1975; Dautzenburg et al., 1978) and an optimal catalyst activity has been shown experimentally to exist for catalysts with intermediate pore size (van Zoonan and Dowes, 1963; Sooter and Crynes, 1975; Spry and Sawyer, 1975; Richardson and Alley, 1975). The catalyst undergoes deactivation as a result of the deposition of metal sulfides and coke, and models to compute the activity and life of the catalyst have been proposed (Newson, 1975; Hughes and Mann, 1978; Dautzenburg et al., 1978). More recently, Rajagopalan and Luss [1979] have also derived simple algebraic equations to calculate *a priori* the pore sizes that would yield the optimal initial activity and the optimal lifetime activity. For all the catalysts with the same surface area and porosity, the largest initial activity is obtained for pellets with a uniform pore size.

4.10.5. Effectiveness Factors for Molecular Sieve Catalysts

The analysis presented in earlier sections was based on the assumption that the effective diffusivity is independent of concentration. In some cases, however, such as in zeolites, the diffusivity is known to be concentration dependent (Ruthven and Loughlin, 1971). Thus the kind of analysis that is used for constant-diffusivity situations would not be applicable to molecular sieve catalysts. Theoretical concentration profiles and the corresponding effectiveness factors have been calculated for diffusion with first-order reaction for systems in which the concentration dependence of D_e arises from the nonlinearity of the Langmuir equilibrium isotherm (Ruthven, 1972). Both the concentration profiles and the effectiveness factors show considerable deviations from the solutions with constant D_e. The following simple approximate expression for the effectiveness factor for zeolites with high diffusional resistance has been derived (Ruthven, 1972):

$$\varepsilon = (2/\hat{C}^*\phi_{p1})\left[-\hat{C}^* - \ln(1 - \hat{C}^*)^{1/2} \right], \qquad \varepsilon < 0.5 \qquad (4.104)$$

where $\hat{C}^*$ is the ratio of sorbate concentration at the surface to that at saturation.

Pereira and Varma (1978) have obtained analytical solutions for effectiveness factors for the case of mildly concentration-dependent diffusion coefficients. They found that the effect of $D(C)$ was minor, being largest at $\phi \to \infty$

Certain peculiar features are associated with zeolite catalysts because of their shape selectivity. For instance, in the catalytic cracking of C_6 to C_9 hydrocarbon on ZSM-5, it has been shown by Haag et al. (1981) that two separate effects on selectivity can be recognized: a mass transfer induced selectivity, and one due to steric inhibition by the size of a reaction complex. The selective cracking of n-paraffins compared to monomethyl paraffins has been shown to be due to the higher intrinsic rate constant of the n-paraffin with diffusional limitation playing practically no part. In contrast, dimethyl paraffin cracking is strongly diffusion limited. The difference between the behavior of the two classes of paraffins is the result of steric constraint on the methyl paraffin carbonium ion reaction complex. It has been shown experimentally by the same authors that the diffusivity (determined under reaction conditions) exceeds the values expected from the Knudsen model. There is thus a need to review the application of the conventional concepts of mass transport to zeolites in general.

4.10.6. Effectiveness Factors for Supported Liquid-Phase Catalysts

Recently, supported liquid-phase catalysts, in which a liquid phase is supported on porous solids, have been shown to be very useful (see Chapter 18). Rony (1968) has theoretically studied the diffusion-limited reaction in supported liquid-phase catalysts. A model, based on the ideal pore concept, has been proposed for a first-order reaction and has been verified experimentally by using the hydroformylation of propylene reaction (Rony, 1969).

Abed and Rinker (1973) have analyzed the problem of diffusion and reaction in supported liquid-phase catalysts for a nonideal porous support. They have proposed a more rigorous model which can predict the effective diffusivity under reaction conditions as well as the effectiveness factor. Livbjerg et al. (1974, 1976) and Chaudhari (1980) have considered diffusion in both the gas and liquid phases. Their results are discussed in Chapter 18.

4.10.7. Effectiveness Factors for Other Heterogeneous Systems (Gas–Liquid, Gas–Liquid–Solid)

Kulkarni and Doraiswamy (1975, 1976) have shown that it is possible to extend the concept of the effectiveness factor to gas–liquid reactions through the postulation of a parameter β_L, defined as

$$\beta_L = \frac{1}{a\delta} \qquad (4.105)$$

which may be regarded as the ratio of the area per unit volume in the film to that in the liquid bulk. The main difference between gas–liquid and gas–solid effectiveness factors is the existence of the bulk phase in gas–liquid systems. The reaction occurring within the catalyst may be likened to the reaction occurring within a film in gas–liquid systems. A general equation has been developed that includes the parameter β_L. When β_L approaches zero the equation reduces to that for the normal effectiveness factor for solid-catalyzed reactions as developed in Section 4.2.

The slurry-bed reactor, where all three phases (gas, liquid, and solid) are present simultaneously, provides perhaps the most general situation for applying the concept of the effectiveness factor. In the general instance, the rate-controlling step can either be the transport across the gas–liquid interface or liquid–solid interface, or may lie within the catalyst particle. Let us define the gas–liquid and liquid–particle effectiveness factors as

$$\varepsilon_G = \frac{k_g}{k_g + k_L} \qquad (4.106)$$

and

$$\varepsilon_L = \frac{\rho_c k_s a_s}{m k_v \varepsilon + \rho_c k_s a_s} \qquad (4.107)$$

where m represents the catalyst loading (in g/cm^3 of slurry); k_L and k_g are the mass transfer coefficients for the liquid side and the gas side, respectively; at the gas–liquid interface, k_s is the liquid–solid mass transfer coefficient; a_s and a_g are the solid–liquid and gas–liquid interfacial area, respectively. The overall effectiveness factor for a slurry-bed reactor can then be obtained as (Sylvester et al., 1975)

$$\varepsilon_{\text{overall}} = \frac{1}{1 + \text{Da}} = 1 - \frac{H r_v m}{C_g k_L a_g \varepsilon_G} \qquad (4.108)$$

where the Damköhler number

$$\mathrm{Da} = \frac{mk_v \varepsilon \varepsilon_\mathrm{L}}{a_g k_\mathrm{L} \rho_c \varepsilon_\mathrm{G}}$$

H is Henry's law constant, and ε is the intraparticle effectiveness factor. The overall effectiveness factor can be estimated by experimentally measuring the terms $Hr_v m$ and $k_\mathrm{L} a_g \varepsilon_\mathrm{G}$. Sylvester et al. (1975) demonstrated the use of this method for calculating the effectiveness factor by employing the data of Satterfield et al. (1968, 1969) and Calderbank et al. (1963).

Effectiveness factors have also been defined for gas–solid reactions by using three main models: moving boundary, diffuse interface, and homogeneous.

REFERENCES

Abed, R., and Rinker, R. G. (1973), *AIChE J.*, **19**, 618.

Abhyankar, S. M., and Doraiswamy, L. K. (1979), *Can. J. Chem. Eng.*, **57**, 481.

Akehata, T., Namkoog, S., Kubotate, H., and Shindo, M. (1961), *Can. J. Chem. Eng.*, **39**, 127.

Aris, R. (1957), *Chem. Eng. Sci.*, **6**, 262.

Aris, R. (1966), *Ind. Eng. Chem.*, **58**(9), 32.

Aris, R. (1971), *J. Catal.*, **22**, 282.

Aris, R. (1975), *The Mathematical Theory of Diffusion and Reaction in Permeable Catalysts*, Vol. 1, Oxford (Clarendon Press), New York and London.

Barnett, L. G., Weaver, R. E. C., and Gilkeson, M. M. (1961), *AIChE J.*, **7**, 211.

Becker, E. R., and Wei, J. (1977), *J. Catal.*, **46**, 365.

Bienert, R., and Gelbin, D. (1967), *Chem. Tech. (Berlin)*, **19**, 207.

Bischoff, K. B. (1965), *AIChE J.*, **11**, 351.

Bischoff, K. B. (1968), *Chem. Eng. Sci.*, **23**, 451.

Blue, R. W., Holm, V. C. F., Regier, R. B., Fast, E., and Heckelsberg, L. F. (1952), *Ind. Eng. Chem.*, **44**, 2710.

Bokhoven, C., and van Raayen, W. (1954), *J. Phys. Chem.*, **58**, 471.

Bond, G. C. (1962), *Catalysis by Metals*, Academic Press, New York.

Boreskov, G. K. (1947), *Khim. Prom.*, p. 256.

Boreskov, G. K., and Slinko, M. G. (1952), *Zh. Fiz. Khim.*, **26**, 235.

Broekhoff, J. C. P., and de Boer, J. H. (1968), *J. Catal.*, **10**, 153.

Butt, J. B., Downing, D. M., and Lee, J. W. (1977), *Ind. Eng. Chem. Fundam.*, **16**, 270.

Calderbank, P. H., Evans, F., Farley, R., Jepson, G., and Poll, A. (1963), *Catalysis in Practice*, Inst. Chem. Eng., London.

Cant, N. W., and Hall, W. K. (1970), *J. Catal.*, **16**, 220.

Carberry, J. J. (1961), *AIChE J.*, **7**, 350.

Carberry, J. J. (1962a), *Chem. Eng. Sci.*, **17**, 675.

Carberry, J. J. (1962b), *AIChE J.*, **8**, 557.

Carberry, J. J. (1976), *Chemical and Catalytic Reaction Engineering*, McGraw-Hill, New York.

Carberry, J. J., and Wendel, M. (1963), *AIChE J.*, **9**, 129.

Chang, C. D., and Silvestri, A. J. (1974), *Ind. Eng. Chem. Process Design Dev.*, **13**, 315.

Chaudhari, R. V. (1980), unpublished work.

Choudhary, V. R., and Doraiswamy, L. K. (1975), *Ind. Eng. Chem. Process Design Dev.*, **14**, 227.

Chu, C., and Chon, K. (1970), *J. Catal.*, **17**, 71.

Chu, C., and Hougen, O. A. (1962), *Chem. Eng. Sci.*, **17**, 167.

Churchill, S. W. (1977), *AIChE J.*, **23**, 208.

Copelowitz, I., and Aris, R. (1970), *Chem. Eng. Sci.*, **25**, 885.

Corbett, W. E., and Luss, D. (1974), *Chem. Eng. Sci.*, **25**, 267.

Coughlin, R. W., and Verykios, X. E. (1977), *J. Catal.*, **48**, 249.

Cresswell, D. L. (1970), *Chem. Eng. Sci.*, **25**, 267.

Damköhler, G. (1943), *Z. Physik. Chem. (Leipzig)*, **A193**, 16.

Dautzenberg, F. M., Van Klinken, J., Pronk, K. M. A., Sie, S. T., and Wijffels, J. B. (1978), *ACS Symp. Ser.*, **65**, 254.

Drott, D. W., and Aris, R. (1969), *Chem. Eng. Sci.*, **24**, 541.

Elnashaie, S. S. E. H., and Mahfouz, A. T. (1978), *Chem. Eng. Sci.*, **33**, 386.

Ferraiolo, G., Peloso, A., Reverberi, A., Del Borghi, M., and Beruto, D. (1973), *Can. J. Chem. Eng.*, **51**, 447.

Finlayson, B. A. (1972), *The Method of Weighted Residuals and Variational Principles*, Academic Press, New York.

Finlayson, B. A. (1974), *Catal. Rev. Sci. Eng.*, **10**, 69.

Forni. L., Zanderighi, L., and Carra, S. (1971), *J. Catal.*, **23**, 38.

Foster, R. N., and Butt, J. B. (1966), *AIChE J.*, **12**, 180.

Friedrichsen, W. (1969), *Chem. Eng. Tech.*, **41**, 967.

Gioia, F. G, Greco, Jr., and Gibilaro, L. G. (1970), *Chem. Eng. Sci.*, **25**, 969.

Goldstein, W., and Carberry, J. J. (1973), *J. Catal.*, **28**, 33.

Gunn, D. J. (1966), *Chem. Eng. Sci.*, **21**, 383.

Gunn, D. J. (1967), *Chem. Eng. Sci.*, **22**, 1439.

Gupta, V. P., and Douglas, W. J. M. (1967), *AIChE J.*, **13**, 883.

Haag, W. O., Lago, R. M., and Weisz, P. B. (1981), *Faraday Disc. Chem. Soc.*, **72**, 317.

Hatfield, B., and Aris, R. (1969), *Chem. Eng. Sci.*, **24**, 1213.

Haynes, H. W., Jr. (1978), *Can. J. Chem. Eng.*, **56**, 582.

Henry, J. P., Chennakeshvan, B., and Smith, J. M. (1961), *AIChE J.*, **7**, 10.

Hlavacek, V., and Kubicek, M. A. (1970), *Chem. Eng. Sci.*, **25**, 1537.

Hlavacek, V., and Marek, M. (1967), *Collect. Czech. Chem. Commun.*, **32**, 4004.

Hlavacek, V., and Marek, M. (1968), *Chem. Eng. Sci.*, **23**, 865.

Hlavacek, V., Marek, M., and Kubicek, M. (1968), *Collect. Czech. Chem. Commun.*, **33**, 718.

Hlavacek, V., and van Rompay, P. (1981), *Chem. Eng. Sci.*, **36**, 1587.

Horvath, L., and Engasser, J. M. (1973), *Ind. Eng. Chem. Fundam.*, **12**, 229.

Hughes, C. C., and Mann, R. (1978), *ACS Symp. Ser.*, **65**, 201.

Hutchings, J., and Carberry, J. J. (1966), *AIChE J.*, **12**, 20.

Ibanez, J. L. (1979), *J. Chem. Phys.*, **71**, 5253.

Irving, J. P. and Butt, J. B. (1967), *Chem. Eng. Sci.*, **22**, 1857.

Jayaraman, V. K., and Kulkarni, B. D. (1981), *Chem. Eng. J.*, **21**, 261.

Jayaraman, V. K., Kulkarni, B. D., and Doraiswamy, L. K., (1982a), *Chem. Eng. J.*, (in press).

Jayaraman, V. K., Kulkarni, B. D., and Doraiswamy. L. K., (1982b), *AIChE J.*, (in press).

Johnson, M. F. L., Kreger, W. E., and Erickson, H. (1957), *Ind. Eng. Chem.*, **49**, 283.

Jouven, J., and Aris, R. (1972), *AIChE J.*, **18**, 402.

Juusola, J. A., Mann, R. F., and Downie, J. (1970), *J. Catal.*, **17**, 106.

Kao, H. S. P., and Satterfield, C. N. (1968), *Ind. Eng. Chem. Fundam.*, **7**, 664.

Karanth, N. G., Koh, H. P., and Hughes, R. (1974), *Chem. Eng. Sci.*, **29**, 451.

Kasaoka, S., and Sakata, Y. (1966), *Kagaku Kogaku*, **30**, 650.

Kasaoka, S., and Sakata, Y. (1968), *J. Chem. Eng. Japan*, **1**, 138.

Kehoe, J. P. G., and Aris, R. (1973), *Chem. Eng. Sci.*, **28**, 2094.

Kehoe, J. P. G., and Butt, J. B. (1972), *AIChE J.*, **18**, 347.

Kesten, A. S. (1969), *AIChE J.*, **15**, 128.

Koh, H. P., and Hughes, R. (1974), *AIChE J.*, **20**, 395.

Kraemer, E. O. (1931), *Treatise on Physical Chemistry*, ed. Taylor, N.S., 2nd ed., Van Nostrand, Princeton, New Jersey.

Krasuk, J. H., and Smith, J. M. (1965), *Ind. Eng. Chem. Fundam.*, **4**, 102.

Kulkarni, B. D., and Doraiswamy, L. K. (1975), *AIChE J.*, **21**, 501.

Kulkarni, B. D., and Doraiswamy, L. K. (1976), *AIChE J.*, **22**, 597.

Kulkarni, B. D., Jayaraman, V. K., and Doraiswamy, L. K. (1981), *Chem. Eng. Sci.*, **36**, 943.

Larsen, A. T., and Tour, R. S. (1922), *Chem. Met. Eng.*, **26**, 647.

Lee, J. C. M., Padmanabhan, L., and Lapidus, L. (1972), *Ind. Eng. Chem. Fundam.*, **11**, 117.

Levy, R. B., and Boudart, M. (1974), *J. Catal.*, **32**, 304.

Lin, K., and Lih, M. M. (1971), *AIChE J.*, **17**, 1234.

Lin, K., and Lih, M. M. (1973), *AIChE J.*, **19**, 832.

Lih, M. M. (1970), *J. Phys. Chem.*, **74**, 2245.

Liu, S. L. (1967), *Chem. Eng. Sci.*, **22**, 871.

Liu, S. L. (1970), *AIChE J.*, **16**, 742.

Liu, S. L., and Amundson, N. R. (1963), *Ind. Eng. Chem. Fundam.*, **2**, 183.

Livbjerg, H., Jensen, K. F., and Villadsen, J. (1976), *J. Catal.*, **45**, 216.

Livbjerg, H., Sorensen, B., and Villadsen, J. (1974), *Adv. Chem. Ser.*, **133**, 242.

Love, K. S., and Emmett, P. H. (1941), *J. Am. Chem. Soc.*, **63**, 3297.

Luss, D. (1968), *AIChE J.*, **14**, 966.

Luss, D. (1977), *Chemical Reactor Theory—A Review*, eds. Lapidus, L., and Amundson, N. R., Prentice-Hall, Englewood Cliffs, New Jersey, Chapter 4.

Luss, D., and Amundson, N. R. (1967), *AIChE J.*, **13**, 759.

Mann, R. S., and Hahn, K. W. (1969), *J. Catal.*, **15**, 329.

Mars, P., and Maessen, J. G. H. (1964), *Proc. 3rd Int. Cong. Catal.*, Amsterdam, **1**, 266.

Maymo, J. A., and Cunningham, R. E. (1966), *J. Catal.*, **6**, 186.

McGinnis, P. H. (1965), *Chem. Eng. Prog. Symp. Ser.*, **61**(55), 2.

McGreavy, C., and Cresswell, D. L. (1969), *Chem. Eng. Sci.*, **24**, 608.

McGreavy, C., and Thornton, J. M. (1970), *Chem. Eng. Sci.*, **25**, 303.

Metcalfe, A., and Rowden, M. W. (1971), *J. Catal.*, **22**, 30.

Miale, J. N., Chen. N. Y., and Weisz, P. B. (1966), *J. Catal.*, **6**, 278.

Michelsen, M. L., and Villadsen, J. (1972), *Chem. Eng. Sci.*, **27**, 751.

Miller, D. N., and Kirk, R. S. (1962), *AIChE J.*, **8**, 183.

Miller, F. W., and Deans, H. A. (1967), *AIChE J.*, **13**, 45.

Mingle, J. O., and Smith, J. M. (1961), *AIChE J.*, **7**, 243.

Minhas, S., and Carberry, J. J. (1969), *Brit. Chem. Eng.*, **14**, 799.

Morbidelli, M., and Varma, A. (1982), *Ind. Eng. Chem. Fundam.*, **21**, 278, 284.

Namjoshi, A., Kulkarni, B. D., and Doraiswamy, L. K. (1982), *AIChE J.* (Communicated).

Newson, E. J. (1975), *Ind. Eng. Chem. Process Design Dev.*, **14**, 27.

Nielsen, A. (1968), *An Investigation on Promoted Iron Catalysts for the Synthesis of Ammonia*, 3rd ed., Gjellerup, Copenhagen.

Nir, A., and Pismen, L. M. (1977), *Chem. Eng. Sci.*, **32**, 35.

Nozik, A. J., and Behnken, D. W. (1965), *J. Catal.*, **4**, 469.

Nystrom, M. (1978), *Chem. Eng. Sci.*, **33**, 379.

Ors, N., and Dogu, T. (1979), *AIChE J.*, **25**, 723.

Östergaard, K. (1963), *Chem. Eng. Sci.*, **18**, 259.

Otani, S., Wakao, N., and Smith, J. M. (1965), *AIChE J.*, **11**, 446.

Pasquon, I., and Dente, M. (1962), *J. Catal.*, **1**, 508.

Patterson, W. R., and Cresswell, D. L. (1971), *Chem. Eng. Sci.*, **26**, 605.

Pereira, C. J., and Varma, A. (1978), *Chem. Eng. Sci.* **33**, 396.

Pereira, C. J., Wang, J. B., and Varma, A. (1979), *AIChE J.*, **25**, 1036.

Petersen, E. E. (1965), Chemical Reaction Analysis, Prentice-Hall, Englewood Cliffs, New Jersey.

Petersen, E. E. (1962), *Chem. Eng. Sci.*, **17**, 987.

Prater, C. D., and Lago, R. M. (1956), *Adv. Catal.*, **8**, 293.

Pratt, K. C., and Wakenham, W. A. (1975), *Chem. Eng. Sci.*, **30**, 444.

Rajadhyaksha, R. A., and Vasudeva, K. (1974), *J. Catal.*, **34**, 321.

Rajadhyaksha, R. A., Vasudeva, K., and Doraiswamy, L. K. (1976), *J. Catal.*, **41**, 61.

Rajagopalan, K., and Luss, D. (1979), *Ind. Eng. Chem. Process Design Dev.*, **18**, 459.

Raja Rao, M., and Smith, J. M. (1963), *AIChE J.*, **9**, 485.

Reddy, K. A., and Doraiswamy, L. K. (1969), *Chem Eng. Sci.*, **24**, 1415.

Rester, S. and Aris, R. (1969), *Chem. Eng. Sci.*, **24**, 793.

Rester, S. Jouven, J., and Aris, R. (1969), *Chem. Eng. Sci.*, **24**, 1019.

Richardson, R. L., and Alley, S. K. (1975), *Preprints Div. Petrol Chem. ACS*, **20**, 554.

Roberts, G. W., and Satterfield, C. N. (1965), *Ind. Eng. Chem. Fundam.*, **4**, 288.

Roberts, G. W., and Satterfield, C. N. (1966), *Ind. Eng. Chem. Fundam.*, **5**, 317.

Rony, P. R. (1968), *Chem. Eng. Sci.*, **23**, 1021.

Rony, P. R. (1969), *J. Catal.*, **14**, 142.

Rozovskii, A. Ya., and Shchekin, V. V. (1960), *Kinet. Catal.*, **1**, 313.

Ruckenstein, E. (1970), *AIChE J.*, **16**, 151.

Ruthven, D. M. (1972), *J. Catal.*, **25**, 259.

Ruthven, M. D., and Loughlin, K. F. (1971), *Trans. Faraday Soc.*, **67**, 1661.

Sadana, A., and Doraiswamy, L. K. (1971), *Chem. Age India*, **22**, 85.

Satterfield, C. N. (1970), *Mass Transfer in Heterogeneous Catalysis*, M.I.T. Press, Cambridge, Massachusetts.

Satterfield, C. N., Ma, Y. H., and Sherwood, T. K. (1968), *Chem. Eng. Prog. Symp. Ser.*, **28**, 22.

Satterfield, C. N., Pelossof, A. A., and Sherwood, T. K. (1969), *AIChE J.*, **15**, 226.

Sato, M., Takayama, N., Kurita, S., and Kwan, T. (1971), *Nippon Kagku. Zasshi*, **92**, 834.

Schilson, R. E. and Amundson, N. R. (1961), *Chem. Eng. Sci.*, **13**, 226.

Schmalzer, D. K. (1969), *Chem. Eng. Sci.*, **24**, 615.

Schneider, P. (1975), *Catal. Rev. Sci. Eng.*, **12**, 201.

Schneider, P., and Mitschka, P. (1966a), *Chem. Eng. Sci.*, **21**, 455.

Schneider, P.,and Mitschka, P. (1966b), *Collect. Czech. Chem. Commun.*, **31**, 3677.

Scott, D. S. (1962), *Can. J. Chem. Eng.*, **40**, 173.

Sinfelt, J. H., and Lucchesi, P. J. (1963), *J. Am. Chem. Soc.*, **85**, 3365.

Smith, N. L., and Amundson, N. R. (1951), *Ind. Eng. Chem.*, **43**, 2156.

Smith, T. G., and Carberry, J. J. (1975), *Can. J. Chem. Eng.*, **53**, 347.

Smith, T. G., Zahradnik, J., and Carberry, J. J. (1975), *Chem. Eng. Sci.*, **30**, 763.

Sohn, H. Y., Merrill, R. P., and Petersen, E. E. (1970), *Chem. Eng. Sci.*, **25**, 399.

Sooter, M. C., and Crynes, B. L. (1975), *Ind. Eng. Chem. Process Design Dev.*, **14**, 199.

Spry, J. C., and Sawyer, W. H. (1975), AIChE 68th ann. meeting, Los Angeles, p. 16.

Sterrett, J. S., and Brown L. F. (1968), *AIChE J.*, **14**, 696.

Suzuki, T., and Uchida, T. (1979), *J. Chem. Eng. Japan*, **12**, 425.

Sylvester, N. D., Kulkarni, A. A., and Carberry, J. J. (1975), *Can J. Chem. Eng.*, **53**, 313.

Tartarelli, R., and Morelli, E. (1967), *Ann. Chim. (Rome)*, **57**, 1316.

Tartarelli, R., Cionis, S., and Caporani, M. (1970), *J. Catal*, **18**, 212.

Thiele, E. W. (1939), *Ind. Eng. Chem.*, **31**, 916.

Tinkler, J. D., and Metzner, A. B. (1961), *Ind. Eng. Chem.*, **53**, 663.

Tsotsis, T. T., and Schmitz, R. A. (1979), *Chem. Eng. Sci.*, **34**, 135.

Uppal, A., and Ray, W. H. (1977), *Chem. Eng. Sci.*, **32**, 649.

Vaidyanathan, K., and Doraiswamy, L. K. (1968). *Chem. Eng. Sci.*, **15**, 326.

Valdman, B. and Hughes, R. (1976), *AIChE J.*, **22**, 192.

van den Bosch, B., and Luss, D. (1977a), *Chem. Eng. Sci.*, **32**, 203.

van den Bosch, B., and Luss, D. (1977b), *Chem. Eng. Sci.*, **32**, 560.

van Zoonen, D., and Dowes, C. T. (1963), *J. Inst. Petrol.*, **49**, 383.

Varghese, P., and Wolf, E. E. (1980), *AIChE J.*, **26**, 55.

Varghese, P., Varma, A., and Carberry, J. J. (1978), *Ind. Eng. Chem. Fundam.*, **17**, 195.

Villadsen, J., and Michelsen, M. L. (1978), *The Solution of Differential Equation models by Polynomial Approximation*, Prentice-Hall, Englewood Cliffs, New Jersey.

Villadsen, J. (1976), *Chem. Eng. Sci.*, **31**, 1212.

Wagner, C. (1943), *Z. Physik. Chem. (Leipzig)*, **A193**, 1.

Wakao, N., Selwood, P. W., and Smith, J. M. (1962), *AIChE J.*, **8**, 478.

Wang, J. B., and Varma, A. (1978), *Chem. Eng. Sci.*, **33**, 1549.

Wang, J. B., and Varma, A. (1980), *Chem. Eng. Sci.*, **35**, 613.

Wedel, S., and Luss, D. (1980), *Chem. Eng. Commun.*, **7**, 245.

Weekman, V. W., Jr. (1966), *J. Catal*, **5**, 44.

Weekman, V. W., Jr., and Gorring, R. L. (1965), *J. Catal.*, **4**, 260.

Weisz, P. B. (1956), *Science*, **123**, 887.

Weisz, P. B. (1957), *Z. Physik. Chem. (Frankfurt)*, **11**, 1.

Weisz, P. B. (1959), *Chem. Eng. Sci. Symp. Ser.*, **55**, 29.

Weisz, P. B., and Hicks, J. S. (1962), *Chem. Eng. Sci.*, **17**, 265.

Weisz, P. B., and Prater, C. D. (1954), *Adv. Catal.*, **6**, 143.

Weisz, P. B., and Swegler, E. W. (1955), *J. Phys. Chem.*, **59**, 823.

Wheeler, A. (1951), *Adv. Catal.*, **3**, 249.

Wicke, E., and Brotz, W. (1949), *Chem. Ing.-Tech.*, **21**, 219.

Wicke, E., Padberg, G., and Arens, H. (1968), *Proc. 4th European Symp. Chem. React. Eng.*, Brussels.

Wong, R. L., Hubbard, G. L., and Denny, V. E. (1976), *Chem. Eng. Sci.*, **31**, 541.

Yazdi, S. F., and Petersen, E. E. (1972), *Chem. Eng. Sci.*, **27**, 227.

Yang, R. T., and Wong, C. (1981), *Chem. Eng. Commun.*, **11**, 317.

Zeldovich, Y. B. (1939), *Z. Fiz. Khim.*, **13**, 163.

CHAPTER FIVE

Role of Pore Diffusion in Complex Reactions

The developments presented in Chapter 4 were restricted to simple (single) reactions in the absence of the complications arising from side reactions. Several reactions of industrial importance are in reality complex reactions that are carried out under conditions such that the desired reaction takes place to the maximum extent possible, thus increasing the yield of the desired product. This is accomplished by employing a catalyst that is highly selective to the reaction of interest. In complex reactions of this type, three main factors need to be considered in determining the yield: (1) the effect of the usual (macropore) diffusion; (2) the effect of diffusion in micropores along with macropore diffusion; and (3) the effect of temperature gradients within the pores. Some of these aspects of pore diffusion were first theoretically considered by Wheeler (1951) and Weisz and Prater (1954) and subsequently by others, for example, Carberry (1962a), Östergaard (1961, 1965), Butt (1966), Wakao and Fujishiro (1966), van der Vusse (1966), Komiyama and Inoue (1968), McGreavy and Thornton (1970), Roberts (1972), Wirges and Rahse (1975), and Varghese et al. (1978). These theories will be discussed in this chapter, along with examples to demonstrate the applicability of the theories.

5.1. EFFECT OF PORE DIFFUSION IN ISOTHERMAL NETWORKS

Three classes of complex reactions will be considered: independent reactions, parallel reactions, and consecutive reactions.

5.1.1. Independent Reactions

Let us first consider the reaction scheme

$$A \xrightarrow{k_{v1}} B, \quad C \xrightarrow{k_{v2}} D \quad \text{(I)}$$

that is, two independent reactions from the same feed stream. If the actual (observed) rate constants are k_{a1} and k_{a2}, and s_a and s represent, respectively, the observed and the intrinsic (true) selectivities for the desired product B, then we have

$$s = \frac{k_{v1}}{k_{v2}} \quad (5.1)$$

and

$$s_a = \frac{k_{a1}}{k_{a2}} \quad (5.2)$$

From the definition of the Thiele modulus and its relationship to the effectiveness factor under conditions of pore diffusional limitation ($\varepsilon = 1/\phi$), we can write

$$s_a = \frac{k_{v1}\varepsilon_1}{k_{v2}\varepsilon_2} = \left(\frac{s}{\alpha_D}\right)^{1/2} \quad (5.3)$$

where α_D represents the ratio of diffusivities:

$$\alpha_D = \frac{D_{eB}}{D_{eA}} \quad (5.4)$$

It may be inferred from Eq. 5.3 that when s is low, the value of s_a can be increased by lowering ε_2. In other words, the second reaction should be relatively curtailed by diffusional resistance.

5.1.2. Parallel Reactions

For a system of parallel reactions

$$A \underset{k_{v2}}{\overset{k_{v1}}{<}} \begin{matrix} B \\ C \end{matrix} \quad \text{(II)}$$

pore diffusion will have no effect on the selectivity when both reactions are first order. Thus:

$$r_{v1} = \varepsilon_1 k_{v1} C_A \tag{5.5}$$

$$r_{v2} = \varepsilon_2 k_{v2} C_A \tag{5.6}$$

Since $\varepsilon_1 = \varepsilon_2$, both representing the effectiveness factor for the same reactant A, we have $s_a = s = k_{v1}/k_{v2}$ at all positions inside the catalyst pellet. If, on the other hand, reaction 1 were first order and reaction 2 second order, then the second reaction would be curtailed to a larger extent inside the pores because of decreased concentration, and as a consequence the selectivity would increase. The effect of diffusion is similar to the effect of pressure, the reaction with the higher order being curtailed to a greater extent with reduction in pressure.

A more detailed analysis of the effect of reaction order has been attempted by Pawlawski (1961) and Roberts (1972). Pawlawski assumed one of the reactions to be first order and examined the effect of changing the order of the other reaction from $\frac{1}{3}$ to 5. He employed a numerical integration method for the asymptotic concentration profile. A better description of the overall behavior of the reaction is obtained, however, by the exact analytical method of Roberts (1972), the results of which are briefly considered below.

For a parallel reaction of (m, n)th order, asymptotic solutions have been worked out for the effectiveness factor as well as for the relative point yield, defined by

$$Y_r = \frac{\text{observed point yield } Y_a}{\text{intrinsic point yield } Y_p} \tag{5.7}$$

where Y_p is given by

$$Y_p = \frac{k_{v1} C_{As}^m}{k_{v1} C_{As}^m + k_{v2} C_{As}^n} = \frac{1}{1 + p_{nm}} \tag{5.8}$$

The term p_{nm} represents the ratio of the intrinsic rate of the nth-order reaction to that of the mth-order reaction. The equations have the final form

$$\varepsilon = \frac{\sqrt{2}\left(\dfrac{1}{m+1} + \dfrac{p_{nm}}{n+1}\right)^{1/2}}{(\phi)_1 (1 + p_{nm})} \tag{5.9}$$

where

and

$$(\phi)_1 = L\left(\frac{k_{v1} C_{As}^{m-1}}{D_{eA}}\right)^{1/2}$$

$$Y_r = \frac{n+1}{2m-n+1} \tag{5.10}$$

For any set of values of m and n, Eq. 5.10 gives the lowest value of the relative yield, and can therefore be used as a quick and approximate guide to the effect of intraparticle diffusion on product distribution.

Exact solutions for specific combinations of m and n have also been worked out by Roberts, who has presented plots for ε vs. $(\phi)_r$ and Y_r vs. $(\phi)_r$ with p_{nm} as the third parameter in both cases. The quantity $(\phi)_r$ is a modified Thiele parameter (see Section 4.2.7) that can be determined directly from observed rate data (without a knowledge of the rate constants) from the relation

$$(\phi)_r^2 = \left(\frac{L^2}{D_{eA} C_{As}}\right) \quad \begin{array}{l}\text{(observed reaction rate per gross}\\ \text{catalyst volume)}\end{array}$$

The following general conclusions can be drawn:

1. The influence of intraparticle diffusion on Y_r increases as the difference between the two reaction orders increases.

2. The decrease in Y_r for a given difference in reaction orders becomes smaller as the reaction order is raised.

3. If the desired product is formed in the lower-order reaction, Y_r is enhanced as a result of intraparticle diffusion. Since, as pointed out earlier in this section, lowering the pressure or diluting the feed has the same effect as a diffusional retardation, operation of such a reaction at reduced pressure will lead to a higher yield.

5.1.3. Consecutive Reactions

For a consecutive reaction scheme

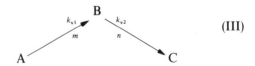

the influence of diffusion is more complicated than in either of the previous cases. The basic differential equation for the conversion of A to intermediate B is given by

$$-\frac{dC_B}{dC_A} = 1 - \frac{1}{s}\left(\frac{C_B^n}{C_A^m}\right) \tag{5.11}$$

Integration of this equation for the case $m = n = 1$ leads to

$$x_{BA} = \frac{s}{s-1}\left[(1-x_A)^{\mathscr{S}} - (1-x_A)\right]$$

where

$$\mathscr{S} = \frac{1}{s} \tag{5.12}$$

This equation expresses the moles of B formed per mole of A in the feed (x_{BA}) as a function of the total conversion of A (x_A).

In order to derive an expression for the selectivity under conditions of diffusional limitation, let us consider a catalyst pellet of any shape on which a consecutive reaction is taking place. The actual yield of B can be expressed as the ratio of the actual rate of formation of B per pellet to the rate of disappearance of A, that is, as the ratio of the diffusive flux of B to that of A at the external surface:

$$Y_a = -\frac{d\hat{C}_B/dt}{d\hat{C}_A/dt} = -\alpha_D\left(\frac{C_{Bs}}{C_{As}}\right)\frac{(d\hat{C}_B/d\hat{L})_{\hat{L}=1}}{(d\hat{C}_A/d\hat{L})_{\hat{L}=1}} \tag{5.13}$$

If there is a net production of intermediate, the signs of the numerator and denominator are opposite and Y_a is positive. If, on the other hand, the second reaction is so fast that there is a net consumption of B, then both the numerator and denominator of Eq. 5.11 are negative and the yield $-(dC_B/dC_A)$ is negative. It will be shown later that in a nonisothermal pellet the yield is so completely annihilated that such a situation does indeed arise.

The following continuity equations can be written for a cylindrical pore, which can be used for any shape provided that L is taken as the ratio of the volume to the surface of the shape:

$$\frac{d^2\hat{C}_A}{d\hat{L}^2} = \frac{L^2 k_{v1}}{D_{eA}}\hat{C}_A = (\phi)_1^2\hat{C}_A \tag{5.14}$$

where $(\phi)_1$ is the Thiele modulus for the first reaction and must be distinguished from ϕ_1, which denotes the modulus for a first-order reaction; and

$$\frac{d^2\hat{C}_B}{d\hat{L}^2} = \frac{1}{\alpha_D}\frac{C_{As}}{C_{Bs}}(\phi^2)_1\hat{C}_A - (\phi^2)_2\hat{C}_B \tag{5.15}$$

where

$$\hat{C}_A = \frac{C_A}{C_{As}}, \qquad \hat{C}_B = \frac{C_B}{C_{Bs}}$$

The boundary conditions are

$$\hat{L} = 0, \qquad \frac{d\hat{C}_A}{d\hat{L}} = \frac{d\hat{C}_B}{d\hat{L}} = 0 \tag{5.16}$$

$$\hat{L} = 1, \qquad \hat{C}_A = \hat{C}_B = 1 \tag{5.17}$$

The solutions are given by

$$\hat{C}_A = \frac{\cosh(\phi)_1\hat{L}}{\cosh(\phi)_1} \tag{5.18}$$

which is identical with the concentration profile equation shown in Table 4.1 for a single reaction, and

$$\hat{C}_B = (1+P)\frac{\cosh(\phi)_2\hat{L}}{\cosh(\phi)_2} - P\frac{\cosh(\phi)_1\hat{L}}{\cosh(\phi)_1} \tag{5.19}$$

where

$$P = \frac{s}{s\alpha_D - 1} = \frac{(\phi^2)_1/\alpha_D}{(\phi^2)_1 - (\phi^2)_2} \tag{5.20}$$

From Eqs. 5.18 and 5.19 the dimensionless fluxes $d\hat{C}_A/d\hat{L}$ and $d\hat{C}_B/d\hat{L}$ can be calculated and the ratio of these fluxes determined to give the instantaneous yield (Carberry, 1962b):

$$-\frac{dC_B}{dC_A} = -\alpha_D\left(\frac{C_{Bs}}{C_{As}} + \frac{s}{\alpha_D s - 1}\right)\mathscr{S} + \frac{\alpha_D s}{\alpha_D s - 1} \tag{5.21}$$

where

$$\mathscr{S} = \frac{(\phi)_2\tanh(\phi)_2}{(\phi)_1\tanh(\phi)_1}$$

With the assumptions that the diffusivities are equal ($\alpha_D = 1$), B is absent in the feed, and the reaction occurs in the diffusion regime [$\tanh(\phi)_1 = \tanh(\phi)_2$], Eq. 5.21 can be integrated to give once again

$$x_{BA} = \frac{s}{s-1}\left[(1-x_A)^{\mathscr{S}} - (1-x_A)\right]$$

where now

$$\mathscr{S} = s^{-\frac{1}{2}} \tag{5.22}$$

Equations 5.12 and 5.22 are of the same form and represent, respectively, the conversion of A to the intermediate B under intrinsic surface reaction conditions and under conditions where pore diffusion is operative. The difference lies only in the definition of $\mathscr{S}$. A comparison of these equations for $s = 3$ is made later in Figure 5.2. This figure is discussed at the end of Section 5.1.5 along with equations corresponding to other situations.

Östergaard (1961) has given analytical expressions for a system of n first-order reactions. van de Vusse (1966) subsequently extended the treatment to zero- and second-order reactions and showed that the yield loss increases with decrease in order. Komiyama and Inoue (1970) further extended the analysis to H–W type rate expressions.

5.1.4. Parallel–Consecutive Networks

The following networks represent typical combinations of parallel and consecutive schemes:

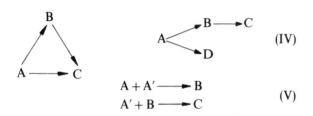

$$(\text{IV})$$
$$(\text{V})$$

Wirges and Rahse (1975) have examined the effect of pore diffusion on the yield of intermediate B in both these networks by writing appropriate mass balance equations and obtaining numerical solutions. Since a large number of parameters are involved, we shall present only their qualitative findings.

Type IV network:

1. The maximum conversion of A to the desired product B is a weak function of the diffusivities ratio $\alpha_D = D_{eB}/D_{eA}$ for given values of the selectivities $s_1 = k_{v\bar{1}}/k_{v\bar{2}}$ and $s_{13} = (k_{v\bar{1}} + k_{v\bar{3}})/k_{v\bar{2}}$, showing a maximum around $\alpha_D = 1$.

2. When B is present in the feed to the reactor, its yield, defined by $(C_{Bs} - C_{B0})/C_{A0}$, shows a maximum at a certain conversion of A (x_A), this maximum falling sharply with increasing amounts of B in the feed (C_{B0}/C_{A0}).

3. For fixed values of s_{13} the maximum conversion of A to the desired product B is a weak function of the Thiele modulus for small values of s_1 (say, 1–3), but falls appreciably with increase in the modulus as s_1 is raised beyond 5.

Type V network:

1. It is known (Kerber and Gestrich, 1966) that even in the absence of diffusion a certain minimum quantity of A′, represented by $C_{A'0}/C_{A0}$, at the reactor entrance is necessary to ensure a maximum in the yield of intermediate B. Above this minimum the yield rises monotonically with conversion x_A. This situation holds even in the presence of a strong diffusional influence, irrespective of the value of the selectivity s_1.

2. When B is present in the feed, a maximum in the yield of B occurs only if the selectivity s_1 is greater than unity.

We shall now define a term to provide a quantitative measure of the loss of yield due to pore diffusion for all the networks. For this purpose a new parameter, Γ, is postulated that is similar, but not equal, to the effectiveness factor (Wirges and Rahse, 1975):

$$\Gamma = \frac{\tanh \phi}{\tanh \phi \, (\alpha_D s_{13})^{1/2}} \tag{5.23}$$

The yield loss (YL) is then

$$\text{YL} = \frac{(C_{Bs}/C_{A0})_{\max}^{\Gamma = (\alpha_D s)^{-1/2}} - (C_{Bs}/C_{A0})_{\max}^{\Gamma = 1}}{(C_{Bs}/C_{A0})_{\max}^{\Gamma = (\alpha_D s)^{-1/2}}} \tag{5.24}$$

where s represents s_1 or s_{13}, depending on whether a Type IV or Type V network is involved.

A plot of YL vs. s, prepared by Wirges and Rahse for Type V, is reproduced in Figure 5.1 for $\Gamma = 1$, $\alpha_D = 1$. It will be noticed that as $C_{A'0}/C_{A0}$ is increased, the Type V network approaches the simpler first-order consecutive scheme (Type I network). On the other hand, as this ratio is decreased, the yield loss also decreases, which is in conformity with van der Vusse's observation referred to earlier.

The treatment presented thus far has dealt with porous particles in which the entire catalytic area was assumed to be internal. Although this may be true for many practical situations, there are instances of reactions wherein the support material used has low internal area. Silver-catalyzed partial oxidation of ethylene is typical of such reactions in which the external catalytic area constitutes a distinct and finite fraction of the total active area, and cannot be neglected in the analysis.

Varghese et al. (1978) have considered such a

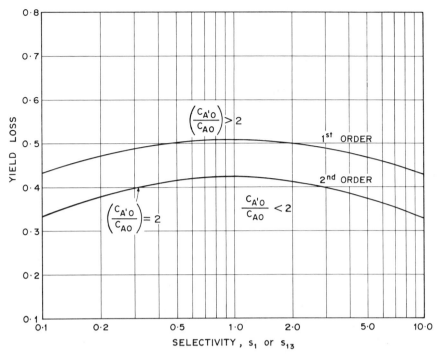

Figure 5.1. Yield loss as a function of the selectivity s for a Type V scheme (Wirges and Rahse, 1975).

catalyst on which a reaction represented by a triangular scheme

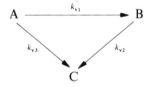

is carried out. The mass balance equation for the species can be written in dimensionless form subject to appropriate accounting of the area available for diffusion:

$$\frac{d^2\hat{C}_A}{d\hat{L}^2} = [(\phi_p)_1^2 + (\phi_p)_3^2](1 - f_e)\hat{C}_A \qquad (5.25)$$

where $(\phi_p)_i$ represents the Thiele modulus based on the ith reaction and f_e the fraction of the total area residing on the external surface.

Equation 5.25 is subject to the boundary conditions

$$\frac{d\hat{C}_A}{d\hat{L}} = 0, \qquad \hat{L} = 0$$

$$\frac{1}{(Bi)_m}\frac{d\hat{C}_A}{d\hat{L}} + [1 + f_e(Da_1 + Da_3)]\hat{C}_A = 1, \qquad \hat{L} = 1$$

$$(5.26)$$

The second boundary condition takes into account the external mass transfer limitation and surface reaction on the fraction f_e of the total area. The parameters Da_1 and Da_3 represent the Damköhler number for reactions 1 and 3, respectively, and are defined as $Da_i = k_{vi}/k_g a_p$, where a_p represents the external surface area per unit volume.

Similar equations can be written for the species B and C, and for linear rate processes they can be solved analytically to obtain the point yield of B. However, because of the cumbersome nature of the solution obtained, the expression is not presented here; only the important conclusions are stated.

The results of the analysis indicate that the effectiveness for species A and the point yield for species B strongly depend on the fraction of the total area that lies on the external surface. As expected, increasing $(Bi)_m$ across the external film has a favorable influence on the yield of B. The analysis also indicates the existence of a finite $(Bi)_m$ above which the curve of the point yield of B vs. the Thiele modulus displays a minimum and then a maximum before dropping off to a value dictated by external mass transfer control.

5.1.5. Effect of Micropore Diffusion on Yield

Carberry (1962b) has extended the treatment of the Type III network to a bidispersed pore model (described in Chapter 4). On the basis of this model, which postulates macropore channels from which micropores branch out in perpendicular directions, it can be shown that the selectivity is not only affected by the macropore effectiveness factor (normally used in monodispersed systems), but also by the micropore effectiveness factor. The development of this concept is restricted in this section to a system of consecutive reactions.

The micropore concentration gradients can be determined by integrating Eqs. 5.14 and 5.15 under the following boundary conditions:

$$\hat{L}_{mi} = 0, \quad \frac{d\hat{C}_{A,mi}}{d\hat{L}_{mi}} = \frac{d\hat{C}_{B,mi}}{d\hat{L}_{mi}} = 0$$

$$\hat{L}_{mi} = 1, \quad \hat{C}_{A,mi} = \hat{C}_{A,ma}, \quad \hat{C}_{B,mi} = \hat{C}_{B,ma} \tag{5.27}$$

Equations 5.14 and 5.15 refer to a monodispersed pore system in a solid of any shape whose characteristic length L represents the ratio of the volume to the surface. If they are integrated under boundary conditions 5.27, then they will specifically apply to the micropores and the dimension L will be replaced by $\dot{L}_{mi}$, the ratio of the volume to the surface of the micropore.

Continuity equations can also be written for transport in the macropores. The final equation derived for point yield would be identical with Eq. 5.21, but the definition of $\mathscr{S}$ would now be different, and would include terms for micropore diffusion. On the assumptions that $\alpha_D = 1$, there is no B in the feed, and reaction occurs in the diffusion regime, this equation can be integrated to give

$$x_{BA} = \frac{s}{s-1}[(1 - x_A)^{\mathscr{S}} - (1 - x_A)] \tag{5.28}$$

where now

$$\mathscr{S} = \frac{(\varepsilon_{mi})_2}{(\varepsilon_{mi})_1} \frac{(\varepsilon_{ma})_2}{(\varepsilon_{ma})_1} = \frac{\varepsilon_2}{\varepsilon_1}$$

$$= \frac{(\phi_{ma})_2 \tanh(\phi_{ma})_2}{(\phi_{ma})_1 \tanh(\phi_{ma})_1} \tag{5.29}$$

$$(\phi_{ma})_1 = L_{ma}\left[\frac{k_{v1}(\varepsilon_{mi})_1}{D_{eA,ma}}\right]^{1/2}$$

$$(\phi_{ma})_2 = L_{ma}\left[\frac{k_{v2}(\varepsilon_{mi})_2}{D_{eB,ma}}\right]^{1/2} \tag{5.30}$$

$$(\phi_{mi})_1 = L_{mi}\left(\frac{k_{v1}}{D_{eA,mi}}\right)^{1/2}$$

$$(\phi_{mi})_2 = L_{mi}\left(\frac{k_{v2}}{D_{eB,mi}}\right)^{1/2} \tag{5.31}$$

ϕ_{ma} may be recognized as the Thiele modulus for macropore diffusion, the term $k_v \varepsilon_{mi}$ representing the rate constant corrected for micropore diffusion. Note that the role of micropore diffusion has been incorporated in the definition of ϕ_{ma} and therefore of $\mathscr{S}$.

As in the previous cases considered, $\mathscr{S}$ is an effective rate parameter that is now inversely proportional to the selectivity s and is dependent on the macro- and microeffectiveness factors of the two reactions. The values of $\mathscr{S}$ under different conditions are as follows

1. Pore diffusion is fully absent:

Equation 5.28 reduces to 5.12, that is, $\mathscr{S} = 1/s$. x_{BA} is therefore given by Eq. 5.12, which defines the conversion to B in the absence of pore diffusion.

2. Macropore diffusion alone is effective:

Equation 5.29 now becomes

$$\mathscr{S} = \frac{(\varepsilon_{ma})_2}{(\varepsilon_{ma})_1}\frac{1}{s} \tag{5.32}$$

Since under diffusional limitation (i.e., for $\phi > 3$) $\varepsilon = 1/\phi$,

$$\mathscr{S} = \left(\frac{s}{\alpha_D}\right)^{-1/2} = \left(\frac{k_{v2} D_{eB,ma}}{k_{v1} D_{eA,ma}}\right)^{1/2} \tag{5.33}$$

or, when $D_{eA,ma} = D_{eB,ma}$, Eq. 5.33 reduces to 5.22, that is, $\mathscr{S} = 1/\sqrt{s}$. The yield of B is then given by Eq. 5.22, which corresponds to diffusional limitation in a monodispersed pore system (macropore diffusion).

3. Both micro- and macrodiffusion are present:

Under diffusional limitation [when $(\phi_{mi})_1$ and $(\phi_{ma})_1 > 3$], on the assumption that

$$\frac{D_{eB,ma}}{D_{eA,ma}} = \frac{D_{eB,mi}}{D_{eA,mi}} = \alpha_D \tag{5.34}$$

we have

$$\mathscr{S} = (\alpha_D s)^{3/4}\frac{1}{s} \tag{5.35}$$

$$= \frac{\alpha_D^{3/4}}{s^{1/4}}$$

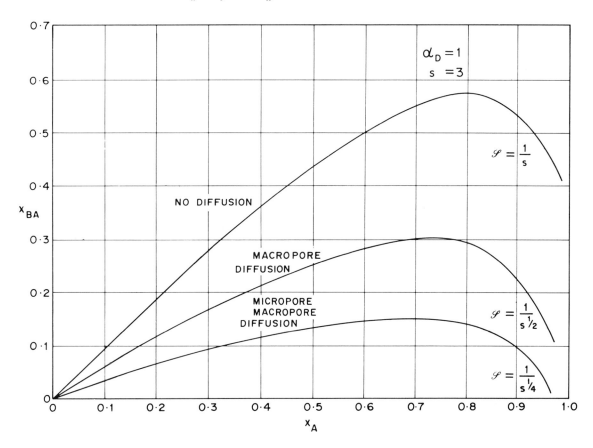

Figure 5.2. Conversion to intermediate B as a function of total conversion of A in a Type III reaction for various models.

or, when $\alpha_D = 1$

$$\mathscr{S} = \frac{1}{s^{1/4}} \qquad (5.36)$$

In Figure 5.2 x_{BA} is plotted as a function of x_A for an intrinsic selectivity (s) of 3 and $\alpha_D = 1$. The effect of pore diffusion on x_{BA} is quite pronounced. When micropore diffusion is neglected the maximum conversion is 30%, as against 57.5% in the absence of diffusion. In a bidispersed pore system (in which micropore diffusion is also considered), the effect of diffusion is even more pronounced, the maximum conversion attainable being only 15%.

The maximum yield is given by $(x_{BA}/x_A)_{max}$ and can be easily obtained from Figure 5.2. The maximum yield thus calculated for different values of $\mathscr{S}$ is plotted as a function of s for all three cases in Figure 5.3. Here again the progressive annihilation of yield with the increasing role of diffusion (macro to micro–macro) is evident.

5.1.6. Example: Dehydrogenation of Butene on Iron Oxide: Calculation of Yield in the Presence of Diffusion

Voge and Morgan (1972) have reported experimental rate data on the dehydrogenation of butene on an industrial iron oxide catalyst (Shell 205) in the presence of steam. The reaction is shown to follow a first-order consecutive scheme with $(k_{v2}/k_{v1}) = 0.9$. Several sizes of catalyst in the form of granules as well as pellets were used in the study. The separate experiments on diffusion at 25°C with oxygen diffusing through the pellet in the presence of a stream of circulating nitrogen indicated that the effective diffusivity is about an order of magnitude lower than the bulk diffusivity. The bulk diffusivity at 620°C for a butene–steam mixture was estimated by the correlation of Hirschfelder et al. (1954) to be 0.720 cm²/sec.

In order to obtain the yield of intermediate, the

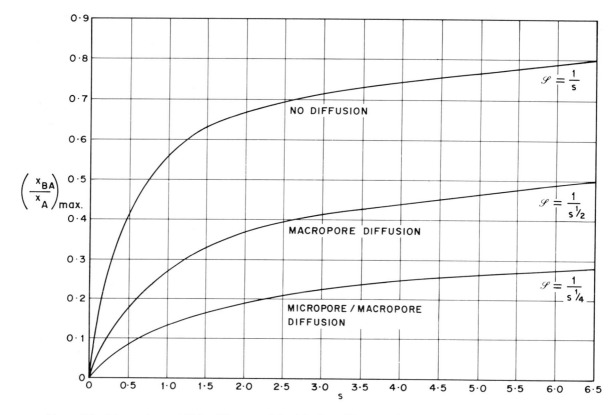

Figure 5.3. The maximum yield for different models of the Type III consecutive scheme as a function of selectivity *s*.

conservation equations for the species in a spherical pellet can be written and solved subject to appropriate boundary conditions to give

$$\frac{r_B}{r_A} = \frac{1 - (\sigma \varepsilon_2 / \varepsilon_1)}{1 - \sigma} - \frac{\varepsilon_2 k_{v2}}{\varepsilon_1 k_{v1}} \frac{C_{Bs}}{C_{As}} \qquad (5.37)$$

where B is the desired intermediate product (butadiene), A is the reactant (butene), and σ and $\varepsilon_2 / \varepsilon_1$ are defined as

$$\sigma = \frac{k_{v2}}{k_{v1}} \left(\frac{D_{eA}}{D_{eB}} \right), \quad \frac{\varepsilon_2}{\varepsilon_1} = \frac{1}{\sigma} \frac{(\phi_{s1})_2 \coth (\phi_{s1})_2 - 1}{(\phi_{s1})_1 \coth (\phi_{s1})_1 - 1} \qquad (5.38)$$

The use of Eq. 5.37, however, requires a knowledge of k_{v1} and k_{v2} independently (to evaluate $\varepsilon_2 / \varepsilon_1$). In the absence of such intrinsic kinetic data, we can make use of Wheeler's modified Thiele modulus defined on the basis of the observed rates (see Section 4.2.7):

$$(\phi)_r^2 = \frac{R^2}{D_{eA}} \frac{r_A}{C_{As}} \qquad (5.39)$$

Using the observed reaction rates for a given size pellet and known D_{eA} (0.072 cm^2/sec), this modulus can be

estimated. It is necessary, however, to use the following relationship between this modified modulus and the normal Thiele modulus to get the value of the latter:

$$(\phi)_r^2 = \varepsilon \phi_{s1}^2 = \frac{3}{\phi_{s1}} (\phi_{s1} \coth \phi_{s1} - 1) \phi_{s1}^2$$

The values of the Thiele moduli $(\phi_{s1})_1$ and $(\phi_{s1})_2$ can therefore be calculated once the appropriate values of the modified moduli are known. These values can then be used to calculate the Thiele modulus for any other particle size.

For a $\frac{3}{16}$-in. pellet at 620°C, Voge and Morgan (1972) have obtained a value of $(\phi_{s1})_1 = 1.5$. Equation 5.37 can now be used to calculate the yield. Relatively good agreement was obtained between experimental and calculated values of yield by using this procedure for several pellet sizes.

5.2. EFFECT OF PORE DIFFUSION IN NONISOTHERMAL NETWORKS

As pointed out in Chapter 4, temperature gradients of up to 50°C or more can exist within a pellet for exothermic reactions on supported catalysts (the supports usually being poor conductors of heat). The yield in complex reactions under these conditions is therefore modified by diffusion not only of mass but also of heat. We shall consider this effect in a system of parallel reactions, and then proceed to the more important case of consecutive reactions.

5.2.1. Parallel Reactions

For a parallel reaction scheme (Type II), the following continuity equations can be written for slab geometry:

$$D_{eA}\left(\frac{d^2 C_A}{dl^2}\right) - (k_{v1} + k_{v2})C_A = 0$$

$$D_{eB}\left(\frac{d^2 C_B}{dl^2}\right) + k_{v1} C_A = 0, \quad D_{eC}\left(\frac{d^2 C_C}{dl^2}\right) + k_{v2} C_A = 0$$

$$(5.40)$$

The conservation equation is

$$k'_e \frac{d^2 T}{dl^2} + \left[(-\Delta H_1)k_{v1} + (-\Delta H_2)k_{v2}\right]C_A = 0$$

$$(5.41)$$

We shall now introduce simplifications so that the set of nonlinear differential Eqs. 5.40 can be solved. First, we shall consider the case where $\Delta H_1 \neq \Delta H_2$ and $E_1 = E_2$, and then the case where $\Delta H_1 = \Delta H_2$ and $E_1 \neq E_2$.

The first case is not significant, since the yield would be no different than for intrinsic kinetics in the absence of heat and mass transport limitations. On the other hand, for the second case the yield is considerably modified. Under the conditions of this case, Eqs. 5.40 and 5.41 become, in dimensionless form,

$$\frac{d^2 \hat{C}_A}{d\hat{L}^2} - (\phi)_{A_f}^2 \hat{C}_A = 0$$

$$\frac{d^2 \hat{C}_B}{d\hat{L}^2} + (\phi)_{A_f}^2 \left[1 + \frac{1}{s_{A_f}} \exp\left(\frac{E_1 - E_2}{R_g T}\right)\right]^{-1} \hat{C}_A = 0 \quad (5.42)$$

$$\frac{d^2 \hat{C}_C}{d\hat{L}^2} + (\phi)_{A_f}^2 \left[\exp\left(\frac{E_2 - E_1}{R_g T}\right) + \frac{1}{s_{A_f}}\right]^{-1} \hat{C}_A = 0$$

$$\frac{d^2 \hat{T}}{d\hat{L}^2} + \beta_m (\phi)_{A_f}^2 \hat{C}_A = 0$$

$$(5.43)$$

where

$$(\phi)_{A_f} = L\left(\frac{A_{f1}}{D_{eA}}\right)^{1/2}\left[\exp\left(-\frac{E_1}{R_g T}\right) + \frac{1}{s_A}\left(-\frac{E_2}{R_g T}\right)\right]^{1/2}$$

$$(5.44)$$

and

$$s_{A_f} = \frac{A_{f1}}{A_{f2}}$$

$$(5.45)$$

Note that in deriving these dimensionless equations, the Arrhenius preexponential factor (A_f) has been used instead of the rate constant in formulating a modified Thiele modulus. Also, a modified selectivity (s_{A_f}) is used, based again on the preexponential factors rather than the rate constants. Employing the usual definition of effectiveness factor and a definition of selectivity similar to that given by Eq. 5.13, Östergaard (1965) has presented numerical solutions for a few specific values of the parameters β_m and E_1/E_2. The following conclusions can be drawn from his analysis:

1. The effectiveness factor plots for different values of β_m behave in much the same way as those for a simple reaction (shown in Figure 4.4). In both cases exothermal reactions are involved; also, in the present case the two reactions combined by the parallel scheme are equally exothermic.

2. The effectiveness factor plots with the ratio E_1/E_2 as a parameter show much less spread. In fact, a single band can be drawn to cover a range from 1.1 to 1.4 with less than 10% error. When $E_1/E_2 = 1$, the reaction can evidently be represented by the simple scheme A → B + C. Since it has been assumed that $\Delta H_1 = \Delta H_2$, the effectiveness factor curve for this case will coincide with the curve for a simple reaction for the same value of β_m.

3. At high values of the Thiele modulus the yield increases sharply with increase in β_m; that is, under conditions of diffusional limitation the influence of β_m is pronounced. The abrupt rise occurs in the same region where the effectiveness factor exhibits a maximum, and then tends to level off. The yield plots for different values of E_1/E_2 show a sharp rise at about the same value of $(\phi)_{A_f}$ and then tend to level off. If the activation energy of the desired reaction is higher than for the wasteful reaction (i.e., $E_1/E_2 > 1$), the best yield is obtained at high values of the

Thiele modulus. Conversely, when $E_1/E_2 < 1$ the best yield is obtained at low values of the Thiele modulus.

5.2.2 Consecutive Reactions

In Chapter 4 we saw that effectiveness factors greater than unity are possible in simple exothermal reactions proceeding within a pellet under nonisothermal conditions. Thus, under certain conditions, diffusional limitations can lead to an enhancement of the reaction rate (since the rate constant is given by the product εk_v). In a complex Type III reaction (a consecutive scheme), however, nonisothermicity of the pellet generally tends to lower the yield of the desired product. As will be shown below, there is often a net *consumption* of intermediate rather than *production*; and since in a degradative reaction system characterized by overall exothermic behavior the maximum value of the effectiveness factor occurs in the region of net consumption, it becomes impossible to exploit the existence of such a maximum for obtaining a higher activity of the catalyst, as can be done in the case of simple reactions.

The Basic Equations

If we consider an infinite slab, the following differential equations can be readily written:

$$D_{eA}\frac{d^2C_A}{dl^2} - k_{v1}C_A = 0 \tag{5.46a}$$

$$D_{eB}\frac{d^2C_B}{dl^2} + k_{v1}C_A - k_{v2}C_B = 0 \tag{5.46b}$$

$$k'_e\frac{d^2T}{dl^2} + (-\Delta H_1)(k_{v1}C_A) + (-\Delta H_2)(k_{v2}C_B) = 0 \tag{5.46c}$$

It has already been pointed out that the rate constant k_v in a nonisothermal system can be eliminated by using the surface rate constant through Eq. 4.39b or 4.39c. In the present case it is convenient to eliminate k_v through Eq. 4.39b. When this is done, and both C_A and l are made dimensionless, the following equations result:

$$\frac{d^2\hat{C}_A}{d\hat{L}^2} - (\phi^2)_1 \exp\left[\alpha_{s1}\left(1 - \frac{1}{\hat{T}}\right)\right]\hat{C}_A = 0 \tag{5.47a}$$

$$\frac{d^2\hat{C}_B}{d\hat{L}^2} + \alpha_D(C_{AB})_s(\phi^2)_1 \exp\left[\alpha_{s1}\left(1 - \frac{1}{\hat{T}}\right)\right]\hat{C}_A$$

$$- (\phi^2)_2 \exp\left[\alpha_{s2}\left(1 - \frac{1}{\hat{T}}\right)\right]\hat{C}_B = 0 \tag{5.47b}$$

$$\frac{d^2\hat{T}}{d\hat{L}^2} + \beta_{m1}(\phi^2)_1 \exp\left[\alpha_{s1}\left(1 - \frac{1}{\hat{T}}\right)\right]\hat{C}_A$$

$$+ \beta_{m2}(\phi^2)_2 \exp\left[\alpha_{s2}\left(1 - \frac{1}{\hat{T}}\right)\right]\hat{C}_B = 0 \tag{5.47c}$$

The various nondimensional parameters appearing in these equations were defined earlier.

The problem now is to calculate the effectiveness factor

$$\varepsilon = -\frac{1}{(\phi^2)_1}\left(\frac{d\hat{C}_A}{d\hat{L}}\right)_{\hat{L}=1} \tag{5.48}$$

and the yield defined by Eq. 5.13. An inspection of Eqs. 5.47 immediately reveals that these constitute a set of highly nonlinear differential equations. Elaborate computer solutions to these have been obtained by Butt (1966) and some of the significant conclusions revealed by these solutions are described below.

Effectiveness Factors

Clearly, when $\beta_{m2} = 0$, the heat effect will be due only to the first reaction; when $\beta_{m1} = 0$, the second reaction alone contributes to nonisothermicity; and when $\beta_{m1} = \beta_{m2} = 0$, the system becomes isothermal. Representative plots of ε vs. $(\phi)_2$ for specific values of β_{m1} and β_{m2} at $s = 1, 4, 8,$ and 12 and $\alpha_{s1} = \alpha_{s2} = 10, 20$ are shown in Figure 5.4. Evidently a multitude of such plots can be prepared. A significant feature of these plots is that the second reaction greatly influences the nature of the curves. When the first reaction is endothermic and the second exothermic, the curve follows the course of an endothermic reaction until the overall effect becomes exothermic, and then it follows the course of an exothermic reaction (passing through a maximum in some cases). This is demonstrated, for instance, by curve B of the figure for $\alpha_{s1} = \alpha_{s2} = 20$ and $s = 1$. Butt (1966) has suggested that the point at which the shift from net endothermic to net exothermic behavior occurs increases roughly in proportion to s.

Yield

It would be instructive to compare the yield under conditions of nonisothermal diffusional limitation with the isothermal yield for the cases (developed earlier) of no diffusion, diffusional limitation in a monodispersed

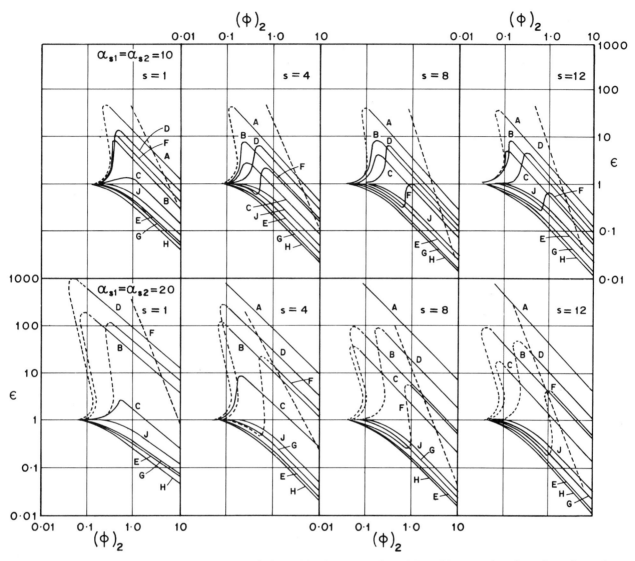

Figure 5.4. Effectiveness factor as a function of the Thiele modulus for a nonisothermal Type III consecutive scheme for various values of selectivity ratio and thermicity (Butt, 1966).

pore system (i.e., macropore diffusion only), and diffusional limitation in a bidispersed pore system (i.e., micropore–macropore diffusion). Such a comparison can best be made in terms of the actual yield Y_a defined by Eq. 5.13. For the three isothermal cases just mentioned Y_a may be calculated from Eqs. 5.11, 5.21, and 5.27, whereas for the nonisothermal case numerical solutions would have to be worked out for specific heat effects.

Figure 5.5 shows Y_a as a function of C_{Bs}/C_{As} for $s = 4$. It may be noted that the annihilation of yield caused by nonisothermicity is far greater than in any other case. In fact, over a large range of conversions, there is a net consumption of intermediate, as can be seen from the positive values of selectivity. No doubt

this will depend on the magnitudes of α_{s1}, α_{s2}, β_{m1}, and β_{m2}, but there is very definite evidence of the destructive role of temperature gradients within the pellet.

Whereas analytical equations exist for estimating the yield in an isothermal system, no such equation is possible for the exact determination of yield in a nonisothermal pellet. The following approximate equation can be used, principally as a guide to the influence of major variables on nonisothermal yield:

$$\text{yield noniso} = \text{yield iso} + \frac{\alpha_{s1} - \alpha_{s2}}{2}$$

$$- \left(\frac{\sqrt{s}}{(1+\sqrt{s})^2} + \frac{C_{Bs}}{C_{As}\sqrt{s}} \right)(\beta_{m1} + \beta_{m2}) \quad (5.49)$$

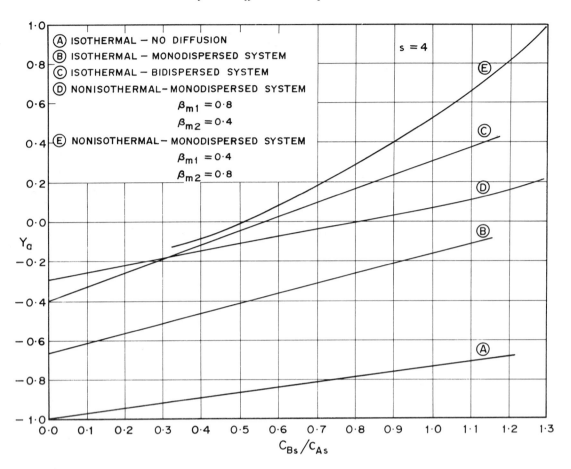

Figure 5.5. Point yield as a function of the ratio C_{Bs}/C_{As} for a nonisothermal Type III consecutive scheme for various diffusion models.

5.2.3. Multiple Solutions in Complex Reactions

The steady-state multiplicity patterns occurring when a single reaction is carried out are well understood at present, and strong criteria for predicting the conditions under which multiplicity occurs are briefly discussed in Chapters 4 and 12 (particularly 12). Most practical problems involve multistep reaction systems, the qualitative behavior of which may be different from that of a single reaction. Thus Luss and Chen (1975) have shown for a parallel (both first-order) endothermic reaction that multiplicity may exist under certain conditions, whereas for a simple first-order endothermic reaction it is known that no multiplicity is ever possible. Such differences warn us against undue generalization of the results from the simple systems, and in general each complex reaction has to be treated separately.

A number of investigators have studied steady-state multiplicity in systems in which several reactions occur simultaneously (Luss and Chen, 1975; Westerterp, 1962; Sabo and Dranoff, 1970; Hlavacek et al., 1972; McGreavy and Thornton, 1973; Butt, 1966; Östergaard, 1965; Michelsen, 1977; Andersen and Michelsen, 1975; Cohen and Keener, 1976). The case of parallel reactions has been satisfactorily investigated (Luss and Chen, 1975; Östergaard, 1965; Michelsen, 1977) and all the possible multiplicity patterns have been determined. Criteria for the prediction of uniqueness and multiplicity are also available.

Pikios and Luss (1979) have investigated the case in which two consecutive or two parallel irreversible first-order reactions occur. Several new and surprising features such as the occurrence of multiplicity for all Damköhler numbers have been reported. The *a priori* criteria for the prediction of the conditions that guarantee steady-state uniqueness or multiplicity in such systems have also been reported, and these are briefly considered in Chapter 12.

5.3. ROLE OF DIFFUSION IN POLYFUNCTIONAL CATALYTIC SYSTEMS

The ability to carry out more than one reaction on a catalyst is often of importance in multistep catalytic processes in which each step requires a different catalyst. This is referred to as polyfunctional catalysis. The very nature of polyfunctional catalysis requires that the overall reaction be a complex network, so that by appropriate choice of catalyst functions the yield of the desired product may be increased. As in the case of complex reactions on monofunctional catalysts considered in the previous sections, diffusion plays an important part in polyfunctional catalysis. Since this problem of diffusion in polyfunctional catalysts is closely connected with reactor performance and optimization, it will be considered in Chapter 12, with only a passing reference to it here.

5.3.1. Polyfunctional Catalysis

Consider the complex reaction†

$$A \rightarrow B \rightarrow C \qquad (1)$$

in which, as in the case of the common monofunctional catalysis, both steps occur on the same catalyst X. The adsorption–reaction scheme for this reaction will be represented by

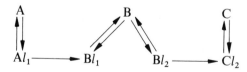

where l_1 is the active site type corresponding to catalyst X. If now the second step is carried out on a second catalyst Y, the adsorption–reaction scheme will be modified to

where l_2 represents the second type of active centers, corresponding to catalyst Y. Note that the intermediate B must now necessarily be desorbed, so that it can be

---

† The numbering of reaction schemes in this section is different from that adopted earlier for complex reactions.

adsorbed again on Y to give the product C. Thus where a bifunctional (or polyfunctional) catalyst is used the intermediate always has a transitory existence in the gas phase.

Examples

We now give some examples illustrating the applications of polyfunctional catalysis.

1. Conversion of hexadecane into a mixture of octanes

Weisz and Swegler (1955) have shown that a catalyst composed of platinum and silica–alumina accomplishes this conversion according to the scheme

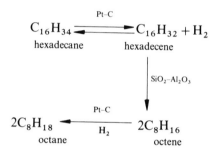

The supported platinum dehydrogenates in step 1, SiO_2–Al_2O_3 cracks in step 2, and the supported platinum hydrogenates to the final product, octane, in step 3.

2. Diversion of the dehydrogenation of cyclohexane (to benzene) to yield methyl cyclopentane

Methyl cyclopentane can be produced by intercepting the platinum-catalyzed dehydrogenation of benzene at the cyclohexene stage and isomerizing to methylcyclopentene on silica–alumina and finally hydrogenating it to methyl cyclopentane by the platinum already present.

3. Conversion of heptane to butane

Heinemann et al. (1953) have reported a bifunctional catalyst that converts heptane into butane in a single operation.

4. Aromatization of methyl cyclopentane

Kinetic studies on reforming reactions have established that the reactions can be represented as pseudo-first-order rate processes. The route that is followed on compounded catalyst pellets containing an active hydrogenation–dehydrogenation agent (X) and an active isomerization agent (Y) can be described (Al-

Samadi et al., 1974) as follows:

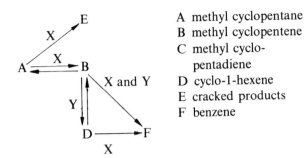

A   methyl cyclopentane
B   methyl cyclopentene
C   methyl cyclo-
      pentadiene
D   cyclo-1-hexene
E   cracked products
F   benzene

On a catalyst pellet containing a discrete mixture of the two active components the route followed is given as

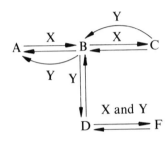

The four examples just cited emphasize the practical importance of polyfunctional catalysis. Not all polyfunctional catalyst systems, however, show advantages over systems in which the constituent reaction steps are carried out individually in separate reactors. Such a system is referred to as a *trivial* polystep reaction. The practical polyfunctional system is the *nontrivial* polystep reaction (Weisz, 1961) in which a composite polyfunctional catalyst gives a higher conversion than if each step of the reaction were carried out separately on the corresponding monofunctional catalyst. Certain conditions must be satisfied for the consequences of a nontrivial reaction to be realized. These are outlined below for the following two nontrivial schemes:

$$A \underset{k_{v1-}}{\overset{k_{v1}}{\rightleftharpoons}} B \xrightarrow{k_{v2}} C \qquad (2)$$

$$A \underset{k_{v1-}}{\overset{k_{v1}}{\rightleftharpoons}} B \begin{matrix} \overset{k_{v2}}{\nearrow} C \\ \underset{k_{v3}}{\searrow} D \end{matrix} \qquad (3)$$

Consider scheme 2, which is limited by equilibrium in step 1. It can be seen that if B is converted to C at a

faster rate on catalyst Y than back to A on catalyst X, the total conversion of A to C (the desired product) would be higher. This can only happen if a composite bifunctional catalyst is used. The conversion to B cannot exceed equilibrium if carried out in a separate reactor, so that the formation of C in the second reactor would also be limited. Similarly, in scheme (3), the reaction on catalyst X giving product C can be intercepted to produce D from intermediate B by using a polyfunctional catalyst.

For a quantitative evaluation of the enhancement of conversion in a nontrivial polystep reaction, we again consider scheme (2). It can be easily shown that if the two steps are carried out separately, the maximum possible conversion of A to C at infinite time would be given by

$$x_{A,max} = \frac{1}{1 + 1/K}, \qquad t = \infty \qquad (5.50)$$

If the equilibrium constant K is small, $x_{A,max}$ will be small. On the other hand, if the catalysts X and Y are intimately mixed in the same reactor, the conversion, no longer limited by equilibrium, will be given by

$$x_A = \frac{m_2}{m_1 - m_2} \exp(m_1 t) + \frac{m_2}{m_1 - m_2} \exp(m_2 t) + 1 \qquad (5.51)$$

where m_1 and m_2 are the roots of

$$m^2 + (k_{v1} + k_{v1-} + k_{v2})m + k_{v1}k_{v2} \qquad (5.52)$$

Equation 5.51 shows that x_A can have a maximum value of unity, thus establishing the superiority of the composite catalyst over the two-reactor system.

In the foregoing treatment, diffusion of B from site X to site Y was considered to be fast in relation to the surface reactions. It is possible that diffusion within a catalyst or between the two catalysts can be controlling. The likelihood of any of these steps influencing the reaction increases when pellets of X and Y are separately prepared and mixed or when composite pellets are prepared from X and Y. These aspects of polyfunctional catalysts are considered in Chapter 12.

5.3.2. Minimum Distance between Active Sites in a Bifunctional Catalyst for Absence of Diffusion

Several criteria to establish the importance of pore diffusion in a catalyst grain have been developed and are discussed in Chapter 9. These criteria require a knowledge of the intrinsic kinetic constant, which at times may not be known. The criterion presented by Weisz (1962), which is in terms of the observed reaction rate, can, however, be used in such instances:

$$\frac{r_a L^2}{D_e C_{As}} \ll 1 \tag{5.53}$$

This can also be used to determine the minimum distance between the two types of active sites in a bifunctional catalyst. Thus consider a reaction

$$A \underset{}{\overset{X}{\rightleftharpoons}} B \overset{Y}{\longrightarrow} C$$

occurring on a bifunctional catalyst. For first-order kinetics the reaction rate for species B can be written on the assumption that the first reaction is rate controlling:

$$r_{aB} = k_1 \left(\frac{1 + K_1}{K_1} \right) (C_B - C_B^*) \tag{5.54}$$

Substituting this observed rate in Eq. 5.53, we get the inequality

$$\frac{r_{aB} L^2}{D_e C_B} \ll 1 \tag{5.55}$$

Also, when the second reaction is rate controlling, we have

$$r_{aC} = k_2 C_B^* \tag{5.56}$$

Substituting this equation in the inequality, we get

$$\frac{r_{aC} L^2}{D_e C_B^*} \ll 1 \tag{5.57}$$

The form of the general inequality 5.57 is the same as that of 5.55. For the observed reaction rates these inequalities can be used to calculate L, the minimum distance (or grain size) that would satisfy the inequality and thus the condition of no diffusional limitation. In the case of *n*-heptane isomerization, Weisz has calculated the distance (from the observed rates at 470°C, corresponding to 40% conversion) as

$$L < 20\text{--}40 \, \mu\text{m}$$

which is in good agreement with the data.

5.4. NONUNIFORM DISTRIBUTION OF CATALYST ACTIVITY

In this section we shall examine the effect of spatially nonuniform catalyst activity on the yield of a consecutive reaction (Type III). The need for this study arises from the fact that many industrially important reactions use catalysts that are deposited on an inert support, resulting in distributed activity. For example, Mars and Gorgels (1964) report the use of such a catalyst for the selective hydrogenation of acetylene, while Friedrichsen (1969) has developed a similar catalyst for the oxidation of *o*-xylene to phthalic anhydride.

5.4.1. Theoretical Analysis

In the theoretical analysis of a consecutive reaction $A \rightarrow B \rightarrow C$, Kasaoka and Sakata (1968) proposed many catalyst distributions, including cases of activity declining or increasing monotonically with distance from the center. As already mentioned in Chapter 4, Yazdi and Petersen (1972) assumed a distribution of catalyst activity of the form $k_v = k_v^0 \hat{R}^q$. Corbett and Luss (1974) have defined a local rate constant $k_v(r) = \bar{k}_v \Omega(r)$, where $\bar{k}_v$ is the volume-averaged rate constant and $\Omega(r)$ is the rate constant density function given by

$$\frac{1}{V_p} \int_{V_p} \Omega(r) \, dr = \frac{\int_0^R 4\pi r^2 \Omega(r) \, dr}{\int_0^R 4\pi r^2 \, dr} = 1 \tag{5.58}$$

Kasaoka and Sakata (1968) have obtained solutions for flat-plate geometry for many catalyst distributions and found that the effectiveness factors for all situations can be brought together by employing a modified Thiele modulus, defined as

$$(\phi_{p1})_m = \frac{\phi_{p1}}{V_p} \int_{V_p} \Omega(1) \, dV \tag{5.59}$$

Smith (1976) has generalized the effectiveness factor plots of Kasaoka and Sakata (1968) to other geometries by incorporating the usual modification of characteristic particle dimension in the Thiele modulus. Further, the activity distribution for a given geometry is expressed in terms of the mathematically equivalent distribution in plane geometry by requiring that the volume-averaged rate constants in the two geometries be equal. This technique (applicable to arbitrary activity distribution in any geometry) eliminates the need for extensive numerical calculations and is illustrated in Section 5.4.2 through a numerical example.

Yazdi and Petersen (1972) have obtained solutions in terms of the instantaneous yield as a function of the Thiele parameter for various activity distributions of the general form $k_v = k_v^0 \hat{R}^q$. Their results suggest that the production of the undesirable product (C) can be suppressed by distributing the catalyst in such a way that the activity decreases toward the center.

Using the rate constant density function defined by Eq. 5.58, Corbett and Luss (1974) have obtained the point yield Y_p (see Eq. 5.21) as a function of the Thiele modulus, the asymptotic form of which is given by

$$\lim_{[(\phi_{s1})_A, (\phi_{s1})_B] \to \infty} Y_p = \frac{1}{1-\Delta} - \sqrt{\Delta}\left(C_{Bs} + \frac{1}{1-\Delta}\right)$$

(5.60)

where

$$\Delta^2 = \frac{(\phi_{s1}^2)_B}{(\phi_{s1}^2)_A}$$

This equation predicts the possibility of a negative yield when a large excess of the desired product (B) is present in the gas feed, as in the case of hydrogenation of traces of acetylene in an ethylene feed. The yield Y_p is plotted as a function of the Thiele parameter in Figure 5.6. The rate constant density function affects the selectivity only for intermediate values of the Thiele modulus, the influence becoming more pronounced with increase in Δ.

The concept of nonuniform distribution of activity can be extended to bifunctional catalysts, where one has to deal with nonuniform distribution of two activity functions.

5.4.2. Example: Calculation of Yield in a Model Consecutive Reaction in a Spherical Pellet with Nonuniform Distribution of Activity

A consecutive reaction A → B → C with the first step 10 times faster than the second is carried out in a spherical pellet with the activity distribution function given by $\Omega(r) = 4\hat{R}^9$ (see Chapter 4). For a feed consisting of pure A, calculate the yield of the desired product B if external diffusion resistance is assumed to be negligible. Assume that $\alpha_D = D_{eB}/D_{eA} = 1$, $s = k_{v1}/k_{v2} = 10$, $C_{B0} = 0$.

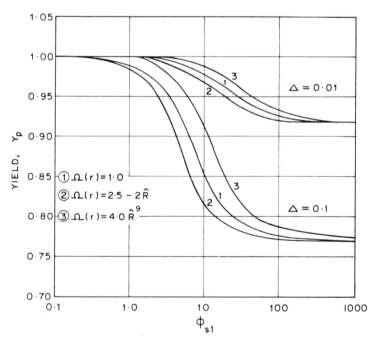

Figure 5.6. Effect of rate constant density function on the intrinsic point yield (Corbett and Luss, 1974).

Solution

The isothermal yield of product B for reaction in a flat plate is given by

$$Y_p = -\alpha_D \left(\frac{C_{Bs}}{C_{As}} + \frac{s}{\alpha_D s - 1} \right) \mathcal{S} + \frac{\alpha_D s}{\alpha_D s - 1}$$

For the conditions of the problem this simplifies to

$$Y_p = -\frac{10\mathcal{S}}{9} + \frac{10}{9} = \frac{10}{9}(1 - \mathcal{S})$$

The quantity $\mathcal{S}$ is given by the expression

$$\mathcal{S} = \frac{(\phi_{p1})_2 \tanh(\phi_{p1})_2}{(\phi_{p1})_1 \tanh(\phi_{p1})_1} = \frac{(\phi_{p1})_2^2}{(\phi_{p1})_1^2} \frac{\tanh(\phi_{p1})_2}{(\phi_{p1})_2}$$
$$\times \frac{(\phi_{p1})_1}{\tanh(\phi_{p1})_1}$$

The Thiele modulus $(\phi_{p1})_2$ can be expressed in terms of $(\phi_{p1})_1$ as follows:

$$(\phi_{p1})_2 = L \left(\frac{k_{v2}}{D_{eA}} \right)^{1/2} = L \left(\frac{k_{v1}}{D_{eA}} \frac{k_{v2}}{k_{v1}} \right)^{1/2}$$
$$= (\phi_{p1})_1 \left(\frac{k_{v2}}{k_{v1}} \right)^{1/2}$$
$$= \frac{(\phi_{p1})_1}{\sqrt{10}} = \frac{(\phi_{p1})_1}{3.16}$$

The quantity $\mathcal{S}$ then becomes

$$\mathcal{S} = \frac{k_{v2}}{k_{v1}} \frac{\tanh[(\phi_{p1})_1/3.16]}{(\phi_{p1})_1/3.16} \frac{(\phi_{p1})_1}{\tanh(\phi_{p1})_1}$$
$$= 0.316 \frac{\tanh[(\phi_{p1})_1/3.16]}{\tanh(\phi_{p1})_1}$$

and the expression for the yield of B

$$Y_p = (1.11 - 0.35) \frac{\tanh[(\phi_{p1})_1/3.16]}{\tanh(\phi_{p1})_1}$$

It should be noted that this equation is valid only for a flat plate with uniform activity distribution.

The modulus $(\phi_{p1})_1$ should now be converted into its equivalent in spherical geometry and corrected for the given activity distribution. For this we shall employ the approximate method proposed by Smith (1976) as outlined in Section 5.4.

The characteristic dimension L in the case of a flat plate can be written as $R/3$ for the case of a sphere. The rate constant is $k_v = k_v^0 \Omega(r) = 4k_v^0 \hat{R}^9$. For the case of uniform distribution, the equivalent rate constant is thus $4k_v^0$. The equivalent Thiele modulus is then obtained as

$$(\phi_{p1})_1 = L \left(\frac{k_v}{D_{eA}} \right)^{1/2} = \frac{R}{3} \left(\frac{4k_v^0}{D_{eA}} \right)^{1/2}$$
$$= \frac{2R}{3} \left(\frac{k_v^0}{D_{eA}} \right)^{1/2} = \frac{2}{3} (\phi_{s1})_1$$

To replace the activity distribution $\Omega(r) = \hat{R}^9$ in spherical geometry by its equivalent in plate geometry, we shall equate the volume-averaged rate constants in the two shapes:

$$\frac{1}{(V)_p} \int_{(V)_p} (k_v)_p(l) \, dV_p = \frac{1}{(V)_s} \int_{(V)_s} (k_v)_s(r) \, d(V)_s = \bar{k}_v$$
$$k_v(r) = \bar{k}_v \Omega(r)$$

Substituting the values

$$\left(\tfrac{4}{3}\pi R^3 \right)^{-1} \int_0^1 \bar{k}_v(r)^9 4\pi r^2 \, dr = \tfrac{1}{4}$$

we get

$$\frac{1}{(V)_p} \int_{(V)_p} (k_v)_p(l) \, d(V)_p = \tfrac{1}{4}$$

Employing the modification due to activity distribution in the rate constant, we have

$$k_v(r) = \bar{k}_v \Omega(r) = \frac{\bar{k}_v}{4}$$

The modified Thiele modulus then becomes

$$(\phi_{p1})_1 = \tfrac{1}{4} \left[\tfrac{2}{3}(\phi_{s1})_1 \right] = \frac{(\phi_{s1})_1}{6}$$

Thus the results are now expressed in terms of the Thiele modulus of the sphere $(\phi_{s1})_1$.

Substituting these results in the expression for the yield leads to

$$Y_p = (1.11 - 0.35) \frac{\tanh[(\phi_{s1})_1/(3.16 \times 6)]}{\tanh[(\phi_{s1})_1/6]}$$

The yield can now be calculated for various values of the Thiele modulus $(\phi_{s1})_1$. The approximate analytical solution and the numerical results of Corbett and Luss (1974) for the sphere (as shown in Figure 5.7) agree quite well except in a narrow region. The approximate solution should therefore be adequate for most computations.

5.5 EXAMPLES OF THE ROLE OF DIFFUSION

5.5.1. Dehydrogenation of Cyclohexane

Herington and Rideal (1947) have shown that the dehydrogenation of cyclohexane occurs via the intermediate cyclohexene as follows:

$$\text{cyclohexane} \rightarrow \text{cyclohexene} \rightarrow \text{benzene}$$
$$\text{(A)} \qquad\qquad \text{(B)} \qquad\qquad \text{(C)}$$

On the assumption that the two reaction steps in the network obey first-order kinetics, the following expression can be derived for reaction in a spherical pellet:

$$-\frac{dC_B}{dC_A} = \left(\frac{s}{1-s} - \frac{C_{Bs}}{C_{As}} \right)$$

$$\left[\frac{\phi_{s1}\sqrt{1/s}\coth(\phi_{s1}\sqrt{1/s})-1}{\phi_{s1}\coth(\phi_{s1})-1} \right] - \frac{s}{1-s} \quad (5.61)$$

It will be noticed that this equation is similar in form to

5.22, with the difference that in the earlier equation cylindrical geometry was considered.

Using a differential reactor, Weisz and Swegler (1955) measured dC_B/dC_A and C_{Bs}/C_{As} for three different particle sizes. Thus for pellets of diameters 0.31, 0.18, and 0.05 cm the values of ϕ estimated by them were 4.2, 2.5, and 0.68, respectively.

Then, for a given pellet size, they prepared plots of dC_B/dC_A vs. C_{Bs}/C_{As} for various values of s using Eq. 5.61. By matching these with the experimental line, the actual value of s was found. This was repeated for the remaining two particle sizes, and it was found that a consistent value of the selectivity (0.071) could be obtained. Then by using the results with the smallest particle size and noting the rate of disappearance of A, k_{v1} was obtained and k_{v2} estimated from the experimental finding, $s = 0.071$.

5.5.2. Hydrogenation of Acetylene

Komiyama and Inoue (1968) studied the hydrogenation of acetylene on a nickel catalyst and found the reaction to proceed according to the scheme

$$\text{acetylene} \rightarrow \text{ethylene} \rightarrow \text{ethane}$$
$$\text{(A)} \qquad\quad \text{(B)} \qquad\quad \text{(C)}$$

They carried out experiments with six different particle sizes, and measured the product distribution under different conditions.

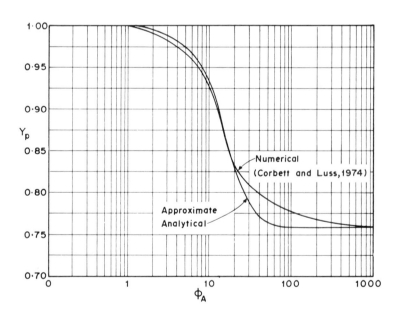

Figure. 5.7. Comparison of approximate and exact yield curves for a spherical pellet.

The selectivity of the reaction as measured by dC_B/dC_A was found to improve with decreasing particle size up to 0.73 mm, below which it becomes independent of particle size. This is in keeping with the theory presented in this chapter, according to which intraparticle diffusion annihilates the selectivity of a reaction.

However, the authors were unable to explain the results quantitatively on the basis of the uniform-pore model. Such a model could not correctly predict the dependence of selectivity on catalyst size. An electron microscope examination of the catalyst showed that it was composed of aggregates of fine nickel crystals inside the carrier block. A modified model, based on micropores inside the aggregates and macropores in the carrier block, explained the dependence adequately. In other words, a modified micropore–macropore model similar to Eq. 5.29 appears to predict correctly the dependence of selectivity on catalyst size.

5.5.3. Oxidation of Toluene

The role of diffusion in the oxydation of toluene to benzaldehyde on a molybdenum oxide–tungstic oxide on alumina catalyst has been studied by Reddy and Doraiswamy (1969). As a first step in the analysis it was established that the oxidation occurs almost exclusively by a parallel (first-order) reaction scheme:

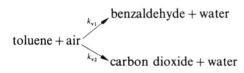

The role of pore diffusion was then studied by increasing both the temperature and the particle size (Reddy, 1968). It was found that, in accordance with the theory outlined in Section 5.1.2, diffusion exercises identical effects on the two steps making up the parallel network. Thus, with increase in temperature, transition to the pore diffusion region was observed to occur at practically the same temperature for both steps, and increase in particle size showed identical effects on the two steps.

5.5.4. Hydrogenation of Acetylene

The process of selective hydrogenation of acetylene in ethylene is used for trace purification of ethylene before polymerization. The gas mixture with excess hydrogen (20–200% of acetylene present) is passed over a partially impregnated Pd-on-alumina catalyst. Mars and Gorgels (1964), using catalysts differing in their Pd content (and hence the depth), have shown that the reaction proceeds in two zones. In the first zone only acetylene is hydrogenated. In the second zone, where no acetylene is present, hydrogenation of ethylene proceeds. Thus in the removal of trace amounts of acetylene from ethylene, the depth of active catalyst is an important design variable and can be manipulated to give a very high selectivity for acetylene hydrogenation. The catalyst particles range in diameter from 0.3 to 0.5 cm with the depth of penetration between 0.01 and 0.03 cm.

5.5.5. Oxidation Reactions

In many oxidation reactions (NH_3 oxidation over Pt or SO_2 oxidation over V_2O_5 or Pt), inter- and intraphase resistances significantly affect the conversion and yield (see also Tajbl et al., 1966). Reduction of the particle size to facilitate the inter- and intraphase transport processes, although fruitful in laboratory research, is not practicable on a large scale (because of excessive pressure drop). Partially impregnated catalysts offer a better alternative in such cases. The simulated data of Minhas and Carberry (1969) on adiabatic SO_2 oxidation over supported platinum compares the performance of a partially impregnated catalyst with that of a totally impregnated support. The results clearly show an effectiveness factor of unity for the partially impregnated catalyst, with definite advantages over the fully impregnated catalyst.

The oxidation of naphthalene to phthalic anhydride over a partially impregnated catalyst, reported by Smith and Carberry (1975), shows an improvement in yield and thermal sensitivity of the process. The partially impregnated catalyst gives better yields, since the diffusion-controlled oxidation of naphthalene is permitted to occur in the outer active zone while the slower degradation of phthalic anhydride is suppressed since no active catalyst is present in the inner zone.

REFERENCES

Al-Samadi, R. A., Luckett, P. R., and Thomas, W. J. (1974), *Adv. Chem. Ser.*, **133**, 316.

Andersen, A. S., and Michelsen, M. L. (1975), *Adv. Chem. Ser.*, **148**, Chap. 7.

Butt, J. B. (1966), *Chem. Eng. Sci.*, **21**, 275.

Carberry, J. J. (1962a), *AIChE J.*, **8**, 557.

Carberry, J. J. (1962b), *Chem. Eng. Sci.*, **17**, 675.

Cohen, D. S., and Keener, J. P. (1976), *Chem. Eng. Sci.*, **31**, 115.

Corbett, W. E., and Luss, D. (1974), *Chem. Eng. Sci.*, **29**, 1473.

Friedrichsen, W. (1969), *Chem. Ing.–Tech.*, **41**, 967.

Heinemann, H., Mill, G. A., Hattman, J. B., and Kirsch, F. W. (1953), *Ind. Eng. Chem.*, **45**, 130.

Herington, E. F. G., and Rideal, E. K. (1947), *Proc. Roy. Soc. (London)*, **A190**, 289.

Hirschfelder, J. O., Curtiss, C. F., and Bird, R. B. (1954), *Molecular Theory of Gases and Liquids*, Wiley, New York.

Hlavacek, V., Kubicek, M., and Visnak, K. (1972), *Chem. Eng. Sci.*, **27**, 719.

Kasaoka, S., and Sakata, Y. (1968), *J. Chem. Eng. Japan*, **1**, 138.

Kerber, R., and Gestrich, W. (1966), *Chem. Ing.–Tech.*, **38**, 536.

Komiyama, H., and Inoue, H. (1968), *J. Chem. Eng. Japan*, **1**, 142.

Komiyama, H., and Inoue, H. (1970), *J. Chem. Eng. Japan*, **3**, 117.

Luss, D., and Chen., G. T. (1975), *Chem. Eng. Sci.*, **30**, 1483.

Mars, P., and Gorgels, M. S. (1964), *Chem. Eng. Sci., Suppl. (Proc. 3d European Symp., Chem. React. Eng.)*, p. 55.

McGreavy, C., and Thornton, J. M. (1970), *Chem. Eng. Sci.*, **25**, 303.

McGreavy, C., and Thornton, J. M. (1973), *Chem. Eng. J.*, **6**, 91.

Michelsen, M. L. (1977), *Chem. Eng. Sci.*, **32**, 454.

Minhas, S., and Carberry, J. J. (1969), *J. Catal.*, **14**, 270.

Östergaard, K. (1961), *Acta Chem. Scand.*, **15**, 2037.

Östergaard, K. (1965), *Proc. 3d Int. Cong. Catal.*, p. 1348.

Pawlawski, J. (1961), *Chem. Eng. Tech.*, **33**, 492.

Pikios, C. A., and Luss, D. (1979), *Chem. Eng. Sci.*, **34**, 919.

Reddy, K. A. (1968), Ph.D. thesis, Bombay University.

Reddy, K. A., and Doraiswamy, L. K. (1969), *Chem. Eng. Sci.*, **24**, 1415.

Roberts, G. W. (1972), *Chem. Eng. Sci.*, **27**, 1409.

Sabo, D., and Dranoff, J. S. (1970), *AIChE J.*, **16**, 211.

Smith, T. G. (1976), *Ind. Eng. Chem. Process Design Dev.*, **15**, 388.

Smith, T. G., and Carberry, J. J. (1975), *Can. J. Chem. Eng.*, **53**, 347.

Tajbl, D., Simons, J. B., and Carberry, J. J. (1966), *Ind. Eng. Chem. Fundam.*, **5**, 171.

van de Vusse, J. G. (1966), *Chem. Eng. Sci.*, **21**, 1239.

Varghese, P., Varma, A., and Carberry, J. J. (1978), *Ind. Eng. Chem. Fundam.*, **17**, 195.

Voge, H. H., and Morgan, C. Z. (1972), *Ind. Eng. Chem. Process Design Dev.*, **11**, 454.

Wakao, N., and Fujishiro, S. (1966), *Kagaku Kogaku*, **30**, 745.

Weisz, P. B. (1961), *Proc. 2nd Int. Conf. Catal.*, **1**, 937.

Weisz, P. B. (1962), *Adv. Catal.*, **13**, 137.

Weisz, P. B., and Prater, C. D. (1954), *Adv. Catal.*, **6**, 143.

Weisz, P. B., and Swegler, E. W. (1955), *J. Phys. Chem.*, **59**, 823.

Westerterp, K. R. (1962), *Chem. Eng. Sci.*, **17**, 423.

Wheeler, A. (1951), *Adv. Catal.*, **3**, 249.

Wirges, H. P., and Rahse, W. (1975), *Chem. Eng. Sci.*, **30**, 647.

Yazdi, S. F., and Petersen, E. E. (1972), *Chem. Eng. Sci.*, **27**, 227.

CHAPTER SIX

Role of External Mass Transfer

In order to apply the catalytic rate equations developed in Chapter 2 it is necessary to formulate procedures whereby the concentrations on the catalyst surface can be estimated from the corresponding values in the ambient fluid and the effect of this external mass transfer resistance included in the design.

In making allowance for external mass transfer it is generally assumed that the chemical transformations occurring on the solid surface do not interfere with the mass transfer process. In fact, the mass transfer coefficients are usually estimated from transfer processes uncomplicated by chemical reaction on the surface, and these are used in estimating the surface concentrations. Another method, which is more fundamental, is to define an effectiveness factor based on the fact that mass transfer and chemical reaction on the external surface are coupled effects, and then obtain the true kinetics of the reaction. This approach is applicable only to those geometric configurations of the fluid–catalyst system for which the hydrodynamics can be precisely described, for example, a rotating disk, a flat plate, the forward stagnation region of a circular cylinder, or a tubular reactor. The entire problem of reaction occurring on the external surface of a catalyst has been discussed in some reviews and recent papers by Chung (1965), Simkins (1965), Fox and Libby (1966), Rosner (1964b), Lindberg and Schmitz (1969, 1970), Mihail (1969, 1970, 1972, 1975), Jeng et al. (1979), and Fournier et al. (1981).

In this chapter, methods for predicting the external mass and heat transfer coefficients are presented. The question of the external effectiveness factor is then considered, including the situation where both internal and external resistances are simultaneously operative.

6.1. THEORETICAL EXPRESSIONS FOR MASS (HEAT) TRANSFER COEFFICIENT

The interaction between the momentum and thermal or mass boundary layers suggests that heat or mass transfer to the surface is influenced by the flow field surrounding the particle. The theoretical analysis of the problem of calculating the mass (heat) transfer coefficient (or the Sherwood, Nusselt, or Biot number in dimensionless form) thus requires an understanding of the existing hydrodynamic conditions. Numerous studies for the case of a single particle with well-defined geometry and for well-defined flows have been reported. These are, however, of limited applicability to the fixed bed, where the flow field is far from being well defined and the presence of other particles affects the transport coefficient in an uncertain way. In the present section we shall therefore focus our attention on the models of mass transfer in multiparticle assemblages; for the problem of single particles the reader is referred to standard textbooks on the topic (e.g., Eckert and Drake, 1972; Reid et al., 1977).

A conceptually elegant model for multiparticle systems is the cell model of Happel (1958). In this "free surface model" the many-body problem of the packed bed is replaced by a simple continuous one involving a single particle. The wall effects and/or entry and exit effects are assumed to be negligible. The model was developed for smooth spherical particles of uniform size at very low Reynolds numbers with the following assumptions: the particles are fixed in space with equal distances separating them; each particle is surrounded by a hypothetical fluid sphere whose radius is related to the actual voidage in the assembly; and the outside surface of each cell so formed is frictionless. Thus, the entire disturbance due to each particle is confined to the cell of fluid with which it is associated. As can be expected, the cell model may be used to predict the bulk or continuum properties in fluid–particle flow.

The cell model has been used by Pfeffer (1964) to obtain heat and mass transport coefficients in multiparticle systems. At high Peclet numbers, where the change in concentration occurs entirely in the thin boundary layer, the "thin boundary layer solution" of the diffusion equation can be combined with the free

TABLE 6.1. Mass (or Heat) Transfer Correlations in Multiparticle Systems

Expression	Range	Remarks	Reference
(1) $\mathrm{Sh}\,(\mathrm{Nu}) = 1.20\left[\dfrac{1-(1-f_B)^{5/3}}{F(f_B)}\right]^{1/3}(\mathrm{Pe}')^{1/3}$ where $F(f_B) = 2 - 3(1-f_B)^{1/3} + 5(1-f_B)^{5/3} - 2(1-f_B)^2$ and $\mathrm{Pe}' = \dfrac{d_p\mu}{D_b}$ for mass transfer $\quad = \dfrac{d_p C_p G}{k'_g}$ for heat transfer	High-Peclet-number–low-Reynolds-number region		Pfeffer (1964)
(2) $\mathrm{Sh} = \dfrac{SF}{6(1-f_B)\mathbf{R}_{cd}}\,\mathrm{Pe}'$ where R_{cd} is the ratio of the average channeling length to the particle diameter	$\mathrm{Pe}' < 10$		Kunii and Suzuki (1967)
(3) $\mathrm{Sh} = \dfrac{2\xi + \left\{\dfrac{2\xi^2(1-f_B)^{1/3}}{[1-(1-f_B)^{1/3}]^2} - 2\right\}\tanh\xi}{\dfrac{\xi}{1-(1-f_B)^{1/3}} - \tan\xi}$ where $\xi = \left[\dfrac{1}{(1-f_B)^{1/3}} - 1\right]\dfrac{\alpha}{2}(\mathrm{Re}')^{1/2}\,\mathrm{Sc}^{1/3}$ and $\alpha \approx 0.6$	$0.08 < \mathrm{Re}' < 1000$	1. As $f_B \to 1$ a relationship similar to that presented for a single sphere by Frossling (1938) and Ranz (1952) is obtained: $\mathrm{Sh}_{f_B\to 1} = 2 + \alpha(\mathrm{Re}')^{1/2}\,\mathrm{Sc}^{1/3}$ 2. The limiting value of the Sherwood number is $\mathrm{Sh}_{\mathrm{Re}'\to 0} = \dfrac{1}{(1-f_B)^{1/3}}\left[\dfrac{1}{(1-f_B)^{1/3}} - 1\right]\dfrac{\alpha^2}{2}\,\mathrm{Re}'\,\mathrm{Sc}^{2/3}$	Nelson and Galloway (1975)

The linear dependence of Sh on Re' is due to the finite radius boundary condition employed. The model is recommended for low Re'.

The model applied to 12 cases of absorption and vaporization in commercial packed beds gives results that are accurate to within 12%. It is recommended for high Re'.

Galloway and Sage (1970)

(4) $j_d = St\, Sc^{2/3} = \dfrac{Sh^{2/3}}{Re'\, Sc^{1/3}} = \dfrac{Sc^{2/3}}{a(HTU)}$ $\qquad Re' > 100$

$j_h = St\, Pr^{2/3} = \dfrac{Nu}{Re'\, Pr^{1/3}} = \dfrac{Pr^{2/3}}{a(HTU)}$

where

$a(HTU) = \dfrac{Re'\, Pr\ (or\ Sc)}{Nu\ (Sh)} = \dfrac{(Re')^{1/2}\, Pr^{2/3}\ (or\ Sc^{2/3})}{Fs + 2/(Re')^{1/2}\, Pr^{1/3}\ (or\ Sc^{1/3})} = \dfrac{1}{St}$

(where St is the Stanton number) and

$Fs = \dfrac{(Nu-2)}{(Re')^{1/2}\, Pr^{1/3}}$ or $\dfrac{(Sh-2)}{(Re')^{1/2}\, Sc^{1/3}}$ for a single sphere

$Fs = \dfrac{Nu}{(Re')^{1/2}\, Pr^{1/3}}$ or $\dfrac{Sh}{(Re')^{1/2}\, Sc^{1/3}}$ for a single cylinder

119

surface model to get an expression for the Sherwood number; this is presented in Table 6.1. The equation, applicable in the low-Reynolds-number–high-Peclet-number region, agrees reasonably well with the available experimental mass transfer data.

The theoretical cell model, limited to the creeping-flow regime, has been extended by El-Kaissy and Homsy (1973) to packed beds at intermediate Reynolds numbers. The inability of the simple cell model to predict the drag at intermediate Reynolds numbers is overcome in this model by altering the shape of the cell. The cell considered has a constant volume but its shape is a function of the Reynolds number. This so-called distorted-cell model has been shown to adequately represent the experimental behavior.

The cell model used by Pfeffer (1964) assumes that the outside surface of each cell is at a constant temperature or concentration. This assumption is valid only at high values of the Peclet number and should be replaced by an impermeable surface condition if the model has to be applied at low Peclet numbers. In an actual packed bed, because of the distribution of voidage, the fluid flows selectively, leading to channeling. Kunii and Suzuki (1967) have developed a model that assumes that there is an equilibrium between the solid and the fluid in the dead zone, that the fluid flows in the channel in plug flow, and heat or mass transfer is restricted to molecular diffusion in the direction perpendicular to the flow. The model uses the ratio of the average channeling length to the particle diameter as a parameter, and the theoretical expression for Nu or Sh obtained is presented in Table 6.1. The equation correlates the data of various workers well in the low Peclet number range, but is of limited use in view of the nature of the parameter values required.

Carberry (1960) and Kusik and Happel (1962) have also proposed models for multiparticle systems, but a particularly general one is that of Galloway and Sage (1970). They developed a boundary layer model for packed and fluidized beds based on an extensive analysis of available data over a range of Pr, Sc, and Re′ for a variety of packing configurations and voidages, and the relationship obtained by them is included in Table 6.1.

The expressions for heat and mass transport coefficients in multiparticle systems derived from the theories of single particles fail to predict the correct values, principally because of the incorrect boundary condition. The constant concentration or temperature boundary condition at the outer spherical shell has the effect of creating a source or sink at infinity. This condition, valid for a single sphere, is inapplicable when a sphere in an assemblage is considered, and should be replaced by a zero gradient condition. Nelson and Galloway (1975) used this condition and combined the boundary layer and Danckwerts's penetration theories to get a more useful relation which correlates much of the data on transfer from dense

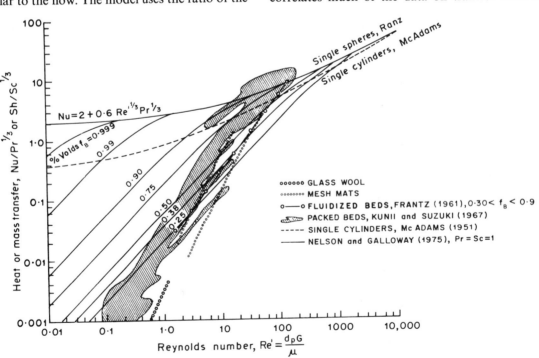

Figure 6.1. Particle-to-fluid heat and mass transfer (Nelson and Galloway, 1975).

suspensions and screen assemblages. The actual expression obtained is included in Table 6.1, and Figure 6.1 shows a comparison of the model with the available experimental data.

No satisfactory correlations are available in the low-Re'–low-Pe region in spite of the extensive work reported (see Wakao, 1976; Gangwal et al., 1977). The main difficulty has been the value of the limiting Sherwood number—whether it goes to zero or takes on a finite constant value as the fluid velocity drops to zero. The limiting value for a single sphere is 2.

The experimental data of Littman et al. (1968) and Kato et al. (1970) show a linear dependence of the Sherwood number on the Reynolds number, whereas Wakao and Tanisho (1974) report a quadratic dependence. On the other hand, several other workers report finite values for the limiting Sherwood number. Thus Padberg and Smith (1968) report a value of 2; Sorensen and Stewart (1974) a theoretical value of 3.89 for a simple cubic array of spheres ($f_B = 0.48$); Gunn and de Souza (1974) a value of about 10; Miyauchi (1972), 8.7; Miyauchi et al. (1975) and Ikeda et al. (1973), 7.41; and Gangwal et al. (1977), 1.

Much of the discrepancy in the reported values of the limiting Sherwood number (particularly values less than 1) can be traced to the influence of axial diffusion, a fact not often recognized in estimating Sh from experimental data. Gangwal et al. (1977) suggest the following correlation for the influence of axial diffusion on Sh (or Nu):

$$\text{Nu} = (\text{Bi})_h = \frac{f_B (\text{Re}' \, \text{Pr})^2}{6(1 - f_B) D_{el}/k'_g} \tag{6.1}$$

The computations of Sorensen and Stewart (1974) show that Nu at low Re' can be two orders of magnitude lower than Nu obtained by allowing for axial dispersion.

6.2. EMPIRICAL CORRELATIONS FOR PARTICLE–FLUID MASS AND HEAT TRANSFER

6.2.1. General Considerations

Since the early studies of Hougen and co-workers (Gamson et al., 1943; Wilke and Hougen, 1945; Taecker and Hougen, 1949), a vast volume of experimental data has accumulated on particle–fluid mass and heat transfer to and from a fixed bed of pellets. In the majority of these correlations the dimensionless factors

j_d and j_h, defined as

$$j_d = \frac{\text{Sh}}{\text{Re}' \, \text{Sc}^{1/3}}, \qquad j_h = \frac{\text{Nu}}{\text{Re}' \, \text{Pr}^{1/3}} \tag{6.2}$$

have been used as the correlating parameters, and the correlations have taken the form

$$j_d \, (\text{or} \, j_h) = a (\text{Re}')^b \tag{6.3}$$

with the exponent b always negative. The definitions of the dimensionless groups are $\text{Re}' = d_p \mu / G$, $\text{Sh} = k_g R / D_e$, $\text{Nu} = h R / k'_e$, $\text{Sc} = \mu / \rho_g D_e$, $\text{Pr} = C_p \mu / k'_g$.

The results have also been presented in the usual form:

$$\text{Sh or } (\text{Bi})_m = a \, (\text{Re}')^b \, \text{Sc}^{1/3}$$

$$\text{Nu or } (\text{Bi})_h = a \, (\text{Re}')^b \, \text{Pr}^{-1/3} \tag{6.4}$$

with the exponent b now being positive. Additional parameters, such as voidage, d_p/d_t or d_p/L, or the Archimedes number, have also been included by some authors to improve the correlations.

The main limitation of these correlations is that they all predict zero transfer in the absence of flow. Thus the role of natural convection is wholly neglected. The most rational way to account for natural convection is to incorporate the Grashof number in the analysis of mass transfer data.

Data uncorrected for axial diffusion should not be used in formulating mass transfer correlations. Most of the early results reported are of doubtful value because this correction was not made. The results of McHenry and Wilhelm (1957), Evans and Kenney (1966), and Gunn (1968) clearly show that the effect of axial diffusion can be quite considerable. Much of the gas-phase and liquid-phase mass transfer data published in the literature have been corrected for axial dispersion by Wakao and Funazkri (1978).

In order to obtain a clearer understanding of the fluid–particle transport process in a packed bed, Thoenes and Kramers (1958) used a single test sphere placed in a regular packed bed of inert spheres (see also Denton, 1951; Wadsworth, 1961; Rhodes and Peebles, 1965). Microwave heating can be employed (Balakrishnan and Pei, 1974, 1979) to eliminate particle–particle conduction and obtain fluid–particle transfer only. Another convenient method is to coat spheres of an inactive solid with a reactive liquid and pass over them a gas that reacts with the liquid. Thus CO_2 can be passed as a pulse in a carrier gas through a bed of spheres coated with caustic potash (Miyauchi et al., 1972). Similarly, a fast mass-transfer-controlled reaction can be carried out in a bed of nonporous catalyst spheres—for example, the decomposition of hydrogen peroxide in a packed bed of silver spheres (Satterfield and Resnick, 1954).

6.2.2. Correlations for Mass Transfer

The more important correlations reported in the literature have been tabulated by many investigators (see, e.g., Wen and Fan, 1975; Upadhyay and Tripathi, 1977; Wakao and Funazkri, 1978). Of these, the following are recommended for the medium and high Reynolds number regions. At very low Reynolds numbers (say, less than 5) none of the correlations seem to be entirely satisfactory, and separate equations are recommended.

Medium and High Reynolds Numbers

The equation of Colquhoun-Lee and Stepanek (1974) is among the best available for medium and high Reynolds numbers. These authors considered the void fraction as a definite parameter in correlating the data. They selected from the literature only such data as were corrected for axial mixing or for which no correction was necessary (owing to the experimental technique used); moreover, they used only data that were obtained in packed beds in which the characteristic dimension of the particle was less than 10% of both bed diameter and length (i.e., wall and end effects were negligible). The final equation is

$$j_d = 1.24 \left(\frac{Re'}{1 - f_B} \right)^{-0.39} \tag{6.5}$$

$$50 < Re' < 5000, \quad 0.3 < f_B < 0.5, \quad 0.6 < Sc < 2000$$

The conditions chosen represent practical situations in industrial reactors; hence, this equation is recommended for design purposes.

Another useful correlation, which is reported to be applicable for Re' as low as 3, is that of Wakao and Funazkri (1978):

$$Sh = 2 + 1.1 \, Sc^{1/3} \, (Re')^{0.6}, \qquad 3 < Re' < 10,000 \tag{6.6}$$

Low Reynolds Numbers

In the low Reynolds number region (less than 3), no completely reliable equation is yet available. Hancil and Mitschka (1969) have compared four typical available

correlations and found that in the Reynolds number range from 10^{-2} to 10, predictions of Sh deviate by a factor that varies from 10 to 10^3. Chambers and Boudart (1966) point out that although the correlation of De Acetis and Thodos (1960) ($j_d = 0.725(Re'^{0.41} - 1.5)^{-1}$) is claimed to be valid for Reynolds numbers as low as 13, actually the exponent on the Reynolds number changes considerably in this region and is much smaller. This observation is even more true as Re' is further decreased.

An acceptable equation for very low Re' appears to be that presented for heat transfer in Section 6.2.4 (Eq. 6.11). After estimating j_h from that equation, j_d may be calculated from the relation $j_h = 0.8 j_d$.

6.2.3. Mass-Transfer-Controlled Fixed-Bed Reactors

The height of a fixed-bed reactor operating under conditions of external mass transfer control would be very small, usually of the order of a few pellet diameters, as compared to the height of a kinetically controlled reactor. If the height is, say, of the order of 100 pellet diameters or more for achieving conversions of the order of 75%, then it is a safe conclusion that external mass transfer is not controlling.

From simple material balance considerations, Satterfield (1970) has derived the following expression for the ratio of the mole fraction of the reactant at the inlet (y_0) to that at a bed height of L (y_L) under conditions of external mass transfer control:

$$\ln \left(\frac{y_0}{y_L} \right) = \frac{k_g a P}{G_M} L \tag{6.7}$$

In terms of the Schmidt and Reynolds numbers, this expression becomes (by combination with 6.5):

$$\ln \left(\frac{y_0}{y_L} \right) = \frac{1.24 \, (Re')^{-0.39} \, Sc^{-2/3} \, a}{(1 - f_B)^{-0.39}} L \tag{6.8}$$

The chief practical difficulty in using these equations is the estimation of the interfacial area a. This can be overcome by employing the results of Sherwood and Pigford (1952), according to which the product $a d_p$ is independent of d_p for a given bed voidage. Thus $a d_p$ is 4.2 for $f_B = 0.3$, 3.6 for $f_B = 0.4$, and 3.0 for $f_B = 0.5$,

6.2.4. Correlations for Heat Transfer

In view of the strong dependence of reaction rate on temperature, correlations for the prediction of gas–solid heat transfer coefficients should provide as accurate an estimate of the solid surface temperature as possible. We shall be concerned here with the total particle–fluid heat transfer, as distinct from convective transfer reported by Bhattacharya and Pei (1975) and Balakrishnan and Pei (1979).

Barker (1965), Kunii and Suzuki (1967), Handley and Heggs (1969), Littman et al. (1968), Gupta et al. (1974), and Balakrishnan and Pei (1979) have presented surveys of the correlations reported in the literature. Bhattacharyya and Pei (1975) give plots of j_h vs. Re′ for 12 selected correlations, but make no specific recommendations.

Gliddon and Cranfield (1970) recommend Ranz's equation (1952) for estimating gas–particle heat transfer for Reynolds numbers greater than 100:

$$\text{Nu or } (\text{Bi})_h = 2.0 + 1.8 \, (\text{Re}')^{1/2} \, \text{Pr}^{1/3} \qquad (6.9)$$

At low Reynolds numbers, as in the case of mass transfer, there is considerable uncertainty in the estimation of the heat transfer coefficient also. It has been shown by Littman et al. (1968) that j_h varies by a factor ranging from 5 to 100 (in general) when estimated from different correlations Gliddon and Cranfield (1970) suggest the following correlation for predicting the heat transfer coefficient in the Reynolds number range from 2 to 100:

$$\text{Nu or } (\text{Bi})_h = 0.36 \, (\text{Re}')^{0.94}$$
$$j_h = 0.41 \, (\text{Re}')^{-0.06} \qquad (6.10)$$

For Re′ less than 2, the equation suggested by Cybulski et al. (1975) may be used:

$$\text{Nu or } (\text{Bi})_h = 0.07 \, \text{Re}' \qquad (6.11)$$

It may be noted that j_h is practically independent of the Reynolds number in both Eqs. 6.10 and 6.11. Also, the Nusselt number becomes much smaller than the theoretical limiting value of 2. Because of the precision of the data involved, this equation is also recommended at low Reynolds numbers for calculating j_d (by assuming that $j_h = 0.8 j_d$, as already mentioned in Section 6.2.3).

Attempts have been made to develop correlations that would be equally applicable to fixed-bed and fluid-bed reactors (see Gupta et al., 1974; Upadhyay and Tripathi, 1975). These equations are less accurate than those developed specifically for each of the two reactor types and are generally not recommended.

6.3. EXTERNAL EFFECTIVENESS FACTORS

6.3.1 Internal and External Flow Configurations

Flow on the surface of a solid can either be restricted to the internal surface of a confined shape or can occur on the surface of an exposed solid. The simplest internal flow system, and no doubt a practically important one, is the tubular reactor with a catalytically active wall; but this is a mathematically complex system. On the other hand, external flow configurations are more easily amenable to theoretical analysis.

Among the external flow configurations, the rotating disk in laminar flow and the forward stagnation region of a blunt-nosed symmetrical body have the rare property that the coefficient of convective transport is uniform over the entire surface. In the more commonly encountered situation of a flat plate (or a sphere), the surface is not uniformly accessible, so that the kinetic and convective terms cannot be uncoupled (as in the case of a uniformly accessible surface). This leads to considerable mathematical difficulties.

6.3.2. Flow around a Catalyst Plate

Consider a catalyst plate immersed in a flowing stream of fluid that contains the greatly diluted reactive species (see Fig. 6.2). This implies that the reaction occurring on and around the catalyst surface does not affect the fluid bulk properties. The reactant concentration in the boundary layer that develops between $y = 0$ and $y = \delta$ is given by

$$u_x \frac{\partial \hat{C}_{Ab}}{\partial x} + u_y \frac{\partial \hat{C}_{Ab}}{\partial y} = D_{bA} \frac{\partial^2 \hat{C}_{Ab}}{\partial y^2} \qquad (6.12)$$

where x and y denote the distance parameters tangential and normal, respectively, to the surface, $\hat{C}_{Ab} = C_A / C_{Ab}$, C_{Ab} is the free stream concentration of A, and D_{bA} is the bulk diffusion coefficient. At the solid surface $y = 0$, and we have

$$D_{bA} \left(\frac{\partial C_{Ab}}{\partial y} \right)_{y=0} = k_S C_{Ab}^n \qquad (6.13)$$

where $k_S C_{Ab}^n$ represents the rate of depletion or of transfer at the solid surface. The problem of obtaining quantitatively the effect of external diffusion on the surface chemical reaction now reduces to solution of the boundary layer equation under the constraint of the kinetic boundary condition 6.13.

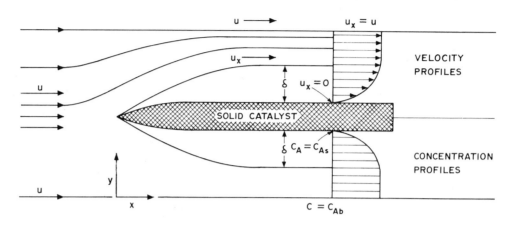

u = FREE STREAM VELOCITY
u_x = VELOCITY TANGENTIAL TO SURFACE
u_y = VELOCITY NORMAL TO SURFACE
δ = BOUNDARY LAYER THICKNESS

Figure 6.2. Hydrodynamic model of a catalyst plate in a flowing stream of reactant.

The ratio of the actual rate of reaction in the presence of diffusion (r_{Sa}) to that in the absence of diffusion (r_S) may be regarded as an external effectiveness factor:

$$\eta = \frac{r_{Sa}}{r_S} \qquad (6.14)$$

where r_S is given by

$$r_S = k_S C_{As}^n \qquad (6.15)$$

Where diffusional resistance is absent, $C_{As} = C_{Ab}$ and

$$r_S = k_S C_{Ab}^n \qquad (6.16)$$

The diffusional rate in the absence of chemical reaction is given by

$$r_d = k_g (C_{Ab} - C_{As}) \;\approx k_g C_{Ab} \quad \text{(for a fast reaction)} \qquad (6.17)$$

The ratio r_S/r_d represents a Damköhler number for a reaction of order n

$$\frac{r_S}{r_d} = \frac{k_S C_{Ab}^{n-1}}{k_g} = Da_n \qquad (6.18)$$

Thus the external effectiveness factor is a function only of the Damköhler number Da_n and reaction order n. This is mathematically formulated below.

6.3.3. Effectiveness Factor by the Quasi-Steady-State Method

Let us first consider the simple case where the diffusional and kinetic terms are uncoupled. This implies that in the usual definition of the rate of mass transfer,

$$r_d = \frac{\text{driving force}}{\text{resistance to mass transfer}} = \frac{\Delta C_A}{1/k_g} \quad (6.19)$$

the driving force for the diffusion of the reactant from the fluid bulk to the catalyst surface is constant over the entire surface; that is, it is independent of hydrodynamic conditions. From a practical point of view, this assumption has found widespread engineering acceptance and has formed the basis of Frank-Kamenetskii's so-called quasi-steady-state solution (1955).

If the local effectiveness factor is defined as

$$\eta' = \left(\frac{C_{As}}{C_{Ab}}\right)^n \qquad (6.20)$$

the following expression can be readily obtained:

$$Da_n' \eta' + (\eta')^{1/n} = 1 \qquad (6.21)$$

where Da_n' is the local Damköhler number for an nth-order reaction.

For the first-order case ($n = 1$), Eq. 6.21 for the local effectiveness factor becomes

$$\eta' = \frac{1}{1 + \mathrm{Da}_1'} \tag{6.22}$$

where the subscript 1 refers to a first-order reaction. Since for a first-order reaction $\mathrm{Da}_1' = k_s/k_g$, Eq. 6.22 can be recast in the form

$$\eta' = \frac{1}{1 + (k_s/k_g)}$$

or

$$\frac{1}{k_s\eta'} = \frac{1}{k_g} + \frac{1}{k_s} \tag{6.23}$$

which can be immediately identified as a simple addition of two independent first-order resistances, one for reaction and the other for mass transfer.

6.3.4. Effectiveness Factor from Quasi–Quasi Steady-State Analysis

Let us refer again to Figure 6.2. The fluid first strikes the inner (sharp) edge of the plate and passes over it. At this inner edge the concentration of the reactive species on the plate is equal to the bulk concentration, but decreases progressively along the length of the plate as a result of an increasing thickness of the boundary layer. Under these conditions the reaction will be chemically controlled near the inner edge of the plate and diffusionally controlled near the outer (or trailing) edge. This problem of nonuniform surface conditions (i.e., of the diffusional driving force) and their influence on the overall effectiveness of a catalyst has been theoretically examined by many workers (Rosner, 1963, 1964a, b; Mihail, 1970).

If the driving force ($C_{Ab} - C_{As}$) increases as some power of the distance x_l along the plate, then the local mass transfer coefficient can be higher than that calculated on the assumption of a constant driving force by a factor of 1.0–1.5 (or more) for laminar flow and of 1.0–1.2 for turbulent flow (Rosner, 1964b). The difficulty in the analysis arises out of the fact that the distribution of the driving force is not known *a priori* and is in fact sought as part of the solution. By writing an expression for the local transfer rate that is valid for an arbitrary distribution of the driving force, the following class of nonlinear Volterra integral equations have been obtained (Rosner, 1964a) for laminar as well as turbulent boundary layers (see also Lighthill,

1950; Tribus and Klein, 1953; Schlichting, 1955; Chambré, 1956):

$$
\begin{aligned}
\eta' &= \left(\frac{C_{As}}{C_{Ab}}\right)^n \\
&= -\frac{1}{\xi}\int_0^\xi \left[1 - \left(\frac{g}{\xi}\right)^p\right]^q \frac{d\hat{C}_A}{dg}\,dg
\end{aligned} \tag{6.24}
$$

where $\xi = k_s C_{Ab}^{n-1}\delta/D_b$ (stream variable), g is a dummy variable, and p and q are constants whose values depend on the magnitude of the Schmidt number and on the nature of the flow (laminar or turbulent). Typical values of p and q have been given by Rosner (1964a).

Equation 6.24 embodies the effect of coupling of the diffusional and kinetic effects, unlike its quasi-steady-state counterpart (Eq. 6.21), in which these effects are uncoupled. In order to quantitatively evaluate the effect of coupling, Eq. 6.24 has to be solved.

Exact solutions for specific cases can be obtained by the procedures outlined by Chambré (1956), Chambré and Acrivos (1956), Acrivos and Chambré (1957), and Chung et al. (1963). An approximate (so-called quasi–quasi-steady-state) solution, which has the advantage of generality, has been provided by Rosner (1963, 1964a) for a reaction of general order on a flat plate and for both the flow regimes, laminar and turbulent. Plots of the overall effectiveness factor η thus prepared for reactions of different orders are shown in Figure 6.3. The similarity between these and the internal effectiveness factor plots is striking. It is interesting to note, however, that for a first-order irreversible reaction, the effectiveness factor (integrated over the region between the forward stagnation point and boundary layer separation) is greatest for the sphere as against the internal effectiveness factor, which is least for the sphere (Mihail, 1972).

6.3.5. Effect of Nonisothermicity

In analogy with Equation 6.14, the local effectiveness factor for the nonisothermal case may be defined as

$$\eta'(T) = \frac{k_s(T_s)C_{As}^n}{k_s(T_b)C_{Ab}^n} \tag{6.25}$$

Under the constraint of isothermicity, it reduces to $\eta' = (C_{As}/C_{Ab})^n$, which may be recognized as Eq. 6.20 for the isothermal case. Since the effect of $k_s(T_s)/k_s(T_b)$ can far outweigh the effect of C_{As}^n/C_{Ab}^n in Eq. 6.25, the external effectiveness factor for a nonisothermal system can exceed unity.

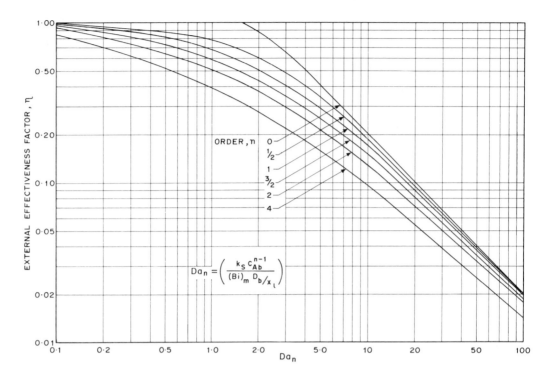

Figure 6.3. Dependence of external effectiveness factor on the Damkohler number for reactions of different orders (Rosner, 1963).

6.3.6. General Comments

In view of the known spatial variation of the boundary layer thickness leading to a surface of unequal accessibility, it would be of practical importance to estimate the magnitude of the error involved in the equiaccessibility approximation.

For this purpose a reaction factor (RF) may be defined as the ratio of the mass transfer coefficient with reaction to that for pure diffusion, k_{gr}/k_g. For an equiaccessible surface this factor must be unity. The theoretical studies of Shirotsuka and Sano (1969) have shown that RF is a function of Da_n. For a second-order reaction, for instance, as shown in Figure 6.4a, RF approaches unity at $Da_2 > 10$. Thus, except at very low values of Da_2 (< 10), the equiaccessible approximation is valid for simple irreversible reactions. A similar plot for the first-order network $A \rightarrow B \rightarrow C$ appears in Figure 6.4b, which shows a much faster approach (from the low Da region) to the equiaccessible approximation. Carra (1972) has compared the two solutions by expressing them in series form and shown that the equiaccessibility approximation is valid to within 5 %. It may therefore be concluded that for most engineering calculations, the essentially crude approximation of an equiaccessible surface can be quite safely used. [A

similar paradoxical situation was noted earlier in an entirely different context: the modeling of catalytic reactions (Chapter 2), where the ideal surface approximation ws found to be surprisingly valid.]

6.4. COMBINED INFLUENCE OF EXTERNAL AND PORE DIFFUSION

We shall now examine the combined effect of external mass transfer (as discussed above) and intraphase diffusion (as presented in Chapters 4 and 5) for a simple reaction. Then, in Section 6.5, we shall extend the analysis to complex networks.

6.4.1. Isothermal Reactions

A simple method, applicable to first-order reactions, is to define an effective rate constant from the principle of the addivity of resistances:

$$\frac{1}{k_{\text{eff}}} = \frac{1}{\varepsilon k_v} + \frac{1}{k_{gv}} \qquad (6.26)$$

where k_{gv} is the mass transfer coefficient based on unit volume and has the dimension reciprocal seconds

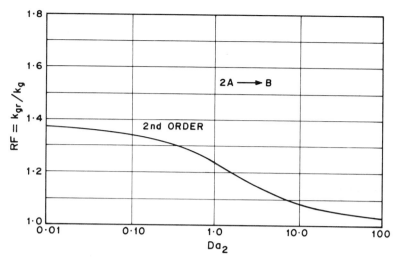

(a) CHEMICAL REACTION FACTOR (RF) FOR A SECOND-ORDER
IRREVERSIBLE REACTION

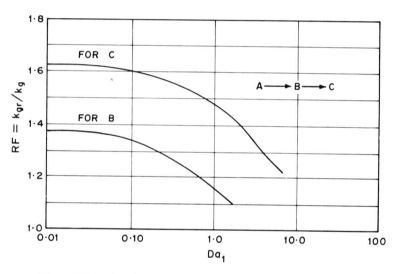

(b) CHEMICAL REACTION FACTOR (RF) FOR A CONSECUTIVE
FIRST-ORDER REACTION

Figure 6.4. Chemical reaction factor (RF) as a function of Damkohler number (Carra, 1972).

(sec^{-1}). It is related to the specific, or phenomenological, mass transfer coefficient by the equation $k_{gv} = k_g a = k_g/L$.

Equation 6.26 can be rearranged to give

$$\frac{k_{\text{eff}}}{k_v} = \frac{\varepsilon}{1 + \varepsilon \, \text{Da}_1} \qquad (6.27)$$

where Da_1 is given by k_v/k_{gv} or $k_v L/k_g$. From the practical point of view, Eq. 6.27 is particularly useful since several reactions of industrial importance are

known to be first order, or can be assumed to be first order without serious loss of accuracy.

A more rational method of determining the combined influence of pore and external diffusion is to redefine the boundary conditions used in Chapter 4 for reaction-cum-pore diffusion so as to include the effect of external mass transfer. Thus

$$\frac{D_{\text{eA}}}{L}\left(\frac{dC_{\text{A}}}{d\hat{L}}\right)_{\hat{L}=1} = k_g(C_{\text{Ab}} - C_{\text{A}}) \qquad (6:28)$$

This can be rearranged to give

$$C_A = C_{Ab} - \frac{1}{(Bi)_m} \left(\frac{dC_A}{d\hat{L}} \right)_{\hat{L}=1} \qquad (6.29)$$

where

$$(Bi)_m = \frac{k_g L}{D_{eA}} \qquad (6.30)$$

The boundary condition at $\hat{L} = 1$ will now be given by 6.29, while that at $\hat{L} = 0$ remains unchanged.

Integrating Eq. 4.1 (for $S = 0$) under these boundary conditions leads to the following equation for the effectiveness factor:

$$\varepsilon = \frac{\tanh \phi_{p1}}{\phi_{p1}(1 + \phi_{p1} \tanh \phi_{p1}/(Bi)_m)} \qquad (6.31)$$

A plot of ε as a function of the Thiele modulus based on Eq. 6.31 is shown in Figure 6.5 for various values of $(Bi)_m$ in the range $0 \leqslant (Bi)_m \leqslant \infty$. At $(Bi)_m = \infty$, the ε–ϕ curve coincides with that for pore diffusion alone. At $(Bi)_m \to 0$, external mass transfer is evidently the controlling influence.

Mehta and Aris (1971) have made a rigorous analysis of the diffusion–reaction problem, taking into account the resistance to external mass transfer. Their analysis is more general than that presented above and includes negative reaction orders in addition to first and higher orders. The asymptotic solution of Mehta and Aris for a slab is of the form

$$A = \frac{(1 - B)^{2/n+1}}{B} \qquad (6.32)$$

where

$$A = \frac{\phi_p}{(Bi)_m} \left(\frac{2}{n+1} \right)^{1/2}$$
$$B = \varepsilon \phi_p \left(\frac{n+1}{2} \right)^{1/2} \qquad (6.33)$$

Their plots (for $n > -1$) are reproduced in Figure 6.6. From a knowledge of the Thiele modulus and Biot number the group A can be calculated, and the effectiveness factor for a reaction of any order can then be extracted from the value of B read from Figure 6.6 for a flat plate.

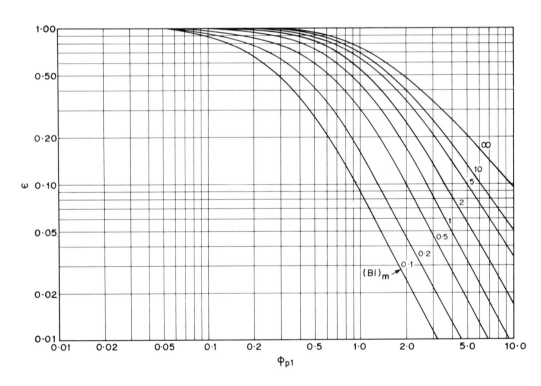

Figure 6.5. Effectiveness factor plot with the effect of interphase mass transfer included through the mass Biot number.

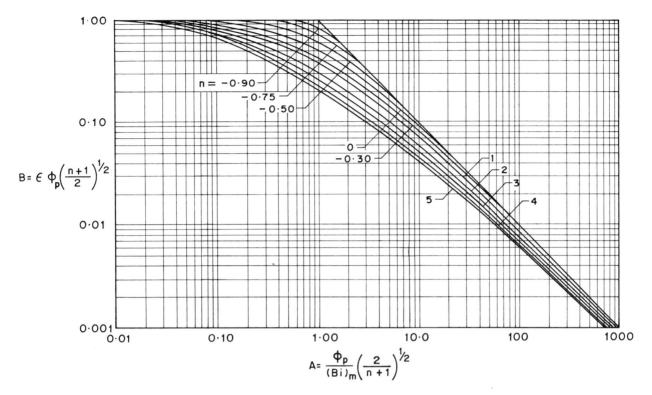

$$B = \epsilon \, \phi_p \left(\frac{n+1}{2} \right)^{1/2}$$

$$A = \frac{\phi_p}{(Bi)_m} \left(\frac{2}{n+1} \right)^{1/2}$$

Figure 6.6. Plot for the combined effect of interphase and intraphase mass transport for reactions of different orders (Mehta and Aris, 1971).

6.4.2. Nonisothermal Reactions: Relative Magnitudes of Interphase and Intraphase Effects; Maximum Temperature Rise

Before proceeding with the question of nonisothermal effectiveness factors we will examine the importance and relative magnitudes of the inter- and intraphase effects due to heat and mass transfer.

The ratio of mass to heat Biot numbers is generally higher than 50. Thus in most real systems heat transfer across the fluid film surrounding the catalyst pellet would appear to be the major resistance, whereas within the pellet mass diffusion is the limiting transport process. More experimental data are needed to uphold the generality of this observation, particularly in view of the contradictory findings that have been reported.

Several investigators (Lee and Luss, 1969; Carberry, 1975; Bischoff, 1976; Smith, 1977; Oh et al., 1978) have presented methods for estimating the upper bounds on inter- and intraparticle temperature differences based on directly observable quantities. The equations derived by Carberry (1975) are given in Table 6.2. Most of these methods are adequate as far as estimation of interphase gradients is concerned, but are rather conservative in the estimation of intraparticle gradients. The equation of Oh et al. (1978), derived for the more

general case of a catalyst of nonuniform activity, allows sharper estimation of intraparticle gradients and is given by

$$\frac{T - T_s}{T_b} = \frac{\beta_{mb}}{6} \frac{(1 - \hat{R}_i)^2 (1 + 2\hat{R}_i)}{(1 - \hat{R}_i)^3} \phi_s \qquad (6.34)$$

where β_{mb} (see Eq. 4.44) and ϕ_s are based on bulk fluid conditions and $\hat{R}_i$ represents the dimensionless extent of the boundary between the unimpregnated inactive core and the outer active shell. For a uniformly impregnated catalyst $\hat{R}_i = 0$ and Eq. 6.34 reduces to $(T - T_s)/T_b = \beta_{mb} \phi_s / 6$.

6.4.3. Nonisothermal Effectiveness Factors for Inter- and Intraphase Effects

Considering a spherical pellet and a simple reaction of the type A → B, the governing continuity and conservation equations derived earlier (Eqs. 4.4 and 4.5) may be recast as

$$\frac{d^2 \hat{C}_{Ab}}{d\hat{R}^2} + \frac{2}{\hat{R}} \left(\frac{d\hat{C}_{Ab}}{d\hat{R}} \right) - \phi_{s1}^2 \, \hat{r}_{vA} = 0$$

$$\frac{d^2 \hat{T}}{d\hat{R}^2} + \frac{2}{\hat{R}} \left(\frac{d\hat{T}}{d\hat{R}} \right) + \beta_m \phi_{s1}^2 \, \hat{r}_{vA} = 0 \qquad (6.35)$$

TABLE 6.2. Relative Importance of External–Internal Temperature Gradients in a Pellet

Equation	Reference	Remarks
$\dfrac{T_s - T_b}{T_{max} - T_b} = \dfrac{(\phi/3)(\text{Bi})_h}{1 + (\phi/3)[1/(\text{Bi})_h - 1/(\text{Bi})_m]}$	Lee and Luss (1969)	Requires knowledge of ϕ, $(\text{Bi})_h$, and $(\text{Bi})_m$.
where $$\phi = \frac{R^2 r_a}{D_e C_0}$$		
$\dfrac{T_s - T_b}{T_{max} - T_b} = \dfrac{[(\text{Bi})_m/(\text{Bi})_h]\varepsilon \, \text{Da}}{1 + \varepsilon \, \text{Da}[(\text{Bi})_m - (\text{Bi})_h]/(\text{Bi})_h}$	Carberry (1975)	$\varepsilon \, \text{Da}$ is an observable quantity.
where $$\varepsilon \, \text{Da} = \frac{r_a}{k_g a C_0} = \varepsilon \left(\frac{k_v C_0^{n-1}}{k_g a} \right)$$		

The reaction rate r_{vA} is now a function of both concentration and temperature. The boundary conditions are

$$\hat{R} = 0, \qquad \frac{d\hat{C}_{Ab}}{d\hat{R}} = 0, \qquad \frac{d\hat{T}}{d\hat{R}} = 0 \qquad (6.36)$$

and

$$\hat{R} = 1, \qquad \frac{d\hat{C}_{Ab}}{d\hat{R}} = (\text{Bi})_m(1 - \hat{C}_{Ab}), \qquad \frac{d\hat{T}}{d\hat{R}} = (\text{Bi})_h(1 - \hat{T}) \qquad (6.37)$$

The Biot numbers are based on pellet diameter; that is $(\text{Bi})_m = k_g d_p/2D_{eA}$ and $(\text{Bi})_h = h_{pf} d_p/2k'_e$. Effectiveness factor plots can be prepared by solving Eq. 6.35 by iteration and employing the familiar relationship $\varepsilon = (3/\phi_{s1}^2)(d\hat{C}_A/d\hat{R})_{\hat{R}=1}$.

Clearly, the catalyst effectiveness can increase severalfold under conditions corresponding to interphase heat transfer limitation at a relatively high thermicity. The effect is considerably less marked at lower thermicity. This confirms (theoretically) the observation that at low finite values of $(\text{Bi})_h$, as exemplified by several real systems, external heat transfer is a major controlling factor.

All situations involving an interphase effect, in addition to the intraphase effect usually associated with the ε–ϕ charts, can be represented by a single curve of the form shown in Figure 6.7. In region 1, $\phi < 1$ and chemical reaction controls; in region 2, mass transfer within the pellet is dominant; and in region 3, heat transfer across the film controls. The last situation results from the inability of the heat of reaction to be dissipated sufficiently rapidly from the pellet surface. Therefore, the surface layers of the catalyst experience very rapid reaction rates, leading to reactant starvation. Inside the pellet the reaction rate is even more rapid, so that the reactant concentration quickly falls to a negligible value. If it is assumed that the concentration gradient across the film is at its maximum, that is, that film transfer controls, then boundary condition 6.37 becomes $(d\hat{C}_{Ab}/d\hat{R})_{\hat{R}=1} = (\text{Bi})_m$, and we have

$$\varepsilon = \frac{3(\text{Bi})_m}{\phi_{s1}^2} \qquad (6.38)$$

This equation for the effectiveness factor corresponds to region 3 of Figure 6.7, a range of practical importance, and can be used for the rapid estimation of the effectiveness factor for strongly exothermic reactions. It can be generalized for different shapes by using the definition $X'_0 (= R/3)$ for the characteristic length instead of R.

6.4.4. Effect of Adsorption

Employing the H–W model for the same reaction $A \rightarrow B$ (with both A and B adsorbed), Hutchings and Carberry (1966) have solved the continuity and conservation equations (Eq. 6.35) under boundary conditions 6.36 and 6.37. Solutions have been provided for $(\text{Bi})_m = (\text{Bi})_h = \infty$; $(\text{Bi})_m$ finite, $(\text{Bi})_h = \infty$; and $(\text{Bi})_m = \infty$, $(\text{Bi})_h$ finite. These situations correspond, respectively, to no interphase resistance, intraphase mass transfer resistance, and interphase heat transfer resist-

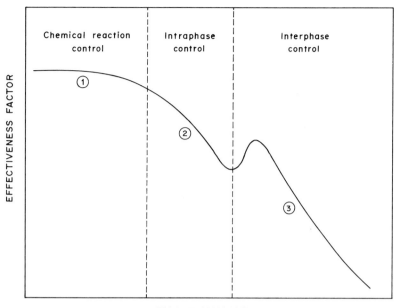

Figure 6.7. A generalized representation of the interphase and intraphase effects.

ance. An inspection of the plots for these three situations shown in Figure 6.8 reveals:

1. When $(Bi)_m = \infty$ and $(Bi)_h = 1$, the effectiveness factor increases by approximately a factor of 30 over the value corresponding to $(Bi)_m = (Bi)_h = \infty$, suggesting that limitation due to interphase heat transport enhances catalyst effectiveness.

2. When $(Bi)_m = 1$ and $(Bi)_h = \infty$, the effectiveness factor is reduced five fold relative to the curve for $(Bi)_m = (Bi)_h = \infty$, suggesting that limitation due to intraphase mass transport leads to a reduction of catalyst effectiveness.

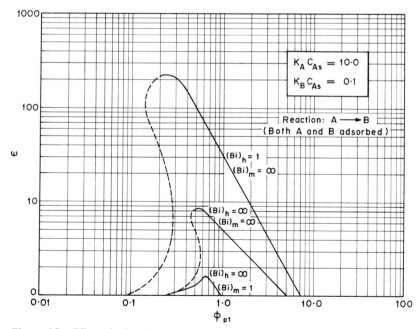

Figure 6.8. Effect of adsorption on interphase and intraphase heat and mass transport (Hutchings and Carberry, 1966).

6.4.5. Effect of Nonuniform External Diffusion

An attempt will be made here to combine the internal and external diffusion problems without uncoupling the diffusion and reaction terms in the latter. Consider a porous flat plate immersed in a fluid, with a boundary layer developed over it. For a simple reaction A → B, the continuity equation for the species A is given by

$$D_{eA} \frac{d^2 C_A(l, x_l)}{dl^2} = k_v C_A \qquad (6.39)$$

where $C_A(l, x_l)$ is the concentration at distance x_l from the leading edge of the plate and depth l. The solution to this equation with the boundary conditions

$$
\begin{aligned}
l = 0, \qquad & C_A = C_A(x_l) \\
l = L, \qquad & C_A \to 0, \qquad \frac{dC_A}{dl} = 0
\end{aligned}
\qquad (6.40)
$$

has been given by Mihail (1970) as

$$C_A(l, x_l) = C_{As}(x_l) e^{-\phi_{p,l}} \qquad (6.41)$$

where $\phi_{p,l}$ is the local Thiele modulus at depth l (for plate geometry) defined as $[l(k_v/D_{eA})^{1/2}]$, and $C_{As}(x_l)$ is the local surface concentration on the plate (at distance x_l). Equation. 6.41 relates this local surface concentration of A to the concentration at any penetration l, and can be solved for the general case of boundary-layer-cum-internal-diffusion-influenced reaction, provided that the local surface concentration $C_{As}(x_l)$ can be obtained from a solution of the boundary layer equation. Thus either Eq. 6.39, developed earlier, can be used to express $C_{As}(x_l)$, or the more general equation for a reaction of general order and arbitrary catalyst shape developed by Mihail (1970) can be used.

From the development presented in Section 6.3 it can be inferred that the local Damköhler number is a measure of the axial distance (on the plate); and since the local Thiele modulus is a measure of the depth of penetration, Eq. 6.41 can be recast in terms of normalized parameters to give

$$\hat{C}_A(\mathrm{Da}_l, \phi_{p,l}) = \hat{C}_{Asb} e^{-\phi_{p,l}} \qquad (6.42)$$

where $\hat{C}_{Asb} = C_{As}/C_{Ab}$. A combined local effectiveness

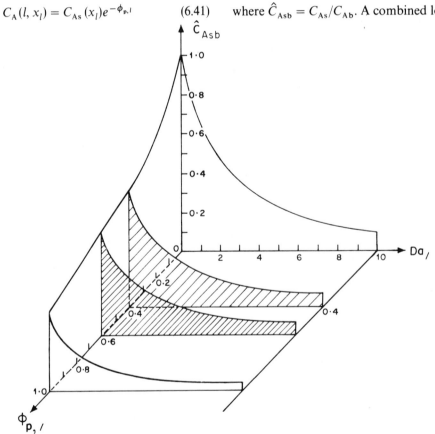

Figure 6.9. Graphical representation of the combined effect of pore diffusion and nonuniform external mass transfer (Mihail, 1970).

factor (ε'_c) for external and internal mass transfer can now be defined, by analogy with the usual definition of the effectiveness factor, as

$$\varepsilon'_c = \frac{r_S[C_A(l, x_l)]}{r_S(C_{Ab})} \quad (6.43)$$

For a first-order irreversible reaction, this becomes

$$\varepsilon'_c = \hat{C}_A(l, x_l) = \hat{C}_A(\text{Da}_l, \phi_{p,l})$$

The integral combined effectiveness factor (ε_c) can be obtained from

$$\varepsilon_c = \frac{1}{A_p} \int_0^{\text{Da}_l} \int_0^{\phi_{p,l}} \hat{C}_A(\text{Da}_l, \phi_{p,l}) \, d\text{Da}_l \, d\phi_{p,l} \quad (6.44)$$

where A_p is the surface area of the catalyst. Equation 6.44 can now be solved in terms of the external mass transfer parameter (Da_l), internal diffusion parameter ($\phi_{p,l}$), and the dimensionless concentration ($\hat{C}_{Asb}$), which comes into the picture through Eq. 6.42. A typical plot is shown in Figure 6.9. Although this plot is not of any direct use in reactor design calculations, it helps us to understand better the combined role of inter- and intraphase transport processes.

6.5. COMBINED EFFECT OF INTRAPHASE AND INTERPHASE DIFFUSION ON YIELD IN COMPLEX REACTIONS

In Chapter 5 we examined the role of intraphase diffusion on selectivity in a complex reaction. In this section we shall extend the treatment to include the effect of external mass transfer.

In the absence of external mass transfer influence, the selectivity in a consecutive reaction of the type $A \rightarrow B \rightarrow C$ is given by Eq. 5.13, while in the presence of external mass transfer it is given by (Carberry, 1962)

$$Y_a = -\alpha_D \frac{m_1(\phi)_2 \tanh(\phi)_2}{m_2(\phi)_1 \tanh(\phi)_1} \left(\frac{C_{Bs}}{C_{As}} + \frac{s}{s\alpha_D - 1} \right) + \alpha_D \frac{s}{s\alpha_D - 1} \quad (6.45)$$

where

$$m_1 = \frac{1 + (\phi)_1 \tanh(\phi)_1}{(\text{Bi})_m}$$

$$m_2 = \frac{1 + (\phi)_2 \tanh(\phi)_2}{(\text{Bi})_m}$$

The effect of external surface on selectivity in the presence of internal and external diffusion has been examined by Goldstein and Carberry (1973) and found to be negligible in the practical ranges of Thiele modulus encountered.

6.6. EFFECT OF SURFACE HETEROGENEITY

In the foregoing analysis of the role of external mass transfer, either by itself or in combination with internal diffusion, the assumption has been made that the surface is *uniformly active*. It is known, however, that activity exists in patches.

Loffler and Schmidt (1975) have postulated a model in which the catalyst surface is made up of active stripes of width $2l_1$ separated by a distance $2L_1$ (Figure 6.10) on an otherwise inert surface. A fluid boundary layer of uniform thickness δ is formed on the surface through which the reactant has to diffuse to react on the surface. Based on this model the following mass balance equation can be written:

$$\frac{\partial^2 C_A}{\partial x^2} + \frac{\partial^2 C_A}{\partial y^2} = 0 \quad (6.46)$$

with the boundary conditions

$$\frac{\partial C_A}{\partial x} = 0, \qquad x = 0, \qquad x = L_1 \quad (6.47a)$$

$$\frac{\partial C_A}{\partial y} = 0, \qquad y = 0, \qquad l_1 < x < L_1 \quad (6.47b)$$

$$\frac{\partial C_A}{\partial y} = r_S = k_S C_A, \qquad y = 0, \qquad x < l_1 \quad (6.47c_1)$$

$$C_A = 0 \quad \text{(for mass transfer control)} \quad (6.47c_2)$$

$$C_A = C_{Ab}, \quad y = \delta, \quad \text{all } x \quad (6.47d)$$

If boundary condition 6.47c₂ were used (instead of 6.47c₁) along with the other boundary conditions, the situation would correspond to a mass-transfer-controlled reaction, whereas with boundary condition 6.47c₁ it would correspond to a finite reaction.

The chief conclusions for a mass-transfer-controlled reation are (1) when $L_1/\delta \ll 1, \eta$ approaches unity; in other words, the entire surface appears to be active even if only a small fraction ($l_1/L_1 \approx 0.1$) is actually active; and (2) as L_1/δ ap-

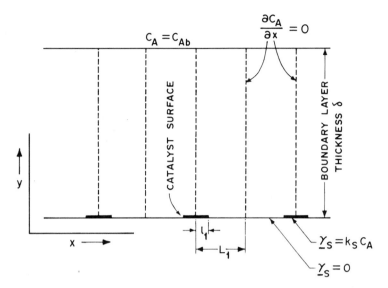

Figure 6.10. Active-stripes model of catalyst surface.

proaches infinity, $\eta = l_1/L_1$; in other words, for a very thin boundary layer, the effectiveness factor equals the fraction of the active surface.

For the case of a finite reaction the conclusions are (1) at low values of Da corresponding to kinetic control, η approaches l_1/L_1—the active patch fraction on the surface; and (2) at high values of Da, say, > 20, the reaction passes into the mass transfer regime.

Loffler and Schmidt have also extended the model to a consecutive reaction scheme and shown that surface inhomogeneity increases the selectivity of the intermediate. This can be understood from the fact that the intermediate can now escape both by vertical movement into the gas phase as well as by lateral movement into the inactive region. On a homogeneous surface, on the other hand, only the former is possible.

6.7. ROLE OF EXTERNAL MASS TRANSFER IN INDUCING MULTIPLE BEHAVIOR

In the previous sections we considered the role of external mass and/or heat transfer in both the absence and presence of intraparticle gradients under isothermal and nonisothermal conditions. In addition to simple reactions, the case of complex reaction where the reaction rate can be a nonlinear function of reactant concentrations has also been considered. In situations of this type under isothermal conditions and in general under semi-isothermal conditions one can expect to get multiplicity of behavior for the same inlet conditions. In the present section we shall briefly comment on the role of external mass and/or heat transfer on the

phenomenon. The role of external mass transfer in inducing multiple behavior of a catalyst pellet can be brought out more clearly by considering a nonporous particle. For this case the governing equations become analogous to the corresponding case of a CSTR with the difference that the flow term (say W/F) in the CSTR now gets replaced by $W/k_g a$ for the particle. This analogy helps us ascertain the role of mass transfer in catalyst particles from the corresponding studies in CSTRs. All these studies relating to CSTRs are considered in greater detail in Chapter 12.

As regards the influence of external mass transfer in the presence of intraparticle gradients, no such generalization seems possible and each case will have to be treated separately. As a general observation it can be stated that the external mass transfer limitations add to the existing number of states. Thus Pereira and Varma (1978) have considered the case of a catalyst particle with substrate-inhibited kinetics. In the range of parameter values where only three solutions are possible, the incorporation of external mass transfer resistance increases the possible number of states to five.

6.8. EXAMPLES OF EXTERNAL MASS TRANSFER CONTROLLED REACTIONS

6.8.1. Decomposition of Hydrogen Peroxide

Satterfield and Resnick (1954) studied the decomposition of H_2O_2 on solid silver spheres in the Reynolds number range 15–61. This reaction has all the desirable characteristics necessary for external mass transfer control. The internal diffusional resistance is eliminated because of the nonporous nature of the catalyst used, and rate data can be correlated through the j_d factor. In fact, Satterfield and Resnick (1954) used this reaction to propose a correlation for j_d.

6.8.2. Dehydrogenation of Cyclohexane

Graham et al. (1968) studied this reaction using platinum supported on alumina as catalyst in the temperature range 470–500°C, pressure range 21–42 atm, and Reynolds number range 20–65. A model based on complete external transport control of the overall reaction has been proposed, with equilibrium of the fluid at the catalyst surface. A very low effectiveness factor is implicit in the equilibrium assumption, the calculated values ranging from 0.025 to 0.068.

6.8.3. Hydrogenation of Phenol

Using platinum on silica gel as catalyst at a temperature of 150°C, Hancil and Mitschka (1969) calculated the effectiveness factors for this reaction at low Reynolds numbers (below 0.1). They also computed the corresponding values of the mass Biot number. They observed that under certain conditions the effectiveness factor is greater than unity. Since the exothermic heat of reaction is relatively high (30 kcal/mol), and this is not transported fast enough to the fluid bulk, there is an overheating of the catalyst surface. Thus, as explained in Section 6.3.5, the effectiveness factor (referred to bulk conditions) can be greater than unity.

6.8.4. Oxidation of Ammonia

Pignet and Schmidt (1974) have shown that the oxidation of ammonia on a Pt–Rh catalyst corresponds to a series–parallel network, which in excess oxygen reduces to the simpler series scheme

$$NH_3 \rightarrow NO \rightarrow N_2$$

Industrially this reaction is carried out by passing 10% ammonia in air over a thin layer of Pt–Rh gauze around 900°C in a conversion range of 95–99%.

The reaction is very fast requiring contact times of the order of only 1 msec to accomplish over 95% conversion; and considering that the catalyst is used in the form of a thin gauze, external mass transfer effects would be expected to be predominant. This should lead to a lowering of the selectivity, while actually there is no such effect observed in practice.

This apparently paradoxical behavior has been explained by Loffler and Schmidt (1975) on the basis of surface nonuniformity explained in Section 6.6. From the data of Nowak (1966), Pignet and Schmidt (1974), and McCabe et al. (1974), they estimated the following parameter values: $L_1/\delta \simeq 0.06$, $l_1/L \simeq 0.3$, $Da_1 \simeq 2.4$ and $Da_2 \simeq 0.8$. For these values the selectivity is over 90%. Thus, abnormally low selectivities will be predicted by assuming a homogeneous surface while in reality the surface happens to be nonhomogeneous.

6.8.5. Oxidation of CO in Automotive Converters

This reaction represents another instance of the role of surface nonuniformity in determining the effect of external mass transfer. CO from automotive exhaust is oxidized in excess air on platinum supported on alumina. As the reaction is very fast, of the order of a fraction of a second, external mass transfer controls the overall reaction. From the kinetic data of Hori and Schmidt (1976) and the conditions normally obtained in a converter, Loffler and Schmidt (1975) estimate the following parameter values: $Da = 25$, $l_1/\delta \simeq 0.02$ and $l_1/L_1 \simeq 0.05$. Under these conditions the effectiveness factor shows a lower value as against that under conditions of surface uniformity. Evidently localized mass transfer resistances at the clusters of platinum particles on the surface are responsible for this situation.

REFERENCES

Acrivos, A., and Chambré, P. L. (1957), *Ind. Eng. Chem.*, **49**, 1025.

Balakrishnan, A. R., and Pei, D. C. T. (1974), *Ind. Eng. Chem. Process Design Dev.*, **13**, 441.

Balakrishnan, A. R., and Pei, D. C. T. (1979), *Ind. Eng. Chem. Process Design Dev.*, **18**, 30.

Barker, J. J. (1965), *Ind. Eng. Chem.*, **57**, 33.

Bhattacharyya, D., and Pei, D. C. T. (1975), *Chem. Eng. Sci.*, **30**, 293.

Bischoff, K. B. (1976), *Ind. Eng. Chem. Fundam.*, **15**, 229.

Carberry, J. J. (1960), *AIChE J.*, **6**, 460.

Carberry, J. J. (1962), *Chem. Eng. Sci.*, **17**, 675.

Carberry, J. J. (1975), *Ind. Eng. Chem. Fundam.*, **14**, 129.

Carra, S. (1972), *Chim. Ind.(Milan)*, **54**, 434.

Chambers, R. P., and Boudart, M. (1966), *J. Catal.*, **6**, 141.

Chambré, P. L. (1956), *Appl. Sci. Res.*, **A6**, 97.

Chambré, P. L., and Acrivos, A. (1956), *J. Appl. Phys.*, **27**, 1322.

Chung, P. M. (1965), *Adv. Heat Transfer*, **2**, 109.

Chung, P. M., Liu, S. W., and Mirels, H. (1963), *Int. J. Heat and Mass Transfer*, **6**, 193.

Colquhoun-Lee, I., and Stepanek, J. (1974), *Chem. Eng. (London)*, **282**, 108.

Cybulski, A., van Dalen, M. J., and van den Berg, P. J. (1975), *Chem. Eng. Sci.*, **30**, 1015.

De Acetis, J., and Thodos, G. (1960), *Ind. Eng. Chem.*, **52**, 1003.

Denton, W. H. (1951), *Proc. General Discussion of Heat Transfer, Inst. Mech. Eng. (London) and ASME.*, p. 370.

Eckert, E. R. G., and Drake, R. M. Jr. (1972), *Analysis of Heat and Mass Transfer*, McGraw-Hill, New York.

El-Kaissy, M. M., and Homsy, G. M. (1973), *Ind. Eng. Chem. Fundam.*, **12**, 82.

Evans, E. V., and Kenney, C. N. (1966), *Trans. Inst. Chem. Eng. (London)*, **44**, T 189.

Fournier, R. L., DeWitt, K. J., and Jeng, D. R. (1981), *AIChE Symp. Ser.*, **77(202)**, 94.

Fox, H., and Libby, P. A. (1966), *Phys. Fluids*, **9**, 33.

Frank-Kamenetskii, D. A. (1955), *Diffusion and Heat Exchange in Chemical Kinetics*, Princeton Univ. Press, Princeton, New Jersey.

Frantz, J. F. (1962), *Chem. Eng.*, **69** (19), 161.

Frossling, N. (1938), *Gerlands Beitr. Geophys.*, **52**, 170.

Galloway, T. R., and Sage, B. H. (1970), *Chem. Eng. Sci.*, **25**, 495.

Gamson, B. W., Thodos, G., and Hougen, O. A. (1943), *Trans. AIChE J.*, **39**, 1.

Gangwal, S. K., Hudgins, R. R., and Silveston, P. L. (1977), "Limiting Sherwood Number for Mass Transfer in Packed Beds," private communication.

Gliddon, B. J., and Cranfield, R. R. (1970), *Brit. Chem. Eng.*, **15**, 481.

Goldstein, W., and Carberry, J. J. (1973), *J. Catal.*, **28**, 33.

Graham, R. R., Vidaurri Jr., F. C., and Gully, A. J. (1968), *AIChE J.*, **14**, 473.

Gunn, D. J. (1968), *Chem. Eng. (London)*, **46**, CE 153.

Gunn, D. J., and de Souza, J. F. C. (1974), *Chem. Eng. Sci.*, **29**, 1363.

Gupta, S. N., Chaube, R. B., and Upadhyay, S. N. (1974), *Chem. Eng. Sci.*, **29**, 839.

Hancil, V., and Mitschka, P. (1969), *Chem. Eng. Sci.*, **24**, 1400.

Handley, D., and Heggs, P. J. (1969), *Int. J. Heat and Mass Transfer*, **12**, 549.

Happel, J. (1958), *AIChE J.*, **4**, 197.

Hori, G. K., and Schmidt, L. D. (1975), *J. Catal.*, **38**, 335.

Hutchings, J., and Carberry, J. J. (1966), *AIChE J.*, **12**, 20.

Ikeda, K., Ohya, H., Kanemitsu, O., and Shimomura, K. (1973), *Chem. Eng. Sci.*, **28**, 227.

Jeng, D. R., DeWitt, K. J., and Lee, M. H. (1979), *Int. J. Heat and Mass Transfer*, **22**, 89.

Kato, K., Kubota, H., and Wen, C. Y. (1970), *Chem. Eng. Prog. Symp. Ser.* (No. 105), 87.

Kunii, D., and Suzuki, M. (1967), *Int. J. Heat and Mass Transfer*, **10**, 845.

Kusik, C. L., and Happel, J. (1962), *Ind. Eng. Chem. Fundam.*, **1**, 163.

Lee, J. C. M., and Luss, D. (1969), *Ind. Eng. Chem. Fundam.*, **8**, 596.

Lighthill, M. J. (1950), *Proc. Roy. Soc. (London).*, **A202**, 359.

Lindberg, R. C., and Schmitz, R. A. (1969), *Chem. Eng. Sci.*, **24**, 1113.

Lindberg, R. C., and Schmitz, R. A. (1970), *Chem. Eng. Sci.*, **25**, 901.

Littman, H., Barile, R. G., and Pulsifer, A. H. (1968), *Ind. Eng. Chem. Fundam.*, **7**, 554.

Loffler, D. G., and Schmidt, L. D. (1975), *AIChE J.*, **21**, 787.

McAdams, W. H. (1951), *Heat Transmission*, McGraw-Hill, London.

McCabe, R. W., Pignet, T. P., and Schmidt, L. D. (1975), *AIChE J.*, **21**, 787.

McHenry, K. W., Jr., and Wilhelm, R. H. (1957), *AIChE J.*, **3**, 83.

Mehta, B. N., and Aris, R. (1971), *Chem. Eng. Sci.*, **26**, 1699.

Mihail, R. (1969), *Int. Chem. Eng.*, **9**, 648.

Mihail, R. (1970), *Chem. Eng. Sci.*, **25**, 461.

Mihail, R. (1972), *Chem. Eng. Sci.*, **27**, 845.

Mihail, R., and Teodorescu, C. (1975), *Chem. Eng. Sci.*, **30**, 993.

Miyauchi, T. (1972), *J. Chem. Eng. Japan*, **5**, 303.

Miyauchi, T., Kikuchi, T., and Kataoka, H. (1972). *Int. Chem. Eng.*, **12**, 373.

Miyauchi, T., Matsumoto, K., and Yoshida, T. (1975), *J. Chem. Eng. Japan*, **8**, 228.

Nelson, P. A., and Galloway, T. R. (1975), *Chem. Eng. Sci.*, **30**, 1.

Nowak, E. J. (1966), *Chem. Eng. Sci.*, **21**, 19.

Oh, S. H., Hegedus, L. L., and Aris, R. (1978), *Ind. Eng. Chem. Fundam.*, **17**, 309.

Padberg, G., and Smith, J. M. (1968), *J. Catal.*, **12**, 172.

Pereira, C. J., and Varma, A. (1978), *Chem. Eng. Sci.*, **33**, 1645.

Pfeffer, R. (1964), *Ind. Eng. Chem. Fundam.*, **3**, 380.

Pignet, T. P., and Schmidt, L. D. (1974), *Chem. Eng. Sci.*, **29**, 1123.

Ranz, W. E. (1952), *Chem. Eng. Prog.*, **48**, 247.

Reid, R. C., Prausnitz, J. N., and Sherwood, T. K. (1977), *Properties of Gases and Liquids.* McGraw-Hill, New York.

Rhodes, J. M., and Peebles, F. N. (1965), *AIChE J.*, **11**, 481.

Rosner, D. E. (1963), *AIChE J.*, **9**, 321.

Rosner, D. E. (1964a), *Chem. Eng. Sci.*, **19**, 1.

Rosner, D. E. (1964b), *AIAA*, **2**, 593.

Satterfield, C. N. (1970), *Mass Transfer in Heterogenous Catalysis*, M.I.T. Press, Cambridge, Massachusetts.

Satterfield, C. N., and Resnick, H. (1954), *Chem. Eng. Prog.*, **50(10)**, 504.

Schlichting, H. (1955), *Boundary Layer Theory*, Pergamon Press, London.

Sherwood, T. K., and Pigford, R. L. (1952), *Absorption and Extraction*, McGraw-Hill, New York.

Shirotsuka, T., and Sano, M. (1969), *Int. Chem. Eng.*, **9**, 155.

Simpkins, G. P. (1965), *Int. J. Heat and Mass Transfer*, **8**, 99.

Smith, T. G. (1977), *Chem. Eng. Sci.*, **32**, 334.

Sorensen, J. P., and Stewart, W. E. (1974), *Chem. Eng. Sci.*, **29**, 827.

Taecker, R. G., and Hougen, O. A. (1949), *Chem. Eng. Prog.*, **45(3)**, 188.

Thoenes, D., Jr., and Kramers, H. (1958), *Chem. Eng. Sci.*, **8**, 271.

Tribus, M., and Klein, J. (1953), *Heat Transfer—A Symposium*, Univ. of Michigan Press, Ann Arbor, Michigan.

Upadhyay, S. N., and Tripathi, G. (1975), *J. Chem. Eng. Data*, **20**, 20.

Wadsworth, J. (1961), *Proc. ASME Heat Transfer Conf.*, Univ. of Colorado.

Wakao, N. (1976), *Chem. Eng. Sci.*, **31**, 1115.

Wakao, N., and Funazkri, T. (1978), *Chem. Eng. Sci.*, **33**, 1375.

Wakao, N., and Tanisho, S. (1974), *Chem. Eng. Sci.*, **29**, 1991.

Wen, C. Y., and Fan, L. T. (1975), *Models for Flow Systems and Chemical Reactors*, (Marcel Dekker, New York.)

Wilke, C. R., and Hougen, O. A. (1945), *Trans. Am. Inst. Chem. Eng. (London)*, **41**, 445.

CHAPTER SEVEN

Role of Transport Limitations in Determining the Controlling Regime

In the presence of finite transport limitations, the catalyst exhibits kinetics falsified by transport processes. This has been referred to as "disguised kinetics" by Wei (1966). Carberry (1969) has examined the implications of this disguise in determining the operating regime of a process, and Rajadhyaksha and Doraiswamy (1976) have presented a detailed analysis of the falsification of kinetics by transport limitations and the role of this falsification in efforts to discern the controlling regime of a reaction.

In this chapter the falsification of kinetic parameters (chiefly activation energy) by various transport limitations, individually or simultaneously, will be considered. The influence of certain other factors, such as adsorption, surface diffusion, catalyst poisoning, and pore structure, will also be examined. Since several industrially important systems involve more than one chemical reaction, falsification of kinetic parameters in complex systems will be discussed. A number of conclusions reached by this analysis have been verified experimentally, and many such instances reported in the literature will be presented. Much of the material described in this chapter is drawn from the review of Rajadhyaksha and Doraiswamy (1976).

7.1. MEASUREMENT OF ACTIVATION ENERGY

7.1.1. True Activation Energy

The expression for the rate constant based on the absolute rate theory is[†]

$$\ln k = \ln\left(\frac{k_b}{h}\right) + \ln T + \frac{\Delta S}{R_g} - \frac{\Delta H}{R_g T} \qquad (7.1)$$

A plot of $\ln k$ vs. $1/T$ should therefore give a curve

[†] The general notation for the rate constant k is used instead of such specific notations as k_v, k_w.

(since neither ΔH nor ΔS is strictly independent of temperature, and of course $\ln T$ also varies with temperature). In most cases, however, the change in $\ln T$ is not significant, and ΔH and ΔS are also reasonably independent of temperature, so that Eq. 7.1 reduces to the familiar Arrhenius form. Arising out of this approximation is the question; What is the difference between the true and apparent Arrhenius values that can occur without a curvature being detected?

Baag (1969) has proposed a method for determining the difference between the true values and the apparent values estimated by applying the least-squares method to the Arrhenius equation. This method is based on resolving the difference into two contributions: one due to a systematic deviation as a result of curvature, and the other due to random experimental error:

$$\Delta E = \Delta E_c \pm \Delta E_e$$
$$\Delta \ln A_f = \Delta \ln A_{fc} \pm \Delta \ln A_{fe}, \qquad (7.2)$$

where the subscripts c and e denote curvature and random error, respectively. On the assumption that the random error is normally distributed about a mean of zero with a variance of σ^2 (in $\ln k$) and employing standard statistical methods (see, e.g., Acton, 1959), Baag (1969) has derived expressions for ΔE and $\Delta \ln A_f$.

Table 7.1 has been prepared on the basis of the conclusions drawn from these equations; this table makes it possible to determine ΔE and $\Delta \log A_f$ at several temperature intervals with minimal input data. Baag (1970) has also reported methods for calculating the activation energies on catalysts with heterogeneous surfaces, but the procedure given above is adequate for most situations encountered in practice.

7.1.2. Point Values of Activation Energy

We now outline a general procedure for obtaining the activation energy of a reaction from raw kinetic data,

TABLE 7.1. Curvature and Random Errors in Arrhenius Parameters[a]

Temperature Interval °K	ΔE_c	$\Delta \log A_{fc}$	No. of Measurements	ΔE_e			$\Delta \log A_{fe}$		
				$\sigma = 0.005$	$\sigma = 0.025$	$\sigma = 0.050$	$\sigma = 0.005$	$\sigma = 0.025$	$\sigma = 0.050$
323–423	0.725	3.00	4	0.080	0.398	0.796	0.024	0.117	0.234
			6	0.073	0.366	0.733	0.018	0.089	0.117
			11	0.059	0.293	0.586	0.016	0.052	0.104
523–623	1.128	3.19	4	0.019	0.950	1.900	0.036	0.181	0.362
			6	0.175	0.873	1.747	0.027	0.145	0.272
			11	0.140	0.697	1.395	0.016	0.080	0.160
723–823	1.528	3.32	4	0.347	1.735	3.471	0.049	0.245	0.489
			6	0.319	1.595	3.190	0.037	0.184	0.370
			11	0.255	1.273	2.546	0.022	0.108	0.217
923–1023	1.928	3.42	4	0.551	2.754	5.509	0.062	0.256	0.617
			6	0.506	2.531	5.062	0.047	0.231	0.505
			11	0.404	2.020	4.039	0.027	0.126	0.273

[a]Curvature error is denoted by subscript c and random error by subscript e.

after which we present a brief description of a simple statistical procedure for determining the activation energy as a function of temperature.

Based on Raw Experimental Data

Plots of $\ln t$ vs. $1/T$ or τ' are prepared for different values of x from experimental plots of x as a function of time. These plots, referred to as isoconversion plots, fall into four categories, as shown in Table 7.2.

Let m kinetic isotherms, corresponding to the reciprocals of the absolute temperatures $\tau'_1, \tau'_2, \ldots, \tau'_m$ be plotted, and n iso-x sections of these isotherms be made at $x = x_1, x_2, \ldots, x_n$. If transformation x_j on the isotherm τ'_j is represented by the matrix $A_{i,j}$, then

$$A_{i,j} = \ln t_{i,j} \qquad (7.3)$$

consisting of n rows (representing x) and m columns (representing τ'). It has been shown by Bliznakov et al. (1970) that another matrix $A'_{i,j}$ can be constructed that can be used as the basis for estimating the point values of E/R_g:

$$A'_{i,j} = \frac{A_{i,j} - A_{i,j-1}}{\tau'_j - \tau'_{j-1}} \qquad (7.4)$$

Based on the values of these elements, the four

categories of the $\ln t$–τ' plots may be interpreted as shown in Table 7.2.

All that is needed to use this method is a tabulation of the results (reaction times or W/F) in rows of x and columns of τ', which would represent the matrix A', as shown in Table 7.3.

If the rate constant is available as a function of temperature, the problem is to determine the activation energy at different temperature levels. For this purpose the Arrhenius equation can be conveniently reparametrized as

$$k = A_f \exp(-E/R_g T') \qquad (7.5)$$

where

$$\frac{1}{T'} = \frac{1}{T} - \frac{1}{\overline{T}} \qquad (7.6)$$

In Eq. 7.6 $1/\overline{T}$ is the mean reciprocal temperature, defined as

$$\frac{1}{\overline{T}} = \frac{1}{N} \sum_{i=1}^{N} \left(\frac{1}{T_i}\right) \qquad (7.7)$$

where N is the total number of temperature points T_i. Thus E can be determined at any temperature $1/T$ in the range represented by an average value $1/\overline{T}$.

TABLE 7.2 Determination of Activation Energy from Raw Kinetic Data

Nature of plots	Matrix Representation	Remarks

Parallel lines

$i \backslash j$	2	3	4	...	m
1	A	A	A	...	A
2	A	A	A	...	A
$\vdots$	...	...	...	$\vdots$	A
n	A	A	A	...	A

Elements constant over both rows and columns

Nonparallel lines

$i \backslash j$	2	3	4	...	m
1	A_1	A_1	A_1	...	A_1
2	A_2	A_2	A_2	...	A_2
$\vdots$	...	...	...	$\vdots$	$\vdots$
n	A_n	A_n	A_n	...	A_n

Elements constant over rows but not over columns

Parallel curves

$i \backslash j$	2	3	4	...	m
1	A_2	A_3	A_4	...	A_m
2	A_2	A_3	A_4	...	A_m
$\vdots$	...	...	...	$\vdots$	$\vdots$
n	A_2	A_3	A_4	...	A_m

Elements constant over columns but not over rows

Nonparallel curves

$i \backslash j$	2	3	4	...	m
1	A_{12}	A_{13}	A_{14}	...	A_{1m}
2	A_{22}	A_{23}	A_{24}	...	A_{2m}
$\vdots$	...	...	...	$\vdots$	$\vdots$
n	A_{n2}	A_{n3}	A_{n4}	...	A_{nm}

Elements not constant over either rows or columns

$\tau'\left(=\dfrac{1}{T}\right) \rightarrow$

TABLE 7.3. Matrix A'

$x \backslash \tau'$	τ'_1	τ'_2	τ'_3	τ'_4
	Reaction time			
x_1	A'_{11}	A'_{12}	A'_{13}	A'_{14}
x_2	A'_{21}	A'_{22}	A'_{23}	A'_{24}
x_3	A'_{31}	A'_{32}	A'_{33}	A'_{34}

7.2. ROLE OF INTRAPARTICLE TRANSPORT PROCESSES

7.2.1. Isothermal Monodispersed Systems

The activation energy of a chemical reaction can be expressed as

$$E = -R_g \frac{d \ln k}{d(1/T)} \tag{7.8}$$

$$E_a = -R_g \frac{d \ln(\varepsilon k)}{d(1/T)} \tag{7.9}$$

where E_a is the apparent activation energy. Equation 7.9 can be recast as

$$E_a = E\left(1 + \frac{1}{2} \frac{d \ln \varepsilon}{d \ln \phi}\right) \tag{7.10}$$

For a first-order reaction in a plate $\varepsilon = \tanh \phi_{p1}/\phi_{p1}$, and

$$E_a = E\left(\frac{1}{2} + \phi_{p1} \frac{1 - \tanh^2 \phi_{p1}}{2 \tanh \phi_{p1}}\right) \tag{7.11}$$

In Figure 7.1 E_a/E is plotted against the effectiveness factor. The apparent activation energy falls to $E/2$ as the temperature is raised, that is, when transition from the kinetic regime to the pore diffusion regime takes place. The $\ln t$ vs. τ' plots will thus result in *parallel*

curves which become linear at both low and high values of τ'.

In the derivation of Eq. 7.10 the effective diffusivity D_e was assumed to be independent of temperature. If D_e also varies exponentially with temperature, with E_d as the activation energy for diffusion, Eq. 7.10 takes the form

$$E_a = E\left[1 + \frac{d \ln \varepsilon}{d \ln \phi}\left(1 - \frac{E_d}{E}\right)\right] \tag{7.12}$$

Gupta and Douglas (1967) report that for the hydration of isobutylene on an ion-exchange resin catalyst, E calculated from Eq. 7.12 is 12.4 as against 16.6 kcal/g mol calculated from Eq. 7.10 (i.e., about 30% lower). However, instances in which such discrepancies are observed must be regarded as rare. In most engineering calculations E_d can be taken to be zero.

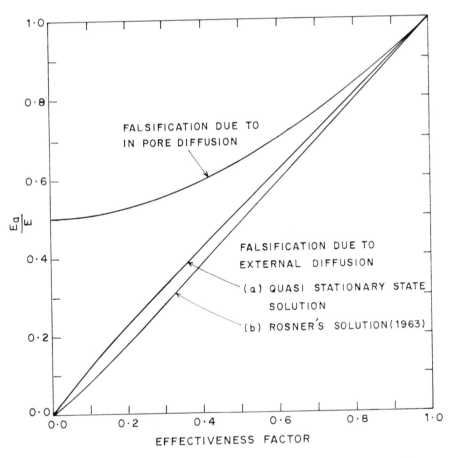

Figure 7.1. Falsification of the activation energy as a function of the effectiveness factor. (Rajadhyaksha and Doraiswamy, 1976)

Pore diffusion also falsifies the order of reaction. For a reaction A → B, the dependence of the apparent reaction order on the parameters of the system would be given by

$$n_a = n + \frac{n-1}{2}\left(\frac{d\ln\varepsilon}{d\ln\phi}\right) \qquad (7.13)$$

For reactions following non-first-order kinetics an analytical expression for the effectiveness factor cannot be obtained. However, as pointed out in Chapter 4, in the asymptotic solution region ε is given by $[2/(n+1)]^{1/2}(1/\phi)$ (Petersen, 1965), so that

$$n_a = \frac{n+1}{2} \qquad (7.14)$$

In the intermediate region, however, n_a will vary with temperature and the point values of E will depend on both concentration and temperature. Thus the $\ln\varepsilon$ vs. τ' plots will give parallel straight lines at either end and nonparallel curves in the transition region.

For a non-first-order reaction, since the reactant concentration is contained in the definition of the Thiele modulus, the apparent reaction order changes during the reaction, and therefore resort to a flow system would be essential. Corrigan et al. (1953) found, for instance, that the cracking of cumene on a silica–alumina catalyst is strongly pressure dependent, a result that Weisz and Prater (1954) showed subsequently to be attributable solely to diffusion effects.

The falsification of reaction order is a rather weak method of discerning the controlling regime, for diffusion within a pellet seldom occurs by a single diffusive mechanism and is usually a complex combination of Knudsen and bulk flow mechanisms (each with its own pressure dependency), with surface diffusion also playing a role in some cases (see Chapter 3). In addition, the method is powerless in the case of first-order reactions.

7.2.2. Isothermal Bidispersed Systems

As pointed out in Chapter 3, when porous particles are pelleted a bidispersed pore system is obtained. Tartarelli and co-workers (Tartarelli and Morelli, 1967, 1968; Tartarelli et al., 1968, 1970) have analyzed this system by defining a modified Thiele modulus. Their results indicate that under conditions of diffu-

sion control,

$$E_a = \frac{E}{4} \qquad \text{and} \qquad n_a = \frac{5}{4} \qquad (7.15)$$

Thus in a bidispersed pore system the falsification caused by pore diffusion is much more drastic than in a monodispersed system.

7.2.3. Combined Influence of Intraparticle Heat and Mass Transport

In the presence of intraparticle heat and mass transport the effectiveness factor depends on two additional parameters, β_m and α_s. It can be shown (Languasco et al., 1972) that

$$\frac{E_a}{E} = 1 + \frac{1}{2}\frac{\partial\ln\varepsilon}{\partial\ln\phi} - \frac{\partial\ln\varepsilon}{\partial\alpha_s} - \frac{\beta}{\alpha_s}\frac{\partial\ln\varepsilon}{\partial\beta_m} \qquad (7.16)$$

Equation 7.16 gives the falsified activation energy in a nonisothermal pellet. Figure 7.2 shows plots of E_a/E vs. ϕ for $\alpha_s = 20$ and different values of β_m. For low values of ϕ, E_a/E is unity. In the multiple-solution region, the effectiveness factor increases rapidly with temperature and values of E_a approaching infinity are obtained. Beyond the maximum, all three terms in Eq. 7.16 tend to reduce E_a, so that E_a/E can be lower than 0.5. For endothermic reactions the influence of β_m and α_s is in the reverse direction, so that the asymptotic value of E_a/E is greater than 0.5.

The expression for the diffusion-disguised order of reaction is

$$n_a = n + \frac{n-1}{2}\frac{d\ln\varepsilon}{d\ln\phi} + \beta_m\frac{d\ln\varepsilon}{d\beta_m} \qquad (7.17)$$

which for a first-order reaction becomes

$$n_a = 1 + \beta_m\frac{d\ln\varepsilon}{d\beta_m} \qquad (7.18)$$

Thus, contrary to the isothermal case, the order of reaction is falsified even for a first-order reaction.

Falsification of the activation energy in a nonisothermal pellet has also been analyzed by Östergaard (1963) using a different approach. He formulated the heat and mass balance equations for the pellet by using

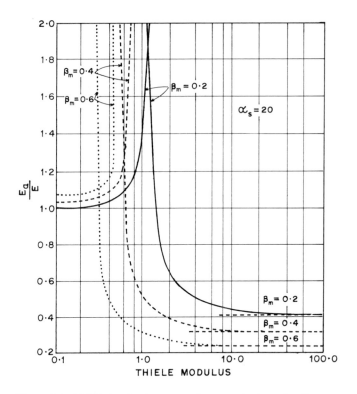

Figure 7.2. Falsified activation energy for a nonisothermal pellet as a function of the Thiele modulus (Languasco et al., 1972).

the following set of parameters:

$$(\phi)_{A_f} = R \left(\frac{A_f}{D_{eA}} \right)^{1/2} \tag{7.19}$$

$$\beta'_m = \frac{D_{eA}(-\Delta H)C_{As}R_g}{Ek'_s} \tag{7.20}$$

It may be noted that the parameter $(\phi)_{A_f}$ is a Thiele modulus based on the Arrhenius frequency factor and is hence temperature independent, while the parameter β'_m is the thermicity factor rendered dimensionless by using E/R_g as reference temperature. Östergaard solved the heat and mass balance equations and obtained the disguised rate constant from the expression

$$\frac{k_a}{A_f} = \frac{3}{(\phi)^2_{A_f}} \left(\frac{d\hat{C}}{d\hat{R}} \right) \tag{7.21}$$

Dimensionless Arrhenius diagrams can be prepared by plotting $\ln(k_a/A_f)$ vs. $E/R_g T_s$. Östergaard (1963) also analyzed the falsification of activation energy for a reaction following zero-order kinetics, and showed

that the falsification is much more drastic than for higher-order reactions.

7.2.4. Effect of Adsorption

A number of gas–solid catalytic reactions intrinsically obey H–W kinetics, in which case the kinetic expression cannot be approximated by a general-order expression. The expression for the apparent activation energy will now include terms for the adsorption equilibrium constant of the component adsorbed:

$$\frac{E_a}{E} = 1 + \frac{1}{2} \frac{\partial \ln \varepsilon}{\partial \ln \phi} - \frac{\partial \ln \varepsilon}{\partial \alpha_s} - \frac{\beta_m}{\alpha_s} \frac{\partial \ln \varepsilon}{\partial \beta}$$
$$+ \frac{1}{E} \sum_i (\Delta H)_{ai} \frac{\partial \ln \varepsilon}{\partial \ln K_i C_{is}} \tag{7.22}$$

where $K_i C_{is}$ and $(\Delta H)_{ai}$ are the adsorption parameter and heat of adsorption, respectively, for the ith species.

Schneider and Mitschka (1969) considered three types of H–W kinetic expressions and showed that the apparent activation energy approaches a value less than $0.5E$ in the asymptotic solution region.

7.2.5. Effect of Surface Diffusion

The phenomenon of surface diffusion has already been explained in Chapter 3. The surface diffusion flux can be expressed as

$$N_S = D_S \frac{dC_{As}}{dl} = K_A D_S \frac{dC_A}{dl} \qquad (7.23)$$

where D_S is the surface diffusivity and C_{As} is the surface concentration. Thus the influence of surface diffusion will be determined by the product of the surface diffusivity and the adsorption equilibrium constant.

The adsorption equilibrium constant decreases with temperature, as given by

$$K_A = K_{A0} \exp\left[-\frac{(\Delta H)_a}{R_g T}\right], \quad (\Delta H)_a \text{ is negative} \qquad (7.24)$$

while

$$D_S = D_0 \exp\left(-\frac{E_S}{R_g T}\right), \quad E_S \text{ is positive} \quad (7.25)$$

Since the activation energy for surface diffusion is normally less than the heat of adsorption (Field et al., 1963; Dacey, 1965; Satterfield, 1970), the overall effect is that the rate of surface diffusion decreases with temperature. As a result of this trend the value of the apparent activation energy will be lowered under conditions where surface diffusion contributes significantly to the total flux, and there can even be a reversal in the slope of the Arrhenius plot.

7.3. ROLE OF INTERPHASE MASS TRANSPORT

In the presence of interphase mass transport limitations the apparent activation energy is given by

$$E_a = -R_g \frac{d \ln (\eta k_v)}{d(1/T)} \qquad (7.26)$$

where η is now the external effectiveness factor defined in Chapter 6, and Da is the Damköhler number, given by $Da = k_v C_{As}^{n-1}/k_g a$. By a procedure similar to that for pore diffusion the following equation can be derived:

$$E_a = E\left(1 + \frac{d \ln \eta}{d \ln Da}\right) \qquad (7.27)$$

The following expression can be readily obtained for the reaction order:

$$n_a = n + (n-1)\frac{d \ln \eta}{d \ln Da} \qquad (7.28)$$

For large values of Da, $(d \ln \eta/d \ln Da)$ attains a value of -1. In this region the reaction order attains a constant value of unity and the apparent activation energy falls to a value corresponding to species transport (1–3 kcal/mol).

Two approaches are possible to the problem of external film transport. The more common of these (the so called quasi-steady-state approach) is based on the assumption that external transport is an independent precursor to the surface reaction. In this case, as already pointed out in Chapter 6, the effectiveness factor is given by $(1 + Da)^{-1}$. In the second approach mass transport and chemical reaction are coupled through the boundary condition at the surface. Rosner (1963) solved the boundary layer equations by taking into account the surface reaction and obtained the values of apparent activation energy and order under the influence of external transport. The results indicate that the value of the apparent activation energy obtained by the simple analysis (uncoupled equations) can deviate by as much as 8 % from that obtained from the boundary layer analysis with a kinetic boundary condition. Plots of E_a vs. η obtained from Rosner's analysis and the quasi-steady-state approach are included in Figure 7.1.

7.4. COMBINED INFLUENCE OF INTERPHASE AND INTRAPARTICLE TRANSPORT

7.4.1. Mass Transport

The effectiveness factor for the combined influence of interphase and intraparticle mass transport for a first-order reaction is given by Eq. 6.31. Mehta and Aris (1971) analyzed this problem for a general-order reaction. The equations are not amenable to analytical solution, but they provided an implicit expression for the effectiveness factor in the asymptotic region (see Chapter 6).

The trend of variation of the apparent activation energy is given by Eq. 7.10. Two transition regions can be realized as the temperature is raised. The apparent activation energy first falls to $E/2$ and then to nearly zero as interphase mass transport becomes controlling.

This transition is also accompanied by a change in reaction order from n to $(n+1)/2$ to unity.

7.4.2. Simultaneous Heat and Mass Transport

Effectiveness factor solutions for a first-order reaction accompanied by all four transport processes have been obtained by McGreavy and Cresswell (1969a, b) and Hatfield and Aris (1969) for a few specific values of parameters. The effect of interphase heat transport is to raise the value of the effectiveness factor. For the same values of the parameters ϕ, β_m, α_s and the same Sherwood number the effectiveness factor increases significantly as the Nusselt number is decreased.

Since the interphase transport parameters Sh and Nu (i.e., mass and heat Biot numbers) are independent of temperature and concentration, the equations for the apparent activation energy and reaction order are the same as those given in Section 7.2.3 (Eq. 7.16 and 7.17) for intraparticle heat and mass transport. However, Sh and Nu have a significant influence on the falsified parameters, since the partial derivatives of ε with respect to ϕ, β_m, and α_s strongly depend on them. The nature of the E_a/E vs. ϕ plot is shown qualitatively in Figure 7.3.

7.5. TRANSITIONS IN COMPLEX NETWORKS

The solutions of mass balance equations yield values of the effectiveness factor for the formation or consumption of species rather than for individual reactions. If a concentration gradient of one of the reactants is set up as a result of only one fast reaction, all the reactions in which that particular species is involved will exhibit diffusional retardation. Thus the kinetics of all such reactions will be falsified even though they may be slow and therefore not subject to diffusional influence if carried out individually. In other words, in a complex reaction scheme the influence of pore diffusion on the different reactions cannot be studied in an isolated manner. The falsification of kinetics in parallel and consecutive reaction schemes will now be considered, and some experimental studies on diffusion with complex reaction will be discussed.

7.5.1. Parallel Reaction

Consider two first-order parallel reactions

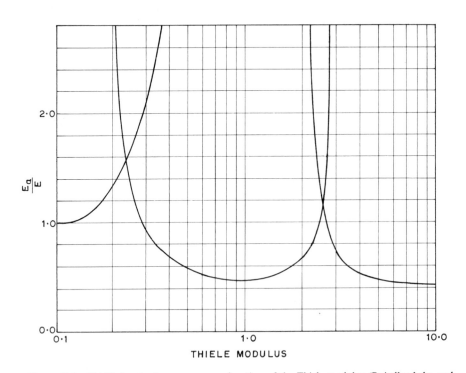

Figure 7.3. Falsified activation energy as a function of the Thiele modulus (Rajadhyaksha and Doraiswamy, 1976).

The effectiveness factor in the asymptotic region is given by

$$\varepsilon = \frac{(D_{eA})^{1/2}}{R} \frac{1}{(k_1 + k_2)^{1/2}} \qquad (7.29)$$

The rates of formation of B and C will be

$$r_B = \frac{(D_{eA})^{1/2}}{R} \frac{k_1}{(k_1 + k_2)^{1/2}} C_A \qquad (7.30)$$

$$r_C = \frac{(D_{eA})^{1/2}}{R} \frac{k_2}{(k_1 + k_2)^{1/2}} C_A \qquad (7.31)$$

The following cases can now be considered:

1. $E_1 \approx E_2 = E$, and k_1 and k_2 are of comparable magnitude:

It can be seen from Eqs. 7.30 and 7.31 that both the activation energies will approach $E/2$ in the asymptotic region.

2. $E_1 \neq E_2, k_1 \gg k_2$:

In this case the activation energy of the reaction A → B will approach $E_1/2$ whereas that for A → C will approach $E_2 - (E_1/2)$ in the asymptotic region.

3. $E_1 > E_2$ and k_1 and k_2 are of comparable magnitude:

Both reactions will exhibit temperature-dependent activation energies, even in the asymptotic region. As the temperature is increased significantly, by virtue of the higher activation energy k_1 will become large as compared to k_2 and this case will reduce to case 2.

Thus the activation energies will be falsified to different extents depending on the values of the rate constants and activation energies for the individual reactions.

7.5.2. Consecutive Reactions

Consider the reaction

$$A \to B \to C$$

where both steps are irreversible and obey first-order kinetics. The following equations can be written:

$$r_A = -\frac{\tanh(\phi_{p1})_1}{(\phi_{p1})_1} k_1 C_A \qquad (7.32)$$

$$r_B = \frac{\tanh(\phi_{p1})_2}{(\phi_{p1})_2} k_2 C_B + \frac{(\phi_{p1}^2)_1 / \alpha_D}{(\phi_{p1}^2)_1 - (\phi_{p1}^2)_2}$$

$$\times \left[\frac{\tanh(\phi_{p1})_2}{(\phi_{p1})_2} - \frac{\tanh(\phi_{p1})_1}{(\phi_{p1})_1} \right] k_2 C_A \qquad (7.33)$$

where

$$(\phi_{p1})_1 = R\left(\frac{k_1}{D_{eA}}\right)^{1/2}, \qquad (\phi_{p1})_2 = R\left(\frac{k_2}{D_{eB}}\right)^{1/2},$$

$$\alpha_D = \frac{D_{eB}}{D_{eA}} \qquad (7.34)$$

From an inspection of Eqs. 7.32 and 7.33 it can be noticed that reaction 1 will be falsified in the same manner as an independent reaction. However, the falsification of the second reaction will be affected by the first as indicated by the presence of the second term in Eq. 7.33.

7.5.3. Examples

The reaction of aniline with ethanol to produce predominantly monoethylaniline is a complex one involving several steps. Goyal and Doraiswamy (1970) made a detailed analysis of this reaction and concluded that some of the steps involved are diffusion controlled while others are not. This fact can be made use of in optimizing the production of the desired compound.

Dehydration of alcohols provides another example of the use of diffusional limitation to enhance the rate of the desired reaction. Abhyankar and Doraiswamy (1979) showed that it is possible to curtail the unwanted reactions by suitable choice of temperature and particle size. It was also observed that the shift to diffusion control in this complex network occurs at progressively lower temperatures as the particle size is increased, thus enlarging the temperature range over which diffusional limitations may be considered to be dominant.

7.6. TRANSITION DUE TO FOULING

In a catalyst that is poisoned at the pore mouth[†] the reaction involves the following steps:

1. diffusion of reacting species through the poisoned core (uncoupled with reaction);
2. diffusion with chemical reaction in the un-poisoned core.

[†] Some of the results presented in Chapter 8 are anticipated at this stage.

By defining a base temperature T_b at which the value of the Thiele modulus is ϕ_b, the rate per pellet can be expressed as (Wheeler, 1955):

$$\frac{\mathscr{R}_f}{\mathscr{R}_{fb}} = \frac{m \tanh\left[(1-p)m\phi_b\right]}{1 + pm\phi_b} \qquad (7.35)$$

where p is the degree of poisoning and

$$m = \exp\left(\frac{E}{2R_g}\right)\left(\frac{1}{T_b} - \frac{1}{T}\right)$$

Equation 7.35 is plotted in Figure 7.4 as $\mathscr{R}_f/\mathscr{R}_{fb}$ vs. $1/T$, using the values $T_b = 500°K$ and $E = 20$ kcal/mol. Five different curves have been drawn, based on different states of the catalyst at $500°K$.

Curves A and B represent catalysts with chemical control in the lower temperature range; but whereas catalyst A is fresh, 80% poisoning has taken place in catalyst B. In the temperature range covered both catalysts show transition to pore diffusional control; but catalyst B deviates from chemical control earlier as a result of poisoning. A hypothetical line ($\phi_b = 0$, $p = 0$) is also drawn in the figure corresponding to chemical control over the entire temperature range.

Catalysts D and E correspond to the requirement of pore diffusional control, and it is in these cases that the effect of poisoning is severe. Curve D is for the

unpoisoned catalyst and the apparent activation energy is 10,000 cal/mol over the entire temperature range. On the other hand, curve E starts with $E_a = 10,000$ cal/mol and levels to $E \to 0$ when diffusion through the poisoned core becomes the controlling resistance.

Catalyst C represents a condition between A, B and D, E and its behavior as brought out in the figure upholds its intermediate character.

These conclusions were confirmed by the results of Narayanan and Doraiswamy (1972) and Sansare and Doraiswamy (1980) on the hydrogenation of nitrobenzene to aniline. Experiments were carried out with (1) catalyst particles of 60–100 mesh, fresh and fouled with thiophene (in nitrobenzene); and (2) catalyst pellets $\frac{1}{2}$-cm in diameter and $\frac{1}{2}$ cm high, fresh and fouled by thiophene; and calculations were made with a base temperature of $500°K$. Any extent of fouling could be achieved by controlling the thiophene added (from 10 to 250 ppm). The results clearly showed that the falsification of activation energy was far greater in catalyst (2), operating in the diffusion regime.

Tartarelli and Morelli (1968) presented a rigorous analysis of the falsification of activation energy and reaction order for a catalyst subject to pore-mouth poisoning. The final expressions obtained by them are

$$\frac{E_a}{E} = \left[2 + (n+1)\left(\frac{1}{\hat{C}_{Ai}} - 1\right)\right]^{-1} \qquad (7.36a)$$

and

$$n_a = 1 + \frac{n-1}{2 + (n+1)(1/\hat{C}_{Ai} - 1)} \qquad (7.36b)$$

where $\hat{C}_{Ai}$ is the normalized reaction concentration C_{Ai}/C_{As} at the poisoned–unpoisoned interface.

Plots of E_a/E vs. p for $\phi = 10$ are shown in Figure 7.5 for orders 1 and 2. Since the equations were derived for the asymptotic solution region, E_a/E is equal to $\frac{1}{2}$ for no poisoning and its value falls to 0 as the extent of poisoning increases, a conclusion also reached by Wheeler's analysis.

A similar analysis based on the macropore–micropore model leads to the expression

$$\frac{E_a}{E} = \left[4 + (n+3)\left(\frac{1}{\hat{C}_{Ai}} - 1\right)\right]^{-1} \qquad (7.37)$$

The curves for this model are also included in Figure 7.5 for $M^{2/3}\phi_{m-m}^{1/3} = 50$. It will be noted that the falsification is considerably more for this model than for the single-pore model.

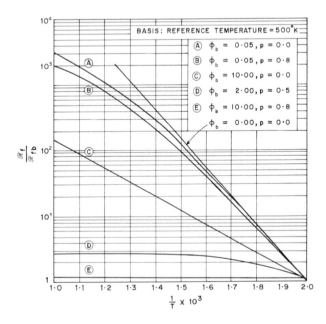

Figure 7.4. Identification of controlling regimes in a fouling catalyst (Rajadhyaksha and Doraiswamy, 1976).

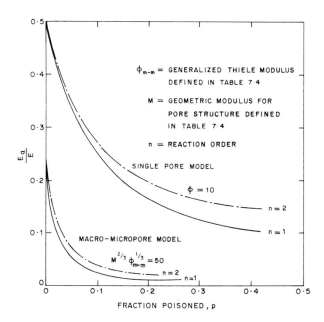

Figure 7.5. Falsification of activation energy due to fouling in monodispersed and bidispersed macropore–micropore models (Tartarelli and Morelli, 1968).

7.7. EXAMPLES OF SHIFT IN CONTROLLING REGIME

7.7.1. Water–Gas Shift Reaction

For pelleted catalysts in which pore diffusional effects are likely to be significant Ruthven (1969) has derived the following expression for the apparent rate constant k_a:

$$k_a = \frac{6}{d_p}\left(\sqrt{kD_e} - \frac{2D_e}{d_p}\right) \qquad (7.38)$$

From the experimental results of Hoogschagen (1955) and Paratella and Sorgato (1964), a value of $D_e = 0.043 \text{ cm}^2/\text{sec}$ has been obtained at $400°C$ and atmospheric pressure. The results obtained with Eq. 7.38 and kinetic data on commercial catalysts (Power Gas, Girdler, and Imperical Chemical Industries) by Marsh and Robsen (1963) are reproduced in Figure 7.6 in the form of Arrhenius plots. It is clear that at higher temperatures the activation energy falls to about half the value at lower temperatures, thus indicating a shift to the pore diffusion regime.

7.7.2. Ammonia Synthesis

The role of pore diffusion in ammonia catalysts was investigated by Dyson and Simon (1968), who solved

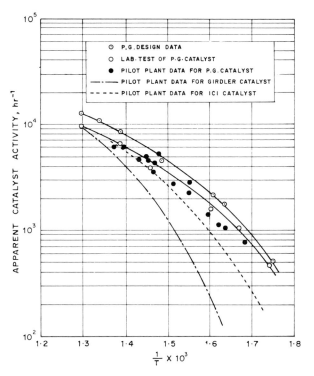

Figure 7.6. Arrhenius plots of apparent activity of the shift catalyst (Marsh and Robson, 1963).

the mass balance equation to obtain the effectiveness factor, using the assumption of Bokhoven and van Raayan (1954) that Knudsen diffusion is absent. The multicomponent bulk diffusivities were calculated from the binary diffusivities reported by Nielson (1956), and the effective diffusivities were then estimated by using a tortuosity factor of 2.

From the results of several calculations, the following expression has been derived for the effectiveness factor as a function of x (conversion of nitrogen) and T for a feed mixture containing 3 parts of H_2 to 1 of N_2 and 12.7% inerts:

$$\varepsilon = b_0 + b_1 T + b_2 x + b_3 T^2 + b_4 x^2 + b_5 T^3 + b_6 x^3 \qquad (7.39)$$

Clearly, for a given pressure and conversion the effectiveness factor decreases rapidly with temperature, so that in the higher-temperature range pore diffusion becomes controlling.

7.7.3. Oxidation of Sulfur Dioxide

Livbjerg and Villadsen (1972) studied the oxidation of sulfur dioxide on kieselgur-supported $V_2O_5–K_2O$

Haldor Topsoe catalyst, using cylindrical pellets 6 mm × 6 mm in size. It was observed that for pellets the activation energy changed from 65 kcal/mol to 16 kcal/mol in the temperature range of 416–484°C. This can be attributed to a shift to pore diffusion control in a catalyst characterized by bidispersed pore structure (in which $E_a/E \approx 1/4$). (Recent findings suggest that V_2O_5 exists in the liquid state.)

7.7.4. Hydrogenation of Ethylene

Fulton and Crosser (1965) studied the hydrogenation of ethylene on a nickel-on-alumina catalyst. The activation energy was found to fall from 9.6 to 1.98 kcal/mol when the particle size was increased from 70–100 mesh to 8–12 mesh. This change can be considered a result to the transition to external mass transport control. Gioia and Green (1967) recalculated the kinetic parameters by properly taking into account the influence of intraparticle and interphase processes, and were able to extract from the same experimental data a constant activation energy value of 11.4 kcal/mol for all particle sizes.

7.8. EXAMPLES OF ANOMALOUS ACTIVATION ENERGY CHANGES

Systems have been reported where a change in activation energy can occur in the kinetic regime itself. An observation is also reported where a high value of activation energy was found in the external mass transfer regime. Examples of such anomalous behavior will be considered in this section.

7.8.1. Activation Energy Changes within the Kinetic Regime

Oxidation of Benzene

Vaidyanathan and Doraiswamy (1968) studied the oxidation of benzene to maleic anhydride on a vanadium catalyst, and observed a fall in E from 19 to 2 kcal/mol with increase in temperature. This lowering of E with temperature can be explained by analyzing the results on the basis of the following scheme:

$$nO_2 + C_6H_6 \xrightarrow{k_1} C_6H_6 \cdot l + nO_2 \xrightarrow{k_2} C_4H_2O_3 \cdot l + CO_2 + H_2O$$

$$\begin{array}{ccc} \downarrow k_4 & \downarrow k_3 & \searrow k_5 \\ CO_2 + H_2O & C_4H_2O_3 & CO_2 + H_2O \end{array}$$

The net rate of formation of the maleic anhydride complex at steady state is

$$r_{C_4H_2O_3(l)} = k_2\theta_{C_6H_6(l)} - k_3\theta_{C_4H_2O_3(l)} - k_5\theta_{C_4H_2O_3(l)} = 0 \tag{7.40}$$

from which the rate of maleic anhydride formation may be obtained as

$$r_{C_4H_2O_3(l)} = \frac{k_2}{1 + k_5/k_3}\theta_{C_6H_6(l)} \tag{7.41}$$

Equation 7.41 can now be analyzed, with the following conclusions. At low temperatures, $E = E_2$; at higher temperatures, $E = E_2 + E_3 - E_5$ leading to a low value of E. To ensure that the lowering of activation energy was not due to diffusional interference, Vaidyanathan and Doraiswamy (1968) carried out experiments with different particle sizes and obtained kinetic data in the particle size range where pore diffusional resistance would be unimportant. The absence of an external mass transfer effect was established by varying the mass velocity.

Hydrogenation of Ethylene

The hydrogenation of ethylene carried out on high-purity alumina shows a fall in activation energy with temperature, but this can be explained purely from kinetic considerations (Sinfelt, 1966). The elementary steps involved are

$$C_2H_4 + l \xrightarrow{k_1} C_2H_4 \cdot l$$

$$C_2H_4 \cdot l + H_2 \underset{k_{-2}}{\overset{k_2}{\rightleftharpoons}} C_2H_5 \cdot l + H \cdot l$$

$$C_2H_5 \cdot l + H \cdot l \xrightarrow{k_3} C_2H_6 + 2l$$

where l represents the active site and $C_2H_4 \cdot l$, $C_2H_5 \cdot l$, and $H \cdot l$ the respective complexes on the surface. On an increase of temperature, the controlling step moves from step 3 to step 2 with consequent change in the activation energy from E_2 to $(E_2 + E_3 - E_{-2})$.

Hydrogenation of Benzene and Cyclohexene

The kinetics of hydrogenation was studied by Voorhoeve and Stuiver (1971) on bulk and alumina-supported Ni–W–S catalyst. Carbon disulfide was added to the feed to maintain the sulfur content of the catalyst. The Arrhenius plots indicate that the activation energy is temperature dependent. On both the bulk and the supported catalysts the activation energy was found to fall to half its value with increase in temperature.

However, it can be said conclusively that the reaction is not diffusion influenced under these conditions. The reaction rate can be varied over an order of magnitude by changing the content of carbon disulfide in the feed. At any given temperature the activation energy is found to be independent of the CS_2 content in the feed. Further, hydrogenolysis of CS_2, which is significantly faster than the other two reactions, shows a constant activation energy over the entire temperature range.

7.8.2. High Activation Energy in the Mass Transfer Regime

Disproportionation of Propylene

Moffat et al. (1970) studied the disproportionation of propylene on molybdate–alumina and tungsten oxide–alumina catalysts. In the case of the tungsten catalyst, experiments showed a strong influence of interphase mass transport. However, the mass transfer rates theoretically calculated using the j_d factor correlations (see Chapter 6) were 100–1000 times greater than the observed rates. This could be attributed to localized mass transport effects at widely separated reaction sites (see Chapter 6).

The temperature dependence of the reaction was

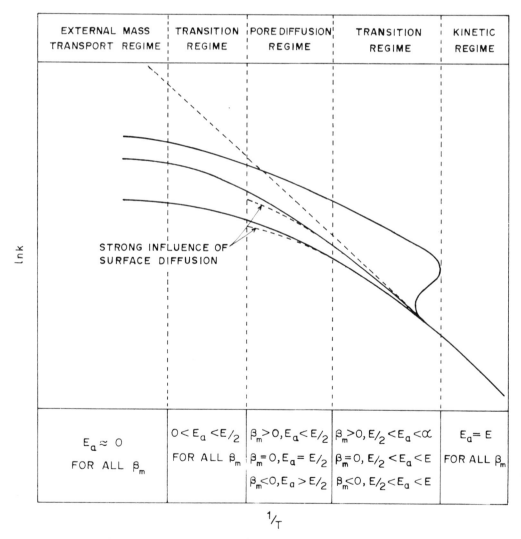

Figure 7.7. Schematic representation of regimes of operation on an Arrhenius diagram. (Rajadhyaksha and Doraiswamy, 1976)

TABLE 7.4 Falsification of Activation Energy and Reaction Order

Reaction Type	Model	E_a/E	n_a	E_a/E under strong diffusional limitation	n_a under strong diffusional limitation
A + B → products	Parallel pore	$1 + \dfrac{1}{2}\dfrac{d\ln\varepsilon}{d\ln\phi}\left(1 - \dfrac{E_d}{E}\right)$	$n + \dfrac{n-1}{2}\dfrac{d\ln\varepsilon}{d\ln\phi}$	$\dfrac{1}{2} + \dfrac{E_d}{2E}$	$\dfrac{n+1}{2}$
2A → products	Macropore–micropore	$2 + \dfrac{3}{4}\dfrac{d\ln\varepsilon}{d\ln\phi_{m-m}}\left(1 - \dfrac{E_d}{E}\right)$	$2 + \dfrac{3}{4}\dfrac{d\ln\varepsilon}{d\ln\phi_{m-m}}$	$\dfrac{1}{4} + \dfrac{3}{4}\dfrac{E_d}{E}$	$\dfrac{5}{4}$ for $n = 2$
A + B → products	Macropore–micropore	$1 + \dfrac{3}{4}\dfrac{d\ln\varepsilon}{d\ln\phi_{m-m}}\left(1 - \dfrac{E_d}{E}\right)$	With respect to A $1 + \left(\dfrac{3}{4}\phi_{m-m} - \dfrac{d\xi}{d\ln\psi}\right)\left(\dfrac{\partial\ln\varepsilon}{\partial\phi_{m-m}}\right)_\psi - \left(\dfrac{\partial\ln\varepsilon}{\partial\ln\psi}\right)_{\phi_{m-m}}$ With respect to B $1 + \dfrac{d\xi}{d\ln}\left(\dfrac{\partial\ln\varepsilon}{\partial\phi_{m-m}}\right)_\psi + \left(\dfrac{\partial\ln\varepsilon}{\partial\ln\psi}\right)_{\phi_{m-m}}$	$\dfrac{1}{4} + \dfrac{3}{4}\dfrac{E_d}{E}$	—

where

$$\xi = \frac{\phi_{m-m}}{(m')^{3/2}}$$

$$\psi = \frac{C_{Bs}}{C_{As}} \quad \text{for } n = 1$$

$$\phi_{m-m} = \sqrt{2}(m')^{3/2}M,$$

$$m' = l_{ma}\left(\frac{2k_s}{r_{p,mi}D_{e,mi}}\right)^{1/2}$$

$$M = \left(\frac{D_{e,mi}l_{ma}^2}{r_{p,ma}D_{e,ma}l_{mi}}\right)^{1/2}$$

aA → products	Parallel pore nonisothermal	$1 + \dfrac{1}{2}\dfrac{\partial \ln \varepsilon}{\partial \ln \phi} - \dfrac{\partial \ln \varepsilon}{\partial \alpha_s} - \dfrac{\beta_m}{\alpha_s}\dfrac{\partial \ln \varepsilon}{\partial \beta_m}$	$n + \dfrac{n-1}{2}\dfrac{\partial \ln \varepsilon}{\partial \ln \phi} + \beta_m \dfrac{\partial \ln \varepsilon}{\partial \beta_m}$	$\dfrac{1}{2} - \dfrac{\partial \ln \varepsilon}{\partial \alpha_s} - \dfrac{\beta_m}{\alpha_s}\dfrac{\partial \ln \varepsilon}{\partial \beta_m}$	$\dfrac{n+1}{2} + \beta_m \dfrac{d \ln \varepsilon}{d \beta_m}$
aA → products	Parallel pore nonisothermal with strong adsorption of A (Langmuir–Hinshelwood kinetics)	$1 + \dfrac{1}{2}\dfrac{\partial \ln \varepsilon}{\partial \ln \phi}\dfrac{\partial \ln \varepsilon}{\partial \alpha_s} - \dfrac{\beta_m}{\alpha_s}\dfrac{\partial \ln \varepsilon}{\partial \beta_m} + \dfrac{\Delta H}{E}\dfrac{\partial \ln \varepsilon}{\partial \ln \omega}$	—	$\dfrac{1}{2} - \dfrac{\partial \ln \varepsilon}{\partial \alpha_s} - \dfrac{\beta_m}{\alpha_s}\dfrac{\partial \ln \varepsilon}{\partial \beta_m} + \dfrac{\Delta H}{E}\dfrac{\partial \ln \varepsilon}{\partial \ln \omega}$	—
aA → products	Parallel pore nonisothermal with strong adsorption of n species (Langmuir–Hinshelwood kinetics)	$1 + \dfrac{1}{2}\dfrac{\partial \ln \varepsilon}{\partial \ln \phi} - \dfrac{\partial \ln \varepsilon}{\partial \alpha_s} - \dfrac{\beta_m}{\alpha}\dfrac{\partial \ln \varepsilon}{\partial \beta_m} + \dfrac{1}{E}\sum_i (\Delta H)_i \dfrac{\partial \ln \varepsilon}{\partial \ln \omega_i}$	—	$\dfrac{1}{2} - \dfrac{\partial \ln \varepsilon}{\partial \alpha_s} - \dfrac{\beta_m}{\alpha_s}\dfrac{\partial \ln \varepsilon}{\partial \beta_m} + \dfrac{1}{E}\sum (\Delta H)_i \dfrac{\partial \ln \varepsilon}{\partial \ln \omega_i}$	—
A + B → products	External mass transport	$1 + \dfrac{d \ln \varepsilon}{d \ln Da}$ where $Da = \dfrac{k C_{As}^{n-1}}{k_g a}$	$n + (n-1)\dfrac{d \ln \varepsilon}{d \ln Da}$	1–3 kcal/mol	

then studied and the system was found to exhibit an activation energy of about 40 kcal/mol in spite of the fact that the experiments conclusively showed mass transport influence. A plausible explanation for this observation is that more and more reaction sites are activated as the temperature is increased, resulting in a significant increase in the reaction rate.

7.9. CONCLUDING REMARKS

The manner in which the activation energy is falsified depends on the transport process involved—interphase, intraphase, or combined. A summary of these various influences on the falsification of the kinetic parameters as tabulated by Rajadhyaksha and Doraiswamy (1976) is reproduced in Table 7.4. This table gives the values of E_a/E and n_a for various types of reactions and diffusion models. The limiting values of these parameters under strong diffusional limitation are also given. The table should be useful in identifying the regime of a particular reaction and thus deciding on the basic design strategy.

A consolidated representation of all the operating regimes in a catalytic reaction is presented in Figure 7.7 in the form of an Arrhenius diagram. It must be noted that the entire gamut of regimes shown may not be realized in a single reaction. The diagram should therefore be regarded purely as indicative.

Several examples have been given in which the falsification of kinetic parameters and the identification of regimes are found to be in conformity with the analysis presented. On the other hand, anomalous activation energy changes have also been reported in several cases, leading to the conclusion that there are many exceptions to the general theory concerning the falsification of kinetic parameters. Thus the methods presented in this chapter for discerning the possible controlling regime should be used with discretion, perhaps primarily as a guide.

REFERENCES

Abhyankar, S. M., and Doraiswamy, L. K. (1979), *Can. J. Chem. Eng.*, **57**, 481.

Acton, F. S. (1959), *Analysis of Straight Line Data*, Wiley, New York.

Baag, J. (1969), *J. Catal.*, **13**, 271.

Baag, J. (1970), *J. Catal.*, **16**, 370.

Bliznakov, G., Bakardjiev, I., and Peshev, O. (1970), *J. Catal.*, **16**, 148.

Bokhoven, C., and van Raayen, W. (1954), *J. Phys. Chem.*, **58**, 471.

Carberry, J. J. (1969), *Catal. Rev.*, **3**, 61.

Corrigan, T. E., Garver, J. C., Rase, H. F., and Kirk, R. S. (1953), *Chem. Eng. Prog.*, **49**, 603.

Dacey, J. R. (1965), *Ind. Eng. Chem.*, **57**, 27.

Dyson, D. C., and Simon, J. M. (1968), *Ind. Eng. Chem. Fundam.*, **7**, 605.

Field, G. J., Watts, H., and Weller, K. R. (1963), *Rev. Pure Appl. Chem.*, **13**, 2.

Fulton, J. W., and Crosser, O. K. (1965), *AIChE J.*, **11**, 513.

Gioia, F., and Green, D. W. (1967), *AIChE J.*, **13**, 395.

Goyal, P., and Doraiswamy, L. K. (1970), *Ind. Eng. Chem. Process Des. Dev.*, **9**, 25.

Gupta, V. P., and Douglas, W. J. M. (1967), *Can. J. Chem. Eng.*, **45**, 119.

Hatfield, B., and Aris, R. (1969), *Chem. Eng. Sci.*, **24**, 1213.

Hoogschagen, H. J. (1955), *Ind. Eng. Chem.*, **47**, 906.

Languasco, J. M., Cunningham, R. E., and Calvelo, A. (1972), *Chem. Eng. Sci.*, **27**, 1459.

Livbjerg, H., and Villadsen, J. (1972), *Chem. Eng. Sci.*, **27**, 21.

Marsh, J. D. F., and Robson, R. (1963), *Proc. Symp. Catalysis in Practice*, Harrogate.

McGreavy, C., and Cresswell, D. L. (1969a), *Chem. Eng. Sci.*, **24**, 608.

McGreavy, C., and Cresswell, D. L. (1969b), *Can. J. Chem. Eng.*, **47**, 583.

Mehta, B. N., and Aris, R. (1971), *Chem. Eng. Sci.*, **26**, 1699.

Moffat, A. J., Johnson, M. M., and Clark, A. (1970), *J. Catal.*, **18**, 345.

Narayanan, T. K., and Doraiswamy, L. K. (1972), NCL report.

Nielson, A. (1956), *Investigation on Promoted Iron Catalysts for the Synthesis of Ammonia*, 2nd ed., Jul Gjellerups, Copenhagen.

Östergaard, K. (1963), *Chem. Eng. Sci.*, **18**, 259.

Paratella, A., and Sorgato, I. (1964), *Proc. 3d European Symp. Chem. React. Eng.* p. 173.

Petersen, E. E. (1965), *Chemical Reaction Analysis*, Prentice-Hall, Englewood Cliffs, New Jersey.

Rajadhyaksha, R. A., and Doraiswamy, L. K. (1976), *Catal. Rev. Sci. Eng.*, **13**, 209.

Rosner, D. E. (1963), *AIChE J.*, **9**, 321.

Ruthven, D. M. (1969), *Can. J. Chem. Eng.*, **47**, 327.

Sansare, S. D., and Doraiswamy, L. K. (1980) *unpublished work*.

Satterfield, C. N. (1970), *Mass Transfer in Heterogeneous Catalysis*, MIT Press, Cambridge, Massachusetts.

Schneider, P., and Mitschka, P. (1969), *Chem. Eng. Sci.*, **24**, 1725.

Sinfelt, J. H. (1966), *Ind. Eng. Chem.*, **58**, 18.

Tartarelli, R., and Morelli, F. (1967), *Ann. Chim. (Rome)*, **57**, 1316.

Tartarelli, R., and Morelli, F. (1968), *J. Catal.*, **11**, 159.

Tartarelli, R., Capovani, M., and Morelli, F. (1968), *Ann. Chim. (Rome)*, **58**, 1050.

Tartarelli, R., Gioni, S., and Capovani, M. (1970), *J. Catal.*, **18**, 212.

Vaidyanathan, K., and Doraiswamy, L. K. (1968), *Chem. En. Sci.*, **23**, 537.

Voorhoeve, R. J. H., and Stuiver, J. C. M. (1971), *J. Catal.*, **23**, 228.

Wei, J. (1966), *Ind. Eng. Chem.*, **58**, 38.

Weisz, P. B., and Prater, C. D. (1954), *Adv. Catal.*, **6**, 143.

Wheeler, A. (1955), *Catalysis*, **2**, 105.

CHAPTER EIGHT

Role of Catalyst Deactivation

Catalyst deactivation often plays an important role in determining the overall kinetics of a catalytic reaction. Just as the intrinsic kinetics can be altered by diffusional effects, catalyst deactivation can also play a decisive role in determining the regime of a reaction in a catalyst pellet (see Chapter 7). As in the case of reactor analysis, the study of catalyst deactivation can be considered to involve three facets. The first is concerned with the chemical nature of catalyst deactivation and is analogous to the kinetics of a chemical reaction. The second involves the behavior of a pellet as a whole under conditions of deactivation, with pore diffusion included in the analysis. This situation corresponds to the internal field problem in reactor analysis. The third is the problem of reactor design, which can be approached by combining the analysis of a deactivating pellet with that of the external flow field.

In this chapter we shall consider the first two aspects of catalyst deactivation, reserving the third for Chapter 16, following a consideration of fixed-bed reactor design for a nondeactivating catalyst in Chapters 11 and 12. The general principles outlined here would also apply to any other type of reactor, such as the fluidized-bed reactor considered in Chapter 14 and the moving-bed reactor considered in Chapter 15.

8.1. THE CHEMICAL NATURE OF DEACTIVATION

In this section we shall briefly consider deactivation as a surface phenomenon in the absence of any diffusional intrusion.

8.1.1. Types of Deactivation

There is considerable overlap of terms in the literature dealing with the chemical nature of catalyst deactivation. In order to avoid confusion, we shall use *deactivation* as a general term to denote loss of catalytic activity (Wojciechowski et al., 1969). The different types of deactivation can then by classified as *fouling, in-hibition, poisoning,* and *independent fouling,* summarized in Table 8.1.

In the first three cases (self- and impurity poisoning) the rate of deactivation would evidently be a function of the reactant, intermediate, or impurity concentration and is therefore subject to the effect of time and temperature in the manner that any reaction would be. On the other hand, in the fourth case (independent fouling), the deactivation is dependent explicity on time and temperature.

Self- and impurity poisoning processes, which are caused by specific chemical reactions, may be represented by the following schemes:

I. Parallel self-poisoning:

$$A \xrightarrow{1} B$$
$$A \xrightarrow{2} C$$

TABLE 8.1. Classification of Catalyst Deactivation

Type of Deactivation	General Features
Fouling (self-poisoning)	Loss of activity due to secondary reaction of reactant or product, corresponding, respectively, to parallel or consecutive fouling (or a combination of the two, referred to as triangular fouling). Practically, the most important is carbon formation.
Inhibition	Loss of activity due to strong product or reactant adsorption; usually the activity falls to a steady nonzero value.
Poisoning (impurity poisoning)	Loss of activity due to strong chemisorption of an impurity from the feed
Decay (aging or independent fouling)	Loss of activity due to structural changes or sintering (leading to a decrease in the active surface area)

II.   Series self-poisoning:

$$A \xrightarrow{\ 1\ } B$$

$$B \xrightarrow{\ 2\ } C$$

III.   Triangular self-poisoning:

$$A \xrightarrow{\ 1\ } B$$
$$2 \searrow \quad \swarrow 3$$
$$C$$

IV.   Impurity poisoning:

$$A \longrightarrow B$$

$$P \longrightarrow \text{poisonous structure}$$

8.1.2.   Coke Formation

Coke formation represents a very common cause of catalyst deactivation. Voorhies (1945) proposed the following empirical equation for the coke content of a catalyst as a function of time:

$$W_c = m t^n \tag{8.1}$$

which was subsequently extensively tested by Blanding (1953). Different values of the exponent n have been reported, but a value of 0.5 appears reasonable (Voorhies, 1945; Rudershausen and Watson, 1954; Wilson and den Herder, 1958; Blue and Engle, 1951; Panchenkov and Lolesnikov, 1959; Crawford and Cunningham, 1956; Eberly et al., 1966; Pozzi and Rase, 1958; Prater and Lago, 1956; Weekman, 1968).

One has to be cautious about too simplistic an approach to the problem of coking. From the data of Dumez and Froment (1976), for instance, it can be shown that coke formation is a strong function of the mechanisms involved in coking (which is to be intuitively expected), the exponent in Eq. 8.1 varying from 0.35 to 1.0. If the coking occurred on distinctive sites, then it would not interfere with the main reaction. On the other hand, competition for a given site would lead to complicated H–W kinetics.

Most of the earlier kinetic studies were restricted to silica–alumina cracking catalysts, but more recent studies show increasing attention to coke formation on zeolites (Eberly and Kimberlin, 1970; Swabb and Gates, 1972; Butt, 1976; Rollmann, 1977; Kokotailo et al., 1978; Chen and Garwood, 1978; Walsh and Rollmann, 1977, 1979; Rollmann and Walsh, 1979). It seems a reasonable conclusion from these studies that coke formation is a spatially demanding reaction directly controlled by the zeolite pore structure.

The study of catalyst deactivation is greatly facilitated by defining a term Ω for catalyst activity:

$$\Omega = \frac{\text{rate of reaction on a deactivating catalyst}}{\text{rate of reaction on a fresh catalyst}} \tag{8.2}$$

The activity is obviously a function of the coke content or poison concentration, namely,

$$\Omega = f(W_c) \quad \text{or} \quad f(C_p) \tag{8.3}$$

Several forms of Eq. 8.3 have been employed, many of which are presented in Table 8.2. One of the forms, based on Eq. 8.1, is $\Omega = \alpha t^{-\beta}$, and holds only when $t > 0$. This form is particularly attractive since it permits the estimation of deactivation order (Szepe and Levenspiel, 1970; see Chapter 9). We shall have occasion to use some of these equations in the developments that follow in the subsequent sections.

From the mechanistic point of view the significant features of deactivation are the following.

1.   Depending on reaction conditions a relatively small amount of coke can completely cover the accessible surface, or coke levels approaching the theoretical limit may be necessary to cause significant blocking of the surface (Levinter et al., 1967). This accounts for the contradictory results on surface area and diffusivity reported in the literature, namely, the findings (a) that coke formation has no effect on these properties (Ozawa and Bischoff, 1968; Toei et al., 1975), and (b) that there is considerable reduction in both (Appleby et al., 1962; Ramser and Hill, 1958; Haldeman and Botty, 1959; Suga et al., 1967; Richardson, 1972; Butt et al., 1975, 1976). An interesting observation (Swabb and Gates, 1972) is that deactivation rates are independent of pore dimensions in the case of methanol dehydration on H–mordenite. Apparently there is no pore blockage in this reaction.

2.   Coke deposition occurs via the formation of condensed nuclei such as benzene, naphthalene,

TABLE 8.2 Forms of the Catalyst Decay Function

Functional Form	Equation[a]	Reference
	$\Omega = 1 - \hat{W}_c$	Masamune and Smith (1966)
Linear	$\Omega = 1 - \beta \hat{C}_P$	Anderson and Whitehouse (1961); Maxted (1951)
Hyperbolic[b]	$\Omega = (1 + \beta \hat{W}_c)^{-1}$	Froment and Bischoff (1961)
	$\Omega = (1 + \beta \hat{C}_P)^{-1}$	Anderson and Whitehouse (1961)
	$\Omega = (1 + \beta \hat{C}_P)^{-1/2}$	Anderson and Whitehouse (1961)
	$\Omega = \dfrac{1}{1 + \phi \hat{W}_c} - \dfrac{\hat{W}_c}{1 + \phi}$	Wheeler (1955)
Exponential	$\Omega = \exp(-\beta \hat{W}_c)$	Froment and Bischoff (1961)
	$\Omega = \exp(-\beta \hat{C}_P)$	Anderson and Whitehouse (1961)
Reciprocal power function	$\Omega = \dfrac{\alpha}{t^\beta}$	Voorhies (1945); Blanding (1953)
Elovich form	$\dfrac{d\Omega}{dt} = \alpha \exp(-\beta x)$	Parravano (1953)

[a]Notation: $\hat{W}_c$ is the relative coke on catalyst, $\hat{C}_P$ the relative poison concentration, and C_P the poison concentration; α and β are constants.
[b]In the distribution of Wheeler (1955), the empirical constant is replaced by the Thiele modulus.

and anthracene. Although low molecular weight straight-chain compounds also lead to carbon formation, aromatic precursors are almost invariably formed (Appleby et al., 1962). Generally, coke deposits consist of highly condensed aromatic structures of low hydrogen content, $C_m H_n$.

8.1.3. Poisoning and Sintering

No general treatment of the chemistry of poisoning is possible, since each poisoning process is distinctive, with its own kinetics and mechanism. An extensive review of the poisoning of catalysts due to metallic and nonmetallic poisons has been presented by Maxted (1951), while a few other aspects of poisoning have been reviewed by Butt (1972). The most common poisons are arsenic, sulfur, nitrogen, phosphorus, nickel, sodium, and copper. The activity of a catalyst subject to impurity poisoning may be expressed in various forms, as in the case of coke formation, and these are included in Table 8.2.

Loss of activity due to sintering has been the subject of relatively few studies. Attention has largely been restricted to Pt–Al$_2$O$_3$ catalysts. Sintering occurs as a result of loss in active surface area and is usually represented by an equation of the form (Maat and

Moscou, 1964)

$$\frac{1}{S_g} = \frac{1}{S_{g0}} + kt \qquad (8.4)$$

where S_g and S_{g0} are the areas at $t = t$ and $t = 0$, respectively.

8.2. DEACTIVATION OF A PELLET

Two approaches have been employed in the analysis of a deactivating pellet: that in which time as an explicit parameter is eliminated, and that based on the use of time as a variable.

In the first approach the degree of deactivation replaces time as a parameter. On the other hand, in the second approach deactivation is considered to be dependent exclusively on time, giving rise to the so-called time-on-stream theory. This approach is essentially a gross kinetic approach in which the effect (if any) of diffusion is included in the deactivation rate constant.

In view of the distinctive nature of the time-on-stream theory, we shall consider it in a separate section; here we shall treat mainly the first approach. An intermediate approach, based on the so-called shell-progressive model, incorporates time as an additional parameter, but is not based exclusively on time. In yet another approach, the concept of time-dependent effectiveness factors is used for a quantitative interpretation of deactivation. These two approaches are also included in this section.

Deactivation of a catalyst is basically a time-dependent phenomenon, and it might therefore seem incongruous that it can be analyzed without employing time as an indispensable variable. However, it is possible to exclude time as an *explicit* variable in the following situations.

1. Part of the catalyst surface is poisoned rapidly, and the catalyst then assumes a stable residual activity. In this situation time is replaced by the degree of poisoning (i.e., the fraction of the surface poisoned).

2. If a reaction is influenced by pore diffusional effects, then clearly the concentration at any plane in the pellet—the center-plane concentration is a convenient general parameter—would be sensitive to the distribution and values of the local Thiele modulus. Thus it is possible to eliminate time and to represent the reaction rate as a function of the center-plane concentration.

8.2.1. Analysis Based on Degree of Deactivation

Three distinct types of deactivation are possible: homogeneous, pore-mouth, and pore-end (or core) poisoning. All concepts and models of deactivation described in subsequent sections of this chapter derive their basic features in one form or another from these three types; hence they are described in some detail.

Homogenous Poisoning

In homogeneous poisoning, the poison and the reactants have equal opportunities for adsorption, so that the poison will be uniformly (or nonselectively) distributed on the pore surface. In a first-order reaction A → products, if the fraction of surface covered by poison is p, then the rate constant decreases to a value $k_v(1-p)$, leading to the modified modulus

$$(\phi_{p1})_f = L \left[\frac{k_v(1-p)}{D_{eA}} \right]^{1/2} = \phi_{p1}(1-p)^{1/2} \quad (8.5)$$

and the modified effectiveness factor for the diffusion regime

$$\varepsilon = \frac{\tanh(\phi_{p1})_f}{(\phi_{p1})_f} = \frac{\tanh\left[\phi_{p1}(1-p)^{1/2}\right]}{\phi_{p1}(1-p)^{1/2}} \quad (8.6)$$

We can now write equations for the reaction rate for the poisoned ($\mathscr{R}_f$) and unpoisoned ($\mathscr{R}$) pores, leading to the following expression for the fraction of activity left in the poisoned pore:

$$\frac{\mathscr{R}_f}{\mathscr{R}} = \frac{\phi_{p1}(1-p)^{1/2}\tanh\left[\phi_{p1}(1-p)^{1/2}\right]}{\phi_{p1}\tanh\phi_{p1}} \quad (8.7)$$

The following limiting values of $\mathscr{R}_f/\mathscr{R}$ may be noted:

Low ϕ_{p1} (slow reaction, entire surface available):

$$\frac{\mathscr{R}_f}{\mathscr{R}} = 1-p \quad (8.8)$$

High ϕ_{p1} (fast reaction, fractional surface available):

$$\frac{\mathscr{R}_f}{\mathscr{R}} = (1-p)^{1/2} \quad (8.9)$$

Evidently, in the chemical control region (say, $\phi_{p1} < 1$) the fractional activity falls in direct proportion to the fractional coverage by poison, but in the diffusion-controlled region (say, $\phi_{p1} > 10$) the activity falls less than linearly with p. As the value of ϕ_{p1} is increased, the effect of poisoning on catalyst effectiveness progressively decreases. This can be understood when it is remembered that in the high-ϕ_{p1} range the reaction is fast and the reaction zone is thus restricted to the pore mouth or external surface of catalyst. If now the catalyst is uniformly poisoned, the reaction at the surface is retarded, but the reaction zone progresses inward, utilizing more of the less active surface. This leads to an increased reaction in comparison with reaction under conditions of chemical control when the entire surface is always available.

Selective or Pore Mouth Poisoning

In the case of pore-mouth poisoning, the poison is preferentially adsorbed at the pore mouth. The Thiele modulus would now be modified through the distance parameter (L or R) and not through the rate constant k_v, as in homogeneous poisoning. Thus

$$(\phi_{p1})_f = \phi_{p1}(1-p) \quad (8.10)$$

On the assumption of a linear concentration profile in the poisoned portion, the following final equation can then be obtained for the fractional activity left $\mathscr{R}_f/\mathscr{R}$ (Wheeler, 1955):

$$\frac{\mathscr{R}_f}{\mathscr{R}} = \frac{\tanh[\phi_{p1}(1-p)]}{\tanh \phi_{p1}} \left(\frac{1}{1+\phi_{p1}p}\right) \quad (8.11)$$

The effect of pore-mouth poisoning is most marked at high values of the Thiele modulus.

The distinction between the different types of poisoning mentioned here is brought out in Figure 8.1, which shows plots of $\mathscr{R}_f/\mathscr{R}$ vs. p for different conditions. Based on these curves, the following classification can be made:

Condition	Type of Poisoning
Low ϕ; $\mathscr{R}_f/\mathscr{R}$ is directly proportional to p.	Homogeneous nonselective (also selective)
High ϕ; $\mathscr{R}_f/\mathscr{R}$ is proportional to $\sqrt{p}$, that is, falls less than linearly with p.	Homogeneous antiselective
High ϕ; $\mathscr{R}_f/\mathscr{R}$ falls sharply with p.	Selective

It will be noticed that whereas at high values of ϕ (i.e., for short pores with large diameters) it is possible to distinguish clearly between nonselective (homogeneous) and selective (pore-mouth) poisoning, in the low-ϕ region (corresponding to kinetic control) no such distinction exists between the types of poisoning involved. In fact, as will be shown in the next section, no distinction between any of the deactivation mechanisms is possible in the kinetic regime.

8.2.2. Analysis Based on Center-Plane Concentration

Recent studies have shown that the center-plane concentration of the catalyst pellet can provide a useful and practical means of characterizing the type of deactivation involved. Since the introduction of this quantity to distinguish between homogeneous and pore-mouth poisoning (Balder and Petersen, 1968a), methods have been proposed for using it as a more general parameter for distinguishing among various types of impurity and self-poisoning (Dougharty, 1970; Hegedus and Petersen, 1973a,b).

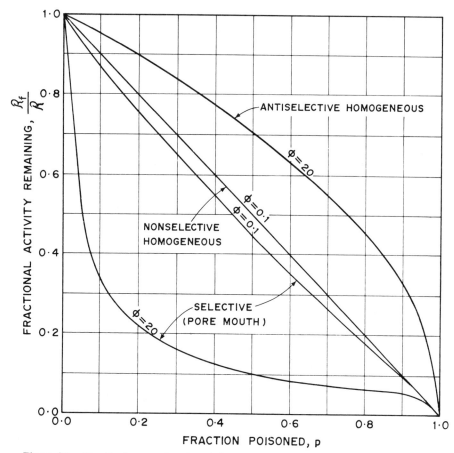

Figure 8.1. Deactivation as a function of degree of poisoning and diffusion (Thiele modulus).

The quantity used is not just the concentration at the center plane, but a normalized relative center-plane concentration, defined as

$$\psi = \frac{(\hat{C}_{cp})_f - \hat{C}_{cp}}{1 - \hat{C}_{cp}} \qquad (8.12)$$

where

$$(\hat{C}_{cp})_f = \frac{(C_{cp})_f}{C_s}, \qquad \hat{C}_{cp} = \frac{C_{cp}}{C_s} \qquad (8.13)$$

In these equations $(C_{cp})_f$ is the center-plane concentration of deactivated catalyst and C_{cp} the center-plane concentration of fresh catalyst.

If the fractional activity left $\mathcal{R}_f/\mathcal{R}$ can be related to the center-plane concentration ψ, then from the nature of these curves a clear distinction can be made between homogeneous and selective poisoning. The final equations for $\mathcal{R}_f/\mathcal{R}$ and for the center-plane concentration (from which ψ can be calculated), as given by Balder and Petersen for a first-order irreversible reaction in a cylindrical pore, are

Homogeneous poisoning:

$$\frac{\mathcal{R}_f}{\mathcal{R}} = \frac{\tanh \phi_{p1}}{\phi_{p1}}, \qquad (\hat{C}_{cp})_f = \frac{1}{\cosh \phi_{p1}} \qquad (8.14)$$

Selective poisoning:

$$\frac{\mathcal{R}_f}{\mathcal{R}} = \frac{\tanh[\phi_{p1}(1-p)]}{\phi_{p1}[1 + p\phi_{p1}\tanh[\phi_{p1}(1-p)]]}$$

$$(\hat{C}_{cp})_f = \{\cosh[\phi_{p1}(1-p)]$$
$$\times [1 + p\phi_{p1}\tanh[\phi_{p1}(1-p)]]\}^{-1} \qquad (8.15)$$

Hegedus and Petersen (1973a) have considered a third kind of poisoning. In this so-called core poisoning, the poison wave starts from the center and moves outward to the surface. This is similar to the reaction starting at the center in a gas–solid system and spreading outward (see Gokhale et al., 1975; Kulkarni and Doraiswamy, 1979). It is thus the reverse of pore-mouth poisoning. Though rather rare, it provides, along with the homogeneous and pore-mouth poisoning mechanisms, three clear-cut bounds within which any real poisoning mechanism must lie. The equations for $\mathcal{R}_f/\mathcal{R}$ and $(\hat{C}_{cp})_f$ for this situation are

$$\frac{\mathcal{R}_f}{\mathcal{R}} = \frac{\tanh[\phi_{p1}(1-p)]}{\tanh \phi_{p1}},$$
$$(\hat{C}_{cp})_f = \cosh[\phi_{p1}(1-p)]^{-1} \qquad (8.16)$$

In any experimental program it is only necessary to determine the rates and center-plane concentrations in the fresh and poisoned catalysts. For this purpose the pellet is mounted in such a manner (Balder and Petersen, 1968a,b) that only one side of it is exposed to bulk flow; then the entire thickness of the pellet constitutes the diffusion length, and the concentration on the other side of the pellet corresponds to the center-plane concentration.

It is possible to extend the treatment to self-poisoning, including all three types (see Section 8.1.1)—parallel, series, and triangular—again by plotting $\mathcal{R}_f/\mathcal{R}$ as a function of ψ. The experimental values will normally fall between the extremes of homogeneous, pore-mouth, and core poisoning. These points are then matched against curves generated for the various models with different values of the fouling parameters. Thus not only the mechanism of deactivation but also the parameter values associated with it may be determined.

Since discrimination among the various mechanisms is based on the behavior of the center-plane concentration, consideration should necessarily be restricted to diffusion-influenced conditions. The analysis is therefore based on the formulation of appropriate transport equations and their solution in terms of the center-plane concentration. The main reaction for all the fouling types considered is A → B. The mass transport equation for A is then

$$D_{eA}\frac{\partial^2 C_A}{\partial l^2} - k_v \Omega C_A^a = 0 \qquad (8.17)$$

and the equations describing the various fouling reactions are

1. Parallel fouling (A $\overset{2}{\rightarrow}$ C):

$$-\frac{\partial \Omega}{\partial t} = k_{v,f} C_A^d \Omega^d \qquad (8.18)$$

2. Series fouling (B $\overset{3}{\rightarrow}$ C):

$$-\frac{\partial \Omega}{\partial t} = k_{v,f} C_B^d \Omega^d \qquad (8.19)$$

3. Triangular fouling (A $\overset{2}{\rightarrow}$ C, B $\overset{3}{\rightarrow}$ C):

$$-\frac{d\Omega}{dt} = C_A \Omega \left(l - \frac{k_{v3,f}}{k_{v2,f}}\right) + \Omega \frac{k_{v3,f}}{k_{v2,f}}(C_A + C_B) \qquad (8.20)$$

4. Impurity fouling:

$$-\frac{\partial \Omega}{\partial t} = k_{v,f} C_p^d \Omega^d \qquad (8.21)$$

5. Independent fouling:

$$-\frac{\partial \Omega}{\partial t} = k_{v,f} \Omega^d \qquad (8.22)$$

The boundary conditions for all the cases considered are identical and may be written in terms of the following nondimensional concentration and distance parameters:

$$\hat{C} = \frac{C(t=t, l=1)}{C(t=0, l=L)} \qquad (8.23)$$

where $C = C_A$, C_B, or C_C.

1. At $t = 0$, $l = 1$:

$$\hat{C}_A = \frac{\cosh(\hat{L}\phi_{p1})}{\cosh \phi_{p1}}; \qquad \hat{C}_B = \hat{C}_C = 0; \qquad \Omega = 1$$

$$(8.24)$$

2. At $t = t$, $l = l$:

$$\hat{C}_A = \hat{C}_B = \hat{C}_C = 1 \qquad (8.25)$$

3. At $t = t$, $l = 0$:

$$\frac{\partial \hat{C}_A}{\partial \hat{L}} = \frac{\partial \hat{C}_C}{\partial \hat{L}} = 0 \qquad (8.26)$$

Hegedus and Petersen (1973a) have solved these equations for impurity, parallel, and series fouling for an intermediate value of the Thiele modulus ($\phi_{p1} = 2.5$) for the main reaction. If $(\phi_{p1})_f$ represents the Thiele modulus for the fouling reaction, then the condition $(\phi_{p1})_f \gg 5$ corresponds to pore-mouth poisoning, while $(\phi_{p1})_f < 0.1$ corresponds to uniform poisoning.

When the results for a first-order main reaction (for $a = 1$) are plotted as the relative rate or activity vs. the center-plane concentration ψ defined by Eq. 8.12, different curves are obtained not only for the different classes of fouling but also for different values of the fouling parameters d and f for a given class of fouling. However, the curves can be grouped in six broad regions, as shown in Figure 8.2, which also contains the

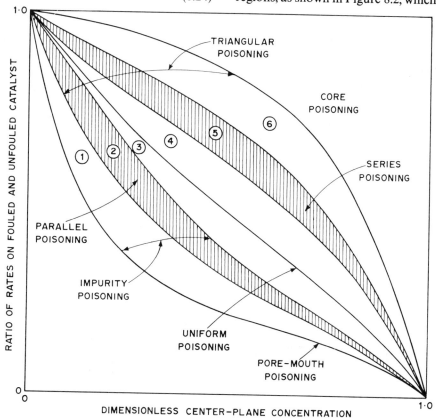

Figure 8.2. Regimes of poisoning.

curves for the three limiting cases: homogeneous (or uniform), pore-mouth, and core poisoning. Analytical solutions for these extreme cases have already been given (Eq. 8.14–8.16).

The fouling mechanisms associated with the various regions are indicated in Table 8.3. Two distinct bands may be observed, one between homogeneous and pore-mouth poisoning and representing parallel fouling (region 2), and the other between homogeneous and core poisoning and representing series fouling (region 5). It is interesting to note that decreasing the Thiele modulus will cause all the regions to shift toward the diagonal from both sides, until eventually, as ϕ_{p1} approaches zero, all of them reduce to a single line, the diagonal.

Since more than one mechanism is represented by some regions (see Table 8.3), a simple experimental strategy can be worked out for obtaining the correct mechanism (Hegedus and Petersen, 1974).

TABLE 8.3. Fouling Mechanisms Associated with the Various Regions of Figure 8.2

Region	Fouling Mechanism
1	Impurity poisoning
2	Parallel self-poisoning
	Triangular self-poisoning
	Impurity poisoning
3	Triangular self-poisoning
	Impurity poisoning
4	Triangular self-poisoning
5	Series self-poisoning
	Triangular self-poisoning
6	Hypothetical case of an impurity-poisoning front progressing from the center plane of the pellet toward its outside boundary

8.2.3. Shell-Progressive Model

It is possible to extend the concept of pore-mouth poisoning by breaking up the overall process in terms of the time required for the various steps involved, thus introducing time as a parameter in the analysis. This is accomplished by idealizing the poison (or carbon) profiles in the pellet into two distinct regions: one in which the carbon is uniformly distributed, and one that is free of carbon. This is called the shell-progressive model (SPM). Weisz and Goodwin (1963), Masamune and Smith (1966), and Murakami et al. (1962) have demonstrated the applicability of the SPM.

In the SPM case, diffusion and reaction must occur in series. It follows therefore that for the assump-

tion of the SPM to be valid the Thiele modulus must be high (reaction should be very fast). Carberry and Gorring (1966) have defined the following approximate criterion for the applicability of the SPM:

$$\phi_{p1} \approx 100$$

The SPM is equivalent to the sharp-interface model characteristic of gas–solid reactions; mathematical treatments of this model have therefore appeared as part of the gas–solid reaction literature and not specifically with reference to catalyst deactivation. Thus, the following criteria derived for the sharp-interface model in general are also applicable:

Calvelo–Cunningham (1970) criterion: $\dfrac{100\hat{R}^2}{(\text{Bi})_m} < 1$

Mantri–Gokarn–Doraiswamy (1976) criterion:
$$\phi_{s1} > 100$$

Consider a flat plate of thickness $2L$ with both faces open. The effective diffusion length to be considered is therefore L. At any time t let a shell of thickness $L - l_i$ be formed, where l_i is the distance of the reaction boundary starting from the center line (where $l = 0$). Each of the three resistances involved (film diffusion, shell diffusion, and reaction) must be equal to the rate of accumulation or consumption of the poison (depending on whether we are considering fouling or regeneration), and can be expressed in terms of a characteristic time. Based on this approach, the following final equation has been derived by Carberry and Gorring (1966) for the total time for the reaction $A + v_B B \rightarrow$ products:

$$t = \frac{C_p L^2}{2v_B D_{eA} C_{Ab}}\left\{6(1 - \hat{L}_i)\left[\frac{1}{(\text{Bi})_m} + \frac{1}{\text{Da}}\right] + (1 - \hat{L}_i)^2\right\} \tag{8.27}$$

where $(\text{Bi})_m = k_g L/D_{eA}$ and $\text{Da} = k_S L/D_{eA}$. For a flat plate, the fraction poisoned p is $1 - \hat{L}_i$. Combining this with Eq. 8.27, and solving the resulting quadratic for p, we have

$$p = -K + (K^2 + 2\hat{t}^2)^{1/2} \tag{8.28}$$

where

$$K = 3\left[\frac{1}{(\text{Bi})_m} + \frac{1}{\text{Da}}\right] \tag{8.29}$$

$$\hat{t} = \frac{1}{L}\left(\frac{v_B D_{eA} C_{Ab} t}{C_p}\right)^{1/2} = \text{dimensionless time} \tag{8.30}$$

The flat-plate equation can be adapted to a sphere merely by replacing L by $R/3$.

This model is applicable up to certain degrees of poisoning only, the actual range depending on the values of $(Bi)_m$ and Da. However, it has the advantage of simplicity in that an explicit p–$\hat{t}$ relationship is possible, and can be used without serious loss of accuracy. It has been found (Masamune and Smith, 1966) that although the SPM is valid for parallel fouling, it is unsuitable for series fouling.

8.2.4. Isothermal Effectiveness Factors

The use of time-dependent effectiveness factors to account for catalyst deactivation brings the analysis in line with that normally employed for catalysts of steady activity.

Slow Fouling. The nondimensional continuity equations for A, B, and carbon in a spherical pellet (see Chapter 4) in the presence of a fouling reaction, and on the assumptions that D_e is unaffected by fouling, $k_{v,f} \ll k_v$, and the accumulation of A or B in the void space of the pellet is negligible, can be written as

Component A:

$$\frac{\partial^2 \hat{C}_A}{\partial \hat{R}^2} + \frac{2}{\hat{R}} \frac{\partial \hat{C}_A}{\partial \hat{R}} - (\phi_{s1}^2)_A \Omega \hat{C}_A = 0 \qquad (8.31)$$

Component B:

$$\frac{\partial^2 \hat{C}_B}{\partial \hat{R}^2} + \frac{2}{\hat{R}} \frac{\partial \hat{C}_B}{\partial \hat{R}} + (\phi_{s1}^2)_A \Omega \frac{D_{eA}}{D_{eB}} \frac{C_{As}}{C_{Bs}} \hat{C}_A = 0 \quad (8.32)$$

Poison P:

$$\frac{\partial^2 \hat{C}_p^2}{\partial \hat{R}^2} + \frac{2}{\hat{R}} \frac{\partial \hat{C}_p}{\partial \hat{R}} - (\phi_{s1}^2)_p \Omega \hat{C}_p = 0 \qquad (8.33)$$

Deactivation:

$$\text{Parallel fouling:} \quad \frac{\partial \Omega}{\partial t_A} = \Omega \hat{C}_A \qquad (8.34)$$

$$\text{Series fouling:} \quad \frac{\partial \Omega}{\partial t_B} = \Omega \hat{C}_B \qquad (8.35)$$

$$\text{Impurity poisoning:} \quad \frac{\partial \Omega}{\partial t_p} = -\Omega \hat{C}_p \qquad (8.36)$$

Boundary conditions:

$$\hat{C}_A = 1, \ \hat{C}_B = 1, \ \hat{C}_p = 0 \quad \text{at} \quad t_A, t_B > 0, \ \hat{R} = 1$$
$$\qquad (8.37)$$

$$\frac{\partial \hat{C}_A}{\partial \hat{R}} = \frac{\partial \hat{C}_B}{\partial \hat{R}} = \frac{\partial \hat{C}_p}{\partial \hat{R}} = 0 \quad \text{at} \quad t_A, t_B > 0, \ \hat{R} = 0$$

$$\Omega = 1 \quad \text{at} \quad t_A, t_B = 0$$

where, as usual

$$(\phi_{s1})_A = R\left(\frac{k_{vA}}{D_{eA}}\right)^{1/2}, \quad (\phi_{s1})_p = R\left(\frac{k_{v,f}}{D_{ep}}\right)^{1/2}$$
$$\qquad (8.38)$$

and

$$t_A = C_{As} k_{v,f} t, \quad t_B = C_{Bs} k_{v,f} t, \quad t_p = C_{ps} k_{v,f} t. \qquad (8.39)$$

Numerical solutions to the self-poisoning equations (parallel and series) were worked out by Masamune and Smith (1966) on the assumption of linear decay due to carbon deposition ($\Omega = 1 - \hat{W}_c$), and the results were then used to obtain the effectiveness factor. The solutions obtained for parallel fouling are shown in Figure 8.3 as plots of ε vs. t_A for different values of $(\phi_{s1})_A$. More recently, Do and Weiland (1981a,b) also provided analytical solutions for the case of series and parallel deactivation in a slab, infinite cylinder, and sphere. They employed the finite Sturm–Liouville integral transforms combined with generalized multitiming techniques to obtain the reactant concentration, activity, and intraphase effectiveness factor valid over the entire life of the catalyst. The solutions obtained in terms of the convergent infinite series compare very well with the numerical solutions of Masamune and Smith (1966). Do and Weiland (1981c) have also solved the case of self-poisoning in finite cylindrical pellets of various aspect ratios (L/d_p), and the computations indicate that infinite cylinder results can be used for low values of the Thiele modulus ($\phi < 1$) without any significant error. However, for pellets with aspect ratio less than unity, the error involved can be substantial, especially at large values of the modulus ($\phi > 5$).

A similar analysis has been carried out by Kam et al. (1975) and by Kulkarni and Ramachandran (1979), who obtained solutions by collocation and modified collocation methods. In view of the analytical nature (and general usefulness) of the solutions obtained by Kulkarni and Ramachandran (1979), their results for different types of fouling are summarized in Table 8.4. Tai and Greenfield (1978) and Lamba and Dudukovic

TABLE 8.4. Analytical Expressions for Effectiveness Factors for Parallel and Series Fouling[a]

Type of Fouling	Expression for Effectiveness Factor

Parallel: $A \begin{smallmatrix} \nearrow B \\ \searrow \text{coke} \end{smallmatrix}$

1. Low values of ϕ

$$\varepsilon = (1 + S)\left[\frac{W_1 \exp(-\psi)}{(\phi^2/B_{12})\exp(-\psi) + 1} + W_2 \exp(-\hat{t}_A)\right]$$

where

$$\hat{t}_A = \frac{\phi^2}{B_{12}}\left[1 - \exp(-\psi)\right] + \psi$$

$S = 0, 1,$ and 2 for slab, cylinder, and sphere, respectively; W_1 and W_2 are the collocation weights; B_{12} is an element of a coefficient matrix; ψ is the cumulative concentration at the collocation point (see Section 19.1.2).

2. High values of ϕ

$$\varepsilon = (1 + S)(1 - \lambda)\int_0^1 \left[u(1 - \lambda) + \lambda\right]^S \left[u^2 + 2u\hat{t}_A\left(\frac{u - 1}{1 - \lambda}\right)\frac{d\lambda}{d\hat{t}_A}\right]\exp(-\hat{t}_A u^2)\, du$$

where u is a dummy variable, and λ is obtained from the solution of the cubic equation

$$\frac{2\hat{t}_A}{(1 - \lambda)^2} + \frac{2S\hat{t}_A u_c}{(1 - \lambda)[u_c(1 - \lambda) + \lambda]} - \phi^2[1 - \exp(-\hat{t}_A u_c^2)] = 0$$

$u_c = 0.464$ and 0.53 for the flat plate and sphere, respectively.

Series: $A \to \nu_B \quad B \to \text{coke}$

1. Low values of ϕ

$$\varepsilon = (1 + S)\left\{\frac{W_1 \exp[-(C_{B0} + \nu)\hat{t}_B + \nu\psi]}{1 + (\phi^2/B_{12})\exp[-(C_{B0} + \nu)\hat{t}_B + \nu\psi]} + W_2 \exp(-C_{B0}\hat{t}_B)\right\}$$

where

$$\hat{t}_B = \frac{1}{C_{B0} + \nu}\ln\frac{[\phi^2(C_{B0} + \nu)/B_{12}C_{B0}][1 - \exp(-C_{B0}\psi)] + 1}{\exp[-\psi(C_{B0} + \nu)]}$$

and C_{B0} is the dimensionless initial concentration of the species B and $\nu = \nu_B D_{eA}/D_{eB}$.

2. High values of ϕ

$$\varepsilon = (1 + S)(1 - \lambda)\int_0^1 [u(1 - \lambda) + \lambda]^S[u^2 + 2\hat{t}_B u\left(\frac{u - 1}{1 - \lambda}\right)\frac{d\lambda}{d\hat{t}_B}]\exp[-(C_{B0} + \nu)\hat{t}_B + \nu\hat{t}_B u^2]\, du$$

where λ is obtained as a solution of the cubic equation

$$\frac{4\hat{t}_B}{(1 - \lambda)^3} + 2\hat{t}_B S\left\{\frac{u_c - 1}{(1 - \lambda)^2[u_c(1 - \lambda) + \lambda]} - \frac{u_c(u_c - 1)}{(1 - \lambda)[u_c(1 - \lambda) + \lambda]^2}\right.$$

$$\left. - \frac{u_c[1 - 2\lambda - 2u_c(1 - \lambda)]}{[u_c(1 - \lambda)^2 + \lambda(1 - \lambda)]^2}\right\} - 2\phi^2\hat{t}_B u_c\left(\frac{u_c - 1}{1 - \lambda}\right)\exp[(-C_{B0} + \nu)\hat{t}_B + \nu\hat{t}_B u_c^2]\frac{d\lambda}{d\hat{t}_B}$$

$$= \phi^2 u_c^2\exp[-(C_{B0} + \nu)\hat{t}_B + \nu\hat{t}_B u_c^2] - \frac{2}{(1 - \lambda)^2} + \frac{2Su_c}{(1 - \lambda)[u_c(1 - \lambda) + \lambda]}$$

[a]Kulkarni and Ramachandran (1979)

162

(1978) have also presented simple methods for obtaining solutions in analytical form.

An examination of the solutions for parallel and series fouling leads to the following useful conclusions.

1. In parallel fouling, as shown in Figure 8.3, the curves for relatively low values of the Thiele modulus cross each other. For fresh catalyst there is a continual decrease in the effectiveness factor as the Thiele modulus is increased, but with increase in time the effectiveness factor first increases and then decreases with the Thiele modulus. Thus a catalyst with some diffusional resistance (say, with $(\phi_{s1})_A$ corresponding to transition from chemical to diffusional control) would appear to be the preferred one in view of its stability over a long process period.

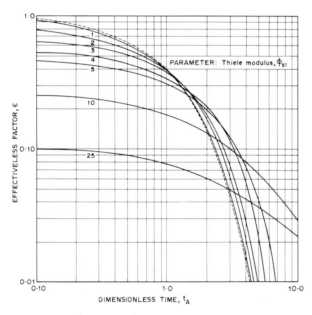

Figure 8.3. Effectiveness factor as a function of time and Thiele modulus for a spherical isothermal pellet subject to parallel fouling (Masamune and Smith, 1966).

2. In series fouling, the extent of fouling increases continuously with the diffusion resistance (ϕ_{s1}) for all values of process time. Thus, a catalyst with the least diffusion resistance is the preferred one.

The analysis has been extended by Kulkarni and Ramachandran (1979) to incorporate the variation of effective diffusivity with coke content. Their results, however, indicate negligible influence of this variation on the effectiveness factor except at large times.

Chu (1968), Gioia et al. (1970), and Wolf and Petersen (1974) have extended the analysis to H–W kinetics. In general, the effect of the introduction of H–W kinetics into the rate expression for the main reaction is to increase the importance of fouling.

Fast Fouling. Murakami et al. (1968) and Kam et al. (1977b) have provided solutions for relatively fast fouling (by including the accumulation term) caused by a parallel or series mechanism. A significant conclusion from their studies is that a catalyst with higher activity or greater diffusional resistance is fouled more rapidly for both parallel and series schemes. According to Kam et al. (1977b), fast fouling is characterized by low values (1–10) of a fouling number FN defined as

$$FN = \frac{D_{eA}}{f_B R^2 k_{v,f} C_{A0}} \tag{8.40}$$

For higher values of FN the pseudo-steady-state assumption (made in slow fouling) is valid.

The extent of carbon deposition inside a pellet is clearly a direct indication of the degree of fouling. The case of regeneration is dependent in some measure on the nature of the carbon profile in the pellet. Thus a pellet with greater carbon deposition on its outer surface can be regenerated more easily. The carbon profile is a function of the fouling mechanism involved, the relative rate of the fouling reaction, and the operative regime of control. An indication of the manner in which carbon deposition occurs under different conditions is given in Table 8.5. The experimental results of Suga et al. (1967) on the dehydrogenation of *n*-butene over an alumina–chromia catalyst suggest, however, that for low values of ϕ uniform deposition can be assumed, whereas in the high-ϕ region the shell model seems to be appropriate.

Reaction of General Order

In the previous section a first-order main reaction with a series or parallel fouling reaction was considered. Greco et al. (1973) extended the analysis to the case of zero-order main reaction and first-order fouling. We shall consider the more important case of general-order main reaction. Thus Eq. 8.31 can be written as

$$\frac{d^2 \hat{C}_A}{d\hat{R}^2} + \frac{2}{\hat{R}}\left(\frac{d\hat{C}_A}{d\hat{R}}\right) - \phi_{s1}^2 \hat{C}_A^n \Omega = 0 \tag{8.41}$$

Following Bischoff (1969) (see Section 16.1) and assum-

TABLE 8.5. Nature of Carbon Deposition in a Decaying Catalyst Pellet

Fouling Mechanism	Catalyst with Small Diffusion Effect (low ϕ_{s1})	Catalyst with Large Diffusion Effect (high ϕ_{s1})
Parallel		
a. Fast	Deposition from outer part of pellet	Deposition from outer part but more nearly uniform
b. Slow	Deposition from outer part of pellet	Deposition from outer part of pellet
Series		
a. Fast	Deposition from center of pellet	Deposition from outer part of pellet
b. Slow	Deposition from center of pellet	Deposition from center of pellet

ing that $\Omega = 1/\hat{W}_c$, we can use a general relationship of the form

$$\hat{C}_A(R, t) \left\{ \left[\frac{\hat{W}_c(R, t) - \hat{W}_c(0)}{\hat{W}_c(0, t) - \hat{W}_c(0)} \right] \hat{C}_A(0) \right\}^{1/m} \quad (8.42)$$

for the reactant concentration $\hat{C}_A$ as a function of carbon content $\hat{W}_c$ (for parallel fouling), where m is some suitably chosen exponent.

Equations 8.41 and 8.42 can be combined to give

$$\frac{d^2\hat{C}_A}{d\hat{R}^2} + \frac{2}{\hat{R}}\left(\frac{d\hat{C}_A}{d\hat{R}}\right) - \phi_{s1}^2 \left[\frac{\hat{C}_A(0)}{\hat{W}_c(0, t)} \right] \hat{C}_A^{n-m} = 0 \quad (8.43)$$

After certain mathematical manipulations, the following asymptotic solution for the effectiveness factor can be obtained for $(n - m) = 1$:

$$\varepsilon = \frac{3}{\phi_{s1}} \left(\frac{\hat{C}_A(0)}{2k_{v,f} C_{A0} t} \right)^{1/4} \quad (8.44)$$

Plots of ε vs. $[2k_{v,f} C_{A0} t / \hat{C}_A(0)]^{-1/4}$ with ϕ_{s1} as a parameter are similar to any normal effectiveness factor plot in the asymptotic region.

We shall now examine the effect of pore–mouth poisoning on catalyst effectiveness. The outer poisoned zone presents a diffusional resistance as a result of which the concentration at the interface (between the poisoned and active zones) represented by R_i is no longer the surface concentration C_{As}, but a modified value C_{Ai}. In dimensionless notation this is expressed as $\hat{C}_{Ai} = C_{Ai}/C_{As}$ and can be calculated on the basis of a mass balance through the diffusion film. The final expression for ε is

$$\varepsilon = \frac{3}{\phi_{s1}^2} \frac{\hat{R}_i(1 - \hat{C}_{Ai})}{1 - \hat{R}_i} \quad (8.45)$$

This is an implicit equation in C_{Ai} that can be solved by trial and error.

Concentration-Independent Deactivation

In the analysis presented in this section so far, it was mainly assumed that the concentration of the reactant and/or product affects the deactivation rate. Although this situation is common, the incorporation of this complex analysis of a single particle system into a macroscopic model for the chemical reactor can become unwieldy. In such instances it is preferable to use concentration-independent deactivation to facilitate easy interpretation and analysis of deactivation data obtained from flow reactors. The expression for the effectiveness factor for a single particle (say, a sphere) is similar to that for a simple first-order system (Eq. 14.18; Krishnaswamy and Kittrell, 1981). The Thiele modulus is, however, modified in this case to take account of deactivation as $(\phi)_m = 3\phi_{s1} \exp(-k_f t/2)$.

The case of concentration-independent poisoning for a simple first-order main reaction has also been analyzed by Angele et al. (1980). The analysis incorporates the diffusional resistance for the poison along with the main reactant, and the model has been used to interpret the kinetic data on the poisoning of a porous noble metal carrier-type catalyst by impurity poisoning.

8.2.5. Nonisothermal Effectiveness Factors

It is logical to expect that the effectiveness factor of a fouling catalyst would be greatly influenced by non-isothermal conditions within a pellet. In order to account for nonisothermicity within a pellet undergoing homogeneous poisoning, the rate constant is expressed in terms of the surface value, as in Eq. 4.43, and the resulting continuity equations for slow poisoning are as follows.

Component A:

$$\frac{\partial^2 \hat{C}_A}{\partial \hat{R}^2} + \frac{2}{\hat{R}} \frac{\partial \hat{C}_A}{\partial \hat{R}} - \phi_{s1}^2 \hat{C}_A \exp\left[\frac{\alpha_s \beta_m (1 - \hat{C}_A)}{1 + \beta_m (1 - \hat{C}_A)}\right] \quad (8.46)$$

Deactivation:

Parallel: $\quad \dfrac{\partial \Omega}{\partial t_A} = \hat{C}_A \Omega \exp\left[\dfrac{\alpha_{s,f} \beta_m (1 - \hat{C}_A)}{1 + \beta_m (1 - \hat{C}_A)}\right] \quad (8.47)$

Series: $\quad \dfrac{\partial \Omega}{\partial t_B} = \hat{C}_B \Omega \exp\left[\dfrac{\alpha_{s,f} \beta_m (1 - \hat{C}_A)}{1 + \beta_m (1 - \hat{C}_A)}\right] \quad (8.48)$

where $\alpha_{s,f} = E_f / R_g T$ and t_A and t_B are dimensionless times defined by Eq. 8.39. The boundary conditions are

$$\begin{aligned} \Omega &= 1 && \text{at} \quad \hat{t} = 0 \quad \text{for} \quad 1 \geqslant \hat{R} \geqslant 0 \\ \hat{C}_A &= 1 && \text{at} \quad \hat{R} = 1 \quad \text{for} \quad \hat{t} > 0 \\ \frac{d\hat{C}_A}{d\hat{R}} &= 0 && \text{at} \quad \hat{R} = 0 \quad \text{for} \quad \hat{t} > 0 \end{aligned} \quad (8.49)$$

Solutions have been worked out by Sagara et al. (1967) for $\Omega = 1 - W_c$ and their results for parallel fouling are presented in Figure 8.4 as plots of ε vs. t_A

for different values of the Thiele modulus at $\alpha_s = 20$, $\alpha_{s,f} = 30$, and $\beta_m = 0.01$. For series fouling an additional parameter involving diffusivities has to be specified. The following conclusions can be drawn:

1. In parallel fouling at a given Thiele modulus, the nonisothermal effectiveness factor $\varepsilon_{\text{noniso}}$ has a higher value than ε_{iso} until a certain time is reached at which the curves cross each other, and thereafter $\varepsilon_{\text{noniso}}$ has a lower value than ε_{iso}. The crossover occurs at progressively larger times as the Thiele modulus is increased. This happens because at shorter times the concentration effect is dominant, whereas at longer times the effect of temperature on the fouling reaction becomes dominant, leading to a lower value of the effectiveness factor.

2. In series fouling, the temperature and concentration effects both tend to increase fouling. In view of the detrimental effect on the concentration of B within the pellet, the crossing of the isothermal and nonisothermal curves will now occur at progressively shorter times as the Thiele modulus is increased (contrary to the situation in parallel fouling).

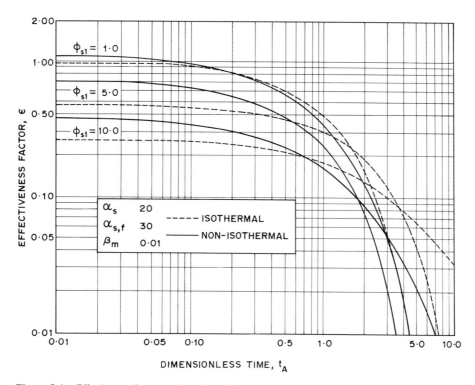

Figure 8.4. Effectiveness factor as a function of time and Thiele modulus for a spherical nonisothermal pellet subject to parallel fouling (Sagara et al., 1967).

3. The opposing influences of reactant concentration and temperature in parallel fouling lead to unusual carbon profiles in the pellet at moderate values of ϕ_{s1}. Thus, the coke concentration rises as one moves from the pellet surface, reaches a maximum at an intermediate point, and then falls toward the center. Evidently, for a situation of this type, the SPM does not hold.

For pore-mouth poisoning caused either by a highly adsorbing poison or through a foulant produced largely near the surface (as, e.g., by a parallel reaction), the following conclusions can be drawn (Ray, 1972):

1. There is an interval in the life of the catalyst pellet during which deactivation may actually improve the conversion of an exothermic reaction as a result of the insulating effect of the deactivated part of the pellet.

2. Small diffusion-limited pellets with pore-mouth poisoning may give higher conversion than pellets under homogeneous or antiselective poisoning.

3. The optimal reactor policy strongly depends on the internal structure of the catalyst poisoning, and hence the optimization of a fixed-bed reactor with catalyst deactivation needs to be examined in much greater detail for the pseudohomogeneous reactor models normally employed (see Chapter 11).

4. The product selectivity must include a detailed analysis of the particle activity profile.

8.2.6. Hougen–Watson Kinetics

The studies of catalyst fouling described thus far have assumed integral-order rate expressions for both the main and fouling reactions. In the present section we shall consider H–W kinetics for the main reaction with either power law or H–W kinetics for the fouling step. Kam et al. (1977a,c) have analyzed the case of first-order fouling under nonisothermal conditions. The work has been extended by Kam and Hughes (1979) to the case of H–W type fouling, and their conclusions may be summarized as follows:

1. For both parallel and series fouling, an increase in ϕ results in an increase in the rate of deactivation. The severity of fouling is, however, much more pronounced for parallel fouling than for series fouling, particularly at low values of ϕ.

2. At low values of ϕ an increase in K_A or K_B results in an increase in the rate of deactivation. This is consistent with the results obtained by Chu (1968) for the corresponding isothermal system. At higher values of ϕ there appears to be a reversal in trend.

8.2.7. Effect of External Film Resistance

Kam et al. (1977a,c) examined the effect of external heat and mass transport limitations (by using the collocation method for solving the heat and mass balance equations) in terms of the usual Biot numbers for heat and mass transfer. Their conclusions for a reaction following H–W kinetics, accompanied by a first-order fouling reaction, may be summarized as follows.

1. The effect of film resistance [i.e., of $(Bi)_m$] is negligible in the case of isothermal fouling; this is also true for nonisothermal fouling accompanying endothermal reactions.

2. In the case of nonisothermal fouling accompanying an exothermic reaction, the extent of fouling increases with increasing $(Bi)_h$, particularly under conditions of diffusion control.

3. Increase in the Arrhenius parameter of the fouling reaction α_f leads to a significant increase in fouling.

4. While the extent of parallel–series fouling depends on the extent of the individual contributions, it is always less than either operating individually.

8.2.8. Example: Fouling of the Chromia–Alumina Catalyst in the Dehydrogenation of *n*-Butane

The dehydrogenation of *n*-butane over a chromia-alumina catalyst is reported to be a series fouling reaction, with the main reaction catalyzed by chromia and coke formation occurring on both chromia and alumina (Toei et al., 1975). The rate of the main reaction is expressed by the H–W type of equation

$$r_1 = \frac{k_1 p_A}{1 + K_A p_A} \qquad (8.50)$$

with $k_1 = 0.45 \times 10^{-4}$ mol/g catalyst sec atm and $K_A = 3.33$ atm^{-1}. Coke formation is found to be first order with respect to *n*-butene, and the effect of adsorption of *n*-butene is negligible. The experimental data on coke formation as a function of the partial pressure of *n*-butene are as follows.

$r_f^0 \times 10^4$	0.70	0.80	1.0	1.25	2.0	3.0
p_A		0.60	0.40	0	0.20	0	0
p_B		0.20	0.20	0.10	0.20	0.20	0.30

Calculate the rate of coke formation, the rate of *n*-butene formation, and the dependence of the activity factor on coke content, assuming $W_{c1,0} = 0.102$, $W_{c2,0} = 0.163$

Solution

The dehydrogenation of *n*-butane can be represented stoichiometrically as[†]

$$n\text{-butane} \xrightarrow{r_1} n\text{-butene} \xrightarrow{r_{f1}} \text{coke chromia catalyst}$$
$$(A) \qquad\qquad (B) \qquad\qquad (C)$$
$$B \xrightarrow{r_{f2}} C \text{ (coke) alumina catalyst}$$

where

$$r_1^0 = \frac{k_1^0 p_A}{1 + K_A p_A}, \qquad r_{f1}^0 = \frac{k_{f1}^0 p_B}{1 + K_A p_A}, \qquad r_{f2}^0 = k_{f2}^0 p_B \tag{8.51}$$

Rate of coke formation:

If the activity is assumed to be linearly proportional to the coke content of the catalyst, these equations can be written as

$$r_1 = k_1^0 \left(1 - \frac{W_{c1}}{W_{c1,0}}\right) \frac{p_A}{1 + K_A p_A}$$
$$r_{f1} = k_{f1}^0 \left(1 - \frac{W_{c1}}{W_{c1,0}}\right) \frac{p_B}{1 + K_A p_A} \tag{8.52}$$
$$r_{f2} = k_{f2}^0 \left(1 - \frac{W_{c2}}{W_{c2,0}}\right) p_B$$

The total amount of coke deposited on the catalyst is the sum of the coke deposited on the chromia (designated 1) and on the alumina (designated 2).

$$W_c = W_{c1} + W_{c2} \tag{8.53}$$

[†] The subscript *w*, used to express the rates and rate constants in terms of unit catalyst weight, is omitted for convenience. Thus $r_{w1} = r_1$, $r_{w,f1} = r_{f1}$, etc.

Integrating Eqs. 8.52 with respect to time at constant p_A and p_B gives

$$1 - \frac{W_{c1}}{W_{c1,0}} = \exp\left[-\frac{k_{f1}^0 p_B t}{W_{c1,0}(1 + K_A p_B)}\right]$$
$$1 - \frac{W_{c2}}{W_{c2,0}} = \exp\left(-\frac{k_{f2}^0 p_B t}{W_{c2,0}}\right) \tag{8.54}$$

Combining Eqs. 8.53 and 8.54 gives

$$W_c = W_{c1,0}\left\{1 - \exp\left[-\frac{k_{f1}^0 p_B t}{W_{c1,0}(1 + K_A p_A)}\right]\right\}$$
$$+ W_{c2,0}\left[1 - \exp\left(-\frac{k_{f2}^0 p_B t}{W_{c2,0}}\right)\right] \tag{8.55}$$

The estimation of the coke content of the catalyst therefore requires a knowledge of four basic parameters: k_{f1}, k_{f2}, $W_{c1,0}$, and $W_{c2,0}$.

From the data of Figure 8.5a the rate constants k_{f1} and k_{f2} are evaluated. At $p_A = 0$ (i.e., when the feed contains *n*-butene only), the rate of carbon formation is given by

$$r_f^0 = r_{f1}^0 + r_{f2}^0 = (k_{f1}^0 + k_{f2}^0)p_B \tag{8.56}$$

This equation is used along with Eq. 8.51 to obtain the values of the rate constants:

$$k_{f1}^0 = 10.27 \times 10^{-4}, \qquad k_{f2}^0 = 0.629 \times 10^{-4}$$

The amount of coke deposited as time tends to infinity is given in the experimental data. As a specific example, let us take

$$p_B = 0.04, \qquad W_{c2,0} = 0.163$$
$$p_A = 0.42, \qquad W_{c1,0} = 0.102$$

Using these values in Eq. 8.55 yields the total carbon content for various values of $p_B t$; this is plotted in Figure 8.5b.

If a single-site mechanism is assumed (i.e. $k_{f1} + k_{f2} = k_f$), then the total carbon deposited on the catalyst will be

$$W_{c1,0} + W_{c2,0} = 0.265 \tag{8.57a}$$

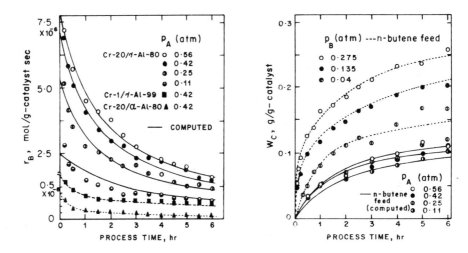

Figure 8.5a. Rate of *n*-butene production and coke content vs. process time (Toei et al., 1975).

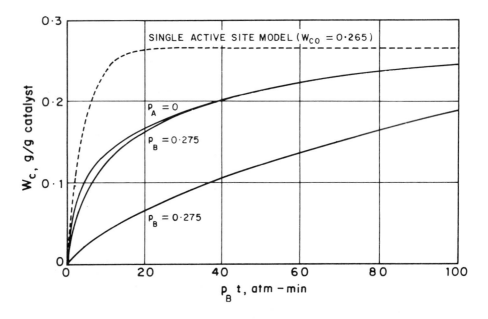

Figure 8.5b. Relation between coke content and $p_b t$ (Toei et al., 1975).

with

$$r_1^0 = \frac{k_1 p_A}{1 + K_A p_A} \quad \text{and} \quad r_f^0 = \frac{k_f p_B}{1 + K_A p_A}$$

$$(8.57b)$$

Following the procedure just outlined enables us to calculate the carbon content as a function of $p_B t$; this curve is also shown in Figure 8.5b (as dotted line).

Let us now calculate the rate of *n*-butene formation

and activity factor dependence on the coke content. $W = 2.6$ g and $F = 32$ liter/hr at 20°C, 1 atm. The reaction is carried out at 580°C, 1 atm.

On the assumption that the concentration (or partial pressure) of the reactant is uniform in the reactor, the mass conservation for *n*-butane (A) and *n*-butene (B) can be written as

$$F(p_{A0} - p_A) = W r_A, \qquad F p_B = W r_B \qquad (8.58)$$

Equation 8.58 can be rearranged to give p_A and p_B as

$$p_A = \left\{ \left(\frac{W}{F} k_1^0 \Omega_1 - K_A p_{A0} + 1 \right) \right.$$
$$+ \left[\left(\frac{W}{F} k_1^0 \Omega_1 - K_A p_{A0} + 1 \right)^2 \right.$$
$$\left. \left. + 4 K_A p_{A0} \right]^{1/2} \right\} \frac{1}{2 K_A} \qquad (8.59)$$

and

$$p_B = \frac{(W/F) k_1^0 p_A \Omega_1}{1 + K_A p_A + (1/v)(W/F) \left[k_{f1}^0 \Omega_1 + k_{f2}^0 \Omega_2 (1 + K_A p_A) \right]}$$
$$(8.60)$$

where

$$\Omega_1 = 1 - \frac{W_{c1}}{W_{c1,0}} \quad \text{and} \quad \Omega_2 = 1 - \frac{W_{c2}}{W_{c2,0}}$$

and the stoichiometric number $v = 4$.

The rate of formation of *n*-butene is given by

$$r_B = r_A - \frac{r_{f1} + r_{f2}}{v} \qquad (8.61)$$

From Eq. 8.52 and the calculated values of k_{f1}^0 and k_{f2}^0, we obtain r_B. Thus for $p_A = 0.56$,

$$k_{f1}^0 = 10.27 \times 10^{-4}, \qquad k_{f2}^0 = 0.629 \times 10^{-4}$$
$$k_1^0 = 0.45 \times 10^{-4} \quad \text{mol/g sec atm}$$

1. Using $p_A = 0.56$ in Eq. 8.60, calculate p_B.
2. Using p_A, p_B, k_{f1}^0, k_{f2}^0, and k_1^0 in Eq. 8.61, calculate r_B.

p_A	0.56	0.56	0.56	0.56
r_B	4.5	3.5	2.5	2.2
t	1	2	3	4

The rate of formation of *n*-butene is shown in Figure 8.5a for different partial pressures of A.

Activity factor dependence on coke content:

From Eq. 8.52 we obtain

$$\frac{dW_{c1}}{dW_{c2}} = \frac{1 - W_{c1}/W_{c1,0}}{1 - W_{c2}/W_{c2,0}} \frac{k_{f1}^0 (1 + K_A p_A)}{k_{f2}^0} \qquad (8.62)$$

For constant partial pressure of *n*-butane, this can be integrated to give

$$1 - \frac{W_{c2}}{W_{c2,0}} = \left(1 - \frac{W_{c1}}{W_{c1,0}} \right)^\alpha \qquad (8.63)$$

where

$$\alpha = \frac{k_{f2}^0}{k_{f1}^0} \frac{W_{c1,0}}{W_{c2,0}} (1 + K_A p_A) \qquad (8.64)$$

Since by definition the activity factor $\Omega = r_A / r_A^0$, from Eqs. 8.63 and 8.64 we obtain

$$W_c = W_{c1,0}(1 - \Omega) + W_{c2,0}(1 - \Omega^z) \qquad (8.65)$$

Equation 8.65 relates the carbon content to the activity of the catalyst based on the experimental data.

8.3. EFFECT OF DEACTIVATION ON SELECTIVITY IN COMPLEX REACTIONS

8.3.1. Deactivation by Feed Stream Poisoning

In the treatment of effectiveness factors in the presence of catalyst fouling presented earlier for simple reactions, a second carbon-generating step was involved, so that the simple reaction was "converted" into a complex one. In contrast to this analysis, in the present case, the reaction being inherently complex, we shall restrict our attention to feed stream poisoning (of the pore-mouth type), that is, class IV presented in Section 8.1.1.

Sada and Wen (1967) derived expressions for the selectivity, defined as the ratio of the fluxes of the desired to the wasteful products (which is similar to the definition employed earlier in Chapter 5). Solutions for all three reaction types considered have been given, based on which the following conclusions can be drawn:

1. For parallel and consecutive schemes, the selectivity decreases monotonically with increase in the modulus $(\phi_{p1})_A$, so that the best results are obtained at very small values of $(\phi_{p1})_A$, corresponding to small, highly porous particles. This behavior is valid at all degrees of poisoning *p*.

2. For independent reactions Type IV (Section 8.1.1), the behavior tends to be quite complex (being a function of several parameters)

and no simple conclusion, as in paragraph 1, appears possible. For instance, at intermediate values of $(\phi_{p1})_A$, the selectivity first increases and then decreases with $(\phi_{p1})_A$, suggesting the existence of an upper limit to the particle size.

8.3.2. Nonuniform Activity Distribution

Consider the case of series self-poisoning $A \rightarrow B \rightarrow C$, where C is a degraded product that is strongly adsorbed on the surface and deactivates the catalyst. If we assume that the local activity decreases linearly with the local concentration of poison (i.e., that $\Omega = 1 - \hat{W}_c$) then the following conservation equations can be written by using Eq. 4.12 for the spatial variation of activity:

$$\frac{d^2\hat{C}_A}{d\hat{R}^2} + \frac{2}{\hat{R}}\frac{d\hat{C}_A}{d\hat{R}} - (\phi_{p1}^2)_A \hat{R}^q(1 - \hat{W}_c)\hat{C}_A = 0 \quad (8.66)$$

$$\frac{d^2\hat{C}_B}{d\hat{R}^2} + \frac{2}{\hat{R}}\frac{d\hat{C}_B}{d\hat{R}} + (\phi_{p1}^2)_A \hat{R}^q(1 - \hat{W}_c)\hat{C}_A$$

$$- (\phi_{p1}^2)_B \hat{R}^q(1 - \hat{W}_c)\hat{C}_B = 0 \quad (8.67)$$

$$\frac{d\hat{W}_c}{dt_B} = \hat{R}^q(1 - \hat{W}_c)\hat{C}_B \quad (8.68)$$

Solutions to this set of equations have been obtained by Yazdi and Petersen (1972) for S = 0.

Nonuniform distribution of activity decreases the rate of deactivation of the catalyst for a self-fouling series reaction. Even though the initial rate of reaction is less for large values of q, after a relatively short time the difference diminishes and finally the rate becomes higher than that corresponding to small values of q. The selectivity in a parallel reaction scheme is not affected significantly by the nonuniform distribution of activity, since the rate of production of both the products will be affected almost equally. It is noteworthy that, since the inner parts of the pellet are not used completely, some decrease in the rate of consumption of A and hence of the production of B is to be expected.

Using the rate constant density function defined by Eq. 5.58, Corbett and Luss (1974) have concluded that the specific deactivation mechanisms and rate, as well as the diffusional resistance, have a marked influence on the desired activity profile. Maximum *selectivity* can be obtained by concentrating the *activity* in the exterior shell; but when the diffusional resistance is small, this results in a minimum resistance to deactivation (i.e.,

low effectiveness) as compared to catalysts with the same volume-averaged activity and different activity profiles. Thus, a compromise between activity and selectivity is necessary for optimal choice of catalyst. On the other hand, when the diffusional resistance is large, both selectivity and effectiveness are improved by concentrating the activity near the surface. Similar qualitative conclusions can be drawn from the results of Becker and Wei (1977); they based their analysis on discontinuous activity distribution, which is practically easier to realize than continuous distribution. A diluted catalyst pellet also manifests a considerably longer lifetime than a uniformly impregnated catalyst under conditions of diffusional limitation and pore-mouth poisoning (Varghese and Wolf, 1980).

8.3.3. Deactivation of Bifunctional Catalysts

The distinguishing feature of the deactivation of a bifunctional catalyst is the possibility of one or both of the functions being deactivated. Webb and Macnab (1972) found, for instance, that in the hydroisomerization of *n*-butene to isobutene on an $Rh-SiO_2$ catalyst, the hydrogenation function (Rh) is preferentially deactivated by small amounts of mercury, while the isomerization function (SiO_2) is practically unaffected. Obviously the two functions operate in parallel, so that one can be selectively suppressed. On the other hand, when the functions operate in series, as in the case of disproportionation of butane over a mechanical mixture of $Pt-Al_2O_3$ and WO_3 (Burnett and Hughes, 1973), the deactivation of one function (in this case, coking or water poisoning of the Pt function) can completely arrest the entire reaction.

Certain aspects of the deactivation of bifunctional catalysts have been computationally explored. Thus for a typical bifunctional network of the type (see Section 5.3.1)

$$A \underset{}{\overset{}{\rightleftharpoons}} B \overset{x}{\underset{y}{\rightleftarrows}} \begin{matrix} C \\ R \end{matrix}$$

with coke formation occurring on the X function and the main reaction on the Y function, Snyder and Mathews (1973) and Lee and Butt (1973) have analyzed the behavior of composite as well as discrete formulations. The chief conclusion appears to be that, for a given catalyst formulation, the presence of a finite diffusional resistance increases both selectivity and catalyst life. It will be recalled (Section 8.2.2) that a similar beneficial effect was noticed for parallel fouling on a monofunctional catalyst.

8.4. TIME-ON-STREAM THEORY OF CATALYST DEACTIVATION

In the analysis of catalyst deactivation presented in the preceding sections, the quantity of deactivating agent always appeared as a primary parameter, either as concentration in the feed stream or as coke on catalyst. In the time-on-stream theory, all deactivation is expressed exclusively in terms of time and the role of the deactivating agent is indirectly recognized by postulating a reduction in the number of active sites. The basic assumption is made that either an active site is completely destroyed or it is present with its full initial activity; intermediate levels of activity of a single site are not permitted in the analysis.

The rate of reaction (say, A → products) at any instant can be expressed as

$$r_v = k_l C_l^n f(C_A) \tag{8.69}$$

where k_l is the rate constant per active site and C_l is the concentration of active sites at time t. According to this definition, the activity of the catalyst Ω (which is the same as the decay function defined by Eq. 8.2) may be expressed as

$$\Omega = \frac{k_l C_l^n f(C_A)}{k_l C_L^n f(C_A)} = \theta^n \tag{8.70}$$

where

$$\theta = \frac{C_l}{C_L} \tag{8.71}$$

Thus there is a direct relationship between the activity of a catalyst and its current concentration of active sites. We shall come back to this equation after a brief consideration of the kinetics of deactivation (or active-site destruction).

In its most general form, the decay kinetics, represented by the rate of change of active centers, can be expressed as (Wojciechowski, 1974).

$$-\frac{dC_l}{dt} = k_{0,f}'' + k_{1,f}'' C_l + k_{2,f}'' C_l^2 + \ldots + k_{n,f}'' C_l^n \tag{8.72}$$

which includes all orders from 0 to n. In terms of the ratio of active centers θ defined by Eq. 8.71, this becomes

$$-\frac{d\theta}{dt} = k_{0,f} + k_{1,f}\theta + k_{2,f}\theta^2 + \cdots + k_{n,f}\theta^n \tag{8.73}$$

where

$$k_{n,f} = k_{n,f}'' C_l^{n-1} \tag{8.74}$$

The integrated form of Equation 8.73 reduces to the linear form $\theta = 1 - k_{0,f}t$ for zero-order decay in which only the first term is involved, and to the exponential form $\theta = \exp(-k_{1,f}t)$ for first-order decay in which only the second term is involved. The most general is the mth-order case represented by $-(d\theta/dt) = k_{m,f}\theta^m$, the integrated form of which is given by

$$\theta = \left[\frac{1}{1 + (m-1)k_{m,f}t}\right]^{1/(m-1)} \tag{8.75}$$

or

$$C_l = C_L(1 + Gt)^{-M} \tag{8.76}$$

where $M = 1/(m-1)$ and $G = (m-1)k_{m,f}$.

By combining Eqs. 8.76 and 8.69 we can obtain the following composite expression for the reaction rate on a decaying catalyst (Wojciechowski et al., 1969):

$$r_v = k_l C_L^n (1 + Gt)^{-N} f(C_A) \tag{8.77}$$

where $N = nM$. Obviously the value of N will determine the behavior of the rate. As pointed out by Pachovsky and Wojciechowski (1972), however, this behavior is also a function of reactor type.

We shall now examine in greater detail the relationship between the residual activity Ω and active site concentration θ as given by Eq. 8.70. Combining this equation with Eq. 8.76 leads to

$$\Omega = \theta^n = (1 + Gt)^{-N} \tag{8.78}$$

while 8.76 itself can be written as

$$\theta = (1 + Gt)^{-M} \tag{8.79}$$

A comparison of these two equations shows that Ω equals θ only when $M = N$ or $n = 1$. In other words, the activity is equal to the relative concentration of active sites only when a single site is involved in the deactivation process. When more than one site is involved, $N > M$, and Ω decreases more than linearly with decreasing θ. The case where $N < M$ (i.e., less than an active site is involved) is purely hypothetical. For the most common first-order decay

$$\Omega = \theta = (1 + Gt)^{-M} \tag{8.80}$$

8.5. CATALYST REGENERATION

The most important form of deactivation is the quick and progressive fouling of cracking catalysts by the deposition of carbonaceous matter (essentially carbon or coke) from the reactants and products. Ausman and Watson (1962) have analyzed the problem of regeneration of this coke-laden catalyst on the assumption that the rate of carbon burning is dependent on oxygen partial pressure alone; but at moderate to low carbon concentrations it does not seem appropriate to disregard the effect of carbon concentration (Bondi et al., 1962; Weisz and Goodwin, 1966). A general mathematical model, whose limiting solutions correspond to chemical and diffusional control, has been proposed by Mickley et al. (1965), and the following treatment is based essentially on this model. A more specific model, applicable to diffusional control alone, was developed earlier by Weisz and Prater (1954).

8.5.1. Isothermal Regeneration

The continuity equation developed in Chapter 4 can be applied to carbon burning by an appropriate modification of the rate term. Considering first the diffusion of oxygen, one can write for a spherical pellet under isothermal conditions:

$$D_{eA} \frac{d^2 C_A}{dr^2} + \frac{2 D_{eA}}{r} \frac{d C_A}{dr} = \frac{(r_v)_{cb}}{v_c} \qquad (8.81)$$

where C_A is the concentration of oxygen, $(r_v)_{cb}$ is the rate of carbon burning in mol/cm^3 sec, and v_c is the moles of carbon reacting per mole of oxygen. The rate of carbon burning may be expressed as

$$\begin{aligned} (r_v)_{cb} &= k_{cb} C_A C_c \\ &= k_{cbp} p_A C_c \end{aligned} \qquad (8.82)$$

where k_{cb} has the units cm^3/mol sec and k_{cbp} the units 1/atm sec; C_c is the concentration of carbon given by $W_{c0}(\rho_c/M_c)\hat{W}_c$. Expressing Eq. 8.82 in dimensionless form and using the ideal gas law to eliminate C_{A0}, we get

$$\frac{\partial^2 \hat{p}_A}{\partial \hat{R}^2} + \frac{2}{\hat{R}} \frac{\partial \hat{p}_A}{\partial \hat{R}} = \mathbf{R}_{cb} \hat{p}_A \hat{W}_c \qquad (8.83)$$

where $\hat{p}_A = C_A/C_{A0}$ and $\mathbf{R}_{cb}$ is a dimensionless diffusion parameter defined by

$$\mathbf{R}_{cb} = \frac{R^2 W_{c0} \rho_c R_g T k_{cbp}}{D_{eA} v_c M_c} \qquad (8.84)$$

High values of $\mathbf{R}_{cb}$ correspond to significant diffusional intrusion, whereas low values would correspond to kinetic control.

In order to develop a similar equation for carbon balance, the time dependence of the carbon concentration must be taken into account. The equation in its normalized form is

$$\frac{\partial \hat{W}_c}{\partial \hat{t}_{cb}} = \hat{p}_A \hat{W}_c \qquad (8.85)$$

where the dimensionless time

$$\hat{t}_{cb} = p_{A0} k_{cbp} t \qquad (8.86)$$

The boundary conditions for the simultaneous solution of Eqs. 8.83 and 8.85 are

$$\hat{p}_A = 1 \quad \text{at} \quad \hat{R} = 1, \qquad \hat{W}_c = 1 \quad \text{at} \quad \hat{t}_{cb} = 1,$$
$$\frac{d\hat{p}_A}{d\hat{R}} = 0 \quad \text{at} \quad \hat{R} = 0 \qquad (8.87)$$

The two limiting solutions are as follows:

Carbon burning rate controlling ($\mathbf{R}_{cb} \to 0$):

$$\hat{W}_{av} = e^{-\hat{t}_{cb}} \qquad (8.88)$$

Pore diffusion controlling ($\mathbf{R}_{cb} \to \infty$):

$$f(W) = \tfrac{1}{3}\hat{W}_{av} - \tfrac{1}{2}\hat{W}_{av}^{2/3} + \tfrac{1}{6} = \frac{\hat{t}_{cb}}{\mathbf{R}_{cb}} \qquad (8.89)$$

This corresponds to the SPM solution of Weisz and Goodwin (1953), in which $\hat{W}_{av} = r_i/r$ where r_i is the radius of the unreacted core. The intermediate region is characterized by dual control. Plots showing oxygen profiles at $\mathbf{R}_{cb} = 5$ and 400 are presented in Figure 8.6.

It can be seen from Figure 8.6 that for $\mathbf{R}_{cb} = 400$ at low values of $\hat{t}_{cb}$ there is still a considerable amount of carbon left on the outside surface of the pellet. As the time is progressively increased, the curve marked A is reached. This curve is different from the other curves of the family in that at $\hat{R} = 1$ (pellet surface) the carbon content is exactly zero, indicating that all the surface carbon has just been oxidized. The burning of carbon up to the time represented by this profile is sometimes referred to as "constant-rate burning" (corresponding to the constant surface evaporation in drying). All the curves to the left of this represent the carbon profiles in the so-called falling-rate period (again with its counter-

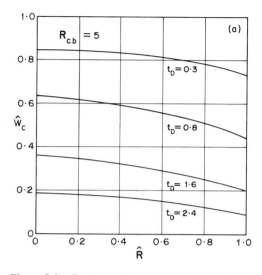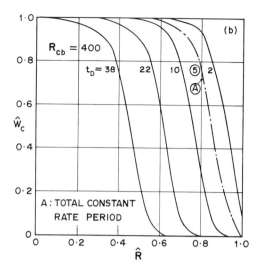

Figure 8.6. Carbon profiles as a function of time for (a) low and (b) high values of the diffusion parameter ($\mathbf{R}_{cb}$).

part in drying). As already pointed out, the constant-rate period corresponds to chemical control, but the converse does not necessarily follow, that the chemical control regime ends abruptly at the value of $\hat{t}_{cb}$ corresponding to curve A. Such a clear demarcation is not possible, and the chemical control regime might well extend a little beyond the constant-rate period.

Ausman and Watsons's analysis, based on the dependence of the burning rate on the oxygen partial pressure alone, has formed the basis of several elaborate models for gas–solid reactions in general (Ishida and Wen, 1968; Tudose, 1970; Mantri et al., 1976). In the case of catalyst regeneration, however, the SPM (or shrinking-core model, as it is also called), as developed by Yagi and Kunii (1955), Levenspiel (1965), and Gokarn and Doraiswamy (1973), corresponding to diffusion control, is generally adequate. At lower temperatures (less than 500°C) the kinetic model represented by Eq. 8.88 may be used. The condition for this is given by

$$\left(\frac{R^2}{D_{eO_2}}\right)\left(\frac{r_{v_{O_2}}}{C_{O_2}}\right) < 1 \qquad (8.90)$$

where A has been replaced by O_2. In other words, if we take order-of-magnitude values ($D_{eO_2} = 10^{-5}$, $R = 10^{-1}$, and $C_{O_2} \approx 10^{-6}$), the rate of oxygen consumption should be less than 10^{-8} mol/cm³ sec if kinetic control is to be operative.

According to Eq. 8.89 the burn-off should be a linear function of time if SPM conditions are operative. Because of this linear relationship, Weisz and Goodwin (1966) define "85% burn-off time" (represented by t_{85}) as a convenient measure of characterizing regeneration.

From Eq. 8.89, a relationship for t_{85} can be readily written by setting $W_{av} = 0.15$:

$$t_{85} = 0.076\left(\frac{R^2 W_{c0}}{\rho_c M_c D_{eO_2} C_{O_2}}\right) \quad \text{min} \qquad (8.91)$$

Several experiments to test the dependence of t_{85} on the different variables of Eq. 8.91 have shown this relationship to be generally valid at the high temperatures normally employed in regeneration.

8.5.2. Nonisothermal Regeneration

In the development presented in Section 8.5.1 no allowance was made for the heat released during oxidation. Incorporation of the heat effect, however, cannot be accomplished by merely writing the energy balance and solving it simultaneously with the oxygen and carbon balances represented by Eqs. 8.83 and 8.85. This is because of the basic assumption that coke is composed entirely of carbon. Actually coke is a condensed product of composition CH_n (see Section 8.1.2) with a low hydrogen content. Whereas the neglect of hydrogen did not pose any serious problem in the analysis of deactivation, in the case of regeneration coke is a participant in the oxidative burning process, so that its composition becomes important. Again, where temperature effects can be neglected, the presence of hydrogen can be ignored without serious consequences (as in the case of isothermal regeneration), but where a steep rise of temperature is expected, an additional oxidative step involving hydrogen becomes necessary to account for this rise.

Massoth (1967) presented an analysis of regeneration on the assumption of a double-interface model for coke removal. Because of the faster rate of oxidation of hydrogen than that of carbon, a hydrogen–oxygen reaction front is first formed, followed by a carbon–oxygen front that moves through the layers of coke depleted of its hydrogen. Neglecting the hydrogen reaction, particularly in the initial stages, can therefore lead to underprediction of the pellet temperature. In fact, none of the general gas–solid models (see Chapter 19) predicts the temperature maximum observed by Shettigar and Hughes (1972) in the initial stages of regeneration. Massoth's model, specific to regeneration, has been modified by Ramachandran et al. (1975); in their model the hydrogen is assumed to react at a sharp interface while carbon reacts in the zone between the interface and the pellet surface. This model shows better conformity with the experimental data of Shettigar and Hughes (1972).

On the assumption of uniform coke deposition, no intrapellet temperature gradient, spherical geometry, irreversible oxidation, and an average carbon concentration in the hydrogen-depleted zone (zone 2 of the model), nondimensional heat and mass balance equations have been developed by Ramachandran et al. (1975). Based on the numerical solutions to these equations using parameter values typical of coke regeneration, the following major observations can be made: The experimentally observed temperature peak in the initial stages of regeneration is predicted by the model with reasonable accuracy, and the best matching with experimental data is obtained when the carbon-to-hydrogen molal ratio in the coke is assumed to be 1 to 0.5 (i.e., the composition of the coke is taken as $CH_{0.5}$).

REFERENCES

Anderson, R. B., and Whitehouse, A. M. (1961), *Ind. Eng. Chem.*, **53**, 1011.

Angele, B., Kirchner, K., and Schlosser, E. G. (1980), *Chem. Eng. Sci.*, **35**, 2093.

Appleby, W. G., Gibson, J. W., and Good, G. M. (1962), *Ind. Eng. Chem. Process Design Dev.*, **1**, 102.

Ausman, J. M., and Watson, C. C. (1962), *Chem. Eng. Sci.*, **17**, 323.

Balder, J. R., and Petersen, E. E. (1968a), *J. catal.*, **11**, 202.

Balder, J. R., and Petersen, E. E. (1968b), *Chem. Eng. Sci.*, **23**, 1287.

Becker, E. R., and Wei, J. (1977), *J. Catal.*, **46**, 372.

Bischoff, K. B. (1969), *Ind. Eng. Chem. Fundam.*, **8**, 665.

Blanding, F. H. (1953), *Ind. Eng. Chem.*, **45**, 1186.

Blue, R. W., and Engel, C. J. (1951), *Ind. Eng. Chem.*, **43**, 494.

Bondi, A., Miller, R. S., and Schlaffer, W. F. (1962), *Ind. Eng. Chem. Process Design Dev.*, **1**, 196.

Burnett, R. L., and Hughes, T. R. (1973), *J. Catal.*, **31**, 55.

Butt, J. B. (1972), *Adv. Chem. Ser.*, **109**, 259.

Butt, J. B. (1976), *J. Catal.*, **41**, 190.

Butt, J. B., Delgado-Diaz, S., and Muno, W. E. (1975), *J. Catal.*, **37**, 158.

Calvelo, A., and Cunningham, R. E. (1970), *J. Catal.*, **16**, 397.

Carberry, J. J., and Gorring, R. L. (1966), *J. Catal.*, **5**, 529.

Chen, N. Y., and Garwood, W. E. (1978), *J. Catal.*, **52**, 453.

Chu, C. (1968), *Ind. Eng. Chem. Fundam.*, **7**, 509.

Corbett, W. E., and Luss, D. (1974), *Chem. Eng. Sci.*, **29**, 1473.

Crawford, P. B., and Cunningham, W. A. (1956), *Petrol. Refiner*, **35**, 169.

Do, D. D., and Weiland, R. H. (1981a), *Ind. Eng. Chem. Fundam.*, **20**, 34.

Do, D. D., and Weiland, R. H. (1981b), *Ind. Eng. Chem. Fundam.*, **20**, 42.

Do, D. D., and Weiland, R. H. (1981c), *Ind. Eng. Chem. Fundam.*, **20**, 48.

Dougharty, N. A. (1970), *Chem. Eng. Sci.*, **25**, 489.

Dumez, F. J., and Froment, G. F. (1976), *Ind. Eng. Chem. Process Des. Dev.*, **15**, 291.

Eberly, P. E., Jr., and Kimberlin, C. N., Jr. (1970), *Proc. 1st Int. Symp. Chem. React. Eng.*, Washington, D.C.

Eberly, P. E., Kimberlin, C. N., Miller, W. H., and Prushel, H. U. (1966), *Ind. Eng. Chem. Process Design Dev.*, **5**, 193.

Froment, G. F., and Bischoff, K. B. (1961), *Chem. Eng. Sci.*, **16**, 189.

Gioia, F., Gibilaro, L. G., and Greco, G., Jr. (1970), *Chem. Eng. J.*, **1**, 9.

Gokarn, A. N., and Doraiswamy, L. K. (1973), *Chem. Eng. Sci.*, **28**, 401.

Gokhale, M. V., Naik, A. T., and Doraiswamy, L. K. (1975), *Chem. Eng. Sci.*, **30**, 1409.

Greco, G., Jr., Alfani, F., and Gioia, F. (1973), *J. Catal.*, **30**, 155.

Haldeman, R. G., and Botty, M. C. (1959), *J. Phys. Chem.*, **63**, 489.

Hegedus, L. L., and Petersen, E. E. (1973a), *Chem. Eng. Sci.*, **28**, 69.

Hegedus, L. L., and Petersen, E. E. (1973b), *Chem. Eng. Sci.*, **28**, 345.

Hegedus, L. L., and Petersen, E. E. (1974), *Catal. Rev. Sci. Eng.*, **9**, 245.

Ishida, N., and Wen, C. Y. (1968), *AIChE J.*, **14**, 311.

Kam, E. K. T., and Hughes, R. (1979), *AIChE J.*, **25**, 359.

Kam. E. K. T., Ramachandran, P. A., and Hughes, R. (1975), *J. Catal.*, **38**, 283.

Kam, E. K. T., Ramachandran, P. A., and Hughes, R. (1977a), *Chem. Eng. Sci.*, **32**, 1307.

Kam, E. K. T., Ramachandran, P. A., and Hughes, R. (1977b), *Chem. Eng. Sci.*, **32**, 1317.

Kam, E. K. T., Ramachandran, P. A., and Hughes, R. (1977c), *J. Catal.*, **48**, 177.

Kokotailo, G. T., Lawton, S. L., Olson, D. H., and Meier, W. M. (1978), *Nature (London)*, **272**, 437.

Krishnaswamy, S., and Kittrell, J. R. (1981), *AIChE J.*, **27**, 120.

Kulkarni, B. D., and Doraiswamy, L. K. (1979), unpublished work.

Kulkarni, B. D., and Ramachandran, P. A. (1979), *Chem. Eng. J.*, **19**, 57.

Lamba, H. S., and Dudukovic, M. P. (1978), *Chem. Eng. J.*, **16**, 117.

Lee, J. W., and Butt, J. B. (1973), *Chem. Eng. J.*, **6**, 111.

Levenspiel, O. (1965), *Chemical Reaction Engineering*, Wiley, New York.

Levinter, M. S., Panchenkov, G. M., and Tanatarov, M. A. (1967), *Int. Chem. Eng.*, **7**, 23.

Maat, E. J., and Moscou, L. (1964), *Proc. 3d Int. Cong. Catal.*, **2**, 1277.

Mantri, V. B., Gokarn, A. N., and Doraiswamy, L. K. (1976), *Chem. Eng. Sci.*, **31**, 779.

Masamune, S., and Smith, J. M. (1966), *AIChE J.*, **12**, 384.

Massoth, F. E. (1967), *Ind. Eng. Chem. Process Design Dev.*, **6**, 200.

Maxted, E. B. (1951), *Adv. Catal.*, **3**, 129.

Mickley, H. S., Nestor, S. W., Jr., and Gould, L. A. (1965), *Can. J. Chem. Eng.*, **43**, 61.

Murakami, Y., Nozaki, F., and Turkevich, J. (1962), *Shokubai* (*Tokyo*), **5**, 262.

Murakami, Y., Kobayashi, T., Hattori, T., and Masuda, M. (1968), *Ind. Eng. Chem. Fundam.*, **7**, 599.

Ozawa, Y., and Bischoff, K. B. (1968), *Ind. Eng. Chem. Process Design Dev.*, **7**, 67.

Pachovsky, R. A., and Wojciechowski, B. W. (1972), *Can. J. Chem. Eng.*, **50**, 306.

Panchenkov, G. M., and Lolesnikov, I. M. (1959), *Izv. Vysshikh, Uchebn. Zavedenii*, **9**, 79.

Parravano, G. (1953), *J. Am. Chem. Soc.*, **75**, 1448.

Pozzi, A. L., and Rase, H. F. (1958), *Ind. Eng. Chem.*, **50**, 1075.

Prater, C. D., and Lago, R. M. (1956), *Adv. Catal.*, **8**, 293.

Ramachandran, P. A., Rashid, M. H., and Hughes, R. (1975), *Chem. Eng. Sci.*, **30**, 1391.

Ramser, J. H., and Hill, P. B. (1958), *Ind. Eng. Chem.*, **50**, 117.

Ray, W. H. (1972), *Chem. Eng. Sci.*, **27**, 489.

Richardson, J. T. (1972), *Ind. Eng. Chem. Process Design Dev.*, **11**, 8.

Rollmann, L. D. (1977), *J. Catal.*, **47**, 113.

Rollmann, L. D., and Walsh, D. E. (1979), *J. Catal.*, **56**, 139.

Rudershausen, C. G., and Watson, C. C. (1954), *Chem. Eng. Sci.*, **3**, 110.

Sada, E., and Wen, C. Y. (1967), *Chem. Eng. Sci.*, **22**, 559.

Sagara, M., Masamune, S., and Smith, J. M. (1967), *AIChE J.*, **13**, 1226.

Shettigar, U. R., and Hughes, R. (1972), *Chem. Eng. Sci.*, **27**, 208.

Snyder, A. C., and Mathews, J. C. (1973), *Chem. Eng. Sci.*, **28**, 291.

Suga, K., Morita, Y., Kunugita, E., and Otake, T. (1967), *Brit. Chem. Eng.*, **7**, 742.

Swabb, E. A., and Gates, B. C. (1972), *Ind. Eng. Chem. Fundam.*, **11**, 540.

Szepe, S., and Levenspiel, O., (1970), *Proc. 4th European Fed. Chem. React. Eng.*, Brussels.

Tai, N. M., and Greenfield, P. F. (1978), *Chem. Eng. J.*, **16**, 89.

Toei, R., Nakanishi, K., Yamada, K., and Okazaki, M. (1975), *J. Chem. Eng. Japan*, **8**, 131.

Tudose, R. Z. (1970), *Bull. Inst. Politech. DIn IASI*, **16**, 241.

Varghese, P., and Wolf, E. E. (1980), *AIChE J.*, **26**, 55.

Voorhies, A., Jr. (1945), *Ind. Eng. Chem.*, **37**, 318.

Walsh, D. E., and Rollmann, L. D. (1977), *J. Catal.*, **49**, 369.

Walsh, D. E., and Rollmann, L. D. (1979), *J. Catal.*, **56**, 195.

Webb, G., and Macnab, J. I. (1972), *J. Catal.*, **26**, 226.

Weekman, V. W., Jr. (1968), *Ind. Eng. Chem. Process Design Dev.*, **7**, 252.

Weisz, P. B., and Goodwin, R. B. (1963), *J. Catal.*, **2**, 397.

Weisz, P. B., and Goodwin, R. B. (1966), *J. Catal.*, **6**, 227.

Weisz, P. B., and Prater, C. D. (1954), *Adv. Catal.*, **6**, 143.

Wheeler, A. (1955), *Catalysis*, **2**.

Wilson, J. L., and den Herder, M. J. (1958), *Ind. Eng. Chem.*, **50**, 305.

Wojciechowski, B. W. (1974), *Catal. Rev. Sci. Eng.*, **9**, 79.

Wojciechowski, B. W., Juusola, J. A., and Downie, J. (1969), *Can. J. Chem. Eng.*, **47**, 338.

Wolf, E., and Petersen, E. E. (1974), *Chem. Eng. Sci.*, **29**, 1500.

Yagi, S., and Kunii, D. (1955), *Proc. 5th Int. Symp. Combustion*.

Yazdi, S. F., and Petersen, E. E. (1972), *Chem. Eng. Sci.*, **27**, 227.

CHAPTER NINE

Experimental Methods in Catalytic Kinetics

One of the primary requirements in the quantitative analysis of catalytic reactors is a rate expression for the reaction concerned. In obtaining the rate data the effect of internal and external heat and mass transport should be considered, and it should further be ensured that the reactor operates under strictly isothermal conditions. Since industrial catalysts are often subject to deactivation, methods should also be formulated for obtaining the kinetics not only of the main reaction but also of the fouling reaction. An important property needed in reactor design, which should preferably be measured experimentally, is the effective diffusivity of the catalyst pellet. The other transport parameters can be estimated from available correlations.

9.1. EVALUATION OF DIFFUSIONAL RESISTANCES IN EXPERIMENTAL REACTORS

Rate data obtained under conditions of diffusional interference cannot be used as such for formulating kinetic expressions. Therefore, either the data have to be suitably corrected, or the experiments must be carried out under conditions where both the diffusional resistances are negligible. In this section we shall first consider the criteria for the absence of the external mass and/or heat transfer effect, and then proceed to a discussion of the in-pore diffusion problem.

9.1.1. Evaluation of the External Mass and Heat Transfer Effect

The principal criterion for the absence of the external mass and heat transfer effects is that the measured rate (or conversion) at a given residence time should be independent of agitation. If the agitation parameter (e.g., flow rate, stirrer speed) is denoted by $\mathscr{A}$ (its units are cm/sec), then the first step in the experimental procedure is to determine the reaction rate (or conversion) as a function of $\mathscr{A}$ at a fixed value of W/F (which is a direct measure of the residence time).

Dependence of Reaction Rate on Flow Rate

From experimental plots of x vs. W/F, as illustrated for the (overall) reaction in Figure 9.1a for different values of W (see Banerjee, 1965), plots of x vs. the velocity can be prepared; the conversion is seen to level off beyond a certain velocity (Figure 9.1b), leading to the conclusion that external mass transfer has no effect at these relatively high velocities. It may be noted that the velocity at which the effect of external mass transfer becomes negligible is a function of W/F. This is shown in Figure 9.1c, from which it is clear that at high contact times, corresponding to low reaction rates, the role of external mass transfer is less pronounced; this is consistent with expectation and should not be lost sight of in experimental measurements of the rate.

A paradoxical situation can arise when, at very low flow rates, corresponding to Reynolds numbers of less than 2–3, the reaction is entirely mass transfer controlled and the mass transfer coefficient is practically independent of Re′. The possibility of this situation can be understood from the fact that at very low Re′ a limiting value of the Sherwood number is obtained (see Chapter 6). Under these conditions, as shown in Figure 9.1d, although the contact time (W/F) is held constant, an increase in the feed rate (and therefore in Re′) would lead to a decrease in conversion until a point is reached when the Sherwood number begins to rise with Re′ and the normal behavior depicted in Figure 9.1b is restored. Puranik and Sharma (1967) have experimentally verified the existence of such a minimum in the dehydration of ethyl alcohol to ethylene by carrying out experiments at Re′ less than 0.6.

In laboratory reactors the Reynolds numbers employed are usually less than 10. Since there seems to be

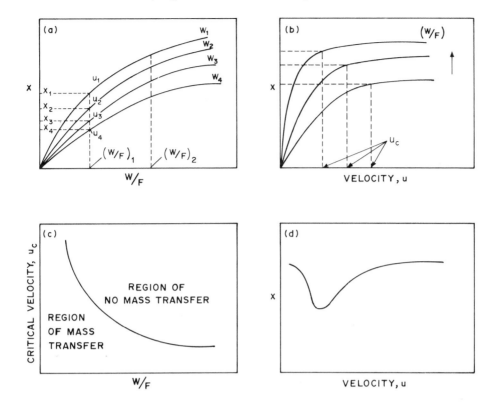

Figure 9.1. Experimental plots for discerning the external mass transfer effect.

considerable uncertainty regarding the validity of the various mass transfer correlations in the low Reynolds number region (see Chapter 6), they should be used with caution to correct for mass transfer in laboratory reactors.

Range of Flow Rates

It has been shown by Ford and Perlmutter (1963) that a yes-or-no decision on the effect of mass transfer based exclusively on the flatness of the curves shown in Figure 9.1b can be misleading. They proposed the following equation for determining the range over which the agitation parameter $\mathscr{A}$ (i.e., the flow rate) must be varied in order to ensure that random fluctuations in the rate (due to experimental error) do not obscure a genuine mass transfer effect:

$$\frac{\mathscr{A}_2}{\mathscr{A}_1} = \frac{G_2}{G_1} > \left[\frac{r_{v,\mathscr{A}_1}}{r_{v,\mathscr{A}_2}}\right]^{1/b(\Delta p/p_{As})} \tag{9.1}$$

where $\Delta p/p_{As}$ is the normalized partial pressure difference across the film, and $b = 0.49$ for $Re' < 350$ and 0.59 for $Re' > 350$. Thus, for fixed values of the experimental error $r_{v,\mathscr{A}_1}/r_{v,\mathscr{A}_2}$ and $\Delta p/p_{As}$, it is possible

to estimate the range over which the agitation parameter has to be varied ($\mathscr{A}_2/\mathscr{A}_1$) in order to ensure that any mass transfer effect is not obscured by the inherent experimental error.

Criteria for the Absence of External Mass and Heat Transfer Effects

Several criteria have been proposed to ensure the absence of external diffusional effects. Ruthven (1968) and Huang and Sather (1970) have developed a simple criterion based on the external effectiveness factor that can be used for the determining whether available kinetic data for a first-order reaction are free from diffusional effects. Mears (1971) has proposed a criterion based on the observed reaction rate for the absence of external heat transfer resistance. Both these criteria are given in Table 9.1.

9.1.2. Evaluation of the Pore Diffusional Effect

Simple Experimental Methods

Differential reactor data can be used to obtain the effectiveness factor directly. By plotting the rate as a

TABLE 9.1 Criteria for the Absence of Mass and Heat Transfer Effects

Interphase Transport	Intraparticle Mass transfer	General[a]	Interparticle Transport	
			Radial Heat Transfer	Axial Mass Transfer

Mass transfer

Isothermal

Isothermal

1. $\eta = 1 - \dfrac{k_a d_p^{1.5}}{11\sqrt{D_b}\,u} > 0.9$

Ruthven (1968)

1. $\left(\dfrac{V'}{a'}\right)^2 \dfrac{r_{vA}}{D_{eA} C_{As}} < 1$

Weisz and Prater (1954); Weisz (1957)

1. $\dfrac{2}{n+1}\left(\dfrac{V'}{a'}\right)^2 \dfrac{\rho S_g k_s C_A^{n-1}}{D_{eA}} \leqslant 0.03$

Guha and Narsimhan (1972)

Nonisothermal

$\dfrac{|\Delta H|\, d_t^2 (1 - f_B) r_{vA}}{4 k'_{el} T}$

$< \dfrac{0.4}{[1 + 8(d_p/d_t)(Bi)_{h_w}]\,\alpha_w}$

$\dfrac{L}{d_p} > \dfrac{20}{Pe_{ml}} \ln \dfrac{C_{A0}}{C_{AL}}$

2. $\dfrac{r_{vA}(V'/a')}{C_A k_g} < \dfrac{0.15}{n}$

2. $\left(\dfrac{V'}{a'}\right)^2 \dfrac{r_{vA}}{D_{eA} C_{As}} < \dfrac{1}{|n|}$, $n \neq 0$

Stewart and Villadsen (1969); Mears (1971)

2. $\alpha_b \beta_{mb} + [0.3 n \alpha_b (-\Delta H) r_{vA} R / h_{fp} T_b] < 0.5n$

3. $\dfrac{r_{vA} d_p^2}{4 C_A D_{eA}} < \dfrac{1 + 0.33\alpha_b \dfrac{(-\Delta H_A) r_{vA} d_p}{2 h_{fp} T_b}}{|n - \alpha_b \beta_m|\left[1 + 0.33n\left(\dfrac{r_{vA} d_p}{2 C_A k_g}\right)\right]}$

Mears (1971)

Volume change

3. $\dfrac{(V''/a')^2 r_{vA}}{C_{As} D_{eA}} < \dfrac{1}{(1+\theta)n}$, $n \neq 0$

Kubota et al. (1969); Mears (1971)

178

Nonisothermal

4. $\dfrac{(V'/a')^2 r_{vA}}{C_{As} D_{eA}} < \dfrac{1}{n - \alpha_s \beta_m}$

Kubota and Yamanaka (1969);
Mears (1971)

5. $\left(\dfrac{V'}{a'}\right) \dfrac{r_{vA}}{D_{eA} C_{As}} \leq \sqrt{\dfrac{2}{n+1}}$,

$n > -1$

Petersen (1965)

Heat transfer

$\left| \dfrac{(-\Delta H) r_{vA}(V'/a')}{T_b h_{fp}} \right| < \dfrac{0.15}{\alpha_b}$

Mears (1971)

Nonisothermal

6. $\dfrac{(-\Delta H) d_p^2}{4 T_s k'_{el}} < 0.75 \left(\dfrac{T_0 R_g}{E}\right)$

Anderson (1963)

---

[a]Interphase and intraparticle mass transfer.

179

function of conversion for both particles and pellets of the catalyst, it is possible to determine the effectiveness factor for a given temperature, conversion, and initial gas composition by merely taking the ratio of the two rates under these conditions. Kadlec et al. (1970) have used this direct method to determine the effectiveness factor for SVD and ICI catalysts in the oxidation of sulfur dioxide.

The effectiveness factor may also be determined from the integrated expression for a single pellet. For this purpose Eq. 4.18 can be used to find the reaction rate per pellet, based on which the following expression can be written for the ratio of the reaction rates for pellets of two different sizes:

$$\frac{r_{p_1}}{r_{p_2}} = \left(\frac{R_2}{R_1}\right)^2 \frac{R_1 (k_v/D_{eA})^{1/2} \coth\left[R_1 (k_v/D_{eA})^{1/2}\right] - 1}{R_2 (k_v/D_{eA})^{1/2} \coth\left[R_2 (k_v/D_{eA})^{1/2}\right] - 1}$$

(9.2)

This equation can be solved by iteration to give k_v/D_{eA} and then ϕ can be obtained from the definition of the modulus. The effectiveness factor can then be calculated from Eq. 4.15. Walker and Mann (1969) have prepared useful plots for determining the effectiveness factor from Eq. 9.2.

Criteria for the Absence of Pore Diffusional Effects

Several criteria have been reported in the literature for the absence of pore diffusional effects both for isothermal and nonisothermal pellets. The more important ones are summarized in Table 9.1. Hudgins (1968) has also proposed a criterion but it is not generally very useful. All these criteria are based on the assumption of a certain permissible difference (usually 5%) between the rates in the presence of transport limitations and under diffusion-free conditions. Such an assumption can be avoided (Petersen, 1968) if the asymptotic solution for the general nonisothermal case given in Chapter 4 is used as the basis for formulating a criterion. Thus we have

$$\frac{(V'/a')^2 r_{vA}}{C_{As} D_{eA}} \leqslant \left(\frac{2}{n+1}\right)^{1/2}$$

(9.3)

It is particularly important to check for pellet isothermicity, since otherwise misleading results can be obtained. For example, in the oxidation of ethylene on a copper oxide–alumina catalyst (Caretto and Nobe, 1969), although the temperature difference was less than 0.30°C, a discrepancy in the rate constant of the order of 4% was observed.

An extension of these criteria to nonisothermal

situations should provide the most general test for the absence of transport effects. Two such criteria are included in Table 9.1.

9.1.3. Evaluation of Interparticle Transport Effects

Thus far we have been concerned with transport limitations within a pellet. In an actual integral reactor, rate data are also influenced by interparticle transport resistances, the consequences of which are manifested as radial and longitudinal diffusion effects in the packed reactor tube. Criteria for the absence of the interparticle effects in both the radial and axial directions are presented in Table 9.1.

9.1.4. A simple Diagnostic Test of the Kinetic Regime

It is likely that changing the pellet size may lead to distortion in the flow field, and it may not always be possible to detect any change in conversion due to a variation in the flow rate at constant space velocity (Chambers and Boudart, 1966).

A simple diagnostic test proposed by Koros and Nowak (1967) overcomes these difficulties. Making use of the fact that the reaction rate is proportional to the number of active sites only in the kinetic regime, these investigators suggest that rates should be measured with particles of catalyst mixed thoroughly with an inert material of the same size and properties. Then, in the kinetic regime, the ratio of the reaction rate using these pellets to the rate using pellets formed from catalyst powder alone should be equal to the fraction of catalyst in the mixed pellet. It must be ensured that the diluting powder is truly inert and possesses the same diffusion characteristics as the catalyst powder.

9.1.5. Recommended Criteria

A summary of the recommended criteria for the absence of various physical effects on the reaction rate is given in Table 9.1, which is an extension of the original table of Doraiswamy and Tajbl (1974).

9.2. ISOTHERMAL REACTORS

9.2.1. Integral Reactors

On the assumption of plug flow a simple material balance over a differential element of catalyst dW gives

$$r_{wA} dW = F dx_A$$

(9.4)

where r_{wA} is the reaction rate (mol/sec g catalyst). Integration of this equation between the limits $W = 0$, $x = 0$ and $W = W$, $x_A = x_{Ae}$ leads to

$$\frac{W}{F} = \int_{x_{A0}}^{x_{Ae}} \frac{1}{r_{wA}} \, dx_A \qquad (9.5)$$

Thus a plot of x_A vs. W/F can be differentiated to provide values of r_{wA}.

The reactor can be heated by molten metal, heat transfer fluid, fused salt, or induction heating, but by far the most convenient and efficient heating is provided by a fluidized solids bath.

A simple equation can be set up to provide a smooth curve through the data points. Since usually there is a considerable scatter of the data points, such a curve cannot be regarded as indicative of the true picture. This is brought out by the curve shown in Figure 9.2a.

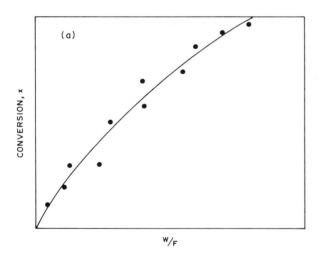

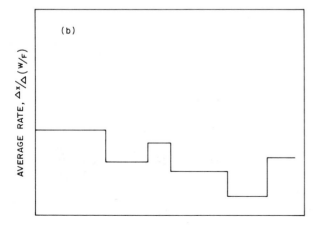

Figure 9.2. "Differential treatment" of integral rate data.

Let us take a pair of data points. The average rate calculated from these two points would be

$$\text{average rate} = \frac{x_2 - x_1}{(W/F)_2 - (W/F)_1} \qquad (9.6)$$

where the subscripts 1 and 2 correspond to the two data points. When average rates are calculated in this manner for successive pairs of experimental points and are plotted as shown in Figure 9.2b, it will be noticed that one would be entirely justified in assuming zero-order, first-order, or H–W kinetics with successive improvements in accuracy. As White and Churchill (1959) have pointed out, such a "differential treatment" of the rate data must be regarded as a useful adjunct to the analysis of integral rate data, since it provides a clear indication of the scope and accuracy of the data.

The main reason for errors in integral data is the lack of isothermicity, which is particularly serious in highly exothermic reactions. Such errors can be minimized by diluting the catalyst bed with inert solids.

Effect of Catalyst Dilution

Rihani et al. (1965) found that merely mixing the catalyst and diluent is not adequate, and that different degrees of dilution are necessary in different parts of the bed to ensure isothermal operation. Calderbank et al. (1969) have employed catalyst dilution as a moderating influence in the design of fixed-bed reactors for exothermic reactions (see Chapter 12), while Weisz and Swegler (1957) and Mosely and Good (1965) have examined the chemical influence of dilution.

Clearly the exit reactant concentration can be influenced by the type of inert distribution in the bed. Although in general this effect of dilution is not significant, under conditions of high experimental precision even a small effect due to dilution (say, of the order of 5 %) can cause avoidable errors. Van den Bleek et al. (1969) have formulated the following criterion for neglecting the effect of catalyst dilution:

$$\frac{b d_p}{l \delta} < 4 \times 10^{-3} \qquad (9.7)$$

where δ is the experimental error, l the undiluted height, and b the dilution ratio.

9.2.2. Differential Reactors

Equation 9.4 shows that if a differential quantity of catalyst is used (say, 1–3 g), then the conversion obtained (dx) would be so small that the compositions of the inlet and outlet streams would be almost identical. Under these conditions the experimental conversion gives the rate directly, corresponding to the partial pressures (average of inlet and outlet values) of the reactants and products in the differential bed:

$$-r_{wA} = \frac{y_{A0} - y_{Ae}}{W/F} \qquad (9.8)$$

where y_{A0} and y_{Ae} represent the inlet and outlet mole fractions of A.

A simple way of ensuring that the feed contains all the reaction components is to use an integral reactor to provide the feed for the differential reactor. The operation of a typical differential reactor has been described by, among others, Pansing and Malloy (1962). The shortcomings of the differential reactor can be overcome by two methods: (1) recycling a large part of the product stream, or (2) using a stirred reactor. These agitated reactors may be regarded as pseudo-differential reactors in that they give the rates directly (like the true differential reactor) but at integral conversion levels.

9.2.3. External Recycle Reactors

If a differential reactor is operated in such a manner that (a) most of the effluent stream is recycled, (b) a small amount of feed is continuously added, and (c) a small net product stream continuously removed, a reactor system is obtained that has several advantages over the conventional differential reactor. Since usually recirculation rates 10–15 times the feed rate are employed, the system would tend to be more nearly isothermal. Also, the resultant high velocities past the catalyst particles would eliminate almost completely any external mass transfer influence. Steady-state recycle reactors of this type have been described by Langer and Walker (1954), Perkins and Rase (1958), Temkin (1962), Satterfield and Roberts (1968), and Carberry (1976). At high rates of recirculation, the reaction rate given by Eq. 9.8 would correspond to the *exit* conditions.

A batch recycle reactor can also be used instead of the steady-state type described above. In this reactor a batch of the reactant is continuously recirculated over a bed of catalyst. If the conversion per pass is kept at a very low value, then this would in effect constitute a batch differential reactor system. By incorporating suitable sampling points fitted with serum stoppers for withdrawing samples through a hypodermic syringe,

for any given run the concentration vs. time curve can be obtained. Reactor systems of this type have been used by Butt et al. (1962) and Cassano et al. (1968). A semibatch recycle reactor has also been used for obtaining kinetic data (Leinroth and Sherwood, 1964), but this is not a particularly useful reactor to employ.

An important aspect of the recycle reactor is that the degree of mixing can be calculated from the recycle rate F_R, and depending on the value of F_R the reactor can be operated at the extremes of zero and complete mixing. Gillespie and Carberry (1966a,b) have given procedures for estimating the effect of the recycle ratio on reactor performance.

9.2.4. Agitated Reactors

Continuous Stirred Gas–Solid Reactions

The principal features of continuous stirred gas–solid reactors are that gas–solid mass transfer resistance is almost eliminated and perfect mixing can be achieved, thus making it possible to operate these reactors under practically gradientless conditions.

Fluidized-Bed Reactor with Agitation. The fluidized bed can be made fully mixed by providing suitable baffles or by using a stirrer. A stirred fluidized-bed reactor of this design has been used by Trotter (1960). The basic disadvantage of this reactor is that the catalyst is subject to disintegration as a result of the mechanical agitation involved. This disadvantage has been overcome by Ramaswamy and Doraiswamy (1973), who employed external agitation through a pulsator operating independently of the feed device. Alternatively, the feed itself can be introduced through a pulse pump. Some useful results have been reported by Massimilla et al. (1966) and Kobayashi et al. (1970).

In view of the availability of other (more reliable) types of reactors, and the inherent uncertainty regarding complete mixing in a pulsed fluidized bed, this reactor is not normally recommended for obtaining intrinsic kinetic data.

Rotating-Basket Reactors. In a rotating-basket reactor the catalyst is held in a wire mesh container (basket) that is attached to a stirrer rotating inside a reaction pot. Variations in design pertain chiefly to basket design and stirring arrangement. This reactor is superior to the recycle reactor in that very high speeds of rotation (say, up to 5000 rev/min) can be employed to ensure the absence of any mass transfer effect. A disadvantage of the basket reactor is that the actual temperature of the surface of the catalyst cannot be directly measured.

Ball-Mill Reactor. A rotating mixer reactor that is based on the concept of the ball mill has been described by Barrett (1971). Mixing of solid and gas is obtained by rotating the reactor; this leads to temperature differences between the particles and the gas as low as 2°C.

The catalyst in the stirred reactors described above is not stationary. Stirred reactors with stationary catalyst are described next.

Reactors with Catalyst Impregnated on Reactor Walls or Placed in an Annular Basket.

A thin layer of the desired catalyst can be directly coated on the inner walls of the reactor (Ford and Perlmutter, 1964) or on a removable liner fitted snugly inside the reactor (Lakshmanan and Rouleau, 1969), or the reactor wall itself can have catalytic properties (Wentzheimer, 1969). In such cases the reactor is provided with an efficient stirrer to ensure thorough mixing; thus the true kinetics can be studied in any desired temperature range. However, since such a reactor has a large ratio of free volume to catalyst volume, its use is limited to systems where noncatalytic homogeneous reactions are not involved. The temperature of the catalyst (the reactor wall) can be directly measured.

In another design the catalyst is in the form of pellets or granules and is placed in an annular basket made of wire mesh fitted close to the reactor wall. The use of such an annular basket type of reactor has been reported by Tajbl et al. (1967), Relyea and Perlmutter (1968), and Lakshmanan and Rouleau (1970).

Reactor with Catalyst Placed in a Stationary Cylindrical Basket.

In certain reactors catalyst of any form and size (or even a single pellet) is placed in a cylindrical basket (provided with a fine wire mesh) fitted at the center of the bottom of the reactor pot, and stirred-tank performance is obtained by rotating a special impeller. Alternatively, the baffle-equipped reactor pot itself rotates around the basket with some clearance (Choudhary and Doraiswamy, 1972). The ratio of the free volume to catalyst volume can be minimized by a proper design of the reactor pot, impeller, and basket. The important feature of this type of reactor is that the actual temperature of the catalyst can be measured directly.

A reactor of this type has been employed by Costa and Smith (1971) for studying a gas–solid noncatalytic reaction hydrofluorination of uranium dioxide using a spherical single pellet.

Internal Recirculation Reactor.

In the internal recirculation reactor mixing is achieved by recirculating the reaction mixture through a stationary catalyst bed by means of a specially designed (turbine type) impeller. Variations in design have been reported by Garanin et al. (1967), Weychert and Trela (1968), Gelanin (1969), Berty et al. (1969), Brown and Bennett (1970), Livbjerg and Villadsen (1971), Bennett et al. (1972), Mahoney (1974) and Baag et al. (1975). The recirculation rate may be determined by the rotational speed of a centrifugal blower mounted at the top or bottom of the catalyst chamber. Two typical designs are shown in Figure 9.3.

The flow rate of the reacting mixture and the degree of recirculation through the catalyst bed are well-defined quantities for these reactors (which is not so for a stirred reactor) and the actual rate of internal recycle can be estimated from a few preliminary experiments. Hence, the mass and heat transfer correlations developed for fixed-bed reactors (Chapter 6) can be successfully applied for estimating the mass and heat transfer rates from gas to solid particles.

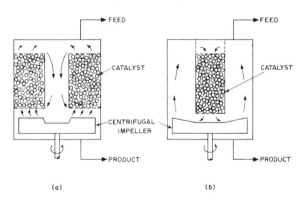

Figure 9.3. Internal recirculation reactors, with catalyst (a) at the wall (Brown and Bennett, 1970) and (b) at the center (Berty et al., 1969). (from Doraiswamy and Tajbl, 1974).

Choice of Stirred Reactor

The rotating-basket type of reactor should be used for studying catalytic reactions involving low heats of reaction only. The use of a reactor with catalyst lined on the wall or placed in an annular basket within it is recommended only when homogeneous reactions do not occur to any appreciable extent. On the other hand, reactors with a stationary cylindrical basket and internal recycle are versatile and may be used for studying any gas–solid reaction.

Mass (and Heat) Transfer Coefficients and Mixing Tests

The most important and basic requirement of the stirred gas–solid reactor is that it should be operated under gradientless conditions. Physical measurements should therefore be carried out to evaluate the reactor behavior with respect to both the mass transfer and mixing characteristics under different conditions of flow rate and stirring speed before undertaking any kinetic study.

Choudhary and Doraiswamy (1972) and Pereira and Calderbank (1975) have studied mass transfer in a spinning-catalyst-basket reactor using naphthalene evaporation, and have developed predictive correlations. Their results show that the mass transfer coefficient depends not only on the rotating speed but also on the radial position of the pellet. While these correlations can be used for reactors of design similar to those for which they were developed separate mass transfer studies should be carried out if the designs are even slightly different.

The surface temperature of the catalyst may be calculated from j_h through the following relationship between j_h and j_d: $j_h = 1.37 j_d$.

Mixing tests may be carried out by studying the kinetics of a simple (first-order) reaction, the true kinetics of which has already been well established, or (more directly) by the tracer technique. Three methods based on the tracer technique may be used: (1) inject a small amount of tracer into a steady stream of an inert gas over a short time (pulse test); (2) switch over from a steady flow of inert gas to a steady flow of tracer gas (step test) or vice versa (purge test); or (3) introduce a steady flow of tracer gas into a steady flow of inert gas (step test) or vice versa (purge test). If the mixing is perfect, the tracer responses to these tests are as follows:

Pulse test:

$$\frac{C}{C_{mp}} = \exp\left(-\frac{tQ}{V_r}\right) \qquad (9.9)$$

Step test:

$$1 - \frac{C}{C_{ss}} = \exp\left(-\frac{tQ}{V_r}\right) \qquad (9.10)$$

Purge test:

$$\frac{C}{C_{ss}} = \exp\left(-\frac{tQ}{V_r}\right) \qquad (9.11)$$

In these equations C_{mp} is the gas concentration in the reactor at the instant the pulse enters, and in an actual test is the maximum gas concentration measured in the effluent; C_{ss} is the steady-state tracer concentration; Q is the volumetric flow rate; and V_r is the reactor volume.

Any of these tests can be used. The easiest is perhaps the step test, in which a steady stream of a tracer (carbon dioxide, ethylene, etc.) is introduced into a steady stream of an inert (usually nitrogen); a plot of $\ln\left[1 - C/C_{ss}\right]$ vs. Qt/V_r must give a straight line with a negative slope of unity provided the reactor is perfectly mixed.

9.2.5. Microreactors

As the name implies, a microreactor is characterized by the use of a very small quantity of catalyst (usually in the range of 0.01–1.0 g). The heat of reaction would therefore be small, and practically isothermal conditions would be more readily obtained than with larger catalyst beds. The nonisothermal character of the internal (or intraparticle) field in the case of exothermal or highly endothermal reactions would, however, remain unaltered. Since such a reactor is usually connected to a gas chromatograph to enable the determination of product composition, and since this involves the use of a carrier gas (usually hydrogen or helium) in relatively large quantities, it is clear that the resultant high velocities would tend to eliminate to a large extent any external mass transfer resistance.

The microreactor can either be operated as a straightforward steady-state reactor ("tail gas" technique), like any of the differential or integral reactors described earlier, or a pulse of reactant can be introduced into the reactor ("slug" technique) and all the basic kinetic parameters extracted from a theoretical analysis of the concentration pulse of the reactant through the reactor. The latter is obviously an unsteady-state operation.

Both techniques have been employed in obtaining kinetic data (see, e.g., Yushchenko et al., 1968; Barbul et al., 1968; Celler, 1970; Luckner and Wills, 1973).

First-Order Irreversible Reactions

For a reaction of order n, the rate equation is

$$r_{vA} = \frac{d\hat{C}_A}{dt} = -k_v C_{A0}^n (1 - x_A)^n \qquad (9.12)$$

so that the conversion in the reactor becomes a function of the inlet concentration and therefore of the input pulse. On the other hand, for a first-order reaction ($n = 1$), the rate is independent of the input pulse.

Bassett and Habgood (1960, 1961) have shown that

$$\ln\left(\frac{1}{1 - x_A}\right) = \frac{V_R}{Q} K_A k_v \qquad (9.13)$$

where V_R is the chromatographic retention volume. This is identical with the equation for conversion in a first-order reaction occurring under steady-state conditions. A plot of V_R/Q, which comprises quantities that are easily measurable, against $\ln[1/(1-x_A)]$ gives the value of the composite constant $K_A k_v$. There is seldom any necessity to determine k_v and the equilibrium constant K_A separately.

Irreversible Reaction of General Order

On the assumption of plug flow and no radial or axial diffusion the following equation can be written:

$$f_B \frac{\partial C_A}{\partial t} + u \frac{\partial C_A}{\partial l} = -r_{VA} \qquad (9.14)$$

where u represents the velocity of the carrier gas, and r_{VA} is the reaction rates per unit volume of the reactor (as against r_{vA}, which represents the unit volume of the pellet).

To solve Eq. 9.14, the nature of the input pulse must be known. Three cases may be considered: square-topped pulse; triangular pulse; and Gaussian pulse. Note that the actual pulse form represents the inlet boundary condition, and its attenuation with time during its transit through the reactor provides a measure of the reaction rate. Thus, for an analysis of these different pulse types, we shall denote the functional dependence of concentration on time by the relation

$$C_A = C_{A0}^0 f(t) \quad \text{or} \quad \hat{C}_A = f(t) \qquad (9.15)$$

where C_{A0}^0 represents the peak input pulse concentration. For the square-topped or rectangular pulse there is no concentration peak, and the constant time-independent value of C_A (giving $C_A = C_{A0}^0$) leads to $f(t) = 1$. This is brought out in Figure 9.4, in which the line representing the rectangular pulse can be readily identified. In view of its time-independent characteristic, the rectangular pulse is similar to a flow reactor operating under steady-state conditions.

The triangular and Gaussian pulses are also shown in the figure. As its name implies, the triangular pulse shows a peak value; in the increasing-concentration region of the pulse, $f(t)$ increases in direct proportion to t until the peak value of 2 is reached. Since the pulse is linear on both sides of the peak, in the falling-concentration region we have $f(t) = 2 - t/t_m$.

The Gaussian distribution is usually employed as the most acceptable form of distribution. Using this distri-

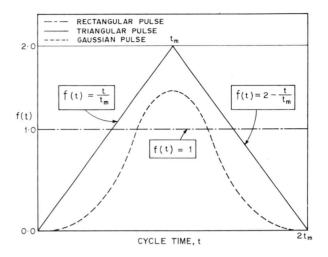

Figure 9.4. Illustration of various pulse forms.

bution and a dimensionless rate constant

$$\hat{k}_v = \left(\frac{L f_B C_{A0}^{0^{n-1}}}{u} \right) k_v = \left(\frac{V_r f_B C_{A0}^{0^{n-1}}}{Q} \right) k_v \qquad (9.16)$$

Blanton et al. (1968) have provided solutions in terms of conversion as a function of the dimensionless rate constant $\hat{k}_v$ for $n = 0.5, 1.0, 1.5,$ and 2.0 (see Figure 9.5).

The procedure for determining the rate constant and reaction order by using the foregoing solutions is as follows:

1. Determine the conversion at different values of C_{A0}^0 and Q.

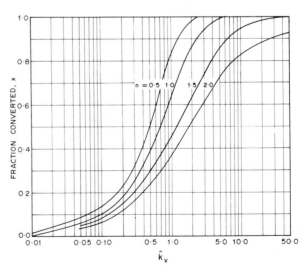

Figure 9.5. Conversion as a function of the dimensionless rate constant for reactions of different orders in a pulsed microreactor (Blanton et al., 1968).

2. Assume a value of the reaction order n; then, if the reaction involved is $A \rightarrow R$, the rate equation is $r_{vA} = -k_v C_A^n$. Plot conversion as a function of the group $V_r f_B C_A^{n-1}/Q$.

3. Superimpose this plot on the curve for the same value of n in Figure 9.5; repeat steps 2 and 3 until good matching is obtained.

4. For a given conversion read the values of $\hat{k}_v$ and the group $V_r f_B C_A^{n-1}/Q$ on the matched curves and thus estimate the rate constant from Eq. 9.16.

5. Repeat the entire procedure for various temperatures and obtain the Arrhenius parameters.

Gaziev et al. (1963) and Deans et al. (1967) have also provided solutions to the problem of non-first-order reactions, while Hightower and Hall (1968) have considered the case of a reaction following the Langmuir–Hinshelwood scheme.

Reversible and Complex Reactions

The behavior of reversible and complex reactions has been examined in considerable detail by Magee (1963), Murakami and Hattori (1967), Phillips et al. (1967), Hattori and Murakami (1968a,b, 1973, 1974) and Sica et al. (1978). For a reversible reaction, the forward rate constant can be obtained by operating at different values of the pulse width and extrapolating to zero width.

Typical Microreactors

Kokes et al. (1955) were among the first to use a microcatalytic reactor for vapor-phase catalytic reactions. Modifications of the original microreactor have been suggested by Hall and Emmett (1957), Keulemans and Voge (1959), Ettre and Brenner (1960), Hartwig (1964), Dutten and Mounts (1964), Danforth and Roberts (1968), and others. Semiautomatic (Hall et al., 1960) and completely automatic (Harrison et al., 1965) precision microreactors have also been reported for obtaining data on reaction kinetics and catalyst behavior. Binter et al. (1969) have described an integrated system comprising a microcatalytic reactor combined with a gas chromatograph. Eliezer et al. (1977) have described a flow microreactor for studying high-pressure (up to 300 atm) catalytic hydroprocessing reactions. Choudhary (1978) has described a simple microcatalytic reactor combined with a dual-column gas chromatograph.

The use of the pulse technique in studying heterogeneous catalysis has been reported by several investigators, and the operating principles have been reviewed by Galeski and Hightower (1970) (see also Varma and Kaliaguine, 1973; Richardson et al., 1975). Richardson and Friedrich (1975) have described a pulsed thermokinetic (PTK) technique, which is basically a combination of a pulsed microreactor and a conventional differential thermal analysis (DTA) apparatus, for studying simultaneous adsorption and reaction on solid catalysts.

Sketches of a few typical microreactors are shown in Figure 9.6. Figure 9.6a represents the simplest microreactor, which is similar to any steady-state integral or differential reactor. The microreactor in Figure 9.6b has separate provisions for introducing the reactant and the carrier gas. The contact time can be varied at constant pressure and flow rate by altering the depth at which the reactant is introduced into the catalyst bed through a long-needled hypodermic syringe.

The more sophisticated design in Figure 9.6c includes a special valve that permits the injection of feed at two points. To start with, the valve position is adjusted such that the carrier gas passes through the inert bed, into which a pulse of the reactant is introduced. Then the valve position is altered so that the carrier gas passes through the catalyst bed, into which a reactant pulse (identical with the earlier pulse) is introduced; the effluent is analyzed in the gas chromatograph. Comparison of the reactant peaks in the effluent in the two cases gives the conversion due to reaction.

Chromatographic Column Reactors

The chromatographic column itself can be used as a reactor by packing it with catalyst. The special feature of this reactor is that, under chromatographic conditions, the products of the reaction are not only separated from the reactants but are also separated from each other. Thus the reverse reaction does not take place to any significant extent, leading to a sharp increase in the conversion. Clearly, then, the use of a chromatographic-column reactor is advantageous only for fast reactions with low equilibrium constants which form at least two reaction products that can be chromatographically separated from each other. This reactor will not be of any special advantage in cases where two reactants are present that will also be separated in the chromatographic column, since in such cases no reaction at all can take place.

One important application of these principles is in studies of kinetics, adsorption, and other rate phenomena (Bassett and Habgood, 1960; Blanton et al., 1968;

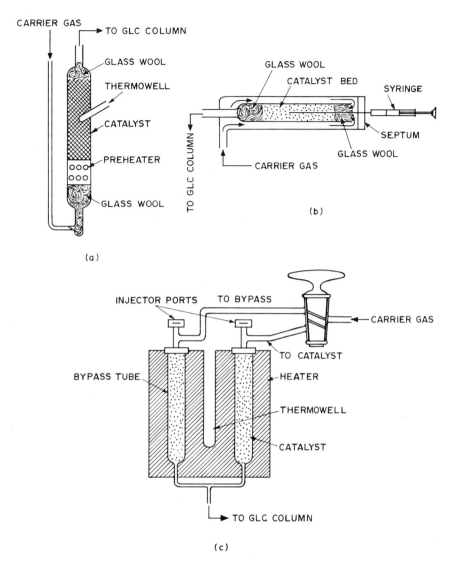

Figure 9.6. (a) Simple microreactor (Kokes et al., 1955); (b) variable-contact microreactor (Dutton and Mounts, 1964); (c) twin-tube injector microreactor (Danforth and Roberts, 1968) (from Doraiswamy and Tajbl, 1974).

Collins and Deans, 1968; Deans et al., 1970; Langer and Yurchak, 1969; Paratella et al., 1968; Roginskii and Rozental, 1964; Schwab and Watson, 1965). Several experimental studies on gas-phase reactions occurring under chromatographic conditions over a solid catalyst have been reported by Matsen et al. (1965); Filinovskii et al. (1966), Gaziev et al. (1963), Roginskii et al. (1962, 1963, 1964), Aliev et al. (1968), Semenenko et al. (1964), and Wetherold et al. (1974). Theoretical analyses for various situations have been presented by many investigators, including Matsen et al. (1965), Klinkenberg (1961), Roginskii et al. (1962), Zimin et al. (1965), Langer et al. (1969), Roginskii and Rozental (1965a,b), Gore (1967), Magee (1963), Saito et al. (1965),

Viswanathan and Aris (1974, 1975) and Cho et al. (1980).

Examples of the Use of Microreactors

A large number of catalysts can be readily evaluated for their activity and selectivity by using any one of the microreactors described in the previous section at different reaction conditions (Hall et al., 1960; Norton, 1962; Steen et al., 1962; Norton and Moss, 1964; Topchieva et al., 1968; Yushchenko and Antipina, 1969). It should be noted, however, that the results obtained by this technique cannot describe the process as a whole, and hence care must be taken in selecting a catalyst on the basis of such data alone.

TABLE 9.2. Some Examples of the Use of Microcatalytic Reactors in Kinetics Studies

Reaction Studied	Catalyst Used	Remarks	Reference
Isomerization of cyclopropane	Linde molecular sieve 13 X	Extent of adsorption under reaction conditions and activation energy for surface reaction step measured	Bassett and Habgood (1960, 1961)
Thiophene hydrodesulfurization	Chromia and cobalt molybdate catalyst	Comparison of steady-state-flow and pulsed microreactors made for non-first-order reaction; differences were more pronounced at lower temperature than higher one	Owens and Amberg (1961)
Dehydrogenation of methane	Silver catalyst	Same reaction order and activation energy obtained by using steady-state-flow and pulsed microcatalytic reactors	Schwab and Watson (1965)
Isomerization of cycloalkanes	Silica–alumina	Results could be explained by Langmuir–Hinshelwood theory	Hightower and Hall (1968)
Hydrogenation of ethylene	Alumina	Rate parameter obtained by pulse technique agreed very closely with the steady-state-flow results	Blanton et al. (1968)
Dehydration of 2-butanol	Hydroxyapatites	Pulsed microreactors and steady-state-flow reactors gave significantly different results (for zero-order reaction)	Bett and Hall (1968)
Dealkylation and disproportionation of cumene	Silica–alumina	Results found consistent with Langmuir–Hinshelwood kinetics	Hattori and Murakami (1968a, b); Murakami and Hattori (1967)
Dehydration of isopropanol Cracking of cumene	γ-Alumina Zeolite NU	Langmuir adsorption isotherm assumed	Yushchenko and Antipina (1969b)
Dehydration of isopropanol	MgO, CaO, and SrO	Both flow and micropulse techniques used in the study	Szabo et al. (1975)
Hydrogenation of cyclopropane	Rhenium powder		Wallace and Hayes (1970)
Oxidation of methanol	MgO_3–$Fe_2(MoO_4)_3$	Kinetic results explained on the basis of catalyst surface heterogeneity and/or of poisoning by products	Liberti et al. (1972)
Disproportionation of propylene	Tungsten oxide on silica catalyst	—	Luckner and Wills (1973)
Hydrogenation of ethylene	Ni–Al_2O_3	Used for studying adsorption of hydrogen and ethylene and for ethylene hydrogenation	Richardson et al. (1975)
Fischer–Tropsch reaction	Ru–Al_2O_3	A microflow reactor used to collect kinetic data under transient operating conditions	Dautzenberg et al. (1977)

The pulsed microreactor has also been used to determine the kinetics of several reactions, and in many instances the results have been compared with those obtained from steady-state reactors. Some significant studies are summarized in Table 9.2.

9.2.6. Reactor Systems with Well-Defined Hydrodynamics

The scientific elegance of conducting kinetic experiments in a well-defined flow field was brought out in Chapter 6. Among the several configurations—internal as well as external—that are possible, the tubular reactor and the forward stagnation region of a circular cylinder offer the best scope for obtaining the intrinsic kinetic parameters of a reaction. Of these two, the tubular reactor (in which the reactor wall is coated with catalyst) presents considerable mathematical difficulties, whereas the stagnation region of a circular cylinder is unique as regards its ability to satisfy the uniform accessibility requirement (see Section 6.3). Although a number of investigations have been reported on the use of the tubular reactor, we shall confine our discussion in this section to the use of the forward stagnation region recently suggested by Balaraman et al. (1980).

Balaraman et al. (1980) have solved the equation for convective diffusion with chemical reaction in the stagnation region of a circular cylinder, and derived an expression for the interaction between reaction and diffusion for an nth-order reaction. For a first-order reaction this reduces to the familiar additivity-of-resistances relation $1/k_a = 1/k_S + 1/k_g$. The mass transfer coefficient k_g in the stagnation region can be evaluated by the exact solution given by Schlichting (1968), which is of the form

$$\frac{1}{2}\left(\frac{Sh}{(Re)^{1/2}}\right) = F(Sc) \qquad (9.17)$$

where $F(Sc)$ is a function of the Schmidt number.

The Balaraman–Mashelkar–Doraiswamy reactor (1980) should prove generally useful in studies of this kind. It comprises a cylindrical galvanized iron duct (inside diameter 5.27 cm) that is 1.8 m long. Suitable entrance and calming section are provided to ensure the development of a perfect laminar flow profile. This is necessary in order to generate a region of constant boundary layer thickness (and therefore of uniform accessibility) around the test cylinder inserted into the duct. As shown in Figure 9.7, this stagnation region corresponds to a segment of angle $60°$ in which the variation of boundary layer thickness is less than 3%.

The test section is so placed that the stagnation region is exactly perpendicular to the upstream flow of gas. The catalyst coating is restricted to this region.

Another useful configuration is the rotating disk, on which a number of studies have been reported. A laboratory reactor consisting of a metal disk coated with catalyst and rotating in a fully mixed environment has been described by Shah and Doraiswamy (1983). The rotating disk ensures a uniform laminar boundary layer, thus enabling extraction of precise kinetic data.

9.3. NONISOTHERMAL REACTORS

9.3.1. Adiabatic Reactors

In an adiabatic reactor preheated feed is passed into a thermally insulated reactor containing the catalyst bed, and the temperature profile is allowed to build up in accordance with the kinetics and thermodynamics of the reaction involved.

The starting point is the conservation equation developed in Chapter 11 (Eq. 11.8, neglecting axial conduction). If the length l from the reactor inlet appearing in this equation is replaced by the catalyst weight W corresponding to this length, then by simple substitutions the following equation can be obtained:

$$GA_c C_p \rho_B \frac{dT}{dW} + r_{VA}(-\Delta H_A) = 0 \qquad (9.18)$$

From the experimental axial profile of T vs. W, numerical values of dT/dW can be obtained, and Eq. 9.18 then solved to provide values of r_{VA} at different temperatures.

The next step is to relate the measured temperature at any point to the mole fractions of the various components of the reaction. Thus, if axial diffusion of heat and mass is neglected, and T_0 is taken as the base temperature, the following energy balance can be written (under adiabatic conditions) for reactant A:

$$y_A = \frac{y_{A0} + M_0(C_p/-\Delta H_A)(T-T_0)}{1 + (\Delta n)_A M_0(C_p/-\Delta H_A)(T-T_0)} \qquad (9.19)$$

where M_0 is the initial molecular weight of the feed and $(\Delta n)_A$ the increase in number of moles per mole of A converted.

Therefore, from a knowledge of the temperature profile, the mole fraction of A at any point in the reactor can be computed. Similar equations can be written for y_B, y_R, and y_S, with $-\Delta H$ being expressed per mole of the concerned component (B, R or S).

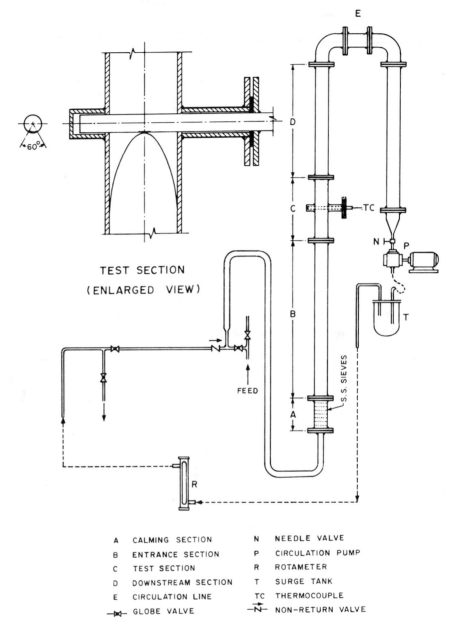

TEST SECTION

(ENLARGED VIEW)

A	CALMING SECTION	N	NEEDLE VALVE
B	ENTRANCE SECTION	P	CIRCULATION PUMP
C	TEST SECTION	R	ROTAMETER
D	DOWNSTREAM SECTION	T	SURGE TANK
E	CIRCULATION LINE	TC	THERMOCOUPLE
—▷◁—	GLOBE VALVE	—N—	NON-RETURN VALVE

Figure 9.7. Experimental setup of Balaraman et al. (1980).

If the reaction order is n, the following rate equation can be written:

$$r_{VA} = A_f (P y_A)^n \exp\left(-\frac{E}{R_g T} \right) \qquad (9.20)$$

Thus a plot of $\ln\left[r_{VA}/(P y_A)^n \right]$ vs. $1/T$ for an assumed value of n should give a straight line of slope E/R_g and intercept A_f if the assumed value of n is correct. Several

trials may be necessary before the final kinetic expression is established.

This method has been tested by Schmidt et al. (1964) in a special multibed adiabatic reactor, and by Chakraborty and Doraiswamy (1971) for the hydrogenation of nitrobenzene to aniline. The catalyst bed is generally divided into several cylindrical sections to enable gas mixing in the intermediate baffled spaces and easy measurement of gas temperatures in these regions.

It must be ensured that the reactor is completely adiabatic. This can be done by insulating it heavily and supplying controlled heat to the insulation by heating coils wound around it. As a test of adiabatic behavior, the product issuing from the reactor may be analyzed and $T - T_0$ calculated from Eq. 9.19; this value should agree with the experimentally determined $T - T_0$ for the catalyst bed as a whole. Note that if this requirement is satisfied, there is no further need for any chemical analysis.

9.3.2. Kinetic Data from Nonisothermal Experiments

Despite the availability of precise techniques for obtaining isothermal rate data, it often becomes necessary to obtain reliable data from a simple packed reactor that can be easily set up in any laboratory. The main difficulty with such a reactor (which is usually 2–3 cm in diameter and 100–400 cm long) is that it is nonisothermal, and temperature variations of the order of $50°C$ are normal. This is more fully discussed in Chapter 11. When such a reactor is used for obtaining kinetic data, it serves the additional purpose of providing a clear insight into the behavior of the fixed-bed reactor system.

9.4. DETERMINATION OF KINETIC PARAMETERS DURING CATALYST DEACTIVATION AND REGENERATION

Attention so far has been restricted to methods for the determination of kinetic parameters in the absence of catalyst deactivation. As pointed out in Chapter 8, there are many catalytic reactions in which rapid fouling occurs, so that an independent determination of the kinetic constants for the main reaction becomes rather involved, and often may not be feasible. An alternative approach is to determine the kinetic parameters simultaneously for both the main and fouling reactions in a single experimental program. A problem of equal importance is the kinetics of carbon burn-off during regeneration, for a knowledge of this is essential in fixing the operating variables for regeneration. The object of this section is to present methods for the independent and simultaneous determination of the kinetic and fouling constants on a rapidly decaying catalyst, as well as to outline briefly the general procedure for determining the kinetics of carbon burn-off.

9.4.1. Independent Determination of the Kinetic and Fouling Parameters

The independent determination of the kinetics of the main reaction alone on a fouling catalyst is possible only if experimental conditions can be so chosen that the influence of fouling is eliminated. A good example is the hydrolysis of chlorobenzene to phenol on tricalcium phosphate in the Raschig process. This catalyst exhibits peak activity for less than a few hours (but sufficiently long for rate measurments) between activation and deactivation periods. Deo and Doraiswamy (1967) studied the kinetics of the main reaction by preparing x vs. time plots for each W/F and noting the steady value of x obtained during every cycle of deactivation and regeneration.

9.4.2. Simultaneous Determination of the Kinetic and Fouling Parameters

Often it is difficult to define conditions under which the kinetic parameters of a reaction carried out on a rapidly fouling catalyst can be independently determined as described above. Procedures for decoupling the intrinsic kinetic and fouling constants for a reaction subject to rapid fouling are described below.

We shall confine the treatment to independent and parallel fouling. A similar procedure can be used for series fouling. The following rate equations can be written.

Main reaction:

$$-r_{vA} = -\frac{dC_A}{dt} = k_v C_A^n \Omega \qquad (9.21)$$

Fouling reactions:

$$-\frac{d\Omega}{dt} = k_{v,f} \Omega^f \qquad \text{(independent)} \qquad (9.22)$$

$$-\frac{d\Omega}{dt} = k_{v,f} C_A^d \Omega^f \qquad \text{(parallel)} \qquad (9.23)$$

$$-\frac{d\Omega}{dt} = k \qquad (9.24)$$

Independent Fouling

When Ω is independent of C_A, if $f = 1$, then an expression for the fouling factor can be found by integrating 9.22:

$$\Omega = \exp(-k_{v,f} t) \qquad (9.25)$$

The fouling equation for a deactivating mixed reactor with $n = 1$ can then be written:

$$\ln\left(\frac{C_{A0}}{C_A} - 1\right) = \ln(k_v \tau_f) - k_{v,f} t \qquad (9.26)$$

where τ_f (a time parameter) $= WC_{A0}/F\rho_c$. A plot of $\ln[(C_{A0}/C_A) - 1]$ as a function of time gives a straight line with slope $-k_{v,f}$ and intercept $\ln(k_v \tau_f)$. Thus the rate and fouling constants can be determined.

This procedure assumes first-order kinetics for both the main and fouling reactions. In the absence of *a priori* knowledge of these orders in practical situations, the following procedure may be adopted:

1. Determine the reaction rate as a function of time in a mixed reactor.
2. Prepare plots of Ω vs. t and Ω vs. C_A (by calculating Ω from any assumed functional relationship given in Table 8.2).
3. Plot $-r_{vA}/\Omega$ against C_A on log–log coordinates and determine the reaction order n and the rate constant k_v.
4. Determine $d\Omega/dt$ from step 2 and plot against Ω on log–log coordinates. The slope and intercept of this line give the exponent f and $k_{v,f}$, respectively.

A relationship analogous to Eq. 9.26 for the mixed reactor can be obtained for the integral reactor. The right-hand side of this equation is exactly the same, while the left-hand side changes to $\ln \ln[C_{A0}/(C_{A0} - C_A)]$. This equation, given by Levenspiel (1972), can be used to interpret data obtained in a plug-flow reactor under conditions of no diffusional limitation. In the event that diffusional limitations are present, an intraparticle effectiveness factor enters the analysis and the simple equation (Eq. 9.26) gets modified to $\ln \ln[C_{A0}/(C_{A0} - C_A)] = \ln(k_{v,f} \varepsilon C_f) - (k_{v,f}/2)t$. A plot of concentration vs. time can thus be used to estimate the deactivation parameter (Krishnaswamy and Kittrell, 1981a,b). The presence of external mass transfer further alters the nature of this relationship. However, the equation is still simple enough to enable us to estimate the parameters.

In case an integral reactor is used for complex reactions, the rate is determined by graphical or analytical differentiation of the concentration time curve. Prasad and Doraiswamy (1974) have used essentially this procedure for obtaining the kinetic and fouling parameters for a consecutive reaction.

Krishnaswamy and Kittrell (1978) characterized the deactivation of immobilized glucose oxidase, which catalyzes the conversion of glucose to lactone, by using an integral reactor.

It is of interest to note that for rapidly deactivating systems equivalent degrees of fit of experimental data and, in some cases, identical equation forms result by a variety of assumptions regarding the kinetics of the primary and deactivation reactions (Krishnaswamy and Kittrell, 1978). More specifically, gas oil cracking data (Paraskos et al., 1976) from a transfer-line reactor may be satisfactorily represented by either a second-order conversion model with concentration-independent deactivation or a first-order conversion model with concentration-dependent deactivation. Similar uncertainty has been demonstrated for the data of Altomare et al. (1974) on the deactivation of immobilized catalase in the decomposition of hydrogen peroxide.

Parallel Fouling I

When Ω is a function of C_A the only way to determine k_v and $k_{v,f}$ is to hold the concentration of A in the reactor constant. In a mixed reactor this can only be achieved by continuously decreasing the feed rate. Thus, the factor affecting the rate of the principal reaction is, in a sense, eliminated, and conditions are created for the isolation and determination of the fouling parameters. Based on Eqs. 9.21 and 9.23, which characterize parallel fouling, the following equation can be derived for fth-order decay:

$$\tau_f^{f-1} = K_1 + K_2 t \qquad (9.27)$$

where

$$K_1 = \left(\frac{C_{A0} - C_A}{k_v C_A^n}\right)^{f-1}$$

$$K_2 = \left(\frac{C_{A0} - C_A}{k_v C_A^n}\right)^{f-1} (f-1)(k_{v,f} C_A^d)$$

Thus a plot of $(\tau_f)^{f-1}$ vs. t should give K_1 and K_2, from which k_v and $k_{v,f}$ can be obtained.

As in the case of independent fouling, here too, unless the deactivation order f is known, Eq. 9.27 cannot be used, except by a laborious trial-and-error procedure. This difficulty can be overcome by adopting the following general procedure.

1. Decrease the feed rate continuously (or in short steps) so that the concentration of A in the reactor remains constant, and plot the feed rate as a function of time (see Prasad and

Doraiswamy, 1974). This can be done by using a calibrated pump.

2. Calculate Ω from $W/F = (C_{A0} - C_A)/C_{A0}K$ (where $K = C_A^n k_v = $ constant) from various feed rates.

3. From a linear plot of Ω vs. $1/\tau_f$ determine K and therefore $k_{v,f}$ (since $\tau_f = (C_{A0} - C_A)/K\Omega$).

4. Differentiate the Ω vs. t plot and prepare a log–log plot of $d\Omega/dt$ vs. Ω; determine $C_A^d k_{v,f}$ and f from the intercept and slope, respectively; $k_{v,f}$ may thus be obtained.

Parallel Fouling II

When (as is often the case) Ω is expressed in terms of the dimensionless coke content of the catalyst $\hat{W}_c$ for reactions in which the fouling reaction results in coke deposition on the catalyst, the procedure formulated by Ozawa and Bischoff (1968a,b) for determining k_v and $k_{v,f}$ by using the linear relationship between Ω and $\hat{W}_c$ (Table 8.2) is appropriate.

On the assumption that the transient coke deposition is slow (i.e., neglecting the accumulation term in Eq. 8.31), and by using Eq. 8.34 for carbon balance (for parallel fouling), the following solution has been obtained by Ozawa and Bischoff:

$$Z\Omega = \frac{1 - \hat{C}_A}{\hat{C}_A} \qquad (9.28)$$

Here Z represents the sum of the rate constants normalized through the catalyst weight W and the volumetric flow rate Q:

$$Z = (k_v + k_{v,f})\frac{W}{Q\rho_c} \qquad (9.29)$$

Since $\hat{C}_A$ is an easily measurable quantity, $Z\Omega$ can be computed at any instant of time. For linear fouling

$$\Omega = 1 - a''\hat{W}_c \qquad (9.30)$$

or

$$Z\Omega = Z - Za''\hat{W}_c \qquad (9.31)$$

Thus a plot of $Z\Omega$ vs. $\hat{W}_c$ must yield a straight line of slope Za'' (where a'' is a constant) and intercept Z.

Having established the value of Z, the next step is to determine the individual rate constants k_v and $k_{v,f}$ from the relation

$$Z\hat{W}_c - \frac{1}{a''}\ln(1 - a''\hat{W}_c) = (k_{v,f}C_A)t \qquad (9.32)$$

Since Z and a'' are known from the experimental plot of

$Z\Omega$ vs. $\hat{W}_c$, the left-hand side of Eq. 9.32 can be calculated and plotted as a function of t to provide the values of $k_{v,f}$. Then, from Eq. 9.29, k_v can be calculated.

The same procedure can be used for any other functional relationship between Ω and $\hat{W}_c$, but the equations get more involved and the determination of Z and $k_{v,f}$ will no longer be as straightforward as for the linear case.

Other investigators have determined the kinetics of deactivation in various reaction systems, mostly by coking, and have then incorporated this information into appropriate models to predict integral reactor performance data. De Pauw and Froment (1975) and Dumez and Froment (1976) studied the isomerization of *n*-pentane on Pt-reforming catalysts and dehydrogenation of 1-butene on Cr_2O_3–Al_2O_3. Reactor modeling dynamics influenced by poisoning have been reported by Blaum (1974). Some experiments in which parallel poisoning is the mechanism of deactivation have been reported by Weng et al. (1975) and Price and Butt (1977). In these studies many kinetic, reactor, and poisoning parameters were determined by separate experimentation.

9.4.3. Fouling Parameters from Center-Plane Concentration

Reactors have recently been designed to produce specific gradients that can be measured and used for parameter estimation. The basis of this method is that measurement of center-plane concentration vs. relative rate of reaction provides data sufficient to distinguish between various types of deactivation and permits parameter estimation (see Section 8.2.2). The experiments based on center-plane concentration are best carried out in a single-pellet reactor.

The single-pellet reactor developed by Hegedus and Petersen (1972, 1973a, b, c, 1974) Balder and Petersen (1968), and Wolf and Petersen (1974b, 1977) is designed for concentration measurements. The single-pellet reactors developed by Butt and co-workers (1962, 1977), Hughes and Koh (1970), Hughes et al. (1974), Koh and Hughes (1974), Benham and Denny (1972), and Trimm et al. (1974) provide temperature profile data.

The measurement of center-plane concentrations for cyclopropane hydrogenolysis by Hegedus and Petersen (1973a,b) resulted in the postulation of a triangular self-poisoning model with the parallel mode predominating at lower temperatures and the series mode at higher temperatures. Measurements of temperature profiles in a single-pellet reactor have also been made experimentally for O_2 poisoning of Ni–SiO_2–Al_2O_3 in ethylene hydrogenation (Hughes and Koh, 1974) and thiophene poisoning of Ni–kieselguhr in benzene hydrogenation (Butt et al. 1977).

9.4.4. Reactors for Fouling and Regeneration Studies

For determining the reaction and fouling parameters for systems in which independent poisoning is involved, either an integral or a mixed reactor can be used. On the other hand, for reactions in which impurity, parallel, series, or triangular fouling is involved, it is necessary to restrict the choice to a mixed reactor, for the rate can be obtained directly in this reactor. Any of the stirred reactors described earlier is recommended after tests have been made to ensure complete mixing: A particularly useful method of operating a reactor has been suggested by Sinfelt (1968) for determining the kinetics of reactions on a fouling catalyst.

The determination of regeneration rates is considerably simpler than that of deactivation rates, because the former are not accompanied by the main reaction. Usually regeneration is resorted to in the case of catalysts fouled by carbon deposition. The coke-laden catalyst is suspended in a reactor by a spring and the loss in weight is continuously recorded in a calibrated weighing arrangement. From the rate of carbon burn-off thus determined, the kinetics of regeneration may be worked out.

9.5. MEASUREMENT OF EFFECTIVE DIFFUSIVITY

Methods of estimating D_e on the basis of hypothetical models for the porous structure of a pellet were described in Chapter 3. Broadly, D_e can be experimentally determined by three techniques: steady-state diffusion; unsteady-state diffusion; and reaction rate measurement.

9.5.1. Steady-State Methods

In steady-state methods the diffusion flux of a fluid flowing through a porous solid under well-defined conditions is measured at steady state. Wicke and Kallenbach (1941) first developed the method, which was subsequently modified by Weisz (1957). Two different pure gases are introduced on opposite sides of a pellet, pressure and temperature gradients across the pellet are eliminated, and the diffusion fluxes are measured. The solid sample is mounted in a slightly undersized tubing of tygon. The choice of H_2 or He as the diffusing gas and the thermal conductivity cell as the detector leads to a very sensitive detection system.

Henry et al. (1961) have suggested a modification of Weisz's method, in which streams of pure N_2 and CO_2 are used as counterdiffusing gases and the rate of diffusion on each side of the pellet is measured by employing two conductivity cells separately. In experiments of this type involving counter diffusing of gases, any of the questions developed in Chapter 3 for calculating D_e from experimental data can be used.

Marginal improvements over this method have been suggested by Cunningham and Geankoplis (1968), Henry et al. (1967), and Satterfield and Cadle (1968). In another modification of the diffusion cell (Gokarn and Doraiswamy, 1972), the catalyst pellet is formed *in situ* by compressing catalyst powder to the strength and pore size corresponding to those used in industrial catalysts.

9.5.2. Unsteady-State Methods (the Chromatographic Method)

As its name suggests, the unsteady-state method employs transient phenomena associated with the adsorption and desorption processes characteristic of porous solid–gas systems. Several methods have been reported (Habgood, 1958; Bolt and Innes, 1959; Gorring and de Rosset, 1964; Gordon, 1945; Testin and Stuart, 1967). The pulse response of single catalyst pellets may also be used to evaluate the effective diffusivity. This technique has been called the *dynamic Wicke–Kallenbach method* (Suzuki and Smith, 1972; Dogu and Smith, 1975; 1976; Shibuya and Kawazoe, 1979). The most important and useful method is that based on the use of the gas chromatographic technique, and is described here.

Effective diffusivities are obtained from gas chromatographic data by relating the pulse spreading to the carrier-gas flow rate in a chromatographic column. The processes involved in the passage of a pulse of gas through a chromatographic column can be characterized in terms of the height equivalent to a theoretical plate (HETP) based on the properties of the effluent pulse. Van Deemter et al. (1956) have shown that for any practical column where the height of a mixing stage is small compared to the column length, the output pulse has a Gaussian distribution. The simplified expression for the height equivalent of a theoretical plate (HETP) is

$$\text{HETP} = H = A + \frac{B}{u_i} + Cu_i \qquad (9.33)$$

Each of the three terms represents an independent contribution to band broadening. The constant A represents eddy diffusion due to turbulence in the gas stream caused by the packing; B represents band broadening as a consequence of the longitudinal diffusion of gas in the pulse; and C is related to the effective diffusivity D_e (Habgood and Hanlan, 1959).

Experimentally the HETP for a given value of u_i is determined by employing the relation suggested by Purnell (1962):

$$\text{HETP} = \frac{LW_e}{t_m t_e} \qquad (9.34)$$

where t_m = retention time at maximum pulse,
   t_e = retention time of pulse at a fraction $1/e$ (= 0.368) of pulse height,
   W_e = width of pulse at a fraction $1/e$ of pulse height.

Davis and Scott (1965) and Leffler (1966) have used this technique to determine D_e. Gangwal et al. (1971) have used an analysis based on Fourier series, while Kelly and Fuller (1972) have proposed a method in which a reactant pulse is passed through a stirred reactor. MacDonald and Habgood (1972) have described a gas chromatographic method based on the theory of Giddings and Schettler (1964) for the determination of intracrystalline diffusivity in zeolite catalysts.

Schneider and Smith (1968a,b) have proposed a method that is particularly attractive from the experimental point of view. This method is based on the theory proposed by Kubin (1965a,b) and Kucera (1965) for relating the moments of the effluent concentration wave from a bed of adsorbent particles to the rate constants associated with the various steps in the overall adsorption process. Schneider and Smith used this method for determining the adsorption equilibrium constants, rate constants, and intraparticle diffusivities for various hydrocarbons on silica gel. The nth absolute moment μ'_n of the function $C(l, t)$ is defined as

$$\mu'_n = \frac{m_n}{m_0} \qquad (9.35)$$

where

$$m_n = \int_0^\infty t^n C(l, t)\, dt \qquad (n = 0, 1, 2, 3, \ldots) \qquad (9.36)$$

and the nth central moment μ_n is defined as

$$\mu_n = \frac{1}{m_0} \int_0^\infty (t - \mu'_1)^n C(l, t)\, dt \qquad (9.37)$$

The first absolute moment μ'_1 characterizes the position of the center of gravity of the chromatographic curve and the second central moment μ_2 depends on the width of this curve.

Schneider and Smith (1968a,b) expressed the first absolute and the second central moments (μ'_1, μ_2) by the following equations:

$$\mu'_1 = \frac{L}{u_i}(1 + \delta_0) + \frac{t_{0A}}{2} \qquad (9.38)$$

$$\mu_2 = \left(\frac{2L}{u_i}\right)\left(\delta_1 + \frac{D_{el}}{f_B}\right)(1 + \delta_0)^2 \frac{1}{u_i^2} + \frac{t_{0A}^2}{12} \qquad (9.39)$$

where

$$\delta_0 = \left(\frac{1 - f_B}{f_B} f_c\right)\left(1 + \frac{\rho_c}{f_c} K_A\right) \qquad (9.40)$$

$$\delta_1 = \left(\frac{1 - f_B}{f_B}\right)\left(\frac{R^2 f_c}{15}\right)\left(1 + \frac{\rho_c}{f_c} K_A\right)^2 \left(\frac{1}{D_{el}} + \frac{5}{k_g R}\right) \qquad (9.41)$$

The moments μ'_1 and μ_2 can be evaluated from the experimental chromatographic curves for the effluent from the column by using Eqs. 9.38 and 9.39. Data at different carrier-gas velocities will give the moments as a function of velocity.

Chromatographic curves are measured in an apparatus that consists of a constant-temperature gas chromatograph with a flame ionization detector. The chromatographic column containing the stationary phase is replaced by a tube containing the solid catalyst particles, and the gas component whose adsorption coefficient and effective diffusivity are to be evaluated is injected into the tube by means of a six-way two-position sampling valve. The effluent peak, which is recorded in the strip chart recorder connected to the flame ionization detector at the column outlet, can be directly correlated to the gas concentration and the integrals in Eqs. 9.36 and 9.37 evaluated numerically by means of Simpson's rule.

Equations 9.38 and 9.40 lead to

$$\frac{[\Delta\mu'_1 - (t_{0A}/2)]}{[(1 - f_B)/f_B] f_c} = \frac{\rho_c K_A}{f_c}\left(\frac{L}{u_i}\right) \qquad (9.42)$$

where t_{0A} is the total injection time of the adsorbable gas, and

$$\Delta\mu'_1 = \mu'_1 - (\mu'_1)_{\text{inert}}. \qquad (9.43)$$

Since the flame ionization detector does not respond

to the inert gas, the first absolute moments of the inert gas $(\mu'_1)_{inert}$ can be calculated from Eq. 9.38 according to which

$$(\mu'_1)_{inert} = \left[1 + \left(\frac{1-f_B}{f_B}\right)f_c\right]\left(\frac{L}{u_i}\right) \qquad (9.44)$$

when

$$K_A = 0 \quad \text{and} \quad t_{0A} = 0$$

As can be seen from Eq. 9.42, a plot of

$$\frac{\Delta\mu'_1 - t_{0A/2}}{f_c(1-f_B)/f_B}$$

vs. L/u_i will yield a straight line passing through the origin, the slope of which will be equal to $\rho_c K_A/f_c$. From the values of ρ_c and f_c the adsorption equilibrium constant K_A and thus the value of δ_0 in Eq. 9.40 can be obtained.

Equations 9.39 and 9.41 can be combined and rewritten in a modified form:

$$\frac{\mu_2 - (t_{0A}^2/12)}{2L/u_i} = (\delta_e + \delta_i) + \left(\frac{D_{el}}{f_B}\right)(1+\delta_0)^2\left(\frac{1}{u_i^2}\right) \qquad (9.45)$$

where

$$\delta_e = \delta_0\left(\frac{R^2 f_c}{15}\right)\left(1 + \frac{\rho_c}{f_c}K_A\right)\left(\frac{5}{k_g R}\right) \qquad (9.46)$$

$$\delta_i = \delta_0\left(\frac{R^2 f_c}{15}\right)\left(1 + \frac{\rho_c}{f_c}K_A\right)\left(\frac{1}{D_e}\right) \qquad (9.47)$$

As can be seen from Eq. 9.45, a plot of $[\mu_2 - (t_{0A}^2/12)]/2L/u_i$ vs. $1/u_i^2$ should yield a straight line with a slope of $(D_{el}/f_B)(1+\delta_0)^2$. Knowing the values of f_B and δ_0, we can calculate the axial dispersion coefficient D_{el}. The straight line in the above plot will have an intercept whose value is equal to $\delta_e + \delta_i$. Schneider and Smith have shown that at low Reynolds numbers, $k_g R$ in Eq. 9.46 is equal to the molecular diffusivity D_b. Thus, if the values of D_b and K_A, and δ_0 from the first absolute moment, are known, δ_e can be calculated and δ_i obtained from the intercept. The effective diffusivity D_e can then be readily calculated from Eq. 9.47.

It may be noted here that Eq. 9.39 is valid when the adsorption isotherm is linear and the rate of adsorption is sufficiently rapid so that equilibrium conditions prevail at the catalyst surface. More general equations can be derived for the case of a finite rate of adsorption. Under certain conditions, it may be possible to evaluate the adsorption rate constant by the method of moments. A second assumption involved in Eq. 9.39 is

that there is no axial dispersion in the chromatographic column and the flow pattern is close to plug flow. This may not be valid for operations with low velocities. A method for correcting for the effects of axial dispersion has been proposed by Suzuki and Smith (1972).

A useful development is the application of the moments method to catalysts with bidispersed porous structure. The method can be used to evaluate both the macro- and micropore diffusivities. The theoretical basis of this method and some experimental results have been presented by Haynes and Sarma (1973), Sarma and Haynes (1974), Hashimoto and Smith (1973, 1974), and Lee and Ruthven (1977).

9.6. EXAMPLES

9.6.1. Evaluation of Diffusional Effects in the Isomerization of *n*-Butenes[†]

n-Butenes can be isomerized to isobutene on fluorinated alumina at 400°C. In order to obtain a rate equation for the reaction, experiments were carried out in a stirred reactor under fully mixed conditions (Choudhary, 1971). A typical rate value reported by him is

$$r_{vA} = 5.25 \times 10^{-6} \quad \text{mol/sec cm}^3$$

It is not known *a priori* whether this run is in the kinetic regime and free from all diffusional influences. By using appropriate criteria, evaluate the influence of interphase and intraparticle resistances under the conditions of the experiment. The following data are available.

Volume of pellets, V'	0.666 cm³/g
External surface area of pellets, a'	67.8 cm²/g
Concentration, C_A	1.37×10^{-5} mol/cm³
Surface reaction rate constant, k_p	1.6×10^{-2} mol/hr g atm
k_S	1.64×10^{-7} cm/sec
Pellet density, ρ_c	1.5 g/cm³
Total surface area, S_g	1.5×10^6 cm²/g
Effective diffusivity, D_{eA}	3.6×10^{-3} cm²/sec
Heat of reaction, $(-\Delta H)$	1400 cal/mol
Activation energy, E	8350 cal/mol
Temperature, T	673°K

[†] This example is selected to specifically demonstrate the absence of diffusion effects; any run which shows these effects is best discarded.

Solution

1. Mass and heat transfer coefficients:

The mass transfer coefficient for the stirred reactor system used can be obtained from (Choudhary and Doraiswamy, 1972)

$$j_d = (2.0 \times 10^{-3})\,(Re')^{-0.70}$$

and the heat transfer coefficient can then be estimated from the assumption that $j_h = 1.37 j_d$.

The following data may be used.

Particle diameter, d_p	0.034 cm
Apparent mass velocity, G	$\pi d_1 N \rho_g$
	$= 0.731$ g/cm^2 sec
Basket diameter, d_i	5.3 cm
Stirrer speed, N	43.3 rev/sec
Gas density, ρ_g	1.014×10^{-3} g/cm^3
Viscosity, μ	0.175×10^{-3} poise
Pressure, P	0.94 atm
Molecular weight, M_m	56
Binary diffusion coefficient, D_{AB}	0.21 cm^2/sec
Heat capacity, C_p	0.65 cal/g °C
Thermal conductivity, k'_g	14.5×10^{-5} cal/cm sec °C

Thus

$$j_d = \frac{k_g P M_m (\mu/\rho_g D_{AB})^{2/3}}{G} = (2.0 \times 10^{-3})\left(\frac{d_p G}{\mu}\right)^{-0.7}$$

or

$$k_g = 9.9 \times 10^{-7} \quad \text{mol/cm}^2 \text{ sec atm}$$

and

$$j_h = 1.37 j_d = \frac{h_{fp}(C_p \mu/k'_g)^{2/3}}{C_p G}$$

or

$$h_{fp} = 4.8 \times 10^{-5} \quad \text{cal/sec cm}^2 \text{ °C}$$

2. Interphase mass transfer:

The following criterion (Mears, 1971) may be used (Table 9.1):

$$\frac{r_{vA}(V'/a')}{C_A k_g} < \frac{0.15}{n}$$

right-hand side $= \dfrac{0.15}{1} = 0.15$

left-hand side $= \dfrac{(5.25 \times 10^{-6})(0.666/67.8)}{0.86 \times 9.9 \times 10^{-7}} = 0.061$

Hence interphase mass transfer effects are absent.

3. Interphase heat transfer:

The following criterion (Mears, 1971) may be used (Table 9.1):

$$\frac{(-\Delta H)r_{vA}(V'/a')}{T_b h_{fp}} < 0.15 \frac{R_g T_b}{E}$$

right-hand side $= \dfrac{0.15 \times 2.0 \times 673}{8350} = 2.42 \times 10^{-2}$

left-hand side $= \dfrac{(1400 \times 5.25 \times 10^6)(0.666/67.8)}{673 \times 4.8 \times 10^{-5}}$

$= 2.230 \times 10^{-3}$

Hence the effects due to interphase heat transfer are negligible.

4. Intraparticle mass transfer (isothermal)

The Weisz–Prater criterion may be used:

$$\left(\frac{V'}{a'}\right)^2 \left(\frac{r_{vA}}{D_{eA} C_{As}}\right) < 1$$

left-hand side $= \left(\dfrac{0.666}{67.8}\right)^2 \times \left(\dfrac{5.25 \times 10^{-6}}{3.6 \times 10^{-3} \times 1.37 \times 10^{-5}}\right)$

$= 1.03 \times 10^{-2} < 1$

Hence intraparticle mass transfer effects are absent.

5. Interphase and intraparticle mass transfer

The following criterion may be used (Table 9.1):

$$\frac{2}{n+1}\left(\frac{V'}{a'}\right)^2 \left(\frac{\rho_c S_g k_S C_A^{n-1}}{D_{eA}}\right) \leqslant 0.03$$

left-hand side $= \left(\dfrac{2}{1+1}\right)\left(\dfrac{0.666}{67.8}\right)^2$

$$\times \left(\frac{1.5 \times 150 \times 10^4 \times 1.64 \times 10^{-7} \times 1}{3.6 \times 10^{-3}}\right)$$

$= 0.0099 < 0.03$

Hence both interphase and intraparticle mass transfer effects are absent. Note that this criterion can be employed only after the rate constant has first been determined from untested data, or is known from other sources.

9.6.2. Decomposition of Hydrogen Peroxide

Satterfield and Resnick (see Section 6.2.1) studied the decomposition of H_2O_2 on solid silver spheres in the Reynolds number range 15–61. This reaction has all the desirable characteristics necessary for external mass transfer control. The internal diffusional resistance is eliminated because of the nonporous nature of the catalyst used, and rate data can be correlated through the j_d factor. The reaction was conducted isothermally at atmospheric pressure (708.8 mm Hg) and at hydrogen feed concentrations in the range 0–3% in air. The reaction was found to be controlled both by diffusion and intrinsic kinetics. The experimental data of Balaraman et al. (1980) are reproduced below.

Run No.	Reaction Temperature (°K)	Reynolds Number Around the Cylinder	Observed or Apparent Rate Constant k_a (mol/hr cm² atm)
1	295	49.20	0.142
2	313	47.06	0.258
3	323	46.04	0.306
4	333	44.99	0.358

In the forward stagnation region of a circular cylinder the Sherwood number (or mass Biot number) is given by Eq. 9.17 as

$$\frac{1}{2}\frac{Sh}{(Re)^{1/2}} = F(Sc)$$

for $Sc = 0.2$, $F(Sc) = 0.3$. Also, the interaction between reaction and diffusion for an nth-order reaction has the form:

$$C_{As}^n + \frac{D_b C_{As}}{k_S \delta} - \frac{D_b C_{Ab}}{k_S \delta} = 0 \qquad (9.48)$$

The following additional data are given:
Bulk diffusion coefficient of hydrogen in air

$$D_b = 4.0517 \times 10^{-5} \times T^{1.75} \qquad \text{at } 708.8 \text{ mm Hg}$$

Kinematic viscosity of air

$$v = 7.780 \times 10^{-6} \times T^{1.75} \qquad \text{at } 708.8 \text{ mm Hg}$$

Schmidt number $Sc = 0.2$.
Diameter of the coated cylinder = 1.25 cm.
Calculate (1) the rate constant k_S, (2) stream variable ξ, and (3) local effectiveness factor η'. Compare the local effectiveness factors obtained at different values of the stream variable with theoretical predictions.

Solution

The local effectiveness factor is given by Eq. 6.23 as

$$\eta' = \frac{1}{1 + k_S/k_g}$$

The stream variable ξ is given by

$$\xi = k_S C_{Ab}^{n-1} \delta/D_b = k_S C_{Ab}^{n-1}/k_g$$

For $n = 1$,

$$\xi = k_S/k_g$$

To estimate the local effectiveness factor η' and the stream variable ξ, we have to determine both the intrinsic rate constant k_S and the mass transfer coefficient k_g.

Equation 9.48 for a first-order reaction reduces to

$$C_{As} = \frac{D_b C_{Ab}}{k_S \delta + D_b} = \frac{C_{Ab}}{k_S(\delta/D_b) + 1} \qquad (9.49)$$

$$= \frac{C_{Ab}}{k_S/k_g + 1} \quad \text{or} \quad \frac{C_{Ab} k_g}{k_g + k_S}$$

We have

$$r_S = k_S C_{As} = \frac{k_S k_g}{k_S + k_g} C_{Ab}$$

$$= k_a C_{Ab}$$

where

$$k_a = \frac{k_S k_g}{k_S + k_g}$$

or

$$\frac{1}{k_a} = \frac{1}{k_S} + \frac{1}{k_g} \qquad (9.50)$$

Calculation of k_S and k_g (specimen calculation for run no. 3):

In the stagnation region, with $Re = 46.04$, $Sc = 0.2$, $F(Sc) = 0.3$, and Sh can be calculated from Eq. 6.48. Thus

$$Sh = 2 \times 0.3 \sqrt{46.04} = 4.071$$

The mass transfer coefficient k_g is then given by

$$k_g = Sh \frac{D_b}{\delta}$$

$$= \frac{4.071 \times 4.0517 \times 10^{-5} \times 323^{1.75}}{1.25} \frac{3600}{R_g T}$$

$$= 0.441 \text{ mol/hr cm}^2 \text{ atm}$$

By the additivity relation 9.50,

$$\frac{1}{k_a} = \frac{1}{0.306} = \frac{1}{k_g} + \frac{1}{0.441}$$

or

$$k_S = 1.007 \, \text{mol/hr cm}^2 \, \text{atm}$$

The local effectiveness factor is then

$$\eta' = \frac{k_g}{k_S + k_g} = 0.3045$$

and the stream variable

$$\xi = \frac{k_S}{k_g} = 2.283$$

The results of all the four runs are tabulated below.

Run No.	Reaction Temperature (°C)	Observed Rate Constant k_a	Mass Transfer Coefficient k_g	Intrinsic Rate Constant k_S	η^1	ξ
1	295	0.142	0.426	0.214	0.665	0.502
2	313	0.258	0.435	0.638	0.405	1.466
3	323	0.306	0.441	1.007	0.304	2.283
4	333	0.358	0.446	1.829	0.196	4.102

The four experimental points agree to within 2% with the predicted values of ξ. This example underlines the importance of accounting for external mass transfer in analyzing a solid catalyzed reaction.

REFERENCES

Aliev, R. R., Berman, A. D., and Yanovskii, M. I. (1968), *Azerb. Khim Zh.*, p. 26.

Altomare, R. E., Greenfield, P. F., and Kittrell, J. R. (1974), *Biotechnol. Bioeng.*, **16**, 1675.

Anderson, J. B. (1963), *Chem. Eng. Sci.*, **18**, 147.

Baag, I. F., Bacon, D. W., and Downie, J. (1975), *J. Catal.*, **38**, 375.

Balaraman, K. S., Mashelkar, R. A., and Doraiswamy, L. K. (1980), *AIChE J.* **26**, 635.

Balder, J. R., and Petersen, E. E. (1968), *J. Catal.*, **11**, 202.

Banerjee, S. C. (1965), Ph.D. thesis, Bombay University.

Barbul, M., Serban, Gh., Ghejan, I., and Fillotti, T. (1968), *Petrol Gaze (Bucharest)*, **19**, 181.

Barrett, D. (1971), *Trans. Inst. Chem. Eng. (London)*, **49**, 80.

Bassett, D. W., and Habgood, H. W. (1960), *J. Phys. Chem.*, **64**, 769.

Bassett, D. W., and Habgood, H. W. (1961), *Chem. Inst. Canada*, **13**, 50.

Benham, C. B., and Denny, V. E. (1972), *Chem. Eng. Sci.*, **27**, 2163.

Bennett, C. O., Cutlip, M. B., and Yang, C. C. (1972), *Chem. Eng. Sci.*, **27**, 2255.

Berty, J. M., Hambrick, J. O., Malone, T. R., and Ullock, D. S. (1969), Paper presented at the 64th national meeting of the AIChE, New Orleans, Paper No. 42E.

Bett, J. A. S., and Hall, W. K. (1968), *J. Catal.*, **10**, 105.

Binter, E. D., Davison, V. L., and Dutton, H. J. (1969), *J. Am. Oil Chem. Soc.*, **46**, 113.

Blanton, W. A., Jr., Byers, C. H., and Merril, R. P. (1968), *Ind. Eng. Chem. Fundam.*, **7**, 611.

Blaum, E. (1974), *Chem. Eng. Sci.*, **29**, 2263.

Bolt, B. A., and Innes, J. A. (1959), *Fuel*, **38**, 333.

Brown, C. E., and Bennett, C. O. (1970), *AIChE J.*, **16**, 817.

Butt, J. B., Bliss, H., and Walker, C. A. (1962), *AIChE J.*, **8**, 42.

Butt, J. B., Downing, D. M., and Lee, J. W. (1977), *Ind. Eng. Chem. Fundam.*, **16**, 270.

Calderbank, P. H., Caldwell, A. D., and Ross, G. L. (1969), *Chim. Ind. Genie Chim.*, **101** (2), 215.

Carberry, J. J. (1976), *Catalytic and Chemical Reaction Engineering*, McGraw-Hill, New York.

Caretto, L. S., and Nobe, K. (1969), *AIChE J.*, **15**, 18.

Cassano, A. E., Matsuura, T., and Smith, J. M. (1968), *Ind. Eng. Chem. Fundam.*, **7**, 655.

Celler, W. (1970), *Prezm. Chem.*, **49**, 68.

Chakraborty, A., and Doraiswamy, L. K. (1971), NCL report.

Chambers, R. P., and Boudart, M. (1966), *J. Catal.*, **6**, 141.

Cho, B. K., Carr, R. W., and Aris, R. (1980), *Chem. Eng. Sci.*, **35**, 74.

Choudhary, V. R. (1971), Ph.D. thesis, Poona University.

Choudhary, V. R. (1978), *J. Chromatog.*, **152**, 208.

Choudhary, V. R., and Doraiswamy, L. K. (1972), *Ind. Eng. Chem. Process Design Dev.*, **11**, 420.

Collins, C. G., and Deans, H. A. (1968), *AIChE J.*, **14**, 25.

Costa, E. C., and Smith, J. M. (1971), *AIChE J.*, **17**, 947.

Cunningham, R. S., and Geankoplis, C. J. (1968), *Ind. Eng. Chem. Fundam.*, **7**, 535.

Danforth, J. D., and Roberts, J. M. (1968), *J. Catal.*, **10**, 252.

Dautzenberg, F. M., Heele, J. N., Van Santen, R. A., and Verbeek, H. (1977), *J. Catal.*, **50**, 8.

Davis, B. R., and Scott, D. S. (1965), AIChE 58th annual meeting, Preprint 48D.

Deans, H. A., Horns, F. J. M., and Klauser, G. (1967), *Perturbative Chromatography*, AIChE., New York.

Deans, H. A., Horn, F. J. M., and Klauser, G. (1970), *AIChE J.*, **16**, 426.

Deo, V. D., and Doraiswamy, L. K. (1967), *Ind. J. Tech.*, **5**, 174.

De Pauw, R., and Froment, G. F. (1975), *Chem. Eng. Sci.*, **30**, 789.

Dogu, G., and Smith, J. M. (1975), *AIChE J.*, **21**, 58.

Dogu, G., and Smith, J. M. (1976), *Chem. Eng. Sci.*, **31**, 123.

Doraiswamy, L. K., and Tajbl, D. G. (1974), *Catal. Rev. Sci. Eng.*, **10**, 177.

Dumez, F. J., and Froment, G. F. (1976), *Ind. Eng. Chem. Process Design Dev.*, **15**, 291.

Dutton, H. J., and Mounts, T. L. (1964), *J. Catal.*, **3**, 363.

Eliezer, K. F., Bhinde, M., Houalla, M., Broderick, D., Gates, B. C., Katzer, J. R., and Olson, J. H. (1977), *Ind. Eng. Chem. Fundam.*, **16**, 380.

Ettre, L. S., and Brenner, N. (1960), *J. Chromatog.*, **3**, 524.

Filinovskii, V. Yu., Gaziev, G. A., and Yanovskii, M. I. (1966), *Dokl. Akad. Nauk SSR*, **167**, 143.

Ford, F. E., and Perlmutter, D. D. (1963), *AIChE J.*, **9**, 371.

Ford, F. E., and Perlmutter, D. D. (1964), *Chem. Eng. Sci.*, **19**, 371.

Galeski, J. N., and Hightower, J. W. (1970), *Can. J. Chem. Eng.*, **48**, 151.

Gangwal, S. K., Hudgins, R. R., Bryson, A. W., and Silveston, P. L. (1971), *Can. J. Chem. Eng.*, **49**, 113.

Garanin, V. I., Kurkchi, V. M., and Minachev, K. M. (1967), *Kinet. Katal.*, **8**, 605.

Gaziev, G. A., Filinovskii, V. Yu., and Yanovskii, M. I. (1963), *Kinet. Katal.*, **4**, 688.

Gelanin, C. (1969), *Chim. Ind. Genie Chim*, **102**, 984.

Giddings, J. C., and Schettler, P. D. (1964), *Anal. Chem.*, **36**, 1483.

Gillespie, B. M., and Carberry, J. J. (1966a), *Ind. Eng. Chem.*, **5**, 164.

Gillespie, B. M., and Carberry, J. J. (1966b), *Chem. Eng. Sci.*, **21**, 472.

Gokarn, A. N., and Doraiswamy, L. K. (1972), *Chem. Eng. Sci.*, **27**, 1515.

Gordon, A. R. (1945), *Ann. N. Y. Acad. Sci.*, **46**, 285.

Gore, F. E. (1967), *Ind. Eng. Chem. Process Design Dev.*, **6**, 10.

Gorring, R. L., and de Rosset, A. J. (1964), *J. Catal.*, **3**, 341.

Guha, B. K., and Narsimhan, G. (1972), *Chem. Eng. Sci.*, **27**, 703.

Habgood, H. W. (1958), *Can. J. Chem.*, **36**, 1384.

Habgood, H. W., and Hanlan, J. F. (1959), *Can. J. Chem.*, **37**, 843.

Hall, W. K., and Emmett, P. H. (1957), *J. Am. Chem. Soc.*, **79**, 2091.

Hall, W. K., MacIver, D. S., and Weber, H. P. (1960), *Ind. Eng. Chem.*, **52**, 421.

Harrison, D. P., Hall, W. K., and Rase, H. F. (1965), *Ind. Eng. Chem.*, **57**, 18.

Hartwig, M. (1964), *Brennstoff-Chem.*, **45**, 234.

Hashimoto, N., and Smith, J. M. (1973), *Ind. Eng. Chem. Fundam.*, **12**, 353.

Hashimoto, N., and Smith, J. M. (1974), *Ind. Eng. Chem. Fundam.*, **13**, 115.

Hattori, T., and Murakami, Y. (1968a), *J. Catal.*, **10**, 114.

Hattori, T., and Murakami, Y. (1968b), *J. Catal.*, **12**, 166.

Hattori, T., and Murakami, Y. (1973), *J. Catal.*, **31**, 127.

Hattori, T., and Murakami, Y. (1974), *J. Catal.*, **33**, 365.

Haynes, H. W., Jr., and Sarma, P. N. (1973), *AIChE J.*, **19**, 1043.

Hegedus, L. L., and Petersen, E. E. (1972), *Ind. Eng. Chem. Fundam.*, **11**, 579.

Hegedus, L. L., and Petersen, E. E. (1973a), *Chem. Eng. Sci.*, **28**, 69.

Hegedus, L. L., and Petersen, E. E. (1973b), *Chem. Eng. Sci.*, **28**, 345.

Hegedus, L. L., and Petersen, E. E. (1973c), *J. Catal.*, **28**, 150.

Hegedus, L. L., and Petersen, E. E. (1974), *Catal. Rev. Sci. Eng.*, **9**, 245.

Henry, J. P., Chennakesavan, B., and Smith, J. M. (1961), *AIChE J.*, **7**, 10.

Henry, J. P., Cunningham, R. S., and Geankoplis, C. J. (1967), *Chem. Eng. Sci.* **22**, 11.

Hightower, J. W., and Hall, W. K. (1968), *J. Phys. Chem.*, **72**, 4555.

Huang, H. I., and Sather, J. F. (1970), *Chem. Eng. Sci.*, **25**, 340.

Hudgins, R. R. (1968), *Chem. Eng. Sci.*, **23**, 93.

Hughes, R., and Koh, H. P. (1970), *Chem. Eng. J.*, **1**, 186.

Hughes, R., Koh, H. P., and Harriot, P. (1974), *Chem. Eng. Sci.*, **29**, 2183.

Kadlec, B., Michalek, J., and Simecek, A. (1970), *Chem. Eng. Sci.*, **25**, 319.

Kelly, J. F., and Fuller, O. M. (1972), *Can. J. Chem. Eng.*, **50**, 534.

Keulemans, A. J. M., and Voge, H., (1959), *J. Phys. Chem.*, **63**, 476.

Klinkenberg, A. (1961), *Chem. Eng. Sci.*, **15**, 255.

Kobayashi, M., Ramaswamy, D., and Brazelton, W. T. (1970), *Chem. Eng. Prog. Symp. Ser.* (No. 105), **66**, 47.

Koh, H. P., and Hughes, R. (1974), *AIChE J.*, **20**, 395.

Kokes, R. J., Tobin, H., and Emmett, P. H. (1955), *J. Am. Chem. Soc.*, **77**, 5860.

Koros, R. M., and Nowak, E. J. (1967), *Chem. Eng. Sci.*, **22**, 470.

Krishnaswamy, S., and Kittrell, J. R. (1978), *Ind. Eng. Chem. Process Design Dev.*, **17**, 200.

Krishnaswamy, S., and Kittrell, J. R. (1981a), *AIChE J.*, **27**, 120.

Krishnaswamy, S., and Kittrell, J. R. (1981b), *AIChE J.*, **27**, 125.

Kubin, M. (1965a), *Collect. Czech. Chem. Commun.*, **30**, 1104.

Kubin, M. (1965b), *Collect. Czech. Chem. Commun.*, **30**, 2900.

Kubota, H., and Yamanaka, Y. (1969), *J. Chem. Eng. Japan*, **2**, 238.

Kubota, H., Yamanaka, Y., and Lana, I. G. D. (1969), *J. Chem. Eng. Japan*, **2**, 71.

Kucera, E. (1965), *J. Chromatog.*, **19**, 237.

Lakshmanan, R., and Rouleau, D. (1969), *Can J. Chem. Eng.*, **47**, 45.

Lakshmanan, R., and Rouleau, D. (1970), *J. Appl. Chem.*, **20**, 312.

Langer, R. M., and Walker, C. A. (1954), *Ind. Eng. Chem.*, **46**, 1299.

Langer, S. H., and Yurchak, J. Y. (1969), *Ind. Eng. Chem.*, **61**, (4), 11.

Langer, S. H., Yurchak, J. Y., and Patton, E. (1969), *Ind. Eng. Chem.*, **61** (4), 10.

Lee, L. K., and Ruthven, D. (1977), *Ind. Eng. Chem. Fundam.*, **16**, 290.

Leffler, A. J. (1966), *J. Catal.*, **5**, 22.

Leinroth, J. P., and Sherwood, T. K. (1964), *AIChE J.*, **10**, 524.

Levenspiel, O. (1972), *Chemical Reaction Engineering*, 2nd ed. Wiley, New York.

Liberti, G., Pernicone, N., and Soattini, S. (1972), *J. Catal.*, **27**, 52.

Livbjerg, H., and Villadsen, J. (1971), *Chem. Eng. Sci.*, **26**, 1495.

Luckner, R. C., and Wills, G. B. (1973), *J. Catal.*, **28**, 83.

MacDonald, W. R., and Habgood, H. W. (1972), *Can. J. Chem. Eng.*, **50**, 462.

Magee, E. M. (1963), *Ind. Eng. Chem. Fundam.*, **2**, 32.

Mahoney, J. A. (1974), *J. Catal.*, **32**, 247.

Massimilla, L., Volpicelli, G., and Razo, G. (1966), *Chem. Eng. Prog. Symp. Ser.*, **62**, 63.

Matsen, J. M., Harding, J. W., and Magee, E. M. (1965), *J. Phys. Chem.*, **69**, 522.

Mears, D. E. (1971), *Ind. Eng. Chem. Process Design Dev.*, **10**, 541.

Mosely, R. B., and Good, G. M. (1965), *J. Catal.*, **4**, 85.

Murakami, Y., and Hattori, T. (1967), *Kogyo Kogaku Zasshi*, **70**, 2098.

Norton, C. J. (1962), *Chem. Ind. (London)*, **6**, 258.

Norton, C. J., and Moss, T. E. (1964), *Ind. Eng. Chem. Process Design Dev.*, **3**, 23.

Owens, P. J., and Amberg, C. H. (1961), *Adv. Chem. Eng. Ser.*, **33**, 182.

Ozawa, Y., and Bischoff, K. B. (1968a), *Ind. Eng. Chem. Process Design Dev.*, **7**, 67.

Ozawa, Y., and Bischoff, K. B. (1968b), *Ind. Eng. Chem. Process Design Dev.*, **7**, 72.

Pansing, W. F., and Malloy, J. B. (1962), *Chem. Eng. Prog.*, **58** (12), 53.

Paraskos, J. A., Shah, Y. T., McKinney, J. D., and Carr, N. L. (1976), *Ind. Eng. Chem. Process Design Dev.*, **15**, 165.

Paratella, A., Orefice, V., and Giordano, N. (1968), *Chem. Eng. Sci.*, **23**, 373.

Perkins, T. K., and Rase, H. F. (1958), *AIChE J.*, **4**, 351.

Pereira, J. R., and Calderbank, P. H. (1975), *Chem. Eng. Sci.*, **30**, 167.

Petersen, E. E. (1965), *Chem. Eng. Sci.*, **20**, 587.

Petersen, E. E. (1968), *Chem. Eng. Sci.*, **23**, 94.

Phillips, C. S. G., Phillips, A. J., Hart-Davis, A. J., Saul, R., and Wormald, J. (1967), *J. Gas Chromatog.*, **5**, 424.

Prasad, K. B. S., and Doraiswamy, L. K. (1974), *J. Catal.*, **32**, 384.

Price, J. H., and Butt, J. B. (1977), Chem. Eng. Sci., **32**, 393.

Puranik, S. A., and Sharma, M. M. (1967), *Ind. Chem. Eng.*, **9**, 124.

Purnell, H. (1962), *Gas Chromatography*, Wiley, New York.

Ramaswamy, V., and Doraiswamy, L. K. (1973), NCL report.

Relyea, D. L., and Perlmutter, D. D. (1968), *Ind. Eng. Chem. Process Design Dev.*, **7**, 261.

Richardson, J. T., and Friedrich, H. (1975), *J. Catal.*, **37**, 8.

Richardson, J. T., Friedrich, H., and McGill, R. N. (1975), *J. Catal.*, **37**, 1.

Rihani, D. N., Narayanan, T. K., and Doraiswamy, L. K. (1965), *Ind. Eng. Chem. Process Design Dev.*, **4**, 403.

Roginskii, S. Z., and Rozental, A. L. (1964), *Kinet. Katal.*, **5**, 104.

Roginskii, S. Z., and Rozental, A. L. (1965a), *Dokl. Akad. Nauk SSSR*, **162**, 621.

Roginskii, S. Z., and Rozental, A. L. (1965b), *Dokl. Akad. Nauk SSSR*, **162**.

Roginskii, S. Z., Yanovskii, M. I., and Gaziev, G. A. (1962), *Kinet. Katal.*, **3**, 529.

Roginskii, S. Z., Semenenko, E. I., and Yanovskii, M. I. (1963), *Dokl. Akad. Nauk SSSR*, **153**, 383.

Roginskii, S. Z., Yanovskii, M. I., and Gaziev, G. A. (1964), *Gaz. Khromatogr. Tr. Vtoroi Vses. Konf., Akad. Nauk SSSR, Moscow*, **1962**, 27.

Ruthven, D. M. (1968), *Chem. Eng. Sci.*, **23**, 759.

Saito, H., Murakami, S. H., and Hattori, T. (1965), *Kagaku Kogaku*, **29**, 585.

Sarma, P. N., and Haynes, H. W. (1974), *Adv. Chem. Ser.*, **133**, 205.

Satterfield, C. N., and Cadle, P. J. (1968), *Ind. Eng. Chem. Fundam.*, **7**, 202.

Satterfield, C. N., and Roberts, G. W. (1968), *AIChE J.*, **14**, 159.

Schlichting, H. (1968), *Boundary Layer Theory*, 6th ed., McGraw-Hill, New York.

Schmidt, J. P., Mickley, H. S., and Grotch, S. L. (1964), *AIChE J.*, **10**, 149.

Schneider, P., and Smith, J. M. (1968a), *AIChE J.*, **14**, 762.

Schneider, P., and Smith, J. M. (1968b), *AIChE J.*, **14**, 886.

Schwab, G. M., and Watson, A. M. (1965), *J. Catal.*, **4**, 570.

Semenenko, E. I., Roginskii, S. A., and Yanovskii, M. I. (1964), *Kinet. Katal.*, **5**, 490.

Shah, R. T., and Doraiswamy, L. K. (1983), *Ind. Chem. Engr.*, (in press).

Shibuya, H., and Kawazoe, K. (1979), *J. Chem. Eng. Japan*, **12**, 33.

Sica, A. M., Valles, E. M., and Gigola, C. E. (1978), *J. Catal.*, **51**, 115.

Sinfelt, J. M. (1968), *Chem. Eng. Sci.*, **23**, 1181.

Steen, K. C., Freeman, J. J., Hofer, L., and Anderson, R. B. (1962), *U.S. Bur. Mines Bull.*, **608**, 19.

Stewart, W. E., and Villadsen, J. (1969), *AIChE J.*, **15**, 28.

Suzuki, M., and Smith, J. M. (1972), *Chem. Eng. (London)*, **3**, 256.

Szabo, Z. G., Jover, B., and Ohmacht, R. (1975), *J. Catal.*, **39**, 225.

Tajbl, D. G., Feldkerchner, H. L., and Lee, A. L. (1967), *Adv. Chem. Ser.*, **69**, 166.

Temkin, M. I. (1962), *Kinet. Katal.*, **3**, 509.

Testin, R. F., and Stuart, E. B. (1967), *Chem. Eng. Prog. Symp. Ser.* (No. 74), **63**, 10.

Topchieva, K. V., Rosolovskaya, E. N., and Shakhnoskaya, O. L. (1968), *Vestn. Mosk. Univ. Ser. 11*, **23** (1), 39.

Trimm, D. L., Corrie, J., and Holton, R. D. (1974), *Chem. Eng. Sci.*, **29**, 2009.

Trotter, I. (1960), Ph.D. thesis, Princeton University.

Van Deemter, J. J., Zuiderweg, F. J., and Klinkenberg, A. (1956), *Chem. Eng. Sci.*, **5**, 271.

Van den Bleek, C. M., van der wiele, K., and van den Berg, P. J. (1969), *Chem. Eng. Sci.*, **24**, 681.

Varma, A., and Kaliaguine, S. (1973), *J. Catal.*, **30**, 430.

Viswanathan, S., and Aris, R. (1974), *Adv. Chem. Ser.*, **133**, 191.

Viswanathan, S., and Aris, R. (1975), *Proc. AMS Symp. Appl. Mater.* (April 1974).

Walker, W. W., and Mann, R. F. (1969), *Can. J. Chem. Eng.*, **47**, 218.

Wallace, H. F., and Hayes, K. E. (1970), *J. Catal.*, **18**, 77.

Weisz, P. B. (1957), *Z. Physik. Chem. (Frankfurt)*, **11**, 1.

Weisz, P. B., and Prater, C. D. (1954), *Adv. Catal.*, **6**, 143.

Weisz, P. B., and Swegler, E. W. (1957), *Science*, **126**, 31.

Weng, H. S., Eigenberger, G., and Butt, J. B. (1975), *Chem. Eng. Sci.*, **30**, 1341.

Wentzheimer, W. W. (1969), Ph.D. thesis, Pennsylvania State University.

Wetherold, R. G., Wissler, E. H., and Bischoff, K. B. (1974), *Adv. Chem. Ser.*, **133**, 181.

Weychert, S., and Trela, M. (1968), *Int. Chem. Eng.*, **8**, 658.

White, R. R., and Churchill, S. W. (1959), *AIChE J.*, **5**, 354.

Wicke, E., and Kallenbach, R. (1941), *Colloid Z.*, **97**, 135.

Wolf, E. E., and Petersen, E. E. (1974), *Chem. Eng. Sci.*, **29**, 1500.

Wolf, E. E., and Petersen, E. E. (1977), *J. Catal.*, **46**, 190.

Yushchenko, V. V., Korneichuk, G. P., Usha-Kova-Stasevich, V. P., and Semenyuk, Yu. V. (1968), *Kinet. Katal.* **4**, 154.

Yushchenko, V. V., and Antipina, T. V. (1969a), *Zh. Fiz. Khim.*, **43**, 540.

Yushchenko, V. V., and Antipina, T. V. (1969b), *Zh. Fiz. Khim.*, **43**, 2932.

Zimin, R. A., Roginskii, S. Z., and Yanovskii, M. (1965), *Metody Issled. Katal. Reakt. Red. Izd. Otd. Silir, Otd. Akad. Nauk SSSR, Novasibirsk*, **3**, 279.

CHAPTER TEN

Estimation of Effective Transport Properties

In the design of a packed-bed reactor one of the most important considerations is heat (and mass) transfer in the direction of heat flux, that is, transverse to the direction of fluid flow. If there is a radial profile, the two-dimensional model must be used. On the other hand, if a uniform radial temperature (and concentration) can be assumed, the simpler one-dimensional model is adequate.

In both these approaches the basic assumption is made that the gas–solid system can be regarded as a single phase through which heat and mass transfer occurs. Thus, property values are assigned to this "continuum" in much the same manner as for a single continuous phase, like a gas, liquid, or solid. Since in reality two phases are involved, the properties are termed "effective" properties which are dependent on the nature of the individual phases constituting the continuum (the packed bed). The property that defines heat transport is designated the effective thermal conductivity of the bed, while that which defines mass transport is called the effective dispersion coefficient (or diffusivity) of the bed.

Radial transport of mass is restricted to the diameter of the reactor tube, whereas that of heat is not, since heat transfer also takes place between the reactor tube and the control fluid outside it. Thus, a knowledge of the heat transfer coefficient at the wall is also necessary in the design of a reactor.

In addition to radial transport, axial diffusion of mass superimposed on bulk flow of the fluid can be of some significance under certain conditions, but axial conduction of heat is generally not of any consequence.

The effective properties are of no real basic value, and are important only to the extent that they permit a simplified mathematical description of the fixed-bed reactor in which effective coefficients of transport are used in place of their basic counterparts for flow in a homogeneous medium. In this chapter we shall outline methods for the prediction of these effective properties

for use in the design of fixed-bed reactors, to be described in Chapter 11. The treatment is based almost entirely on the review of Kulkarni and Doraiswamy (1980). Other reviews consulted are those of Hlavacek and Votruba (1978) and Balakrishnan and Pei (1979).

10.1. EFFECTIVE THERMAL CONDUCTIVITY

The effective thermal conductivity of a fluid–solid heterogeneous system, exemplified by a dispersion or packing of discrete solid particles in a fluid that may be either static or in a state of flow, is a gross property that can be defined as the overall heat flux density divided by the overall temperature difference in the direction of the flux. It should be regarded purely as a convenient engineering concept with no thermodynamic significance.

In a tubular packed bed, the effective thermal conductivity in the radial direction alone is important, since this corresponds to the direction of heat flux. In a flowing system, an effective thermal conductivity in the direction of fluid flow can theoretically exist as conduction superimposed on bulk flow, but is so small relative to heat transfer by bulk flow that it can usually be neglected. In this section, therefore, we shall primarily be concerned with transport in the direction of heat flow, and shall refer to axial conduction only briefly in the next section.

10.1.1. Radial Transport

A reference to Figure 10.1 shows that if the radial temperature profile in a packed tube were uniform from the axis of the bed right through to the wall (curve a), the entire radial heat transfer process could be described by a single parameter, namely, the effective thermal conductivity of the bed, and the problem of a

PACKED TUBE

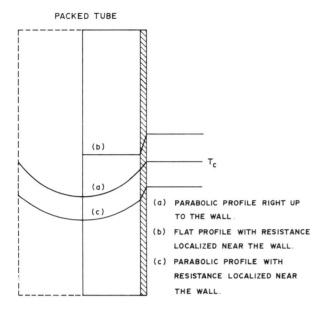

(a) PARABOLIC PROFILE RIGHT UP
    TO THE WALL.

(b) FLAT PROFILE WITH RESISTANCE
    LOCALIZED NEAR THE WALL.

(c) PARABOLIC PROFILE WITH
    RESISTANCE LOCALIZED NEAR
    THE WALL.

Figure 10.1. Patterns of radial temperature distribution.

separate heat transfer coefficient at the wall does not arise. Actually, however, there is a break in the profile near the wall (curve b or c), so that two parameters—an effective thermal conductivity and a heat transfer coefficient in the vicinity of the wall—are required to describe the heat transfer process. In both the cases a temperature profile is assumed in the direction of heat flow (i.e., transverse to the direction of fluid flow). Thus they represent two-dimensional models requiring one or two parameters to describe the heat transfer process, the latter being more generally used in reactor design (Chapter 11).

If the temperature profile in the main part of the cross section were flat, curve b, then the entire resistance to heat transfer would be localized in the vicinity of the wall. Thus the question of an effective thermal conductivity does not arise, and the entire heat transfer process can be described by an overall heat transfer coefficient at the wall. In this one-dimensional model whatever resistance to heat transfer exists within the bed is incorporated in the heat transfer coefficient.

Thus the following variations are possible in the analysis of radial heat transfer data:

1. Assume a one-dimensional model and calculate an overall heat transfer coefficient at the wall.

2. Assume a two-dimensional model with two parameters, and calculate an effective thermal conductivity for the bed and a heat transfer coefficient at the wall.

3. Assume a two-dimensional model with one parameter, and calculate an effective thermal conductivity for the bed on the basis of a continuously varying temperature from the axis to the wall.

The third variation is seldom used in design, for there is strong evidence in support of a break in the temperature profile near the wall. We shall therefore be primarily concerned with the first two approaches.

In an actual experimental program, where measurements are made under nonreactive conditions, the basic two-dimensional pseudohomogeneous model for the packed bed is solved in order to obtain the transport coefficients. While the thermal conductivity is a basic parameter of the equation itself, the wall heat transfer coefficient appears as a boundary condition at the wall. Analytical solution, in terms of an infinite series, is possible in the absence of a reaction term, and for sufficiently long beds the first term of this solution can be used to compute the transport coefficients. It will be noticed that this procedure is the inverse of simulating the reactor, where the temperature profiles are generated from a knowledge of the transport parameters.

Among the methods used by different workers the following may be mentioned: (1) four-point temperature measurement (Cybulski, 1966, 1970); (2) axial temperature measurement (Yagi and Wakao 1959); (3) mixing-cell and center-line temperature profile measurement (Olbrich et al., 1966; Agnew and Potter, 1970); and (4) radial temperature profile measurement (Felix, 1951; Coberly and Marshall, 1951; Maeda, 1952; Valstar, 1969; De Wasch and Froment, 1972). Most of these methods yield reasonable estimates of transport coefficients only for sufficiently long beds. Tsang et al. (1976) have used the orthogonal collocation method to obtain more reliable values. A relatively fast computational technique based on modified orthogonal collocation for estimation of heat transfer parameters in a packed bed from radial temperature measurements has been proposed by Michelsen (1980).

The radial temperature profile is usually measured at several bed depths with thermocouples placed just above the packing. Li and Finlayson (1977) have commented on the methods based on such measurements, and conclude that the transfer coefficients obtained by these methods are not always the asymptotic values (the ones to be used for design purposes) and that such data should be compared with caution.

The experimental data of some typical investigations on effective thermal conductivity are plotted in Figure 10.2. The lines shown in the figure can be fitted

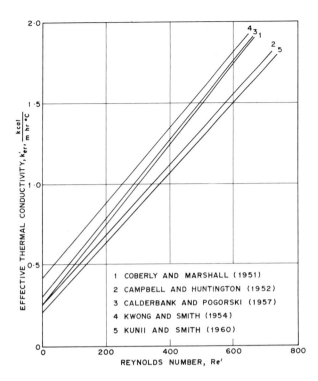

Figure 10.2. Some reported correlations for effective radial thermal conductivity (Kulkarni and Doraiswamy, 1980).

into a linear equation:

$$k'_{er} = A + B \, Re' \qquad (10.1)$$

Equation 10.1 is significant when viewed in relation to the values of A and B as provided by the intercepts and slopes of the different lines shown in the figure. Evidently the effective thermal conductivity has a definite value when there is no flow, rising sharply as flow is commenced and then increased. Thus there are two distinct contributions to the radial effective thermal conductivity in a packed-bed catalytic reactor static (k'^0_e) and dynamic (k'^d_{er}). While the static contribution can be predicted from theoretical models, the larger dynamic contribution is a function of the Reynolds number and the model employed in its determination. Methods for estimating the two contributions are outlined below.

Static Contribution

The total effective thermal conductivity of a packed bed (in the direction of heat flow) can be treated as being made up of additive contributions for natural convection, radiation, and conduction:

$$k'^0_e = k'^0_{ea} + k'^0_{eb} + k'^0_{ec} \qquad (10.2)$$

The heat transfer due to convection can usually be neglected in relation to that caused by radiation and conduction (Gorring and Churchill, 1961; Gopalrathnam et al., 1961). Radiant heat transfer can be expressed as

$$k'_{eb} = h_b d_p = 4\sigma g d_p T^3 \qquad (10.3)$$

Except at high temperatures and for larger particles, k'_{eb} is also usually negligible. Of the many methods reported for estimating the effective radial thermal conductivity, the following are particularly useful.

First, Eq. 10.2 can be simplified to $k'^0_e = k'^0_{ec}$ without serious loss of accuracy. Gorring and Churchill (1961) proposed a model based on conduction alone and obtained the following expression:

$$k'^0_{ec} = 1.92 \left(\frac{k'^2_s / k'_g}{C^2_r} \right)^{1/3} \qquad (10.4)$$

where

$$C_r = \left[\frac{5}{3} - \left(\frac{20 f_B}{\pi} \right)^{1.5} \right]^{-1}$$

In a more elaborate model (Yagi and Kunii, 1957; Kunii and Smith, 1960), a single comprehensive relationship has been proposed that accounts for all modes of heat transfer in the packed bed. It is assumed that the particles are surrounded by a stagnant fluid film, and heat transfer (in the direction of heat flux) occurs (1) through the stationary fluid in the void spaces of the packed bed and (2) through the solid phase. The following final expression for the effective thermal conductivity can be derived by using this model:

$$\frac{k'^0_e}{k'_g} = f_B \left(1 + \beta'' \frac{h_{bv} d_p}{k'_g} \right) + \frac{\beta'' (1 - f_B)}{[(1/\delta)(d_p h_{bs}/k'_g)]^{-1} + \frac{2}{3}(k'_g/k'_s)} \qquad (10.5)$$

where β'' is the ratio of the effective distance between particles to the particle diameter ($\Delta L / d_p$), being equal to unity for loose packing and to 0.895 for close packing; δ is a measure of the stagnant fluid thickness, also normalized by d_p, values of which can be read from Figure 10.3 for both loose and dense packings. The coefficients h_{bv} and h_{bs} represent radiant transfer between voids and between solid surfaces, respectively, and can be estimated from the following equations (Kulkarni and Doraiswamy, 1980):

$$h_{bv} = (1.952 \times 10^{-7}) \frac{T^3}{1 + \frac{f_B}{2(1 - f_B)} \left(\frac{1 - \varepsilon_M}{\varepsilon_M} \right)} \qquad (10.6)$$

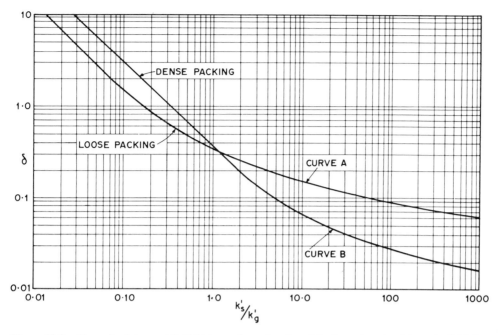

Figure 10.3. Normalized stagnant fluid thickness δ as a function of the thermal conductivity ratio k'_s/k'_g for loose and dense packing (Kunii and Smith, 1960).

and

$$h_{bs} = (1.952 \times 10^{-7}) \left(\frac{\varepsilon_M}{2 - \varepsilon_M} \right) T^3 \qquad (10.7)$$

where ε_M is the emissivity.

If radiation between voids and between solid surfaces alone is considered, the following equation for the radiant contribution can be obtained by using the Kunii–Smith model:

$$k'^0_{eb} = (\beta'' f_B d_p h_{bv}) + \frac{1 - f_B}{(1/k'_s) + (\beta'' d_p h_{bv})^{-1}} \qquad (10.8)$$

where β'' has a value of 0.67 for loose packing and 0.7 for dense packing.

Diessler and Boegli (1958) considered the packed bed as a cubic lattice of spheres ($f_B = 0.476$) and obtained the solutions for heat transport by conduction. Subsequently the method was extended to the orthorhombic model ($f_B = 0.395$) for heat transport by conduction and radiation by Wakao and Kato (1968). The following working equation has been recommended for predicting k'^0_e to within about 15%:

$$\frac{k'^0_e}{k'_g} = \left(\frac{k'^0_e}{k'_g} \right)_{\mathrm{Bi}=0} + 13\,(\mathrm{Bi})_b \frac{k'_s}{k'_g} \qquad (10.9)$$

for

$$10 \leqslant \frac{k'_s}{k'_g} \leqslant 500 \qquad \text{and} \qquad (\mathrm{Bi})_b = \frac{h_b R}{k'_e} \leqslant 0.1$$

The following empirical equation has been proposed by Krupiczka (1966):

$$\frac{k'^0_e}{k'_g} = \left(\frac{k'_s}{k'_g} \right)^{0.28 - 0.757 \log f_B - 0.057 \log(k'_s/k'_g)} \qquad (10.10)$$

An alternative expression is that of Beveridge and Haughey (1971).

Dynamic Contribution

We shall be concerned primarily with the two-dimensional model with two parameters k'^d_{er} and h_w. Methods of estimating h_w are presented in Section 10.2. De Wasch and Froment (1972) have obtained elaborate experimental data, based on which they suggest the following equation for k'^d_{er}:

$$k'^d_{er} = \frac{0.0025}{1 + 46\,(d_p/d_t)^2} \mathrm{Re}' \qquad (10.11)$$

in which the effect of the ratio of particle diameter to tube diameter is also included. Since the experimental data are represented by this equation with a deviation of less than 1.5 % and it includes the influence of d_p/d_t, it may be regarded as the equation of choice for estimating the dynamic contribution.

De Wasch and Froment also suggest an equation for the two-dimensional model with one parameter:

$$k_{er}'^d = \frac{0.0022}{1 + 120(d_p/d_t)^2} \, \text{Re}' \qquad (10.12)$$

Dixon and Cresswell (1979) have presented the following theoretical relationship for predicting the effective radial thermal conductivity:

$$k_{er}' = k_g' + k_s' \frac{1 + (8k_g'/h_w d_t)}{1 + \dfrac{(16/3)k_s'(h_{fp}d_p)^{-1} + (0.1/k_p)}{(1 - f_B)(d_t/d_p)^2}} \qquad (10.13)$$

In the limit as $d_t/d_p \to \infty$ this equation permits us to treat the effective thermal conductivity as the sum of the two conductivities operating in parallel.

10.1.2. Model Based on Effect of Fluid Flow on Both Conductive and Convective Transport

In the studies of Yagi and Kunii (1957), Kunii and Smith (1960), and others (Chan and Tien, 1973; Wakao and Vortmeyer, 1971; Yovanovich, 1973), the static (or conductive) contribution is assumed to be uninfluenced by flow. Thus, for obtaining experimental values of the static contribution, the lines of Figure 10.2 were extrapolated to zero Reynolds number. Actually, however, the total heat transfer consists of a conductive mode, represented by particle-to-particle heat transfer, and a convective mode, represented by fluid-to-particle heat transfer.

Bhattacharya and Pei (1975) suggest three modes of transfer: (1) conduction alone with motionless fluid; (2) effect of fluid flow on conduction; and (3) convective heat transfer due mainly to flow. In other words, Eq. 10.1 based on static and dynamic contributions will now be written as

$$k_{er}' = k_{ec}'^0 + k_{ec}'^d + k_{ea}'^d \qquad (10.14)$$

where a new term $k_{ec}'^d$ has been introduced to account for the dynamic contribution to conduction, and $k_{ea}'^d$ is due to convection alone. Clearly, $k_{er}'^d = k_{ec}'^d + k_{ea}'^d$. By obtaining experimental data on the contribution due to convection alone (i.e., $k_{ea}'^d$) by using microwave power for heating the bed and subtracting the values from the earlier reported results on $k_{er}'^d$, Bhattacharya and Pei (1975) obtained values of the flow contribution to conduction $k_{ec}'^d$.

The static contribution $k_{ec}'^0$ is then recovered by plotting $k_{ec}'^d$ vs. Re' and extrapolating to Re' = 0. Note that in earlier studies $k_e'^d$ was extrapolated to Re' = 0, so that the resulting value of the "static" contribution also contained the flow contribution to conduction.

The correctness of the approach outlined above is demonstrated by the fact that the values of $k_e'^0/k_g'$ obtained by this method closely match those predicted by the Kunii–Smith model for the static contribution. It will be recalled that this model does not account for flow, and hence comparisons with $k_e'^0/k_g'$ estimated by extrapolating the curves of Figure 10.2 (which really give the combined effects of static and dynamic contributions to conduction) would not be correct, and in fact show wide disparities. The final correlations of Bhattacharya and Pei (1975) are summarized in Table 10.1.

10.1.3. Axial Transport

Normally the effect of axial thermal conductivity is ignored in reactor design calculations. It is, however, desirable to develop a procedure that will indicate the relative importance of axial conductivity and that can thus be used for justifying its neglect.

Useful correlations for axial effective thermal conductivity are very meagre. Among the significant contributions are those of Argo and Smith (1953), Yagi et al. (1960), Kunii and Smith (1961), Bischoff (1962, 1967), de Ligny (1970), Votruba et al. (1972), and Dixon and Cresswell (1979).

A useful correlation appears to be that of Votruba et al. (1972):

$$\frac{1}{\text{Pe}_{hl}'} = \frac{k_{el}'^0/k_g'}{\text{Pr Re}'} + \frac{14.5}{d_p[1 + (C/\text{Pr Re}')]} \qquad (10.15)$$

where C is a constant whose value varies between 0 and 5; data for a few typical systems are listed in Table 10.2. For any system under consideration, the value for the one closest to it in the table can be chosen.

The static conductivity term appearing in Eq. 10.15 can be estimated by any of the methods presented in Section 10.1.1, preferably from the equation recommended later in Table 10.5.

TABLE 10.1. Equations for Radial Heat Transfer Suggested by Bhattacharya and Pei (1975)

Type of transport	Equation
Conductive heat transport in the motionless fluid	$\dfrac{k_e'^0}{k_g'} = f_B\left(1 + \beta'' \dfrac{h_b d_p}{k_g'}\right) + \dfrac{\beta''(1 - f_B)}{[1/\delta + (d_p/k_g')h_{bs}]^{-1} + \frac{2}{3}(k_g'/k_s')}$
Conductive heat transport due to fluid flow (i.e., effect of fluid flow on conduction)	$(j_h)_{fc} = \left(\dfrac{h_{fc}}{C_{pf}G_f}\right)\text{Pr}^{2/3} = (2.05 \times 10^{-5})\left(\dfrac{\rho_c C_{pc}}{\rho_g C_{pg}}\right)$
Heat transfer due to convection (i.e., fluid–particle heat transfer)	$(j_h)_{fp} = \left(\dfrac{h_{fp}}{C_{pf}G_f}\right)\text{Pr}^{2/3} = 0.018(\text{Ar})^{0.25}\left(\dfrac{\text{Re}'}{1 - f_B}\right)^{-0.5}$

where Ar is the Archimedes number, given by

$$\text{Ar} = \frac{d_p^3 g \rho_g (\rho_c - \rho_g)}{\mu_g^2}$$

TABLE 10.2. Values of Constant C of Eq. 10.15 (Votruba et al., 1972)

System	d_p (mm)	$\dfrac{d_t}{d_p}$	$\dfrac{k_e'^0}{k_g'}$	$\dfrac{14.5}{d_p}$	C	f_B	Interval of Re'
Glass spheres–oxygen	3.90	6.60	38.8	3.92	4.9	0.38	2.8–93.5
Glass spheres–oxygen	2.25	11.55	8.3	7.93	2.9	0.33	0.4–90.3
Lead spheres–nitrogen	2.25	11.55	29.8	4.14	0.0	0.33	1.7–22.7
Iron spheres–nitrogen	5.15	5.05	23.3	2.27	0.0	0.40 0.33	3.5–86.7
Raschig rings (ceramics)–air–nitrogen	6.5 × 3	4.00	19.6	2.07	11.4	0.49	5.1–134.8
Glass spheres–nitrogen–air	0.45	57.77	27.5	22.41	0.0	0.27 0.35	0.31–3.4
Iron spheres–air	5.00	5.20	25.0	2.17	5.1	0.42 0.43 0.38 0.36	0.7–192.2
Alumina spheres–air	3.40	7.65	4.5	4.95	0.5	0.36 0.51	0.5–115.8
Alumina cylinders–air	5.60	4.64	7.3	2.61	3.2	0.49 0.52	1.5–84.8
Duracryl particle–air	1.32	19.7	34.2	11.35	2.9	0.40	0.9–8.8
Sand–air	0.25	104.0	11.2	63.04	0.03	0.42	0.2–1.03
Glass spheres–air	6.50	3.85	228	1.92	70.8	0.40	260–1000

Dixon and Cresswell's (1979) theoretical relationship can also be conveniently used:

$$k_{el}' = k_g' + \frac{k_s'}{1 + \dfrac{(16/3)k_s'(h_{fp}d_p)^{-1} + (0.1/k_p)^2}{(1 - f_B)(d_t/d_p)^2}} \qquad (10.16)$$

In the limit as $d_t/d_p \to \infty$, this equation reduces to the simple form $k_{el}' = k_g' + k_s'$, conforming to the additive contributions from the constituent phases.

10.2. WALL HEAT TRANSFER COEFFICIENT

In this section we shall confine the treatment to procedures for estimating the effective heat transfer coefficient (h_e) for the one-dimensional model and the wall heat transfer coefficient (h_w) for the two-dimensional model.

10.2.1. Effective Heat Transfer Coefficient

Empirical Correlations

Several empirical correlations for predicting h_e have been proposed, notably by Colburn (1951), Maeda (1952), Verschoor and Schuit (1950), and Leva (1947) and co-workers (Leva and Grummer, 1948; Leva et al., 1948). A simple correlation involving readily available input data is that of Leva and Grummer (1948), but the equation of De Wasch and Froment (1972) provides a very accurate estimate:

$$(\text{Bi})_h = (\text{Bi})_h^0 + 0.024\, \text{Re}' \qquad (10.17)$$

where $(\text{Bi})_h$ is the Biot number for wall heat transfer:

$$(\text{Bi})_h = h_e d_p / 2 k_{er}' \qquad (10.18)$$

and $(\text{Bi})_h^0$ represents the static contribution given by $(h_e^0 d_p / 2 k_{er}')$. The term h_e is the effective coefficient in the vicinity of the wall (and not the heat transfer coefficient at the wall h_w for a two-dimensional model). The static effective coefficient h_e^0 cannot be predicted from theory, but the following equation can be used for its estimation from a knowledge of $k_e'^0$:

$$h_e^0 = 6.15\,(k_e'^0 / d_t) \qquad (10.19)$$

The effective heat transfer coefficient h_e must also be distinguished from the overall coefficient U, which includes heat transfer outside the wall:

$$\frac{1}{U} = \frac{1}{h_e} + \frac{1}{h_C} \qquad (10.20)$$

where h_C is the heat transfer coefficient for the control fluid in the shell.

Whether a model is one-dimensional or two-dimensional, the mechanism of heat transfer at the wall in a packed bed involves the following static and dynamic contributions (see Figure 10.4). The static contribution comprises (1a) heat transfer through the solid phase by contact conduction, (1b) radiation between solid phases, (2a) heat transfer in the void spaces arising out of the molecular thermal conduction, and (2b) radiation between voids. The dynamic contribution is brought in as a result of (3a) the lateral mixing

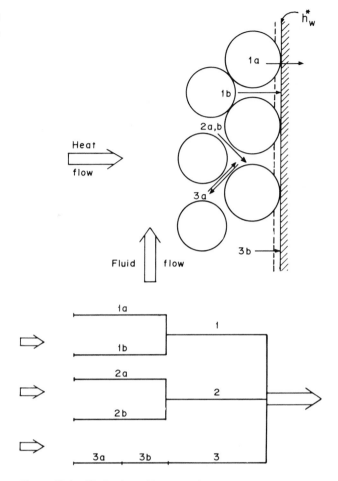

Figure 10.4. Mechanism of heat transfer at the wall in a packed bed.

of the flowing fluid and (3b) transport through a free boundary film. The total heat transfer may be written as the sum of the static and dynamic contributions:

$$h_e = h_e^0 + h_e^d \quad \text{and} \quad h_w = h_w^0 + h_w^d \qquad (10.21)$$

The experimental results of several investigators (Yagi and Kunii, 1960; Felix, 1951; Plautz and Johnstone, 1955) suggest that the value of $(\text{Bi})_h^0$ varies between 1 and 8.

Yagi and Kunii (1960) proposed another model for heat transfer at the wall in a packed bed by incorporating transport across a so-called true fluid boundary layer that acts in series with transport due to lateral mixing. Thus the total dynamic contribution h_{wt}^d now becomes

$$\frac{1}{h_{wt}^d} = \frac{1}{h_w^*} + \frac{1}{h_w^d} \qquad (10.22)$$

leading to the final relationship

$$(Bi)_h = (Bi)_h^0 + \left[\frac{1}{(Bi)_h^*} + \frac{1}{\alpha_w \, Pr \, Re'} \right]^{-1} \quad (10.23)$$

where

$$(Bi)_h^* = \frac{h_w^* d_p}{k_g'} \quad (10.24)$$

Since there are no data available from which to calculate $(Bi)_h^*$, Eq. 10.23 can only be taken as hypothetical.

Correction for Parabolic Profile

The overall heat transfer coefficient determined by any of the methods listed above is based on an average bed temperature. The existence of such an average temperature at any axial position depends, however, on the relative magnitudes of the heat transfer coefficient at the wall (h_w) and the radial effective thermal conductivity of the solid–fluid system involved (k_e'). If $k_e' \gg d_p h_w$, then it may be assumed that the bed has a flat profile and the resistance to heat transfer resides exclusively in the vicinity of the wall. On the other hand, if $k_e' > d_p h_w$, we might reasonably expect a radial temperature profile in the bed. Evidently, then, the Biot number for wall heat transfer defined by an equivalent form of Eq. 10.18, $(Bi)_h = h_w d_p / k_e'$, would be a direct measure of the existence of a radial profile.

The presence of a radial temperature profile affects the temperature driving force in the vicinity of the wall, and a correction has to be applied to the heat transfer coefficient in order to get reliable estimates. If a parabolic profile is assumed, the following equation can be easily obtained:

$$h_e = \frac{h_w}{1 + (Bi)_h / 8} \quad (10.25)$$

Crider and Foss (1955) have shown that a much better approximation is obtained when 8 is replaced by 6.2. As already anticipated, $h_e \simeq h_w$ for low values of $(Bi)_h$. It is therefore suggested that the one-dimensional approximation be used only when $(Bi)_h < 1$. For $(Bi)_h = 2$, for instance, such an approximation would lead to an error of 20% if a correction is not made in accordance with Eq. 10.25.

10.2.2. Heat Transfer Coefficient at the Wall

The heat transfer coeffiient at the wall (h_w) is a more meaningful coefficient than the effective coefficient (h_e)

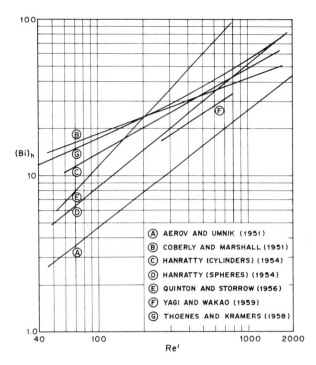

Figure 10.5. Correlations for heat transfer coefficient at the wall (Kulkarni and Doraiswamy, 1980).

of the previous section. This can be calculated by the procedure outlined in Section 10.1. The coefficient thus determined has been correlated as a function of Reynolds number by several investigators, and their results are plotted in Figure 10.5. Each of the lines can be correlated by a general equation of the form

$$(Bi)_h = a \, (Re')^b \quad (10.26)$$

Based on an evaluation of these equations, the following are recommended by Hougen (1961).

Spheres: (Yagi and Wakao, 1959):

$$(Bi)_h = 0.186 \, (Re')^{0.8} \quad (10.27)$$

Cylinders: (Hanratty, 1954):

$$(Bi)_h = 0.95 \, (Re')^{0.5} \quad (10.28)$$

Beek (1962) has included the effect of the Prandtl number on h_w and suggests the following equations for spheres. From Hanratty's data:

$$(Bi)_h = 0.203 \, (Re')^{1/3} \, Pr^{1/3} + 0.220 \, (Re')^{0.8} \, Pr^{0.4} \quad (10.29)$$

If it is assumed that $h_w = 0.8\,k'_g$, then

$$(Bi)_h = 2.58\,(Re')^{1/3}\,Pr^{1/3} + 0.094\,(Re')^{0.8}\,Pr^{0.4} \tag{10.30}$$

These equations are based on the particle-to-fluid mass transfer data of Thoenes and Kramers (1958) and have been adapted to heat transfer at the wall on the assumption that the same correlation holds for heat transfer and that a correction factor of 0.8 enables the equation to be used for heat transfer at the wall.

Olbrich and Potter (1972) obtained data over a wide range of Reynolds numbers and particle sizes. The results were fitted to an equation of the form represented by 10.26. Although this correlation was developed for mass transfer from the wall, it can also be used for heat transfer at the wall by employing a correction factor of 0.8.

In the correlations presented above the packed bed operates as a heat exchanger with no reaction involved. The values of h_w predicted by these equations are always less than those measured under reacting conditions. Further, the heat fluxes involved are of the order of 1500 kcal/hr m², whereas in commercial reactors they can be as high as 100,000 kcal/hr m². For instance, in the steam-reforming reaction the heat flux is 55,000 kcal/hr m². Hawthorne et al. (1968) suggest the following relationship, which was tested with data obtained for the dehydrogenation of methyl cyclohexane:

$$(Bi)_h = 0.28\,(Re')^{0.77}\,Pr^{0.4} \tag{10.31}$$

Chao et al. (1973) suggest an alternative (less useful) equation.

It must be noted that the forgoing equations are not applicable for stagnant systems, since h_w does not become zero at $Re' = 0$. De Wasch and Froment (1972) propose the following equation (as in the case of the one-dimensional model), which includes a stagnant contribution:

$$h_w = h_w^0 + 0.0115\,\frac{d_t}{d_p}\,Re' \tag{10.32}$$

where h_w^0 can be obtained from a knowledge of the stagnant thermal conductivity:

$$h_w^0 = \frac{20\,k_e'^0}{d_t} \tag{10.33}$$

An extensive discussion of the heat transfer coefficient at the wall is presented by Aerov and Todes (1968).

Dixon and Cresswell (1979) have presented the theoretical relation

$$(Bi)_h\,(d_p/r_0) = 3\,Re^{-0.25} \quad (Re > 40) \tag{10.34}$$

which adequately predicts the wall heat transfer coefficients over a wide range of conditions.

10.2.3. Effect of Axial Dispersion

The influence of axial dispersion on the effective radial thermal conductivity and heat transfer coefficient at the wall is known to be small. The following simple expression gives an *a priori* estimate of the error involved in neglecting it (Li and Finlayson, 1977):

$$\frac{\Delta k'_{er}}{k'_{er}} = \frac{\Delta h_w}{h_w} = -\frac{1}{Pe_{hr}}\left(\frac{d_p}{r_0}\right)^2 \frac{3(Bi)_w}{(Bi)_w + 3} \tag{10.35}$$

A similar expression has been derived by Young (1975) with an additional term $d_p/2L$ on the right-hand side. This equation shows the dependence of the error on the length of the reactor and for very long reactors can usually be neglected. In small reactors such as are used in laboratory-scale experiments the heat transfer rates are appreciably altered, suggesting that such data should be used with caution. The experimental results of Gelbin et al. (1976) also convincingly prove this point.

10.2.4. Asymptotic Values of Heat Transfer Parameter and Choice of Equations for the Design of Industrial Reactors

De Wasch and Froment (1972) and Paterson (1975) found that k'_e and h_w decrease with increasing bed length. It is thus necessary to examine the effect of bed length on the values of these parameters, particularly since considerable disagreement exists between predictions from the equations reported above (Beek, 1962; Hlavacek, 1970). By reevaluating all the heat transfer data published in the literature, Li and Finlayson (1977) obtained expressions for asymptotic values of the two parameters (which would be independent of length), and these are reproduced later in Table 10.5.

Typical L/r_0 values of industrial fixed-bed reactors are in the range of 100–200. For such reactors the asymptotic correlations given in Table 10.6 should be used, but for cases where the tube length is small the correlations given in earlier sections can be used. For a list of all recommended equations, reference may be made to Section 10.4.

10.3. EFFECTIVE DIFFUSIVITY

10.3.1. General Considerations

In the case of thermal conductivity, a physical model based on the different modes of heat transfer and bed geometry could be built up and tested with experimental data. For effective diffusivity, on the other hand, such an approach is not possible. Mass transport in a packed bed is the resultant of molecular and turbulent phenomena, but the present state of the theory of turbulence in packed beds does not permit the *a priori* prediction of eddy (or turbulent) diffusivity as a separate additive contribution. As in the case of thermal conductivity, since the packed bed is anisotropic, different diffusivity values exist in the radial and axial directions.

In fully developed flow, the lateral Peclet number Pe'_{mr}, defined as $d_p u / D_{er}$, has a value of 12, whereas the longitudinal Peclet number Pe'_{ml} is a function of Re' (see Section 11.2). At low Reynolds numbers both Pe'_{mr} and Pe'_{ml} are functions of Re'. Equations will be presented in this section for estimating radial and longitudinal Peclet numbers as functions of the Reynolds number.

The effective diffusivity (D_{el} or D_{er}) can be regarded as the sum of the molecular and turbulent contributions, analogous to the static and dynamic contributions, respectively, of thermal conductivity. Thus

$$\frac{1}{Pe''} = \frac{1}{(Pe'')_t} + \frac{1}{(Pe'')_M} \tag{10.36}$$

where the subscripts t and M represent the turbulent and molecular contributions, respectively.

The molecular Peclet number is directly related to the Reynolds number based on interstitial velocity u/f_B, and to the molecular Schmidt number:

$$(Pe'')_M = Re'' Sc = \frac{Re' Sc}{f_B} \tag{10.37}$$

where Re' is the Reynolds number based on superficial velocity u. Actually the molecular diffusion contribution given by Eq. 10.37 is slightly lower because of the tortuosity effect and should be modified as $\tau Re' Sc/f_B$. Equation 10.36 can then be written as

$$\frac{1}{Pe''} = \left(\frac{1}{Pe''}\right)_t + \frac{f_B}{\tau Re' Sc} \tag{10.38}$$

For the parameter values normally encountered in practice, namely, $\tau = 1.5$, $f_B = 0.37$, and $Sc = 0.7$, Eq. 10.38 can be approximated as

$$\frac{1}{Pe''} = \left(\frac{1}{Pe''}\right)_t + \frac{0.38}{Re'} \tag{10.39}$$

Equation 10.38 or 10.39 can be used to estimate the radial or axial Peclet numbers, and the results are usually expressed as Peclet number vs. Reynolds number. Although several workers have reported data for gaseous as well as liquid systems, the use of different definitions of the Reynolds number and Peclet number has caused considerable confusion. For instance, the Peclet number has been based on the interstitial velocity (u_i) as well as on the superficial velocity (u), whereas the Reynolds number has been calculated by using the particle diameter (d_p) or an effective diameter (d_e) analogous to the hydraulic diameter in flow through pipes, the two being equal for large values of d_t/d_p.

It should be mentioned that Eq. 10.36 is a general equation applicable in both the axial and radial directions. It is only necessary to substitute the proper values for the turbulent and molecular Peclet numbers. The molecular Peclet number is the same for both the axial and the radial direction, but the turbulent component differs.

10.3.2. Radial Dispersion

The behavior of the three Peclet numbers of Eq. 10.36 is shown in Figure 10.6 as a function of the Reynolds number. The molecular Peclet number is represented by a straight line, while the total effective and turbulent Peclet numbers are represented by curves 2 and 3, respectively, in the figure. It will be noted that beyond $Re' \geq 100$ the total Peclet number coincides with that for turbulent flow, whereas in the low Reynolds number region, particularly below 20, the rate of molecular diffusion becomes significant. The turbulent Peclet number decreases sharply as the Reynolds number goes below 10, but the eddy effect is present down to very low Reynolds numbers, and only in the neighborhood of zero is the Peclet number governed by the true laminar conditions, corresponding to molecular diffusion.

Estimation of the total effective radial Peclet number (Pe''_{mr}) requires a knowledge of molecular ($Pe''_{mr})_M$ and turbulent Peclet numbers ($Pe''_{mr})_t$ in the radial direction. The molecular Peclet number can be obtained by using Eq. 10.37, while the turbulent Peclet number may be obtained by experimental determination of Pe''_{mr} and

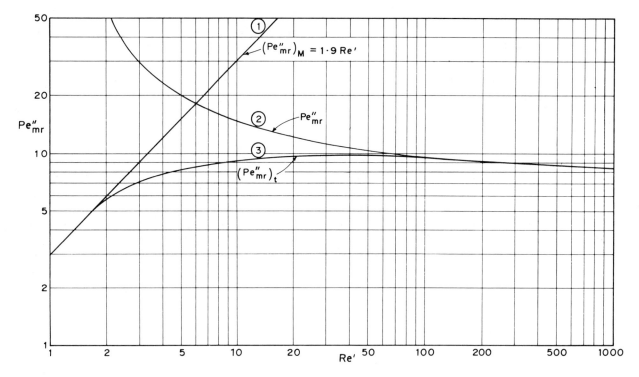

Figure 10.6. The total effective, turbulent, and molecular Peclet numbers as functions of the Reynolds number (Kulkarni and Doraiswamy, 1980).

use of Eq. 10.36. Many methods are available for determining D_{er} (and hence Pe''_{mr}), a simple and convenient one being that of Bernard and Wilhelm (1950). The turbulent Peclet number can also be estimated by using the stochastic model proposed by Gunn (1969) for radial dispersion in a packed bed. Here the turbulent Peclet number, also called the fluid-mechanical Peclet number, is related to the probability P of a molecule at the entrance traveling a distance x_p. This probability, and hence the fluid-mechanical Peclet number, is related to the Reynolds number, as shown in Table 10.3. Thus $(Pe''_{mr})_t$ can be estimated.

TABLE 10.3. Dependence of $(Pe''_{mr})_t$ and Probability P on Reynolds Number (Gunn, 1969)

Reynolds Number Re'	$(Pe''_{mr})_t$	Probability P
1	40	0.17
5	32	0.18
10	25	0.20
20	22	0.27
40	18	0.36
100	14	0.43
200	12	0.47
400	11	0.48
1000	11	0.50

For large Reynolds numbers, as can be seen from Table 10.3, the fluid-mechanical Peclet number approaches a value equal to 11. Equation 10.38 can therefore be written as

$$\frac{1}{Pe''_{mr}} = \frac{1}{11} + \frac{f_B}{\tau \, Re' \, Sc} = \frac{1}{11} + \frac{0.38}{Re'} \quad (10.40)$$

Based on data from a number of reported investigations, Wen and Fan (1975) have developed the following empirical correlation:

$$\frac{1}{Pe''_{mr}} = \frac{0.09}{1 + (10/Re' \, Sc)} + \frac{0.4}{(Re' \, Sc)^{0.8}} \quad (10.41)$$

for $0.4 < Re' < 500$, $0.77 < Sc < 1.2$.

The radial Peclet number is actually a point property that increases with radial position (toward the tube wall). The results of Schwartz and Smith (1953) have shown that interstitial velocity varies with radial position. Thus an empirical correlating parameter

$$\frac{Pe''_{mr}}{1 + 19.4 \, (d_p/d_t)^2} \quad (10.42)$$

can be used (Fahien and Smith, 1955) instead of Pe''_{mr} to bring together the data of different particle sizes as a function of Reynolds number. Froment (1967) has replotted the results of Bernard and Wilhelm (1950), Hiby (1962), Plautz and Johnstone (1955), Fahien and Smith (1955), and Dorrweiler and Fahien (1959) using the parameter 10.42. In assigning any significance to the effect of d_p/d_t, however, it should be remembered that the scatter of experimental points may be due to hydrodynamic factors alone, unconnected with d_p/d_t (Roemer et al., 1962; Gunn, 1969). In any case, the effect of d_p/d_t appears to be negligible below a value of about 0.08, and at higher values the following relationship is reasonably accurate:

$$Pe''_{mr} = 11\left[1 + 19.4\left(\frac{d_p}{d_t}\right)^2\right] \qquad (10.43)$$

This equation represents satisfactorily all the available data at high Reynolds numbers, including the Peclet number for an empty tube (for which the ratio d_p/d_t is obviously unity), and can therefore be used for estimating Pe''_{mr} in reactor design calculations at high Reynolds numbers where the effect of d_p/d_t alone is significant. At low Reynolds numbers Pe''_{mr} may be estimated from Eq. 10.37.

10.3.3. Axial Dispersion

The total displacement of a fluid element in a packed tube can be regarded as the sum of several individual steps, each of which is determined by the combined effect of the mean interstitial velocity and a random component. The net result of these steps is a diffusionlike process superimposed on convective flow. The so-called mixing-cell model has been used to arrive at the theoretical value of 2 for the axial Peclet number (Bernard and Wilhelm, 1950), but it has been shown by Gunn and Pryce (1969) that mixing in a packed bed is better described by a diffusionlike process than by the mixing-cell concept. This would indicate that mixing is basically a concentration-dependent phenomenon. On the other hand, the study of Sundaresan et al. (1980) shows that mixing in a packed bed is caused more by mechanical mixing due to fluid flow than by molecular or turbulent diffusion or Taylor dispersion. If this conclusion is correct, it would be difficult to sustain the concept that mixing in a packed bed is determined by a concentration gradient.

In spite of its limitations, the effective transport model offers a convenient and practical method of characterizing mixing through an axial diffusion coef-

ficient D_{el}. This has been fully discussed by Aris and Amundson (1957), Carberry and Bretton (1958), Levenspiel and Smith (1957), Bischoff and Levenspiel (1962), Edwards and Richardson (1968), and Gunn (1969). The experimental results of a number of investigators are presented later in Figure 11.7.

The general equation 10.38 can therefore be used to calculate the axial dispersion coefficient. As in the case of radial dispersion, the fluid-mechanical Peclet number $(Pe''_{mr})_t$ approaches a constant value (in this case, 2) at high Reynolds numbers and Eqs. 10.38 and 10.39 can be approximated as

$$\frac{1}{Pe''_{mr}} = \frac{1}{2} + \frac{f_B}{\tau\,Re'\,Sc} = \frac{1}{2} + \frac{0.38}{Re'} \qquad (10.44)$$

Although this equation can be used for a quick and approximate estimate of the axial dispersion coefficient, a more accurate method would be to incorporate the effect of the Reynolds number on the fluid-mechanical component. The fluid-mechanical component is related to the Reynolds number as

$$\frac{1}{(Pe''_{ml})_t} = \frac{1/Pe''_{ml}}{1 + (\beta/\tau\,Re'\,Sc)} \qquad (10.45)$$

where (Pe''_{ml}) refers to the limiting value (usually 2) of the axial Peclet number and β, the radial dispersion factor, is equal to 8 using the highly turbulent random walk model. The term $\beta/\tau\,Re'\,Sc$ accounts for the effect of radial dispersion on the concentration gradient caused by axial dispersion. Using this definition of the fluid-mechanical Peclet number in Eq. 10.38, we can write the axial dispersion coefficient as

$$\frac{1}{Pe''_{ml}} = \frac{1/Pe''_{ml}}{1 + (\beta/\tau\,Re'\,Sc)} + \frac{f_B}{\tau\,Re'\,Sc} \qquad (10.46)$$

When plotted as axial Peclet number vs. Reynolds number, this equation gives the correct asymptotic values at low and high Reynolds numbers.

The Peclet number as given by this equation exhibits a maximum value in the range of the Reynolds numbers rather than just smoothly interpolating between the values at the two extremes. Some of the experimental data reported (e.g., McHenry and Wilhelm, 1957) also indicates such a trend. The simpler expressions like 10.44, which do not show any maximum, therefore cannot represent the behavior over the entire range. Using the Bischoff–Levenspiel (1962) equation for the prediction of axial dispersion from the radial dispersion and the appropriate value for the tortuosity factor, Bischoff (1969) has reported the following equation:

$$\frac{1}{Pe''_{ml}} = \frac{f_B}{\tau \, Re' \, Sc} + \frac{0.45}{1 + (7.3/Re' \, Sc)} \qquad (10.47)$$

which closely resembles Eq. 10.46 with specific values for the parameters (Pe_{ml}) and β.

Edwards and Richardson (1968) have also obtained the value for the fluid-mechanical component as

$$\left(\frac{1}{Pe''_{ml}}\right)_t = \frac{0.5}{1 + (0.5/Re')}.$$

Recently Wen and Fan (1975) have recorrelated a wide range of reported experimental data and obtained the following expression using $\tau = 1.5$:

$$\frac{1}{P''_{ml}} = \frac{0.5}{1 + (3.8/Re' \, Sc)} + \frac{0.3}{Re' \, Sc} \qquad (10.48)$$

for $0.008 < Re' < 400$, $0.28 < Sc < 2.2$.

Gunn (1969) has derived a more rigorous equation for Pe''_{ml} in terms of the probability P (already referred to). A similar expression has been obtained by Miyauchi and Kikuchi (1975). At high Reynolds numbers, Gunn's equation reduces to

$$\frac{1}{Pe''_{ml}} = \left(\frac{1}{Pe''_{ml}}\right)_t = \frac{2P}{1 - P} \qquad (10.49)$$

and at low Reynolds numbers to $(f_B/Re' \, Sc)$. The functional dependence of P on the Reynolds number is shown in Table 10.3.

The molecular component of effective diffusivity is evidently the same for both radial and axial directions; thus the anisotropy is provided only by the turbulent component. In both the cases the molecular Peclet number is given by $Re' \, Sc \, \tau/f_B$. Equation 10.37 was obtained by assuming τ to be unity, whereas actually it is clear from Eq. 10.39, in which it has been assigned a value of 1.50, that $(Pe''_{ml})_M = 2.67 \, Re'$.

The empirical constants in Eq. 10.46 $[\tau, \beta, (Pe_{ml})]$ have been evaluated by fitting this equation to the experimental results of the various authors by Langer et al. (1978). Analysis of their results shows that except for the value of τ, all other empirical constants vary over a wide range. Based on the analysis of reported experimental data (Suzuki and Smith, 1972; Kawazoe et al., 1974; Urban and Gomezplata, 1969; Evans and Kenney, 1966; Scott et al., 1974), Langer et al. (1978) conclude that the limiting value of the Peclet number at high flow rates increases linearly with particle diameter $d_p \leqslant 0.25$ cm and only reaches the generally assumed value of 2 when $d_p \geqslant 0.3$ cm. The empirical equation obtained by these authors is

$$(Pe''_{ml}) = 6.7 d_p \qquad (10.50)$$

In general the effects of axial mass transport are larger for small packing diameter and low L/d_t columns. Equation 10.46, applicable for industrial equipment, becomes restrictive when analyzing the data from a bench-scale apparatus, and it is advisable in such cases to conduct separate runs for evaluating axial dispersion.

10.3.4. A Unified Equation for Radial and Axial Dispersion

In the forgoing sections we saw that radial transport occurs essentially through a random walk (or sidestepping of particles) and axial transport by a coupling of diffusive and fluid-mechanical forces. An attempt was made in Section 10.3.3 to explain transport in both directions through a stochastic diffusive fluid-mechanical model based on the probability of a molecule at the entrance moving to a given position (Gunn,

TABLE 10.4. Constants C_1 and C_2 of Eq. 10.51 (de Ligny, 1970)

	Liquid Fluid		Gaseous Fluid	
Constant	Spherical Packings	Irregular Packings	Spherical Packings	Irregular Packings
C_1 (longitudinal dispersion)	2.5	2.5	0.7	4.0
C_2 (radial dispersion)	0.08	0.08	0.12	0.12
C_1 (longitudinal dispersion)	8.8	7.7	5.8	5.1
C_2 (radial dispersion)	—	—	78 ± 20	—
C_3		0.7		

TABLE 10.5. Recommended Correlations for Estimating Effective Transport Parameters (Kulkarni and Doraiswamy, 1980)

Transport Parameter	Contributions	Recommended Correlation
Effective radial thermal conductivity for the two-dimensional model	Static contribution due to conduction	$\dfrac{k_{ec}^{\prime 0}}{k_g^{\prime}} = f_B + (1 - f_B)\left[\dfrac{\beta^{\prime\prime}}{\delta + \frac{2}{3}(k_g^{\prime}/k_s^{\prime})}\right]$ where $\beta^{\prime\prime} = 1$, δ from curve A of Figure 10.3 for loose packing $\beta^{\prime\prime} = 0.895$, δ from curve B of Figure 10.3 for dense packing
	Static contribution due to radiation	$k_{eb}^{\prime 0} = \beta^{\prime\prime} f_B d_p h_{bv} + \dfrac{(1 - f_B)}{1/k_s^{\prime} + (d_p h_{bv} \beta^{\prime\prime})^{-1}}$ where $h_{bv} = (1.952 \times 10^{-7}) T^3 \left[1 + \dfrac{f_B}{2(1 - f_B)}\left(\dfrac{1 - \varepsilon_M}{\varepsilon_M}\right)\right]^{-1}$
	Total static contribution	$k_e^{\prime 0} = k_{ec}^{\prime 0} + k_{eb}^{\prime 0}$
	Dynamic contribution	$k_{er}^{\prime d} = \dfrac{0.0025}{1 + 46(d_p/d_t)^2}\,\text{Re}^{\prime}$
	Total effective value (static + dynamic)	$k_e^{\prime} = k_e^{\prime 0} + k_{er}^{\prime d}$
Wall heat transfer coefficient for the two-dimensional model		$h_w = h_w^0 + 0.0115\left(\dfrac{d_t}{d_p}\right)\text{Re}^{\prime}$ where $h_w^0 = \dfrac{20 k_e^{\prime 0}}{d_t}$
Overall wall heat transfer coefficient for the one-dimensional model		1. $(\text{Bi})_h = (\text{Bi})_h^0 + 0.024\,\text{Re}^{\prime}$ where $(\text{Bi})_h = \dfrac{h_e d_p}{k_e^{\prime}}$, $\quad (\text{Bi})_h^0 = \dfrac{h_e^0 d_p}{k_{er}^{\prime}}$ $h_e^0 = \dfrac{6.15 k_e^{\prime 0}}{d_t}$ 2. $\dfrac{1}{h_e} = \dfrac{1}{h_w} + \dfrac{4 d_t}{k_{er}^{\prime}}$
Effective axial thermal conductivity	Static plus dynamic contribution	$\dfrac{1}{\text{Pe}_{hl}^{\prime}} = \dfrac{k_{el}^{\prime 0}/k_g^{\prime}}{\text{Re}^{\prime}\,\text{Pr}} + \dfrac{14.5}{d_p\left(1 + \dfrac{C}{\text{Re}^{\prime}\,\text{Pr}}\right)}$ where C is a constant (see Table 10.2), and $k_{el}^{\prime 0}$ is the effective axial static contribution (same as that for radial conduction).

1969). However, different forms of equations were obtained for the radial and the axial direction.

It has been shown by de Ligny (1970) that an equation of the same form based on a coupling of diffusive and convective mechanisms can be derived for predicting both radial and axial transport coefficients. The equation derived by him is similar to 10.47, proposed by Bischoff (1969) for axial dispersion:

$$\frac{1}{Pe''_{ml}} = \frac{C_1}{1 + (C_2/Re' \, Sc)} + \frac{C_3}{Re' \, Sc} \qquad (10.51)$$

According to the theory, the values of the coefficient C_1 would be different for longitudinal and radial dispersion, whereas C_2 should have the same value for both. Actually, however, C_2 is also found to have different values for the two modes of transport. The recommended values of C_1 and C_2 for flow of liquids as well as gases through spherical and irregular particles are listed in Table 10.4. For C_3 de Ligny suggests an approximate value of 0.7, but it seems more logical to assign a value of f_B/τ with f_B in the range 0.32–0.45 and τ in the range 1.2–1.5.

TABLE 10.6. Recommended Equations for Asymptotic Values of Heat Transfer Parameters (Kulkarni and Doraiswamy, 1980)

	One-Dimensional Model	Two-Dimensional Model
	Overall heat transfer coefficient	*Wall heat transfer coefficient*
Spherical packing	$\dfrac{h_e d_t}{k'_g} \exp\left(\dfrac{6d_p}{d_t}\right) = 2.03 (Re')^{0.8}$	$\dfrac{h_w d_p}{k'_g} = 0.17 (Re')^{0.79}$
	$20 < Re' < 7600$	$20 < Re' < 7600$
	$0.05 < d_p/d_t < 0.3$	$0.05 < d_p/d_t < 0.3$
Cylindrical packing	$\dfrac{h_e d_t}{k'_g} \exp\left(\dfrac{6d_p}{d_t}\right) = 1.26 (Re')^{0.95}$	$\dfrac{h_w d_p}{k'_g} = 0.16 (Re')^{0.93}$
	$20 < Re' < 800$	$20 < Re' < 800$
	$0.03 < d_p/d_t < 0.2$	$0.03 < d_p/d_t < 0.2$
		$d_p = 6V_p/A_p$
		Effective thermal conductivity
		$Bi\left(\dfrac{d_p}{r_0}\right)\dfrac{f_B}{1 - f_B} = 0.27$
		$0.05 < \dfrac{d_p}{d_t} < 0.15$
		$500 < Re'' < 6000$
		$R'' = \dfrac{G d_p}{\mu(1 - f_B)}$
		where $Bi = \dfrac{h_w r_0}{k'_e}$

10.4. RECOMMENDED EQUATIONS; RANGES OF PARAMETER VALUES

In the forgoing sections several equations for estimating the effective transport parameters of the fixed-bed reactor were presented. The principal effective parameters considered were the effective radial thermal conductivity for the two-dimensional model, the overall heat transfer coefficient for the one-dimensional model, the effective axial thermal conductivity, the effective radial diffusivity, and the effective axial diffusivity.

From the point of view of design certain equations appear to be more useful than others for estimating these properties. The recommended equations for short tubes (< 1 m) for all the effective heat transport parameters are summarized in Table 10.5. These predict average values and should be used for simulating single-tube fixed-bed reactors in a laboratory. The recommended equations for designing industrial fixed-bed tubular reactors are summarized in Table 10.6 for two distinct types of packing, spherical and cylindrical. The recommended equations for the effective diffusional transport parameters are given in Table 10.7.

It is always useful to know the order of magnitude of the parameter values. Unfortunately very few values are reported for specific reactions systems. The correlations developed are for nonreactive gases passing through a bed of solids. Even so, the ranges of values of the important parameters, as reported in a variety of studies, are summarized in Table 10.8.

TABLE 10.7. Recommended Correlations for Estimating Effective Diffusional Transport Parameters (Kulkarni and Doraiswamy, 1980)

Transport Parameter	Recommended Correlation
Effective radial diffusivity	1. Where $d_p/d_t > 0.1$ $$\frac{1}{Pe''_{mr}} = \frac{1}{(Pe''_{mr})_t} + \frac{f_B}{\tau\,Re'\,Sc}$$ 2. Where $d_p/d_t < 0.1$, multiply the value obtained above, which will be around 11 for a high Reynolds number, by $$1 + 19.4\left(\frac{d_p}{d_t}\right)^2$$ 3. $\dfrac{1}{Pe''_{mr}} = \dfrac{0.09}{1 + \dfrac{10}{Re'\,Sc}} + \dfrac{0.4}{(Re'\,Sc)^{0.8}}$ for $\quad 0.4 < Re' < 500, \qquad 0.77 < Sc < 1.2$
Effective axial diffusivity	$\dfrac{1}{Pe''_{ml}} = \dfrac{0.5}{1 + \dfrac{3.8}{Re'\,Sc}} + \dfrac{0.3}{Re'\,Sc}$ for $\quad 0.008 < Re' < 400, \qquad 0.28 < Sc < 2.2$

TABLE 10.8. Approximate Ranges of Values of Effective Transport Parameters in a Packed Bed (Kulkarni and Doraiswamy, 1980)

Effective thermal conductivities

1. effective radial thermal conductivity, k'_{er} $1\text{--}10 \dfrac{\text{kcal}}{\text{m hr } ^\circ\text{C}}$

2. $\dfrac{\text{effective axial thermal conductivity}}{\text{thermal conductivity of fluid}}, \dfrac{k'_{el}}{k'_g}$ $1\text{--}300$

3. $\dfrac{\text{effective radial thermal conductivity}}{\text{thermal conductivity of fluid}}, \dfrac{k'_{er}}{k'_g}$ $1\text{--}12$

4. static contribution, k'^0_e $0.14\text{--}0.32 \dfrac{\text{kcal}}{\text{m hr } ^\circ\text{C}}$

Heat transfer coefficients

1. heat transfer coefficient for the one-dimensional model, h_e $15\text{--}75 \dfrac{\text{kcal}}{\text{m}^2 \text{ hr } ^\circ\text{C}}$

2. wall coefficient for the two-dimensional model, h_w $100\text{--}250 \dfrac{\text{kcal}}{\text{m}^2 \text{ hr } ^\circ\text{C}}$

3. static contribution (one-dimensional model), h^0_e $5\text{--}20 \dfrac{\text{kcal}}{\text{m}^2 \text{ hr } ^\circ\text{C}}$

4. static contribution to wall coefficient (two-dimensional model), h^0_w $15\text{--}85 \dfrac{\text{kcal}}{\text{m}^2 \text{ hr } ^\circ\text{C}}$

Effective diffusivities

1. radial and axial Peclet numbers, Pe''_{mr}, Pe''_{ml} $6\text{--}20, 0.01\text{--}10$
2. radial fluid-mechanical Peclet number, $(Pe''_{mr})_t$ $10\text{--}40$
3. axial fluid-mechanical Peclet number, $(Pe''_{ml})_t$ $0.1\text{--}5$
4. tortuosity factor, τ $1\text{--}1.5$

REFERENCES

Aerov, M. E., and Todes, O. H. (1968), *Hydraulic and Thermal Principles of the Operation of Apparatus with Fixed and Fluidized Beds*, Khimiya, Leningrad.

Aerov, M. E., and Umnik, N. N. (1951), *J. Tech. Phys. (USSR)*, **21**, 1351.

Agnew, J. B., and Potter, O. E. (1970), *Trans. Inst. Chem. Eng. (London)*, **48**, T 15.

Argo, W. B., and Smith, J. M. (1953), *Chem. Eng. Prog.*, **49**, 443.

Aris, R., and Amundson, N. R. (1957), *AIChE J.*, **3**, 281.

Balakrishnan, A. R., and Pei, D. C. T. (1979), *Ind. Eng. Chem. Process Design Dev.*, **18**, 4.

Beek, J. (1962), *Adv. Chem. Eng.*, **3**, 203.

Bernard, R. A., and Wilhelm, R. H. (1950), *Chem. Eng. Prog.*, **46**, 233.

Beveridge, G. S. G., and Haughey, D. P. (1971), *Int. J. Heat and Mass Transfer*, **14**, 1093.

Bhattacharya, D., and Pei, D. C. T. (1975), *Chem. Eng. Sci.*, **30**, 293.

Bischoff, K. B. (1962), *Can. J. Chem. Eng.*, **40**, 161.

Bischoff, K. B. (1967), *Ind. Eng. Chem.*, **59**, 18.

Bischoff, K. B. (1969), *Chem. Eng. Sci.*, **24**, 607.

Bischoff, K. B., and Levenspiel, O. (1962), *Chem. Eng. Sci.*, **17**, 245.

Calderbank, P. H., and Pogorski, L. A. (1957), *Trans. Inst. Chem. Eng. (London)*, **35**, 195.

Campbell, J. M., and Huntington, R. L. (1952), *Hydrocarbon Process. Petrol. Refiner*, **31**, 123.

Carberry, J. J., and Bretton, R. H. (1958), *AIChE J.*, **4**, 367.

Chan, C. K., and Tien, C. L. (1973), ASME–AIChE Heat Transfer Conference, Atlanta, Paper No. 73-HT-1.

Chao, R. E., Caban, R. A., and Irizarry, M. M. (1973), *Can. J. Chem. Eng.*, **51**, 67.

Coberly, C. A., and Marshall, W. R. (1951), *Chem. Eng. Prog.*, **47**, 141.

Colburn, A. P. (1951), *Ind. Eng. Chem.*, **23**, 190.

Crider, J. E., and Foss, A. S. (1955), *AIChE J.*, **11**, 1012.

Cybulski, A. (1966), *Bull. Acad. Pol. Sci.*, **14**, 801.

Cybulski, A. (1970), *Bull. Acad. Pol. Sci.*, **18**, 109.

Diessler, R. G., and Boegli, J. S. (1958), *Trans. ASME*, **80**, 1417.

de Ligny, C. L. (1970), *Chem. Eng. Sci.*, **25**, 1177.

De Wasch, A. P., and Froment, G. F. (1972), *Chem. Eng. Sci.*, **27**, 567.

Dixon, A. G., and Cresswell, D. L. (1979), *AIChE J.*, **25**, 663.

Dorrweiler, V. P., and Fahien, R. W. (1959), *AIChE J.*, **5**, 139.

Edwards, M. F., and Richardson, J. F. (1968), *Chem. Eng. Sci.*, **23**, 109.

Evans, E. V., and Kenney, C. N. (1966), *Trans. Inst. Chem. Eng. (London)*, **44**, T 189.

Fahien, R. W., and Smith, J. M. (1955), *AIChE J.*, **1**, 28.

Felix, T. R. (1951), Ph.D. thesis, University of Wisconsin.

Froment, G. F. (1967), *Ind. Eng. Chem.*, **59**, 27.

Gelbin, D., Radeke, K. H., Rosahl, B., and Stein, W. (1976), *Int. J. Heat and Mass Transfer*, **19**, 987.

Gopalrathnam, C. D., Hoelscher, H. E., and Laddha, G. S. (1961), *AIChE J.*, **7**, 249.

Gorring, R. L., and Churchill, S. W. (1961), *Chem. Eng. Prog.*, **57**, 53.

Gunn, D. J. (1968), *Chem. Eng. (London)*, **219**, 153.

Gunn, D. J. (1969), *Trans. Inst. Chem. Eng. (London)*, **47**, T 351.

Gunn, D. J., and Pryce, C. (1969), *Trans. Inst. Chem. Eng. (London)*, **47**, T 341.

Hanratty, T. J. (1954), *Chem. Eng. Sci.*, **3**, 209.

Hawthorne, R. D., Ackerman, G. R., and Nixon, A. C. (1968), *AIChE J.*, **14**, 69.

Hiby, J. W. (1962), *Proc. Symp. Interaction between Fluids and Particles, Inst. Chem. Eng. (London)*, p. 312.

Hlavacek, V., and Votruba, J. (1978), In *Chemical Reactor Theory— A Review*, eds. Lapidus, D., and Amundson, N. R., Prentice-Hall, Englewood Cliffs, New Jersey.

Hlavacek, V. (1970), *Ind. Eng. Chem.*, **62** (7), 8.

Hougen, O. (1961), *Ind. Eng. Chem.*, **53**, 509.

Kawazoe, K., Suzuki, M., and Chihara, K. (1974), *J. Chem. Eng. Japan*, **7**, 151.

Krupiczka, R. (1966), *Chim. Ind. Genie Chim.*, **95**, 1393.

Kulkarni, B. D., and Doraiswamy, L. K. (1980), *Catal. Rev. Sci. Eng.*, **22**, 431.

Kunii, D., and Smith, J. M. (1960), *AIChE J.*, **6**, 71.

Kunii, D., and Smith, J. M. (1961), *AIChE J.*, **7**, 29.

Kwong, S. S., and Smith, J. M. (1954), *Ind. Eng. Chem.*, **49**, 894.

Leva, M. (1947), *Ind. Eng. Chem.*, **39**, 857.

Leva, M., and Grummer, M. (1948), *Ind. Eng. Chem.*, **40**, 415.

Leva, M., Weintraub, M., Grummer, M., and Clark, E. L. (1948), *Ind. Eng. Chem.*, **40**, 747.

Langer, G., Roethe, A., Rothe, K. P., and Gelbin, D. (1978), *Int. J. Heat and Mass Transfer*, **21**, 751.

Levenspiel, O., and Smith, W. K. (1957), *Chem. Eng. Sci.*, **6**, 227.

Li, C. H., and Finlayson, B. A. (1977), *Chem. Eng. Sci.*, **32**, 1055.

Maeda, S. (1952), *Tech. Rep. Tohoku Univ.*, **16**, 4.

McHenry, Jr. K. W., and Wilhelm, R. H. (1957), *AIChE J.*, **3**, 83.

Michelsen, M. L. (1980), Personal communication.

Miyauchi, T., and Kikuchi, T. (1975), *Chem. Eng. Sci.*, **30**, 343.

Olbrich, W. E., and Potter, O. E. (1972), *Chem. Eng. Sci.*, **27**, 1723.

Olbrich, W. E., Agnew, J. B., and Potter, O. E. (1966), *Trans. Inst. Chem. Eng. (London)*, **44**, T 207.

Paterson, W. R. (1975), Ph.D. thesis, University of Edinburgh.

Plautz, D. A., and Johnstone, H. F. (1955), *AIChE J.*, **1**, 193.

Quinton, J. H., and Storrow, J. A. (1956), *Chem. Eng. Sci.*, **5**, 245.

Roemer, G., Dranoff, J. B., and Smith, J. M. (1962), *Ind. Eng. Chem. Fundam.*, **1**, 284.

Schwartz, C. E., and Smith, J. M. (1953), *Ind. Eng. Chem.*, **45**, 1209.

Scott, D. S., Lee, W., and Papa, J. (1974), *Chem. Eng. Sci.*, **29**, 2155.

Sundaresan, S., Amundson, N. R., and Aris, R. (1980), *AIChE J.*, **26**, 529.

Suzuki, M., and Smith, J. M. (1972), *Chem. Eng. J.*, **3**, 256.

Thoenes, Jr. D., and Kramers, H. (1958), *Chem. Eng. Sci.*, **8**, 271.

Tsang, T. H., Edgar, T. F., and Hougan, F. O. (1976), *Chem. Eng. J.*, **11**, 57.

Urban, J. C., and Gomezplata, A. (1969), *Can. J. Chem. Eng.*, **47**, 353.

Valstar, J. M. (1969), Ph.D. thesis, Delft, Netherlands.

Verschoor, H., and Schuit, G. C. A. (1950), *Appl. Sci. Res.*, **A2**, 97.

Votruba, J., Hlavacek, V., and Marek, M. (1972), *Chem. Eng. Sci.*, **27**, 1845.

Wakao, N., and Kato, K. (1968), *J. Chem. Eng. Japan*, **2**, 24.

Wakao, N., and Vortmeyer, D. (1971), *Chem. Eng. Sci.*, **26**, 1753.

Wen, C. Y., and Fan, L. T. (1975), *Models for Flow Systems and Chemical Reactors*, Marcel Dekker, New York.

Yagi, S., and Kunii, D. (1957), *AIChE J.*, **3**, 373.

Yagi, S., and Kunii, D. (1960), *AIChE J.*, **6**, 97.

Yagi, S., and Wakao, N. (1959), *AIChE J.*, **5**, 79.

Yagi, S. Kunii, D., and Wakao, N. (1960), *AIChE J.*, **6**, 543.

Young, L. C. (1975), private communication.

Yovanovich, M. M. (1973), Paper presented at ASME–AIChE Heat Transfer Conference, Atlanta, Paper No. 73-HT-43.

CHAPTER ELEVEN

Modeling of Fixed-Bed Multitubular Reactors

11.1. INTRODUCTION

11.1.1. The Rationale of Design

The tubular fixed-bed reactor is among the commonest reactor types used in industry. Its construction is similar to that of a shell-and-tube heat exchanger. The reacting fluid flows through the tubes, which are packed with catalyst pellets, while the heat transfer medium (oil, flue gas, molten salt, or condensing vapor) flows through the shell side.

While designing such equipment, the objective of the designer is to arrive at the set of system parameters that result in optimal operation. In the case of fixed-bed reactors these parameters are temperature, pressure, composition, and flow rate of the feed stream; the dimensions of the reactor tubes and the catalyst pellets; and the details of the heat transfer system. The objective of the design is to obtain the desired conversion with the specified selectivity.

To carry out this exercise we need an equation or a system of equations relating the outlet concentration to the system parameters, as in the case of a homogeneous reactor. Unfortunately this is not possible because of the complex hydrodynamics of the packed bed. Hence we have to resort to a system of equations such that the behavior of the reactor is reasonably well represented by the solutions of these equations. Many such systems of equations (i.e., mathematical models) have been proposed and techniques developed for their solution. These models differ in their complexity and in their ability to predict the reactor behavior under different conditions.

11.1.2. Reactor Models

In a widely accepted approach to the modeling of fixed-bed reactors, the solid particles and reacting gas are assumed to form an anisotropic continuum, and the gross concentration and temperature profiles in the bed are consequently assumed to be smooth functions of axial and radial coordinates. Actually, however, three types of transport processes are involved in the operation of a fixed-bed reactor: (1) interparticle transport through the voids between the catalyst pellets (i.e., in the external field); (2) intraparticle transport within the catalyst pellet (i.e.,in the internal field); and (3) interphase transport between the mainstream of the fluid and the surface of the catalyst pellet.

The reactor model consists of a combination of models for all three transport processes. In the *quasi-homogeneous model*, the intraparticle and interphase processes are unimportant and the concentration and temperature in the internal field are the same as those in the external field; the global reaction rate is based on these values. On the other hand, in the *heterogeneous model*, transport processes play a finite role, and the equations for the internal field have to be solved in order to obtain the global rate.

Two other entirely different classes of models—mixing-cell and channel models—have been proposed as alternatives to these more common differential balance or quasicontinuum models (see Wilhelm, 1962). In the cell model, each pellet along with its environment is considered as a small reactor. A sequence of such cells connected in the direction of flow is assumed to approximate the heterogeneous one-dimensional model. In the two-dimensional model, the cells are additionally considered to be laterally connected. The so-called channel models (Kondelik et al., 1968; Marivoet et al., 1974) are based on the observed distribution of local void volume in the bed. The void volume passes through alternate zones of maxima and minima in the radial direction. The channel model in its simplest form assumes that the packed bed is built up of coaxial cylindrical surfaces passing through regions of minimum voidage.

The three classes of models are shown schematically

220

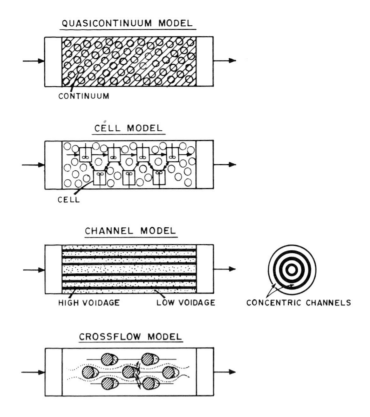

Figure 11.1. Models for describing the packed-bed reactor.

in Figure 11.1. We shall be primarily concerned with the quasicontinuum models in our treatment. Studies in fixed-bed reactor behavior using these reactor models have uncovered several significant features of this reactor, such as multiplicity of solutions, temperature runaway, and parametric sensitivity. Some of these aspects of fixed-bed behavior will also be considered. We shall make a brief reference to the cell model. The channel model, as yet inadequately developed, will not be considered at all. More recently, another model has been proposed (see Section 11.5.5) that divides the fluid into stagnant and flowing zones with exchange between the zones. This so-called crossflow model, sketched in Figure 11.1, is also not considered.

Of the reviews and books published so far on this subject, those by the following are particularly noteworthy: Beek (1962), Wilhelm (1962), Denbigh (1965), Levenspiel (1974), Smith (1956), Froment (1967, 1974), Hlavacek (1970), Karanth and Hughes (1974), Hlavacek and Votruba (1977), Froment and Bischoff (1979), and Butt (1980).

11.2. THE ROLE OF MIXING; FLOW IN PACKED REACTORS

11.2.1. Ideal and Nonideal Reactors: The Role of Mixing

Reactors in which the fluid elements travel in pistonlike motion without any backmixing of elements represent the ideal plug-flow limit. Our concern here is primarily with this class of reactors in which any mixing effects resulting in nonideality are accounted for by incorporating an axial diffusion parameter (see Section 11.2.2). At the other extreme we have the ideal fully mixed reactor, represented by the continuous stirred-tank reactor (CSTR) or the batch-operated stirred-tank reactor (STR).

The degree (and quality) of mixing therefore becomes an important parameter in reactor analysis. In any reactor the elements of the reactant fluid travel through it via different paths, so that the fluid leaving the reactor at any instant comprises elements of various ages; this gives rise to the concept of residence time distribution (RTD) or an age distribution function (E). The fraction of the stream with age between t and $t + dt$ is given by $E\,dt$, so that for the entire stream we have $\int_0^\infty E\,dt = 1$. The concept of RTD, explained at length in many books (e.g., Levenspiel, 1974), has now become an indispensable tool in the analysis of mixed flow.

Every pattern of mixing is characterized by a definite RTD, but every RTD does not correspond to a definite flow pattern. This is because mixing can occur early or late during the fluid's transit through the reactor. Thus an additional parameter, "earliness of mixing", also comes into the picture.

Mixing itself has been analyzed in terms of two distinct postulations: macromixing and micromixing. Macromixing is characterized by clumps of fluid elements, fully segregated from each other, flowing through the vessel. Micromixing, on the other hand, implies mixing at a molecular level where fluid packets or clumps no longer exist and there is no segregation.

Based on the foregoing concepts we shall now summarize a few important aspects of mixing (which can sometimes be exploited in reactor optimization) and then proceed to the question of flow in a packed reactor.

1. Macromixing is entirely characterized by the RTD of the system, whereas micromixing may be viewed as a transition from segregation by age to segregation by life expectancy (Weinstein and Adler, 1967; Nishimura and Matsubara; 1970; Kattan and Adler, 1972).

2. Reactor performance depends on kinetics, RTD, degree of segregation, and earliness of mixing. Thus segregation and late mixing improve conversion for $n > 1$ and decrease conversion for $n < 1$. The effect of various levels of micromixing on an nth-order reaction is negligible, but can be quite pronounced in multistep reactions.

3. A state of "maximum mixedness" (Zwietering, 1959) corresponding to complete micromixing (or no segregation) is assumed in the analysis of a fully mixed CSTR.

4. Various models of partial segregation corresponding to different levels of micromixing have been postulated. The model of Danckwerts (1958) defines an intensity of segregation (I) as the ratio of the variation in average ages of aggregates to variation in ages of the individual molecules. Curl (1963) has proposed a coalescence model, in analogy with the behavior of

dispersed-phase droplets, in which a coalescence parameter I is defined that represents the number of coalescences experienced by a droplet. For macromixing, $I = 0$; for micromixing, it equals ∞.

Other models proposed are the consecutive and parallel segregated–maximum-mixedness models (Weinstein and Adler, 1967; Nishimura and Matsubara, 1970), and the generalized recycle model (Dohan and Weinstein, 1973; Dudukovic and Lamba, 1975). As the names imply, the first two models are based on consecutive and parallel combinations of the segregated and maximum-mixedness states.

11.2.2. Flow in a Packed Reactor

The analysis of diffusion in both radial and axial directions is greatly facilitated by the use of the dimensionless Peclet numbers,

$$\text{Pe}'_{mr} = \frac{ud_p}{D_{er}}, \qquad \text{Pe}_{mr} = \frac{uR}{D_{er}} \qquad (11.1)$$

$$\text{Pe}'_{ml} = \frac{ud_p}{D_{el}}, \qquad \text{Pe}_{ml} = \frac{uL}{D_{el}} \qquad (11.2)$$

These can be written as $(L^2/D_e)/(L/u)$, which represents the ratio of the diffusion time constant to the average residence time. The magnitude of this ratio is obviously a measure of the role of diffusion. In the ideal plug-flow limit, $\text{Pe}'_{mr} \to 0$ and $\text{Pe}'_{ml} \to \infty$.

The velocity distribution in the direction perpendicular to the direction of flow, which causes radial diffusion, is also responsible for diffusion in the axial direction (Taylor diffusion). It can be shown from a frequency response analysis of fully mixed and plug-flow reactors that the longitudinal Peclet number for mass transfer has a value of 2 in the high-velocity regime (Kramers and Alberta, 1953; McHenry and Wilhelm, 1957). On the other hand, the value for Pe′ for heat transfer is approximately 10 (Votruba et al., 1972).

Transport in the radial direction occurs as a result of the sideways deflection (or sidestepping) of the fluid molecules flowing through a fixed bed of particles. Based on the three-dimensional tetrahedral structure of the packing, Ranz (1952) has shown that $\text{Pe}'_{mr} = 11.2$. On the other hand, it has been shown that $\text{Pe}'_{hr} \simeq 8\text{--}10$ for heat transfer.

The methods of estimating the various *effective transport* parameters mentioned above have already been described in Chapter 10 and summarized in Tables

10.4–10.7. Based on these properties we are now in a position to undertake the design of a multitubular fixed-bed reactor using the quasicontinuum model, with varying degrees of complexity.

11.3. MODELS FOR THE EXTERNAL FIELD: INTERPARTICLE TRANSPORT

11.3.1. Isothermal Plug-Flow Reactor

The following steady-state material balance can be written by considering an element dl of a packed tubular reactor:

$$u\left(\frac{dC_A}{dl}\right) + r_{VA} = 0 \tag{11.3}$$

with the boundary condition $C_A = C_{A0}$ at $l = 0$. Note that r_{VA} is based on reactor volume. For a first-order irreversible reaction Eq. 11.3 takes the following form:

$$u\left(\frac{dC_A}{dl}\right) + (1-f_B)k_v C_A = 0 \tag{11.4}$$

where k_v is now based on catalyst volume. The solution of this equation is

$$\ln\left(\frac{C_{Ae}}{C_{A0}}\right) = (1-f_B)k_v t = (1-f_B)\frac{k_v L}{u} \tag{11.5}$$

where t is a measure of the residence time and is given by L/u.

11.3.2. Nonisothermal Plug-Flow Reactor

The energy balance analogous to Eq. 11.3 for mass balance is given in Table 11.1. Upon introduction of the dimensionless variables†

$$c = \frac{C_A}{C_{A0}}, \qquad \tau = \frac{T}{T_0}, \qquad z' = \frac{l}{d_p} \tag{11.6}$$

the following equations result:

$$-\frac{dc}{dz'} = \frac{d_p}{uC_{A0}} r_{VA} \tag{11.7}$$

---

† Note that the method used in earlier chapters of representing a normalized variable by the symbol $\hat{}$ has not been adopted in the case of *reactor coordinates*.

$$\frac{d\tau}{dz'} + \frac{4Ud_p}{u\rho_g C_p d_t}(\tau - \tau_w) - \frac{(-\Delta H)d_p}{u\rho_g C_p T_0} r_{VA} = 0 \tag{11.8}$$

with the boundary conditions

$$\tau = 1, \quad c = 1 \quad \text{at} \quad z' = 0 \tag{11.9}$$

Equations 11.7–11.9 constitute a model of the non-isothermal plug-flow reactor. The integration of such a system of coupled nonlinear equations is not feasible by analytical methods, but numerical solution on a digital computer can be obtained by replacing the derivatives in the equation by their difference approximations. Solutions can also be obtained using an analog computer (Östergaard, 1965), or by the method of orthogonal collocation (Villadsen and Stewart, 1967; Finlayson, 1974).

11.3.3. Reactors with Axial Dispersion

The basic continuity equation is given in Table 11.1. In dimensionless form this becomes

$$\frac{1}{Pe_{ml}}\left(\frac{d^2c}{dz^2}\right) - \frac{dc}{dz} - \left(\frac{L}{uC_{A0}}\right)r_{VA} = 0 \tag{11.10}$$

where $Pe_{ml} = uL/D_{el}$ (see Eq. 11.2) and is based on the reactor length L and not on d_p, and $z = l/L$.

The boundary conditions can be obtained as follows. At $l = 0$, the condition is

$$-D_{el}\frac{dC_A}{dl} = u(C_{A0} - C_A) \tag{11.11}$$

At $l = L$, Danckwerts (1953) reasoned that dC_A/dl can be neither negative nor positive and should therefore be zero; thus

$$\frac{dC_A}{dl} = 0 \tag{11.12}$$

In dimensionless form, these conditions become

$$-\frac{1}{Pe_{ml}}\frac{dc}{dz} = (1-c) \text{ at } z = 0, \quad \frac{dc}{dz} = 0 \text{ at } z = 1 \tag{11.13}$$

Several other boundary conditions have been postulated (see Levenspiel and Bischoff, 1963), of which those of Wehner and Wilhelm (1956) appear to be the most rigorous. They postulated two zones, one before and

TABLE 11.1. Models for Interparticle Transport: The Reactor or External Field Equations

Model	Equation[a]	Boundary Conditions	Method of Solution
1. Isothermal plug-flow reactor	$u\dfrac{dC_A}{dl} + r_{VA} = 0$	$C_A = C_{A0}$ at $l = 0$	Equations can be solved analytically.
2. One-dimensional model: nonisothermal plug-flow reactor	$u\dfrac{dC_A}{dl} + r_{VA} = 0$ $u\rho_g C_p \dfrac{dT}{dl} + \dfrac{4U}{d_t}(T - T_w) - (-\Delta H)r_{VA} = 0$	$C_A = C_{A0}$ at $l = 0$ $T = T_0$ at $l = 0$	Equations require numerical solution. Since they constitute an initial value problem, digital solution presents no difficulty. The equations can also be solved on an analogue computer.
3. Isothermal reactor with axial dispersion	$D_{el}\dfrac{d^2C_A}{dl^2} - u\dfrac{dC_A}{dl} - r_{VA} = 0$	$-D_{el}\dfrac{dC_A}{dl} = u(C_{A0} - C_A)$ at $l = 0$ $\dfrac{dC_A}{dl} = 0$ at $l = L$	The equations constitute a boundary value problem. An analytical solution can be obtained for a first-order reaction. Numerical solution becomes necessary for nonlinear kinetic expressions, and an iterative procedure has to be followed to satisfy the boundary conditions.
4. Nonisothermal reactor with axial dispersion of heat and mass	$D_{el}\dfrac{d^2C_A}{dl^2} - u\dfrac{dC_A}{dl} - r_{VA} = 0$ $k'_{el}\dfrac{d^2T}{dl^2} - u\rho_g C_p\dfrac{dT}{dl} - \dfrac{4U}{d_t}(T - T_w)$ $\qquad\qquad + (-\Delta H)r_{VA} = 0$	$-D_{el}\dfrac{dC_A}{dl} = u(C_{A0} - C_A)$ at $l = 0$ $\dfrac{dC_A}{dl} = 0$ at $l = L$ $-k'_{el}\dfrac{dT}{dl} = u\rho_g C_p(T_0 - T)$ at $l = 0$ $\dfrac{dT}{dl} = 0$ at $l = L$	An iterative numerical solution is necessary.

5. Two-dimensional model: nonisothermal plug-flow reactor with radial gradients

$$D_{er}\left(\frac{\partial^2 C_A}{\partial r^2} + \frac{1}{r}\frac{\partial C_A}{\partial r}\right) - u\frac{\partial C_A}{\partial l} - r_{VA} = 0$$

$$k'_{er}\left(\frac{\partial^2 T}{\partial r^2} + \frac{1}{r}\frac{\partial T}{\partial r}\right) - u\rho_g C_p \frac{\partial T}{\partial l}$$
$$+ (-\Delta H)r_{VA} = 0$$

$\dfrac{\partial C_A}{\partial r} = 0$ at $l > 0, r = 0$

$\dfrac{\partial C_A}{\partial r} = 0$ at $l > 0, r = 0$

$C_A = C_{A0}$ at $l = 0, r > 0$

$\dfrac{\partial T}{\partial r} = 0$ at $l > 0, r = 0$

$-k'_{er}\dfrac{\partial T}{\partial r} = h_w(T - T_w)$ at $l > 0, r = R$

$T = T_0$ at $l = 0, r > 0$

The system of equations constitutes a coupled parabolic nonlinear partial differential equation. The solution is obtained numerically. An analog solution is possible by converting the system into difference equations.

6. Two-dimensional model with axial dispersion of heat and mass

$$D_{er}\left(\frac{\partial^2 C_A}{\partial r^2} + \frac{1}{r}\frac{\partial C_A}{\partial r}\right) + D_{el}\frac{\partial^2 C_A}{\partial l^2}$$
$$- u\frac{\partial C_A}{\partial l} - r_{VA} = 0$$

$$k'_{er}\left(\frac{\partial^2 T}{\partial r^2} + \frac{1}{r}\frac{\partial T}{\partial r}\right) - u\rho_g C_p \frac{\partial T}{\partial l}$$
$$+ k'_{el}\frac{\partial^2 T}{\partial l^2} + (-\Delta H)r_{VA} = 0$$

$-D_{el}\dfrac{\partial C_A}{\partial l} = u(C_{A0} - C_A)$ at $l = 0, r > 0$

$\dfrac{\partial C_A}{\partial l} = 0$ at $l = L, r > 0$

$\dfrac{\partial C_A}{\partial r} = 0$ at $l > 0, r = 0$

$-k'_{el}\dfrac{\partial T}{\partial l} = u\rho_g C_p(T_0 - T)$ at $l = 0, r > 0$

$\dfrac{\partial T}{\partial l} = 0$ at $l = L, r > 0$

$-k'_{er}\dfrac{\partial T}{\partial r} = h_w(T - T_w)$ at $l > 0, r = R$

$\dfrac{\partial T}{\partial r} = 0$ at $l > 0, r = 0$

An iterative numerical solution is necessary

---

[a]The equations are given in dimensional form and can be easily recast in nondimensional form.

one after the reaction zone, on the hypothesis that diffusion must also occur away from the inlet and exit planes of the reactor.

Various aspects of the model have been studied with particular reference to homogeneous reactors (Adler and Vortmeyer, 1963, 1964; Bischoff, 1968). For a first-order reaction the rate term (Lr_{VA}/uC_{AO}) in Eq. 11.10 will be replaced by $k_v(1-f_B)tc$, and the solution for a closed vessel is given by (Danckwerts, 1953; Wehner and Wilhelm, 1956).

$$c = (1 - x_A) =$$

$$\frac{4a \exp(\mathrm{Pe}_{ml}/2)}{\left[(1+a)^2 \exp(a\,\mathrm{Pe}_{ml}/2) - (1-a)^2 \exp\left(\frac{-a\,\mathrm{Pe}_{ml}}{2}\right)\right]}$$

(11.14)

where $a = 1 + 4tk_v(1+f_B)(1/\mathrm{Pe}_{ml})$. Levenspiel and Bischoff (1963) have plotted this equation in a useful form. It may be noticed from their plots, reproduced in Figure 11.2, that the effect of mixing is far more pronounced at higher than at lower conversions. In order to compute the reactor volume for non-first-order reactions, the charts prepared by Fan and Bailie (1960) may be used. For a consecutive reaction of the type $A \rightarrow B \rightarrow C$, the fractional decrease in the maximum yield of the intermediate B can be approximated by the value of the Peclet number itself provided that $\mathrm{Pe}_{ml} > 25$, that is, the reaction is only mildly nonideal (Tichacek, 1963).

The basic energy balance equations are given in Table 11.1. These can be written in dimensionless form as

$$\frac{1}{\mathrm{Pe}_{hl}} \frac{d^2\tau}{dz^2} - \frac{d\tau}{dz} + \frac{(-\Delta H)L}{u\rho_g C_p T_0} r_{VA} - \frac{4UL}{u\rho_g C_p d_t}(\tau - \tau_W)$$

(11.15)

with the boundary conditions

$$\frac{1}{\mathrm{Pe}_{hl}} \frac{d\tau}{dz} = \tau - 1 \quad \text{at} \quad z = 0$$

$$\frac{d\tau}{dz} = 0 \qquad \text{at} \quad z = 1$$

(11.16)

where

$$\mathrm{Pe}_{hl} = \frac{u\rho_g C_p L}{k'_{el}}$$

(11.17)

and represents the Peclet number for axial conduction. Note that here too, as in the case of mass diffusion, the Peclet number is based on reactor length.

Equations 11.10 and 11.15 with boundary conditions 11.13 and 11.16 constitute the model for a nonisothermal reactor with axial mixing and conduction. The numerical solution of these equations presents certain difficulties. For a concise description of the various methods of solution, reference may be made to Hlavacek and Votruba (1977). Our brief treatment will be confined to the method of difference approximations. These approximations for the first and second derivatives are written in the same manner as for the ideal reactor, but since the difference approximations for the second derivatives involve at least three points, concentrations and temperatures at two preceding points have to be known in order for calculations to be made for a new point. This creates difficulties at the outset of the computation. The concentration and temperature at the inlet as well as the next nodal point are required to start the solution. If both the concentration and the first derivative of concentration are known at the inlet, the solution can be started because, from a knowledge of the derivative at the inlet, the concentration at the first nodal point can be evaluated by using the relation

$$C_1 = C_0 + \left(\frac{dC}{dl}\right)\Delta l$$

(11.18)

This also applies to the determination of temperature.

In the present case, however, the derivative is known only at the outlet (and not at the inlet) together with a relation between the first derivative at the inlet and the inlet value. Hence, to start the solution, values of the inlet concentration and temperature are assumed. The first derivatives are then evaluated by using the boundary conditions at the inlet, and calculations are carried out through the whole bed. If the boundary conditions at the exit are satisfied, the solution is accepted; otherwise new values of inlet concentration and temperature are assumed and the calculations are repeated. To save computation time, initially calculations are done by taking large grid intervals. When the exit boundary conditions are approximately satisfied the calculations are repeated with smaller grid intervals. Sometimes the values of the derivatives at the exit are very sensitive to the values of concentration and temperature at the inlet. In that case the solution does not converge unless the inlet values are specified up to a large number of digits. Thus the addition of the second derivative in the model significantly increases the effort required for computation.

An alternative method of solving Eqs. 11.10 and 11.15 is by using an analog computer. Hofmann and

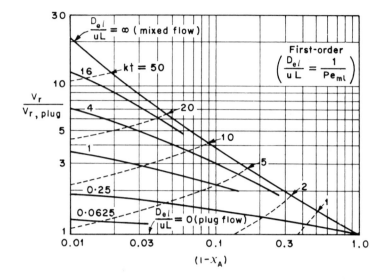

(a) FIRST-ORDER REACTION

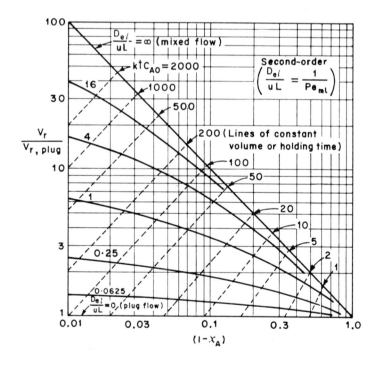

(b) SECOND-ORDER REACTION

Figure 11.2. Effect of longitudinal diffusion on conversion (Levenspiel, 1974).

Astheimer (1963) generated the concentration and temperatures profiles for a homogeneous reaction by analog computation.

11.3.4. Reactors with Radial Gradients: Two-Dimensional Models

There are two major factors that cause radial variation of temperature and concentration, namely, the presence of a velocity profile normal to the direction of flow, and heat generation and dissipation. In packed beds the hydrodynamic effect is not significant, and the radial temperature profile is determined largely by the rates of heat generation due to reaction and of heat transfer at the wall.

Formulation of Equations

The mass and heat balance equation can be obtained by considering the balances over an element of length dl of a thin cylindrical shell of thickness dr at a distance r from the axis. These are included in Table 11.1. The mass balance equation alone is not of any serious consequence for the purpose of design. In conjunction with its heat transfer analog, however, it provides a powerful basis for the rigorous design of a fixed-bed reactor.

These equations can be recast in dimensionless form with the following additional reduced coordinate:

$$\omega' = \frac{r}{d_p} \tag{11.19}$$

with the result

$$\frac{1}{Pe'_{mr}}\left(\frac{\partial^2 c}{\partial(\omega')^2} + \frac{1}{\omega'}\frac{\partial c}{\partial \omega'}\right) - \frac{\partial c}{\partial z'} - R_M r_{VA} = 0 \tag{11.20}$$

$$\frac{1}{Pe'_{hl}}\left[\frac{\partial^2 \tau}{\partial(\omega')^2} + \frac{1}{\omega'}\frac{\partial \tau}{\partial \omega'}\right] - \frac{\partial \tau}{\partial z'} + R_H r_{VA} = 0 \tag{11.21}$$

The boundary conditions would be

$$c = 1, \quad \tau = 1 \quad \text{at} \quad z' = 0, \quad \omega' > 0$$

$$\frac{\partial c}{\partial \omega'} = \frac{\partial \tau}{\partial \omega'} = 0 \quad \text{at} \quad z' \geqslant 0, \quad \omega' = 0 \tag{11.22}$$

$$\frac{\partial c}{\partial \omega'} = 0, \quad \frac{\partial \tau}{\partial \omega'} = (Bi)_h(\tau - \tau_w) \quad \text{at} \quad z' \geqslant 0,$$

$$\omega' = \omega'_{r_0} = \frac{r_0}{d_p}$$

where τ_w represents the value of τ at the wall, $(Bi)_h$ is the Biot number for heat transfer $(h_w d_p / 2k'_{er})$, and

$$R_M = \frac{d_p}{uC_{A0}}, \qquad R_H = \frac{d_p(-\Delta H)}{u\rho_g C_p T_0} \tag{11.23}$$

R_M and R_H are mass and heat transfer factors with the dimensions of reciprocal volume rate (i.e., cm³ sec/mol). Thus the terms $R_M r_{VA}$ and $R_H r_{VA}$ appearing in the mass and heat balance equations 11.20 and 11.21, respectively, are nondimensional, like the rest of the terms in these equations. The Peclet numbers Pe' are now based on the particle diameter d_p.

Numerical Solution

The set of equations 11.20 and 11.21 with the boundary conditions 11.22 can be solved by numerical methods. The novel graphical procedure suggested by Baron (1952) is cumbersome and therefore not recommended. A worked-out example is presented in Section 11.3.6. For a discussion of the various numerical methods, reference may be made to several theoretical treatments of chemical reactors, notably that of Finlayson (1971, 1974), and Hlavacek and Votruba (1977).

11.3.5. Semi-Two-Dimensional Models

Let us consider a reaction

The axial and radial concentration and temperature profiles of A for this reaction taking place in a fixed-bed reactor can be generated by using the two-dimensional model, that is, by solving Eqs. 11.20–11.22. From the parameter values listed in the example of Section 11.3.6 [which are the same as those used by Froment (1967) in his simulation of the phthalic anhydride reactor] radial concentration profiles were generated; the results are displayed in Figure 11.3. It will be noticed that there is practically no variation in the concentration of A. A similar conclusion was reached by McGreavy and Cresswell (1968), Hawthorn et al. (1968), and Zahner (1972).

On the other hand, radial variations in temperature are quite pronounced and cannot be neglected in any design procedure. It would therefore seem reasonable to develop a design procedure based on the existence of a radial temperature profile and a uniform radial concentration at any axial position.

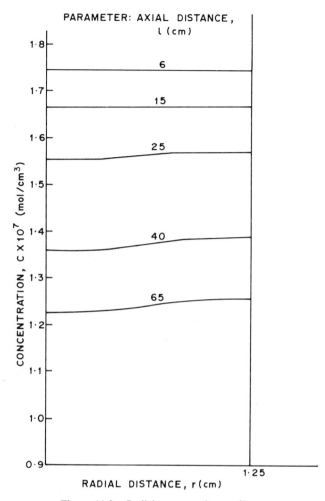

Figure 11.3. Radial concentration profiles.

The heat and mass balance equations for this simplified two-dimensional model can thus be written as

$$k'_{er}\left(\frac{\partial T}{\partial r^2} + \frac{1}{r}\frac{\partial T}{\partial r}\right) - u\rho_g C_p \frac{\partial T}{\partial l}$$

$$+ (-\Delta H)r_{VA}[C_A(l), T(r, l)] = 0 \qquad (11.24)$$

$$u\frac{dC_A}{dl} + r_{VA}(l) = 0 \qquad (11.25)$$

with the boundary conditions

$$
\begin{aligned}
l = 0, & \quad & C_A &= C_{A0} \\
l = 0, \quad r > 0, & \quad & T &= T_0 \\
l > 0, \quad r = 0, & \quad & \frac{\partial T}{\partial r} &= 0 \qquad (11.26) \\
l > 0, \quad r = r_0, & \quad & k'_{er}\frac{\partial T}{\partial r} &= h_w(T_w - T)
\end{aligned}
$$

It may be noted that the concentration term appearing in the heat balance equation is independent of radial position. The rate term can be expressed in two different ways, as shown here:

Semi-two-dimensional model 1 (STD-1):

$$\bar{r}_{VA} = \frac{1}{\pi r_0^2} \int_0^{r_0} (2\pi r)r_{VA}[C_A, T(r)]\, dr \qquad (11.27)$$

Semi-two-dimensional model 2 (STD-2):

$$\bar{r}_{VA} = r_{VA}(C_A, \bar{T}) \qquad (11.28)$$

Equation 11.27 obtains the rate at a given cross section by evaluating the average rate ($\bar{r}_{VA}$) over the cross section, while Eq. 11.28 uses the average temperature ($\bar{T}$) over the cross section, both at a given axial position.

The results of typical computations (see the example that follows) indicate that the computer time requirement for STD-2 is approximately 50% of that required for the two-dimensional model.

11.3.6. Example: Oxidation of *o*-Xylene to Phthalic Anhydride: Comparison of Various Models

Consider the oxidation reaction

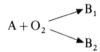

where A represents phthalic anhydride, carried out in a fixed-bed reactor. The reactions obey the following kinetic expressions:

$$r_{v1} = k_1 p_{O_2} p_A, \qquad r_{v2} = k_2 p_{O_2} p_A$$

The kinetic constants and the other parameters of the reactor are

k_1	$\exp(-27000/R_g T + 19.837)$
k_2	$\exp(-28600/R_g T + 18.980)$
$-\Delta H_1$	307000 cal/mol
$-\Delta H_2$	1090000 cal/mol
Superficial mass velocity	4684 kg/m² hr
Heat capacity	0.25 cal/g °C
o-Xylene feed partial pressure	0.00924 atm

Partial pressure of oxygen	0.208 atm
Diameter of tube	2.5 cm
Density of bed	1.3 g/cm^3
Radial Peclet number (heat)	5.25
Radial Peclet number (mass)	10.00
Wall heat transfer coefficient	134.0 kcal/cm^2 °C
Overall heat transfer coefficient	92.7 kcal/m^2 hr °C

Calculate the temperature and concentration profiles in the reactor using (1) the one-dimensional model, (2) the semi-two-dimensional model based on average temperature, (3) the semi-two-dimensional model based on average rates, and (4) the two-dimensional model. Assume that the inlet temperature is 631 °K.

Solution

1. One-dimensional model:

The reactor equations for the one-dimensional model can be written as

$$-\frac{dC}{dl} = J(C, T), \qquad \frac{dT}{dl} = F(C, T)$$

where

$$J(C, T) = (1 - f_B)\frac{r_v}{u}$$

$$F(C, T) = \frac{(-\Delta H)r_v}{u\rho_g C_p} - \frac{4U}{u\rho_g C_p d_t}(T - T_w)$$

$$r_v = (k_{p1} + k_{p2})p_{O_2}p_A$$

These equations constitute coupled nonlinear first-order ordinary differential equations that can be conveniently solved by the Runge–Kutta–Gill procedure.

In the present problem all the quantities involved in the evaluation of $J(C, T)$ and $F(C, T)$ are known. The fluid density may be taken to be constant and evaluated at 631 °K. The computed concentration and temperature profiles are given in Figures 11.4 and 11.5, respectively.

2. Semi-two-dimensional model based on average temperature:

The model equations are given by 11.24 and 11.25. To obtain a difference analog of Eq. 11.24 the

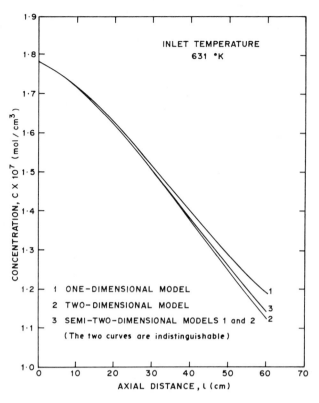

Figure 11.4. Comparison of the axial concentration profiles obtained from different models.

Crank–Nicholson method may be used. Equation 11.25 is then solved simultaneously by the simple forward difference scheme.

The average temperature required for the solution of the mass balance equation is evaluated as follows:

$$(\pi r_0^2)r_0 = \Pi(\Delta r)^2 \tfrac{1}{2}(T_{m,1} + T_{m,2})$$
$$+ \sum_{n=2}^{N-1} (n\,\Delta r)^2 - (n-1)^2(\Delta r)^2 \tfrac{1}{2}(T_{m,n} + T_{m,n+1})$$

which simplifies to

$$\overline{T} = \frac{1}{(N-1)^2}0.5(T_{m,1} + T_{m,2})$$
$$+ \sum_{n=2}^{N-1}(n-0.5)(T_{m,n} + T_{m,n+1})$$

The Thomas algorithm (Lapidus, 1962) is used to solve the set of algebraic equations involved in the solution of the heat balance equation. The resultant concentration and temperature profiles are included in Figures 11.4 and 11.5.

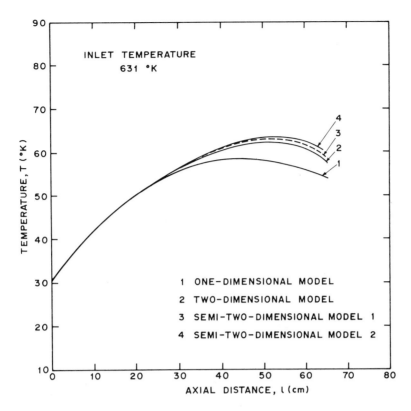

Figure 11.5. Comparison of the axial temperature profiles obtained from the different models.

3. Semi-two-dimensional model based on the average rate:

The difference approximations used for the model equations are the same as in the last case. The average rate appearing in the mass balance equation is obtained by a procedure similar to that followed for the evaluation of the average temperature at each cross section. The expression so obtained is

$$\bar{r}_v = \frac{1}{(N-1)^2} 0.5 [r_v(C_m, T_{m,1}) + r_v(C_m, T_{m,2})]$$

$$+ \sum_{n=2}^{N-1} (n-0.5)[r_v(C_m, T_{m,n}) + r_v(C_m, T_{m,n+1})]$$

The computed concentration and temperature profiles are included in Figures 11.4 and 11.5, respectively.

4. Two-dimensional model:

In the two-dimensional model both mass and energy balance equations are parabolic differential equations. These equations can be solved by means of the

Crank–Nicholson procedure. The Thomas algorithm is used to solve the simultaneous algebraic equations. The results of the computations are included in Figures 11.4 and 11.5.

The computer time required for STD-2 is found to be approximately 50 % of that for the two-dimensional model (TD), whereas STD-1 involves practically the same computational effort as TD. Hence STD-1 is of no practical importance, since it incorporates an approximation without presenting any advantage.

Figure 11.6 shows the radial temperature profiles computed from the TD and STD models. Clearly, when the profiles are not severe, the results of the STD models are the same as those of the TD model.

11.4. INTERNAL FIELD EQUATIONS: EVALUATION OF THE GLOBAL RATE

The rate of reaction term appearing in the models discussed in the previous sections was the global rate based on the unit volume of the reactor. The intrinsic reaction rate per unit reactor volume (i.e., the rate when

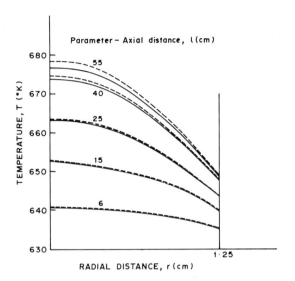

Figure 11.6. Comparison of radial temperature profiles (inlet temperature 631°K) obtained from the two-dimensional and semi-two-dimensional (Std– – –); models. (Curves for STD-1 and STD-2 are very close.)

no external or internal gradients are present) is given by

$$r_{vA} = \rho_c(1 - f_B)r_{wA}(C_A, T) \qquad (11.29)$$

where r_{wA} is the intrinsic rate (mol/sec g catalyst). We shall now impose diffusional limitations arising out of the presence of solid particles and derive expressions for the global rate for use in the various model equations.

11.4.1. External (Interphase) Gradients Only

In the case of external gradients only, the catalyst pellet is at a uniform temperature T_s (corresponding to the catalyst surface) and the concentration of the reactant is also uniform within the pellet. The steady-state heat and mass balance equations for the pellet can be written as

$$k_g A_p(C_A - C_{As}) = V_p \rho_c r_{wA}(T_s, C_{As}) \qquad (11.30)$$

where A_p and V_p are the area and volume, respectively, of a single pellet. Similarly,

$$-h_{fp} A_p(T - T_s) = \rho_c V_p(-\Delta H)r_{wA}(T_s, C_{As}) \qquad (11.31)$$

In Eqs. 11.30 and 11.31 C_{As} and T_s are the two unknowns which can be evaluated by solving, the equations simultaneously. Then the global rate can be

written as[†]

$$\rho_c r_{wA} = r_{vA} = \frac{A_p}{V_p} k_g(C_A - C_{As}) \qquad (11.32)$$

11.4.2. Internal Gradients Only

In the case of internal gradients only the fluid concentration and temperature at the surface of the pellet are identical with those of the fluid, but vary with position within the pellet. To evaluate the global rate it therefore becomes necessary to solve the mass and heat balance equations for the pellet. These equations have been presented (Eqs. 4.4 and 4.5) and their solutions discussed in Chapter 4. The solutions yield the concentration and temperature profiles in the pellet. The global rate can then be expressed as

$$r_{vA}(C_A, T) = \rho_c \varepsilon r_{wA}(C_A, T) \qquad (11.33)$$

where

$$\varepsilon = -\frac{D_{eA} A_p (dC_A/dr)_{r=R}}{V_p r_{wA}(C_A, T)\rho_c} \qquad (11.34)$$

It should be noted that Eqs. 4.4 and 4.5 are nonlinear and hence require numerical solution. Since the fluid concentration and temperature that appear as boundary conditions in these equations vary from point to point in the reactor, they will have to be solved at every mesh point in the reactor for the evaluation of the global rate while integrating the heat and mass balance equations for the fluid phase (i.e., for the reactor as a whole) by means of any of the models described in Section 11.3.

The calculations can be considerably simplified if a uniform value of ε can be assumed for the entire reactor. This is possible for a thermoneutral reaction or for one with a low heat effect, provided the reaction is first order.

11.4.3. Both Internal and External Gradients

When there are both internal and external gradients the heat and mass balance equations for the pellet are the same as for the previous case. However, the boundary conditions should be modified to include the effect of external gradients (Eq. 6.29). The solution of Eqs. 4.4 and 4.5 would now be more involved because the concentration C_{As} and temperature T_s are not known initially and are required to be assumed before starting the solution. After the solution is obtained, the global

---

[†] The term $\dfrac{A_p}{V_p}$ is also denoted by a (area per unit volume).

rate is given by Eq. 11.33. For further details regarding the equations for estimating the effectiveness factor when both internal and external gradients are present, see Chapter 6.

11.4.4. Summary of Models and Rate Equations

A summary of all the models with the corresponding equations and their methods of solution is presented in Table 11.1, and the rate equations to be used with these models under various diffusional limitations are summarized in Table 11.2. Thus, if in a particular design a nonisothermal ideal plug-flow model is to be used and both internal and external resistances in the solid phase are to be accounted for, the corresponding equations would be 2 of Table 11.1 and 4 of Table 11.2.

11.4.5. Simplified (Reformulated) Reactor Equations

The mass and heat balance equations for the pellet in the presence of interphase gradients (Eqs. 11.30 and 11.31) get modified to

$$k_g A_p (C_A - C_{As}) = \varepsilon V_p \rho_c r_{wA}(C_{As}, T_s) \quad (11.35)$$

$$-h_{fp} A_p (T - T_s) = \varepsilon V_p \rho_c (-\Delta H) r_{wA}(C_{As}, T_s) \quad (11.36)$$

These equations can be reduced to a single equation, since C_{As} and T_s are interrelated by the expression

$$k_g (C_A - C_{As})(-\Delta H) = h_{fp}(T_s - T) \quad (11.37)$$

Thus the evaluation of the global rate for given values of C_{As} and T_s involves the solution of a nonlinear implicit equation by a trial-and-error procedure. The reactor equations can, however, be reformulated to eliminate trial and error (Rajadhyaksha, 1976).

Equations 11.35 and 11.36 can be recast in the following form:

$$C_A = C_{As} + \psi(C_{As}, T_s) \quad (11.38)$$

$$T = T_s - \lambda \psi(C_{As}, T_s) \quad (11.39)$$

where

$$\psi(C_{As}, T_s) = \frac{V_p \rho_c r_{wA}(C_{As}, T_s)}{k_g A_p}$$

$$\lambda = \frac{k_g(-\Delta H)}{h_{fp}} \quad (11.40)$$

On substituting these equations for C_A and T in Eqs. 3 and 4 of Table 11.1 for the quasihomogeneous plug-flow nonisothermal model and rearranging, the following expressions result:

$$\frac{dC_{As}}{dl} = \frac{\left(1 - \lambda \dfrac{\partial \psi}{\partial T_s}\right) G(C_{As}, T_s) - \dfrac{\partial \psi}{\partial T_s} H(C_{As}, T_s)}{\lambda \dfrac{\partial \psi}{\partial T_s}\dfrac{\partial \psi}{\partial C_{As}} + \left(1 - \lambda \dfrac{\partial \psi}{\partial T_s}\right)\left(1 + \dfrac{\partial \psi}{\partial C_{As}}\right)}$$

$$(11.41)$$

$$\frac{dT_s}{dl} = \frac{\lambda \dfrac{\partial \psi}{\partial C_{As}} G(C_{As}, T_s) + \left(1 + \dfrac{\partial \psi}{\partial C_{As}}\right) H(C_{As}, T_s)}{\lambda \dfrac{\partial \psi}{\partial T_s}\dfrac{\partial \psi}{\partial C_{As}} + \left(1 - \lambda \dfrac{\partial \psi}{\partial T_s}\right)\left(1 + \dfrac{\partial \psi}{\partial C_{As}}\right)}$$

$$(11.42)$$

where

$$G(C_{As}, T_s) = \frac{\rho_c(1 - f_B)\varepsilon r_{wA}(C_{As}, T_s)}{u} \quad (11.43a)$$

$$H(C_{As}, T_s) = \frac{\rho_c(1 - f_B)(-\Delta H)\varepsilon r_{wA}(C_{As}, T_s)}{u\rho_g C_p}$$

$$-\frac{4U}{d_t}\frac{[T_s - \lambda \psi(C_{As}, T_s) - T_w]}{u\rho_g C_p} \quad (11.43b)$$

The initial conditions are given by Eqs. 11.35 and 11.36, in

$$k_g A_p (C_{A0} - C_{As0}) = \varepsilon r_{wA}(C_{As0}, T_{s0}) V_p \rho_c \quad (11.44)$$

$$-h_{fp} A_p (T_{s0} - T_0) = (-\Delta H)\varepsilon r_{wA}(C_{As0}, T_{s0}) V_p \rho_c$$

$$(11.45)$$

where C_{As0} and T_{s0} are the values of C_{As} and T_s at the inlet. These can be obtained by solving Eqs. 11.44 and 11.45 by trial and error.

Equations 11.41 and 11.42 can now be readily integrated by the usual numerical techniques for first-order ordinary differential equations (which do not involve trial and error). The partial derivatives $\partial \psi / \partial T_s$ and $\partial \psi / \partial C_{As}$ can be obtained by differentiating the function ψ numerically. If, however, the effectiveness factor can be assumed to be unity, the rate term in Eq. 11.40 for an nth-order reaction becomes $\rho_c(1 - f_B) A_f \exp(-E/R_g T_s) C_{As}^n$, leading to the following

TABLE 11.2. Models for the Global Reactions Rate: The Pellet or Internal Field Equations

Model	Model Equation	Boundary Conditions	Method of Solution
1. No transport limitations: quasihomogeneous model	$r_v = \rho_c r_w(C_A, T)$	—	The rate can be calculated explicitly.
2. Interphase gradients only: heterogeneous model	$k_g A_p(C_A - C_{As}) = V_p \rho_c r_w(C_{As}, T_s)$ $-hA_p(T - T_s) = V_p \rho_c(-\Delta H)r_w(C_{As}, T_s)$	—	An iterative solution of a nonlinear algebraic equation is involved.
3. Interphase gradients only: heterogeneous model a. Intraphase concentration gradients only	$D_{eA}\left(\dfrac{d^2 C_A}{dr^2} + \dfrac{S}{r}\dfrac{dC_A}{dr}\right) - \rho_c r_w = 0$ $S = 0$ for slab, 1 for cylinder, 2 for sphere	$C_A = C_{As}$ at $r = R$ $\dfrac{dC_A}{dr} = 0$ at $r = 0$	Analytical solution can be obtained for first-order reactions. Numerical solution is necessary for nonlinear kinetics. Since the equations form a boundary value problem, and iterative solution is involved.
b. Both concentration and temperature gradients	$D_{eA}\left(\dfrac{d^2 C_A}{dr^2} + \dfrac{S}{r}\dfrac{dC_A}{dr}\right) - \rho_c r_w = 0$ $k'_{er}\left(\dfrac{d^2 T}{dr^2} + \dfrac{S}{r}\dfrac{dT}{dr}\right) + (-\Delta H)\rho_c r_w = 0$	$C_A = C_{As}$ at $r = R$ $\dfrac{dC_A}{dr} = 0$ at $r = 0$ $T = T_s$ at $r = R$ $\dfrac{dT}{dr} = 0$ at $r = 0$	A nonlinear boundary value problem is involved, and iterative numerical solution is necessary.
4. Both interphase and intraphase gradients: heterogeneous model	$D_{eA}\left(\dfrac{d^2 C_A}{dr^2} + \dfrac{S}{r}\dfrac{dC_A}{dr}\right) - \rho_c r_w = 0$ $k'_{er}\left(\dfrac{d^2 T}{dr^2} + \dfrac{S}{r}\dfrac{dT}{dr}\right) + (-\Delta H)\rho_c r_w = 0$	$\dfrac{dC_A}{dr} = 0$ at $r = 0$ $-D_{eA}\dfrac{dC_A}{dr} = k_g(C_A - C_{As})$ at $r = R$ $\dfrac{dT}{dr} = 0$ at $r = 0$ $-k'_{er}\dfrac{dT}{dr} = h(T - T_s)$ at $r = R$	An iterative numerical solution is necessary.

analytical expressions for $\partial \psi / \partial C_{As}$ and $\partial \psi / \partial T_s$:

$$\frac{\partial \psi}{\partial C_{As}} = \frac{V_p \rho_c (1 - f_B) A_f \exp(-E/R_g T_s) n C_{As}^{n-1}}{k_g A_p}$$

$$(11.46)$$

$$\frac{\partial \psi}{\partial T_s} = \frac{V_p \rho_c (1 - f_B) A_f \exp(-E/R_g T_s) C_{As}^n}{k_g A_p} \left(\frac{E}{R_g T_s^2} \right)$$

$$(11.47)$$

Rajadhyaksha (1976) has calculated the concentration and temperature profiles using (1) the conventional reactor equations, and (2) the reformulated reactor equations. The profiles obtained by the two methods coincide almost completely with each other, as shown in Figure 11.7.

It is clear from the forgoing example that by invoking some approximations dictated by practical considerations, it is possible to simplify the rigorous model without serious loss of accuracy. Some of the approximations usually invoked include isothermality

of the catalyst pellet, negligible interphase mass transfer resistance $[(Bi)_m \gg 1]$, approximation of the concentration at the center of the pellet by the concentration corresponding to a pseudo-first-order reaction, and approximation of (k_{vs}/k_v) by the first-order term in the Taylor series expansion. In addition to these assumptions, Lee (1981) has invoked the generalized effectiveness factor of Bischoff (1965) to obtain the reactor point effectiveness factor, which is then used in the macroscopic equations describing the chemical reactor. The approximations involved are valid for most practical situations, and the method of Lee can be used for the approximate design and analysis of fixed-bed catalytic reactors.

It is possible to convert the two-dimensional model into an almost equally rigorous one-dimensional model if the radial temperature and concentration profiles are represented by equations whose parameters are functions of axial position. Such an approach has been followed by Ahmed and Fahien (1980a), and the resulting model has been used to simulate the SO_2

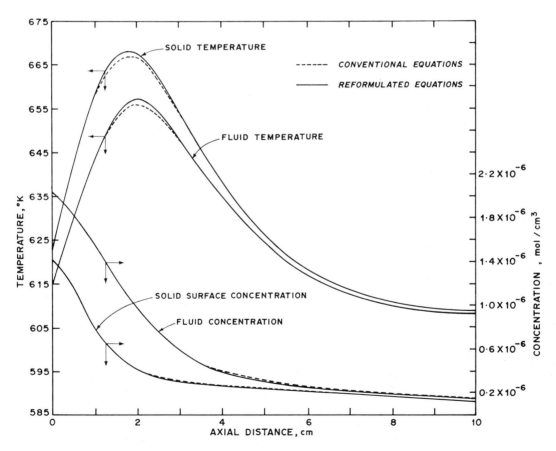

Figure 11.7. Comparison of concentration and temperature profiles computed by the conventional and reformulated equations (Rajadhyaksha, 1976).

oxidation reactor. Alternatively, it is also feasible to consider the radial variation of D_{er}, k'_{er}, and u (Ahmed and Fahien, 1980b). This approach is equivalent to the third approach mentioned in Section 10.1.1, and avoids the use of a separate heat transfer coefficient at the wall, h_w. The model applied to the data of SO_2 oxidation gave almost quantitative predictions.

11.5. SOME CONSIDERATIONS IN THE SELECTION OF A MODEL

Since fixed-bed reactors are almost without exception nonisothermal in character, we shall confine the question of model selection to those models for which the continuity and conservation equations have to be solved simultaneously.

The one-dimensional plug-flow model involves solution of an initial value problem governed by ordinary differential equations. The addition of axial mixing converts it into a boundary value problem. The two-dimensional models involve solution of partial differential equations. If all these models are assumed to be quasihomogeneous, the global rate is obtained by direct substitution of the values of temperature and concentration in the rate expression at each mesh point. On the other hand, for a heterogeneous model involving interphase gradients, implicit equations have to be solved at each mesh point, whereas for one involving intrapellet gradients solution of a nonlinear boundary value problem governed by two coupled differential equations at every mesh point is involved. Presence of both interphase and intrapellet gradients makes the evaluation of the global rate still more complicated.

11.5.1. Models Involving Axial Mixing

The effect of axial mixing is generally negligible for values of L/d_p greater than 50. This condition is usually if not always satisfied in industrial reactors; this is also true for axial conduction. (A mathematical criterion for neglecting axial mixing has already been presented in Table 9.1.) It is instructive to note that the computations of Young and Finlayson (1973) based on the data of Carberry and Wendel (1963) showed that inclusion of radial and axial diffusion was necessary to obtain good agreement with experimental results. It would therefore be dangerous to neglect axial diffusion without first gauging its role through preliminary calculations.

11.5.2. Heterogeneous Models

Since the intraparticle problem involves solution of a nonlinear boundary value problem at every mesh point, an extensive computational program becomes necessary. Three simpler alternatives are possible in this situation.

1. Assume that the entire heat transfer resistance is located in the film at the catalyst surface and that the pellet is isothermal (McGreavy and Thornton, 1970 a,b). Equations for calculating the effectiveness factor by this method have already been derived in Section 4.3.

2. Use an approximate procedure to solve the heat and mass balance equations. Thus the technique employing an effective reaction zone (Cresswell and Paterson, 1971), can be used.

3. Use an empirical expression relating the effectiveness factor to system parameters. Liu (1970) has solved the reactor simulation equations using such correlations.

11.5.3. Radial Gradients: One-Dimensional vs. Two-Dimensional Models

Since the thermal conductivity of industrial catalysts is usually low, the one-dimensional assumption is violated in the majority of systems. A simple way to check the validity of the one-dimensional model is to analyze the same system by both models and compare the results. This exercise, however, is not very straightforward, since the two models involve different sets of parameters (h_e in one case and h_w and k'_{er} in the other). The following relationship between these parameters (Froment, 1967):

$$\frac{1}{h_e} = \frac{1}{h_w} + \frac{r_0}{4k'_{er}} \tag{11.48}$$

provides a common basis for comparison since it builds up h_e from h_w and k'_{er}. As already pointed out in Section 11.3.5, the effect of radial concentration gradients is usually negligible. Other equations as alternatives to 11.48 are also available: $1/h_e = 1/h_w + D_{er}/8k'_e$ (Beek and Singer, 1951), and $1/h_e = 1/h_w + D_{er}/12.266k'_{er}$ (Crider and Foss, 1965). In all these equations the overall heat transfer coefficient U can be introduced in place of h_e if a term for control fluid heat transfer ($1/h_C$) is added on the right-hand side.

In the one-dimensional model the rate at any cross section is evaluated at the average temperature.

McGreavy and Turner (1970) have evaluated this discrepancy and observed that the ratio of the average rate to the rate at the average temperature lies between 1.1 and 1.3 for the system investigated by them.

It can generally be said that for the majority of systems the one-dimensional plug-flow model satisfactorily predicts the performance of the reactor. Only in cases where severe spots occur does the two-dimensional or the semi-two-dimensional model present an advantage (see Froment, 1967; Paterson and Cresswell, 1970). The concentrations and temperatures predicted by the one-dimensional model in such cases are usually low. Since the one-dimensional model does not give any information about the detailed concentration and temperature distributions, the ultimate design calculation may be carried out according to the two-dimensional model; but for reactor optimization and transient studies the one-dimensional model is most commonly used.

Examples of Simulation

The results of reactor modeling for a number of industrially important systems reported in the literature are summarized in the Table 11.3. These include both adiabatic and multitubular nonadiabatic reactor systems. The former will be considered in Chapter 12.

11.5.4. The Relative Importance of Kinetic and Transport Parameters

The sensitivity of the performance predicted by reactor models to various model parameters has been examined by different investigators. Froment (1967) showed that with a small change in the value of the effective thermal conductivity, there can be a significant change in the predicted performance. Richarz and Lattmann (1968) and Polidor et al. (1970) observed that the reactor performance can also be sensitive to kinetic parameters.

Rajadhyaksha (1976) simulated a reactor for the hydrogenation of nitrobenzene to aniline. He estimated the actual uncertainty in the experimental measurement of the model parameters. The uncertainty in the wall heat transfer coefficient was estimated from the scatter in experimental data, while that in the kinetic parameters was estimated by obtaining the joint confidence region. The computations indicate that the uncertainty in the predicted performance due to that in kinetic parameters can be very significant, whereas that due to the wall heat transfer coefficient is relatively unimportant. Thus the rate measurements in laboratory reactors (as described in Chapter 9) should be

carried out with a high degree of precision, and there is really nothing like an "engineering accuracy" in these experiments.

11.5.5. Alternative Models of Dispersion in Packed-Bed Reactors

In the models considered thus far, the dispersion phenomenon has been accounted for by appropriately incorporating the axial and radial Peclet numbers for mass and heat. While this way of representing dispersion in packed reactors appears quite sound, the resulting equations are invariably rather cumbersome in that they involve solution of two-point boundary value problems. Also, this class of models predicts that the injected pulse of tracer spreads at an infinite velocity in all directions, which has never been observed experimentally. Chiefly because of these disadvantages, a new class of models that requires the solution of only initial value problems has been proposed. The simple mixing-cell models exemplify this situation.

In the simplest case of this class of models, the bed is modeled as an interconnected set of perfectly mixed regions. The model gives rise to a recurrence relation rather than to differential equations. In the limiting case of a large number of such cells, or when the size of each cell decreases, the difference equations reduce to differential equations for the plug-flow model. Clearly, dispersion cannot be described by this simple model. In addition to the simply connected set of cells, if suitable recycle flows between successive cells are incorporated, the model becomes capable of describing axial dispersion. Although this model has the advantage of being an initial value problem, in the limiting case of a large number of cells this advantage is clearly lost.

More recently, Hinduja et al. (1980) have proposed an alternative class of models (the so-called crossflow models) based on the idealization of the flow pattern in the bed. They classified the whole void region into stagnant and flowing parts. The stagnant parts correspond to the wakes of packing elements, and dispersion is introduced by assuming that there is an exchange of fluid between the stagnant and flowing regions throughout the bed. The physical situation gives rise to first-order differential equations as against the recurrence relations in mixing-cell models and second-order differential equations in the Fickian diffusion models. The models are described as an initial value problem and avoid all the complications inherent in the existing models. Further, this class of models predicts that the tracer spreads at a finite bounded velocity.

In the following section, despite the advantages of

TABLE 11.3. Some Examples of Fixed-Bed Reactor Simulation

System	Type of Reactor	Model Used	Special Remarks	Reference
Dehydrogenation of methyl cyclohexane to toluene	Nonadiabatic	Two-dimensional quasihomogeneous plug-flow model	Predictions are compared with the experimental results.	Hawthorn et al. (1968)
Synthesis of methanol	Multibed adiabatic	One-dimensional quasihomogeneous plug-flow model	Complex kinetic expression is used; optimization study is presented.	Bakemier et al. (1970)
Synthesis of methanol	Multibed adiabatic	One-dimensional plug-flow model	Predictions are compared with experiments.	Cappelli et al. (1972)
Oxidation of sulfur dioxide	Multibed adiabatic	One-dimensional plug-flow model	Reactor optimization methods are compared.	Chartland and Crowe (1969)
Dehydrogenation of ethyl benzene	Single-bed adiabatic	One-dimensional quasihomogeneous plug-flow model	A scheme of six reactions is considered; results are compared with the experiment.	Sheel and Crowe (1969)
Hydrochlorination of acetylene	Nonadiabatic	Two-dimensional quasihomogeneous with axial mixing		Yablonski et al. (1968)
Hydrogenation of benzene	Nonadiabatic	One-dimensional plug-flow model	Predictions are compared with experimental results.	Pexidr et al. (1970).
Reduction of nitrobenzene	Nonadiabatic	One-dimensional plug-flow model	Predictions are compared with experiments.	Rajadhyaksha (1976)
Synthesis of ammonia	Multitubular with continuous heat exchange	Quasihomogeneous plug-flow model with feed exchanging heat with reacting gases	Predictions of the model are compared with plant data.	Baddour et al. (1965)
Oxidation of hydrocarbon	Multitubular	Two-dimensional heterogeneous model	Nonporous catalyst used. The predictions from one- and two-dimensional homogeneous and heterogeneous models are compared.	De Wasch and Froment (1971)

238

Reaction	Reactor type	Model	Remarks	Reference
Oxidation of o-xylene	Nonadiabatic	One-dimensional homogeneous plug-flow model	Effect of uncertainty in model parameters is investigated.	Froment (1967)
Oxidation of o-xylene	Nonadiabatic	Two-dimensional heterogeneous plug-flow model	Effect of uncertainty in model parameters is investigated.	Cresswell and Paterson (1970)
Oxidation of o-xylene	Multitubular	Quasihomogeneous plug-flow model	Influence of feed composition, wall temperature on hot spot, and sensitivity are reported.	Van Welsenaere and Froment (1970)
Oxidation of o-xylene	Multitubular	Two-demensional homogeneous model with porosity and velocity profiles taken into account	Predictions are compared with experimental results.	Lerou and Froment (1977)
Dehydrogenation of 1-butene to butadiene	Adiabatic reactor diluted with inert solids	One-dimensional heterogeneous model with heat exchange between gas and solid particles properly accounted	The kinetics of the main reaction along with the deactivation step is experimentally determined. The intraparticle resistances were incorporated and the industrial reactor was simulated and optimized.	Dumez and Froment (1976)
Oxidation of carbon monoxide	Single-bed adiabatic	Heterogeneous axial dispersion model	Steady states as well as transient experiments are performed. Transient perturbations include step changes in concentration and temperature. The model used is found to adequately represent steady-state behavior.	Sharma and Hughes (1979)
Oxidation of ethylene	Single-bed adiabatic	A modified two-phase cell model including axial dispersion and axial bed conduction	Experimental data are analyzed in terms of the cell model to extract the heat and mass transfer parameters under reaction conditions. The values thus obtained are considerably different from those under non-reactive conditions.	Paspek and Varma (1980)

the crossflow model, for reasons of conceptual simplicity the cell model is described in some detail.

The Finite-Stage Model

The finite-stage model is based essentially on building up material and heat balance equations for a one-dimensional model and extending them to the more realistic two-dimensional model. The following combinations of cells are possible in the absence of radial profiles: (1) an array of stirred cells with no backflow; (2) an array of stirred cells with axial backflow; (3) an array of stirred cells with gas–solid heat and mass transfer; (4) an array of stirred cells with gas–solid heat and mass transfer and axial backflow, and (5) an array of interconnected mixers.

We consider the third combination of cells (with no transport limitations). Equation 11.10 can be recast to give (with $Pe'_{ml} = 2$)

$$\frac{1}{2}\frac{\partial^2 c}{\partial z'^2} - \frac{\partial c}{\partial z'} = \frac{\partial c}{\partial \hat{t}} \tag{11.49}$$

where $\hat{t}$ is a dimensionless time given by

$$\hat{t} = \frac{t}{d_p/u}$$

This is a one-dimensional equation that includes axial diffusion but neglects radial diffusion, contrary to the simplified form used in the differential balance model, which neglects axial diffusion but includes radial diffusion. For details of derivation and analysis for various combinations of cells and their interconnections, reference may be made to Amundson (1969), Kunii and Furusawa (1972), Vanderveen et al. (1968), Sinkule et al. (1974), Olbrich et al. (1972), and Levic et al. (1967).

The performance of a packed bed can be regarded as equivalent to that of a series of stirred vessels of one-particle diameter each. The residence time in each of these well-mixed stages is given by d_p/u. The material balance for any stage in such a model, bounded by two parallel planes at distances id_p and $(i-1)d_p$ from the inlet, is given by

$$Q(C_{i-1} - C_i) = V\frac{dC_i}{dt} \tag{11.50}$$

Since $V/Q = d_p/u$ = average residence time,

$$C_{i-1} - C_i = \frac{dC_i}{d\hat{t}} \tag{11.51}$$

In extending the treatment to the two-dimensional case (Deans and Lapidus, 1960; Deans, 1963), a single stage i, j is considered to be bounded by two horizontal planes at distances id_p and $(i-1)d_p$ from the inlet end and two concentric disks of radius jd_p and $(j-1)d_p$. Thus, by defining both the vertical and radial coordinates in multiples of the particle diameter, a unit stage is defined that corresponds to the voids associated with a single particle.

A closer examination shows that the radial bounding surfaces are actually Kjd_p and $K(j-1)d_p$ where K represents the multiple of half-particle diameters that the fluid has to sidestep in the bed. If M is the number of stages in a radial row, then

$$2Md_p K = d_t \tag{11.52}$$

or

$$K = \frac{d_t}{2Md_p}$$

In the regular geometric arrangement shown in Figure 11.8, consecutive rows of particles are assumed to be in staggered orientation. In this offset arrangement, the radial multiple j takes on the values $\frac{1}{2}, \frac{3}{2}, \frac{5}{2}, \ldots, M$ when i is an odd integer, and the values $1, 2, 3, \ldots, M$ when i is an even integer.

In the two-dimensional model the feed concentration to the i, jth stage may be arbitrarily designated $\phi_{i-1,j}$ [since the ith horizontal position must emanate from the $(i-1)$th position] so that the material balance equation becomes

$$\tilde{\phi}_{i-1,j} - C_{i,j} = \frac{dC_i}{d\hat{t}}\, dC_{i,j} \tag{11.53}$$

From geometric considerations Deans and Lapidus (1960) have derived the following expression for estimating $\tilde{\phi}_{i-1,j}$:

$$\tilde{\phi}_{i-1,j} = \frac{(j-3/4)C_{i-1,j-1/2} + (j-1/4)C_{i-1,j+1/2}}{(2j-1)} \tag{11.54}$$

Thus Eqs. 11.53, 11.54 together replace the material balance equation 11.51 of the one-dimensional model.

By a similar procedure an energy balance can be written for the ith stage for the one-dimensional case and the development extended to the two-dimensional model for the i, jth stage. The final expressions obtained are

$$\tilde{A}_{i-1,j} - \tau_{i,j} = \tilde{B}\frac{dT_{i,j}}{d\hat{t}} \tag{11.55}$$

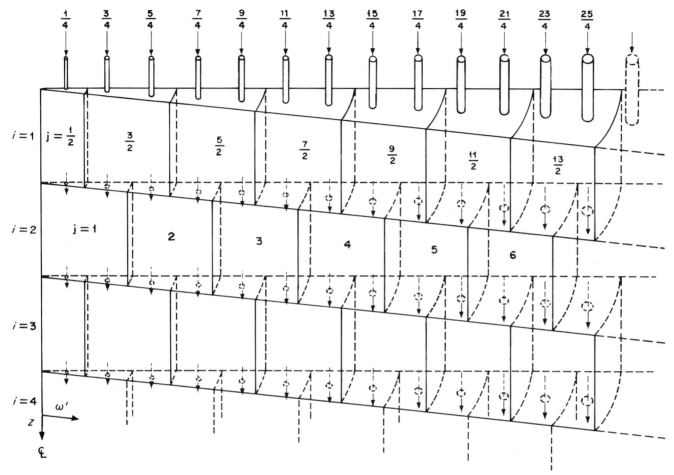

Figure 11.8. Schematic representation of the arrangement of stages and the configuration of the flow pattern in the basic stage model (Deans and Lapidus, 1960).

where

$$\tilde{A}_{i-1,j} = \frac{(j-3/4)\tau_{i-1,j-1/2} + (j-1/4)\tau_{i-1,j+1/2}}{(2j-1)}$$

(11.56)

and

$$\tilde{B} = 1 + \left(C_{ps}\rho_c \frac{1-f_B}{C_p\rho_g f_B}\right)$$

In the foregoing paragraphs the development of the model was explained for a nonreactive system. This model can now be extended to a reactive system (fixed-bed reactor) on the assumption of (1) constant density, (2) constant property values, and (3) absence of pore diffusion and film transport. A coupling term will have to be introduced in Eqs. 11.53 and 11.55 to account for chemical reaction. The final equations obtained by Deans and Lapidus (1960) for a first-order reaction are

$$\tilde{\phi}_{i-1,j} - C_{i,j}(1 + \hat{k}e^{-\alpha_0\tau_{i,j}}) = \frac{dC_{i,j}}{d\hat{t}}$$

(11.57)

$$\tilde{A}_{i-1,j} - \tau_{i,j} + \lambda' C_{i,j}\hat{k}e^{-\alpha_0\tau_{i,j}} = \tilde{B}\frac{d\tau_{i,j}}{d\hat{t}}$$

(11.58)

where

$$\hat{k} = \frac{d_p A_f}{u}, \qquad \alpha_0 = \frac{E}{RT_0}, \qquad \lambda' = \frac{(-\Delta H)C_0}{C_p\rho_g T_0}$$

For $j = M$, Eq. 11.58 must include the effect of heat transfer at the wall. By a treatment entirely analogous to that for a nonreactive system, it can be shown that (assuming St_w to be constant)

$$\frac{\tilde{A}_{i-1,M} + St_w\tau_{wi}}{1 + St_w} - \tau_{i,M} + \left(\frac{\lambda'}{1+St_w}\right)C_{i,M}\hat{k}e^{-\alpha_0\tau_{i,M}}$$

$$= \left(\frac{\tilde{B}}{1+St_w}\right)\left(\frac{d\tau_{i,M}}{d\hat{t}}\right)$$

(11.59)

where St_w is the Stanton number at the wall, given by. $(h_w/G_M C_p)(2M/2M-1)$.

The solution of these equations involves a marching technique requiring only the initial values. The need for fixing the boundary conditions is therefore eliminated, thus enabling the inclusion of axial diffusion in the computations. The model has been extended by McGuire and Lapidus (1965) to include the effect of fluid–particle heat and mass transfer.

11.6. PARAMETRIC SENSITIVITY

An important factor that is not considered in the optimization procedures employed (see, e.g., Horn and Klein, 1971) is the effect of small perturbations in the operating parameters, such as feed temperature, coolant temperature, and feed composition. Bilous and Amundson (1956) have shown that such small changes can severely affect the operation of the reactor by causing a large increase in the hot-spot temperature. The reactor is then considered to be "parametrically sensitive."

11.6.1. Sensitivity in Quasihomogeneous Models

Barkelew (1959) rearranged the usual one-dimensional heat and mass balance equations and solved them numerically for a large number of different cases by varying the ratio of the inlet temperature to coolant temperature and the nature of the kinetic expression. The results of these extensive calculations were presented as N/S vs. J_m/S curves for specific values of S, where $N, S,$ and J are given by

$$N = \frac{4U}{d_t C_p A_{fp} \rho_{g0} f_B}, \qquad S = \frac{(-\Delta H') y_0 (E/R_g T_C^2)}{C_p}$$

$$J = \frac{E}{R_g T_C^2}(T - T_C), \quad J_m = \frac{E}{R_g T_C}(T_m - T_C)$$

$$(11.60)$$

The quantity N/S represents the ratio of the rate of heat transfer to the rate of heat generation, and J_m represents the value of the parameter J at the maximum temperature, that is, at $T = T_m$.

It can be seen from the typical set of curves shown in Figure 11.9 that the family of curves is bounded from below by an envelope. For every value of S the curve turns sharply at the envelope, so that any further decrease in N/S at that point causes a sharp increase in the hot-spot temperature. The point of tangency [which gives the critical value $(N/S)_{cr}$] therefore marks the onset of the region of parametric sensitivity. Since the parameters S, N and J involve three design variables,

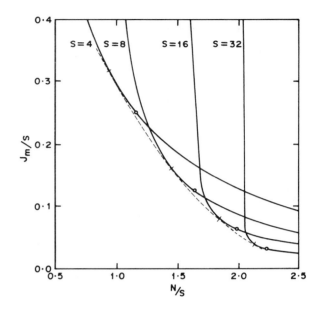

Figure 11.9. Barkelew plot for the parametric sensitivity of an ideal tubular reactor (Barkelew, 1959).

namely, $d_t, y_0,$ and T_C, if any two of them are given, the third can be computed from a plot of this type.

Van Welsenaere and Froment (1970) also developed criteria for a first-order reaction based on the intrinsic properties of the partial-pressure–temperature trajectories of a fixed-bed reactor (shown in Figure 11.10). The locus of the maxima of these curves is referred to as the "maximum curve." Note that this curve also exhibits a maximum. Van Welsenaere and Froment obtained the following equation[†] for the maxima curve:

$$p_m = \frac{(4U/\rho_g C_p d_t)(T_m - T_w)}{[(-\Delta H)\rho_B/\rho_g C_p] A_{fp} \exp(-E/R_g T_m)}$$

$$(11.61)$$

The trajectories corresponding to low inlet partial pressures (p_{01}, p_{02}) intersect the maxima curve (Eq. 11.61) before its maximum (i.e., before M), and those corresponding to high initial partial pressure (i.e., p_{04}) intersect beyond M. It can be seen that after the point of intersection crosses the maximum of the maxima curve the intersection occurs at considerably higher values of T, that is, the hot spot grows very fast. Hence the trajectory passing through M can be considered critical. It represents the locus of the critical inlet partial pressure (p_{03}) and temperature for a given wall temperature. Van Welsenaere and Froment obtained the

[†] Partial pressure units are used here for convenience, and if the rate is expressed per unit weight of catalyst, then A_{fp} (in this section) has the units mol/g catalyst sec atm.

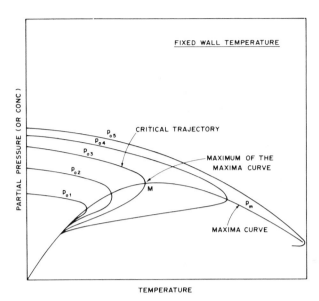

FIXED WALL TEMPERATURE

PARTIAL PRESSURE (OR CONC)

TEMPERATURE

p_{o5}
p_{o4}
p_{o3}
p_{o2}
p_{o1}

CRITICAL TRAJECTORY

MAXIMUM OF THE MAXIMA CURVE

M

p_m

MAXIMA CURVE

Figure 11.10. Diagram showing fixed-bed reactor sensitivity on a partial-pressure–temperature plane.

following expression for the critical temperature T_{cr}:

$$T_{cr} = \frac{1}{2}\left\{\frac{E}{R_g} - \left[\frac{E}{R_g}\left(\frac{E}{R_g} - 4T_w\right)\right]^{1/2}\right\} \quad (11.62)$$

The critical partial pressure can be obtained by substituting $T_m = T_{cr}$ in Eq. 11.61.

The criterion discussed above gives the critical conditions inside the reactor (i.e., at the maximum point). A method is now required to compute the inlet conditions that result in these critical conditions. An extrapolation procedure is used to evaluate the lower and upper bounds of the inlet partial pressure. These bounds are given by:

Lower bound:

$$p_0^l = p_{cr} + \frac{A''}{B''}(T_{cr} - T_0) \quad (11.63)$$

provided that $T_w < T_{cr} < T_0$.

Upper bound:

$$p_0^u = \frac{A''}{B''}(T_{cr} - T_w)\left\{\left[\frac{C''}{A''}A_{fp}\exp\left(\frac{-E}{R_g T_{cr}}\right)\right]^{-1/2} + 1\right\}^2 \quad (11.64)$$

where, as before in this section, A_{fp} has the units of mol/g catalyst sec atm, and

$$A'' = \frac{MP\rho_B}{\rho_g}, \qquad B'' = \frac{(-\Delta H)\rho_B}{\rho_g C_p}, \qquad C'' = \frac{4U}{\rho_g C_p d_t}$$

The lower limit is always safe, and the upper limit always unsafe. It has been found that the mean of the upper and lower bounds represents an excellent approximation of the true critical value. Thus, using the criterion developed earlier and setting $T_0 = T_w$, we get

$$(p_0)_{mean} = \frac{1}{2}(p_0^l + p_0^u) \quad (11.65)$$

Criteria have also been presented by Thomos (1961), Dente and Cappelli (1966), Hlavacek et al. (1970), Hlavacek (1970), and Oroskar and Stern (1979). More recently, based on the method of isoclines, Morbidelli and Varma (1982) have provided exact criteria that can be readily applied for all positive-order exothermic reactions, using the full Arrhenius temperature dependence of the reaction rate, and for all values of the inlet temperature. The procedure to evaluate critical values of heat of reaction and heat transfer parameters is simple and does not involve any trial and error. The results of their analysis indicate that, all other parameters being fixed, chances of reactor runaway increase as the reaction order decreases, the reaction activation energy increases, or the inlet temperature of reactants increases.

Example: Synthesis of Vinyl Acetate (A Complex Reaction)

No criteria are available for complex systems. The complex kinetic expressions will have to be approximated to pseudo-first-order kinetics for the application of these criteria. If such an approximation is not possible, Barkelew-type diagrams will have to be prepared by integration of the model equation for different values of the parameters. Because of the large number of parameters involved in complex systems, no general criteria are possible, and each case has to be dealt with individually. An experimental study on the runaway of a catalytic fixed-bed reactor for the synthesis of vinyl acetate has been reported by Emig et al. (1980) and provides an example of the analysis of a complex reaction. Their data have been compared with the Barkelew (1959), McGreavy and Adderley (1973), and Agnew and Potter (1966) diagrams for parametric sensitivity. The simple Barkelew diagram reasonably explains the observed experimental results. McGreavy and Adderley's diagram shows a better fit to the experimental data, but requires more experimental information.

11.6.2. Sensitivity in the Presence of Transport Limitations: The Rajadhyaksha–Vasudeva–Doraiswamy (RVD) Analysis

Transport limitations can drastically alter the global rate of a reaction; hence, the critical inlet conditions can be significantly different from those obtained on the quasihomogeneous assumption. For a first-order irreversible reaction in an isothermal pellet (at T_p) in the presence of external and internal diffusion, Rajadhyaksha et al. (1975) obtained the following expressions for T_m and p_m:

$$T_m = T_p - \frac{\dfrac{(-\Delta H)A_{fp}}{h_{fp}a'}\exp\left(\dfrac{-E}{R_g T_p}\right)p_m \varepsilon}{1 + \dfrac{\varepsilon A_{fp}\exp\left(-E/R_g T_p\right)}{k_{gp}a'}} \quad (11.66)$$

$$p_m = \frac{\dfrac{4U}{\rho_g C_p d_t}(T_p - T_w)\left[1 + \dfrac{\varepsilon A_{fp}\exp\left(-E/R_g T_p\right)}{k_{gp}a'}\right]}{\varepsilon A_{fp}\exp\left(\dfrac{-E}{R_g T_p}\right)\left[\dfrac{(-\Delta H)\rho_B}{\rho_g C_p} + \dfrac{4U}{\rho_g C_p d_t}\left(\dfrac{-\Delta H}{h_{fp}a'}\right)\right]} \quad (11.67)$$

where $k_{gp}a'$ is expressed as mol/g catalyst sec atm. These equations together constitute the equation of the maxima curve (with the pellet temperature T_p appearing as a parameter).

The maximum of the maxima curve can be obtained by setting $dp_m/dT_m = 0$ (as in the case of the quasihomogeneous model); this reduces to

$$\frac{dp_m}{dT_p} = 0 \quad (11.68)$$

since dT_p/dT_m is always nonzero. The critical point can thus be obtained by solving Eq. 11.68. The properties of the maxima curve, however, are now profoundly dependent on the extent of the transport limitations. Thus, in working out solutions to Eq. 11.68, it is convenient to divide the parameter space into four regimes with different transport limitations. Conditions for discerning the regimes are outlined in Table 11.4.

Regime 1: Very Slow Reactions

In regime 1 the transport processes do not influence the global rate, so that $\varepsilon \to 1$, $h_{fp} \to \infty$, $k_{gp}a' \to \infty$. With these conditions Eqs. 11.66 and 11.67 reduce to 11.61 for the quasihomogeneous model.

Regime 2: Slow Reactions

Regime 2 is characterized by the absence of interphase mass transport resistance in the entire temperature range of interest. The interphase heat transport and intraparticle mass transport resistances can have finite effects on the global rate in the whole or part of the operating range. The conditions for a system to lie in this regime may be written as in Table 11.4 (column 3). The equations of the maxima curve can then be obtained by setting $k_{gp}a' \to \infty$, with the result

$$p_m = \frac{\dfrac{4U}{\rho_g C_p d_t}(T_p - T_w)}{A_{fp}\varepsilon \exp\left(\dfrac{-E}{R_g T_p}\right)\left[\dfrac{(-\Delta H)\rho_B}{\rho_g C_p} + \dfrac{(-\Delta H)}{h_{fp}a'}\dfrac{4U}{\rho_g C_p d_t}\right]} \quad (11.69)$$

$$T_m = T_p - \frac{\varepsilon(-\Delta H)A_{fp}\exp\left(-E/R_g T_p\right)p_m}{h_{fp}a'} \quad (11.70)$$

The nature of this maxima curve is shown in Table 11.4. While obtaining the solution to Eq. 11.66 for this regime two cases should be considered.

1. Negligible intraparticle gradients:

When there are no intraparticle gradients, $\varepsilon = 1$ and the temperature-dependent part of p_m becomes $(T_p - T_w)/e^{(-E/R_g T_p)}$ (assuming ρ_g, C_p, ΔH to be temperature independent), and Eq. 11.62 for T_{cr} results. The critical fluid pressure and temperature can then be obtained by substituting $T_p = T_{pcr}$ in Eqs. 11.69 and 11.70.

2. Significant intraparticle gradients:

In the presence of intraparticle gradients Rajadhyaksha et al. (1975) defined a term

$$\theta = R\left(\frac{A_{fp}\rho_B R_g' E}{D_e R_g}\right)^{1/2} \quad (11.71)$$

which constitutes the temperature-independent part of the Thiele modulus and is constant for a system. It is related to the usual Thiele modulus by

$$\phi = \theta\sqrt{\frac{1}{\alpha}e^{-\alpha}} \quad (11.72)$$

Equation 11.68 for this case may be solved by differentiating the function $(T_p - T_w)/\varepsilon \exp(-E/R_g T_p)$ numerically with assumed values of T_p and finding the value of T_w so that the equation is satisfied. The results

TABLE 11.4. Identification of Sensitivity Regimes[a]

	Regime 1	Regime 2	Regime 3	Regime 4
Conditions	1.1 $r_w(p_{A0}, T_u) < 0.005\, k_a a' p_{A0}$	2.1 $\varepsilon r_w(p_{A0}, T_u) < 0.05\, k_a a' p_{A0}$	3.1 $\varepsilon r_w(p_{A0}, T_u) < 0.05\, k_a a' p_{A0}$	4.1 $\varepsilon r_w(p_{A0}, T_0) > 20\, k_a a' p_{A0}$
	1.2 $\phi^2(T_u) < 1$	2.2 $\phi^2(T_u) > 1$	3.2 $r_w(p_{A0}, T_0) < 20 k_a a' p_{A0}$	
		and/or		
	1.3 $\dfrac{r_w(p_{A0}, T_u)}{r_w(p_{A0}, T_{fu})} - 1 < 0.05$	2.3 $\dfrac{r_w(p_{A0}, T_u)}{r_w(p_{A0}, T_{fu})} - 1 > 0.05$		
	where	where		
	1.4 $T_{fu} = T_u - \dfrac{r_w(p_{A0}, T_u)(-\Delta H)}{h_{fp} a'}$	2.4 $T_{fu} = T_u - \dfrac{r_w(p_{A0}, T_u)(-\Delta H)}{h_{fp} a'}$		
Equation of maxima curve	11.61	11.69	11.66	11.78
Nature of the maxima curve				
Type of sensitivity	Global	Global	Global and local	No sensitivity

[a]Units: k_a (actual value of k_p) is in mol/cm^2 sec atm; a' in cm^2/g; r in mol/g catalyst sec.

245

of this computation are presented in Figure 11.11 as plots of $\Delta(1/\alpha)$ vs. $1/\alpha_w$. The curve for the quasihomogeneous model is the plot of Eq. 11.62, which can be rearranged to the form

$$\Delta\left(\frac{1}{\alpha}\right) = \frac{1}{2}\left[1 - \left(1 - \frac{4}{\alpha_w}\right)^{1/2}\right] - \frac{1}{\alpha_w} \quad (11.73)$$

The curves for different values of θ coincide with the quasihomogeneous curve at low values of $1/\alpha_w$. At larger values of $1/\alpha_w$ the curves for all values of $\theta > 0$ again merge into a single curve. This limiting curve (represented by curve B in Figure 11.11) in the asymptotic region will be given by

$$\Delta\left(\frac{1}{\alpha}\right) = \frac{1}{2}\left\{\frac{1}{2} - \left[\frac{1}{2}\left(\frac{1}{2} - \frac{4}{\alpha_w}\right)\right]^{1/2}\right\} - \frac{1}{\alpha_w} \quad (11.74)$$

The curves for different values of θ will be contained between the two curves given by Eqs. 11.73 and 11.74.

By means of the curves given in Figure 11.11 the value of T_{pcr} can be obtained for a given set of values of E, T_w, and θ. The fluid conditions (partial pressure and temperature) can then be obtained by using Eqs. 11.69 and 11.70.

After determining the critical conditions the corresponding inlet conditions should be estimated. For a given inlet temperature the lower limit of the inlet partial pressure can be estimated by assuming adiabatic operation between the critical point and the inlet of the reactor.

There does not appear to be any rigorous mathematical procedure for obtaining an equation for the upper limit. However, an equation has been obtained by intuitively modifying 11.64 to the following form:

$$p_0^u = \frac{A''}{B''}(T_{cr} - T_0)\left\{\left[\frac{C''}{A''}A_f \exp\left(\frac{-E}{R_g T_{cr}}\right)\varepsilon\right.\right.$$
$$\left.\left. \times \left(1 + \frac{4U}{\rho_B d_t h_{fp} a'}\right)\right]^{-1/2} + 1\right\}^2$$
$$(11.75)$$

Regime 3: Fast Reactions

As we move from very slow reactions in regime 1 to very fast reactions in regime 4, it is necessary to consider a regime in which all the resistances are operative, namely, regime 3, which represents the most general case. The conditions for a system to lie in this regime are given in Table 11.4 (column 4).

Since all the transport processes are to be taken into account, Eqs. 11.66 and 11.67 without any simplification describe the maxima curve for this regime. This curve will exhibit a minimum along with a maximum, in contrast to the maxima curves in regimes 1 and 2, which exhibit a monotonically decreasing trend after the maximum over a wide range of temperatures. Although in theory this observation is true for regimes 1 and 2 also, for practical purposes it comes into the picture approximately with the addition of interphase mass transport resistance to regime 2.

The solution to Eq. 11.68 for $\varepsilon = 1$ is given by

$$1 - \frac{E}{R_g T_{cr}^2}(T_{cr} - T_w) + Da' \exp\left(\frac{-E}{R_g T_{cr}}\right) = 0$$
$$(11.76)$$

where $Da' = A_{fp}/k_{gp}a'$. Figure 11.12 is a graphical representation of this equation for $Da' = 10^6$ for $\theta = 0$. The figure indicates that to the left of the point p there are two solutions to the equation for a given value of $1/\alpha_w$. These two solutions correspond to the maximum and minimum of the maxima curve (the lower corresponding to the maximum). As the wall temperature is increased, the maximum and the minimum approach each other and coincide at the point p. In other words, at the wall temperature corresponding to $1/\alpha_w^*$, the maxima curve will exhibit a point of inflection. To the right of the point p, there is no solution to Eq. 11.68 and the maximum partial pressure increases monotonically.

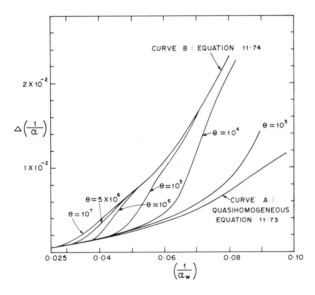

Figure 11.11. Analysis of regime 2: $\Delta(1/\alpha)$ vs. $1/\alpha_w$ curves (Rajadhyaksha et al., 1975).

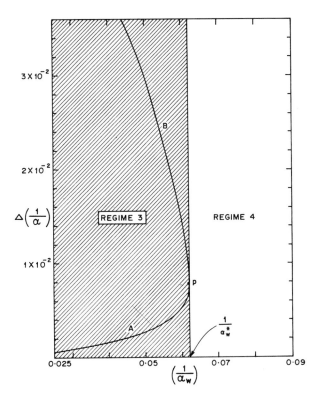

Figure 11.12. Analysis of a typical $\Delta(1/\alpha)$ vs. $1/\alpha_w$ curve: Da$'$ = 10^6, $\theta = 0$ (Rajadhyaksha et al., 1975).

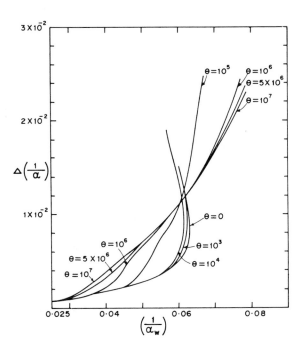

Figure 11.13. Analysis of regime 3: $\Delta(1/\alpha)$ vs. $1/\alpha_w$) curves for Da$'$ = 10^6 (Rajadhyaksha et al., 1975).

Rajadhyaksha et al. also obtained solutions at finite values of θ for different values of Da$'$. Some typical plots are presented in Figure 11.13. The critical conditions can be evaluated from these figures by following a procedure similar to that for regime 2. Because of the high rates of reaction in this regime the operation before the hot spot is close to adiabatic. Hence the lower limit obtained on the basis of the adiabatic reactor equation gives a good estimate of the critical inlet partial pressure.

Regime 4: Very Fast Reactions

In regime 4 the operation is interphase mass transport controlled at all temperatures above the inlet temperature, and the equation of the maxima curve can be obtained by setting the kinetic rate very large as compared to the mass transfer rate in Eqs. 11.66 and 11.67; that is,

$$\frac{A_{fp}\exp(-E/R_g T_p)}{k_{gp}a'} \gg 1 \qquad (11.77)$$

with the result

$$p_m = \frac{\dfrac{4U}{\rho_g C_p d_t}(T_p - T_w)}{k_{gp}a'\left[\left(\dfrac{(-\Delta H)\rho_B}{\rho_g C_p}\right) + \left(\dfrac{4U}{\rho_g C_p d_t}\right)\dfrac{(-\Delta H)}{h_{fp}a'}\right]} \qquad (11.78)$$

$$T_m = T_p - \left(\frac{(-\Delta H)k_{gp}}{h_{fp}}\right)p_m \qquad (11.79)$$

It is evident from these equations that the maxima curve is a straight line with no indication of global sensitivity. The unshaded region in Figure 11.12 bounded on the left by the ordinate through p refers to operation in this regime where there is no solution to Eq. 11.68. Thus point p marks the transition from regime 3 to regime 4.

11.6.3. Local Sensitivity

McGreavy and Adderly (1973) studied the variation of the pellet temperature with the fluid temperature. Curve A of Figure 11.14 is a plot of the fluid temperature vs. pellet temperature. It can be noticed that initially the pellet temperature rises gradually as the fluid temperature increases, but when the latter exceeds a certain critical value the pellet temperature increases drastically with a small increase in T_b. Even though the sensitivity is not reflected in the temperature of the main

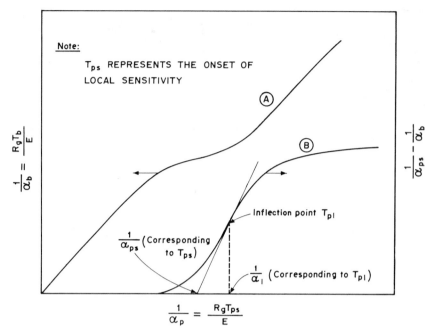

Figure 11.14. Local parametric sensitivity in fixed-bed reactors.

fluid stream, it has all the undesirable effects of temperature runaway, such as loss of catalyst activity and selectivity. McGreavy and Adderly termed this phenomenon *local sensitivity* or *instability*, as distinct from the sensitivity considered in Section 11.6.2, which may be termed *global sensitivity*.

McGreavy and Adderly (1973) have also described a general method to identify this sensitivity (in a heterogeneous system). It consists in plotting $T_p - T_b$ as a function of T_p, as shown in curve B of Figure 11.14. A point of inflection T_{pi} can be observed; the tangent from this point intersects the pellet-temperature axis at the point T_{ps}, which marks the onset of the parametrically sensitive region. To obtain an analytical expression for the temperature at the inflection point the second derivative $d^2(T_p - T_b)/dT_p^2$ is set equal to zero; the final equations obtained are

$$\frac{1}{\alpha_I} = \frac{A_I'(1-g_I)(2A_I'g_I-1)+g_I}{4[g_I-(1-g_I^2)]} - \frac{A_I'[r(1-g_I^2)+1]}{2(rg_I+A_I')}$$

(11.80)

$$\frac{1}{\alpha_{ps}} = \frac{1}{\alpha_I} - \frac{4}{\alpha_I^2}\left\{\frac{(A_I'-g_I)(rg_I+A_I')}{\text{Sh}(A_I')^2[g_I^2-(A_I')^2]+A_Ig_I}\right\}$$

(11.81)

where $A_I' = \phi_{A_f}\exp(-\alpha_I/2)$ and $g_I = \tanh A_I'$.

To obtain the runaway limit, T_{ps} is estimated by

solving Eqs. 11:80 and 11.81 at the given values of the Sherwood number Sh and ϕ_{A_f}. With this value of T_{ps} the corresponding fluid temperature can be obtained from Eq. 12.20. Hence, the method specifies an upper limit for the fluid temperature for a given set of values of Sh and ϕ_{A_f}. It may be noted that T_{pi} and T_{ps} are independent of B_I', which represents the value of B' in Eq. 12.24 at the inflection point.

Rajadhyaksha et al. (1975) investigated the role of this local sensitivity in fixed-bed reactor operation in the light of the operating regimes postulated by them. They showed that such a sensitivity can be observed only in systems operating in regime 3. Because of the limitation on the rate of heat removal, the fixed-bed reactor has to be operated at lower values of B' in order to restrict the hot spot. Local sensitivity becomes important, however, at relatively high values of B', but these are rarely realized in actual systems.

11.6.4. Examples: Operating Regimes of Some Industrial Reactions

The analysis presented above has evolved a classification of systems into four regimes of operation. In order to relate these regimes to physical reality, some industrially important reactions that are carried out in fixed-bed reactors are considered below with the object of identifying their regimes of operation. The conditions for different regimes may be applied to these systems by using the kinetic and transport data reported in the literature. The systems considered are the

1. oxidation of benzene to maleic anhydride;
2. oxidation of *o*-xylene to phthalic anhydride;
3. hydrogenation of nitrobenzene to aniline;
4. hydrochlorination of acetylene to vinyl chloride;
5. hydrogenation of phenol to cyclohexanol.

The interphase heat and mass transfer coefficients may be evaluated by using the correlations presented in Chapter 6, taking the particle Reynolds number to be 100. Effective diffusivities may be taken to be 10% of the molecular diffusivities (Eq. 3.37).

The conditions stated in Table 11.5 are independent of the kinetic expression but are applicable to a single reaction only. Approximations are required to be made when the reaction follows a complex scheme. In the case of systems 1 and 2 the reaction follows a triangular

network. In the analysis of 1 the reaction of anhydride with oxygen may be neglected, since the concentration of the anhydride is low in the region close to the inlet.[†] Also, the rate constants for the two parallel reactions, that is,

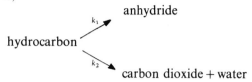

can be lumped into a single rate constant, which represents the rate of disappearance of the hydrocarbon. Such a lumping is possible only because the activation energies for the two parallel reactions are almost identical. Since the two reactions occur in a constant proportion, an effective heat of reaction can be calculated as

$$-\Delta H = \frac{k_1(-\Delta H_1) + k_2(-\Delta H_2)}{k_1 + k_2} \quad (11.82)$$

TABLE 11.5. Operating Regimes for Some Industrial Systems (Rajadhyaksha et al., 1975)

	System				
	Oxidation of Benzene	Oxidation of *o*-Xylene	Hydrogenation of Nitrobenzene	Hydrochlorination of Acetylene	Hydrogenation of Phenol
Reference for the kinetic expression	Vaidyanathan and Doraiswamy (1968)	Froment (1967)	Rihani et al. (1965)	Wesselhoft et al. (1959)	Hancil and Mitschka (1969)
Temperature range, °C	350–450	325–400	275–350	75–125	150–200
Partial pressure range, atm	0–0.015	0–0.015	0–0.1	0–0.5	0–0.2
$-\Delta H$, cal/mol	537,000	395,000	152,000	60,000	30,000
D_e, cm²/sec	0.0431	0.0208	0.088	0.0247	0.0615
h, cal/sec °C cm²	1.53×10^{-2}	1.53×10^{-2}	1.47×10^{-2}	1.55×10^{-2}	1.47×10^{-2}
k_g, cm/sec	30.0	30.0	26.2	7.6	13.8
Kinetic rate, T_u, mol/sec g	5.9×10^{-8}	5.1×10^{-7}	4.6×10^{-6}	5.9×10^{-6}	1.1×10^{-4}
Mass transfer rate, mol/sec g	5.8×10^{-5}	5.4×10^{-5}	1.9×10^{-4}	8.9×10^{-4}	4.6×10^{-4}
$\phi^2(T_u)$	0.521	12.8	6.27	1.42	47.4
$H^{*\,a}$	0.005	0.04	0.131	0.09	—
Regime	1	2	2	2	3

[a] $H^* = [r(T_u)/r(T_{fu})] - 1$

[†] A more correct scheme is a triangular network (see Section 2.6.11)

The results of the computations are summarized in Table 11.5. It will be noticed that reaction 1 falls in regime 1, reactions 2–4 in regime 2, and reaction 5 in regime 3. It follows therefore that it would be incorrect to carry out a sensitivity analysis of these reactions on the assumption of a quasihomogeneous model (i.e., that they all fall in regime 1).

An example is presented next to illustrate the methods suggested in this section for calculating the upper concentration bound of reactant for different regimes. Hydrocarbon oxidation reactions are used with several assumed parameter values.

11.6.5. Example: Oxidation of Hydrocarbons: Role of Transport Resistances in Fixing the Maximum Allowable Inlet Partial Pressure in Regime 2

The following data may be used: $d_p = 0.9$ cm, $\rho_B = 1.3$ g/cm^3, $U = 100$ kcal/m^2 hr °C, $h_{fp} = 4.16 \times 10^{-3}$ cal/cm^2 sec °C, $\rho_g = 0.524 \times 10^{-3}$ g/cm^3, $C_p = 0.25$ cal/g °C, $A_{fp} = 51000$ mol/sec g atm; $E = 27.8$ kcal/mol, $d_t = 1.25$ cm, $a' = 5.13$ cm^2/g, $-\Delta H = 197.5$ kcal/mol, $k_g = 30$ cm/sec.

The upper limit of the inlet partial pressure set by flammability considerations is 0.075 atm. The operation of the catalyst should be restricted to a temperature range of 600–660°K for satisfactory performance. Intraparticle gradients are negligible. If the inlet temperature and the wall temperature are maintained at 615°K, calculate the maximum allowable partial pressure at the inlet (1) by disregarding the influence of the transport limitations, and (2) by taking into account the influence of transport processes.

Solution

1. Calculation of critical inlet partial pressure by Van Welsenaere and Froment's criterion:

The critical temperature is given by (Eq. 11.62)

$$T_{cr} = \tfrac{1}{2}[a - \sqrt{a(a-4T_w)}]$$
$$= 645°K$$

(Note: $a = E/R_g$.)

$$A'' = \frac{MP\rho_B}{\rho_g} = 6.89 \times 10^4$$

$$B'' = \frac{(-\Delta H)\rho_B}{\rho_g C_p} = \frac{(1.97 \times 10^5)(1.3)}{(0.524 \times 10^{-3})(0.25)} = 1.96 \times 10^9$$

U is 100 kcal/hr m^2 °C or 2.78×10^{-3} cal/sec cm^2 °C.

Therefore

$$C'' = \frac{4U}{\rho_g C_p d_t} = \frac{4(2.78 \times 10^{-3})}{(0.524 \times 10^{-3})(0.25)(1.25)} = 68$$

The critical partial pressure is given by Eq. 11.61.

$$p_{cr} = \frac{68(645-615)}{(1.96 \times 10^9)\left[(5.1 \times 10^4)\exp\left(-\dfrac{27800}{1.98 \times 645}\right)\right]}$$
$$= 0.06 \text{ atm}$$

The lower limit of partial pressure at the inlet is given by Eq. 11.63:

$$p_0^l = 0.06 + \frac{6.89 \times 10^4}{1.96 \times 10^9}(645-615) = 0.061 \text{ atm}$$

and the upper limit can be obtained from Eq. 11.64:

$$p_0^u = \frac{6.89 \times 10^4}{1.96 \times 10^9}(645-615)$$
$$\left[\left(\frac{6.89 \times 10^4}{68} 1.79 \times 10^{-5}\right)^{-1/2} + 1\right]^2$$
$$= 0.071 \text{ atm}$$

Hence the critical inlet partial pressure will be given by

$$(p_0)_{mean} = \tfrac{1}{2}(p_0^u + p_0^l) = 0.066 \text{ atm}$$

2. Calculation of the critical inlet partial pressure in the presence of transport limitations using the RVD method:

 a. Identification of regime:

 Taking

$$p_0 = 0.075 \text{ atm}, \quad k_g a' p_0 = 2.14 \times 10^{-4} \quad \text{mol/sec g}$$

we have

$$r(p_0, T) = r(0.075, 660) = 2.22 \times 10^{-6} \quad \text{mol/sec g}$$

Hence

$$r(p_0, t) < 0.05\, k_g a' p_0$$

Therefore the system can be in regime 1 or 2.

From Eq. 2.4 of Table 11.4

$$T_{\text{fu}} = T - \frac{r(p_0, t)(-\Delta H)}{h_{\text{fp}}a'} = 660 - 20.4 = 639.6°\text{K}$$

$$\frac{r(p_0, T)}{r(p_0, T_{\text{fu}})} - 1 = 1.1 > 0.05$$

Hence the system belongs to regime 2.

b.   Calculation of critical conditions:

Since the intraparticle gradients are negligible, the critical pellet temperature is given by

$$T_{\text{pcr}} = \tfrac{1}{2}\big[a - \sqrt{a(a - 4T_{\text{w}})}\big] = 645°\text{K}$$

and (from Eq. 11.69)

$$p_{\text{cr}} = 0.043 \text{ atm}$$

So also (from Eq. 11.70)

$$T_{\text{cr}} = 637.8°\text{K}$$

Hence (from Eq. 11.63)

$$p_0^l = 0.0438 \text{ atm}$$

and (from Eq. 11.75)

$$p_0^u = 0.061 \text{ atm}$$

Therefore the critical inlet partial pressure, which is given by the average of p_0^l and p_0^u, is 0.0523 atm.

It will be noted that the partial pressure computed by neglecting the transport limitation (0.066 atm) is over 25% higher than the correct value of 0.0523 atm. This emphasizes the danger of neglecting the transport limitation in calculating the maximum inlet partial pressure.

REFERENCES

Adler, J., and Vortmeyer, D. (1963), *Chem. Eng. Sci.*, **18**, 99.

Adler, J., and Vortmeyer, D. (1964), *Chem. Eng. Sci.*, **19**, 413.

Agnew, J. B., and Potter, O. E. (1966), *Trans. Inst. Chem. Eng. (London)*, **44**, T216.

Ahmed, M., and Fahien, R. W. (1980a), *Chem. Eng. Sci.*, **35**, 889.

Ahmed, M., and Fahien, R. W. (1980b), *Chem. Eng. Sci.*, **35**, 897.

Amundson, N. R. (1969), *Mathematical Models of Fixed Bed Reactors* (Diskussion stagung der DBG), Konigstein, Taunus, West Germany.

Baddour, R. F., Brian, P. L. T., Logeais, B. A., and Eymery, J. P. (1965), *Chem. Eng. Sci.*, **20**, 281.

Bakemeier, H., Laurer, P. R., and Schroder, W. (1970), *Chem. Eng. Prog. Symp. Ser.* (No. 78), **66**, 1.

Barkelew, C. H. (1959), *Chem. Eng. Prog. Symp. Ser.* **55** (25), 37.

Baron, T. (1952), *Chem. Eng. Prog.*, **48**, 118.

Beek, J., and Singer, E. (1951), *Chem. Eng. Progr. Symp. Ser.*, **47**, 534.

Beek, J. (1962), *Adv. Chem. Eng.*, **3**, 204.

Bilous, R., and Amundson, N. R. (1956), *AIChE. J.*, **2**, 117.

Bischoff, K. B. (1965), *AIChE J.*, **11**, 351.

Bischoff, K. B. (1968), *AIChE J.*, **14**, 820.

Butt, J. B. (1980), *Reaction Kinetics and Reactor Design*, Prentice-Hall, New Jersey.

Cappelli, A., Collina, A., and Dente, M. (1972), *Adv. Chem. Ser.*, **109**, 35.

Carberry, J. J., and Wendel, M. (1963), *AIChE J.*, **9**, 129

Chartland, C., and Crowe, C. M. (1969), *Can. J. Chem. Eng.*, **47**, 296.

Cresswell, D., and Paterson, W. R. (1970), *Chem. Eng. Sci.*, **25**, 1405

Crider, J. E., and Foss, A. S. (1965), *AIChE J.*, **11**, 1012.

Curl, R. L. (1963), *AIChE J.*, **9**, 175.

Danckwerts, P. V. (1953), *Chem. Eng. Sci.*, **2**, 1.

Danckwerts, P. V. (1958), *Chem. Eng. Sci.*, **8**, 93.

Deans, H. A., and Lapidus, L. (1960), *AIChE J.*, **6**, 656.

Deans, H. A. (1963), *Soc. Petrol. Eng. J.*, **228**, 3.

Denbigh, K. G. (1965), *Chemical Reactor Theory*, Cambridge Univ. Press, New York and London.

Dente, M., and Cappelli, A. (1966), *Ing. Chim. Ital.*, **48**, 702.

De Wasch, A. P., and Froment, G. F. (1971), *Chem. Eng. Sci.*, **26**, 629.

Dohan, L. A., and Weinstein, H. (1973), *Ind. Eng. Chem. Fundam.*, **12**, 55.

Dudukovic, M. P., and Lamba, H. S. (1975), AIChE 68th ann. meeting.

Dumez, F. J., and Froment, G. F. (1976), *Ind. Eng. Chem. Process Design Dev.*, **15**, 291.

Emig, G., Hofmann, H., Hoffmann, V., and Fiand, V. (1980), *Chem. Eng. Sci.*, **35**, 249.

Fan, L. T., and Bailie, R. C. (1960), *Chem. Eng. Sci.*, **13**, 63.

Finlayson, B. A. (1974), *Catal. Rev. Sci. Eng.*, **10**, 69.

Finlayson, B. A. (1971), *Chem. Eng. Sci.*, **26**, 1081.

Froment, G. F. (1967), *Ind. Eng. Chem.*, **59** (2), 18.

Froment, G. F. (1974), *Proc. 7th European Symp. Computer Application in Process Development*, Erlanger, Dechema.

Froment, G. F., and Bischoff, K. B. (1979), *Chemical Reactor Analysis and Design*, Wiley, New York.

Gray, P., and Lee, P. R. (1965), *Combust. Flame*, **9**, 201.

Hancil, V., and Mitschka, R. (1969), *Chem. Eng. Sci.*, **24**, 1400.

Hawthorn, R. D., Ackerman, G. H., and Nixon, A. C. (1968), *AIChE J.*, **14**, 69.

Hinduja, M. J., Sundaresan, S., and Jackson, R. (1980), *AIChE J.*, **26**, 274.

Hlavacek, V. (1970), *Ind. Chem. Eng.*, **62** (7), 8.

Hlavacek, V., and Votruba, J. (1977), *Chemical Reactor Theory—A Review*, eds. Lapidus, L., and Amundson, N. R., Prentice-Hall, Englewood Cliffs, New Jersey, p. 314.

Hlavacek, V., Kubicek, M., and Jelinek, J. (1970), *Chem. Eng. Sci.*, **25**, 1441.

Hofmann, H., and Astheimer, H. S. (1963), *Chem. Eng. Sci.*, **18**, 643.

Horn, F., and Klein, J. (1971), *Proc. 1st Int. Symp. Chem. React. Eng.*, Washington, D.C.

Karanth, N. G., and Hughes, R. (1974), *Catal. Rev. Sci. Eng.*, **9**, 169.

Kattan, A., and Adler, R. J. (1972), *Chem. Eng. Sci.*, **27**, 1013.

Kondelik, P., and Boyarinov, A. I. (1969), *Collect. Czech. Chem. Commun.*, **34**, 3852.

Kramers, H., and Alberta, G. (1953), *Chem. Eng. Sci.*, **2**, 173.

Kunii, D., and Furusawa, T. (1972), *Chem. Eng. J.*, **4**, 268.

Kuo, J. C. W., Morgan, C. R., and Lassen, H. G. (1971), *Mathematical Modeling of CO and HC Catalytic converter Systems*, SAE Report No. 710289, Detroit.

Lapidus, L. (1962), *Digital Computations for Chemical Engineers*, McGraw-Hill, New York.

Lee, H. H. (1981), *AIChE J.*, **27**, 558.

Lerou, J. J., and Froment, G. F. (1977), *Chem. Eng. Sci.*, **32**, 853.

Levenspiel, O. (1974), *Chemical Reaction Engineering*, 2nd ed., Wiley Eastern Pvt. Ltd., New Delhi.

Levenspiel, O., and Bischoff, K. B. (1963), *Adv. Chem. Eng.*, **4**, 95.

Levich, V. G., Pismen, L. M., and German, E. D. (1967), *Theoret. Found. Chem. Tech.*, **1**, 366.

Liu, S. L. (1970), *AIChE J.*, **16**, 742.

Marivoet, J., Teodoroiu, P., and Wajc, S. J. (1974), *Chem. Eng. Sci.*, **29**, 1836.

Marivoet, J., Teodoroiu, P. and Wajc, S. J. (1974), *Chem. Eng. Sci.*, **29**, 1836.

McGreavy, C., and Cresswell, D. L. (1968), *Proc. 4th European Symp. Chem. React. Eng.*

McGreavy, C., and Adderly, C. I. (1973), *Chem. Eng. Sci.*, **28**, 577.

McGreavy, C., and Thornton, J. M. (1970a), *Chem. Eng. Sci.*, **25**, 303.

McGreavy, C., and Thornton, J. M. (1970b), *Can. J. Chem. Eng.*, **48**, 187.

McGreavy, C., and Turner, K. (1970), *Can. J. Chem. Eng.*, **48**, 200.

McGuire, M. L., and Lapidus, L. (1965), *AIChE J.*, **11**, 85.

McHenry, Jr. K. W., and Wilhelm, R. H. (1957), *AIChE J.*, **3**, 83.

Morbidelli, M., and Varma, A. (1982), *AIChE J.* (in press).

Nishimura, Y., and Matsubara, M. (1970), *Chem. Eng. Sci.*, **25**, 1785.

Olbrich, W. E., Agnew, J. B., and Potter, O. E. (1972), *Chem. Eng. Sci.*, **27**, 1723.

Oroskar, A., and Stern, S. A. (1979), *AIChE J.*, **25**, 903.

Östergaard, K. (1965), *Brit. Chem. Eng.*, **10**, 449.

Östergaard, K. (1971), *Brit. Chem. Eng.*, **26**, 605.

Paspek, S. C., and Varma, A. (1980), *Chem. Eng. Sci.*, **35**, 33.

Paterson, W. R., and Cresswell, D. (1971), *Chem. Eng. Sci.*, **26**, 605.

Pexidr, V., Hlubucek, V., and Pacek, J. (1970), *Int. Chem. Eng.*, **10**, 188.

Polidor, J., Pexidr, V., Jiracek, F., and Horak, J. (1970), *Int. Chem. Eng.*, **10**, 124.

Rajadhyaksha, R. A. (1976), Ph.D. (Tech.) thesis, University of Bombay.

Rajadhyaksha, R. A., Vasudeva, K., and Doraiswamy, L. K. (1975), *Chem. Eng. Sci.*, **30**, 1399.

Ranz, W. E. (1952), *Chem. Eng. Prog.*, **48**, 247.

Raskin, A. Ya., Sokolinskii, Yu. A., Mukosei, V. I., and Aerov, M. E. (1968), *Theoret. Found. Chem. Eng. USSR (Engl. Transl.)*, **2**, 220.

Richarz, R., and Lattmann, S. (1968), *Proc. 4th European Symp. Chem. React. Eng.*, p. 123.

Rihani, D. N., Narayanan, T. K., and Doraiswamy, L. K. (1965), *Ind. Eng. Chem. Process Design Dev.*, **4**, 403.

Sharma, C. S., and Hughes, R. (1979), *Chem. Eng. Sci.*, **34**, 613.

Sheel, J. G. P., and Crowe, C. M. (1969), *Can. J. Chem. Eng.*, **47**, 183.

Sinkule, J., Hlavacek, V., Votruba, J., and Tvrdik, I. (1974), *Chem. Eng. Sci.*, **29**, 689.

Smith, J. M. (1956), *Chemical Engineering Kinetics*, McGraw-Hill, New York.

Thomas, P. H. (1961), *Proc. Roy. Soc. (London)*, **A262**, 192.

Tichacek, L. J. (1963), *AIChE J.*, **9**, 394.

Vaidyanathan, K., and Doraiswamy, L. K. (1968), *Chem. Eng. Sci.*, **23**, 537.

Vanderveen, J. W., Luss, D., and Amundson, N. R. (1968), *AIChE J.*, **14**, 636.

Van Welsenaere, R. J., and Froment, G. F. (1970), *Chem. Eng. Sci.*, **25**, 1503.

Villadsen, J., and Stewart, W. E. (1967), *Chem. Eng. Sci.*, **22**, 1483.

Votruba, J., and Hlavacek, V. (1972), *Chem. Eng. J.*, **4**, 91.

Votruba, J., Hlavacek, V., and Marek, M. (1972), *Chem. Eng. Sci.*, **27**, 1845.

Wehner, J. F., and Wilhelm, R. H. (1956), *Chem. Eng. Sci.*, **6**, 89.

Weinstein, H., and Adler, R. J. (1967), *Chem. Eng. Sci.*, **22**, 65.

Wesselhoft, R. D., Woods, J. M., and Smith, J. M. (1959), *AIChE J.*, **5**, 361.

Wilhelm, R. H. (1962), *Pure Appl. Chem.*, **5**, 403.

Yablonski, G. S., Kamenko, B. L., Gel'bshtein, A. I., and Slinko, M. G. (1968), *Int. Chem. Eng.*, **8**, 6.

Young, L. C., and Finlayson, B. A. (1973), *Ind. Eng. Chem. Fundam.*, **12**, 412.

Zahner, J. C. (1972), *Adv. Chem. Ser.*, **109**, 201.

Zwietering, Th. N. (1959), *Chem. Eng. Sci.*, **11**, 1.

CHAPTER TWELVE

Fixed-Bed Reactors: Some Design and Operational Considerations

The ultimate design and optimization of a fixed-bed reactor is strongly dependent on the type of reactor proposed to be used. The reactor can be adiabatic, in which case interstage heat exchange is often involved; or tubular nonadiabatic, in which case provision for heat exchange within the reactor is necessary. Both these aspects of fixed-bed reactor design and operation have been considered in detail by Aris (1965).

A catalytic reactor can be optimized with respect to the feed composition, initial temperature, temperature profile in the reactor, feed rate, mixing pattern of the reactant fluid, and tube diameter. The most significant variable is temperature, and several studies have been reported in the literature on the temperature optimization of complex catalytic reactions with respect to the yield of the desired species (e.g., Denbigh, 1958; Horn and Troltenier, 1961a, b; Coward and Jackson, 1965; Jaspan et al., 1972).

Optimization methods can be classified as follows (Horn and Klein, 1972): (1) constructive methods, (2) indirect methods based on the Euler equations and the maximum principle of variational calculus (Hestenes, 1966; Denn, 1969), and (3) direct-search methods. In the first method the optimum is obtained in a finite number of steps; the most useful example of this method in reaction engineering is dynamic programming (Bellman and Dreyfus, 1962; Aris, 1961, 1964). The second method is based on certain necessary conditions that should be satisfied for optimality, while in the direct-search method the optimum is reached by successive improvements that often require prohibitive computer time. The most important examples of this method are those based on steepest ascent (Wilde, 1964; Kelley, 1962).

In this chapter we shall consider several aspects of fixed-bed reactor design and optimization for both adiabatic and nonadiabatic operation. Some practical considerations in design and operation will be highlighted, after which a discussion of reactor types and a strategy for reactor choice and design will be presented.

12.1. ADIABATIC REACTORS

Adiabatic reactors, by their very nature, are not required to be multitubular in construction. In principle there is no restriction on the diameter that can be used, since there is no radial transport of heat involved. The old German reactor for aniline had a diameter of 2.5 m, whereas styrene reactors are known to be about 4 m in diameter. Depending on the exothermicity of the reaction, a single-stage reactor or a multistage reactor with interstage cooling can be used.

12.1.1. Basic Design Principles

For a simple adiabatic reaction $A \rightleftharpoons B$, on the assumption of constant property values, the following relationship can be written:

$$C_p(T - T_0) = (-\Delta H')(x - x_0)$$

or

$$T = T_0 + H(x - x_0) \qquad (12.1)$$

where

$$H = \frac{(-\Delta H')}{C_p} \qquad (12.2)$$

and T_0 and x_0 represent the inlet conditions of temperature and conversion, respectively. On an $x - T$ plane Eq. 12.1 represents the adiabatic path of the reaction—a straight line of slope $1/H$, terminating at the equilibrium position at which the rate is zero. The rate exhibits a maximum at some point x_{max}, T_{max} on the adiabatic

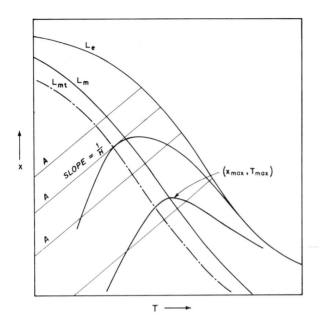

Figure 12.1. Reaction paths in an adiabatic reactor.

path.[†] The locus of the maxima provides a rational starting point for design, since the reaction should logically be localized in a region close to it. Figure 12.1 shows the equilibrium line L_e and the adiabatic paths A, as well as the locus of maximum rates L_m.

The conversion–temperature plot for the oxidation of SO_2 based on the data of Collina et al. (1971) has been prepared by Froment and Bischoff (1979). The equilibrium curve L_e and the locus of the maximum rates L_m are also included in their figure, along with the rate contours and adiabatic reaction paths of slope $1/H$.

For a single-bed reactor, from a knowledge of the production capacity envisaged, the feed rate can be tentatively fixed and the following simple material balance written (see Section 9.2):

$$F\, dx = r_{wA}\, dW$$

or

$$F\, dx = r_{wA}\rho_B A_c\, dl = r_{VA} A_c\, dl \qquad (12.3)$$

This equation may be expressed in the form

$$\Delta l = \alpha_1 \Delta x \qquad (12.4)$$

[†] Strictly, the maximum would be located at $x_{max,t}$, $T_{max,t}$ where the adiabatic line is tangent to the rate curve (with a slope of $1/H$); but the locus of these maxima L_{mt} is usually very close to L_m.

Similarly Eq. 12.1 can be rewritten as

$$\Delta T = \alpha_2 \Delta x \qquad (12.5)$$

These two equations can be readily solved by trial and error to give the conversion and temperature as functions of height l.

12.1.2. Multibed Reactors: Optimization by Dynamic Programming

Multibed reactors are characterized by interstage cooling, so that the feed state to every bed can be controlled—optimized if possible. This cooling can be accomplished by (1) indirect cooling between stages of (2) mixing with a cold shot of unreacted feed.

We shall now describe a procedure for calculating the number of stages for case 1. Let there be N stages in the reactor, with the last stage representing stage 1 (see Figure 12.2). The number of stages can be calculated as follows (starting from stage N):

1. Assume inlet temperature T'_N and outlet conversion x_N.
2. Since x_{N+1} (feed condition) is known, obtain x'_N from $x'_N = x_{N+1}$.
3. Calculate S_N from $S_N = T'_N - Hx'_N$ (constant for the stage).
4. Obtain holding time and therefore reactor volume from

$$t_N = \int_{x'_N}^{x_N} \frac{dx}{r_{V,N}}$$

5. Calculate outlet temperature T_N from $T_N = T'_N + H(x_N - x'_N)$.

Thus the inlet conditions x'_N, T'_N and outlet conditions x_N, T_N for stage N are fixed. We are now in a position to proceed to stage $N-1$ by assuming T'_{N-1} and x_{N-1}. In practice it is simpler to start from stage 1 (outlet) and proceed to stage N (inlet).

It will be noticed that for any stage n two values (T'_n and x_n) have to be assumed for the calculations. Thus for N stages $2N$ decision variables are involved. These decisions determine the reaction boundary on one side and the cooling limit on the other. Aris (1965) has shown how each of these decisions can be optimally made by the method of dynamic programming. According to this method (Bellman, 1957), the maximum profit is a function only of the initial state (or

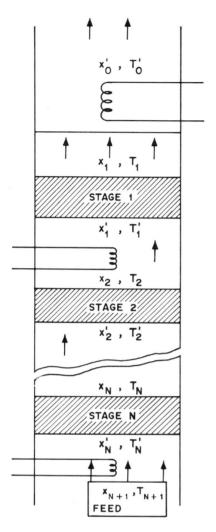

Figure 12.2. The multibed adiabatic reactor.

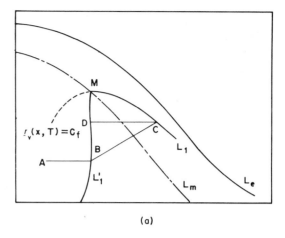

(a)

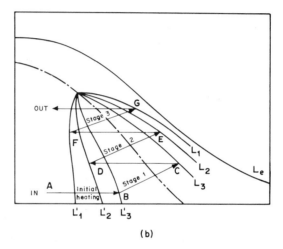

(b)

Figure 12.3. Design of (a) a single-stage adiabatic reactor and (b) a three-stage adiabatic reactor with interstage cooling.

decision) x_{N+1}, T_{N+1}. If it is assumed that the cost of cooling from T_{N+1} to T'_n is negligible in relation to the cost of catalyst volume for achieving a conversion $x_n - x'_n$, and if each decision is made optimally, then the maximum value of the profit P is a function only of x_{N+1}.

The end point of reaction is obviously that at which any further reaction would be economically unattractive because of the rapidly decreasing rate (the limit being the equilibrium curve L_e at which $r_v = 0$). The locus of all such points is given by the curve L_1 in Figure 12.3a. Similarly, L'_1 to the left of the maximum curve represents the locus of the entrance temperatures. Thus, if the feed is initially at state A, it should be heated to B and reacted along the adiabatic path to C. The optimal location of the loci L_1 and L'_1 maximizes the profit from the operation of a single-stage adiabatic

reactor. Aris (1965) has defined a profit function based on the cost of holding time (in units of rate), by maximizing which the optimal paths L_1 and L'_1 can be determined. Without going into the mathematical details, we outline the procedure in its practical essence as follows.

1. In order to determine the locus L_1 of outlet states for a single stage, the following profit function should be maximized:

$$\mathbf{P}_1 = \int_{x_2}^{x_1} 1 - \left(\frac{\mathbf{C}_f}{r_{v,1}(x)}\right) dx \qquad (12.6)$$

where x_2 is the feed to the stage, or outlet from the preceding (second) stage (see Figure 12.2), $r_{v,1}(x_1)$ is the rate corresponding to the exit conversion x_1, and C_f is a cost factor. Differentiating with respect to x_1 and T'_1 (the parameters to be optimized) and setting the results to zero, we obtain

$$\frac{\partial \mathbf{P}_1}{\partial x_1} = 1 - \frac{\mathbf{C}_f}{r_{V,1}(x_1)} = 0 \qquad (12.7)$$

$$\frac{\partial \mathbf{P}_1}{\partial T'_1} = \mathbf{C}_f \int_{x_2}^{x_1} \frac{(dr_{V,1}/dT)}{r_{V,1}^2(x)} dx = 0 \quad (12.8)$$

Equation 12.7 sets the limit to the rate contours on the x–T plane. Thus if $r_{V,1}(x,T) > C_f$ the operation would be uneconomical, leading to the limit $r_{V,1}(x,T) = \mathbf{C}_f$. It is also obvious that only the portion to the right of the point M of this rate contour (see Figure 12.3a) can be considered, for the portion to the left would be uneconomical. The portion to the right then represents the locus L_1 of exit states from stage 1.

2. To get the locus of inlet states, we make use of Eq. 12.8. What we look for is the inlet point x'_1, T'_1 on an adiabatic path such that the integral in 12.8 just vanishes, that is, the positive part just cancels the negative part. Note that in this integral $(dr_{V,1}/dT)$ can be readily calculated as a function of x; x_1 is fixed by the exit requirement; and $r_{V,1}(x_1) = \mathbf{C}_f$. Thus the value of x_2 (or x'_1) at which the integral vanishes can be determined by graphical integration. The calculations are repeated for different values of $r_{V,1}(x_1) < \mathbf{C}_f$ to provide the locus of inlet states L'_1.

3. To extend the design to a multistage system let us first consider two stages, 1 and 2. The profit function to be maximized is now

$$\mathbf{P}_2 + \mathbf{P}_1 = \int_{x'_2}^{x_2} \left(1 - \frac{\mathbf{C}_f}{r_{V,2}(x)}\right) dx$$

$$+ \int_{x'_1}^{x_1} \left(1 - \frac{\mathbf{C}_f}{r_{V,1}(x)}\right) dx \qquad (12.9)$$

Differentiating this with respect to x_2 and T'_2 and setting the derivatives to zero, we have

$$\frac{\partial(\mathbf{P}_2 + \mathbf{P}_1)}{\partial x_2} = \left[1 - \frac{\mathbf{C}_f}{r_{V,2}(x_2)}\right] dx$$

$$- \left[1 - \frac{\mathbf{C}_f}{r_{V,1}(x'_1)}\right] dx = 0 \qquad (12.10)$$

$$\frac{\partial(\mathbf{P}_2 + \mathbf{P}_1)}{\partial T'_2} = \mathbf{C}_f \int_{x_3}^{x_2} \frac{1}{r_{V,2}^2(x_2)} \left(\frac{\partial r_{V,2}}{\partial T}\right) dx = 0 \qquad (12.11)$$

It is clear from 12.10 that for maximum profit $r_{V,2}(x_2) = r_{V,1}(x'_1)$; that is, the rate at the exit of bed 2 should be equal to the rate at the inlet to bed 1. Such a condition is possible, as indicated by the points A and B in Figure 12.4. The point B corresponds to the outlet from bed 2 and the locus of all such points L_2 can be readily drawn.

The locus of inlet states L_2 can be drawn from Eq. 12.11 by noting the points at which the integral vanishes (as in step 2).

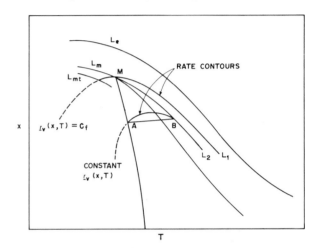

Figure 12.4.   Location of inlet states.

4. The procedure can be repeated to obtain the loci L_3 and L'_3, L_4 and L'_4, and so on. The loci for a three-stage reactor are shown in Figure 12.3b together with the graphical procedure for fixing the inlet and outlet states of each stage for a given feed condition.

In method 2, involving cooling by cold-shot mixing, an additional consideration is the determination of the optimum recycle fraction (Aris, 1965; Narsimhan, 1969). Figure 12.5a shows that the fraction of flow through a stage is simply the ratio OM/ON where point N lies on the ray connecting the exit state to the origin. The locus of optimum inlet states should pass through the points M on different rays. A procedure can be evolved (see Aris, 1965) for determining the loci of the optimum inlet and outlet states based on the scheme sketched in Figure 12.5b, in which the recycle fraction $F_{R,n}$ is included in the definition of the profit function.

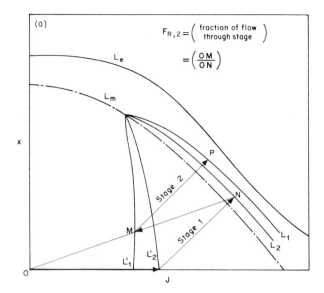

(a)

$F_{R,2} = \left(\begin{array}{c}\text{fraction of flow} \\ \text{through stage}\end{array}\right)$

$= \left(\dfrac{OM}{ON}\right)$

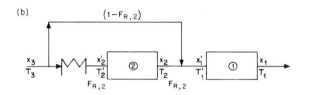

(b)

Figure 12.5. Adiabatic bed with cold-shot cooling.

12.1.3. Example: Optimization of a Multibed Adiabatic Reactor for the Catalytic Oxidation of Sulfur Dioxide: Comparison of Methods

Studies have been reported by Chartrand and Crowe (1969) on the optimum conversion obtainable in an existing four-bed converter of Canadian Industries Ltd. for SO_2 oxidation with both direct and indirect cooling between beds. Also, the efficiencies of several methods of optimization—dynamic programming, gradient search, Hooke and Jeeves's pattern search, and the discrete maximum principle—have been compared. Some of these methods were mentioned earlier in this chapter, and have been dealt with in detail by Wilde and Beightler (1967).

The reactor beds were essentially adiabatic and vanadia catalyst was employed. The following fixed data for the optimization process were available: reactor bed diameter 5.18 m; total volume of catalyst available 35.94 cm³; total air available to add to the reactor 159.8 kg mol/hr; reactor feed rate 1328 kg mol/hr; reactor feed composition 9.5 % SO_2,

11.5 % O_2, 79 % N_2; reactor feed pressure 1.2 kg/cm² absolute. The decision variables were the temperature into each bed, the catalyst distribution among the beds, and air addition before each bed. The differential equations describing the individual beds were based on the assumption of plug flow, negligible axial dispersion, uniform radial temperature and concentration profiles, adiabatic reactor walls, and ideal gas behavior. The equations are

$$\frac{dF_{SO_3}}{dl} = A_c r_{vSO_3} \tag{12.12}$$

$$\frac{dT}{dl} = \frac{A_c r_{vSO_3}(-\Delta H)}{\sum_i F_i C_{pm,i}} \tag{12.13}$$

The kinetic equation used was that of Kubota et al. (1959), namely

$$r_{vSO_3} = \frac{\varepsilon k_p \left[p_{O_2} - (p_{SO_3}/K p_{SO_2})^2 \right]}{D^2} \tag{12.14}$$

with

$$D = \begin{cases} \dfrac{1 + K_a p_{SO_3}}{p_{SO_2}} & \text{for} \quad T < 450°C \\[3mm] \dfrac{1 + K_b p_{O_2}^{1/2}}{p_{SO_2}} & \text{for} \quad T > 500°C \end{cases}$$

where k_p is the rate constant and K the equilibrium constant. The values of k_p, K_a, and K_b are given as functions of temperature. An approximation of constant effectiveness factor ε for each bed was made because of paucity of data.

The gradient search technique employed was similar to that of Zellnick et al. (1962). Here the gradient at a point was found numerically and the maximum along the gradient direction was found to within $\pm 1°C$ by golden section (Wilde and Beightler, 1967). The random search was done around a stationary point since none was reached. The gradient search is inefficient, since inaccuracies in the numerical determination of the gradient and the maximum in the gradient direction can lead to problems when a ridge exists in the response surface (as was the case in the study), and the computer time involved is excessive.

The discrete maximum principle was equivalent to finding zeros of the first partial derivatives of the objective function with respect to the decision variables, and hence it had the disadvantage that it could locate only a stationary point and not an optimum. When the second derivative was to be evaluated numerically this test could be somewhat inconclusive. In the present study the stationary point obtained did happen to be the optimum.

There was a wide variation possible in the choice of T_i with very little effect on either the optimum inlet temperatures to the remaining beds or an optimum conversion. This corresponded to a resolution ridge, and Hooke and Jeeves's method could move along this ridge, giving its shape along the way, whereas the gradient search could not progress very quickly to the optimum. Thus the prominence of the optimum peak could be revealed by this method, wherein perturbation of each independent variable was carried out about successively shifting base points.

Dynamic programming was the only method capable of finding a global maximum rather than local maxima, as was the case with the other methods. Generally, if more than one maximum exists, the global maximum can be found only by comparing local ones.

For the optimization problem given here:

Number of beds	Four
Bed inlet temperature	Varied
Catalyst bed depths	Fixed
An addition to beds	Fixed
Constraint on temperature	a. None
	b. $T \leqslant 600°C$.

a comparison of the methods of optimization yielded the results given in Table 12.1.

The most useful method of optimization for the generalized problem was found to be that of Hooke and

TABLE 12.1. Comparison of Optimization Methods

Method	Conversion to SO_3 (%)	Computing time[a]
Dynamic programming	a. 97.78	a. 20 min 26 sec
	b. 97.76	b. 21 min 19 sec
Gradient search	a. 96.84	a. 23 min[b]
	b. 96.84	b. 23 min
Hooke and Jeeves	a. 97.80	a. 7 min 53 sec
	b. 97.78	b. 9 min 33 sec
Discrete maximum	a. 97.78	a. 5 min 6 sec
principle	b. —	b. —

[a]On an IBM 7040.
[b]Optimum conversion not reached within the computer time limit.

Jeeves. The discrete maximum principle was the fastest but did not show the contours near the optimum. Dynamic programming was most useful for studying a range of values of the initial conversion.

The optimum conversion, or values very close to it, could be obtained with a range of air addition rates and inlet temperatures to the first bed as well as a range of catalyst distributions. With the imposition of a temperature constraint of 600°C the range of feasible inlet temperatures to the first bed was greatly reduced. The temperatures into the third and fourth beds were always very close to 452°C and 440°C, respectively, no matter what the catalyst distribution and the temperature of the first bed. Within the limitations of this model the conversion of the industrial plant studied could be increased by 0.25 % (to 97.84 %).

12.1.4. Design by a Simpler Approximate Method

In practice the method outlined above is difficult to use since the loci of outlet points often tend to coincide with each other. This is clearly seen in the solution of a problem involving the design of a multistage adiabatic reactor system for the oxidation of SO_2 (Lee and Aris, 1963). A simpler method, in which the optimization is not as rigorous as in the dynamic programming method, can be usefully employed (Levenspiel, 1977). In this method, constant rate contours are drawn on a concentration–temperature plot and the number of stages is determined by making use of the criterion demanding the equivalence of rates between the outlet of a given stage and the inlet to the next stage (see step 4).

12.1.5. A Shortcut Method Based on Constant Conversion Plots

Hougen and Watson (1947) suggested a simple practical method of attempting a preliminary design of an adiabatic reactor. A plot of rate vs. temperature is prepared for various values of conversion. A representative plot for SO_2 oxidation is shown in Figure 12.6. The maxima in the curves are characteristic of reversible exothermal reactions as well as reactions following H–W kinetics in which the decrease in the adsorption equilibrium constant with temperature is greater than the increase in the rate constant.

Adiabatic lines are then superimposed on these constant conversion curves and the average rate in passing from one conversion level to another (close to it) along this path is directly obtained from the plot. The

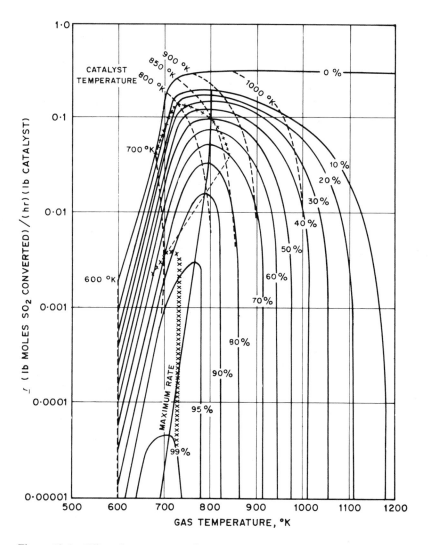

Figure 12.6. Effect of temperature and conversion on the rate of catalytic oxidation of sulfur dioxide (composition of entering gas: 7.8% SO_2, 10.8% O_2, 81.4% N_2); (Hougen and Watson, 1947).

quantity of catalyst required for this incremental conversion is

$$W = F \frac{\Delta x}{r_{av}} \qquad (12.15)$$

from which the total weight of catalyst for accomplishing a specified conversion can be obtained:

$$W = F \sum \frac{\Delta x}{r_{av}} \qquad (12.16)$$

The main feature of this plot is that the operations can be visualized, which is a distinct advantage in arriving at preliminary decisions regarding the number of beds

that may have to be used. More rigorous designs can then be undertaken using any of the methods described in the earlier sections.

Another quick method calls for operation of the reactor at an optimum exit temperature given by the curve L_m in Figure 12.1. An approximate value of this temperature can be calculated by the following equation (Denbigh, 1944):

$$\frac{T_{eq} - T_{opt}}{T_{eq} T_{opt}} = \frac{R_g}{(-\Delta H)} \ln\left[\frac{E + (-\Delta H)}{E}\right] \qquad (12.17)$$

where T_{eq} represents the equilibrium temperature, which lies on the curve L_e. Beyond the optimum locus, the temperature would have to be lowered (for an exothermic reaction) to maximize the rate. T_{opt} calculated from Eq. 12.17 gives the temperature for maximizing the outlet rate, and since it is influenced by a number of parameters namely, pressure, reactants ratio, and complexity of the reaction, it can only be regarded as providing an approximate preliminary basis for locating the true optimum. By combining this procedure with that described in Section 12.1.2, the optimum strategy, such as the number of beds or cold shots, can be estimated. However, the actual quantity of catalyst required for each bed for carrying out the desired conversion has still to be estimated from either kinetic data or pilot plant trials.

12.2. MULTIPLE STEADY STATES[†]

There are many situations where a reactor tends to operate at more than one steady state, jumping from one to another with even the slightest perturbation in the operating conditions. In simple terms this multiplicity of steady states is the "number of different sets of state variables at which the time rate of change of all state variables is identically zero for a fixed set of conditions or parameters".

A number of papers dealing with multiplicity and stability in chemical reactors have appeared in the literature over the past few years. They have been condensed into five significant books on the subject (Oppelt and Wicke, 1964; Gavalas, 1968; Perlmutter, 1972; Aris, 1975; Denn, 1975). In addition, the reviews of Ray (1972), Uppal et al., (1974), Schmitz (1975, 1978), Luss (1977), Varma and Aris (1977), Luss (1981), Hlavacek and van Rompay (1981), and Pismen (1980) are noteworthy.

In the present section we shall first consider multiplicity in a CSTR (a lumped-parameter system) and then in the homogeneous and heterogeneous models of the fixed-bed reactor (distributed-parameter systems).

12.2.1. Multiplicity in a CSTR

The concept of multiplicity of steady states can be understood by considering a simple exothermic reaction carried out adiabatically in a batch reactor. The heat generation curve is of the form represented by B in Figure 12.7; but if there is no depletion of the reactant, the effect of decreasing concentration does not show up and a shape similar to the sigmoidal curve A is obtained. For an exothermic reversible reaction A $\rightleftharpoons$ B the net rate of reaction, and therefore of heat generation, can

[†]As this is a vast subject by itself we present only a brief review in this section

diminish beyond a certain temperature if the acceleration due to temperature is greater for the reverse step than for the forward step.

If the system is not adiabatic, curve B will still be of the same type, but the actual temperature attained will now depend on the balance of heat generation (represented by curve B) and heat removal. The latter is given by the sum of the sensible heat and heat removal terms: $\rho_g C_p (T - T_0) + U A_h (T - T_C)$. In Figure 12.7 heat removal lines for three situations are shown. In the case of line 1 the intersection with the heat generation curve occurs at a very low temperature, whereas for line 3 it occurs at a high temperature. For line 2, on the other hand, there are three points of intersection, a, b, and c. Point b represents metastable operation. Point a, corresponding to a very low temperature, at which practically no reaction occurs, is called the extinction point; and c is referred to as the ignition point. This multiplicity of steady states is possible only in the region bounded by the dotted lines 4 and 5.

In this connection the studies of van Heerden (1953, 1958), Bilous and Amundson (1955), and Aris and Amundson (1958) have established that even for a simple exothermic reaction, multiple steady states, unstable states, and even limit cycles (sustained oscillatory outputs) are possible. Schmitz and Amundson (1963) and Gilles and Hofmann (1961) demonstrated that states other than the intermediate state could be unstable in nonadiabatic systems.

The static and dynamic conditions of Gilles and Hofmann (1961) assure the uniqueness of the steady state. The static condition demands that the slope of the heat removal curve exceed that of the heat generation curve at the steady state, while the dynamic condition demands that the trace of the coefficient matrix be negative. For a general-order reaction ($n \geqslant 0$) van den Bosch and Luss (1977) have proposed a criterion for the adiabatic CSTR to have a unique solution.

The multiplicity problem in CSTRs has been extended to include consecutive reaction schemes (Hlavacek et al., 1972; Cohen and Keener, 1976), parallel reaction schemes (Luss and Chen, 1975; Andersen and Michelsen, 1975; Michelsen, 1977; Pikios and Luss, 1979), reactors in series (Horak et al., 1971; Varma, 1981), and multiphase systems (Elnashaie and Yates, 1973; Bukur et al., 1974; Bukur, 1978) under both isothermal and nonisothermal conditions. The existence of multiplicity under isothermal conditions owes its existence to the presence of some form of autocatalytic feedback (Matsuura and Kato, 1967; Higgins, 1967; Sheintuch and Schmitz, 1977, 1978; Slinko and Slinko, 1978; Endo et al., 1978). The simplest example of isothermal multiplicity is provided by reacting systems

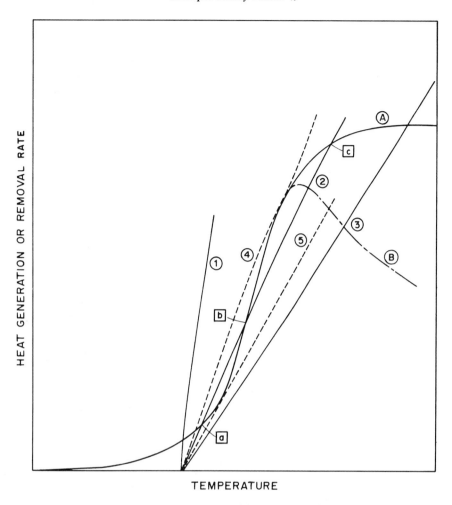

Figure 12.7. Existence of multiple solutions.

exhibiting Langmuir–Hinshelwood bimolecular kinetics. A more complicated situation involving two reactant species has been analyzed by Kulkarni and Ramachandran (1980a). Although the nonlinear dependence of the rate on the concentration of species (as evident in the studies cited) generates bistability of states, it still does not provide an explanation for certain phenomena, such as periodic behavior.

More recently, Ravi Kumar et al. (1983) have proposed an alternative form of autocatalysis to explain the occurrence of periodic behavior in isothermal systems. It has been assumed that the intermediate species has an autocatalytic influence on the rate constant of the previous step, and the influence is gauged in terms of an autocatalytic parameter α. For low values of α the reactor transients quickly approach a stable state, whereas for higher values long transients develop. There exists a critical value of the parameter α at which a stable node becomes an unstable focus with

a limit cycle surrounding it. Further increase of α converts the unstable focus into a stable focus. In the limit cycle region, the reactor behavior depends on the inlet feed composition, with amplitude and period of oscillations decreasing with decrease in concentration of the main reactant.

Most studies in CSTRs assume homogeneous conditions in the reactor and ignore any interphase or interparticle resistances. Kulkarni and Ramachandran (1982) have analyzed the case of systems that include these resistances and obtained an approximate analytical criterion for the occurrence of multiplicity.

12.2.2. Fixed-Bed Reactor: Multiplicity in Homogeneous Models with Axial Mixing

As pointed out by van Heerden (1953), multiple steady states can occur in a fixed-bed reactor as a result of axial diffusion of heat and mass. In other words, some kind of thermal feedback is necessary. Such multiplicity is observed only in certain regions of the parameter space. Attempts have been made to locate them and to obtain criteria for the uniqueness of the solution.

The continuity and conservation equations for the one-dimensional model with axial mixing and conduction are given by Eqs. 5 and 6 of Table 11.1. Raymond and Amundson (1964) solved these equations for a simple first-order reaction A → B carried out adiabatically. The existence of multiple solutions can be explained through plots of inlet vs. outlet temperatures. Figure 12.8 shows such plots for two different sets of parameter values. In Figure 12.8a there is a unique outlet temperature for any given inlet temperature. Hence the reactor will always have a unique solution. But in Figure 12.8b there exists a range of inlet temperatures in which three outlet temperatures are possible for a given inlet value. This indicates the presence of multiple steady states. The outer two are stable to small perturbations, whereas the middle one is unstable. Various other types of instabilities, including sustained oscillations for a nonadiabatic reactor, have also been demonstrated by Amundson (1965), Kuo and Amundson (1967a, b, c), Luss and Amundson, (1967a, b, 1968), and Varma and Amundson (1972a, b, 1973a, b).

A number of criteria have been developed to ensure the uniqueness of the steady state. They have been derived for adiabatic as well as nonadiabatic quasi-homogeneous reactors and are summarized in Table 12.2. Han and Agarwal (1973) and Varma and Amundson (1972a, b) have worked out sufficient conditions for uniqueness for the nonadiabatic case with

$Pe_m = Pe_h$. In general, the common feature of these criteria is that they are satisfied for large values of the Peclet number. Hlavacek and Hofmann (1970) obtained, after extensive numerical computations, the limits of the region of parameter space wherein multiple states are observed. The parameters that determine the operation of the reactor are β_m, α_s, the Damköhler number (Da), and the Peclet number (Pe) based on reactor length. Hlavacek and Hofmann (1970) have shown that the region of multiple solutions is very narrow; also, the range of Peclet numbers over which multiplicity is observed lies far below the range of Peclet numbers observed in industrial reactors (Froment, 1972). The length of the reactor tubes in industrial reactors is so large that the possibility of observing multiple solutions is usually negligible.

At very low Peclet numbers the performance equations in the homogeneous model reduce to those of a CSTR. The qualitative features of the CSTR are thus expected to be reproduced at smaller values of the Peclet number. The analysis by Cohen (1972) using the singular perturbation technique and that by Hlavacek and Hofmann (1970) using a linearization method indeed show similar trends and parameter effects. However, the CSTR analogy seems to fail under certain conditions, such as for a first-order nonadiabatic reaction (McGowin and Perlmutter, 1971; Hlavacek et al., 1971; Varma and Amundson, 1973a, b).

Besides the simple exothermic reaction using the homogeneous model with axial mixing, more complex reactions like autocatalytic reactions have also been investigated for multiplicity (Chen, 1972; Lin, 1979).

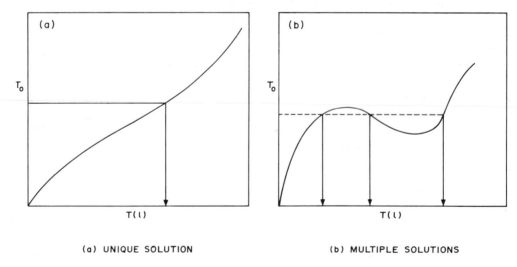

(a) UNIQUE SOLUTION   (b) MULTIPLE SOLUTIONS

Figure 12.8. Multiple solutions due to axial mixing.

TABLE 12.2. Criteria for Uniqueness of Solution for a Quasihomogeneous Reactor with Axial Mixing

Reactor Type	Kinetic Expression	Criterion	Remarks	Reference
Adiabatic	First order	$\alpha_0 \dfrac{T_{ad} - T_0}{T_0} \leqslant 1$	$Pe_m = Pe_h$	Luss and Amundson (1967)
Adiabatic	First order	$\alpha_0 \dfrac{T_{ad} - T_0}{T_0} \leqslant 1$	$Pe_m = Pe_h$	Luss (1968)
Adiabatic	First order	$\alpha_0 \dfrac{T_{ad} - T_0}{T_0} \leqslant \dfrac{4}{1 + [4/(E/R_g T_0)}$	$Pe_m = Pe_h$	Hlavacek and Hofmann (1970)
Nonadiabatic	First order	$\sqrt{2}\, Da \exp\left(\dfrac{B}{1 + \alpha_0 B}\right)$ $\left\{\left(\dfrac{4}{Pe_m}\right)^2 + \left(\dfrac{Pe_h B}{(Pe_h^2/4) + Pe_h B \rho_g/\alpha_0}\right)^2\right\}^{1/2}$ $\leqslant 1$	$Pe_m \neq Pe_h$	Han and Agrawal (1973)

Replace B by P for $P < B$ where

$$B = \frac{(-\Delta H)C_{A0}\alpha_0^2}{C_p T_0} > \alpha_0 = \frac{E}{R_g T_0}$$

P is the dimensionless maximum reactor temperature, defined as

$$\sup_{T^*}\left[\max_{0 \leqslant z \leqslant 1} T^*(z)\right] \leqslant P$$

and

$$T^* = \frac{\alpha_0(T - T_0)}{T_0}$$

12.2.3. Fixed-Bed Reactor; Multiplicity in Heterogeneous Models

The explicit recognition of the catalyst phase complicates the analysis, since the catalyst particle itself can exhibit multiple states. It is likely, too, that for a given value of the fluid temperature in a fixed-bed reactor the particle temperature will assume more than one value, again leading to multiple solutions. Thus an important design requirement is a criterion for avoiding such multiple solutions. In this section we shall be concerned with the formulation of criteria for the uniqueness of the steady state for the following situations: interphase gradients only; intraparticle gradients (the catalyst particle problem); both inter- and intraphase gradients;

and isothermal conditions. Finally, examples of multiplicity under various conditions will be cited.

Interphase Gradients Only

Liu and Amundson (1962, 1963), Wicke (1961), and Wicke and Vortmeyer (1959) investigated the reactor behavior under conditions of plug flow with interphase gradients but no intraphase gradients. The heat and mass balance equations for a catalyst pellet in such circumstances are given in Section 11.4. In addition to these equations, we have

$$-h_{fp}(T_g - T_s) = (-\Delta H)k_g(C_{Ab} - C_{As}) \quad (12.18)$$

Substituting this in the mass balance equation, we get

$$h_{fp} A_p (T_s - T_b) = (-\Delta H) V_p A_f \exp\left(\frac{-E}{R_g T_s}\right)$$

$$\times \left[C_{Ab} - \frac{h_{fp}}{k_g(-\Delta H)} (T_s - T_b) \right] \quad (12.19)$$

This is an implicit equation in T_s. The left- and right-hand sides of the equation correspond to the straight line and sigmoid curve, respectively, of Figure 12.7. It can be seen that, depending on the system parameters, the intersection of the two curves can occur at one or three points, leading, respectively, to unique or multiple solutions. The situation considered here is, however, rather uncommon, because in industrial reactors an intraphase concentration gradient always accompanies an interphase concentration gradient.

Intraparticle Gradients Only: The Catalyst Particle Problem

The governing differential equations for mass and energy conservation have already been stated in Chapter 4 (see also Section 11.4). In studies concerning multiplicity, stability, and sensitivity, these equations are used with varying assumptions. Frequently a regular geometry is assumed, the mass and thermal diffusivities are assumed to be equal, and interphase gradients are ignored. These assumptions yield results

that are unrealistic for most practical systems, since $(Bi)_m/(Bi)_h$ is normally greater than 50. Also, the ratio of mass to thermal diffusivities always exceeds unity, with an average value of about 30.

The analogy of the catalyst particle problem with the CSTR becomes evident when the diffusion terms are replaced by the flow terms. Multiplicities are therefore to be expected under certain conditions in this case also (see Chapter 4). Criteria for uniqueness of the steady state have been reported by Weisz and Hicks (1962), Drott and Aris (1969), and Luss (1971).

Weisz–Hicks: $\alpha_s \beta_m < 5$ (12.20)

Drott–Aris: $\alpha_s \beta_m < 4.074 \beta_m + 4.132$ (12.21)

Both are restricted to a first-order reaction. As can be seen from the values listed in Table 12.3, these conditions for uniqueness are satisfied by many industrial systems.

Jouven and Aris (1972) have presented a critical plot of β_m vs. α_s, reproduced in Figure 12.9, that represents the interface between uniqueness and multiplicity. It is clear from this plot that for the normal values of α_s encountered (20–40), β_m would have to exceed 0.2 for the appearance of multiplicity. Since usually $\beta_m < 0.05$, the chances of multiplicity are remote. Hence multiplicity due to intraphase gradients is not of any practical importance.

TABLE 12.3. Intraparticle Transport Parameters for Some Common Systems (Hlavacek et al., 1969).

System	α_s	$\alpha_s \beta_m$
Synthesis of ammonia	29.4	0.0018
Synthesis of higher molecules from CO and H_2	28.4	0.024
Oxidation of methanol to formaldehyde	16.0	0.175
Hydrochlorination of acetylene to vinyl chloride	6.5	1.65
Hydrogenation of ethylene	23–27	1–2.7
Oxidation of hydrogen	6.75–7.52	0.21–2.3
Oxidation of ethylene to ethylene oxide	13.4	1.76
Decomposition of nitrous oxide	22	1–2
Hydrogenation of benzene	11–16	1.7–2.0
Oxidation of sulfur dioxide	14.8	0.175
Decomposition of ammonia	24.3	0.26
Oxidation of naphthalene	30	0.45

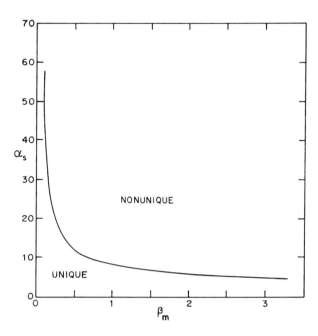

Figure 12.9. Plot showing demarcation between regions of unique and nonunique solution (Jouven and Aris, 1972).

A number of investigators have emphasized the analogy between the catalyst problem and the CSTR, including Kuo and Amundson (1967a, b, c), Jackson (1973), Wei (1965), Hlavacek and Marek (1971), Lee and Luss (1970), Yang and Lapidus (1974), Gavalas (1966), and Wedel et al. (1977). Instances have been reported, however, where the intraparticle gradient problem deviates from the analogous CSTR problem (Copelowitz and Aris, 1970; Michelsen and Villadsen, 1972; Hlavacek and Marek, 1968).

The problem of intraparticle gradients has also been investigated for more complex kinetics. In general, the number of steady states is an odd number $(2m + 1)$ with at least m unsteady states (Gavalas, 1968).

As in the case of a CSTR, multiplicity of states has been shown to exist in the case of a single particle under isothermal conditions. Becker and Wei (1977) have analyzed the case of H–W bimolecular kinetics to show that the effectiveness factor can indeed exceed unity under certain conditions, leading to multiplicity for certain ranges of parameter values (typically for KC_{As} greater than 8). The case of a finite cylinder for the same reaction scheme has been analyzed by Ho (1976) and a criterion for uniqueness has been proposed. More complex reaction schemes have also been considered (Cresswell and Patterson, 1970; McGreavy and Thornton, 1970a, b; 1972; Pareira and Varma, 1978; Hegedus et al., 1977; Kulkarni and Ramachandran,

1980a; Sadana et al., 1980) under both nonisothermal and isothermal conditions.

Both Interphase and Intraphase Gradients

The system equations that take into account both inter- and intraphase gradients are given in Tables 11.1 and 11.2. Solutions to these equations have been presented by Hatfield and Aris (1969), McGreavy and Cresswell (1969), and Hutchings and Carberry (1966). Imposition of interphase gradients results in multiple solutions in the region where the intraparticle problem gives a unique solution. In certain ranges of parameters two distinct regions of multiple solutions (each giving three solutions) have been observed. In some cases these two regions merge to give a region of five multiple solutions. Hlavacek et al. (1968) have proposed a method to locate the region of multiple solutions when the Biot numbers for heat and mass transfer are equal.

The effect of axial conduction in the solid phase has been taken into account by Eigenberger (1972) and his model describes the wandering or creeping profiles first observed experimentally by Wicke and Vortmeyer (1959).

Cresswell (1970) analyzed the problem of multiplicity by assuming the pellet to be isothermal (see Section 4.6.2) and using the following ranges of the parameter values.

α_s	β_m	$(Bi)_h$	$(Bi)_m$
10–40	0–0.1	0–10	100–500

Since in the region where multiple solutions are observed the reaction is restricted to the layers close to the pellet surface, flat geometry can be used; this also justifies the use of asymptotic solution in the analysis. On the basis of these assumptions the following criterion may be obtained for the uniqueness of the solution:

$$\alpha_s \beta_m < 8 \left[\frac{(Bi)_h}{(Bi)_m} + \beta_m \right] \qquad (12.22)$$

Since $(Bi)_h / (Bi)_m$ in most cases is less than 0.05 and α_s is greater than 8, a number of systems listed in Table 12.3 fail to satisfy this criterion. Hence multiplicity of this type can be observed in real systems.

The analysis of multiplicity given above makes use of the parameters α_s, β_m, and ϕ, which vary from point to point in the reactor. McGreavy and Thornton (1970b) restated the problem through the following relationship, using a new set of parameters:

$$\frac{1}{\alpha_b} = \frac{1}{\alpha_p} - B'(\text{Bi})_m \frac{A' - g}{\{[(\text{Bi})_m/2] - 1\}^{g+A}} \quad (12.23)$$

where

$$\alpha_b = \frac{E}{R_g T_b}, \qquad \alpha_p = \frac{E}{R_g T_p}$$

$$B' = \frac{(-\Delta H) C_{\text{Ab}} D_e R_g}{2 R h_{\text{fp}} E}, \qquad A' = \phi_{A_f} \exp\left(-\frac{\alpha_p}{2}\right)$$

$$g = \tanh A', \qquad \phi_{A_f} = R\left(\frac{A_f}{D_e}\right)^{1/2}, \qquad (\text{Bi})_m = \frac{2R k_g}{D_e}$$

$$(12.24)$$

where ϕ_{A_f} is a modified Thiele modulus based on the Arrhenius preexponential factor A_f. The use of the modified form of the Thiele modulus greatly generalizes the treatment and makes the equations applicable throughout the reactor.

For fixed values of $(\text{Bi})_m$ and the modified Thiele

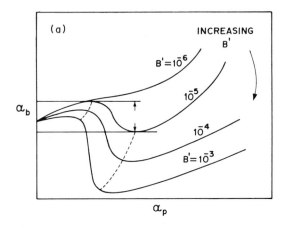

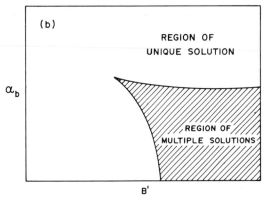

Figure 12.10. Multiplicity of solution due to interphase and intraphase gradients.

parameter, Eq. 12.23 gives curves of the form shown in Figure 12.10a with B' as parameter and the dimensionless particle and fluid temperatures as ordinates. It will be noticed that as B' increases, multiple solutions make their appearance. The region of multiple solutions can be precisely represented by noting that $d\alpha_b/d\alpha_p = 0$ at two points on each of the curves shown in Figure 12.10a; multiple solutions are possible only between these two extremes for given values of B'. This situation can be represented on an α_b–B' plane by plotting the lower and upper limits of the fluid temperature (i.e., the loci of the maxima and minima) as functions of B'. The region bounded by the resulting curves, as shown by the shaded area in Figure 12.10b, represents the region of nonunique solutions.

The chief advantage of this treatment is that the operation of the reactor can be represented on the α_b–B' plane by a series of trajectories corresponding, say, to different radial positions or to adiabatic operation. It can be shown, for instance, that the adiabatic operation line passes through the nonunique region, so that the adiabatic reactor must be regarded as a potentially unstable one.

12.2.4. Multiplicity Induced by Kinetics Coupled or Uncoupled with Diffusion

In the discussion presented above a change in temperature was usually involved in generating a curve with a maximum and with the necessary properties for producing multiplicity (Figure 4.4). It is possible, however, to obtain curves of this type purely as a result of kinetics, coupled or uncoupled with diffusion. Thus negative-order reactions coupled with isothermal intraphase diffusion resistance can lead to multiple solutions (Roberts and Satterfield, 1966; Schneider and Mitschka, 1966). Multiplicity can also be induced by multicomponent diffusion (Lowe and Bub, 1976). Reactions following Hougen–Watson models and exhibiting a rate maximum (such as a self-inhibited rate expression) can also exhibit multiple solutions (Denbigh et al., 1948; Matsuura and Kato, 1967; O'Neill et al., 1971; Bruns et al., 1973; Pareira and Varma, 1979). Uppal and Ray (1977) have demonstrated clearly the existence of multiplicity in H–W type kinetics.

Multiplicity can also be induced by volume changes occurring during reaction or by surface diffusion. These physical factors can alter the effectiveness factor curve to either include a region of multiple solutions or shift (or expand) an existing region of multiplicity. Micromixing effects can also induce multiplicity for reactions with a rate maximum. In fact, certain patterns of micromixing may give rise to unexpected multiple steady states at intermediate degrees of segregation when only a single state is possible at the condition of maximum mixedness (Dudukovic, 1977). In the treatments of the CSTR, maximum mixedness was assumed.

The kind of multiplicity described above arose as a result of the interaction between transport and kinetic effects. It is of interest that multiplicity may also occur because of kinetic factors alone. Heterogeneously catalyzed surface reactions are usually described by multistep mechanisms involving chemisorption, reaction, and desorption of the reacting species. The complete consideration of these steps frequently leads to complex kinetic models with implicit multiple solutions. Thus Eigenberger (1978) has considered the simple three-stage mechanism

$$A + l \rightleftharpoons Al$$

$$B + l \rightleftharpoons Bl$$

$$AZ + Bl \rightarrow AB + 2l$$

to show the existence of multiplicity. His analysis shows that competing chemisorption of A and B upon the same active site can lead to rate multiplicities. A similar conclusion emerges from the work of Bykov et al. (1976a, b). Besides the three-stage mechanism just discussed, multiplicity may also occur as a result of the generation or annihilation of active centers during the course of reaction, as suggested by Barelko and Volodin (1976) and Ravi Kumar et al. (1981). It is interesting to note at this point that such generation or annihilation of active centers has been experimentally recorded for some systems in recent years.

These models which explain the rate multiplicity are, however, inadequate to explain the experimentally observed instabilities (see, e.g., Beusch et al., 1972; Hugo and Jakubith, 1972; Belyaev et al., 1976; Cant and Hall, 1971; Dauchot and van Cakenberghe, 1973; Plichta and Schmitz, 1979; Zuniga and Luss, 1978; Schmitz et al., 1977, 1979; Rossler and Wegmann, 1978; Yamazaki et al., 1978). Belyaev et al. (1976) and Pikios and Luss (1977) explain these instabilities by accounting for the variation of the activation energy of the surface reaction step with surface coverage of the reactant species. Eigenberger (1978), on the other hand, favors the assumption of the presence of a slow buffer step along with the other main steps to explain these instabilities. Daggonier and Nuyts (1976) suggest the existence of a

putative temperature on the surface as a possible reason for the occurrence of instability, while in a series of papers Takoudis et al. (1981a, b) explored the possibility that vacant sites were responsible for this phenomenon.

The studies just mentioned explain the occurrence of one-peak-per-cycle oscillations. Several experimental systems, however, indicate the existence of multipeak-per-cycle oscillations. Schmitz et al. (1980) have provided an explanation for this type of behavior by invoking the interaction between external diffusion and nonlinear kinetics of the reaction. Jensen and Ray (1980) have presented a more detailed analysis that takes into account the metal crystallite size distribution and invokes the coupling of internal diffusion with nonlinear kinetics of the reaction. More recently, Prasad and Kulkarni (1982) provided an analysis that does not depend on the interaction of diffusion and reaction. The true kinetic explanation for the occurrence of chaotic behavior provided by them assumes different heterogeneous patches on the catalyst surface. When the patches are out of phase with each other, they are shown to generate chaotic behavior. The recent papers by Hlavacek and van Rompay (1981), Pismen (1980), and Kehlert et al. (1981) review the situation in regard to these complex situations.

12.2.5. Examples of Multiplicity

Although multiple solutions are not a normal occurrence under conditions of industrial operation, several experimental studies have been reported which demonstrate that multiplicities can be observed if the reaction is carried out within the predicted limits for the appearance of multiple solutions. The most commonly studied reaction appears to be the oxidation of carbon monoxide on different catalysts. This reaction has been shown to exhibit isothermal multiplicity as well as multiplicity under adiabatic operation. It is interesting to note that multiple steady states, which may sometimes appear in smaller reactor lengths, tend to disappear as the length is increased (Cresswell, 1970). The possibility of multiple steady states occurring in industrial nonadiabatic fixed-bed reactors appears to be very remote, even in the presence of film resistance (Smith, 1977; Hlavacek and Votruba, 1977).

Several examples of multiplicity reported in the literature are summarized in Table 12.4. The type of reactor used (CSTR, fixed-bed adiabatic, or fixed-bed nonadiabatic) is also indicated in the table, along with the type of multiplicity observed.

TABLE 12.4. Examples of Observed Steady-State Multiplicity and Instability

System	Reactor	Remarks	Reference
Polymerization of ethylene	Nonadiabatic tubular reactor	Oscillatory states	Volter (1963)
Oxidation of ethane on Pd–Al$_2$O$_3$ catalyst	Fixed-bed reactor	Multiple steady states observed, probably due to influence of both axial dispersion and interphase transport	Wicke et al. (1968)
Decomposition of N$_2$O on CuO catalyst	Adiabatic circulating reactor	Oscillatory states	Hugo (1968)
Vapor-phase chlorination of methyl chloride	Nonadiabatic CSTR	Oscillatory states	Bush (1969)
Mildly exothermic water–gas shift reaction	Autothermal reactor with internal countercurrent heat exchange	Multiple states; both one-dimensional and two-dimensional pseudohomogeneous models employed to predict the behavior	Ampaya and Rinker (1977)
Reduction of NO with CO over fiberglass-supported catalyst	Nonadiabatic fixed-bed reactor	Instability	Schleppy and Shah (1977)
Oxidation of CO on Pt–Al$_2$O$_3$ catalyst	Fixed-bed reactor	Multiple steady states observed, probably due to influence of both axial dispersion and interphase transport	Padberg and Wicke (1967); Wicke et al. (1968)
Isothermal oxidation of CO on Pt catalyst	Recirculating reactor	Oscillatory states and multiple steady states	Hugo (1970); Hugo and Jakubith (1972)
Oxidation of CO on Pt–Al$_2$O$_3$ catalyst	Adiabatic fixed-bed reactor	Multiple steady states	Fieguth and Wicke (1971)
Oxidation of CO on CuO–Al$_2$O$_3$ catalyst	Adiabatic circulating reactor	Oscillatory and multiple states	Eckert et al. (1973a, b)
Oxidation of CO on Pd–Al$_2$O$_3$ and CuO–Al$_2$O$_3$ catalyst	Adiabatic fixed-bed reactor	Three stable states	Votruba and Hlavacek (1973; 1974)
Oxidation of CO on Pt–Al$_2$O$_3$	Isothermal integral reactor	Several stable steady states observed; catalyst aging widens the region of multiplicity; reaction kinetics (H–W model) and intraparticle diffusion successfully employed to interpret the phenomenon	Hegedus et al. (1977)
Oxidation of CO on supported platinum Pt–α–Al$_2$O$_3$	CSTR spinning-basket reactor	Sustained isothermal oscillations	McCarthy et al. (1975)
Oxidation of CO on Pt–SiO$_2$ pellets	Pellet	Oscillating states	Dagonnier and Nuyts (1976)
Oxidation of CO on Pd–Al$_2$O$_3$ catalyst	Diluted adiabatic reactor		Votruba et al. (1976)
Simultaneous CO and 1-butene oxidation over supported Pt catalyst	Gradientless recyle reactor	Multiple steady states and oscillations	Cutlip and Kenney (1978)

Table 12.4. (continued)

System	Reactor	Remarks	Reference
Isothermal oxidation of H_2 on Pt wires	Continuous-flow reactor	Sustained multipeak relaxation oscillations observed; steady-state multiplicity leading to two different oscillatory states at the same operating conditions also observed	Zuniga and Luss (1978)
Isothermal CO oxidation by supported Pt on α-Al_2O_3	CSTR	Spurious limit cycles resulting from trace impurities (e.g., N_2, CH_4, Ar, H_2O) in the O_2 stream observed	Varghese et al. (1978)
Isothermal CO oxidation on Pt foil	Gradientless CSTR	Varying catalyst activities yielding unusual dynamic behavior reported	Plichta and Schmitz (1979)
Ethylene oxidation on polycrystalline Pt films	CSTR	Self-sustained multipeak oscillations reported	Vayenas et al. (1980)
Isothermal oxidation of H_2 on nickel plates	Continuous-flow reactor	Periodic oscillations observed when either significant mass or heat transfer limitations existed. Aperiodic behavior was noted when transport resistances were small	Kurtanjek et al. (1980)
Propylene oxidation on Pt wires	Continuous-flow reactor	Aperiodic behavior and hysteresis effects for variations in oxygen concentrations observed	Sheintuch and Luss (1981)
CO oxidation on Pt–alumina catalyst	Recycle reactor	"Soft" and "hard" complex oscillations observed; soft oscillations arise around steady states of high parametric sensitivity, with small changes in exit conversion; hard oscillations are more or less symmetrical, characterized by large changes in exit conversions.	Rathouský et al. (1981)

12.3. HEAT EXCHANGE IN FIXED-BED REACTORS

12.3.1. Adiabatic Versus Nonadiabatic Operation

One of the first decisions in the design of a fixed-bed reactor is the mode of heat exchange to be employed. An important consideration in this decision is whether to operate the reactor adiabatically or nonadiabatically. Two parameters are useful in making a preliminary choice: adiabatic temperature change at complete conversion, and temperature sensitivity. Several other practical considerations would also contribute to the final decision.

The adiabatic temperature change (ATC) is the calculated temperature change accompanying a reaction when it is allowed to go to completion, even though complete conversion may be impossible because of equilibrium limitations. It is therefore a measure of the maximum temperature change, and for a simple reaction A → products, is given by

$$\text{ATC} = \frac{(-\Delta H_A) y_{AO}}{C_{pm}} \qquad (12.25)$$

For a complex reaction scheme the ATC may be calculated for the individual steps, and the highest value considered in evaluation. Obviously the relative rates of the individual reactions would have to be computed for a firmer estimate of the temperature change. A parameter of primary importance in estimating ATC is the heat of reaction. This can be calculated from the heats of formation of the reactants and products obtained either from standard handbooks (Perry, 1973) or from group contributions (Verma and Doraiswamy, 1965). The heat capacity of an organic compound as a function of temperature can be obtained by the method of Rihani and Doraiswamy (1965).

The temperature sensitivity of a reaction is an equally important parameter in assessing its thermal behavior. It is a measure of the effect of temperature on the reaction rate and therefore on the rate of heat generation. For a simple reaction, the rate equation is $-r_{wA} = A_f \exp(-E/R_g T) f(C_A)$. Differentiating this with respect to temperature at constant conversion (concentration) leads to

$$\left| \frac{d(-r_{wA})}{dT} \right|_{C_A} = (-r_{wA}) \frac{E}{R_g T^2} \qquad (12.26)$$

The term $E/R_g T^2$ (with units of reciprocal temperature) is a measure of the temperature sensitivity (TS) of the reaction:

$$\text{TS} = \frac{E}{R_g T^2} \qquad (12.27)$$

Multistage adiabatic reactors are more common industrially than single-stage reactors. As already pointed out, these multistage units involve interstage cooling or heating. Values of the ATC can be as high as about $-1000°$K for ethane cracking and about $-680°$K for steam reforming, with corresponding TS values of 0.029 and 0.036°K^{-1}. In view of the very high values of the reaction temperature and ATC for both these reactions, they are carried out in a furnace. On the other hand, milder endothermic reactions with ATC in the range from -30 to $-150°$K and TS in the range 0.01–0.6 are well suited for adiabatic operation. A typical example is the styrene reaction.

In the case of exothermic reactions the choice is more critical, since temperature excursions leading to runaway or undesired side products can occur more easily. For the aniline reaction, where side reactions are not very significant, either adiabatic or nonadiabatic operation can be employed. Present-day practice, however, favors nonadiabatic operation exclusively. Well-known

examples of adiabatic exothermic reactions are oxidation of SO_2 (ATC = 180°K, TS = 0.03°K^{-1}); ammonia synthesis (ATC = 630°K, TS = 0.04°K^{-1}); shift reaction (ATC = 90°K, TS = 0.012°K^{-1}). In spite of the high temperature sensitivity of the ammonia synthesis reaction, it is carried out adiabatically, but an elaborate method of internal heat exchange is used.

12.3.2. Types of Heat Exchange in a Fixed-Bed Reactor

In formulating procedures for the design of multitubular reactors in Chapter 11, a constant wall temperature was assumed and no attention was paid to the nature of heat exchange outside the catalyst tube. Two distinct methods of heat exchange are possible: (1) through an external control fluid, and (2) through internal heat exchange with the product. Considering the possibility of using cocurrent or countercurrent operation in each case, four different modes of heat exchange can be employed, as shown in Figure 12.11.

The heat balance equations for the four modes can be readily written by considering the following: sensible heat of the reactant stream $uA_c C_p \rho_g (T - T_0)$; sensible heat of the coolant $u_C A_c \rho_C C_{pC} (T_C - T_{C0})$; and rate of heat generation $(-\Delta H')uA_c x \rho_g$. From these quantities the control fluid temperature at any position can be expressed in terms of its inlet value T_{C0}, the temperature and conversion at that position within the reactor, and the parameter H:

$$T_C = T_{C0} + \xi(Hx - T + T_0)$$

where

$$\xi = \frac{uC_p \rho_g}{u_C C_{pC} \rho_C} \qquad (12.28)$$

The term ξ represents the ratio of the heat capacities of the reactant and control fluid streams, and has the following values:

Value	Condition
Positive	Cocurrent operation
Negative	Countercurrent operation
Zero	Isothermal jacket
Infinity	Adiabatic operation
Unity	Reactant is used as control fluid cocurrently $(+1)$ or countercurrently (-1).

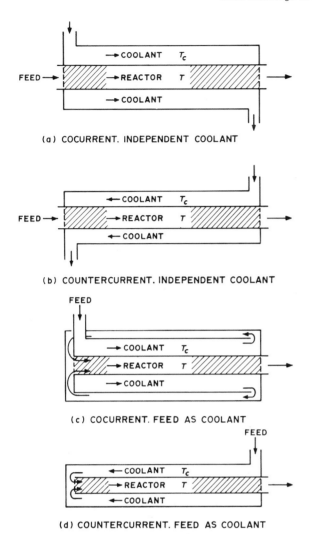

(a) COCURRENT. INDEPENDENT COOLANT

(b) COUNTERCURRENT. INDEPENDENT COOLANT

(c) COCURRENT. FEED AS COOLANT

(d) COUNTERCURRENT. FEED AS COOLANT

Figure 12.11. Arrangements for cooling the tubular reactor.

Heat balance expressions for the different situations mentioned have been derived by Aris (1965).

Figure 12.12a depicts the situation for an exothermal reaction with a constant wall temperature (the case considered in Chapter 11). If point A represents a relatively high initial temperature, the reaction path would be given by $A A' A''$ (as against the adiabatic path) with a temperature maximum at A'. As the initial temperature is decreased, an unfavorable situation may arise, as shown by curve $C C' C''$, when the temperature falls monotonically. From the practical point of view it is best to operate the reactor at an initial temperature T_{02} (corresponding to point B), which ensures maximum rate almost throughout the reactor (by following a path closest to L_{m}). Thus an optimum temperature is likely to exist.

Consider now the case of countercurrent cooling with fresh feed. The reaction cannot progress beyond the line $T - Hx = T_{\mathrm{f}}$, shown by OP in Figure 12.12b (which is the adiabatic line through the feed temperature). In other words, for this type of heat exchange, the conversion can never exceed the adiabatic equilibrium value for the given feed. As in the case of constant temperature cooling, a value of T_0 exists at which the reaction path follows the maximum rate line L_{m} for the greater part of the reactor length. This corresponds to the optimum inlet temperature.

The figure also shows three curves with different values of γ drawn through B, representing a single initial temperature T_{02}. At a high value of γ there is practically no heat exchange, whereas at a low value there is good heat exchange. Here again an optimum value of γ can be visualized that permits reactor operation close to L_{m}.

It will thus be seen that optimum values of the initial temperature and heat exchange parameter γ exist. These values, along with those of the other operating parameters, can be determined by the method of steepest ascent provided an appropriate reactor model has been selected.

12.3.3. Autothermal Reactions

Several reactions of practical interest produce large quantities of heat that would be sufficient to raise the temperature of the feed stream to such levels that the desired conversions can be obtained in the reactors without extra addition of energy. Such a mode of operation is referred to as autothermic operation, and processes such as the synthesis of ammonia, water–gas shift reaction, and oxidation reactions provide common examples of reactions that can be operated this way. In many instances the heat liberated as a result of chemical reaction is much larger than what is required for maintaining the best operating temperature, thus necessitating its removal (recovery, if possible). One simple way to recover this extra heat is to preheat the feed by means of the reacting fluid and/or the effluent in a suitable heat exchanger. This procedure, however, brings about a changed behavior of the reactor that now needs careful analysis. One feature that immediately becomes evident is that the presence of the exchanger provides a thermal feedback, thus providing the basic condition for multiplicities of states to occur under suitable conditions. The problem of identifying the region of multiplicity, the transient behavior of the reactor, and the stability of the steady states therefore becomes important. This would, however, be specific to the mode of reactor operation employed, method of

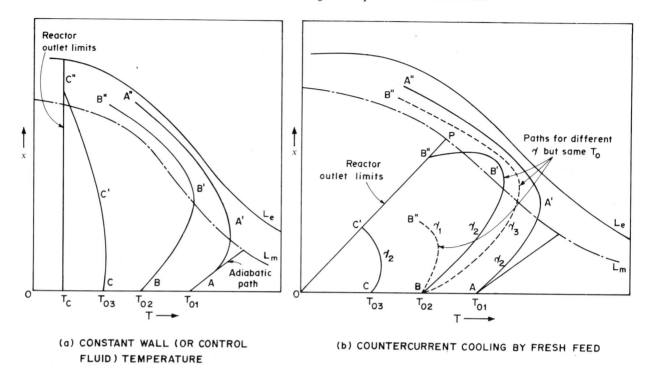

Figure 12.12. Plots of x vs. T for two typical heat exchange situations in tubular reactor operation.

heat removal, and nature of the chemical reaction, necessitating a separate analysis of each system. Examples analyzing a few typical industrial systems are presented later in this section. Figure 12.13 shows some of the important reactor–heat-exchanger configurations used in industrial practice.

Figure 12.13a represents a single adiabatic bed with preheating of reactants by the effluent gases. Even a simple configuration like this gives rise to a two-point boundary value problem, with the boundary conditions at the exit from the exchanger and the inlet conditions to the reactor unspecified. The problem can therefore be solved only by trial and error. Several modifications of this simple mode of operation seem desirable from the point of view of better operation. Thus in certain instances it may not be necessary to preheat the entire feed to the reactor, and part of it can be made to bypass the heat exchanger. Such a mode of operation is beneficial in NH_3 synthesis, and is shown in Figure 12.13b. Figure 12.13c shows a multibed adiabatic reactor with bypass; Shah (1967) has numerically integrated the system of differential equations describing an ammonia synthesis reactor of this type. Fodder (1971) has described a multibed adiabatic reactor with interstage cooling.

Multitubular reactors with internal heat exchange by cocurrent and countercurrent flows are shown in Figure 12.13d. The countercurrent reactor, which is a typical representation of the Nitrogen Engineering Co. (NEC) and Tennessee Valley Authority (TVA) ammonia reactors, has been analyzed by Baddour et al. (1965) and Murase et al. (1970).

12.3.4. Example 1: Simulation of an Adiabatic Reactor for High-Pressure Methanol Synthesis

The Fauser–Montecatini reactor for the synthesis of methanol from CO and H_2 consists of four adiabatic beds, a nonadiabatic bed, and an internal heat exchanger. A ZnO–Cr_2O_3 catalyst with a surface area of $70 \, m^2/g$ is used.

Since the synthesis gas contains about $2\% \, CO_2$, the reactions involved may be written as

$$CO + 2H_2 \rightleftharpoons CH_3OH$$

$$CO_2 + H_2 \rightleftharpoons CO + H_2O$$

Capelli et al. (1972) used a one-dimensional heterogeneous model with external and internal gradients to

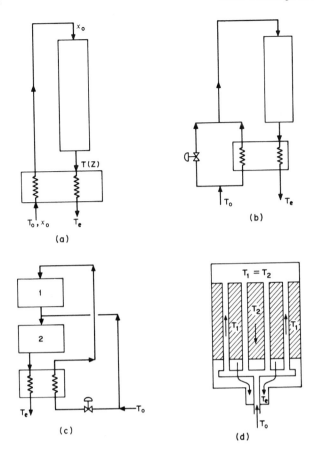

Figure 12.13. Modes of heat exchange between feed and effluent or reacting gas.

simulate this reactor. The conservation equations for mass and heat in the fluid and solid phases can be written, in accordance with the model, as follows.

Gas Phase (Methanol)

$$-\frac{1}{A_c}\frac{dF_m}{dl} = k_g A_p (p_m - p_{ms}) \qquad (12.29)$$

The feed rate of methanol F_m can be expressed in terms of the total flow rate F_t as $F_m = p_m F_t / p_t$. Equation 12.29 can therefore be written as

$$-\frac{1}{A_c}\frac{p_m}{p_t}\frac{dF_t}{dl} - \frac{F_t}{A_c p_t}\frac{dp_m}{dl}$$
$$= k_g A_p (p_m - p_{ms}) \qquad (12.30)$$

where p_m and p_t represent methanol and total pressure, restectively. It is important to note from the reaction stoichiometry that the total flow rate changes with

position, so that

$$\frac{dF_t}{dl} = 2k_g A_p (p_m - p_{ms}) \qquad (12.31)$$

Similar equations can be written for other gaseous species—for example, CO_2:

$$\frac{p_{CO_2}}{A_c p_t}\frac{dF_t}{dl} - \frac{F_t}{A_c p_t}\frac{dp_{CO_2}}{dl} = k_{gCO_2} A_p (p_{CO_2} - p_{CO_2,s})$$
$$(12.32)$$

The heat balance equation in the gas phase can be written as

$$-\frac{1}{A_c}\frac{d}{dl}(F_t C_p T) = k_g A_p (p_m - p_{ms})(-\Delta H)_1$$
$$+ k_{gCO_2} A_p (p_{CO_2} - p_{CO_2,s})$$
$$(-\Delta H)_2 - \frac{4U}{d_t}(T - T_{ex}) \qquad (12.33)$$

The last term of Eq. 12.33 is zero for adiabatic beds. For nonadiabatic beds T_{ex} represents the temperature of the feed that is preheated inside the tubes. It is important to note that T_{ex} varies with position and can be obtained as a solution of

$$\frac{F_t C_p}{A_c}\frac{dT_{ex}}{dl} = 4\frac{U}{d_t}(T - T_{ex}) \qquad (12.34)$$

The set of equations 12.29–12.34 along with the initial conditions

$$p_m = p_{mo}, \quad p_{CO_2} = p_{CO,0}, \quad T = T_0, \quad F_t = F_{t0}$$
$$\text{at} \quad l = 0 \qquad (12.35)$$

completely describes the external field situation.

Solid Phase

The internal field equations for mass and heat transfer for the species methanol and CO_2 are

$$\frac{D_{ei}}{R_g T}\frac{1}{r^2}\left(r^2\frac{dp_i}{dr}\right) = r_{vi} \qquad (12.36)$$

where the subscript i denotes methanol or CO_2. The corresponding heat balance equation is

$$-\frac{k_e'}{r^2}\frac{d}{dr}\left(r^2\frac{dT_s}{dr}\right) = r_{vm}(-\Delta H)_1 - r_{v,CO_2}(-\Delta H)_2$$
$$(12.37)$$

Equations 12.36 and 12.37 together with the boundary conditions

$$r = R, \qquad p_m = p_{ms}, \qquad P_{CO_2} = p_{CO_2, s}, \qquad T = T_s$$

$$r = 0, \qquad \frac{dp_m}{dr} = 0, \qquad \frac{dp_{CO_2}}{dr} = 0, \qquad \frac{dT}{dr} = 0$$

$$(12.38)$$

completely define the internal field problem. The rate equation proposed by Cappelli and Dente (1965) is presented below (note that $r_{vm} = r_{wm} \rho_c$):

$$r_{wm} = \frac{f_{CO} p_{CO} f_{H_2}^2 p_{H_2}^2 - f_m (p/K)}{A^3 [1 + B f_{CO} p_{CO} + C f_{H_2} p_{H_2} + D f_m p_m + E f_{CO_2} p_{CO_2}]}$$
$$\times \left(\frac{\text{kg mol}}{\text{kg catalyst hr}} \right)$$

where f represent the fugacity coefficients and A, B, C, D, E are given by

$$A = (2.78 \times 10^5) \exp \left(\frac{-8280}{R_g T} \right)$$

$$B = (1.33 \times 10^{-10}) \exp \left(\frac{23850}{R_g T} \right)$$

$$C = (4.72 \times 10^{-14}) \exp \left(\frac{30500}{R_g T} \right)$$

$$D = (5.05 \times 10^{-12}) \exp \left(\frac{31250}{R_g T} \right)$$

$$E = (3.33 \times 10^{-10}) \exp \left(\frac{23850}{R_g T} \right)$$

Cappelli et al. (1972) have solved Eqs. 12.36–12.38 by the method of successive approximations and obtained the parameter values that are needed in the gas-phase equations. The external field problem is then solved by the Runge–Kutta procedure. The results of the simulation have been compared with industrial results and are reproduced in Figure 12.14. The agreement between the two is remarkably good.

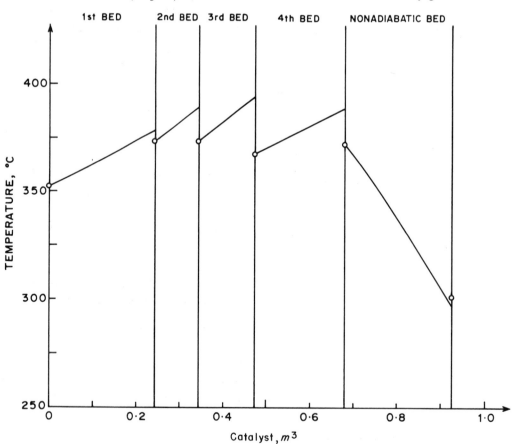

Figure 12.14. Comparison of the industrial and simulated results for a methanol synthesis reactor (Cappelli et al., 1972).

12.3.5. Example 2: Simulation of Ammonia Synthesis Reactors

The ammonia synthesis reaction produces a large quantity of heat that has to be removed in the interests of safe operation. Several ways of removing this heat have been used in practice: In autothermic reactors the heat is removed throughout the bed volume, whereas other designs have adiabatic beds (two or three) between which cooling by either external heat exchangers or quench-gas mixing is effected. Figure 12.15 shows schematic diagrams of (a) a three-stage adiabatic reactor with external cooling of the reaction mixture and (b) an autothermal reactor.

Three-Stage Adiabatic Reactor

On the assumption of uniform composition and temperature at any cross section of the bed, and when the axial diffusion of mass and heat is ignored, the conservation equations for mass and heat for a three-stage adiabatic reactor can be written as

$$\frac{dx}{dV_c} = \frac{\varepsilon r_{NH_3}(x, T, P)}{FM} \qquad (12.39)$$

$$\frac{dT}{dV_c} = \frac{(-\Delta H)}{\bar{C}_p} \frac{\varepsilon r_{NH_3}(x, T, P)}{FM} \qquad (12.40)$$

where x represents the extent of reaction, defined as

$$x = \frac{y_j - y_{j0}}{v_j} \qquad (12.41)$$

and V_c represents the volume of catalyst in the bed. Note that the effectiveness factor ε may be evaluated from the equations given in Section 4.3.

Singh and Saraf (1979) have solved the set of equations 12.39–12.41 for a given initial feed condition (flow rate, feed composition, volume of catalyst in each bed, total pressure of operation). Three different values of total feed flow at three different pressures were used to check the computational results with the data available from the running plant. In general, the agreement of the simulated results with the plant data seems extremely good, verifying the simple assumptions made in the analysis.

Shah (1967) has considered a two-stage adiabatic bed reactor with an intermediate quench by means of cold feed. His results, which are of particular relevance to reactor stability, are considered in Section 12.3.6.

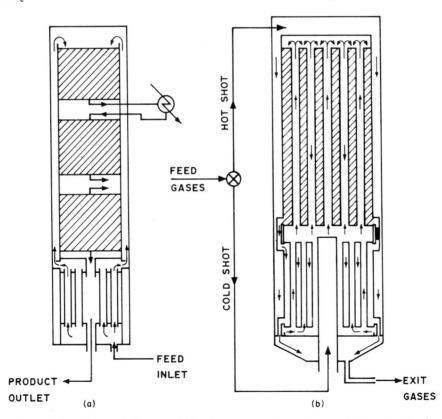

Figure 12.15. Schematic diagrams of (a) a three-stage reactor and (b) an autothermal reactor for the synthesis of ammonia.

Nonadiabatic Bed (*Autothermal Reactor*)

The mass conservation equation for this mode of operation is the same as mentioned in the earlier case. The energy balance equation, however, should include the loss in heat due to cooling and becomes

$$\frac{dT}{dV_c} = \frac{(-\Delta H)}{\bar{C}_p M}\frac{\varepsilon r_{NH_3}}{F} - \frac{U A_{sc}(T - T_c)}{F\bar{C}_p M} \quad (12.42)$$

where T_c is the temperature of the feeding gas in the embedded cooling tubes and A_{sc} is the net exchange area per unit catalyst volume. An additional equation around the exchanger completes the description of the process:

$$\frac{dT_C}{dV} = \frac{U A_{sc}(T - T_c)}{F\bar{C}_p M} \quad (12.43)$$

The set of initial and boundary conditions to this problem at the inlet can be formulated as

$$T = T_C \quad x = x_0 \quad (12.44a)$$

and at the outlet as

$$T_C = T_0 \quad (12.44b)$$

The equations present a two-point boundary value problem. Singh and Saraf (1979) have solved these equations for several inlet feed conditions (flow rate, composition, pressure of operation, total heat transfer surface, volume of catalyst, etc.). Figure 12.16 shows

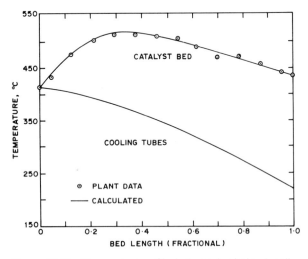

Figure 12.16. Temperature profiles in the catalyst bed and cooling tubes along the length in an autothermal reactor (Singh and Saraf, 1979).

the temperature profiles in the catalyst bed and cooling tubes along the bed height together with the available plant data. The remarkable closeness of the actual and calculated data upholds the validity of the model used for simulation.

Baddour et al. (1965) have also solved this set of equations for the TVA reactor and observed that under certain conditions three steady states are possible. The linearized dynamic analysis of the problem indicated that the upper and lower steady states are asymptotically stable, while the middle one is unstable to small temperature perturbations (around 5°C).

12.3.6. Example 3: Multiplicity in the Ammonia Synthesis Reactor

The kinetic equations for the different catalysts used in ammonia synthesis are available, and an example illustrating the simulation of the synthesis reactor was presented in Section 12.3.5. We shall now see how multiple steady states can seriously interfere with the operation of the synthesis reactor.

Shah (1967) has numerically integrated the system of differential equations describing a typical ammonia synthesis reactor. A schematic representation of the two-stage adiabatic reactor with intermediate quench of cold feed considered by him is presened in Figure 12.13c. For such a reactor system it has been found that when the hydrogen conversion is plotted as a function of the fraction of the feed preheated or the inlet temperature, multiple steady states appear. This is brought out in Figure 12.17. The figure shows, for instance, that when the fraction of the feed preheated exceeds 0.7, two exit conditions are possible; when all the feed is preheated, conversions of 18.5% and 15.5% are obtained, corresponding respectively to 848°K and 760°K.

Figure 12.17 also shows that when the inlet temperature is reduced to a value below 380°K the reactor is extinguished. In other words, the stable point of Figure 12.17 is crossed and there is a sudden drop in the reaction. Shah (1967) has carried out further simulation studies to show that the reactor can be extinguished when the pressure is decreased from 240 to 160 atm, or the inert fraction increased from 9 to 18%. It follows therefore that the inlet temperature should be kept sufficiently above the temperature at which the reactor is extinguished, in order to avoid the possibility of an increase in inert content leading to instability. It can also be shown that a multitubular reactor with heat exchange can lead to multiple steady states in the ammonia synthesis reactor.

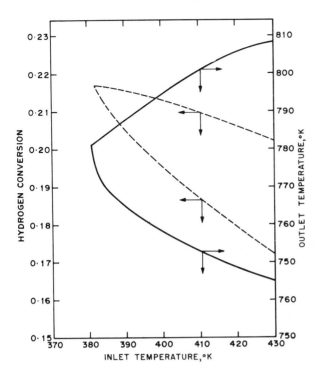

Figure 12.17. Plot of hydrogen conversion vs. inlet temperature for a two-stage adiabatic reactor (Shah, 1967).

12.3.7. A Simple Method for Estimating Tube Diameter and Flow Rate in a Multitubular Reactor

In a tubular reactor, the ability to remove completely at the wall heat generated by the reaction is a primary consideration in design. This would ensure radial uniformity of temperature and justify the use of the one-dimensional model (see Section 11.5.3). Brötz (1965) has suggested an approximate criterion for this purpose. Plots have been prepared (for reactors of different shapes) of a parameter

$$A_{max} = \left[\frac{d_t}{2} \left(\frac{\alpha_w q_v}{T_w k'_{er}} \right)^{1/2} \right]_{max}$$

vs. a parameter

$$B = \left(\frac{k'_{er} \alpha_w q_v}{U^2} \right)^{1/2}$$

that give the maximum values of A for a given value of B (Figure 12.18). In these definitions q_v represents the heat generated per unit volume, calculated at wall conditions, and the other quantities have the usual meaning. Values of A higher than A_{max} indicate that the

heat generated by the reaction cannot be removed at the wall. It would then be necessary to choose another value of tube diameter and of the flow rate (thus indirectly altering the heat transfer coefficient U) to satisfy the criterion.

An approximate design can be carried out based on Figure 12.18 (Rase, 1977). By assuming a value of tube diameter, wall temperature, and flow rate (which determines U), the parameter B can be calculated, and the corresponding value of A_{max} can be read from Figure 12.18. This value is then compared with the actual value of A_{max} calculated from its defining equation. If the calculated value is larger, the tube diameter or q_v must be reduced. The (heat generated per unit volume) q_v can be reduced by diluting the bed with inert packing (see Section 12.5). A quantitative measure of the extent of dilution can be obtained from $q_v = (-\Delta H) r_{wA} \rho_B / (1 + b)$. After the calculated value is matched with the value read from Figure 12.18, the reactor tube length can be readily calculated from the following equation on the assumption of isothermicity:

$$L = \frac{G_m y_{A0}}{\rho_B} \int_0^{x_A} \frac{dx_A}{(r_{wA})} \qquad (12.45)$$

The best isothermal temperature to take would be the mean value of the circulating heat transfer fluid temperature (based on inlet and outlet values).

12.4. OPTIMUM TEMPERATURE PROFILES IN A TUBULAR REACTOR

It is frequently required to find a temperature profile in a tubular reactor of finite dimensions such that the yield of the desired product is maximum. Seeking an optimum temperature profile in a chemical reactor is a problem in the area of optimal control. The formulation of the problem is a typical variational problem and the desired solution can be obtained by using the classical calculus of variations. However, the calculus of variations is used less frequently than the methods of dynamic programming and the maximum principle. This is because the variational technique is of little use in problems in which limitations resulting from the physical interpretation of the process are imposed on the function sought. For example, a limitation of the type $T_{min} \leqslant T \leqslant T_{max}$ is often imposed in constructing the optimum temperature profile. Such limitations, however, do not substantially interfere in a consideration of the maximum principle or dynamic programming.

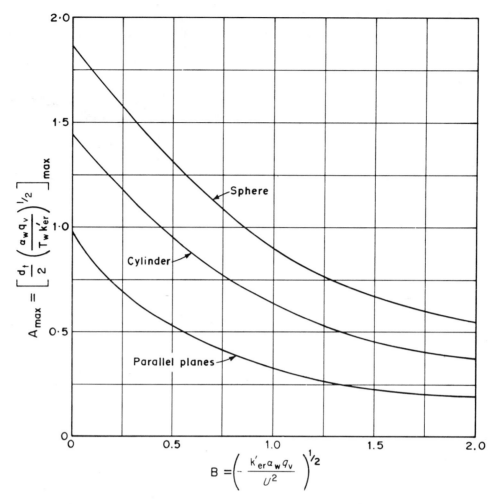

Figure 12.18. Maximum allowable values of $[(d_t/2)(\alpha_w q_v/T_w k'_{er})^{1/2}]_{max}$ (Brotz, 1965).

Obviously the method of dynamic programming is better suited for multistage multidecision processes. Its use involves partial differential equations, for which analytical solutions are usually not possible. The method based on the maximum principle (Pontryagin, 1962; Athens and Falb, 1969) has also been applied to the problems of optimization (Lee, 1964; Coward and Jackson, 1965). The formulation results in ordinary differential equations and avoids the usual two-point boundary value problems. This is in contrast with the method of dynamic programming, which does not offer such possibilities of prediction. Both these methods along with the calculus of variations are employed in seeking an optimum temperature profile.

12.4.1. Simple Reversible Reaction

In the case of an irreversible reaction the optimum temperature will be the maximum temperature, since a higher reaction rate is synonymous with greater product yield. However, for a single reversible reaction

$$A \underset{E_2}{\overset{E_1}{\rightleftharpoons}} B$$

the temperature accelerates the reaction in both directions and the equilibrium degree of conversion is altered. If $E_2 > E_1$, temperature will have a more pronounced effect on the backward reaction than on the forward reaction. The solution then is not a constant temperature profile but a decreasing one.

We seek the temperature at any cross section of a tubular reactor that will result in a minimum volume of the reactor for a predetermined amount of product. Thus for the reaction

$$A + B \underset{k_2}{\overset{k_1}{\rightleftharpoons}} C + D$$

the following equation can be derived (Denbigh, 1944):

$$\frac{T_{opt}}{T^*} = \left[1 + \frac{R_g T^*}{E_2 - E_1} \ln\left(\frac{E_2}{E_1}\right)\right]^{-1} \quad (12.46)$$

where T^* represents the equilibrium temperature. For small difference between the temperature T^* and T_{opt}, this can be approximated to

$$\frac{T^* - T_{opt}}{T^*} \simeq \frac{R_g T^*}{(-\Delta H)} \ln\left(\frac{E_2}{E_1}\right) \quad (12.47)$$

For an equimolar feed of A and B, this expression can be recast in terms of the mole fraction (y) as

$$\frac{A_{f1} \exp(-E_1/R_g T_{opt})(E_1 - 2R_g T_{opt})}{A_{f2} \exp(-E_2/R_g T_{opt})(E_2 - 2R_g T_{opt})} = \frac{y^2}{(1-y)^2} \quad (12.48)$$

A trial-and-error solution of this equation leads to the optimum temperature for any value of the degree of conversion.

12.4.2. Consecutive Reactions

Several studies have been reported on the optimal temperature profile in a system of consecutive reactions (e.g., Bilous and Amundson, 1956; Aris, 1961, 1965; Coward and Jackson, 1965; Lee, 1964; Denn and Aris, 1965; Fine and Bankhoff, 1967). Most of them, however, are restricted to a first-order consecutive scheme. This class of reactions was first considered by Bilous and Amundson (1956) in order to find the temperature profile that would maximize the yield of intermediate in a tubular reactor of given length. The results of Bilous and Amundson (1956) showed the optimal temperature profile to be independent of the activation energies E_1 and E_2. The profile decreased monotonically from inlet to outlet for $E_1 < E_2$.

The governing differential equations for a consecutive reaction

$$A \xrightarrow[\;E_1\;]{k_{v1}} B \xrightarrow[\;E_2\;]{k_{v2}} C$$

are as follows:

$$\frac{dC_A}{dt} = -k_{v1} C_A, \quad \frac{dC_B}{dt} = k_{v1} C_A - k_{v2} C_B \quad (12.49)$$

The Hamiltonian H defined as a function of the concentrations and the corresponding impulse functions I_A and I_B may be minimized by using the combined random search and integration procedures to get the solution to this problem. The results of Lee (1964) correspond to those of Bilous and Amundson (1956). The temperature and concentration profiles obtained are sketched in Figure 12.19a.

12.4.3. Parallel Reactions

Let us consider a first-order parallel reaction scheme

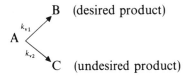

taking place in an ideal tubular reactor. This scheme has been analyzed by Horn (1961), Horn and Troltenier (1961a, b), and Skrzypek (1974). Horn and coworkers used the classical calculus of variations and consequently could not consider the temperature limitations. Skrzypek (1974) analyzed the problem by means of the Pontryagin maximum principle. The governing differential equations can be written as

$$\frac{dc_A}{dz} = -(\hat{k}_{v1} + \hat{k}_{v2})c_A, \quad \frac{dc_B}{dz} = \hat{k}_{v1} c_B \quad (12.50)$$

with the initial condition

$$c_A(0) = 1, \quad c_B(0) = 0$$

and the modified rate constants defined as

$$\hat{k}_{v1} = \frac{L}{u} k_{v1} = \hat{A}_{f1} \exp\left(\frac{-E_1}{R_g T}\right)$$

$$\hat{k}_{v2} = \frac{L}{u} k_{v2} = \hat{A}_{f2} \exp\left(\frac{-E_2}{R_g T}\right) \quad (12.51)$$

where $\hat{A}_f = (L/u)A_f$. The objective now is to find the optimum temperature profile $T(z)$, so that the yield of the product B will be maximal at the reactor outlet ($z = 1$). The temperature limitation $T_{min} < T < T_{max}$ will be imposed on the problem and the changes in viscosity occurring during the reaction will be ignored.

Using the maximum principle, we introduce the function H (Hamiltonian), which depends on c_A and the

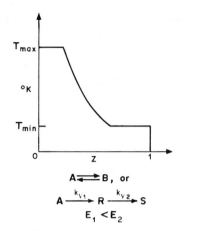

(a) REVERSIBLE AND CONSECUTIVE
REACTIONS

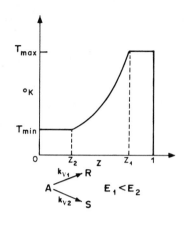

(b) PARALLEL REACTIONS

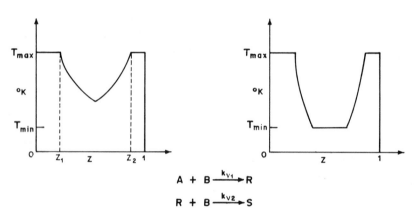

(c) COMPLEX REACTION

Figure 12.19. Temperature profiles for certain typical complex reactions.

related function I_A through the expression

$$H = -I_A(\hat{k}_{v1} + \hat{k}_{v2})c_A + \hat{k}_{v1}c_A I_B \qquad (12.52)$$

The maximum principle uses a temperature selection such that for every value of z the Hamiltonian will be maximal with respect to T. When the temperature selection is carried out in accordance with the principle of maximality, the Hamiltonian will remain constant along the optimal temperature profile. Construction of an optimal temperature profile for the most general case indicates the existence of three sections: (1) the lower horizontal segment, where the temperature remains constant at $T = T_{min}$ ($z = z_2$); (2) a section where the temperature rises until the point $z = z_1$ is reached, where the temperature corresponds to $T = T_{max}$; and

(3) the upper horizontal segment, where the temperature remains constant at $T = T_{max}$ until the end of the reactor. These three segments are shown in Figure 12.19b (see also Section 12.4.5). The distance at which the maximum temperature is reached can be calculated by using the following equation, derived by Skrzypek (1974):

$$z = \frac{E_2}{E_1 \hat{A}_{f1}} \left\{ \frac{E_2 \hat{A}_{f2}}{R_g \hat{A}_{f1}} \frac{1}{T} - \exp\left(\frac{E_1}{R_g T}\right) \right.$$
$$+ \frac{E_2 \hat{A}_{f2}}{E_1 \hat{A}_{f1}} \ln\left[\frac{E_1 \hat{A}_{f1}}{E_2 \hat{A}_{f2}}\right.$$
$$\left.\left. + \exp\left(\frac{E_1}{R_g T}\right)\right]\right\} + \text{constant} \qquad (12.53)$$

At $z = z_1$, $T = T_{max}$, so that the constant can be evaluated and the equation rearranged to give

$$z - z_1 = \frac{E_2}{E_1 \hat{A}_{f1}} \left\{ \frac{E_2 \hat{A}_{f2}}{R_g \hat{A}_{f1}} \left(\frac{1}{T} - \frac{1}{T_{max}} \right) - \exp \left(\frac{E_1}{R_g T} \right) \right.$$

$$+ \exp \left(\frac{E_1}{R_g T_{max}} \right) + \frac{E_2 \hat{A}_{f2}}{E_1 \hat{A}_{f1}}$$

$$\left. \times \ln \left[\frac{E_1 \hat{A}_{f1}/E_2 \hat{A}_{f2} + \exp(-E_1/R_g T)}{E_1 \hat{A}_{f1}/E_2 \hat{A}_{f2} + \exp(-E_1/R_g T_{max})} \right] \right\} \quad (12.54)$$

The distance $z = z_2$ can be obtained from Eq. 12.53 with $T = T_{min}$.

The concentrating profiles for the reactant A and the desired product B can be obtained by integrating the basic differential equations 12.50. As stated earlier, the temperature profile shows three different zones, so that substituting for the temperature profile in these equations and integrating leads to expressions for the concentration profiles in these three zones. The results are presented in Table 12.5 for both the reactant A and product B.

12.4.4. Complex Networks

A system of reactions in which a reactant and its product both react with the second reactant is a matter of common occurrence in industrial practice. Let us then consider a reaction scheme (Burghardt and Skrzypek, 1974):

$$A + B \xrightarrow[k_{v2}]{k_{v1}} R \quad \text{(desired product)}$$

$$R + B \xrightarrow{k_{v2}} S \quad \text{(undesired product)}$$

where the rate constants follow the Arrhenius law and $E_2 > E_1$.

It can be shown that for a plug-flow reactor, the optimum temperature profile will begin as the isotherm $T = T_{max}$, as in the case of a consecutive reaction. This temperature $T = T_{max}$ will remain constant up to the point at which $z = z_1$, where the necessary condition for the local maximum of the Hamiltonian ($\partial H/\partial T = 0$) will be satisfied. At the point where $z = z_1$ the curve will begin to fall, and after passing through a minimum it will rise until it again reaches the isotherm $T = T_{max}$. A sketch of the profile is shown in Figure 12.19c. The optimum profile for this system of reactions combines the features of the profile for consecutive reactions (falling curve) and that for the parallel reactions (rising curve).

12.5. MODULATION OF NONISOTHERMICITY BY CATALYST DILUTION

As already pointed out, the design of a fixed-bed reactor becomes critical only in the region where temperature runaway is experienced. In the case of highly exothermic reactions, for which the critical region cannot be ignored in design, it is possible to cut down the temperature rise by diluting the catalyst with a catalytically inert material. This will ensure that the desired steady-state temperature is obtained from the competing rates of heat generation and heat transfer, since the rate of a catalytic reaction per unit volume is controlled only by the amount of catalyst present in the volume.

For a fully mixed isothermal reactor the degree of dilution is limited to a single uniform value for the whole reactor. On the other hand, for a fixed-bed isothermal plug-flow reactor, since the concentration is a function of axial position, the rate per unit volume also becomes a function of position. A dilution profile will therefore have to be worked out to match the concentration (i.e., rate) profile in the reactor in order to ensure isothermal operation. In the case of an exothermic reaction with an optimum temperature profile in the reactor, the extent of dilution at each position will have to be calculated at the concentration *and* temperature at that position.

The presence of inert pellets tends to flatten the radial temperature gradients. Thus, unless a reaction is extremely exothermic, for tubes of the size normally used in fixed-bed reactors radial temperature gradients can be ignored in a diluted catalyst bed; that is, the one-dimensional model can be safely used.

12.5.1. Dilution in an Isothermal Reactor

If the total weight of the diluted bed (catalyst + inerts) is W_t, then a dilution factor DF can be defined as

$$\text{DF} = \frac{W_t}{W} \quad (12.55)$$

The following equation can be written:

$$\frac{dT}{dx_A} = \frac{y_A}{C_p} \left[(-\Delta H) + \frac{4U(T_w - T)(\text{DF})}{r_{wA} d_t \rho_B} \right] \quad (12.56)$$

TABLE 12.5. Optimum Concentration Profiles for the Reactant and Product in a Parallel Reaction Scheme[a]

Zone	Profile for Reactant Species	Profile for Product Species
$0 < z < z_2$ $T = T_{\min}$	$c_A = \exp[-(\hat{k}_{1\min} + \hat{k}_{2\min})z]$	$c_B = -\dfrac{\hat{k}_{1\min}}{\hat{k}_{1\min} + \hat{k}_{2\min}}\left\{\exp[-(\hat{k}_{1\min} + \hat{k}_{2\min})z] - 1\right\}$
$z_2 < z < z_1$	$c_A = c_A(z_2)\dfrac{E_1\hat{A}_{f1}\exp(E_2/R_gT) + E_2\hat{A}_{f2}\exp(E_1/R_gT)}{E_1\hat{A}_{f1}\exp(E_2/R_gT_{\min}) + E_2\hat{A}_{f2}\exp(E_1/R_gT_{\min})}$	$c_B = -\dfrac{c_A(z_2)E_1\hat{A}_{f1}}{E_1\hat{A}_{f1}\exp(E_2/R_gT_{\min}) + E_2\hat{A}_{f2}\exp(E_1/R_gT_{\min})}$ $\times\left[\exp\left(\dfrac{E_2}{R_gT}\right) - \exp\left(\dfrac{E_2}{R_gT_{\min}}\right)\right] + c_B(z_2)$
$z_1 < z < 1$	$c_A = c_A(z_1)\exp[-(\hat{k}_{1\max} + \hat{k}_{2\max})(z - z_1)]$	$c_B = \dfrac{\hat{k}_{1\max}}{\hat{k}_{1\max} + \hat{k}_{2\max}}c_A(z_1)\{\exp[-(\hat{k}_{1\max} + \hat{k}_{2\max})](z - z_1) - 1\} + c_B(z_1)$

[a]Note that $\hat{k}_{\max}$ and $\hat{k}_{\min}$ in these profiles correspond to $T_{\max}$ and $T_{\min}$; the subscript v has been dropped for convenience.

For isothermal operation this reduces to

$$\frac{r_{wA}\rho_B}{DF} = \frac{4U(T - T_w)}{(-\Delta H)d_t} = \text{constant} \qquad (12.57)$$

In other words, the catalyst must be diluted in such a manner that the rate of reaction per unit volume is maintained constant throughout the length of the reactor.

The following expressions have been derived by Caldwell and Calderbank (1969) for estimating the dilution factor as a function of position for a given exit conversion x_{AL} for a first-order reaction:

$$DF = \frac{(1/x_{AL}) - (l/L)}{(1/x_{AL}) - 1} \qquad (12.58)$$

The average value of the dilution factor is then obtained from

$$\overline{DF} = \frac{L}{\displaystyle\int_0^L \frac{dl}{DF}} = \frac{\dfrac{x_{AL}}{1 - x_{AL}}}{-\ln(1 - x_{AL})}. \qquad (12.59)$$

12.5.2. Dilution in a Nonisothermal Reactor

For a predetermined constraint on the temperature at each point, catalyst volume can be determined by first calculating the optimum temperature at each point from Eq. 12.46 and then estimating the dilution required to obtain this temperature. This is a much better and more practical way of imposing a temperature profile than any other method, such as permitting the heat dissipation capacity to vary along the length of the reactor in an appropriate manner.

The recommended procedure is briefly as follows:

1. Obtain the optimal temperature profile by the method described in Section 12.4.
2. Prepare a plot of T_{opt} as a function of x_{opt} and determine the derivative dT_{opt}/dx_{opt} (denoted simply by dT/dx), which will be a function of position.
3. Rewrite Eq. 12.56 as

$$\frac{dT}{dx_A} = \frac{y_{A0}}{C_p}\left[(-\Delta H) + \frac{4U(T_w - T_{opt})(DF)}{r_{wA}d_t\rho_B}\right] \qquad (12.60)$$

Then from the value of dT/dx_A at every position,

a knowledge of $(-\Delta H)$ and U, and with fixed values of T_{opt} (at that position) and T_w, calculate DF.

Narsimhan (1976) has proposed a more elaborate method for calculating the catalyst (or dilution) profile in isothermal, adiabatic, and nonisothermal non-adiabatic reactors. Thus a dilution profile can be established that dispenses with the need to impose a varying wall temperature. It may so happen that the reactor volume now becomes too large, so that any decision on the adoption of this dilution policy will depend on the relative costs of the catalyst and reactor.

It is reported (Doraiswamy et al., 1970) that the oxidation of carbon monoxide could be carried out under essentially isothermal conditions by using a diluted bed of palladium catalyst. Similar results have been obtained on a pilot plant reactor for aniline by Doraiswamy et al. (1970), who used a multitubular reactor (containing seven tubes 3.5 cm in diameter and 300 cm long) filled with pellets of a supported copper catalyst diluted with unimpregnated pellets.

Another interesting feature of the diluted bed is that it can be used as a recycle reactor (see Section 12.6). All that is necessary to do is to insert a coaxial cylinder, and fill the larger inner space with dilutent only and the annular space with the active catalyst. Recycling is accomplished by the lateral diffusion of the reactant from the inert core.

12.6. OPTIMIZATION THROUGH MIXING: COMBINED AND RECYCLE REACTORS

In this section we describe two typical reactor systems—combined and recycle—in which the concept of mixing has been employed in two different ways to optimize reactor performance.

12.6.1. Combined Reactors

A useful method of optimizing a chemical reactor from the point of view of mixing is to combine the fully mixed and tubular reactors in such a manner that the output would be maximum for a given total volume or the total volume would be minimum for a given output. Such a combination of the mixed and tubular reactors is normally referred to as the MT or TM reactor, depending on which of the reactors comes first in the combination.

In the case of an adiabatic reactor two opposing influences, those of decreasing concentration and increasing temperature, can be visualized as the reaction progresses. In the initial stages the effect of temperature is dominant, whereas in the later stages that of concentration becomes more significant. Since in a fully mixed reactor the outlet conditions determine the performance of the reactor, it would obviously be reasonable to carry out the first part of the reaction, which is associated with an increasing trend in the rate, in a mixed reactor. The second part, characterized by a decreasing rate, can be more advantageously carried out in a tubular reactor. The first studies on these mixed–tubular (MT) and tubular–mixed (TM) reactor combinations were carried out by Cholette and Cloutier (1959), Cholette et al. (1960), and Cholette and Blanchet (1961). These studies were essentially concerned with the effect of an MT or TM combination in an isothermal system.

We now examine the effect of combining the mixed and tubular reactors under adiabatic conditions. A plot of conversion vs. residence time for the mixed and tubular reactors is shown in Figure 12.20. It is possible to carry out the reaction in a CSTR up to a residence time t_m as represented by point D, then continue the reaction under plug-flow conditions (along the broken line). It will be seen that for the same total residence time t a conversion represented by B is obtained, which is much higher than in either reactor operated singly.

A similar conclusion can be reached by the method of Aris (1962), in which the reciprocal rate is plotted as a function of conversion as shown in Figure 12.21. The residence time (W/F) is given by

$$t = \frac{W}{F} = \int_0^{x_A} \frac{1}{r_{wA}} dx_A \tag{12.61}$$

Let us consider a conversion represented by point A. If the reaction is carried out under fully mixed conditions, the residence time is given by the rectangle $OABD$, whereas if it is carried out under plug-flow conditions it is given by $OABC$. Evidently the residence time is lower in the former case. This situation continues until point E is reached. On the other hand, for a desired conversion corresponding to point P, the tubular reactor has a smaller holding time than a mixed reactor. However, it is now clear that an MT combination of a mixed reactor converting to an intermediate extent corresponding to L followed by a tubular reactor will be optimal. It is also clear that the TM combination will never be optimal.

Douglas (1964) also presented criteria for optimal combinations under adiabatic conditions for a reaction in which a single reactant species is involved. He put forward a condition according to which the point at which the slopes of the two curves in Figure 12.20 are equal represents the optimum.

Trambouze and Piret (1959) have discussed the use of combined reactors for a system of three competing parallel reactions under isothermal conditions. King (1965) has extended the treatment to include heat exchange and described methods for the calculation of the optimal combination of operating conditions.

12.6.2. Optimality Criteria

Criteria for optimal combinations of the mixed and tubular reactors for different classes of reactions have been reported by Aris (1962), Douglas (1964), and Babu Rao and Doraiswamy (1967), and the final expressions are given below.

Simple Reaction with a Single Reactant

The criterion for a simple reaction with a single reactant

$$A \longrightarrow products$$

is given by

$$\frac{\alpha M}{(1 + Mx_m)^2} = \frac{m}{1 - x_m} \tag{12.62}$$

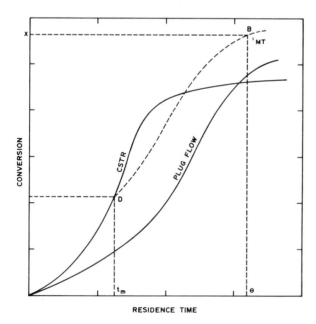

Figure 12.20. Plot of conversion vs. residence time for (adiabatic) mixed and tubular reactors.

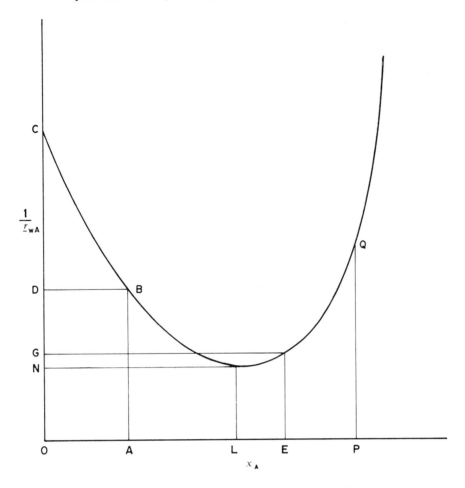

Figure 12.21. Nature of the $1/r_{wA}$ vs. x_A plot for an adiabatic exothermal reaction.

Simple Reaction with Two Reactants

For a simple reaction with two reactants

$$A + v_B B \xrightarrow{m, n} \text{products}$$

the criterion is given by

$$\frac{m}{1 - x_m} + \frac{n v_B}{(C_{B0}/C_{A0} - v_B x_m)} = \frac{\alpha M}{(1 + M x_m)^2} \quad (12.63)$$

It will be noticed that in the absence of the second component B this reduces to the criterion for a simple reaction involving a single reactant.

Consecutive Reactions

The criterion for consecutive reactions

$$A \xrightarrow{m} R \xrightarrow{n} S$$

is given by

$$\frac{k_{10}}{k_{20}} \frac{(\exp \alpha_1 Z)}{(\exp \alpha_2 Z)} = \frac{\alpha_2 (x - y)^n}{\alpha_1 (1 - x)^m - \dfrac{m H^2 (1 - x)^{m-1}}{M_1 + M_2}}$$

$$\times \frac{1}{C_{A0}^{m-n}} \quad (12.64)$$

where $M = \Delta T/T_0$ (ΔT is the total adiabatic temperature rise) and

$$H = 1 + M_1 x + M_2 y, \qquad Z = \frac{M_1 x + M_2 y}{1 + M_1 x + M_2 y}$$

$$\alpha_1 = \frac{E_1}{R_g T_0}, \qquad\qquad \alpha_2 = \frac{E_2}{R_g T_0}$$

12.6.3. Example: Oxidation of Benzene in an MT Reactor

Babu Rao and Doraiswamy (1967) carried out the oxidation of benzene on a vanadium catalyst in tubular as well as fluidized-bed reactors. It was ensured that the fluidized-bed reactor operated under conditions close to full mixing. The results obtained by them at 380°C with a benzene-to-air ratio of 1:170 are graphically displayed in Figure 12.22. Given the rate equation

$$r_{wB} = (1.537 \times 10^{-3})p_B$$

determine the optimum conversion to which the mixed reactor should be operated in a combined MT adiabatic reactor by the following methods: (1) plot of $1/r_{wB}$ vs. W/F; and (2) the criterion equation. The following data are available: adiabatic temperature rise (for total conversion) $\Delta T = 480°C$, activation energy $E = 2900$ cal/mol, initial temperature $T_0 = 380°C$, initial benzene partial pressure is 0.0059 atm.

Solution

1. From the plot of 1/r vs. x:

Since the reaction is adiabatic, the temperature can be expressed in terms of the conversion and adiabatic temperature rise ΔT. Thus the rate equation becomes

$$r_{wB} = k_{p0} \exp(\alpha_0)\left(\frac{Mx}{1 + Mx}\right)$$

where

$$\alpha_0 = \frac{E}{R_g T_0} = \frac{2900}{2 \times 653} = 2.22$$

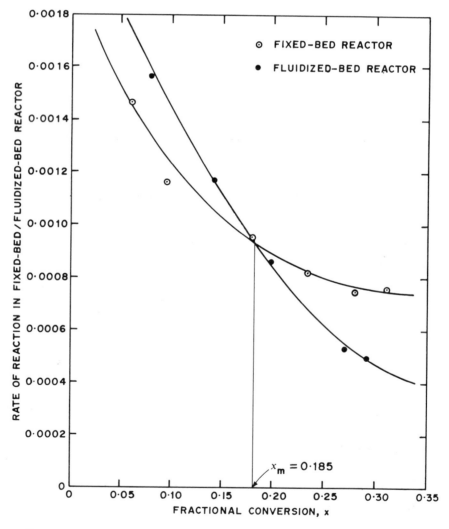

Figure 12.22. Calculation of the optimal conversion in the mixed stage in an MT reactor.

and

$$M = \frac{\Delta T}{T_0} = \frac{480}{653} = 0.735$$

Since the conversion x is given by

$$x = \frac{p_{B0} - p_B}{p_{B0}}$$

where $p_{B0} = 0.0059$ (corresponding to the ratio $1:170$), x can be calculated for different values of p_B and the corresponding rates can also be estimated from the rate equation. Thus for $p_B = 0.003$ we have $x = 0.492$ and the rate 9.193×10^{-6} mol/g catalyst hr.

Table 12.6 records the results obtained for different conversions. A plot of $1/r_{wB}$ vs. x can be prepared, from which it can be noticed that the reciprocal rate shows a minimum at

$$(x_m)_{max} = 0.185$$

TABLE 12.6. Calculated Values of the Rate of Benzene Disappearance at Different Partial Pressures under Adiabatic Conditions

p_B	$x = \dfrac{p_{B0} - p_B}{p_{B0}}$	r_{wB}
0.0015	0.7457	5.596×10^{-6}
0.0020	0.6610	7.0261×10^{-6}
0.0030	0.4915	9.1927×10^{-6}
0.0040	0.3220	1.0399×10^{-5}
0.0045	0.2373	1.0637×10^{-5}
0.0050	0.1525	1.0631×10^{-5}
0.0055	0.0678	1.0389×10^{-5}

2. From the criterion equation:

The criterion for obtaining the optimum value of conversion in the mixed reactor (for a simple first-order reaction) is given by Eq. 12.62 (with $m = 1$):

$$\frac{\alpha M}{(1 + M x_m)^2} = \frac{1}{1 - x_m}$$

Substituting the values of α, M in this expression, we have

$$(x_m)_{max} = 0.195$$

It will be noticed that the optimum values of conversion obtained by the two methods agree closely

with each other. The criterion given by Eq. 12.62, which is simple to use and based only on kinetic data, can therefore be safely used for calculating x_m.

In order to test the correctness of the value of x_m Babu Rao and Doraiswamy (1967) carried out the oxidation of benzene in an MT reactor proposed by them earlier based on the principle of semifluidization (Babu Rao et al., 1965). The fraction of the total volume corresponding to the mixed portion (m') at the bottom was expressed as

$$m' = \frac{W_f/F}{W/F} = \frac{W_f}{W} \tag{12.65}$$

The results for $T_0 = 380°C$ and a benzene-to-air ratio of $1:170$ are reproduced in Figure 12.23 as total conversion x_{mt} (obtained in the combined reactor) vs. m'. It may be noticed that beyond a certain value of the total residence time W/F, the curves show a maximum, thus distinctly bringing out the superiority of the MT combination.

From the value of m' corresponding to the maximum, the conversion x_m in the mixed portion can be calculated from

$$\frac{W_f}{F} = \frac{x_m}{r_{wA}} \tag{12.66}$$

The value of $(x_m)_{max}$ thus determined has been shown to agree closely with the optimum value calculated earlier.

12.6.4. MT Reactor with Heat Exchange

Since the use of MT reactors is restricted to exothermic reactions, the question of heat exchange assumes considerable importance. A practically useful scheme is to provide for unrestricted heat exchange in the mixed stage and an intercooler between stages. We thus have three parameters to be optimized: conversion x_m and temperature T_m in the mixed zone; and inlet temperature T_0 to the tubular zone.

Based on the developments presented in the preceding sections, the following equations can be written for the total, mixed, and tubular reactor holding times:

$$t = t_m + t_t \tag{12.67}$$

$$t_m = \frac{C_0 (x_m - x_0)}{r_V (x_m, T_m)} \tag{12.68}$$

$$t_t = C_0 \int_{x_m}^{x_e} \frac{dx}{-r_V (x, T)} \tag{12.69}$$

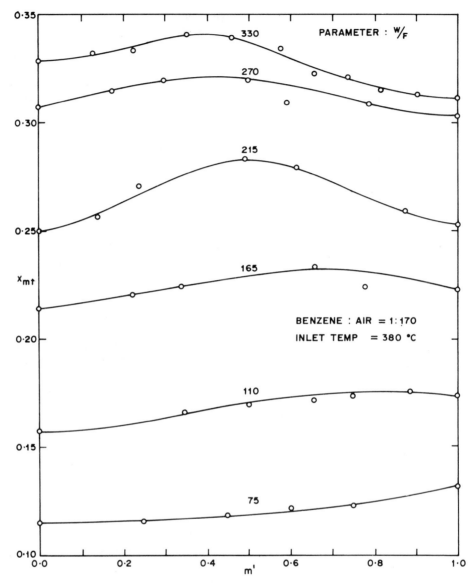

Figure 12.23. Plot of x_{mt} vs. m' for semifluidized MT reactor at 380°C (Babu Rao and Doraiswamy, 1967).

The problem now is to minimize the total time t. According to the optimality principle of the method of dynamic programming, the operating variable T in the second stage must be chosen to make this stage optimal with respect to its feed composition x_m. Since in an adiabatic reactor T is restricted by Eq. 12.1, any temperature variation can be effected only through T_0. Thus t_{min} may be expressed as

$$t_{min} = \min_{x_m} \left\{ \min_{T_m} \left[C_0(x_m - x_0) \frac{1}{-r_V(x_m, T_m)} \right] \right.$$
$$\left. + \min_{T_0} C_0 \int_{x_m}^{x_e} \frac{dx}{-r_V(x, T)} \right\} \quad (12.70)$$

where the first minimization is done through choice of x_m, which applies to both steps; the second through T_m, which applies only to the first (mixed) stage; and the third through T_0, which applies only to the second (tubular) stage. T_m is controlled by cooling in the mixed stage and T_0 through cooling in the interstage cooler.

For minimizing t_m, t_t, and t, the following conditions should be satisfied: $\partial t_m / \partial T_m = 0$, $\partial t_t / \partial T_0 = 0$, and $\partial t / \partial x_m = \partial t_m / \partial x_m + \partial t_t / \partial x_m$. King (1965) has suggested methods of satisfying these conditions both by simultaneous solution and by the method of dynamic programming. In the former the three equations are solved simultaneously by assuming various values of x_m, T_m, and T_0. In the dynamic programming method, an optimum is sought for each variable in turn, as brought out in Eq. 12.70. Thus the equations $\partial t_m/\partial T_m = 0$ and $\partial t_t/\partial T_0 = 0$ are solved independently for different values of x_m and a maximum is then sought along the x_m coordinate.

The existence of a minimum in t can be visualized by reference to Figure 12.24. Part (a) of this figure shows the usual T–x plot for an adiabatic reactor (see Section 12.1.2) with X representing the adiabatic path in the tubular reactor. The $(1/r)$–x portrait of the adiabatic line is shown as curve Y in Figure 12.24b. The rectangle A represents the holding time t_m for achieving a conversion x_m from x_0 at T_m (point a) in the mixed reactor. The products are then cooled to T_0 (point b) and reacted along curve z. The area B under this curve represents the holding time t_t. The values of x_m, T_m, and T_0 at which the sum of the areas (A + B) is minimum represent the optimum values of these parameters.

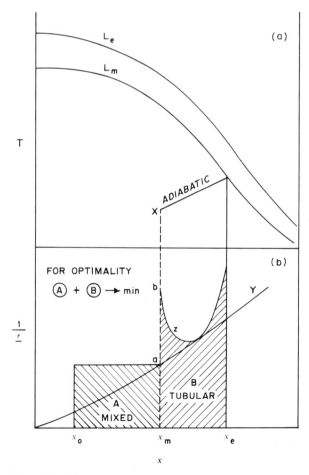

Figure 12.24. Optimal design of an adiabatic MT reactor with interstage cooling.

12.6.5. Recycle Reactor

For a simple first-order reaction occurring in a plug-flow reactor whose outlet fluid stream is partly recycled the following equation can be derived:

$$k_v \frac{V_r}{F} = -\ln\left[\frac{C_A(R+1)}{RC_A + 1}\right] \qquad (12.71)$$

For large values of the recycle ratio $q/Q = R$, by expanding the log term and retaining only the first term we have

$$C_A = \frac{1}{1 + k_v(V_r/Q)} \qquad (12.72)$$

This is easily recognized as the expression for a CSTR. Depending on the recycle ratio, the plug-flow performance can be made to approach that of a CSTR. This also brings out the possibility of describing the intermediate levels of mixing in the reactor by using the recycle ratio as a convenient parameter (besides the conventional axial dispersion or tanks-in-series model).

In the case of complex reactions the recycle ratio can have a profound influence on the yield of the desired product. The general conclusion appears to be (Carberry and Gillespie, 1966) that for an isothermal consecutive reaction, the ultimate product (and for a simultaneous reaction network, the lower-order reaction) is favored at higher levels of mixing. This brings us to a paradoxical situation when a series–parallel network is involved. Van de Vusse (1964) has treated such a system for the two extremes in the level of mixing. Carberry and Gillespie (1966) have extended the analysis to intermediate levels of mixing and arrived at the rather significant conclusion that neither plug flow nor a CSTR leads to a maximum in the intermediate concentration. An intermediate level of mixing depending on the values of the rate constants involved gives the maximum yield.

12.7. REACTORS FOR POLYFUNCTIONAL CATALYSTS

A brief description of bifunctional catalysis was presented in Chapter 5. We shall now examine its implications in reactor design. Consider catalysts X and Y constituting a bifunctional system. These can be combined in three ways:

1. Intimate but unconsolidated mixture of X and Y.

2. Mixture of pellets of X and Y.
3. Composite pellets formed from a mixture of X and Y.

Clearly the role of diffusion will depend on the method of admixture. Case 1 involves intimate mixing of particles of X and Y, and since the role of diffusion in particles is less significant than in pellets, it can be assumed that for all practical purposes the performance of case 1 can be evaluated from Eq. 5.51, derived under conditions where diffusional resistance is absent. We shall therefore be concerned here with only cases 2 and 3.

Mixture of Pellets of Single Catalysts

For reaction scheme 2 (Section 5.3.1), for which the application of polyfunctional catalysis has been shown to be advantageous, the following continuity equations can be written:

$$(D_e)_X\left(\frac{\partial^2 C_A}{\partial r^2} + \frac{2}{r}\frac{\partial C_A}{\partial r}\right) - k_{v1}C_A + k_{v1-}C_B = 0$$

$$(D_e)_X\left(\frac{\partial^2 C_B}{\partial r^2} + \frac{2}{r}\frac{\partial C_B}{\partial r}\right) + k_{v1}C_A - k_{v1-}C_B = 0$$

$$(12.73)$$

$$(D_e)_Y\left(\frac{\partial^2 C_B}{\partial r^2} + \frac{2}{r}\frac{\partial C_B}{\partial r}\right) - k_{v2}C_B = 0$$

where $(D_e)_X$ and $(D_e)_Y$ represent the effective diffusivities in catalysts X and Y, respectively. The boundary conditions are

$$r = R_X, \qquad C_A = C_{As}, \qquad C_A \text{ finite at } r = 0$$

$$r = R_Y, \qquad C_B = C_{Bs}, \qquad C_B \text{ finite at } r = 0$$

where R_X and R_Y are the radii of pellets of X and Y, respectively. The rate of flow of A in pellets of X is given by

$$r_A(X) = V_X(1 - f_c)\frac{3}{R_X}(D_{eA})_X\left(\frac{\partial C_A}{\partial r}\right)_{r = R_X} \qquad (12.74)$$

Similarly, the rates for B in pellets of X are given by

$$r_B(X) = V_X(1 - f_c)\frac{3}{R_X}(D_{eB})_X\left(\frac{\partial C_B}{\partial r}\right)_{r = R_X} \qquad (12.75)$$

and

$$r_B(Y) = (1 - V_X)(1 - f_c)\frac{3}{R_Y}(D_{eB})_Y\left(\frac{\partial C_B}{\partial r}\right)_{r = R_Y}$$

$$(12.76)$$

For the complete formulation of equations, we shall write the material balances for all three components over a differential length dl of a plug-flow reactor:

$$u\frac{\partial C_{As}}{\partial l}dl = -r_A(X)dl$$

$$u\frac{\partial C_{Bs}}{\partial l}dl = -r_B(X)dl \qquad (12.77)$$

$$u\frac{\partial C_{Cs}}{\partial l}dl = -r_B(Y)dl$$

where u is the superficial velocity of the reactant.

Gunn and Thomas (1965) have solved these equations to give the concentrations of A, B, and C as functions of V_X as well as of reactor length for the conditions $(D_e)_X = (D_e)_Y$ and $R_X = R_Y$. Our primary concern here is with the determination of an optimum value of V_X, that is, an optimum catalyst composition. The existence of such an optimum can be easily rationalized, since with catalyst X alone the product C cannot form, and with catalyst Y alone also it cannot form, since X is required to produce the intermediate B. Evidently X and Y can be combined in an optimum proportion to give the highest conversion to C.

Gunn and Thomas (1965) have also extended this analysis to reaction schemes 1 and 3. Their numerical solutions for all three cases using specific values of the kinetic parameters are reproduced in Figure 12.25. Clearly, an optimal catalyst composition exists for each of the reaction schemes. For scheme 1 almost equal parts of X and Y are indicated; for scheme 2 the proportion of X is considerably lower; and for scheme 3, in which Y intrudes to produce D, the optimum is displaced even more to the left (the composition being less than 15% X).

Composite Pellets

In the case of composite pellets, since each pellet is prepared from a mixture of catalysts X and Y, the governing continuity equations will be different from those characterizing pellets of single catalysts. Again considering reaction scheme 2 and a composite pellet

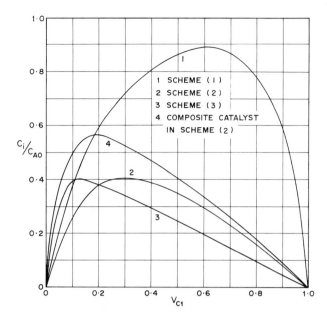

Figure 12.25. Concentration of desired product at the outlet of a plug-flow reactor as a function of the volume fraction of catalyst X for various schemes of operation of a reactor with polyfunctional catalysts.

of spherical geometry, we can write

$$D_e\left(\frac{\partial^2 C_A}{\partial r^2} + \frac{2}{r}\frac{\partial C_A}{\partial r}\right) - V_X(k_{v1}C_A - k_{v1\text{-}}C_B) = 0$$

$$D_e\left(\frac{\partial^2 C_B}{\partial r^2} + \frac{2}{r}\frac{\partial C_B}{\partial r}\right) + V_X(k_{v1}C_A - V_X k_{v1\text{-}}C_B)$$

$$\tag{12.78}$$
$$- (1 - V_X)k_{v2}C_B = 0$$

Note that the same effective diffusivity is operative in all the pellets. Equations for the fluxes of A, B, and C are similar to those derived for the previous case with the difference that the term V_X will not appear, since each pellet is composed of X and Y, and $(D_e)_X$ and $(D_e)_Y$ are each replaced by D_e. Equations identical to 12.77 can be written for the material balance over an infinitesimal section of the reactor. Solutions to these equations have been provided by Gunn and Thomas. The variation of the concentration of C in reaction scheme 2 as a function of catalyst composition as worked out by these investigators is included in Figure 12.25. The composite catalyst requires a lesser proportion of X at its optimum value than does the mixture of individual pellets of X and Y. It has also been established (through plots of concentration vs. reactor length) that the output of a reactor is greater with composite pellets than with mixtures.

From the forgoing discussion it is clear that for a given set of conditions an optimum catalyst composition exists for reaction on a polyfunctional catalyst. In view of the changing conditions along the length of a catalytic reactor, the possibility of an optimum catalyst profile can be readily conceived for such a system.

12.8. PRACTICAL CONSIDERATIONS

It is not the purpose of this section to describe in detail the various aspects of fixed-bed reactor operation, but a few pertinent observations will be made concerning pressure drop, gas mixing, and flammability and explosion limits. Start-up and shut-down procedures, methods of catalyst reduction and activation *in situ*, and safety precautions to be observed depend almost entirely on the type of reaction or catalyst employed.

12.8.1. Pressure Drop

A particularly important consideration during the charging of catalyst in a tubular reactor is the uniformity of pressure drop in all the tubes. If the pressure drop is nonuniform, uneven flow of the reacting gas occurs in the different tubes, leading to variations in conversion and a lower final conversion.

In order to avoid this situation it is necessary to fill each tube with great care so that the pressure drop at any given height is the same in all the tubes (with a variation of less than 5%). The following steps are normally recommended to ensure uniform packing.

1. A weighed quantity of catalyst is taken for charging in each tube.
2. About 10% of this quantity is charged in each tube and the pressure drop is noted. Tubes should be tapped or vibrated for a specific period before the second 10% is charged. The pressure drop is again noted, and the extent of tapping or vibration in the different tubes is varied to bring the pressure drop to within 5% of the mean value.
3. This procedure is continued until the tubes are filled to the top. The pressure drop should again be measured. If the value is higher in one of the tubes, then probably a certain degree of attrition has taken place; on the other hand, a lower pressure drop indicates loose packing. Appropriate corrections should be made in either case.

For purposes of estimation, the pressure drop can be obtained from the following equation of Mehta and Hawley (1969):

$$\frac{\Delta P}{L} = \frac{G^2}{\rho_g d_p} \left(\frac{1 - f_B}{f_B^2} \right) \left(1 + \frac{4d_p}{6(1 - f_B)d_t} \right)^2$$
$$\times \left[\frac{150(1 - f_B)}{Re'} + \frac{1.75}{1 + (4d_p/6(1 - f_B)d_t)} \right]$$

$$(12.79)$$

Several other equations have also been reported, such as those of Ergun (1952), Chilton and Colburn (1931), Carman (1937), and Brownell and Katz (1947). Many of these have been reviewed by Brown (1950). In Ergun's equation, the first term represents viscous loss while the second represents the kinetic energy loss. At low Reynolds numbers (Re' < 20) viscous losses are dominant and only the first term remains. On the other hand, at Re' > 100 only the first term remains. On the other hand, at Re' > 100 only the second term need be considered.

One of the practical ways of ensuring uniform pressure drop in all the tubes is to introduce a plate with a single nozzle at the bottom of each tube, with a pressure drop that is considerably higher than through the granular material within the tube. Thus, changes in pressure drop within the tube will not be very important, so that the mode of packing or charging would also not be significant. This practice, however, is not generally followed, since with adequate care it is possible to ensure uniform pressure drop in all the tubes. The introduction of a nozzle plate at the bottom is nevertheless essential in the case of multitubular reactors in which the solids are allowed to fluidize.

12.8.2. Premixing of Gases

When the reactions involved are very fast and highly exothermic, it is often necessary to mix the reactant gas with a stream of diluent gas, or with part of the product gas that is recycled to the reactor. This is done to moderate the reaction and to increase the heat capacity of the gas stream in order to achieve better control on the temperature of the bed. Also, in cases where two reactant gases are involved, mixing of the gas streams must be ensured before introduction into the reactor.

Jet mixers, compressors (or blowers), baffled flow mixers, and static mixers are some of the equipment usually employed for gas–gas mixing. The former two can be used only when one of the gases is available at a sufficiently high pressure to overcome the loss in pressure during mixing. In the case of jet mixers, jets of one gas are introduced into the mainstream of the other either perpendicularly or at a slight angle (10–20°). Baffled flow mixers in general, and disk and doughnut baffled mixers in particular, are ideally suited for premixing of two gases when high pressure drops cannot be tolerated.

Spargers can be conveniently used when a small quantity of one reactant is to be added to a flowing stream of the major reactant. Mixing **T** or **Y** can also be used; the best results are obtained when the mass velocity of the added gas is sufficiently higher than that of the mainstream (generally 1.5–3.0 times higher). Other devices used are orifices and on-line mixers. The fraction ΔP recovered can be estimated from the approximate relationship

$$\Delta P_{recovered} = \left(\frac{\text{orifice diameter}}{\text{conduit diameter}} \right)$$

derived from the graphical correlation of Bennet and Myers (1962). This is valid for orifice-to-conduit ratios greater than 0.2.

12.8.3. Explosion and Flammability

Oxidation reactions normally carried out in fixed-bed reactors are potentially explosive, and care must be taken to avoid compositions that would lead to flaming and eventually to explosion. A convenient measure of the explosion potential of a system is its adiabatic reaction temperature at constant volume. (Note that this is different from the adiabatic temperature at constant pressure discussed in Section 12.1.1.) From a knowledge of the adiabatic temperature at constant volume for complete conversion, the corresponding pressure can be calculated from the simple gas law $P = zn\rho_0 R_g T$, where n is the total moles per unit mass, z the compressibility, and ρ_0 the initial density of the reactor gases. The term $R_g Tn$ is called the explosion potential of the reaction.

Values of some typical explosion potentials have been tabulated by Steffesen et al. (1966). For example, for the oxidation of ethylene, the adiabatic temperature is 3732°K and the explosion potential is 13,000 atm cm^3/g.

Flammability diagrams can be prepared for explosive mixtures, and the operating conditions so chosen that the flammability region is avoided. For complex mixtures, lower flammability limits can be obtained from the following equation (Coward and Jones, 1952;

Zabatakis, 1965): $\Sigma \, y_i/\tilde{L}_i = 1$, where i represents the components A, B, C, . . . ; $\tilde{L}$ the flammability limit; and y the mole fraction at the flammability limit of the mixture. In a mixture of A, B, and C, if y_A and y_B and $\tilde{L}_A$ and $\tilde{L}_C$ are known, the value of y_C can be readily calculated.

From the practical point of view, it would be desirable to have ratings of different substances and processes as a measure of hazard. Shabica (1963) has provided exhaustive tables of ratings of chemical substances as well as chemical processes. The degree of hazard is categorized in six levels, starting from the most hazardous chemicals at level 1. Chemical processes are classified into three categories—a, b, and c—again in decreasing order of hazard. Shabica (1963) has prepared a plot showing the relationship between these two hazards that can form a preliminary basis for hazard rating.

12.9. SOME SPECIAL REACTOR TYPES

There are at least two important reactor types in which the catalyst is stationary but the methods of design are different from those for the more common fixed-bed reactors described earlier. These are the catalytic wire-gauze reactors and radial-flow reactors. In both types, practice seems to be much ahead of theory. In the last few years, however, some significant attempts have been made toward a more rational approach to the design of these reactors. Two other reactor types, the direct-fired fixed-bed reactor and the pebble reactor, are also noteworthy. The latter is noncatalytic but heterogeneous and is therefore briefly considered. Although hot pebbles are continuously added to replace the cold pebbles that are withdrawn, the reactor acts essentially as a fixed bed of constant depth.

12.9.1. Catalytic Wire-Gauze Reactors

In certain catalytic reactors the catalyst is used in the form of a wire gauze or filament. Many of these catalysts are precious metals like platinum or silver. Oxidation of ammonia is one of the most important processes carried out on a wire-gauze catalyst. Other major reactions of this class are

1. Andrussow process for manufacturing HCN on Pt gauze with NH_3, air, and CH_4.
2. Oxidation of methanol to formaldehyde on Ag–Cu screen.
3. Oxidation of some hydrocarbon on Pt screen.

4. Oxidation of ethylene to ethylene oxide on Ag gauze.

The following considerations are important in the design of wire-gauze reactors: (1) selection of operating conditions for L/d_t ratios of the order of $1/200$ (characteristic of these reactors); (2) study of mass and heat transfer from wire-gauze catalysts and the phenomenon known as catalyst flicker; (3) stability and bounds of operating variables, and (4) loss prevention due to vaporization of the precious metal catalyst.

Choice of operating conditions becomes critical since the object of design in these reactors is to increase the time cycle of a catalyst charge. It is well known that while the catalyst surface affects the reaction between the reactants, the gas phase causes significant changes in the structure of the surface. Since these changes are responsible for an initial activation period during commercial use (Schmidt and Luss, 1971), selection of optimum operating conditions can be a very critical factor.

It has been found that the loss of catalyst is a major cost consideration and may account for as much as 85% of the operating cost of the process. Since the loss is a function of the operating conditions, failure to control these conditions will increase the volatilization of the precious metal catalyst, leading to uneconomical operation (Gillespie and Kenson, 1971).

Whereas earlier studies on wire-gauze catalysts were restricted almost entirely to empirical selection of operating conditions and prevention of catalyst loss, more recent studies have thrown light on heat and mass transfer to and/or from catalyst wires and on the phenomenon of catalyst flicker (e.g., Schmidt and Luss, 1971; Edwards et al., 1973, 1974; Luss and Erwin, 1972).

Exothermic reactions occurring on catalytic wires and gauzes lead to nonuniform surface temperatures. Even on industrial Pt gauze one can observe hot spots that grow and decay in time. This is the phenomenon referred to as "flicker." The numerical computations of Luss and co-workers indicate that appreciable temperature fluctuations could be caused by fluctuating transport coefficients. They define a ratio of two characteristic times, namely,

$$a_F = \frac{\text{characteristic time for changes in wire temperature}}{\text{characteristic time for changes in surface concentration}}$$

and suggest that flicker occurs when the value of this ratio is $a_F < 1$. However, the estimated value of 9900 for ammonia oxidation was found to be considerably higher than the approximate value of 100 indicated by experiments on a 0.025-mm Pt gauze. It would therefore appear that considerably more work needs to be done to understand and interpret the phenomenon of catalyst flicker.

Studies on mass and heat transfer to and/or from catalytic wires and gauzes have also been reported (e.g., Satterfield and Cortez, 1970; Shah and Roberts, 1974; Rader and Weller, 1974). The most significant correlations appear to be those of Satterfield and Cortez (1970): for one to three screens,

$$0.4 < Re'' < 9, \qquad f_s j_d = 0.865 \left(\frac{Re''}{f_s}\right)^{-0.648} \tag{12.80}$$

where f_s is the porosity of a single screen and Re'' is based on the interstitial velocity.

Shah and Roberts (1974) report, for from one to five screens,

$$3 < Re'' < 107, \qquad \gamma_f j_d = \left(\frac{0.644}{Re''}\right)^{-0.570} \tag{12.81}$$

where γ_f is the minimal fractional opening of a single screen and Re'' is based on the maximum interstitial velocity.

As already mentioned, the most important application of wire-gauze reactors is in the oxidation of ammonia to give nitric oxide, which is subsequently converted to nitric acid. The catalyst employed consists of 20–60 layers of shining Pt wire screen that takes on a blistered appearance after being in operation for a few hours. Details of the process have been described by Powell (1969), Chilton (1960), and Gillespie and Kenson (1971). The use of random pack and Degussa Getter packings has greatly improved the efficiency of this reactor. A typical reactor design is shown in Figure 12.26.

Hydrocyanic acid is also produced in a gauze reactor by reacting oxygen and methane over Pt screen. This process is covered chiefly by the patents of Andrussow (1935) and Luckey et al. (1958).

Process improvements (Longman, 1956) in the production of formaldehyde from methanol since the operation of the first plant in the thirties allow operation at 60–65% methanol in a mixture of air and steam. The catalyst used is electrolytically refined Ag laid to a depth of about 1 cm on a fine-mesh Cu gauze.

12.9.2. Radial-Flow Reactors

Certain essential products like ammonia are required in very large tonnages. The reactor design strategy for such products should specifically take into account this aspect of the requirement. The conventional fixed-bed tubular reactors discussed earlier would be inadequate to accommodate such high orders of flow (50–100 tonnes/hr). One obvious way of persisting with conventional design is to pack the tubes with large quantities of catalyst, but this would result in increased pressure drop and operating cost. The pressure drop can be reduced by using large pellets, but this would have the effect of enhancing the intraparticle diffusional resistance, leading to decreased reaction rates. Typically, when the particle size is increased fivefold (say, from 2 mm to 10 mm), the reaction rate falls to almost half the value.

A completely different reactor design is therefore necessary to meet this demanding situation. The Haldor Topsøe Company has proposed a reactor system that seeks to accommodate such large throughputs. In this design the catalyst is placed between coaxial cylinders and the reacting gas is allowed to flow either from or to the center, as shown in Figure 12.27. Clearly the pressure drop is expected to be low in such a configuration, owing to the short length of the catalyst bed. A spherical configuration that is entirely analogous to the cylindrical radial-flow reactor has also been suggested (Cimbalnik et al., 1962). Studies on radial-flow reactors have been very sparse (Raskin et al., 1968a, b; Hlavacek and Kubicek, 1972; Strauss and Budde, 1978; Calo,

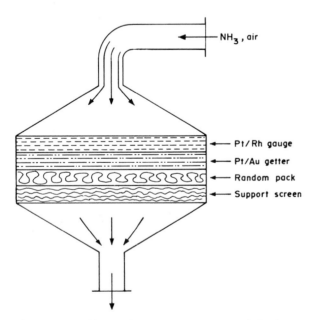

Figure. 12.26. Sketch of a wire-gauze converter (Gillespie and Kenson, 1971).

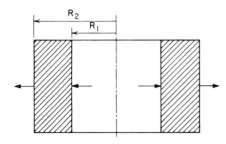

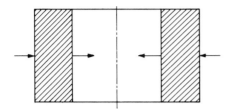

Figure 12.27. Schematic representation of radial-flow reactors.

1978), and have been concerned essentially with the problem of radial profiles and the effect of mixing.

The two configurations—outward flow and inward flow—are sometimes referred to as centrifugal and centripetal flow, respectively. Consider outward flow in a cylindrical reactor, and make the following assumptions: Heat transfer in the radial direction can be described by the dispersion model; intraparticle effects and axial gradients are absent; and a simple reaction $(A \rightarrow B)$ of the power law type is involved. Then the following mass and heat balance equations (for component A) can be written (Hlavacek and Votruba, 1977):

$$\frac{d}{dr}(uC_A r) - \frac{d}{dr}\left(r D_{er} \frac{dC_A}{dr}\right) + k_g a r(C_A - C_{As}) = 0$$

(12.82)

$$k_g a(C_A - C_{As}) - A_f C_{As}^n \exp\left(-\frac{E}{R_g T_s}\right) = 0$$

(12.83)

$$\frac{d}{dr}(\rho_g T C_p u r) - \frac{d}{dr}\left(r k_{er}' \frac{dT}{dr}\right) + h_{fp} a r(T - T_s) = 0$$

(12.84)

$$h_{fp} a(T - T_s) + (-\Delta H) A_f C_{As}^n \exp\left(-\frac{E}{R_g T_s}\right) = 0$$

(12.85)

with the boundary conditions

$$r = R_1, \quad u\rho_g C_p(T_0 - T) = -k_{er}' \frac{dT}{dr}$$

$$u(C_{A0} - C_A) = -D_{er}\frac{dC_A}{dr}$$

(12.86)

$$r = R_2, \quad \frac{dC_A}{dr} = \frac{dT}{dr} = 0$$

(12.87)

where R_1 and R_2 are the inner and outer radii of the annulus containing the catalyst. These equations can be rendered dimensionless by using the following dimensionless groups (some of which have already been defined):

$$\text{Pe}_{hr} = \frac{R_2 u\rho_g C_p}{k_{er}'}, \quad \text{Pe}_{mr} = \frac{R_2 u}{D_{er}}, \quad \theta = \frac{E(T - T_0)}{R_g T_0^2}$$

$$D_m = \frac{R_2 k_g a}{u}, \quad D_h = \frac{R_2 h_{fp} a}{u\rho_g C_p}, \quad \theta_s = \frac{E(T_s - T_0)}{R_g T_0^2}$$

$$\text{Da}_R = \frac{A_f C_0^{n-1} R_2 \exp(-\alpha_0)}{u}, \quad \beta_M = \frac{(-\Delta H)C_{A0}}{\rho_g C_p}\frac{E}{R_g T_0}$$

$$C_A' = \frac{C_{A0} - C_A}{C_{A0}}, \quad C_{As}' = \frac{C_{A0} - C_{As}}{C_{A0}}$$

(12.88)

The dimensionless equations in their final form are

$$\frac{dC_A'}{d\omega} - \frac{1}{\text{Pe}_{mr}}\frac{d}{d\omega}\left(\omega\frac{dC_A'}{d\omega}\right) - D_m\omega(C_{As}' - C_A') = 0$$

(12.89)

$$\frac{d\theta}{d\omega} - \frac{1}{\text{Pe}_{hr}}\frac{d}{d\omega}\left(\omega\frac{d\theta}{d\omega}\right) - D_h\omega(\theta_s - \theta) = 0$$

(12.90)

$$D_m(C_{As}' - C_A') - \text{Da}_R(1 - C_{As}')^n \exp\left(\frac{\theta_s}{1 + \theta_s/\alpha_0}\right) = 0$$

(12.91)

$$D_h(\theta_s - \theta) - \text{Da}_R\beta_M(1 - C_{As}')^n \exp\left(\frac{\theta_s}{1 + \theta_s/\alpha_0}\right) = 0$$

(12.92)

with the boundary conditions

$$\omega = \omega_0, \quad \text{Pe}_{mr}C_A' = \omega_0\frac{dC_A'}{d\omega}, \quad \text{Pe}_{hr}\theta = \omega_0\frac{d\theta}{d\omega}$$

(12.93)

$$\omega = 1, \quad \frac{d\theta}{d\omega} = \frac{dC_A'}{d\omega} = 0$$

These equations can be solved by the shooting techniques and radial profiles of concentration and temperature obtained.

Equations 12.89–12.92 can be considerably simplified if plug flow is assumed (i.e., if the Peclet numbers Pe_{mr} and Pe_{hr} have very high values). A simple and approximate criterion for plug flow is

$$Pe_{hr}(1 - \bar{\omega}) > 50 \qquad (12.94)$$

In the ammonia reactor the catalyst size is about 2 mm, and the left-hand side of inequality (12.94) is about 700. Hence plug flow may be assumed and the simplified equations used.

In an elaborate analysis of the behavior of radial flow reactors, Balakotaiah and Luss (1981) derived mass and heat balance equations similar to those presented here, but they considered a reaction involving a change in volume and described by various types of kinetics. They examined the implications of both types of flow, outward and inward, for cylindrical as well as spherical reactors, and their chief conclusions for the more important cylindrical reactors may be summarized as follows.

1. For reaction with no volume change, outward flow yields a higher conversion than inward flow for positive-order kinetics, whereas inward flow is superior for negative-order kinetics.
2. For a first-order reaction with increase in volume, outward flow gives higher conversion, whereas for a reaction with decrease in volume, inward flow gives better conversion.
3. For the ideal plug-flow condition, the direction of flow makes no difference; the difference in behavior between the two directions of flow is therefore attributable to the dispersion effect. (Thus for the ammonia reactor, change of direction of flow will have no effect.)

Equations for estimating heat and mass transfer coefficients in radial reactors have been given by Hlavacek and Votruba (1977). These are

$$Pe_{hr} = \frac{k'_{er} \ln(R_1/R_2)}{(u\rho_g C_p)(R_1 - R_2)} \qquad (12.95)$$

$$Pe_{mr} = \frac{D_{er} \ln(R_1/R_2)}{u(R_1 - R_2)} \qquad (12.96)$$

12.9.3. Catalytic Monoliths

An important class of reactors in the control of automobile emissions are the so-called catalytic monoliths characterized by a honeycomb structure. These composite and/or monolith structures can be used for control of air pollution, notably from automobile exhausts, and in the elimination of nitrogen oxides from air (Searles, 1973) and from ammonia plants, as well as from operations involving nitration in

TABLE 12.7. Some Distinguishing Features of Monoliths as Compared to Conventional Packed-Bed Reactors

Packed Bed	Monolith
Useful basically for conventional designs	Readily adaptable to new concepts in reactor design
Pelleted catalysts are prone to attrition and shrinkage, which lead to catalyst fines and reactor plugging.	Production of fines is not a problem.
Pelleted catalyst systems have at least one or two orders of magnitude higher pressure drop.	Reactor throughout can be maximized without significantly increasing the pressure drop.
Pore diffusion is almost always significant.	Reactions occur on the catalyst surface; hence pore-diffusional influences are practically absent, implying a higher rate.
External reactor geometry strongly influences mass-transfer-controlled conversion in particulate catalysts.	Monolith system is almost independent of mass transfer phenomena.
Particulate catalyst systems are subject to radial gas flow and radial heat transfer, and the reactor can either be isothermal or have axial and/or radial temperature gradients.	The typical monolith does not have radial gas flow. For ceramic monoliths with low thermal conductivity this implies an adiabatic mode of operation; with metallic monoliths, however, a quasi-isothermal mode of operation seems possible.

general (McDermott, 1971; Balgord et al., 1972). The monoliths, made of metal or ceramic honeycomb structures, have been shown to be very effective in these operations, but the design of monolith systems is still entirely proprietary. In comparison with the packed bed, the monolith possesses much higher geometric surface area. This is important in reactions which are so fast that they are confined to the external surface area. Other advantages are the low pressure drop and uniform distribution of flow. A comparison between the two is presented in Table 12.7.

One type of metal monolith is a matrix prepared from crimped Nichrome ribbon, giving rise to matlike structures (see Johnson et al., 1961). Among the ceramic structures the most common is the so-called Oxycat configuration. Each monolith consists of a pair of end plates between which rodlike elements of airfoil cross section are fixed. Flow through this structure is similar to flow through a bank of staggered tubes. Such a composite can be coated with a film of catalyst, such as activated alumina impregnated with finely divided platinum. Tucci and Thomson (1979) have compared the characteristics of metal with those of ceramic monoliths for the methanation reaction and concluded that metallic monoliths, because of their higher thermal conductivity, ease of reactivation, and so on, are far superior to ceramic monoliths.

It has been shown by Carberry and Kulkarni (1973) that the effects of mass and heat transfer in the body of the matrix are negligible. The main resistance to be considered is therefore transport between the fluid bulk and the active catalytic layer. The usual method of crushing the catalyst obviously cannot be adopted here (since the structure has to be preserved). A recycle reactor into which the matrix is placed appears to be a very satisfactory way of obtaining kinetic data. Alternatively, a Berty reactor may be employed.

Heat and mass transfer studies on composites have been summarized by Hlavacek and Votruba (1977). Mathematical models describing monoliths have been summarized by Wei (1975). In view of the scanty information published on monoliths, no acceptable design procedure has been proposed.

12.9.4. Direct-Fired Fixed-Bed Tubular Reactors

Empty tube reactors (or coils) heated in a furnace are commonly used for the cracking of petroleum feedstocks and some organic chemicals (e.g., ethylene dichloride to vinyl chloride). On the other hand, tubular fixed-bed reactors that are directly fired in a furnace are relatively rare. The most striking example is the cata-

lytic steam reforming of methane (or higher hydrocarbons).

The steam-reforming process carried out on a nickel catalyst in a number of vertical tubes directly heated in a furnace involves two main reactions:

$$CH_4 + H_2O \rightarrow CO + 3H_2 \quad (\Delta H = 54 \text{ kcal/mol})$$
$$CO + H_2O \rightarrow CO_2 + H_2 \quad (\Delta H = -8.21 \text{ kcal/mol})$$

The first reaction, which constitutes the actual reforming process, is endothermic and is therefore favored by high temperatures. The second reaction, commonly known as shift conversion, is exothermic and is therefore favored by lower temperatures. Thus, if the shift reaction is to be suppressed and mixtures of CO and hydrogen produced for the oxo process, the entire reactor has to be operated at temperatures of the order of 900°C. On the other hand, if CO is to be converted to CO_2 by the shift reaction, a falling temperature profile has to be introduced to ensure that the second reaction occurs at the exit end of the reactor system.

Although the methods used in thermal cracking can also be employed for direct-fired fixed-bed reactors, pilot plant operation seems inescapable in the design of commercial reactors. Several types of furnaces are used, the main points of difference being in the geometry and orientation of the tube rows and burner rows.

Contact times in these reactors are usually less than 1 sec, and the very high temperatures involved call for robust mechanical design and maintenance during operation. Daily monitoring of the tubes is necessary in order to detect and forestall failure.

12.9.5. Pebble Reactors

In addition to the use of direct-fired reactors, high temperatures can be obtained by passing the reactants through a chamber of refractory material with heat stored from a previous heating cycle. A more elaborate and effective method is to feed into a refractory-lined reactor pebbles (also made of refractory material) that have been previously heated; they are fed into the reactor in such a way that they come into contact with the reactant gases moving countercurrently to the pebbles. The reactor temperature can be controlled through the ratio of the feed rate and the pebble circulation rate, while the contact time is determined by the pebble depth and the feed rate.

The main principle behind this reactor is that heat is transferred from the hot pebbles to the fluid in a very short time. Normally the contact time is of the order of a fraction of a second for pebbles ranging in diameter from 0.5 to 1 cm. The chief requirements for the pebbles are that they have a high softening point, high heat capacity, and good abrasion resistance. Details of the operation of pebble reactors have been described by Findlay and Goins (1959).

Although pebble reactors are not currently very popular, the basic advantage that they offer—attainment of very high temperatures within a fraction of a second—can be exploited for certain industrial reactions. The design of pebble reactors has to be based almost entirely on pilot plant data, although some basic design equations (which neglect the heat generated as a result of reaction) have been developed by Munro and Amundson (1951). Fluid-to-pebble heat transfer coefficients are of the order of 8.2 for air heating and 20 for hydrogen heating (Norton, 1946).

REFERENCES

Ampaya, J. P., and Rinker, R. G. (1977), *Ind. Eng. Chem. Process Design Dev.*, **16**, 63.

Amundson, N. R. (1965), *Can. J. Chem. Eng.*, **43**, 49.

Andersen, A. S., and Michelsen, M. L. (1975), *Adv. Chem. Ser.*, **148**, Chapter 1.

Andrussow, L. (1935), U.S. Patent 1, 934, 838.

Aris, R. (1961), *The Optimal Design of Chemical Reactors—A Study in Dynamic Programming*, Academic Press, New York.

Aris, R. (1962), *Can. J. Chem. Eng.*, **40**, 87.

Aris, R. (1964), *Discrete Dynamic Programming*, Blaisdell, Massachusetts.

Aris, R. (1965), *Introduction to the Analysis of Chemical Reactors*, Prentice-Hall, Englewood Cliffs, New Jersey.

Aris, R. (1975), *The Mathematical Theory of Diffusion in Permeable Catalysts*, Vol. 1, Oxford (Clarendon Press), New York and London.

Aris, R., and Amundson, N. R. (1958), *Chem. Eng. Sci.*, 7, 121, 132, 148.

Athens, M., and Falb, P. L. (1969), *Optimal Control*, WHT, Warsaw.

Babu Rao, K., and Doraiswamy, L. K. (1967), *AIChE J.*, **13**, 397.

Babu Rao, K., Mukherjee, S. P., and Doraiswamy, L. K. (1965), *AIChE J.*, **11**, 741.

Baddour, R. F., Brian, P. L. T., Logeais, B. A., and Eymery, J. P. (1965), *Chem. Eng. Sci.*, **20**, 281.

Balakotaiah, V., and Luss, D. (1981), *AIChE J.*, **27**, 442.

Balgord, W. D., Strickland, R. W., and Wang, K. W. K. (1972), *AIChE Symp. Ser.* (No. 126), **68**, 102.

Barelko, V. V., and Volodin, Yu. E. (1976), *Kinet. Katal.*, 593.

Becker, E. R., and Wei, J. (1977), *J. Catal.*, **46**, 372.

Bellman, R. (1957), *Dynamic Programming*, Princeton Univ. Press, Princeton, New Jersey.

Bellman, R., and Dreyfus, S. E. (1962), *Applied Dynamic Programming*, Princeton Univ. Press, Princeton, New Jersey.

Belyaev, V. D., Slinko, M. M., and Slinko, M. G. (1976), *Proc. Int. Cong. Catal.*, **B**, 15.

Bennett, C. O., and Myers, J. E. (1962), *Momentum, Heat and Mass Transfer*, McGraw-Hill, New York.

Beusch, H., Fieguth, P., and Wicke, E. (1972), *Chem Ing.–Tech.*, **44**, 445.

Bilous, O., and Amundson, N. R. (1955), *AIChE J.*, **1**, 513.

Bilous, O., and Amundson, N. R. (1956), *Chem. Eng. Sci.*, **5**, 81, 115.

Brötz, W. (1965), *Fundamentals of Chemical Reactor Engineering*, Addison-Wesley, Reading, Massachusetts.

Brown, G. G., ed. (1950), *Unit Operations*, Wiley, New York.

Brownell, L. E., and Katz, D. L. (1947), *Chem. Eng. Prog.*, **43**, 537.

Bruns, D. D., Bailey, J. E., and Luss, D. (1973), *Biotech. Bioeng.*, **15**, 1131.

Bukur, D. B., Wittmann, C. V., and Amundson, N. R. (1974), *Chem. Eng. Sci.*, **29**, 1173.

Bukur, D. (1978), *Chem. Eng. Sci.*, **33**, 1055.

Burghardt, A., and Skrzypek, J. (1974), *Chem. Eng. Sci.*, **29**, 1311.

Bush, S. F. (1969), *Proc. Roy. Soc. (London)*, **A309**, 1.

Bykov, V. I., Elokhin, V. I., and Yablonskii, G. S. (1976a), *React. Kinet. Catal. Lett.*, **4**, 191.

Bykov, V. I., Chumakov, G. A., Elokhin, V. I., and Yablonskii, G. S. (1976b), *React. Kinet. Catal. Lett.*, **4**, 397.

Caldwell, A. D., and Calderbank, P. H. (1969), *Brit. Chem. Eng.*, **14**, 470.

Calo, J. M. (1978), *ACS Symp. Ser.*, **65**, 550.

Calo, J. M., and Chang, H. C. (1978), Paper presented at CHISA, Prague.

Cant, N. W., and Hall, W. K. (1971), *J. Catal.*, **22**, 310.

Cappelli, A., and Dente, M. (1965), *Chim. Ind. (Milan)*, **47**, 1068.

Cappelli, A., Collina, A., and Dente, M. (1972), *Ind. Eng. Chem. Process Design Dev.*, **11**, 184.

Carberry, J. J., and Gillespie, B. M. (1966), *Chem. Eng. Sci.*, **21**, 472.

Carberry, J. J., and Kulkarni, A. A. (1973), *J. Catal.*, **31**, 41.

Carman, P. C. (1937), *Trans. Inst. Chem. Eng. (London)*, **15**, 150.

Chartrand, C., and Crowe, C. M. (1969), *Can. J. Chem. Eng.*, **47**, 296.

Chen, M. S. K. (1972), *AIChE J.*, **18**, 849.

Chilton, T. H. (1960), *CEP Monograph Ser.* (No. 3), **56**.

Chilton, T. H. and Colburn, A. P. (1931), *Trans Am. Inst. Chem. Eng. (New York)*, **26**, 178.

Cholette, A., and Blanchet, J. (1961), *Can. J. Chem. Eng.*, **39**, 192.

Cholette, A., Blanchet, J., and Cloutier, L. (1960), *Can. J. Chem. Eng.*, **38**, 1.

Cholette, A., and Cloutier, L. (1959), *Can. J. Chem. Eng.*, **37**, 105.

Cimbalnik, Z., Prchal, J., and Vermouzek, L. (1962). Czech. Patent 117, 150.

Cohen, D. A. (1972), *Proceedings of the Summer Institute of Nonlinear Mathematics*, eds. Stackgold, I., Joseph, D. D., and Sattinger, D. H. Springer-Verlag, Berlin.

Cohen, D. S., and Keener, J. P. (1976), *Chem. Eng. Sci.*, **31**, 115.

Collina, A., Corbetta, D., and Capelli, A. (1971), *Proc. European Symp. Use of Computers in Design of Chemical Plants*, Firenze.

Copelowitz, I., and Aris, R. (1970), *Chem. Eng. Sci.*, **25**, 885.

Coward, I., and Jackson, R. (1965), *Chem. Eng. Sci.*, **20**, 911.

Coward, H. F., and Jones, G. W. (1952), *U.S. Bur. Mines Bull.*, **503**, 155.

Cresswell, D. L. (1970), *Chem. Eng. Sci.*, **25**, 267.

Cresswell, D. L., and Paterson, W. R. (1970), *Chem. Eng. Sci.*, **25**, 1405.

Cutlip, M. B., and Kenney, C. N. (1978), *ACS Symp. Ser.* **65**, 475.

Dagonnier, R., and Nuyts, J. (1976), *J. Chem. Phys*, **65**, 2061.

Dauchot, J. P., and van Cakenberghe, J. (1973), *Nature Phys. Sci.*, **246**, 61.

Denbigh, K. G. (1944), *Trans. Faraday Soc.*, **40**, 352.

Denbigh, K. G. (1958), *Chem. Eng. Sci.*, **8**, 125.

Denbigh, K. G., Hicks, M. J., and Page, F. M. (1948), *Trans. Faraday Soc.*, **44**, 479.

Denn, M. M. (1969), *Optimization by Variational Methods*, McGraw-Hill, New York.

Denn, M. M. (1975), *Stability of Reaction and Transport Processes*, Prentice-Hall, Englewood Cliffs, New Jersey.

Denn, M. M., and Aris, R. (1965), *Ind. Eng. Chem. Fundam.*, **4**, 213.

Doraiswamy, L. K., Venkitakrishnan, G. R., and Sadasivan, N. (1970), NCL report.

Douglas, J. M. (1964), *Chem. Eng. Progr. Symp. Ser.* (No. 48), **61**, 1.

Drott, D. W., and Aris, R. (1969), *Chem. Eng. Sci.*, **24**, 541.

Dudukovic, M. P. (1977), *Chem. Eng. Sci.*, **32**, 985.

Dudukovic, M. P., and Lamba, H. S. (1975), paper presented at 80th AIChE national meeting, Boston.

Eckert, E., Hlavacek, V., and Marek, M. (1973a), *Chem. Eng. Commun.*, **1**, 89.

Eckert, E., Hlavacek, V., Marek, M., and Sinkule, J. (1973b), *Chem. Ing.–Tech.*, **45**, 83.

Edwards, W. M., Worley, F. L., Jr., and Luss, D. (1973), *Chem. Eng. Sci.*, **28**, 1479.

Edwards, W. M., Zuniga-Chaves, J. E., Worley, F. L., Jr., and Luss, D. (1974), *AIChE J.*, **20**, 571.

Eigenberger, G. (1972), *Chem. Eng. Sci.*, **27**, 1909, 1917.

Eigenberger, G. (1978), *Chem. Eng. Sci.*, **33**, 1255, 1263.

Elnashaie, S., and Yates, J. G. (1973), *Chem. Eng. Sci.*, **28**, 515.

Endo, I., Furusawa, T., and Matsuyama, H. (1978), *Catal. Rev. Sci. Eng.*, **18**, 297.

Ergun, S. (1952), *Chem. Eng. Prog.*, **48**(2), 89.

Fieguth, P., and Wicke, E. (1971), *Chem. Ing.–Tech.*, **43**, 604.

Findlay, R. A., and Goins, R. R. (1959), *Adv. Petrol. Chem. Refining*, **4**, 126.

Fine, F. A., and Bankhoff, S. G. (1967), *Ind. Eng. Chem. Fundam.*, **6**, 288.

Fodder, L. (1971), *Chimique*, **104**, 1002.

Froment, G. F. (1972), *Adv. Chem. Ser.*, **119**, 1.

Froment, G. F., and Bischoff, K. B. (1979), *Chemical Reactor Analysis and Design*, Wiley, N.Y.

Gavalas, G. R. (1966), *Chem. Eng. Sci.*, **21**, 477.

Gavalas, G. R. (1968), *Non-Linear Differential Equations of Chemically Reacting Systems*, Springer-Verlag, New York.

Geiringer, P. L. (1962), *Handbook of Heat Transfer Media*, Chapman & Hall, London.

Gilles, E. D., and Hofmann, H. (1961), *Chem. Eng. Sci.*, **15**, 328.

Gillespie, G. R., and Kenson, R. E. (1971), *Chem. Tech.*, **1**, 627.

Gunn, D. J., and Thomas, W. J. (1965), *Chem. Eng. Sci.*, **20**, 89.

Han, C. D., and Agrawal, S. (1973), *Chem. Eng. Sci.*, **28**, 1617.

Hatfield, B., and Aris, R. (1969), *Chem. Eng. Sci.*, **24**, 1213.

Hegedus, L. L., Oh, S. H., and Baron, K. (1977), *AIChE J.*, **23**, 632.

Hestenes, M. R. (1966), *Calculus of Variation and Optimal Control Theory*, Wiley, New York.

Higgins, J. (1967), *Ind. Eng. Chem.*, **59**(5), 19.

Hlavacek, V., and Hofmann, H. (1970), *Chem. Eng. Sci.*, **25**, 173, 1517.

Hlavacek, V., and Kubicek, M. (1972), *Chem. Eng. Sci.*, **27**, 1770.

Hlavacek, V., and Marek, M. (1968), *Chem. Eng. Sci.*, **23**, 865.

Hlavacek, V., and Marek, M. (1971), *Proc. 4th European Symp. Chem. React. Eng.*, p. 1070.

Hlavacek, V., and Votruba, J. (1977), *Chemical Reactor Theory— A Review*, eds. Lapidus, L., and Amundson, N. R., Prentice-Hall, Englewood Cliffs, New Jersey, p. 314.

Hlavacek, V., Marek, M., and Kubicek, M. (1968), *Chem. Eng. Sci.*, **23**, 1083.

Hlavacek, V., Marek, M., and John, T. M. (1969), *Collect. Czech. Chem. Commun.* **34**, 3664.

Hlavacek, V., Hofmann, H., and Kubicek, M. (1971), *Chem. Eng. Sci.*, **26**, 1629.

Hlavacek, V., Kubicek, M., and Visnak, K. (1972), *Chem. Eng. Sci.*, **27**, 719.

Hlavacek, V., and van Rompay, P. (1981), *Chem. Eng. Sci.*, **36**, 1587.

Ho, T. C. (1976), *Chem. Eng. Sci.*, **31**, 235.

Horak, J., Jiracek, F., and Krausova, L. (1971), *Chem. Eng. Sci.*, **26**, 1.

Horn, F. (1961), *Chem. Eng. Sci.*, **14**, 77.

Horn, F., and Klein, J. (1972), *Adv. Chem. Ser.*, **109**, 141.

Horn, F., and Troltenier, U. (1961a), *Chem. Ing.–Tech.*, **32**, 382.

Horn, F., and Troltenier, U. (1961b), *Chem. Ing.–Tech.*, **33**, 413.

Hougen, O. A., and Watson, K. M. (1947), *Chemical Process Principles*, Vol. 3, Wiley, New York.

Hugo, P. (1968), *Proc. 4th European Symp. Chem. React. Eng.*, p. 459.

Hugo, P. (1970), *Ber. Bunsenges. Phys. Chem.*, **74**, 121.

Hugo, P., and Jakubith, M. (1972), *Chem. Ing.–Tech.*, **44**, 383.

Hutchings, J., and Carberry, J. J. (1966), *AIChE J.*, **12**, 20.

Jackson, R. (1973), *Chem. Eng. Sci.*, **28**, 1355.

Jaspan, R. K., Coull, J., and Andersen, T. S. (1972), *Adv. Chem. Ser.*, **109**, 160.

Jensen, K. F., and Ray, W. H. (1980), *Chem. Eng. Sci.*, **35**, 241.

Johnson, L. L., Johnson, W. P., and O'Brien, D. L. (1961), *Chem. Eng. Prog. Symp. Ser.* **35**, 55.

Jouven, J., and Aris, R. (1972), *AIChE J.*, **18**, 402.

Kehlert, C., Rössler, O. E., and Varma, A., (1981), *Springer Ser. Chem. Phys.*, **18**, 355.

Kelley, H. S. (1962), *Methods of Gradients in Optimization Techniques*, ed. Leltman, G., Academic Press, New York.

King, R. P. (1965), *Chem. Eng. Sci.*, **20**, 537.

Kubota, H., Ishizawa, M., and Shindo, M. (1959), *Sulphuric Acid Japan*, **12**, 243.

Kulkarni, B. D., and Ramachandran, P. A. (1980a), *Chem. Eng. Commun.*, **4**, 353.

Kulkarni, B. D., and Ramachandran, P. A. (1980b), *Biotech. Bioeng.*, **22**, 1759.

Kulkarni, B. D., and Ramachandran, P. A. (1982), *AIChE J* (in press).

Kuo, J. C. W., and Amundson, N. R. (1967a), *Chem. Eng. Sci.*, **22**, 49.

Kuo, J. C. W., and Amundson, N. R. (1967b), *Chem. Eng. Sci.*, **22**, 443.

Kuo, J. C. W., and Amundson, N. R. (1967c), *Chem. Eng. Sci.*, **22**, 1185.

Kurtanjek, Z., Sheintuch, M., and Luss, D. (1980), *J. Catal.*, **66**, 11.

Lee, E. S. (1964), *AIChE J.*, **10**, 309.

Lee, K. U., and Aris, R. (1963), *Ind. Eng. Chem. Process Design Dev.*, **2**, 300, 306.

Lee, J. C. M., and Luss, D. (1970), *AIChE J.*, **16**, 621.

Levenspiel, O. (1977), personal communication.

Lin, K. F. (1979), *Can. J. Chem. Eng.*, **57**, 476.

Liu, S. L., and Amundson, N. R. (1962), *Ind. Eng. Chem. Fundam.*, **1**, 200.

Liu, S. L., and Amundson, N. R. (1963), *Ind. Eng. Chem. Fundam.*, **2**, 183.

Longman, C. W. (1956), *Ind. Chemist*, **32** (328), 307.

Lowe, A., and Bub, G. (1976), *Chem. Eng. Sci.*, **31**, 175.

Luckey, G. W., Robinett, J. M., and Stiles, A. B. (1958), U.S. Patent 2, 831, 752.

Luss, D. (1968). *Chem. Eng. Sci.*, **23**, 1249.

Luss, D. (1971), *Chem. Eng. Sci.*, **26**, 1713.

Luss, D. (1977), *Chemical Reactor Theory—A Review*, eds. Lapidus, L., and Amundson, N. R., Prentice-Hall, Englewood Cliffs, New Jersey, Chapter 4.

Luss, D. (1981), in *Multiphase Chemical Reactors*, Vol. I: *Fundamentals*, eds. Rodrigues, A. E., Calo, J. M., and Sweed, N. H., Springer-Verlag, New York.

Luss, D., and Amundson, N. R. (1967a), *Chem. Eng. Sci.*, **22**, 253.

Luss, D., and Amundson, N. R. (1967b), *Can. J. Chem. Eng.*, **45**, 341.

Luss, D., and Amundson, N. R. (1968), *Can. J. Chem. Eng.*, **46**, 424.

Luss, D., and Chen, G. T. (1975), *Chem. Eng. Sci.*, **30**, 1483.

Luss, D., and Erwin, M. A. (1972), *Chem. Eng. Sci.*, **27**, 315.

Matsuura, T., and Kato, M. (1967), *Chem. Eng. Sci.*, **22**, 171.

McCarthy, E., Zahradnik, J., Kuczynski, G. C., and Carberry, J. J. (1975), *J. Catal.*, **39**, 28.

McDermott, J. (1971), *Catalytic Conversion of Automobile Exhaust* (*Pollution Control Review*, No. 2).

McGowin, C. R., and Perlmutter, D. D. (1971), *Chem. Eng. Sci.*, **26**, 275.

McGreavy, C., and Cresswell, D. (1969), *Chem. Eng. Sci.*, **24**, 608.

McGreavy, C., and Thornton, J. M. (1970a), *Can. J. Chem. Eng.*, **48**, 187.

McGreavy, C., and Thornton, J. M. (1970b), *Chem. Eng. Sci.*, **25**, 303.

McGreavy, C., and Thornton, J. M. (1972), *Adv. Chem. Ser.*, **109**, 607.

Mehta, D., and Hawley, M. C. (1969), *Ind. Eng. Chem. Process Design Dev.*, **8**, 280.

Michelsen, M. L. (1977), *Chem. Eng. Sci.*, **32**, 454.

Michelsen, M. L., and Villadsen, J. (1972), *Chem. Eng. Sci.*, **27**, 751.

Munro, W. D., and Amundson, N. R. (1950), *Ind. Eng. Chem.*, **42**, 1481.

Murase, A., Roberts, H. L., and Converse, A. O. (1970), *Ind. Eng. Chem. Process Design Dev.*, **9**, 503.

Narsimhan, G. (1969), *Brit. Chem. Eng.*, **14**, 1402.

Narsimhan, G. (1976), *Ind. Eng. Chem. Process Design Dev.*, **15**, 302.

Norton, C. L., Jr. (1946), *Chem. Met. Eng.*, **53**, 116.

O'Neill, S. P., Lilly, M. D., and Rowe, P. N. (1971), *Chem. Eng. Sci.*, **26**, 173.

Oppelt, W., and Wicke, E. (1964), *Adv. Chem. Ser.*, **109**, 122.

Padberg, G., and Wicke, E. 1967), *Chem. Eng. Sci.*, **22**, 1035.

Pareira, C., and Varma, A (1978), *Chem. Eng. Sci.*, **33**, 1645.

Pareira, C., and Varma, A (1979), *Chem. Eng. Sci.*, **34**, 1187.

Perlmutter, D. D. (1972), *Stability of Chemical Reactors*, Prentice-Hall, Englewood Cliffs, New Jersey.

Perry, R. H. (1973), *Chemical Engineering Handbook*, McGraw-Hill, New York.

Pikios, C. A., and Luss, D. (1977), *Chem. Eng. Sci.*, **32**, 191.

Pikios, C. A., and Luss, D. (1979), *Chem. Eng. Sci.*, **34**, 919.

Pismen, L. M. (1980), *Chem. Eng. Sci.*, **35**, 1950.

Plichta, R. T., and Schmitz, R. A. (1979), *Chem. Eng. Commun.*, **3**, 387.

Pontryagin, L. S. (1962), *The Mathematical Theory of Optimal Processes*, Wiley-Inter Science, New York.

Powell, R. (1969), *Chem. Process. Rev. 30* (*Nitric Acid Technology, Recent Developments*).

Prasad, S. D., and Kulkarni, B. D. (1982), *Chem. Eng. Sci.*, **36**, 1731.

Rader, C. G., and Weller, S. W. (1974), *AIChE J.*, **20**, 515.

Rase, H. F. (1977), *Chemical Reactor Design for Process Plants*, Wiley, New York.

Raskin, A., Ja., et al. (1968a), *Theoret. Found. Chem. Tech.*, **2**, 220.

Raskin, A., Ja., et al. (1968b), *Chim. Ind. (Milan)*, **44**, 199.

Rathousky, J., Kira, E., and Hlavacek, V. (1981), *Chem. Eng. Sci.*, **36**, 776.

Ravi Kumar, V., Jayaraman, V. K., and Kulkarni, B. D. (1981), *Chem. Eng. Sci.*, **36**, 945.

Ravi Kumar, V., Kulkarni, B. D., and Doraiswamy, L. K. (1982), *AIChE J.* (in press).

Ray, W. H. (1972), *Proc 5th European/2nd Int. Symp. Chem. React. Eng.*, p. A8.

Raymond, L. R., and Amundson, N. R. (1964), *Can. J. Chem. Eng.*, **42**, 173.

Rihani, D. N., and Doraiswamy, L. K. (1965), *Ind. Eng. Chem. Fundam.*, **4**, 17.

Roberts, G. W., and Satterfield, C. N. (1966), *Ind. Eng. Chem. Fundam.*, **5**, 317.

Rossler, O. E., and Wegmann, E. (1978), *Nature*, **271**, 89.

Sadana, A., Kulkarni, B. D., and Ramachandran, P. A. (1980), *Chem. Eng. Commun.*, **7**, 389.

Satterfield, C. N., and Cortez, D. H. (1970), *Ind. Eng. Chem. Fundam.*, **9**, 613.

Schleppy, R., Jr. and Shah, Y. T. (1977), *Chem. Eng. Sci.*, **32**, 881.

Schmidt, L. D., and Luss, D. (1971), *J. Catal.*, **22**, 269.

Schmitz, R. A. (1975), *Adv. Chem. Ser.*, **148**, 156.

Schmitz, R. A. (1978), *Proc. Joint Automatic Control Conf.*, **2**, 21.

Schmitz, R. A., and Amundson, N. R. (1963), *Chem. Eng. Sci.*, **18**, 265, 391, 415.

Schmitz, R. A., Graziani, K. R., and Hudson, J. L. (1977), *J. Chem. Phys.*, **67**, 3040.

Schmitz, R. A., Renola, G. T., and Garrigan, P. C. (1979), *Ann. N.Y. Acad. Sci.*, **316**, 638.

Schmitz, R. A., Renola, G. T., and Zioudas, A. P. (1980), In *Dynamics and Modeling of Reactive Systems*, Academic Press, New York, p. 179.

Schneider, P., and Mitschka, P. (1966), *Collect. Czech. Chem. Commun.*, **31**, 3677.

Searles, R. A. (1973), *Plat. Met. Rev.*, **17**, 57.

Shabica, A. C. (1963), *Chem. Eng. Prog.*, **59**, 57.

Shah, M. A., and Roberts, D. (1974), *Adv. Chem. Ser.* **133**, 259.

Shah, M. J. (1967), *Ind. Eng. Chem.*, **59**, 72.

Sheintuch, M., and Luss, D. (1981), *J. Catal.*, **68**, 245.

Sheintuch, M., and Schmitz, R. A. (1977), *Catal. Rev. Sci. Eng.*, **15**, 107.

Sheintuch, M., and Schmitz, R. A. (1978), *ACS Symp. Ser.*, **65**, 487.

Singh, C. P. P., and Saraf, D. N. (1979), *Ind. Eng. Chem. Process Design Dev.*, **18**, 364.

Skrzypek, J. (1974), *Int. Chem. Eng.*, **14**, 214.

Slinko, M. G., and Slinko, M. M. (1978), *Catal. Rev. Sci. Eng.*, **17**, 119.

Smith, T. G. (1977), *Chem. Eng. Sci.*, **32**, 334.

Strauss, A., and Budde, K. (1978), *Chem Tech. (Berlin)*. **30**, 73.

Steffesen, R. J., Agnew, J. T., and Olsen R. A. (1966), *Tables for Adiabatic Gas Temperature and Equilibrium Composition of Six Hydrocarbons* (Engineering Extensions Ser. No. 122), Purdue Univ., Lafayette, Indiana.

Takoudis, C. G., Schmidt, L. D., and Aris, R. (1981a), *Chem. Eng. Sci.*, **36**, 377.

Takoudis, C. G., Schmidt, L. D. and Aris, R. (1981b), *Chem. Eng. Sci.*, **36**, 1795.

Trambouze, P. J., and Piret, E. L. (1959), *AIChE J.*, **5**, 384.

Tucci, E. R., and Thomson, W. J. (1979), *Hydrocarbon Process. Petrol. Refiner*, **59** (2), 123.

Uppal, A., Ray, W. H., and Poore, A. B. (1974), *Chem. Eng. Sci.*, **29**, 967.

Uppal, A., and Ray, W. H. (1977), *Chem. Eng. Sci.*, **32**, 649.

van de Vusse, J. G. (1964), *Chem. Eng. Sci.*, **19**, 994.

van den Bosch, and Luss, D. (1977), *Chem. Eng. Sci.*, **32**, 203.

van Heerden, C. (1953), *Ind. Eng. Chem.*, **45**, 1242.

van Heerden, C. (1958), *Chem. Eng. Sci.*, **8**, 133.

Varghese, P., Carberry, J. J., and Wolf, E. E. (1978), *J. Catal.*, **55**, 76.

Varma, A., and Amundson, N. R. (1972a), *Chem. Eng. Sci.*, **27**, 907.

Varma, A., and Amundson, N. R. (1972b), *Can. J. Chem. Eng.*, **50**, 470.

Varma, A., and Amundson, N. R. (1973a), *Can. J. Chem. Eng.*, **51**, 206.

Varma, A., and Amundson, N. R. (1973b), *Can J. Chem. Eng.*, **51**, 459.

Varma, A., and Aris, R. (1977), *Chemical Reactor Theory—A Review*, eds. Lapidus, L., and Amundson, N. R., Prentice-Hall, Englewood Cliffs, New Jersey, Chapter 2.

Varma, A. (1981), *Ind. Eng. Chem. Fundam.*, **19**, 316.

Vayenas, C. G., Lee, B., and Michaels, J. (1980), *J. Catal.*, **66**, 36.

Verma, K. K., and Doraiswamy, L. K. (1965), *Ind. Eng. Chem. Fundam.*, **4**, 389.

Vautrain, L. H. (1977), U.S. Patent 4, 033, 727.

Volter, B. V. (1963), *Proc. Int. Fed. Automatic Control Cong.*, p. 507/1.

Votruba, J., and Hlavacek, V. (1973), *Chem. Prumysl*, **11**, 541.

Votruba, J., and Hlavacek, V. (1974), *Int. Chem. Eng.*, **14**, 461.

Votruba, J., Hlavacek, V., and Sinkule, J. (1976), *Chem. Eng. Sci.*, **31**, 971.

Wedel, S., Michelsen, M. L., and Villadsen, J. (1977), *Chem. Eng. Sci.*, **32**, 179.

Wei, J. (1965), *Chem. Eng. Sci.*, **20**, 729.

Wei, J. (1975), *Adv. Chem. Ser.*, **148**, 1.

Weisz, P. B., and Hicks, J. S. (1962), *Chem. Eng. Sci.*, **17**, 265.

Wicke, E. (1961), *Z. Elektrochem.*, **65**, 2670.

Wicke, E., and Vortmeyer, D. (1959), *Z. Elektrochem.*, **63**, 145.

Wicke, E., Padberg, G., and Arens, A. (1968), *Proc. 4th European Symp. Chem. React. Eng.*, p. 425.

Wilde, D. (1964), *Optimal Seeking Methods*, Prentice-Hall, Englewood Cliffs, New Jersey.

Wilde, D., and Beightler, C. S. (1967), *Foundations of Optimization*, Prentice-Hall, Englewood Cliffs, New Jersey.

Yamazaki, H., Yoshidsgu, U., and Hirakawa, K. (1978), *J. Phys. Soc. Japan*, **44**, 335.

Yang, R. Y. K., and Lapidus, L. (1974), *Chem. Eng. Sci.*, **29**, 1567.

Zabatakis, M. G. (1965), *U.S. Bur. Mines Bull.*, **627**.

Zuniga, J. E., and Luss, D. (1978), *J. Catal.*, **53**, 312.

Zellnick, H. E., Sondak, N. E., and Davis, R. S. (1962), *Chem. Eng. Prog.*, **58**, 35.

CHAPTER THIRTEEN

Design of Fluidized-Bed Reactors I: Properties of the Fluidized-Bed

13.1. THE FLUIDIZED BED

Fluidization is a technique of gas–solids contact in which a bed of solid particles is brought to a state of contained or uncontained motion by the gas flowing through the bed. The gas–solid system in such a situation may be likened to a fluid, since it exhibits fluidlike properties such as viscosity; thus, in a gross sense, the solid is "fluidized."

The fluid-bed reactor, by its very nature, can allow for continuous recirculation of solids. This is the basic advantage it offers over fixed beds for reactions in which the catalyst is subject to rapid deactivation. In petroleum cracking, for instance, significant reduction of activity occurs in a matter of seconds as a result of deposition of carbon on the catalyst. The answer to this situation is a fluid-bed reactor with provision for carbon burn-off in a second reactor (called the regenerator) through which the catalyst is circulated. Another major advantage of a fluid-bed reactor is that a uniform temperature is established as a result of the very process of fluidization.

As the velocity of a fluid through a packed bed is progressively increased, a stage is reached when the pressure drop across the bed becomes equal to the weight of the solids. This situation is referred to as the onset of fluidization, and the corresponding velocity is called the velocity for minimum or incipient fluidization (u_{mf}). A bed in this state exhibits the properties of a fluid. In fact, the viscosities of a number of fluidized systems have been experimentally determined, and a correlation has also been developed (Grace, 1970), based on the shape of the bubbles rising in real liquids, for estimating viscosities (which usually range between 4 and 20 poise).

As the velocity is further increased, the situation becomes similar to the introduction of a gas in a liquid. The additional gas flows through this gas–solid phase in the form of bubbles. There is no longer any increase in the pressure drop, and the fluid-mechanical behavior of a fluidized bed under these conditions can be predicted from the same principles as govern gas–liquid systems. In this two-phase concept of fluidization the bubble and the surrounding gas–solid "continuous" medium are referred to as the *bubble* phase and the *emulsion* phase, respectively.

A considerable volume of literature has accumulated over the last 20 years on several aspects of fluidization; these have been summarized in several significant books on the subject, including those by Zenz and Othmer, 1960; Davidson and Harrison, 1963, 1971; Kunii and Levenspiel, 1969; Botterill, 1975; and Keairns, 1976 (see also Davidson et al. and Caram et al. in Lapidus and Amundson, 1978; Grace and Matsen, 1980). Our main concern here is with the use of the fluidized bed as a chemical reactor. Thus a brief survey of the main features of the fluidized bed, to the extent necessary for a rational design of a fluid-bed reactor, is presented in this chapter.

13.2. BASIC CONSIDERATIONS

13.2.1. Classification of Particles

For the sake of broad classification the fluidization characteristics of powders can be described in terms of the mean particle size and density of particles. Geldart (1973) has defined four groups of particles, as shown in Figure 13.1. Group A particles are aeratable and give small bubbles with an expanded dense phase and are therefore particularly suitable as catalyst particles. When these particles are fluidized there exists a region in which the bed expands without bubble formation; this distinguishes group A particles from group B particles. Group B particles, the most commonly encountered, are sandlike.

Most of the measurements of bubble properties and

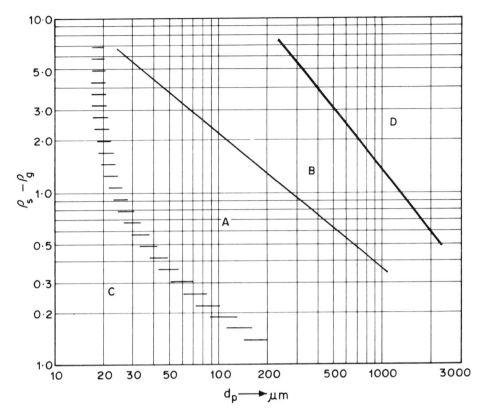

Figure 13.1. Classification of fluidization properties according to the size and density of powders (Geldart, 1973).

correlations that have so far been reported are for group B particles. In recent years measurements of bubble properties for other groups of particles have also been attempted, but the information is scanty. In what follows we shall mainly be concerned with correlations for group B particles, and wherever possible information on other groups of particles will be given.

13.2.2. Minimum Fluidization Velocity

Wen and Yu (1966) have developed an expression for the whole range of Reynolds numbers normally encountered. Saxena and Vogel (1977) have modified this equation based on data for coarse dolomite particles at various temperatures and pressures:

$$\mathrm{Re}'_{mf} = \left[(25.28)^2 + 0.0571 \frac{d_p^3 \rho_g (\rho_s - \rho_g) g}{\mu^2} \right]^{1/2} - 25.28$$

(13.1)

Another useful correlation is that of Broadhurst and

Becker (1975):

$$\frac{g(\rho_s - \rho_g)d_p}{\rho_g u_{mf}^2} = (2.42 \times 10^5) \left[\frac{\mu^2}{\rho_g g(\rho_s - \rho_g)d_p^3} \right]^{0.85}$$

$$\times \left(\frac{\rho_s}{\rho_g} \right)^{0.13} + 33.7$$

$$0.01 < \mathrm{Re}'_{mf} < 1000, \qquad 500 < \frac{\rho_s}{\rho_g} < 50{,}000,$$

$$1 < \frac{\rho_s g(\rho_s - \rho_g)d_p^3}{\mu^2} < 10 \qquad (13.2)$$

This is based on a statistical analysis of the experimental data (95% confidence interval) over a wide range of parameter values. A number of other correlations have also been proposed but will not be presented here.

The minimum fluidization velocity decreases with increase in temperature. However, the quantitative influence of temperature is different for small and large particles. For small particles (i.e., when $\mathrm{Re}'_{mf} < 1$) the second term in Ergun's equation for pressure drop in a fixed bed is negligible, so that u_{mf} would be inversely proportional to μ and will therefore decrease with increase in temperature. On the other hand, for large particles ($Re'_{mf} > 1000$) the first term is negligible, and u_{mf} would be inversely proportional to $\rho_g^{0.5}$ and will again decrease with increase in temperature but in a different way. For a close range of particle sizes, Desai et al. (1977) have proposed a linear relation between G_{mf} and T on a log–log scale. Botterill and Toeman (1980) and Pattipati and Wen (1981) have measured the minimum fluidization velocities of small and large particles at high temperatures. McKay and McLain (1980) have provided experimental data for cuboid large particles. The experimental data have been satis-factorily correlated by using Wen and Yu's (1966) equation.

For a polydispersed bed, Kondukov and Sosna (1965) and Gelperin et al. (1967) have reported data on the minimum velocities at which fluidization starts and the velocities at which all the particles are fluidized. Chen and Keairns (1975) have reported further work on this aspect of fluidization. It has been observed that the range of velocities within which the fluidization begins and is completed is generally narrower than the range of velocities for minimum fluidization of the finer and coarser particles measured independently. This is clearly attributable to the interaction between the finer and coarser particles. Invoking the similarity of the solid–liquid–vapor state of a substance to the packed–fluidized–dilute state of the bed, Kondukov and Sosna (1965) have presented the data in the form of a phase equilibrium diagram. Such diagrams may be useful when polydispersed beds are involved.

13.2.3. Two-Phase Theory of Fluidization

It was mentioned at the beginning of the chapter that all the gas in excess of that required for the onset of fluidization travels through the bed in the form of bubbles. This two-phase theory, first proposed by Toomey and Johnstone (1952), has been the subject of several early investigations (see Davidson et al., 1978). Two significant limitations of this theory are the following:

1. Comparison of the emulsion phase with a liquid is not strictly correct. When a bubble travels through a liquid there is no exchange of gas between the bubble and liquid phases; this is not true in a fluidized bed, where the emulsion is permeable to the gas and promotes its circu-lation through the bubbles.

2. If the two-phase theory were strictly valid, then the bubble-phase velocity should be

$$\frac{Q_b}{A_c} = (u - u_{mf}) \qquad (13.3)$$

Actually, observed values of Q_b/A_c are less than the velocity difference (Geldart, 1968; 1971; Grace and Harrison, 1969; McGrath and Streatfield, 1971; Geldart and Cranfield, 1972). This is so because experiments were carried out to measure directly (or estimate) the *visible* bubble flow rate, which does not account for throughflow in the bubbles (Lockett and Harrison, 1967).

The discrepancy between the observed and actual values of Q_b/A_c has been empirically accounted for by the so-called n theory (Grace and Harrison, 1969), in which the right-hand side of Eq. 13.3 is multiplied by a factor of $1 + nf_{mf}$. The n theory, however, found few applications, for no single value of n that would satisfy most data could be obtained. In fact, the value of n covers a wide range, between -8 and $+140$ (Grace and Clift, 1974). Also, the observed variation of the visible flow is sometimes incompatible with the modified equation (see Yacono, 1977).

An alternative explanation for the discrepancy between the observed and experimental values of Q_b/A_c has also been advanced (Baumgarten and Pigford, 1960; Pyle and Harrison, 1967; Godard and Richardson, 1969; Rowe and Everett, 1972). Here it is assumed that the average superficial velocity in the dense phase is not constant at u_{mf} but varies like Q_b/A_c along the height. This implies a variability of the dense-phase porosity that has not been proved experimen-tally, although it is known that the voidage in the dense phase may be different from that at minimum fluidization.

A large proportion of the bubbles near the distribu-tor are small, slow moving, and without cloud, and this proportion decreases sharply with increasing height. The analysis of Yacono et al. (1979) distinguishes between the bubbles with and without clouds and provides a comprehensive theory for describing the division of gas between the bubble and interstitial phases of the fluid bed.

Although the points just mentioned show up the weakness of the two-phase theory and bring out some alternative explanations, it can still be used as a reasonable basis for design.

13.2.4. Quality of Fluidization

Beyond the minimum fluidization velocity, gas-fluidized beds (such as catalytic reactors) are characterized by bubble formation and rise, giving the appearance of a boiling liquid. Such a bed is said to be aggregatively fluidized. A liquid-fluidized bed, on the other hand, which generally precludes bubble formation and is smoothly fluidized, is referred to as a particulately fluidized bed.

Two approximate criteria (Wilhem and Kwauk, 1948; Romero and Johanson, 1962) are available for distinguishing between the two regimes. Both are based on the magnitude of the Froude group at incipient fluidization, $Fr_{mf} = u^2_{mf}/gd_p$; the latter is reasonably reliable:

$$(Fr_{mf})(Re'_{mf})\left(\frac{\rho_s - \rho_g}{g}\right)\left(\frac{L_{mf}}{d_t}\right) < 100, \quad \text{particulate}$$

$$> 100, \quad \text{aggregative}$$

The practically useful regime of aggregative fluidization is bounded by bubbling at low superficial fluid velocities on one side, and by the onset of slugging at high velocities on the other. Correlations have been developed by Broadhurst and Becker (1975) for the minimum bubbling velocity (onset of aggregative fluidization) and minimum slugging velocity (end of aggregative fluidization). An expression for minimum bubbling velocity for fine powder has been proposed by Geldart and Abrahamsen (1978).

13.3. THE BUBBLE PHASE

13.3.1. Bubble Formation at a Single Nozzle

It has been shown by Davies and Taylor (1950) that the rise velocity of the bubble formed from a single nozzle in an inviscid liquid is related to the bubble size by the following equation:

$$u_{bs} = 0.711 (gd_b)^{1/2}$$
$$= 22.26d_b^{1/2} \quad \text{cm/sec} \tag{13.4}$$

where u_{bs} is the rising velocity of a single bubble and d_b its diameter. This equation is recommended for design (see Rowe and Matsuno, 1971), notwithstanding the observations of Godard and Richardson (1969) and Chiba et al., (1972).

The bubble diameter or volume can be calculated by assuming that bubble formation occurs at a point source and that the detachment of this bubble from the source takes place when the center of the bubble has covered a distance equal to its radius (Davidson and Schuler, 1960). Based on these assumptions the following equation has been derived for the bubble volume:

$$V_b = 1.4\left(\frac{Q^{1.2}}{g^{0.6}}\right) \tag{13.5}$$

The experimental results of Harrison and Leung (1961) and of Bloore (1961) show that Eq. 13.5 is also applicable to the formation of bubbles from a single nozzle in an incipiently fluidized bed. Several more correlations have been proposed for estimating the bubble size (diameter or volume), and the more important ones are summarized in Table 13.1. The most useful from the practical point of view are those of Miwa et al., (1971). Ramakrishnan et al. (1969), Satyanarayanan et al., (1969), and Kumar and Kuloor (1970) suggest the separation of the formation process into two stages, expansion and detachment, which makes it possible to solve the equations independently.

Caution must be exercised in the use of the correlations in Table 13.1, however. This is evident from the findings of Whitehead (1979), who has shown that an increase in bed depth or gas flow can significantly alter the bubble coalescence near the distributor and therefore the bubble diameter. A clearer understanding of the solids movement pattern in large systems is desirable before any fully reliable predictive expression can be developed.

13.3.2. Bubbles in a Fluidized Bed

An important finding from the design point of view is that Eq. 13.5 for a single orifice is also valid for multiple orifices in an evenly bubbling bed (Geldart, 1967). While dealing with bubbles from single orifices an equation for the rise velocity was presented. In the case of a fluidized bed in which bubbles are continuously generated at the distributor plate, it is necessary to develop an equation for the rise velocity (u_b) of the bubbles that also takes into account the motion of the bed. Thus we have (Nicklin 1962; Davidson and Harrison, 1963):

$$u_b = (u - u_{mf}) + u_{bs}$$
$$= (u - u_{mf}) + 0.711 (gd_b)^{1/2} \tag{13.6}$$

TABLE 13.1. Correlations for Bubble Diameter in Fluidized Beds

Correlation for Bubble diameter (cm)	Reference[a]
$d_{\mathrm{b}} = 1.6\rho_{\mathrm{s}}d_{\mathrm{p}}\left(\dfrac{u}{u_{\mathrm{mf}}} - 1\right)^{0.63} l$	Yasui and Johanson (1958)
$d_{\mathrm{b}} = 9.76\left(\dfrac{u}{u_{\mathrm{mf}}}\right)^{0.33} (0.032l)^{0.54}$	Whitehead and Young (1967)
$d_{\mathrm{b}} = 1.44\rho_{\mathrm{s}}d_{\mathrm{p}}\left(\dfrac{u}{u_{\mathrm{mf}}}\right) l + \left(\dfrac{6G}{\pi}\right)^{0.4} g^{-0.2}$ where $G = \dfrac{u - u_{\mathrm{mf}}}{N_{\mathrm{or}}}$	Kato and Wen (1969)
$d_{\mathrm{b}} = 33.3 d_{\mathrm{p}}\left(\dfrac{u}{u_{\mathrm{mf}}} - 1\right)^{0.77} l$	Park et al. (1969)
$d_{\mathrm{b}} = \left(\dfrac{6G}{\pi}\right)^{0.4} g^{-0.2} + 0.027(u - u_{\mathrm{mf}})^{0.94} l$ where $G = \dfrac{u - u_{\mathrm{mf}}}{N_{\mathrm{or}}}$	Geldart (1971)
$d_{\mathrm{b}} = -A_1 + A_2 l + A_3\left(\dfrac{u}{u_{\mathrm{mf}}}\right) + A_4 l\left(\dfrac{u}{u_{\mathrm{mf}}}\right) + A_5\left(\dfrac{u}{u_{\mathrm{mf}}}\right)^2$	Rowe and Everett (1972)

where A_1, A_2, A_3, A_4, A_5 are constants determined by the properties of the soild particles.

$\dfrac{d_{\mathrm{bm}} - d_{\mathrm{b}}}{d_{\mathrm{bm}} - d_{\mathrm{b0}}} = \exp\left(\dfrac{-0.3l}{d_{\mathrm{t}}}\right)$ where d_{b0} (initial d_{b}) $= 0.347\left[\dfrac{A_{\mathrm{c}}(u - u_{\mathrm{mf}})}{n_{\mathrm{or}}}\right]^{0.4}$ for a perforated plate, $d_{\mathrm{b0}} = 0.00376(u - u_{\mathrm{mf}})^2$ for a porous-plate distributor, and d_{bm} (maximum d_{b}) $= 0.652[A_{\mathrm{c}}(u - u_{\mathrm{mf}})]^{0.4}$	Mori and Wen (1976)
$d_{\mathrm{b}} = d_{\mathrm{b0}} + 1.4\rho_{\mathrm{s}}d_{\mathrm{p}}h\left(\dfrac{u}{u_{\mathrm{mf}}}\right)$	Kobayashi et al. (1966)
$d_{\mathrm{b}} = d_{\mathrm{b0}}\left(1 + 1.245\dfrac{l - l_{\mathrm{b0}}}{d_{\mathrm{b0}}}\right)^{2/7}$ for $l < l_{\mathrm{k}}$ where $d_{\mathrm{b0}} = g^{-1/5}\left(\dfrac{6Q_{\mathrm{b}}}{\pi A_{\mathrm{c}}}\right)^{0.4}$	Chiba and Kobayashi (1973)

TABLE 13.1. (continued)

Correlation for Bubble diameter (cm)	Reference[a]

$$l_k = l_{b0} + \left(\frac{2^{1.167s} - 1}{1.245}\right) d_{b0}$$

$$s = \frac{3 \ln(1 - s_{wb})(p/d_{b0})}{\ln 2}$$

$$Q_b = (u - u_{mf})\frac{A_c}{n_{or}}, \qquad l_{b0} = l_b + \frac{6d_{b0}}{7}$$

l_b is the height of the spout emanating from each perforation and p is the pitch of the distributor-plate holes.

$$d_b = (u - u_{mf})^{1/2} g^{1/4}(l + h_0)^{3/4}$$ 　　　　　　Rowe (1976)

where l is the height above the distributor plate and h_0 a constant characterizing the distributor plate. The equation has been rearranged as

$$\frac{d_b}{l + h_0} = (Fr')^{1/4} = \frac{(u - u_{mf})^2}{g(l + h_0)}$$

where analogy is evident with the equation for maximum bubble diameter of Harrison et al. (1961):

$$\frac{(d_p)_{max}}{d_p} \propto Fr$$ 　　　　　　Epstein (1976)

$$d_e = 0.54(u - u_{mf})^{0.4}\frac{(l + 4\sqrt{A_0})^{0.8}}{g^{0.2}}$$

The equation accounts for bubble growth due to coalescence; d_e represents the effective diameter of the bubble, A_0 the catchment area for the bubble stream at the grid plate given by area of plate per orifice, and l the height above the distributor plate. 　　　　　　Darton et al. (1977)

$$d_b = 0.853[1 + 0.272(u - u_{mf})]^{1/3}(1 + 0.684l)^{1.21}$$ 　　　　　　Werther (1978)

[a]See also Hayes et al. (1964); Kupferberg and Jameson (1969); McCann and Prince (1969); Potter (1969); Lanauze and Harris (1972, 1974); Chiba et al. (1973).

In writing Eq. 13.6 the assumption has been made that the rise velocity of a single bubble in a swarm is the same as that of an isolated bubble. This is not strictly true, since the coalescence of bubbles leads to an increased mean velocity in the swarm (Grace and Harrison, 1968). For practical purposes, we may either ignore any difference between the two, or use a value $1.2u_{bs}$ in the swarm, that is, $1.2u_{bs}$ for u_{bs} in Eq. 13.6 (Toei et al., 1966). The first alternative is usually adequate.

13.3.3. Bubble Frequency

The bubble frequency enables us to determine many other parameters, and is also the most attractive from the experimental point of view. Three types of frequencies can be identified: point frequency n_p, total frequency n_t, and area frequency n_a.

Park et al. (1974) have shown that the commonly used point-probe method adequately represents the other two only for small bubble sizes, that is, less than one tenth the column diameter. For larger bubbles the probe data should be used with caution. As pointed out by Geldart and Cranfield (1972), these frequencies are used interchangeably in the literature, which can lead to confusion.

The results of Kunii and Levenspiel (1969) show that beyond a distance of 15–20 cm from the distributor the point frequency is independent of both height and flow rate. On this assumption they derived the following useful relationship between bubble diameter and point frequency for a bed with large bubbles:

$$n_{p} = \frac{1.5(u - u_{mf})}{d_{b}} \tag{13.7}$$

Other equations for bubble frequency are those of Toor and Calderbank (1967), Matsuno and Rowe (1970), and Rowe (1973).

Davidson and Harrison (1963) suggest that the condition for maximum bubble size d_{bm} is $u_{b} = u_{bm} = u_{t}$. However, experimental evidence (Hardebol, 1961; Harrison, 1961; Matsen, 1968; Young, 1967) indicates that the maximum bubble size is approximately twice that predicted from this theory. Based generally on this evidence Grace (1971a) suggests the following simple equation for u_{bm}:

$$u_{bm} = 2\sqrt{u_{t}} \tag{13.8}$$

Thus a reasonably reliable estimate of the maximum bubble size can be obtained by inserting the value of u_{bm} calculated from Eq. 13.8 in the following expression for u_{t} corrected for larger bubble size:

$$u_{b} = 0.711 g^{1/2} d_{b}^{1/2} \left[\frac{1 - f_{mf}}{\rho_{s}/(\rho_{s} - \rho_{g}) - f_{mf}} \right]^{1/2} \tag{13.9}$$

From the practical point of view a good procedure for obtaining the maximum bubble diameter is to determine the initial diameter and multiply by 1.87 (Rowe, 1973). Thus

$$d_{bm} = 1.87 d_{bo} \tag{13.10}$$

13.4. THE EMULSION PHASE

The emulsion phase is important because it provides the particulate medium in which the reaction occurs. As the bubbles move through a fluid bed, the solid particles in the emulsion phase circulate according to certain patterns and cause various degrees of solids mixing. Gas mixing in the emulsion (which can vary from zero to complete mixing) is also important, since it influences the conversion.

13.4.1. Solids Mixing

In addition to purely theoretical studies on the movement of fluidized particles (see, e.g., Gabor, 1967, 1972), essentially two procedures have been followed in analyzing solids mixing:

1. The mixing phenomenon is considered through an axial dispersion coefficient, as in the case of gas mixing (May, 1959; Sutherland, 1961; Rowe and Sutherland, 1964; Lewis et al., 1962; Hayakawa et al., 1964).
2. The mixing of solids is related to the turbulence caused by the bubbles rising in the bed (Rowe and Partridge, 1962, 1965).

Experimental studies on solids mixing have generally been carried out by:

1. studying trajectories of particles either by taking motion pictures or by tagging with radioactive substances (Massimilla and Westwater, 1960; Kondukov et al., 1964; Borlai et al., 1967);
2. considering the intermixing of solids between two sections of the bed (Leva and Grummer, 1952; Talmore and Benenati, 1963); or
3. freezing a fluidized bed of particles coated with thermosetting resins of different colors, sectioning the "cake" so formed, and analyzing each section (Budkov et al., 1970).

Motion picture, tagged-particle, and frozen-bed studies show that generally particles move up the center and down along the walls, although theoretical analyses of single-particle trajectories indicate that the loop is not complete (Gabor, 1971). A model has been suggested similar to the elementary theory of gases for calculating particle velocities (Meissner and Kusik, 1970). Models have also been proposed by Todes et al., (1966), Gibilaro and Rowe (1974), and Fan and Chang (1979).

13.4.2. Gas Mixing

Residence time measurements cannot be used to determine gas mixing in the emulsion, since in a vigorously fluidizing bed a fraction of the bed is always occupied by bubbles, which may not contain any solids. On the other hand, results obtained under conditions of incipient fluidization (with no bubbles) would not be realistic.

A simple method would be to carry out a reaction in a fluidized bed; then by drawing out samples from the emulsion phase through a filter-tipped sample probe inserted at various positions to determine the extent of conversion as a function of position. Calderbank et al. (1967) carried out such a study by using the ozone reaction, and concluded that plug-flow conditions prevail for a short distance above the distributor, and that an internal circulation loop is set up with downflow of gas at the center and upflow at the wall.

13.5. GAS MIXING IN THE BED

Quantitative accounting of mixing in a fluidized bed is normally done in two ways (see Kunii and Levenspiel, 1969; Atimtay and Cakloz, 1978):

1. An effective axial and radial diffusion coefficient for the bed as a whole is defined, as in the case of fixed-bed reactors.
2. The bed is considered to be composed of two regions, an emulsion region and a gas-phase region, with continuous exchange of gas between the two regions. The models based on this concept are referred to as Type I or arbitrary models (Grace, 1971b; Rowe, 1972).

A modification of the Type I models would be to consider a bubble with its accompanying cloud (see Section 13.6) as an isolated element, rather than to consider the gas phase as a "region" that ignores the distinctive presence of bubbles. These are the so-called Type II models. Since the bubble is a part of the fluidized bed, however, some exchange must occur between the bubble–cloud unit and the surrounding emulsion. Several theoretical and experimental studies have been reported to estimate the exchange coefficient, for example, those of Lewis et al., (1959), Szekely (1962), Davies and Richardson (1966), Toei and Matsuno (1967), Rowe et al., (1971), Drinkenburg and Rietema (1972, 1973), Fontaine and Harriot (1972), Rietema and Hoebink (1976), and Chavarie and Grace (1976).

For our present purpose we shall define an inter-change coefficient K_{be}:

$$K_{be} = \frac{\text{volume of gas interchanging between bubble and emulsion}}{(\text{volume of bubbles in the bed})(\text{time})}$$

(13.11)

which can be readily related to the axial dispersion coefficient of the first approach or the crossflow coefficient of the second approach.

13.6. MODELS OF THE FLUID-BED REACTOR

13.6.1. Davidson's Simple Two-Phase Model and Its Extension

The chief features of the two-phase Davidson model are described below.

1. Bubbles of uniform size are assumed to move through the bed with an absolute rise velocity given by Eq. 13.6. An important parameter is the ratio of the rise velocity u_{bs} to the interstitial velocity u_i of the gas passing through the emulsion,

$$\alpha = \frac{u_{bs}}{u_i}$$

(13.12)

For values of $\alpha < 1$, the gas merely enters at the bottom of the bubble and leaves at the top. On the other hand, if $\alpha > 1$, the bubble carries with it some of the gas entering through its lower region. This gas is returned around the bubble to the lower part of it, forming the so-called cloud phase. The thickness of the cloud (represented by its radius r_c) is infinite when $u_{bs} = u_i$ (i.e., $\alpha = 1$) and decreases with increasing bubble velocity. It is given by

$$\frac{r_c}{r_b} = \left(\frac{\alpha + 2}{\alpha - 1}\right)^{1/3}$$

(13.13)

A more rigorous fluid-mechanical analysis of the bubble and cloud phases has been attempted by Jackson (1963), Reuter (1963), Murray (1965), and Pyle and Rose (1965). Figure 13.2 compares the cloud boundaries calculated from the Davidson and Murray

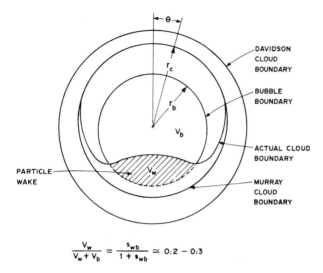

$$\frac{V_w}{V_w + V_b} = \frac{s_{wb}}{1 + s_{wb}} \simeq 0.2 - 0.3$$

Figure 13.2. Representation of the bubble–cloud–wake unit at a relatively low value of α (around 2). Note that both the Davidson and Murray models assume the bubble to be a full circle, whereas actually it is not.

models with the actual cloud boundary (Rowe, 1971). It will be seen that the circular nature of the boundaries predicted by both these models is not correct. The reason for this is the formation of a wake region at the bottom due to a suctionlike action.

Another limitation of the Davidson model is the assumption of a constant voidage in the emulsion phase, which is one of the basic postulates of the two-phase theory (Section 13.2.2). If allowance is made for voidage variations around a bubble, gas flow through the bubbles would actually be less than $3u_{mf}$ (Leung et al., 1970).

From the point of view of design, no general equation is yet available for predicting the cloud volume that also accounts for interaction between bubbles, and the equations for a single bubble appear to be the best available at present. Brief reference will be made in Section 13.6.4 to a mathematical model for coalescence developed by Orcutt and Carpenter (1971).

2. Gas in excess of u_i passes through the bubble phase. If the cross-sectional area of the bed is A_c and the actual gas velocity is u, then the gas passing through the bubble phase is (Eq. 13.3)

$$Q_b = A_c(u - u_i)$$

with a velocity of $u - u_i$.

3. The diffusive interchange between bubble and emulsion is governed by a mass transfer coef-

ficient defined by (Davidson and Harrison, 1963)

$$k_g = 0.975 D_b^{1/2} \left(\frac{g}{d_b}\right)^{1/4} \quad \text{cm/sec} \quad (13.14)$$

If q is the volume flow rate between the bubble and emulsion phases induced by circulation and A_{be} is the surface area of the bubble, then the total volumetric flow rate between the phases may be written as

$$Q_{be} = k_g A_{be} + q \qquad (13.15)$$

The simple picture of the fluid-bed reactor visualized above, together with the basic equations presented, can be used to calculate the bed height required to achieve a stated conversion for a reactor of fixed diameter (Section 14.3).

13.6.2. Assumptions Concerning Flow between the Bubble and Cloud Phases

The Davidson model assumes complete mixing (and therefore a uniform concentration) within the bubble–cloud system and a diffusional resistance inside the bubble at its interface with the emulsion phase. Actually, however, several mechanisms of gas flow are possible between the cloud and the bubble (Rowe, 1964; Partridge and Rowe, 1966; Rose, 1965; Lancaster, 1965; Hovmand and Davidson, 1971):

1. the flow is streamline in both the bubble and cloud phases;

2. the bubble gas is fully mixed, and flow between the bubble and the cloud is streamline;

3. the gas is completely mixed in the bubble and in the cloud, with flow from one region to the other;

4. the gas is completely mixed in the entire bubble–cloud system (with cloud-phase particles assumed to be equally distributed), which behaves like a single region of uniform concentration.

Mechanism 1 is the most rigorous (see Pyle and Rose, 1965). Mechanism 2 appears to be the closest to reality, and has been considered by Hovmand and Davidson (1971). Mechanism 4 (used in the Davidson model among others) is the simplest; it is a reasonable mechanism for large values of α and low circulation rates. Mechanism 3 has been assumed in the Kunii–Levenspiel model described next.

13.6.3. Kunii and Levenspiel's Bubbling-Bed (or Countercurrent-Flow) Model

In the bubbling-bed model (Kunii and Levenspiel, 1968a, b, 1969), the bubble phase consists of large fast-moving bubbles of uniform size surrounded by their clouds and wakes. At the velocities normally employed, there is countercurrent flow of the fluid in the bubble and emulsion phases. The model may therefore be referred to as a countercurrent-flow model. The main features of the model are the following.

1. The bubble size can be represented by a constant effective diameter. For a cloud thickness less than about 10% of the bubble diameter, it is postulated that

$$\frac{u_b}{u_i} > 5 \qquad (13.16)$$

The bubble fraction can be calculated from

$$\delta = 1 - \frac{1 - f_f}{1 - f_{mf}} = 1 - \frac{L_{mf}}{L_f} \qquad (13.17)$$

In the intermediate range $1 < u_b/u_i < 5$, where bubbles have appreciable clouds, interpolation of results from the two equations is necessary.

2. Every bubble is accompanied by a wake at its base. The solid particles in the wake are carried upward along with the bubble right up to the surface and are then returned through the emulsion phase (at a velocity u_s). If u_s exceeds u_i, as can happen in a vigorously fluidizing bed, then the emulsion gas will flow downward. Kunii and Levenspiel show that for $f_{mf} \simeq 0.5$ and $s_{wb} = 0.2$–0.6, such a flow reversal occurs when

$$\frac{u}{u_{mf}} > 6\text{--}11 \qquad (13.18)$$

3. Gas interchange between the bubble and emulsion is assumed to occur in two consecutive steps: (1) from bubble to cloud, characterized by a coefficient k_{bc}, and a bulk flow term q; and (2) from cloud to emulsion, characterized by a coefficient k_{ce}. Then, on the assumption that mechanism 3 listed in Section 13.6.2 for mixing

in the bubble–cloud system is operative, the following equation can be written for the overall coefficient of interchange K_{be} between the bubble and emulsion:

$$\frac{1}{K_{be}} = \frac{1}{K_{bc}} + \frac{1}{K_{ce}} \quad \sec^{-1} \qquad (13.19)$$

The coefficients k_{bc} and K_{bc} are given by

$$k_{bc} = \tfrac{3}{4} u_{mf} + 0.975 D_g^{1/2} \left(\frac{g}{d_b} \right)^{1/4} \quad \text{cm/sec} \qquad (13.20)$$

$$K_{bc} = 4.5 \frac{u_{mf}}{d_b} + 5.85 \left(\frac{D_g^{1/2} g^{1/4}}{d_b^{5/4}} \right) \quad \sec^{-1} \qquad (13.21)$$

For the coefficients k_{ce} and K_{ce}, Kunii and Levenspiel have derived the following equations based on the penetration model of Higbie (1935) and on the assumption that there is no bulk flow between cloud and emulsion:

$$k_{ce} = \frac{2}{\sqrt{\pi}} \left[f_{mf} D_e \frac{(u_{bs} - u_{fr})}{d_b} \right]^{1/2} \quad \text{cm/sec} \qquad (13.22)$$

$$K_{ce} \simeq 6.78 \left(\frac{f_{mf} D_e u_b}{d_b^3} \right)^{1/2} \quad \sec^{-1} \qquad (13.23)$$

where u_{fr} refers to the upward velocity of the gas with respect to the emulsion and D_e is the effective diffusivity of the gas, approximated by $D_e = f_{mf} D_b / \tau$. Usually, $f_{mf} \simeq 0.5$ and the tortuosity $\tau \simeq 1.5$, so that $D_e \simeq D_g / 3$.

4. Distribution of solids between phases is an important consideration. Thus Kunii and Levenspiel define the volume fractions of the solids in the bubble, cloud–wake, and emulsion as follows:

$$s_{bb} = \frac{\text{volume of solids in the bubble}}{\text{volume of the bubble phase}}$$

$$s_{cb} = \frac{\text{volume of solids in the cloud–wake phase}}{\text{volume of the bubble phase}}$$

$$s_{eb} = \frac{\text{volume of solids in the emulsion phase}}{\text{volume of the bubble phase}}$$

These are interrelated through

$$\delta[s_{bb} + s_{cb} + s_{eb}] = 1 - f_t = (1 - f_{mf})(1 - \delta)$$
(13.24)

Usually the solids fraction in the bubble s_{bb} is less than 0.01 and can therefore be neglected. The solids fraction in the cloud–wake s_{cb} can be estimated from the relation

$$s_{cb} = (1 - f_{mf})\left[\frac{3u_i}{0.711(gd_b)^{1/2} - u_i} + s_{wb}\right]$$
(13.25)

where s_{wb} (the volume fraction of the wake solids with respect to the bubble volume) is usually of the order of 25%.

Based on the postulates outlined above, the conversion of the fluidizing gas can be calculated either by making some simplifying assumptions or in a more rigorous manner (Chapter 14).

13.6.4. Models Based on Distributed Bubble Size

Partridge and Rowe (1966) have proposed a model in which experimental information on bubble size distribution is assumed to be available from X-ray photography (see also Rowe, 1964). The chief feature of this model is that a stepwise procedure is used in which the bubble size in each step is assumed to have a constant average value. This average value is determined by measuring the bubble flow rate (at any level in the bed), which is made up of a range of bubble sizes rising at different velocities. This range of bubble sizes is associated with a range of α values. Thus, in essence, the problem reduces to one of determining an average value of α (i.e., $\bar{\alpha}$) for every step.

Toor and Calderbank (1967) have modified the simple Davidson model to account for the increase in bubble size with height due to coalescence by using the relationship $n_a = ae^{-bl}$.

Another mathematical model for coalescence has been proposed (Orcutt and Carpenter, 1971) that is far more rigorous than that of Toor and Calderbank. It is assumed that the rise velocity of a trailing bubble is enhanced as a result of the "wake effect" of the leading bubble by a factor that is proportional to the rise velocity of the latter and the distance between the two bubbles.

Several other models of the fluidized bed have also been proposed, based essentially on different mass exchange mechanisms, different methods of computing the bubble size variation, and incorporation of end effects. The important models are discussed in Chapter 14, with particular reference to reactor design.

13.7. HEAT AND MASS TRANSFER IN FLUIDIZED BEDS

In any fluidized-bed reactor, the following three important cases of heat transfer may be identified:

1. heat transfer from fluidizing solids to the containing wall;
2. heat transfer from fluidizing solids to an immersed tube;
3. fluid–particle heat transfer within the bed.

As can be readily visualized, h_w increases with u slowly at first, and then at the onset of fluidization the turbulence caused by fluidization results in a marked increase in h_w. This increasing trend does not continue, however; and at the velocity corresponding to the terminal velocity of the solids, h_w falls even below the value for a fixed bed. Thus a maximum value of the heat transfer coefficient h_{max} exists, corresponding to an optimum velocity u_{opt}, at which the heat transfer properties of a fluidized bed can be exploited to best advantage.

Two empirical approaches have generally been followed in correlating heat transfer data: one that relates h_{max} to u_{opt}, and another in which correlations are developed for the rising or falling branches of the h–u curve. Before we outline the recommended equations, a brief reference will be made to some theories of fluidized bed heat transfer.

13.7.1. Theories of Heat Transfer between Fluidizing Solids and Surfaces

A fluidized bed at incipient fluidization behaves like a gas of high heat capacity passing through the reactor at a high Reynolds number (Callahan, 1971). This causes a large gain in the capacity of the medium to transfer heat. The heat transfer behavior of a fluidized bed is governed essentially by happenings in the vicinity of the transfer surface. Unlike in packed or empty tubes, these happenings are largely the result of particle–film interactions. Basically, theories pertaining to these interactions have followed four lines of reasoning:

1. The fluid film at the wall is scoured by the solid particles, thus enhancing the heat transfer coefficient.

2. Heat transfer occurs directly through the emulsion; this can happen by a continual renewal of pockets of emulsion at the surface or through an emulsion film similar to a pure fluid film.

3. Both the fluid and emulsion films operate simultaneously in the manner of gas and liquid films in gas–liquid systems.

4. Heat transfer occurs across a string of spheres or alternate slabs adjacent to the wall.

Mathematical formulations based on these theories (see Kunii and Levenspiel, 1969; Gelperin and Einstein, 1971; Botterill, 1975; Xavier and Davidson, 1978; Xavier et al., 1980), though useful in understanding the probable mechanism of heat transfer in a fluidized bed, do not provide accurate estimates of the heat transfer coefficient. It is therefore necessary to resort to empirical correlations.

13.7.2. Fluid–Particle Heat Transfer

Fluid–particle heat transfer studies, the results of most of which are presented as plots of $Nu \, Pr^{-0.33}$ vs. Re'/f_f, have been carried out by several investigators. Another way of expressing heat transfer coefficients (as in the case of fixed-bed reactors) is through plots of j_h vs. Reynolds number. As found by McConnachie and Thodos (1963), Malling and Thodos (1967), and Rowe and Claxton (1965), the bed voidage plays a distinct role in determining the heat transfer coefficient. Gupta et al. (1974) propose the following expression, which is independent of the type of bed:

$$f j_h = \frac{2.876}{Re'} + \frac{0.3023}{(Re')^{0.35}} \quad (13.26)$$

where f equals f_B for the fixed bed and f_f for the fluid bed. This equation is reported to correlate the data in the Reynolds number range of 10–10000. It is generally not advisable to use a single expression for the entire range of Reynolds numbers, and expression 13.26 is therefore recommended for Reynolds numbers greater than 50. More recently, Naud et al. (1981) have also reported data on heat transfer between gas and particulate solids.

Extreme caution is desirable in extending correlations obtained from laboratory reactors to industrial-size reactors, for which the dynamics are known to be different. Fortunately, fluid–particle heat transfer is

generally not the limiting factor in most fluid-bed operations, primarily because of the very large surface area exposed by the particles (30–450 cm^2/cm^3).

13.7.3. Heat Transfer between a Fluidizing Bed and Surface

Einstein and Gelperin (1965), Gelperin et al. (1967), and Botterill (1975) have presented detailed analyses of heat transfer to the containing wall, particularly for the falling portion of the *h–u* curve. Of the many correlations reported, Wender and Cooper's (1958) correlation based on selected data has been recommended for use by Geldart (1970). Wender and Cooper define a dimensionless quantity H:

$$H = \frac{h_w d_p / [(1 - f_{mf})(C_{ps}\rho_s / C_{pg}\rho_g)k_g']}{1 + 7.5 \exp[(-0.44 L/d_t)(C_{pg}/C_{ps})]} \quad (13.27)$$

which is a function of the Reynolds number based on the initial velocity $d_p u_0 \rho_g / \mu$. The functional relationship between H and the Reynolds number is shown in Figure 13.3.

It is noteworthy that the results of different investigators differ widely. As shown in Table 13.2, even major trends appear to be different. This is probably due to a wide variation in the scale of operation. For instance, in laboratory equipment it is possible to achieve a very high bed-to-surface heat transfer coefficient compared to that in industrial units (radiation heat transfer being neglected).

TABLE 13.2. Variations in the Reported Dependence of Heat Transfer Coefficient on Different Parameters

Parameter	Heat Transfer Coefficient	Reference
Particle size	$h \propto d_p^{-0.23}$	Dow and Jakob (1951)
	$h \propto d_p^{-0.96}$	Miller and Logwinuk (1951)
Gas conductivity	$h \propto (k_g')^{0.33}$	Mickley et al. (1961)
	$h \propto k_g'$	Leva et al. (1949)
Particle heat capacity	$h \propto C_{ps}^{0.25}$	Dow and Jakob (1951)
	$h \propto C_{ps}^{0.8}$	Wender and Cooper (1958)

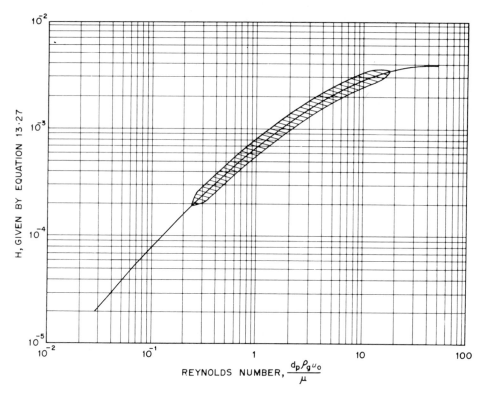

Figure 13.3. Correlation for heat transfer at container walls (Eq. 13.27) (Wender and Cooper, 1958).

Bed-to-wall heat transfer involves three modes of transport: convective particulate transport, convective gas-phase transport, and radiation transport. Most of the theories of bed-to-wall heat transfer really pertain to different models for particulate convective transport. Except in certain special situations, convective gas transport and radiation transport are negligible. Convective particulate transport occurs by virtue of solids circulation, which in turn influences the conduc-

tion through the fluid film at the surface. There is convincing evidence (Ziegler et al., 1964) that 80–95 % of the heat transfer is by particulate convection in normal fluid beds.

Part (a) of Figure 13.4 shows the influence of particle diameter on the particulate heat transfer coefficient, while part (b) brings out the regions of influence of the particulate and gas-phase convective heat transport processes.

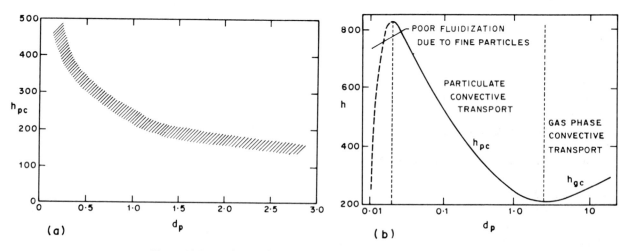

Figure 13.4. Variation of heat transfer coefficients with particle diameter.

Clearly, it is the packing density of the particles close to the wall and the frequency at which these particles are exchanged from the bulk of the bed, along with the gas conductivity, which limit the attainable heat transport rates in a given operation. Botterill (1975) recommends the following correlation of Zabrodsky (1966) for heat transfer by particulate motion:

$$h_{pc} = 35.8\rho_s^{0.2}(k_g')^{0.6}d_p^{-0.36} \quad \text{(mks units)} \quad (13.28)$$

Both Zabrodsky's equation and Eq. 13.27 of Wender and Cooper appear suitable for estimating bed-to-wall heat transfer coefficients. Since Wender and Cooper's correlation seems to have been more widely tested, however, it is recommended for use.

13.7.4. Heat Transfer to Immersed Tubes

Heat from a fluidized bed can be extracted either by tubes immersed vertically or by a bank of horizontal tubes of different configurations. The relatively inefficient behavior of horizontal (Lese and Kermode, 1972) tubes is attributed to the fact that only a narrow region on the sides is contacted by vigorously agitating solids, while the top of the tube is covered by stagnant solids and the bottom by gas film. On the other hand, the results of Korotjanskaja et al. (1967) show that there is hardly 5–6 % rise in the heat transfer coefficient (h_{max}) as the angle of the tube is varied from zero (horizontal) to 90° (vertical). A more detailed account of heat transfer to immersed surfaces has been provided by Staub (1980).

Vertical Tubes

For vertical tubes the best equation appears to be that of Wender and Cooper (1958):

$$\frac{h_w d_p}{k_g'} = 0.01844(CF)(1-f_f)\left(\frac{C_{pg}\rho_g}{k_g'}\right)^{0.43}$$

$$\left(\frac{d_p\rho_g u}{\mu}\right)^{0.23}\left(\frac{C_{ps}}{C_{pg}}\right)^{0.8}\left(\frac{\rho_s}{\rho_g}\right)^{0.66} \quad (13.29)$$

for

$$\frac{d_p\rho_g u}{\mu} = 10^{-2}\text{–}10^2$$

where CF represents the correction factor for nonaxial location of the immersed tube. This is a function of position and can be estimated from the correlation of Vreedenberg (1960).

Horizontal Tubes

Gelperin et al. (1958) and Vreedenberg (1952, 1958) have studied heat transfer to horizontal tubes. The correlation of Vreedenberg can be conveniently used for design purposes:

$$\frac{h_w d_0}{k_g'} = 0.66\left(\frac{C_{pg}\mu_g}{k_g'}\right)^{0.3}\left[\left(\frac{d_0 u\rho_g}{\mu}\right)\left(\frac{\rho_s}{\rho_g}\right)\left(\frac{1-f_f}{f_f}\right)\right]^{0.44}$$

$$(13.30)$$

for $d_0 u\rho_g/\mu < 2000$; and

$$\frac{h_w d_0}{k_g'} = 420\left[\left(\frac{C_{pg}\mu}{k_g'}\right)\left(\frac{d_0 u\rho_g}{\mu}\right)\left(\frac{\rho_s}{\rho_g}\right)\left(\frac{\mu^2}{d_p^3\rho_s g}\right)\right]^{0.3}$$

$$(13.31)$$

for $d_0 u\rho_g/\mu > 2500$.

More recent work on heat transfer between immersed surfaces and gas–solid fluidized beds of large particles is presented by Catipovic et al. (1980, 1981), Wood et al. (1980), and Grewal and Saxena (1981). The simple model of Catipovic et al. (1980) seems more useful for design purposes.

13.7.5. Example: Heat Removal from Vinyl Acetate Reactor

It is desired to design a fluidized-bed reactor for the production of vinyl acetate from acetylene and acetic acid. As a first step in this design it is necessary to examine the possibilities of heat removal by jacket cooling and by vertical tubes immersed in the bed. Thus (a) calculate the bed-to-wall heat transfer coefficient in a reactor 578 cm in diameter, that is 915 cm high, using the correlation of Wender and Cooper; (b) calculate the heat transfer coefficient between the fluidized bed and a vertical tube (10 cm outer diameter) positioned at the bed axis.

The following data are given:

$$d_p = 0.05 \text{ cm}, \quad \rho_s = 2 \text{ g/cm}^3, \quad C_{ps} = 0.2 \text{ cal/g °C}$$

$$C_{pg} = 0.55 \text{ cal/g °C}, \quad \rho_g = 8 \times 10^{-4} \text{ g/cm}^3$$

$$k_g' = 9 \times 10^{-5} \text{ cal/g sec °C}, \quad \mu = 1.6 \times 10^{-4} \text{ g/cm sec}$$

$$u_{mf} = 10 \text{ cm/sec}, \quad f_{mf} = 0.7$$

$$u = 30 \text{ cm/sec}, \quad f_f = 0.6$$

Solution

1. Bed-to-wall heat transfer:

The Reynolds number is first calculated from the given data:

$$\text{Re}' = \frac{d_p \rho_g u}{\mu}$$

$$= \frac{0.05 \times 8 \times 10^{-4} \times 30}{1.6 \times 10^{-4}} = 7.5$$

From Figure 13.3, $H = 2.3 \times 10^{-3}$. Using Wender and Cooper's correlation (Eq. 13.27)

$$H\{1 + 7.5 \exp[(-0.44 L/d_t)(C_{pg}/C_{ps})]\}$$

$$= \frac{h_w d_p}{k_g'} \frac{1}{[(1 - f_{mf})(C_{ps}e_s/C_{pg}e_g)]}$$

we calculated h_w. Thus

$$(2.3 \times 10^{-3})\left[1 + 7.5 \exp\left(-0.44 \times \frac{915}{578} \times \frac{0.55}{0.2}\right)\right]$$

$$= \frac{h_w \times 0.05}{9 \times 10^{-5}} \times \frac{1}{0.4 \times \dfrac{0.2}{0.55} \times \dfrac{2}{8 \times 10^{-4}}}$$

or

$$h_w = 0.00316 \quad \text{cal/cm}^2 \text{ sec } ^\circ\text{C}$$

$$= 113.76 \quad \text{kcal/m}^2 \text{ hr } ^\circ\text{C}$$

2. Bed-to-immersed-tube heat transfer:

To calculate the heat transfer coefficient to an immersed tube we use Eq. 13.29:

$$\frac{h_w d_p}{k_g'} = 0.01844(\text{CF})(1 - f_f)\left(\frac{C_p \rho_g}{k_g'}\right)^{0.43}\left(\frac{d_p \rho_g u}{\mu}\right)^{0.23}$$

$$\left(\frac{C_{ps}}{C_{pg}}\right)^{0.8}\left(\frac{\rho_s}{\rho_g}\right)^{0.66}$$

Since the tube is located at the axis, the correction factor (CF) is unity, and we have

$$\frac{h_w \times 0.05}{9 \times 10^{-5}} = 0.01844 \times 1 \times 0.4$$

$$\times \left(\frac{0.55 \times 8 \times 10^{-4}}{9 \times 10^{-5}}\right)^{0.43}$$

$$\times \left(\frac{0.05 \times 8 \times 10^{-4} \times 30}{1.6 \times 10^{-4}}\right)^{0.23}$$

$$\times \left(\frac{0.2}{0.55}\right)^{0.8} \times \left(\frac{2}{8 \times 10^{-4}}\right)^{0.66}$$

or

$$h_w = 16.14 \quad \text{kcal/m}^2 \text{hr} \quad ^\circ\text{C}$$

In view of the large particles used (0.05 cm), the heat transfer coefficient to an immersed tube is very low. Note that the diameter of the tube (10 cm) does not come into the picture during the calculation, indicating that the heat transfer coefficient is independent of the tube diameter employed. When the relatively low heat of reaction for this reaction is also considered, jacket cooling appears not only adequate but desirable for this system.

13.7.6. Fluid–Particle Mass Transfer

A large number of correlations have been reported for estimating the fluid–particle mass transfer coefficient in fluidized beds. Based on the data of McCune and Wilhelm (1949), Ricetti and Thodos (1961), Petrovic and Thodos (1967), and Snowdon and Turner (1967), Beek (1971) has proposed the following correlations:

$$\text{St Sc}^{2/3} \propto k_g f_f \text{Sc}^{2/3}$$

$$= a\left(\frac{u d_p \rho_g}{\mu}\right)^b \qquad (13.32)$$

where

$$\left.\begin{array}{l} a = 0.81 \pm 0.05 \\ b = -0.5 \end{array}\right\} \quad \text{for} \quad 5 < \frac{u d_p \rho_g}{\mu} < 500$$

and

$$\left.\begin{array}{l} a = 0.60 \pm 0.10 \\ b = -0.43 \end{array}\right\} \quad \text{for} \quad 50 < \frac{u d_p \rho_g}{\mu} < 2000$$

This equation is valid for all values of Sc and for the practical range of Re' normally involved.

13.8. DESIGN OF DISTRIBUTOR

The manner of gas distribution greatly influences the quality and overall performance of the fluidized bed, and is in fact one of the most important factors in the design of a fluid-bed reactor. The main considerations involved in the design of the distributor are (1) uniform distribution of gas in the bed; (2) prevention of solids leakage (also called "weeping") from the distributor; (3) prevention of solids attrition at the gas-injection nozzles; (4) complete fluidization of the bed on start-up, with no stagnant or semistagnant pockets in the vicinity of the distributor; and (5) control of jet formation during gas injection through the distributor holes.

13.8.1. Types of Distributors

Several types of distributors have been used in laboratory and industrial reactors (Kunii and Levenspiel, 1969; Whitehead, 1971; Botterill, 1975; Rigby et al., 1977; Mori and Wen, 1977; Sathiyamoorty and Rao, 1978).

13.8.2. Pressure Drop across Distributor

To ensure uniformity of gas distribution several studies have been carried out on the pressure drop across a distribution plate expressed as a fraction of the pressure drop in the fluidized bed (namely, that required to support the solids). For design purposes Zuiderweg (1967) recommends a maximum value of 0.1. On the other hand, important fluidized-bed reactors, particularly catalytic crackers, are known to operate at ratios of the order of 0.4 (Gregory, 1967).

For the case of perforated plates and nozzles, Hiby's (1967) data suggest the following equation:

$$\frac{\Delta P_d}{\Delta P} = \begin{cases} 0.15 & \left(\dfrac{u}{u_{mf}} \approx 1\text{--}2\right) \\ 0.015 & \left(\dfrac{u}{u_{mf}} \gg 1\right) \end{cases} \quad (13.33)$$

Mori and Moriyama (1978) have developed the following criteria for uniform fluidization of nonaggregative particles.

Uniform distributor such as perforated or porous plates:

$$\frac{\Delta P_d}{\Delta P} = \frac{\delta}{1-\delta} \frac{1}{1 - u_{mf}/u} \quad (13.34a)$$

Nonuniform distributor such as a bubble cap:

$$\frac{\Delta P_d}{\Delta P} = \beta \frac{\delta}{1-\delta} \frac{1}{(1 - u_{mf}/u)^2} \quad (13.34b)$$

where $1 - \delta$ is the volume fraction of the fluidized part in the stationary section.

No specific recommendation seems possible regarding the pressure drop except that a value between the low and high pressure extremes, say 0.2, should be quite acceptable. In the chloromethanes reactor, for instance, the pressure drop is about 80 cm of water, corresponding to a ratio of 0.3.

A simple procedure (Kunii and Levenspiel, 1969) is to assume a pressure drop ratio of 0.3 (or $\Delta P_d = 70\text{--}80$ cm of water) and a discharge coefficient of 0.65 and to calculate the orifice velocity u_{or} from

$$u_{or} = 0.7 \left(\frac{2.0 g_c \Delta P_d}{\rho_g}\right)^{1/2}, \quad Re' > 100 \quad (13.35)$$

If now the number of holes per unit area n_{or}/A_c is fixed, then the orifice diameter d_{or} can be calculated from

$$u = \frac{\pi}{4} d_{or}^2 u_{or} \left(\frac{n_{or}}{A_c}\right) \quad (13.36)$$

The use of the orifice equation (Eq. 13.35) for the design of the grid has been questioned by Behie et al. (1978), primarily because of the assumption that solids over the grid exercise no influence on grid pressure drop, which is in reality incorrect. In the absence of any other quantitative expression this equation can, however, be used as a first approximation.

Several factors associated with the grid plate influence the uniformity of distribution that can be achieved. These are the plate diameter (D_p), thickness (t_p), diameter and number of openings (d_{or}, n_{or}), spacing between openings (s), and the ratio of the total area of the openings to the area of the grid (i.e., the free area of the grid, f_a). These parameters exercise opposing effects on the performance of the grid and the only way to predict them is by grouping the parameters into a single quantity and examining the effect of this quantity on a selected response.

The most acceptable response is the average solids-phase concentration in the bed. The solids concentration changes from point to point in the bed, depending on the distribution efficiency of the grid. The upper and lower limits of the confidence interval (ΔC) associated with an average value of the concentration appears to be a sound statistical parameter for determining the essential validity of the average—in other words, of the uniformity of the bed. A low value would suggest good uniformity.

Razumov et al. (1972) have obtained data on a 1-m column using granulated alumina–silicate catalyst and have related the statistical quantity $\overline{\Delta C}$ to a parameter **A** defined as follows:

$$\mathbf{A} = \frac{st_p d_{or}}{D_p^3} \tag{13.37}$$

The final expression obtained by them is

$$\log \overline{\Delta C} = -0.692\,(\log \mathbf{A})^2 - 6.472 \log \mathbf{A} - 12.817 \tag{13.38}$$

In any design calculation it is only necessary to calculate $\mathbf{A}$ for various combinations of parameters (mainly number and diameter of holes) and choose that combination which gives the lowest possible value of $\overline{\Delta C}$ within the ranges of parameters that can be practically used.

At the orifice velocities employed in some industrial reactors (50–200 cm/sec) the fluid enters the bed as multiple jets, which penetrate a certain distance above the distributor plate before collapsing and giving rise to bubbles. In fact, bubble behavior in the vicinity of the distributor is quite different from that in the main part of the bed. A much larger amount of gas than at minimum fluidization flows through the continuous phase in this region than in the rest of the bed (Toei et al., 1974).

13.8.3. Formation and Dissipation of Jets

Several workers have proposed expressions for calculating the depth of jet penetration in a fluidized bed, but these are generally restricted to the conditions of their experiments (Zenz, 1968; Basov et al., 1969; Shakhova, 1968). Markhevka et al. (1971) studied the motion pictures of vertical jets in a cylindrical bed and observed the pulsating nature of the jet. The maximum jet length was found to vary by $\pm 20\%$ about the mean as a result of these pulsations. Based on this study and other experimental data, Merry (1975) correlated the jet half-angle θ with the properties of the fluid and particle and the nozzle diameter:

$$\cot \theta = 10.4 \left(\frac{\rho_s d_p}{\rho_g d_{or}}\right)^{-0.3} \tag{13.39}$$

and also proposed the following equation for jet penetration:

$$\frac{J_p}{d_{or}} = 5.2 \left(\frac{\rho_s d_p}{\rho_g d_{or}}\right)^{-0.3} \left[1.3\left(\frac{u^2}{gd_{or}}\right)^{0.2} - 1\right] \tag{13.40}$$

Jet penetrations can be as high as 60 cm under certain conditions. However, the jets dissipate faster in a fluidized bed than in a single-phase (air) system.

More recent studies on jets have also been reported by Wen et al. (1977) in a two-dimensional bed and by Yang and Keairns (1978a) in a semicircular unit. Wen et al. (1977) have proposed the following equation for jet penetration:

$$\frac{J_p}{d_{or}} = 814.2 \left(\frac{\rho_s d_p}{\rho_g d_{or}}\right)^{-0.585} \left(\frac{\rho_g d_{or} u_{or}}{\mu}\right)^{-0.654}$$
$$\times \left(\frac{u_{or}^2}{g d_{or}}\right)^{0.47} \tag{13.41}$$

while Yang and Keairns (1978a) employed a two-phase Froude number $(\rho_s u_{or}^2/(\rho_s - \rho_g)gd_{or})^{0.5}$ to correlate the data. Yang and Keairns (1979) have tested several of these correlations proposed for jet penetration against the experimental data of Basov et al. (1969), Behie et al. (1971), and Wen et al.. (1977) (see also Yang and Keairns, 1978b) and proposed the following correlation, based on the two-phase Froude number that correlated the data to within $\pm 40\%$:

$$J_p = 15\,d_{or} \left(\frac{\rho_s}{\rho_s - \rho_g}\frac{u_{or}^2}{gd_{or}}\right)^{0.187} \tag{13.42}$$

Equation 13.42 has been developed based on single-jet and multiple-jet data obtained in both two-dimensional and three-dimensional beds. More recently Wen and Chen (1982) have proposed another correlation for jet penetration. The correlation based on the data of Ghadiri and Clift (1980), Tanaka et al. (1980), and Wen and Chen (1982) has a maximum deviation of $\pm 30\%$ and can also be reliably used.

In none of the foregoing expressions has the effect of fluid viscosity and of particle size range been included. The fluid viscosity may not affect the jet penetration depth significantly; but the effect of particle size distribution is uncertain and further experiments are needed to correct the present correlations so as to render them applicable to a much wider range of particle sizes.

The dissipation of the momentum along the axis of the jet is obviously a clear parameter for determining the jet zone in a fluidized bed. Behie et al. (1970, 1971) have prepared (from experimental data) plots of the normalized momentum flux M/M_0 (where M_0 is the initial momentum) as a function of the axial distance.

All the data for different nozzle sizes have been brought together by replacing the axial distance by a modified Froude group defined as $Fr' = (\rho_g/\rho_\beta)(u_{or}^2/l^2 g)d_{or}$. The normalized momentum flux is plotted as a function of this modified Froude group in Figure 13.5 for different velocities.

This figure is of considerable importance in design, for it would be safe to place the internal at a distance corresponding to a very low value of M/M_0. On the other hand, in the case of very fast reactions, in which jets of one of the reactants are introduced inside a bed fluidized by the other reactant, as in the case of the chloromethanes reactor, the jets must be broken before natural dissipation occurs under the environmental conditions involved. Thus jet breakers will have to be provided.

13.8.4. Operating and Nonoperating Modes

An important consideration in the operation of a fluid bed is the requirement that all nozzles or tuyeres perform equally well. Whitehead and Dent (1967) have reported some results of practical significance using a multituyere distributor in a relatively large reactor. As the velocity is increased beyond u_{mf}, a stage is reached when all the tuyeres operate at an average freeboard velocity u_f. In other words, there would be no stagnant pockets in the vicinity of any tuyere, and *all* the tuyeres are said to be in the operating mode. If the velocity is reduced, a critical value u_m is reached at which pockets momentarily appear in the vicinity of some tuyeres, which are now said to be in the nonoperating mode. But they quickly change back to the operating mode. Further reduction in velocity increases the number of nonoperating tuyeres until more and more of them become permanently nonoperating. The velocity u_m therefore represents a critical minimum velocity that ensures operation of all the tuyeres, and may be estimated from a relationship given by Whitehead and Dent (1967).

In another analysis of distributor operation, Fakhimi and Harrison (1970) obtained the following relationship for the fraction of active (or operating) orifices for a multiorifice distributor of n orifices:

$$\frac{n_{oper}}{n} = \frac{u - u_{mf}}{[u_{mf}^2 + (2\phi^2/\rho_g)(\Delta P_1 - \Delta P_2)]^{1/2} - u_f}$$

(13.43)

where ϕ is the ratio of the total orifice area to the distributor area, ΔP_1 is the pressure drop over the fixed bed of height l_0 over the nonoperating orifice, and ΔP_2 the pressure drop over the spouted bed of height l_0' over the operating orifice given by $2\rho_s(1 - f_{mf})gl_0'/\pi$. The height l_0' represents the entrance region above the plate and is approximated by $l_0' = 2s$ where s is the center-to-center spacing between adjacent orifices (Hovmand et al., 1971).

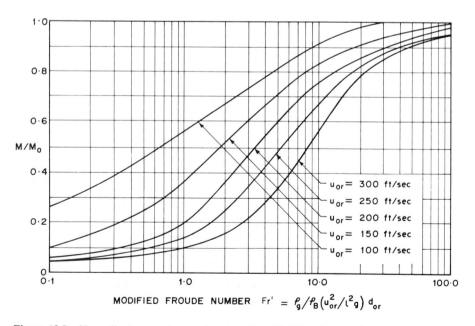

Figure 13.5. Normalized momentum as a function of modified Froude group for different velocities of gas at the nozzle (Behie et al., 1970).

13.9. ENTRAINMENT

The particle size range and the concentration change with height in the freeboard (the region above the bed) until a height is reached when both attain constant values. This height is referred to as the transport disengagement height, or TDH, and the solids concentration above the TDH is called the saturation-carrying capacity (SCC) of the gas. In order to account for entrainment in the overall design, equations must be available for estimating (1) terminal velocity; (2) entrainment above the TDH; and (3) cyclone dip-leg diameter.

13.9.1. Terminal Velocity

Zenz and Othmer (1960) have provided a graphical correlation for determining u_t, but the recommended equations are the Stoke's law for laminar flow $(d_p u_t \rho_g / \mu < 2.0)$

$$u_t = \frac{(\rho_s - \rho_g) g d_p^2}{18 \mu} \qquad (13.44)$$

and the modified expression

$$u_t = \frac{0.153 d_p^{1.14} g^{0.71} (\rho_s - \rho_g)^{0.71}}{\mu^{0.43} \rho_g^{0.29}} \qquad (13.45)$$

for $d_p u_t \rho_g / \mu$ between 2.0 and 500.

Several alternative methods are also reported in the literature for estimating the terminal velocity. For the purpose of design calculations, however, the simple procedure described above should be adequate. A table of terminal velocities is also available (Heywood, 1962).

13.9.2. Entrainment Rates

For the purpose of design, entrainment above the TDH is far more important than below it, and is given by the SCC. This can be estimated from the correlation of Zenz and Weil (1958), which is redrawn in Figure 13.6. In this correlation the group $u^2 / g d_p \rho_s^2$ is plotted against the entrainment group $E' / A_c \rho_g u$ where E' is the entrainment rate in grams per second; thus E' can be estimated for any given velocity.

Figure 13.6 is restricted to particles of a very narrow size range (essentially a single size). According to Zenz and Weil (1958), however, it can also be used for solids with a size distribution. It is only necessary to divide the size range into several narrow cuts and determine the terminal velocity corresponding to the average size of each cut from the figure. Obviously only those solids whose terminal velocities are lower than the actual velocity used will be entrained. The total entrainment E' through all the cuts is then given by the summation of the product (fraction of range) × (entrainment rate of the range).

Studies have also been reported on the entrainment of solid particles in a polydispersed fluidized bed

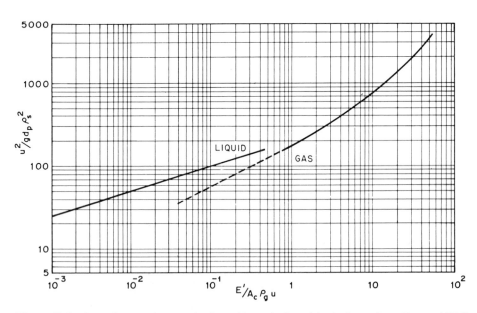

Figure 13.6. Saturation carrying capacity for uniform-sized particles (redrawn from Zenz and Weil, 1958).

(Sycheva and Donat, 1975). Correlations have been proposed for the concentration (y) of particles of a given fraction at the outlet as a function of its concentration (x) in two- and three-component fluidized beds at varying gas velocities. Most of the data could be represented by straight lines with a break. Thus, for a polydispersed system consisting of n fractions,

$$y = \frac{x}{x^*} y^* \tag{13.46}$$

where y^* is the limiting yield concentration, given by

$$y^* = 9.8 \times 10^{-0.093n} \times 10^{-0.085/u_t} - \frac{62.10^{-0.063n} u_t^{1.25}}{u} \tag{13.47}$$

and x^* is the limiting concentration corresponding to y^* and is given by

$$x^* = \frac{(u - u_t)^{0.52} n^{0.474}}{(1.1 \times 10^{-2})n^{0.168}} \tag{13.48}$$

These equations are valid for relatively coarse materials ($u_t = 0.3-1.0$ m/sec); for finer particles ($u_t = 0.3-0.03$ m/sec) the following correlation is suggested for y^*:

$$\log y^* = 1.266 - 0.1675n + 2.7(u - u_t) \tag{13.49}$$

Devices to reduce entrainment mostly follow the simple principle of breaking larger bubbles. The larger bubbles entrain a substantial fraction of particles, and reducing the bubble size would help in reducing the entrainment loss. Such devices, however, have to be located within the dense phase (bed). Thus the internals used in the bed indirectly help in reducing the entrainment.

Other devices, also commonly employed, are located in the dilute phase above the bed. Perforated-grid and tubular arrays (Blyakher and Pavlov, 1966), vertical arrays of slanted baffles (Harrison et al., 1974, Harrison, 1973), and louvers (Martini et al., 1976) have been successfully employed in the dilute phase. Most of these devices are more effective when located close to the bed surface.

13.9.3. Cyclone and Dip-Leg Design

In evaluating the performance of a cyclone, or in the design of a cyclone for that matter, a characteristic cyclone number (CN) is used as a measure of its performance. The number is defined (Rietema and Verver, 1961) as

$$CN = \frac{\bar{d}_p^2(\rho_s - \rho_g)}{\mu} L \frac{\Delta P}{\rho_g q_c} \tag{13.50}$$

where $\bar{d}_p$ is the diameter of the particles, of which 50 % is separated in the cyclone; ΔP is the total pressure differential in a cyclone; and q_c is the capacity of a single cyclone. For efficient operation of a cyclone, CN should be as small as possible.

The performance data for cyclones are always based on operation with a sealed dip-leg. The dip-leg is conveniently sealed by simply immersing it in the fluidized bed. Doing this, however, permits gas backflow, so that the quantity of solids escaping from the cyclone becomes a function not only of the solids concentration in the gas stream but also of the quantity of backflow gas. It is therefore advantageous to keep the size of the dip-leg as small as possible. In calculating the capacity of the cyclone dip-leg it is necessary to remember that a dip-leg is nothing but a free downpipe immersed in the bed. Thus the correlations available for the flow of solids through orifices can be used to calculate the capacity. One such correlation is that of Rausch (1948). If w_A is the solids loading, β_A the angle of repose of the solid, and d_{d1} the diameter of the dip-leg, we get (see Zenz, 1976)

$$w_A = \frac{g}{\tan(\beta_A + 27)}(\rho_s - \rho_g)(1 - f_f)(d_{d1})_e^{1/2} \tag{13.51}$$

from which the diameter of the dip-leg can be estimated.

In a multistage cyclone assembly, the latter stages usually have low solids loading, and a balance should be made between dip-leg size, solids downflow capacity, and backflow losses. Several means for avoiding the gas backflow in dip-legs have been suggested. A simple method is to immerse the dip-leg in a small seal pot having its own aeration gas supply (Zenz and Othmer, 1960).

REFERENCES

Atimtay, A., and Cakloz, T. (1978), *Powder Technol.*, **20**, 1.

Basov, V. A., Markhevka, V. I., Akhnazarov, T. Kh. M., and Orochko, D. I. (1969), *Int. Chem. Eng.*, **9**, 263.

Baumgarten, P. K., and Pigford, R. L. (1960), *AIChE J.*, **6**, 115.

Beek, W. J. (1971), *Fluidization*, eds. Davidson, J. F., and Harrison, D., Academic Press, London, p. 431.

Behie, L. A., Bergougnou, M. A., Baker, C. G. J., and Buloni, W. (1970), *Can. J. Chem. Eng.*, **48**, 158.

Behie, L. A., Bergougnou, M. A., Baker, C. G. J., and Base, T. E. (1971), *Can. J. Chem. Eng.*, **49**, 557.

Behie, L. A., Voegelin, B. E., and Bergougnou, M. A. (1978), *Can. J. Chem. Eng.*, **56**, 404.

Bloore, P. D. (1961), Ph.D. thesis, Birmingham University.

Blyakher, I. G., and Pavlov, V. M. (1966), *Int. Chem. Eng.*, **6**, 47.

Borlai, O., Hodany, L., and Blickle, T. (1967), *Proc. Int. Symp. Fluidization*, p. 433.

Botterill, J. S. M. (1975), *Fluid Bed Heat Transfer*, Academic Press, London.

Botterill, J. S. M., and Toeman, Y. (1980), Fluidization, eds. J. R. Grace and J. M. Matsen, Plenum Press, New York, p. 93.

Broadhurst, T. E., and Becker, H. A. (1975), *AIChE J.*, **21**, 238.

Budkov, V. A., Maslovskii, M. F., and Prozorov, E. N. (1970), *Khim. Prom.*, **46**, 216.

Calderbank, P. H., Toor, F. D., and Lancaster, F. H. (1967), *Proc. Int. Symp. Fluidization*, p. 652.

Callahan, J. T. (1971), *J. Basic Eng.*, **93**, 165.

Caram, H. S., Bukur, D., and Amundson, N. R. (1978), *Chemical Reactor Theory—A Review*, eds. Lapidus, L., and Amundson, N. R., Prentice-Hall, Englewood Cliffs, New Jersey, p. 686.

Catipović, N. M., Jovanović, G. N., Fitzgerald, T. J., and Levenspiel, O. (1980), *Fluidization*, eds. J. R. Grace and J. M. Matsen, p. 225, Plenum Press, New York.

Catipović, N. M., Jovanović, G. N., and Fitzgerald, T. J. (1981), *Ind. Eng. Chem. Fundam.*, **20**, 82.

Chavarie, C., and Grace, J. R. (1976), *Chem. Eng. Sci.*, **31**, 741.

Chen, J. L. P., and Keairns, D. L. (1975), *Can. J. Chem. Eng.*, **53**, 395.

Chiba, T., and Kobayashi, H. (1973), *Proc. Int. Symp. Fluidization*, p. 468.

Chiba, T., Terashima, K., and Kobayashi, H. (1972), *Chem. Eng. Sci.*, **27**, 965.

Chiba, T., Terashima, K., and Kobayashi, H. (1973), *J. Chem. Eng. Japan*, **6**, 78.

Darton, R. C., Lanauze, R. D., Davidson, J. F., and Harrison, D. (1977), *Trans. Inst. Chem. Eng.*, **55**, 274.

Davidson, J. F., and Harrison, D., (1963), *Fluidized Particles*, Cambridge Univ. Press, New York and London.

Davidson, J. F., and Harrison, D. (1971), *Fluidization*, Academic Press, London.

Davidson, J. F., and Schuler, B. O. G. (1960), *Trans. Inst. Chem. Eng. (London)*, **38**, 335.

Davidson, J. F., Harrison, D., Darton, R. C., and Lanauze, R. D. (1978), *Chemical Reactor Theory—A Review*, eds. Lapidus, L., and Amundson, N. R. Prentice-Hall, Englewood Cliffs, New Jersey, p. 583.

Davies, L., and Richardson, J. F. (1966), *Trans. Inst. Chem. Eng. (London)*, **44**, T 293.

Davies, R. M., and Taylor, G. (1950), *Proc. Roy. Soc. (London)*, **A200**, 375.

Desai, A., Kikukawa, H., and Pulsifer, A. H. (1977), *Powder Technol.*, **16**, 143.

Dow, W. M., and Jakob, M. (1951), *Chem. Eng. Prog.*, **47**, 637.

Drinkenburg, A. A. H., and Rietema, K. (1972), *Chem. Eng. Sci.*, **27**, 1765.

Drinkenburg, A. A. H., and Rietema, K. (1973), *Chem. Eng. Sci.*, **28**, 259.

Einstein, V. G., and Gelperin, N. I. (1965), *Khim. Prom.*, **6**, 416.

Epstein, N. (1976), *Chem. Eng. Sci.*, **31**, 852.

Fakhimi, S., and Harrison, D. (1970), *Inst. Chem. Eng. Symp. Ser.*, **33**, 29.

Fan, L. T., and Chang, Y. (1979), *Can. J. Chem. Eng.*, **57**, 88.

Fontaine, R. W., and Harriot, P. (1972), *Chem. Eng. Sci.*, **27**, 2189.

Gabor, J. D. (1967), *Proc. Int. Symp. Fluidization*, p. 230.

Gabor, J. D. (1971), *Chem. Eng. Sci.*, **26**, 1247.

Gabor, J. D. (1972), *Chem. Eng. J.*, **4**, 118.

Geldart, D. (1967), *Chem. Ind. (London)*, p. 1474.

Geldart, D. (1968), *Powder Technol.*, **1**, 355.

Geldart, D. (1970), *Powder Technol.*, **4**, 41.

Geldart, D. (1971), Ph.D. thesis, University of Bradford.

Geldart, D. (1973), *Powder Technol.*, **7**, 285.

Geldart, D., and Cranfield, R. R. (1972), *Chem. Eng. J.*, **3**, 211.

Geldart, D., and Abrahamsen, A. R. (1978), *Powder Technol.*, **19**, 133.

Gelperin, N. I., and Einstein, V. G. (1971), *Fluidization*, Academic Press, London, Chapter 10.

Gelperin, N. I., Kruglikov, V. I., and Einstein, V. G. (1958), *Khim. Prom.*, **6**, 358.

Gelperin, N. I., Einstein, V. G., and Kwasha, V. B. (1967), *Fluidization Technique Fundamentals*, Izd. Khimia, Moscow.

Ghadiri, M., and Clift, R. (1980), *Ind. Eng. Chem. Fundam.*, **19**, 440.

Gibilaro, L. G., and Rowe, P. N. (1974), *Chem. Eng. Sci.*, **29**, 1403.

Godard, K., and Richardson, J. F. (1969), *Chem. Eng. Sci.*, **24**, 663.

Grace, J. R. (1970), *Can. J. Chem. Eng.*, **48**, 30.

Grace, J. R. (1971a), *AIChE Symp. Ser.* (No. 116), **67**, 159.

Grace, J. R. (1971b), *Chem. Eng. Sci.*, **26**, 1955.

Grace, J. R., and Clift, R. (1974), *Chem. Eng. Sci.*, **29**, 327.

Grace, J. R., and Harrison, D. (1968), *Tripartite Chem. Eng. Conf., Inst. Chem. Eng., London.*

Grace, J. R., and Harrison, D. (1969), *Chem. Eng. Sci.*, **24**, 497.

Grace, J. R., and Matsen, J. M. (1980), *Fluidization*, Plenum Press, New York.

Gregory, S. A. (1967), *Proc. Int. Symp. Fluidization*, p. 751.

Grewal, N. W., and Saxena, S. C. (1981), *Ind. Eng. Chem. Process Des. Dev.*, **20**, 108.

Gupta, S. N., Chaube, R. B., and Upadhyay, S. N. (1974), *Chem. Eng. Sci.*, **29**, 839.

Hardebol, J. (1961), *Trans. Inst. Chem. Eng. (London)*, **39**, 229.

Harrison, D. (1961), *Trans. Inst. Chem. Eng. (London)*, **39**, 238.

Harrison, D. (1973), *Chem. Eng. Prog. Symp. Ser.* (No. 128), **69**, 14.

Harrison, D., and Leung, L. S. (1961), *Trans. Inst. Chem. Eng. (London)*, **39**, 409.

Harrison, D., Davidson, J. F., and De Kock, J. W. (1961), *Trans. Inst. Chem. Eng. (London)*, **39**, 202.

Harrison, D., Ashpinall, P. N., and Elder, J. (1974), *Trans. Inst. Chem. Eng. (London)*, **52**, 213.

Hayakawa, T., Graham, W., and Osberg, G. L. (1964), *Can. J. Chem. Eng.*, **42**, 99.

Hayes, W. B., Hardy, B. W., and Holland, C. D. (1964), *AIChE J.*, **24**, 749.

Heywood, H. (1962), *Proc. 1st Symp. Interaction between Fluids and Particles* (Inst. Chem. Eng., London), pp. A1–A8.

Hiby, J. W. (1967), *Chem. Ing.–Tech.*, **39**, 1125.

Higbie, R. (1935), *Trans. Am. Inst. Chem. Eng.*, **31**, 365.

Hovmand, S., and Davidson, J. F. (1971), *Fluidization*, eds. Davidson, J. F., and Harrison, D., Academic Press, London.

Hovmand, S., Freedman, W., and Davidson, J. F. (1971), *Trans. Inst. Chem. Eng. (London)*, **49**, 149.

Jackson, R., (1963), *Trans. Inst. Chem. Eng. (London)*, **41**, 13.

Kato, K., and Wen, C. Y. (1969), *Chem. Eng. Sci.*, **24**, 1351.

Keairns, D. L., (1976), ed. *Fluidization Technology*, Vols. 1 and 2, Hemisphere, Washington, D.C.

Kobayashi, H., Arai, F., and Chiba, T. (1966), *Kagaku Kogaku*, **4**, 147.

Kondukov, N. B., and Sosna, M. Kh. (1965), *Khim. Prom.*, **6**, 402.

Kondukov, N. B., Kornilaev, A. N., Skachko, I. M., Akhromenkov, A. A., and Kruglov, A. S. (1964), *Int. Chem. Eng.*, **4**, 43.

Korotjanskaja, L. A., Gelperin, N. I., Einstein, V. G., and Makotkin, A. V. (1967), *Process and Equipment for Chemical Technology*, Inst. Fine Chem. Technol., Moscow, p. 175.

Kumar, R., and Kuloor, N. R. (1970), *Adv. Chem. Eng.*, **8**, 256.

Kunii, D., and Levenspiel, O. (1968a), *Ind. Eng. Chem. Fundam.*, **7**, 446.

Kunii, D., and Levenspiel, O. (1968b), *Ind. Eng. Chem. Process Design Dev.*, **7**, 481.

Kunii, D., and Levenspiel, O. (1969), *Fluidization Engineering*, Wiley, New York.

Kupferberg, A., and Jameson, G. J. (1969), *Trans. Inst. Chem. Eng. (London)*, **47**, T 241.

Lanauze, R. D., and Harris, I. J. (1972), *Chem. Eng. Sci.*, **27**, 2102.

Lanauze, R. D., and Harris, I. J. (1974) *Chem. Eng. Sci.*, **29**, 1663.

Lancaster, F. H. (1965), Ph.D. thesis, University of Edinburgh.

Lese, H. K., and Kermode, R. I. (1972), *Can. J. Chem. Eng.*, **50**, 44.

Leung, L. S., Sandford, I. C., and Mak, F. K. (1970), *Chem. Eng. Sci.*, **25**, 220.

Leva, M., and Grummer, M. (1952), *Chem. Eng. Prog.*, **48**, 307.

Leva, M., Weintraub, M., and Grummer, M. (1949), *Chem. Eng. Prog.*, **45**, 563.

Lewis, W. K., Gilliland, E. R., and Glass, W. (1959), *AIChE J.*, **5**, 419.

Lewis, W. K., Gilliland, E. R., and Girouard, H. (1962), *Chem. Eng. Prog. Symp. Ser.* (No. 38), **58**, 87.

Lockett, M. J., and Harrison, D. (1967), *Proc. Int. Symp. Fluidization*, p. 257.

Malling, G. F., and Thodos, G. (1967), *Int. J. Heat and Mass Transfer*, **10**, 489.

Markhevka, V. I., Basov, V. A., Melik-Akhnazarov, T.Kh., and Orochko, D. I. (1971), *Theoret. Found. Chem. Eng.*, **5**, 80.

Martini, Y., Bergougnou, M. A., and Baker, G. G. J. (1976), *Fluidization Technology*, Vol. 2, ed. Keairns, D. L., Hemisphere, Washington, D. C., p. 29.

Massimilla, L., and Westwater, J. W. (1960), *AIChE J.*, **6**, 134.

Matsen, J. M. (1968), *Ind. Eng. Chem. Process Design Dev.* **7**, 159.

Matsuno, R., and Rowe, P. N. (1970), *Chem. Eng. Sci.*, **25**, 1587.

May, W. G. (1959), *Chem. Eng. Prog.*, **55**, 49.

McCann, D. J., and Prince, R. G. H. (1969), *Chem. Eng. Sci.*, **24**, 801.

McConnachie, J. T. L., and Thodos, G. (1963), *AIChE J.*, **9**, 60.

McCune, L. K., and Wilhelm, R. H. (1949), *Ind. Eng. Chem.*, **41**, 1124.

McGrath, L., and Streatfield, R. E. (1971), *Trans. Inst. Chem. Eng. (London)*, **49**, 70.

McKay, G., and McLain, H. D. (1980), *Ind. Eng. Chem. Process Design Dev.*, **19**, 712.

Meissner, H. P., and Kusik, C. L. (1970), *Can. J. Chem. Eng.*, **48**, 349.

Merry, J. M. D. (1975), *AIChE J.*, **21**, 507.

Mickley, H. S., Fairbanks, D. F., and Hawthorn, R. D. (1961), *Chem. Eng. Prog. Symp. Ser.* (No. 32), **57**, 51.

Miller, C. O., and Logwinuk, A. K. (1951), *Ind. Eng. Chem.*, **43**, 1220.

Mori, S., and Moriyama, A. (1978), *Int. Chem. Eng.*, **18**, 245.

Mori, S., and Wen, C. Y. (1976), *Fluidization Technology*, Vol. 1, ed. Keairns, D. L., Hemisphere, Washington, D.C., p. 179.

Mori, S., and Wen, C. Y. (1977), *AIChE Symp. Ser.* (No. 161), **73**, 121.

Murray, J. D. (1965), *J. Fluid Mech.*, **22**, 57.

Naud, M., Large, J. F., and Bergougnou, M. A. (1981), *Ind. Eng. Chem. Process Design Dev.*, **20**, 00.

Nicklin, D. J. (1962), *Chem. Eng. Sci.*, **17**, 693.

Orcutt, J. C., and Carpenter, B. H. (1971), *Chem. Eng. Sci.*, **26**, 1049.

Park, W. H., Kang, W. K., Capes, C. E., and Osberg, G. L. (1969), *Chem. Eng. Sci.*, **24**, 851.

Park, W. H., Lee, N. G., and Capes, C. E. (1974), *Chem. Eng. Sci.*, **29**, 339.

Partridge, B. A., and Rowe, P. N. (1966), *Trans. Inst. Chem. Eng. (London)*, **44**, T 335.

Pattipati, R. R., and Wen, C. Y. (1981), *Ind. Eng. Chem. Process Design Dev.*, **20**, 705.

Petrovic, J. J., and Thodos, G. (1967), *Proc. Int. Symp. Fluidization*, p. 586.

Potter, O. E., (1969), *Chem. Eng. Sci.*, **24**, 1733.

Pyle, D. L., and Harrison, D. (1967), *Chem. Eng. Sci.*, **22**, 1199.

Pyle, D. L., and Rose, P. L. (1965), *Chem. Eng. Sci.*, **20**, 25.

Ramakrishnan, S., Kumar, R., and Kuloor, N. R. (1969), *Chem. Eng. Sci.*, **24**, 731.

Rausch, J. M. (1948), Ph.D. thesis, Princeton University, Princeton, New Jersey.

Razumov, I. M., Manshilin, V. V., and Nemets, L. L. (1972), *Khim. Tekhnol. Topl. Masel.*, **17**, 31.

Reuter, H. (1963), *Chem. Ing.–Tech.*, **35**, 98.

Ricetti, R. E., and Thodos, G. (1961), *AIChE J.*, **7**, 442.

Rietema, K., and Hoebink, J. (1976), *Fluidization Technology*, Vol. 1, ed. Keairns, D. L., Hemisphere, Washington, D.C., p. 279.

Rietema, K., and Verver, C. G. (1961), *Cyclones in Industry*, Elsevier, Amsterdam.

Rigby, G. R., Callcott, T. G., Singh, B., and Evans, B. R. (1977), *Trans. Inst. Chem. Eng.*, **55**, 68.

Romero, J. B., and Johanson, L. N. (1962), *Chem. Eng. Prog. Symp. Ser.* (No. 38), **58**, 28.

Rose, P. L. (1965), Ph.D. thesis, Cambridge University.

Rowe, P. N. (1964), *Chem. Eng. Prog.*, **60**, 75.

Rowe, P. N. (1971), *Fluidization*, eds. Davidson, J. F., and Harrison, D., Academic Press, New York, p. 121.

Rowe, P. N. (1972), *Proc. 2nd Int. Symp. Chem. React. Eng.*, A 9–1.

Rowe, P. N. (1973), *AIChE Symp. Ser.* (No. 128), **69**, 23.

Rowe, P. N. (1976), *Chem. Eng. Sci.*, **31**, 285.

Rowe, P. N., and Claxton, K. T. (1965), *Trans. Inst. Chem. Eng. (London)*, **43**, 321.

Rowe, P. N., and Everett, D. S. (1972), *Trans. Inst. Chem. Eng. (London)*, **50**, 42.

Rowe, P. N., and Matsuno, R. (1971), *Chem. Eng. Sci.*, **26**, 923.

Rowe, P. N., and Partridge, B. A. (1962), *Proc. Symp. Interaction between Fluids and Particles*, p. 135.

Rowe, P. N., and Partridge, B. A. (1965), *Trans. Inst. Chem. Eng. (London)*, **43**, T 157.

Rowe, P. N., and Sutherland, K. S. (1964), *Trans. Inst. Chem. Eng. (London)*, **42**, T 55.

Rowe, P. N., Evans, T. J., and Middletone, J. C. (1971), *Chem. Eng. Sci.*, **26**, 1943.

Satyanarayanan, A., Kumar, R., and Kuloor, N. R. (1969), *Chem. Eng. Sci.*, **24**, 749.

Sathiyamoorthy, D., and Rao, C. S., (1978), *Powder Technol.*, **20**, 47.

Saxena, S. C., and Vogel, G. J. (1977), *Trans. Inst. Chem. Eng. (London)*, **55**, 184.

Shakhova, N. A. (1968), *Inzhenernyi Fiz. Zh.*, **14**, 61.

Snowdon, C. B., and Turner, J. C. R. (1967), *Proc. Int. Symp. Fluidization*, p. 599.

Staub, F. W. (1980), *Proc. NSF Workshop on Fluidization*, Rensselaer Polytechnic Institute, Troy, New York.

Sutherland, K. S. (1961), *Trans. Inst. Chem. Eng. (London)*, **39**, 188.

Sycheva, T. N., and Donat, E. V. (1975), *Int. Chem. Eng.*, **15**, 346.

Szekely, J. (1962), *Proc. Symp. Interaction between Fluids and Particles*, p. 197.

Talmor, E., and Benenati, R. F. (1963), *AIChE J.*, **9**, 536.

Tanaka, I., Ishikura, T., Hiromasa, S., Yoshimura, Y. and Shinohara, H. (1980), *J. Powder Technol. (Japan)*, **17**, 21.

Todes, O. M., Bondareva, A. K., and Greenbaum, M. B. (1966), *Khim. Prom.*, **6**, 408.

Toei, R., Matsuno, R., Oichi, M., and Yamamoto, K. (1974), *J. Chem. Eng. Japan*, **7**, 447.

Toei, R., and Matsuno, R. (1967), *Proc. Int. Symp. Fluidization*, p. 271.

Toei, R., Matsuno, R., Kojima, H., Nagai, Y., Nagakawa, K., and Yu, S. (1966), *Kagaku Kogaku*, **4**, 142.

Toomey, R. D., and Johnstone, H. F. (1952), *Chem. Eng. Prog.*, **48**, 220.

Toor, F. D., and Calderbank, P. H., (1967), *Proc. Int. Symp. Fluidization*, p. 373.

Vreedenberg, H. A. (1952), *J. Appl. Chem.*, **2**, Suppl. 1, 526.

Vreedenberg, H. A. (1958), *Chem. Eng. Sci.*, **9**, 52.

Vreedenberg, H. A. (1960), *Chem. Eng. Sci.*, **11**, 274.

Wen, C. Y., and Yu, Y. H. (1966), *AIChE J.*, **12**, 610.

Wen, C. Y., and Chen, L. H. (1982), *AIChE J.*, **28**, 348.

Wen, C. Y., Horio, M., Krishna, R., Khosravi, R., and Rengrajan, P. (1977), *Proc. 2nd Pacific Chem. Eng. Conf.*, p. 1182.

Wender, L., and Cooper, G. T. (1958), *AIChE J.*, **4**, 15.

Werther, J. (1978), *Chem. Ing.–Tech.*, **50**, 1978.

Whitehead, A. B. (1979), *Chem. Eng. Sci.*, **34**, 751.

Whitehead, A. B. (1971), *Fluidization*, eds. Davidson, J. F., and Harrison, D., Academic Press, London, p. 781.

Whitehead, A. B., and Young A. D. (1967), *Proc. Int. Symp. Fluidization*, p. 284, 294.

Whitehead, A. B., Dent, D.C., and Bhat, G. N. (1967), *Powder Technol.*, **1**, 143, 149.

Whitehead, A. B., and Dent, D. C. (1967), *Proc. Int. Symp. Fluidization.*, p. 802.

Wilhelm, R. H., and Kwauk, M. (1948), *Chem. Eng. Prog.*, **44**, 201.

Wood, R. T., Kuwata, M., and Staub, F. W. (1980), *Fluidization*, eds. Grace, J. R., and Matsen, J. M., Plenum Press, New York, p. 235.

Xavier, A. M., and Davidson, J. F. (1978), *Fluidization*, eds. Davidson, J. F., and Keairns, D. L., Cambridge Univ. Press, New York and London.

Xavier, A. M., King, D. F., Davidson, J. F., and Harrison, D. (1980), *Fluidization*, eds. Grace, J. R., and Matsen, J. M., Plenum Press, New York, p. 209.

Yacono, C. (1977), Thèse de Doctorat des Sciences, Institut National Polytechnique de Toulouse, France.

Yacono, C., Rowe, P. N., and Angelino, H. (1979), *Chem. Eng. Sci.*, **34**, 789.

Yang, W. C., and Keairns, D. L. (1978a), *Fluidization*, eds. Davidson, J. F., and Keairns, D. L., Cambridge Univ. Press, New York and London, p. 208.

Yang, W. C., and Keairns, D. L. (1978b), *AIChE Symp. Ser.* (No. 175), **74**, 218.

Yang, W. C., and Keairns, D. L. (1979), *Ind. Eng. Chem. Fundam.*, **18**, 317.

Yasui, G., and Johanson, L. N. (1958), *AIChE J.*, **4**, 445.

Young, A. D. (1967), *Proc. Int. Symp. Fluidization*, p. 305.

Zabrodsky, S. S. (1966), *Hydrodynamics and Heat Transfer in Fluidized beds*, MIT Press, Cambridge, Massachusetts.

Zenz, F. A., and Othmer, D. F. (1960), *Fluidization and Fluid Particle Systems*, Reinhold, New York.

Zenz, F. A. (1968), *Fluidization*, ed. Pierie, J. M., Institution of Chemical Engineers, London, p. 136.

Zenz, F. A. (1976), *Fluidization Technology*, Vol. 2, ed. Keairns, D. L., Hemisphere, Washington, D.C., p. 239.

Zenz, F. A., and Weil, N. A. (1958), *AIChE J.*, **4**, 472.

Ziegler, E. N., Koppel, L. B., and Brazeltom, W. T. (1964), *Ind. Eng. Chem. Fundam.*, **3**, 324.

Zuiderweg, F. J. (1967), *Proc. Int. Symp. Fluidization*, p. 739.

CHAPTER FOURTEEN

Design of Fluidized-Bed Reactors II: Models and Applications

Calculation of the conversion based on an appropriate model constitutes a major step in the complete design of a fluidized-bed reactor. Among the other factors to be considered are catalyst fouling and regeneration, heat transfer, choice of reactor internals, loss of catalyst due to entrainment, choice of particle size, probability of maloperation (such as slugging, described in Section 14.12.1), and the need for large-scale operation using cold models. In this chapter we shall refer to these and a few other practical considerations as part of the overall strategy for the design and development of a fluid-bed reactor. We shall briefly describe two early models (axial dispersion and two-region models), and then proceed to the more rigorous bubbling-bed models. Among the major books and reviews in the area are those of Davidson and Harrison (1963), Kunii and Levenspiel (1969), Davidson and Harrison (1971), Yates (1975), Keairns (1975), Horio and Wen (1977), Davidson and Keairns (1978), Potter (1978), and Grace and Matsen (1981).

14.1. AXIAL DISPERSION MODEL

As in the case of the fixed-bed reactor, the axial dispersion model assumes that the bed is pseudohomogeneous. It will be noticed that in this approach the bubble and emulsion phases are treated as one and the single pseudohomogeneous phase is assumed to be characterized by effective properties. Such an assumption in the case of fixed-bed reactors led to the postulation of effective properties for the fixed-bed. Similarly effective properties for the fluidized bed are necessary in the present approach.

The basic continuity equation for a pseudohomogeneous bed of this type (one-dimensional isothermal) is given by Eq. 11.10. The effect of mixing in this equation is given by the Peclet number. In the complete absence of mixing, namely, under plug-flow conditions, the Peclet number approaches infinity, and the solution reduces to $C_A/C_{A0} = \exp(-K_0)$, whereas under fully mixed conditions it reduces to $C_A/C_{A0} = 1/(1 + K_0)$, where K_0 is a dimensionless rate constant defined as $k_v L_0/u$. The partial mixing model, represented by Eq. 11.10, has been found to fit fluid-bed reactor data at low gas velocities (close to G_{mf}) quite satisfactorily (Ishii and Osberg, 1965; Brammer et al., 1967).

Limitations of the Model

The axial diffusion model ignores the consequences of the two-phase mode of fluid bed operation, and assumes that the reactor must operate between the limits of plug flow and full mixing. Several experimental studies have clearly shown that fluid bed performance can be worse than that of a fully mixed reactor, which indicates the inadequacy of the axial mixing model (Gilliland and Knudsen, 1971; Hovmand and Davidson, 1968; Hovmand et al., 1971).

The explanation is provided by the two-phase theory of fluidization, according to which a large portion of the gas passes through the bed as bubbles. If this entire quantity were to pass unreacted, the conversion would be extremely poor. It is known, however, that mass exchange occurs between the bubble and emulsion phases (see Section 13.5), leading to a considerable rise in the level of conversion. Even so, at high velocities, it remains lower than in a fully mixed reactor.

14.2. TWO-REGION MODELS

Two-region models, also referred to as Type I models, are represented by the following independent parameters: (1) volume of the emulsion (or bubble) phase; (2) volumetric flow rate through the emulsion (or bubble) phase; (3) interchange between phases; (4) longitudinal dispersion coefficient in the emulsion phase; (5) longitudinal dispersion coefficient in the bubble phase; and (6) solids content of the bubble phase *m*.

The diffusivity terms account for mixing in the two phases independently, in contrast to the axial dispersion model described above, in which a single effective diffusivity was used for describing the mixing as a whole, without recognizing the separate existence of the bubble and emulsion phases. Thus the two-region model may be regarded as an improvement over the lumped axial dispersion model. Models have been proposed that account for from one to five of the parameters listed above. Depending on the number of parameters in the model, the mathematical complexity increases.

Van Deemter (1961) has proposed a modified version of the two-region model, which may be regarded as a countercurrent-flow model. He assumed that gas pockets together with the surrounding particles constitute the upflowing phase, and that the other particles and gas (i.e., the emulsion) constitute the downflowing phase. The upflowing particles (along with the associated gas) that accompany the bubbles may possibly

be likened to the cloud–wake phase of the more rigorous bubbling-bed (or Type II) models to be described later. The portion of the cloud that becomes detached in the wake of the bubble may be regarded as representing the transfer of material from the upflow to the downflow stream. Van Deemter's model may therefore be considered as providing the link between the two-region and the bubbling-bed models to be described in subsequent sections.

14.3. DAVIDSON'S TWO-PHASE BUBBLING-BED MODEL

The basic fluid-mechanical equations for the bubble phase were formulated in Section 13.3. We shall now present equations for the conversion of the fluidizing gas A, according to the reaction A → products, as it passes through the bed. (The subscript A will be omitted.)

Let us consider Figure 14.1a, which is an idealized representation of the fluid bed, with complete mixing in the emulsion phase and within a single bubble. Material

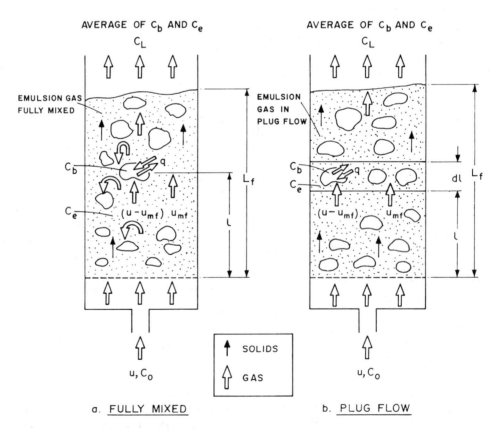

Figure 14.1. The bubble model of Davidson and Harrison (1963).

balance equations for reactant A can be written for both the bubble and emulsion phases. Then, by employing the mixing cup principle for calculating the composition of the gas leaving the top of the bed, the following final expression may be obtained for the conversion of A:

$$\frac{C_{AL}}{C_{A0}} = (1 - x_A) = Ue^{-Y} + \frac{(1 - Ue^{-Y})^2}{K_{0f} + (1 - Ue^{-Y})} \quad (14.1)$$

$$U = 1 - \frac{u_{mf}}{u} \quad (14.2)$$

$$K_{0f} = \frac{k_v L_{mf}}{u} = \frac{k_p PW}{F} \quad (14.3)$$

$$Y = \frac{Q_{be} L_f}{u_b V_b} \quad (14.4)$$

where k_p is the rate constant, expressed as mol/sec g catalyst atm; P is the pressure; W the catalyst weight; and F the molal feed rate.

The dimensionless quantities U and K_{0f} appearing in Eq. 14.1 can be calculated from readily obtainable fluidization and kinetic data, namely, the velocities involved, bed height, and rate constant. The quantity Y, on the other hand, requires a knowledge of the bubble diameter, which is not readily available. This quantity therefore constitutes the basic uncertain factor in using Eq. 14.1.

If the emulsion-phase gas is assumed to be in piston flow (Figure 14.1b), the following equation for conversion results:

$$\frac{C_L}{C_0} = 1 - x_A = \frac{1}{m_1 - m_2}\left[m_1 \exp(m_2 L_f)\left(1 + m_2 \frac{L_f u_{mf}}{uY}\right)\right.$$
$$\left. - m_2 \exp(m_1 L_f)\left(1 + m_1 \frac{L_f u_{mf}}{uY}\right)\right]$$

$$(14.5)$$

where m_1 and m_2 are the roots of the quadratic

$$m^2 + \frac{Y + K_0}{L_f(1 - U)}m + \frac{K_0 Y}{L_f^2(1 - U)} = 0 \quad (14.6)$$

or

$$2L_f(1 - U)m = (Y + K_0) \pm [(Y + K_0)^2$$
$$- 4K_0 Y(1 - U)]^{1/2} \quad (14.7)$$

where $U = 1 - (u_{mf}/u)$ and $m = m_1$ with the positive sign and $m = m_2$ with the negative sign. Comparisons of

calculated and experimental conversions for the ozone reaction reported by Davidson and Harrison (1963) and Toor and Calderbank (1967) indicate that the simpler Eq. 14.1 is adequate.

Normally we would expect that for very fast reactions there would be no gas bypassing as unreacted bubbles. However, Eqs. 14.1 and 14.5 for the complete-mixing and plug-flow models show that in both cases C_L/C_0 has a finite asymptote Ue^{-Y} even at very large values of the reaction parameter K_{0f}.

14.4. COUNTERCURRENT-FLOW MODELS

14.4.1. Kunii and Levenspiel's Simplified Approach

Kunii and Levenspiel (1968a, b, 1969) postulate that the amount of gas passing upward through the emulsion phase decreases as the gas velocity is raised; at sufficiently high velocities there is actually a flow reversal in this phase, with a net downward flow of gas. The conditions for distinguishing between the two flow regimes (as given in Section 13.6.3) are

$$\frac{u}{u_{mf}} < 6\text{--}11, \quad \text{upflow of emulsion}$$

$$\frac{u}{u_{mf}} > 6\text{--}11, \quad \text{downflow of emulsion} \quad (14.8)$$

In the case of downward flow of emulsion gas, since the bubbles (with their clouds) always move upward, we have a countercurrent flow of gas in the two phases.

Consider the reaction

$$A \rightarrow B$$

Assuming isothermal plug flow, we can write the following equation (see Chapter 11):

$$\frac{C_{AL}}{C_{A0}} = 1 - x_A = e^{-K_0} \quad (14.9a)$$

or

$$\ln\frac{C_{A0}}{C_{AL}} = K_0 = \frac{k_v L_0}{u} \quad (14.9b)$$

Note that the rate parameter K_0 appearing here is identical with that used in the Davidson model (Eq. 14.3), except that L_0 is replaced by L_{mf}.

For a fully mixed reactor, on the other hand, the corresponding equation is

$$\frac{C_{AL}}{C_{A0}} = 1 - x_A = \frac{1}{1 + K_0} \quad (14.10)$$

It will be noticed that the conversion is related directly to a dimensionless rate constant K_0. The central feature of this model is the definition of an equivalent nondimensional quantity for a fluidized bed. This quantity must account for reaction in the cloud and emulsion phases as well as for mass interchange between bubble and cloud and between cloud and emulsion.

The first main assumption of the model is that the concentration of the reactant varies with height only in the bubble phase, while the concentrations in the cloud–wake and emulsion phases are independent of position. In other words, bulk flow between the bubble and emulsion phases—an important postulate of the Davidson model—is neglected. The situation involved is represented in Figure 14.2. With the overall disappearance of A taken into consideration, the following material balances can be written:

$$-\frac{dC_b}{dt} = -u_b\frac{dC_b}{dl} = s_{bb}k_vC_b + K_{bc}(C_b - C_c)$$

$$(14.11a)$$

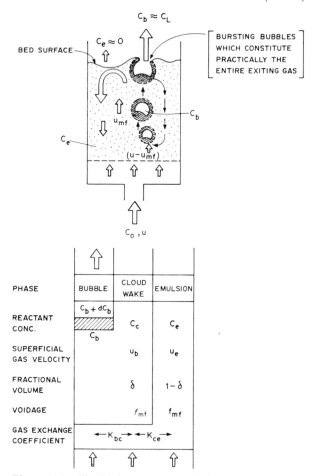

Figure 14.2. The simple countercurrent backmixing model of Kunil and Levenspiel (1969).

(amount transferred to cloud–wake);

$$K_{bc}(C_b - C_c) = s_{cb}k_vC_c + K_{ce}(C_c - C_e)$$

$$(14.11b)$$

(amount transferred to emulsion);

$$K_{ce}(C_c - C_e) \simeq s_{eb}k_vC_e \qquad (14.11c)$$

The overall rate of disappearance can also be written as

$$-\frac{dC_b}{dt} = -u_b\frac{dC_b}{dl} = K_f'C_b \qquad (14.12)$$

where K_f' may be regarded as an overall rate constant, which includes the effects of all the steps visualized in Eqs. 14.11 and is similar to the rate constant k_v. Upon elimination of the concentration terms in the cloud and emulsion phases (C_c and C_e), a nondimensional rate parameter K_f can be defined that is similar to the parameter K_0 (or K_{of}) used in the fixed-bed or Davidson model:

$$K_f = \frac{K_f'L_f}{u_b}$$

$$= \frac{L_fk_v}{u_b}\left(s_{bb} + \cfrac{1}{\cfrac{k_v}{K_{bc}} + \cfrac{1}{s_{cb} + \cfrac{1}{k_v/K_{ce} + 1/s_{eb}}}}\right) \qquad (14.13)$$

Integration of Eq. 14.12 for the inlet condition $l = 0$ leads to the following expression for the concentration of the reactant at any level l in the bubble phase:

$$C_b = C_0 \exp\left[-K_f\left(\frac{l}{L_f}\right)\right] \qquad (14.14)$$

If it is now assumed that the reaction taking place in the emulsion is negligible in relation to that occurring in the cloud–wake phase (the second major assumption of the model), then at the end of the reactor the concentration in the bubble or the total conversion achieved may be written as

$$\frac{C_L}{C_0} = 1 - x_A = \exp(-K_f) = \exp(-k_vE_c)$$

$$(14.15)$$

where E_c is the efficiency of contact defined as

$$E_c = \frac{1}{(1-f_{mf})} \frac{u}{u_{bs}} \left(s_{bb} + \cfrac{1}{\cfrac{k_v}{k_{bc}} + \cfrac{1}{s_{cb} + \cfrac{1}{k_v/K_{ce} + 1/s_{eb}}}} \right)$$

The similarity of this equation to that for the simple plug-flow reactor is apparent.

Usually several simplifications are possible. Thus the amount of solids in the bubble phase can be assumed to be zero, that is, $s_{bb} = 0$. This assumption is not valid for highly exothermic reactions, in which case the entrapped solids can attain such high temperatures as to contribute significantly to the overall reaction (Kunii, 1973). If the reaction is further assumed to be very slow and bubbles small, then $k_v \ll K_{bc}$ and K_{ce}. Thus Eq. 14.13 reduces to $K_f = K_0$, and Eq. 14.9 for a fixed-bed plug-flow reactor will apply.

When the flow of gas in the emulsion phase is upward, then obviously the gas that leaves the bed will be a mixture of the bubble-phase gas and the emulsion-phase gas. According to Kunii and Levenspiel, the amount of gas passing through the emulsion phase would be only a small fraction (between 5 and 10%) of the total gas passing through the fluidized bed, and therefore the reaction occurring in the emulsion phase may be neglected.

14.4.2. Rigorous Approach

In the simplified countercurrent backmixing model of Kunii and Levenspiel, two basic assumptions were made: (1) that the reactants are transported through the bed solely in the bubble phase, and (2) that the contribution of the emulsion-phase gas to the total reaction is negligible. Fryer and Potter (1972a, b) have proposed a rigorous model in which both these restrictions are removed.

In the Kunii–Levenspiel model the velocity of the bubble–cloud phase was described by a single parameter u_b, whereas in the present model we shall use separate velocities for the bubble and cloud phases (u_b and u_c, respectively). The three velocities u_b, u_c, and u_e are given, respectively, by

$$
\begin{aligned}
u_b &= u - u_{mf}[1 - \delta(1 + s_{wb})]\\
u_c &= s_{wb} f_{mf} u_b\\
u_e &= u_{mf}[1 - \delta(1 + s_{wb})](1 + f_{mf} s_{wb}) - u_{mf} s_{wb}
\end{aligned}
$$

(14.16)

It may be recalled that in the simpler Kunii–Levenspiel model, a differential element was considered for the bubble phase only, there being no throughflow through the cloud and emulsion phases. With this restriction removed and the concentration allowed to change in all three phases, the following equations can be written:

$$-\frac{dC_b}{dl} = \frac{K_{bc}(C_b - C_c)}{u_b} \tag{14.17a}$$

$$-\frac{dC_c}{dl} = \frac{\delta[K_{ce}(C_c - C_e) - K_{bc}(C_b - C_c) + k_v C_c s_{wb}]}{u_b} \tag{14.17b}$$

$$-\frac{dC_e}{dl} = \frac{\delta K_{ce}(C_e - C_c) + k_v C_e[1 - \delta(1 + s_{wb})]}{u_b} \tag{14.17c}$$

The boundary conditions are (a) at $l = 0$

$$C_b = C_0 \tag{14.18a}$$

and

$$-u_e C_e + (u - u_b)C_0 = u_c C_c \tag{14.18b}$$

Boundary condition 14.18a follows from the postulation that the entire bubble-phase gas is provided by the incoming gas, and 14.18b from the postulation that the remaining incoming gas combines with the downflowing emulsion gas to constitute the gas for the cloud–wake phase.

(b) At $l = L_f$

$$C_c = C_e \tag{14.19}$$

The exit concentration of the gas is obtained as

$$uC_L = u_b C_b + (u - u_b)C_c \tag{14.20}$$

Equation 14.20 removes the approximation of the simpler Kunii–Levenspiel model that the exit gas is derived entirely from the bubble-phase gas. It is now considered to be a mixture of the bubble-phase gas and a part of the cloud–wake gas. The rest of the cloud–wake gas provides the downflowing emulsion gas, which accounts for boundary condition 14.20.

Fryer and Potter (1972a) have provided an analytical solution in dimensionless form to the above set of equations. In an extension of this model, Fryer and Potter (1972b) and Fryer (1974) allowed for the variation of bubble size with height but at the sacrifice of analytical solution. The subsequent elaborate experimental studies of Fryer and Potter (1973) and Nguyen et al. (1977) showed that beyond a certain gas velocity all profiles show a minimum at some axial position in the bed, a situation well accounted for by their model. The experiments were chosen to minimize the effect of radial nonuniformity and adsorption. The inclusion of adsorption as an additional mechanism for gas exchange accounts for the enhanced gas mixing observed in some experiments (Nguyen et al., 1977).

14.4.3. A Simplified Initial Value (Jayaraman–Kulkarni–Doraiswamy) Model

The chief drawback of the Fryer–Potter model appears to be the extensive computations necessary. The model visualizes the bed as a two-point boundary value problem, which is often difficult to solve even for simple reactions. Besides, even numerically the problem of convergence and stability is very severe. A model that retains all the rigor of the Fryer–Potter model and yet

TABLE 14.1. Equations for Jayaraman–Kulkarni–Doraiswamy Model

$$\hat{C}_b = R_1 e^{\lambda_1 Z} + R_2 e^{\lambda_2 Z} + R_3$$

$$\hat{C}_c = R_1 \alpha_1 e^{\lambda_1 Z} + R_2 \alpha_2 e^{\lambda_2 Z} + R_3$$

$$\hat{C}_e = A_7 A_8 R_1 + A_7 A_9 R_2$$

where λ_1 and λ_2 are the roots of the equation

$$D^2 \hat{C}_b + (A_1 + A_2 + A_3 + A_4) D \hat{C}_b + A_1 (A_3 + A_4) \hat{C}_b - A_1 A_3 \hat{C}_e = 0$$

and R_1 and R_2 can be obtained by solving

$$
\begin{bmatrix}
1 + \dfrac{A_3 A_7 A_8}{A_3 + A_4} & 1 + \dfrac{A_3 A_7 A_9}{A_3 + A_4} \\[2ex]
B_2 \alpha_1 + A_8 \left(\dfrac{B_2 A_3 A_7}{A_3 + A_4} - B_1 A_7 \right), & B_2 \alpha_2 + A_9 \left(\dfrac{B_2 A_3 A_7}{A_3 + A_4} - B_1 A_7 \right)
\end{bmatrix}
\begin{bmatrix}
R_1 \\[2ex]
R_2
\end{bmatrix}
=
\begin{bmatrix}
1 \\[2ex]
1
\end{bmatrix}
$$

Parameters Definitions

$$\alpha_1 = \frac{\lambda_1 + A_1}{A_1} \qquad \alpha_2 = \frac{\lambda_2 + A_1}{A_1} \qquad B_1 = -\frac{u_e}{u - u_b}$$

$$B_2 = \frac{u_c}{u - u_b}$$

$$A_1 = \frac{\delta K_{bc} L_f}{u_b} \qquad A_2 = \frac{\delta K_{bc} L_f}{u_c} \qquad A_3 = \frac{\delta K_{ce} L_f}{u_c}$$

$$A_4 = \frac{k f_w \delta L_f}{u_c} \qquad A_5 = \delta K_{bc} L_f / u_e$$

$$A_6 = k[1 - \delta(1 + f_w)] \frac{L_f}{u_e}$$

$$A_7 = \left(1 + A_6 - A_5 + \frac{A_5 A_3}{A_3 + A_4} - \frac{A_3}{A_3 + A_4} \right)^{-1}$$

$$A_8 = \alpha_1 e^{\lambda_1} + \frac{A_5 \alpha_1}{\lambda_1} (1 - e^{\lambda_1})$$

$$A_9 = \alpha_2 e^{\lambda_2} + \frac{A_5 \alpha_2}{\lambda_2} (1 - e^{\lambda_2})$$

$$R_3 = \frac{A_3}{A_3 + A_4} \hat{C}_e$$

simplifies the calculation procedure is therefore desirable. Such a model has been formulated by Jayaraman et al., (1980) by assuming the emulsion phase to be completely mixed (instead of assuming plug flow). Although there exists an indirect justification for plug flow in the emulsion, the model predictions are relatively insensitive to the extent of mixing in the emulsion phase. This is particularly true of industrial beds, where the diameter of the bubbles involved is fairly large. The rate of mass transport across the bubble–cloud and cloud–emulsion phases is therefore almost always restrictive. The assumption of a completely mixed emulsion phase considerably simplifies the problem, which now assumes the form of an initial value problem. The solutions of Jayaraman et al., (1980) are summarized in Table 14.1.

The question of the extent of mixing in the emulsion phase is trivial from the practical standpoint. Although indirect verification of plug flow in the emulsion exists, it must be noted that all these studies really pertain to laboratory-scale beds. Very little information on industrial-size beds is available. Also, the presence of internals, generally used in industrial beds, ensures sufficient mixing in the bed, so that the assumption of complete mixing in the bed may be appropriate.

Both the Fryer–Potter and the simpler Jayaraman–Kulkarni–Doraiswamy models predict practically identical conversions at the reactor exit. Use of the simpler model may therefore be advantageous from the computational point of view.

14.5. DISTRIBUTED-BUBBLE MODELS

We shall now consider models in which allowance is made for the distribution of bubble size. First we shall briefly refer to the Partridge–Rowe model, in which bubble size distribution is measured by X-ray photography, and then proceed to models in which bubble coalescence (rather than random distribution) is included in the analysis.

14.5.1. Partridge–Rowe Model

As mentioned in Section 13.6.4, in the Partridge–Rowe model the reactor is divided into several sections and allowance is made for bubble size distribution in every section. Based on this an average value of α is calculated for each section, from which average values of the various other parameters of the model are computed. Complete mixing in the bubble–cloud phase and plug flow in the dense phase constitute the basic assumptions

of the model. X-ray data on bubble size distribution at various levels replace the single effective diameter of the previous models.

14.5.2. The Bubble-Assemblage Model

In the bubble-assemblage model (Kato and Wen, 1969) the variation of bubble size with distance from the distributor plate is allowed for by the following empirical equation of Kobayashi et al. (1966), which has already been listed in Table 13.1:

$$d_{b} = d_{b0} + 1.4 \rho_{s} d_{p} l \left(\frac{u}{u_{mf}} \right) \qquad (14.21)$$

The fluidized bed is then represented by a number of compartments in series, the height of each compartment being equal to the bubble diameter at the corresponding bed height. In each of these compartments the reactant is assumed to be perfectly mixed, with no backmixing of the gas between compartments. The interphase mass transfer coefficient is calculated from the correlation of Kobayashi et al. (1967b); $K_{be} = 11/d_{b}$.

A material balance for the reactant around the nth compartment gives

$$C_{b,n} = \frac{C_{b,n-1} u A_{c}}{K_{be,n} V_{b,n} - \dfrac{(K_{be,n} V_{b,n})^{2}}{k_{v} V_{e,n} + K_{be,n} V_{b,n}} + k_{v} V_{b,n} + A_{c} u}$$

$$(14.22a)$$

$$C_{e,n} = \frac{C_{b,n} K_{be,n} V_{b,n}}{k_{v} V_{e,n} + K_{be,n} V_{b,n}} \qquad (14.22b)$$

Based on these equations the concentration profiles of the reactant in the bubble and dense phases can be calculated. There are no adjustable parameters in this model, but it leans heavily on an empirical equation for bubble size distribution.

14.5.3. Coalescence and Bubble-Flow Models

In the coalescence model of Toor and Calderbank (1967), allowance is made for bubble coalescence as a function of height through an empirical equation. An elaborate mathematical procedure has also been proposed by Orcutt and Carpenter (1971) for including the effect of coalescence through a bubble size simulation method. In other models (see Table 13.1) d_{b} is estimated as a function of height.

14.6. DISTRIBUTED-CATALYST MODELS

In all the bubbling-bed models considered so far, continuous exchange of particles between the cloud and emulsion was assumed. Provided the bubble size is constant, this assumption would imply uniform solids distribution in the bed. Since, however, a distribution of bubble size is known to exist, uniform solids distribution would be highly unlikely.

Further, a dilute catalyst phase exists above the emulsion, and in all the analyses presented so far this phase has been ignored. In physical terms the solids content of the dilute phase is equivalent to that of the bubble phase, and accounts for direct contact between the bubble and catalyst phases. Models will now be presented in which these two situations are explicitly accounted for.

14.6.1. Distributed-Catalyst Model

Large gas bubbles travel faster, with residence times less than the mean; they contain practically no catalyst and exchange relatively little gas with the emulsion. On the other hand, small bubbles travel slowly, with considerable exchange of catalyst and gas with the emulsion. Gilliland and Knudsen (1971) account for this distribution of catalyst in the bed by postulating that the amount of catalyst in contact with a gas fraction during its passage through the bed $(u \rho_i)$ is an exponential function of the residence time of that fraction, that is,

$$u \rho_i = C \theta_i^d \qquad (d > 0) \qquad (14.23)$$

In this equation θ_i represents the dimensionless residence time of the gas fraction in a hypothetical plug-flow tube i in a bed assumed to be divided into an infinite number of such parallel tubes. Based on this concept, Gilliland and Knudsen (1971) derived the following expression for the exit concentration:

$$\frac{C_L}{C_0} = \int_0^\infty \exp(-K_f'') f(\theta) \, d\theta \qquad (14.24)$$

where

$$K_f'' = \frac{K_0' \theta^{1+d}}{\int_0^\infty \theta^{1+d} f(\theta) \, d\theta} \qquad (14.25)$$

and K_0' is given by

$$K_0' = \frac{k_p R_g T W}{Q} = \frac{k_v V_r}{Q} = \frac{k_v L_0}{u} \qquad (14.26)$$

Note that when $d = 0$ (for uniform catalyst distribution), $f(\theta) d\theta = 1$, giving $K_f'' = K_0' \theta = K_0$. Under these conditions Eq. 14.24 reduces to 14.9 for the plug-flow reactor.

The chief disadvantage of this model is that an *a priori* value of the catalyst distribution factor d is not available. Thus an experimental plot of conversion vs. θ must be compared with solutions of Eq. 14.24 for different values of d until the two curves match.

14.6.2. Successive Contact Model

In a modification of the distributed-catalyst model, Miyauchi (1974) suggests that instead of directly accounting for the presence of solids in the bubble phase, it would be more logical to include the dilute phase (above the emulsion) in the analysis. The basis of the analysis is the experimental finding (see, e.g., Lewis et al., 1962; Fan et al., 1962) that an axial distribution of bed density exists. Miyauchi (1974) and Miyauchi and Furusaki (1974) postulate that bubbles of gas are confined to the central region of the emulsion, which moves upward along with the bubbles. The bubbles then enter the dilute phase (probably by a process of bursting on the emulsion surface), while a bubble-free emulsion flows downward peripherally.

Based on this model an equation for the conversion has been derived that is composed of three nondimensional terms: a reaction group K_0 (for the emulsion phase) defined by Eq. 14.9, a mass transfer group K_m, a group for reaction in the bubble K_b, and a group for reaction in the dilute phase K_d, defined by

$$K_0 = \frac{k_v L_0}{u}, \qquad K_m = \frac{k_{ob} a_b L_f}{u} = \frac{K_{ob} L_f}{u}$$

$$K_b = \frac{(s_{bb} \delta) k_v L_f}{u}, \qquad K_d = \frac{k_v L_f}{u} \int_1^{z_{ft}} (1 - \delta) \, dz_f \qquad (14.27)$$

In these expressions $(s_{bb} \delta)$ represents the solids content of the bubbles, z_f is the nondimensional height l/L_f, and z_{ft} the total nondimensional height L_t/L_f.

The mass transfer group K_m is calculated from the overall mass transfer coefficient k_{ob}, which may be estimated from the equation of Miyauchi and Morooka (1969):

$$\frac{1}{k_{ob}} = \frac{1}{k_b} + \frac{1}{\beta_r k_e} \qquad (14.28)$$

where

$$k_b = \text{bubble-side} = \frac{2}{\sqrt{\pi}} \left(\frac{D_g u_b}{d_b} \right)^{0.5} \qquad (14.29)$$
$$\text{coefficient}$$

$$k_e = \text{emulsion-side coefficient} = \frac{2}{\sqrt{\pi}} \left[\frac{(PF)D_e u_b}{d_b} \right]^{0.5} \quad (14.30)$$

$$\beta_r = \beta_H - \left(\frac{\delta_{fe}}{PF} \right) J \quad (14.31)$$

In Eq. 14.31 δ_{fe} is the volume fraction of the fluid in the emulsion phase; PF is a partition function for solids distribution between the solids and fluid in the emulsion: $PF = (C_{s,eq}/C_{f,eq})\delta_{pe} + \delta_{fe}$; β_H is the Hatta number for unsteady-state gas absorption and is given by Danckwerts (1952) as a function of $m_H = \sqrt{k_v D_e/k_e}$; J is an integral associated with the cloud and is shown in

Figure 14.3 as a function of m_H (given by $\sqrt{k_v D_e/k_e}$) and a Peclet number defined as

$$\text{Pe} = \frac{PF(r_c - r_b)^2 u_b}{4 d_b D_e} \quad (14.32)$$

The concentration at the end of the dilute phase C_{Lt} can then be expressed in terms of the mass transfer group and the other two groups K_b and K_d as follows:

$$\frac{C_{Lt}}{C_0} = \exp\left[-(K' + K_b + K_d) \right] = \exp\left(-K_R' \right) \quad (14.33)$$

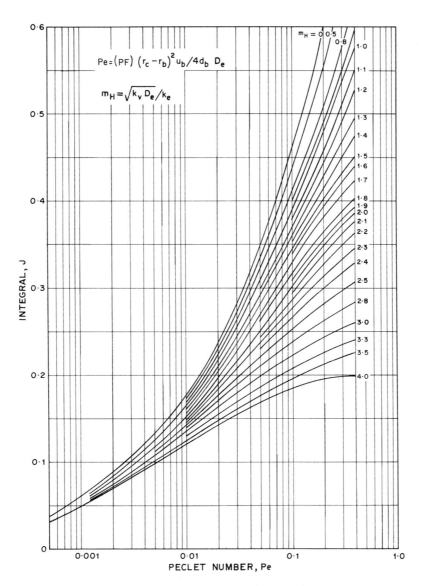

Figure 14.3. Integral J of Eq. 14.31 as a function of Pe and m_H

where K' is an overall constant given by the sum of the resistances for mass transfer in the bubble phase and reaction in the emulsion:

$$\frac{1}{K'} = \frac{1}{K_m} + \frac{1}{K_0(1-\delta)} \qquad (14.34)$$

Equation 14.34 is valid only for the case where the diffusional resistance in the emulsion phase is negligible. Note that this equation reduces to 14.9 for the plug-flow condition, namely, $C_L/C_0 = \exp(-K_0)$.

14.7. NONISOTHERMAL MODELS

All the models discussed so far were based on an isothermal reactor operating under steady-state conditions. The solids carried along with the bubble in the form of a wake (or entrapped in the bubble) can, however, act as sources of nonisothermicity in the fluidized bed, particularly when a high heat of reaction is involved (Kunii, 1973). As in the case of fixed-bed reactors, multiple steady states are also possible in fluid-bed reactors because of the nonlinear dependence of reaction rate on temperature (Westerterp, 1962; Luyben and Lamb, 1963; Luss and Amundson, 1968; Hatfield and Amundson, 1971; Nashaie and Yates, 1973; Lee and Kugelman, 1973; Bukur et al., 1974; Bukur, 1978; Sheplev and Luss, 1979). Another type of multiplicity, "concentration multiplicity," first reported by Matsuura and Kato (1967) and demonstrated for the fluid-bed reactor by Bukur (1978), occurs as a result of a nonlinear dependence of reaction rate on reactant concentration.

The approach of Bukur et al. (1974) is distinctive in that they have modified the common approach by postulating a model in which the reactor is divided into three phases: a dilute phase, the interstitial gas phase, and a solid phase. The last two phases are combined into a single phase (emulsion or dense phase) in the classical two-phase theory. In the development of the three-phase analysis it has been assumed that the particle temperature and the concentration of reactant within the particle depend on its age. The numerical results (Bakur et al., 1974) indicate that the solid phase is completely mixed and uniform. In a modified approach, Bukur and Amundson (1975a) assume uniform catalyst temperature and concentration, which leads to a considerable simplification in the mathematical treatment.

Numerical calculations for various parameter values

indicate that unique solutions with a high conversion are more likely to occur as the circulation rate of the catalyst increases. The effect of interchange coefficients on the multiplicity of steady states is quite significant. When there is no interchange between the two gas phases the steady-state solution is unique. For the other limiting case where the interchange coefficients tend to infinity (i.e., single-gas-phase model) multiple solutions are shown to be possible for all values of bubble diameter beyond a certain value.

The models described above were based on the experimental finding (Kunii and Levenspiel, 1969; Mireur and Bischoff, 1967) that mixing increases with bed diameter, and hence it was assumed that the catalyst particles were completely mixed. As was pointed out in Section 13.5, however, provision can be made for the mixing of solids and emulsion-phase gas by using an axial dispersion model. Based on this model, a more general axial dispersion model for a continuous fluidized-bed reactor under nonisothermal conditions has been proposed by Bukur and Amundson (1975b). In this model the dilute phase is assumed to be in plug flow with axial dispersion. Numerical computations indicate that the solution becomes unique for smaller values of superficial gas velocity and the region of multiplicity expands as the axial dispersion coefficient of solids decreases. In the limiting case of zero dispersion coefficient, the solution becomes unique for all values of u.

Bukur et al., (1977) have considered several models based upon the simple two-phase theory and demonstrated the importance of inclusion of the particles as a separate phase (in the emulsion) in the analysis. The high heat capacity of the particles as compared to that of the gas necessitates the treatment of particles as a separate phase, at least when nonisothermal conditions are involved. From the practical viewpoint, however, one is more interested in approximately locating the region of multiplicity, so that use of simple models should be adequate.

A simple criterion that would give sufficient indication of the onset of multiplicity for a given set of parameter values should therefore be useful from the practical standpoint, especially in multiparameter problems like the fluid bed. Kulkarni et al. (1981) have presented such a criterion, based on the two-phase version of the bed, which should prove useful in a quick estimation of the region of multiplicity and indicate whether multiplicity would occur at all. It appears from their results that the parameter values at which multiplicity would occur are far beyond the region of practical interest.

14.8. REACTOR MODELS INCORPORATING END EFFECTS

The models discussed in the forgoing sections do not distinguish between the region in the vicinity of the grid and the rest of the reactor. Also, the freeboard region is altogether neglected. Some of the recent experimental results indicate that these end effects may also be important. In the present section we describe reactor models that take into account these end effects. The grid models are discussed in Section 14.8.1, while the models that incorporate the freeboard region are discussed in Section 14.8.2.

14.8.1. Grid Models

The design of the grid plate significantly influences the performance of the reactor. This is evident from the data of Cooke et al. (1968), who observed a drop in conversion with an increase in grid hole size; also, Hovmand et al. (1971) showed that increasing the number of holes causes a marked improvement in conversion. The grid region thus plays a critical role in determining reactor performance, particularly for fast reactions in large beds.

The gas entering the reactor as high-speed jets penetrates the bed to a certain height h before breaking up into bubbles. This distance h from the grid plate corresponds to the grid region and contains no bubbles. The emulsion phase in this region can be assumed to be completely mixed because of the highly turbulent jets. These jets can be assumed to be perfectly mixed radially with plug flow in the axial direction. Mass transfer occurs between the jets and the emulsion phase and represents a significant quantity in the case of fast reactions. The bubbling region above the grid region can also be included in the model for a complete analysis of the bed starting from the grid region. For the sake of simplicity we may use the model proposed by Orcutt (1960) to describe the bubbling region; in this model the bubble gas is in plug flow and the emulsion is perfectly mixed.

The jets issuing from the orifices entrain the catalyst particles. However, the catalyst concentration within a jet is negligibly small, so that the reaction in the jet can be neglected in writing the mass balance equations. The material balance for this model can then be written for the jet, bubble, and emulsion phases, and solved for a first-order reaction to give (Behie and Kehoe, 1973):

$$x = \frac{(K''V_e\rho_g/Q)[1-(1-u_{mf}/u)\exp(-\gamma)]}{K''V_e\rho_g/Q+1-(1-u_{mf}/u)\exp(-\gamma)} \quad (14.35)$$

where K'' is the effective rate constant $k_v\rho_sW_e$; W_e the catalyst loading per unit volume of emulsion, cm^3/cm^3; V_e the volume of the emulsion; and γ a dimensionless group $P_1+P_2(L_f/J_p-1)$; $P_1=k_{je}a_jA_cJ_p/Q$; $P_2=k_{be}a_bA_cJ_p/Q$. Areas a_j and a_b are the jet and bubble areas, respectively, per unit emulsion volume, and k_{je} is the jet-to-emulsion mass transfer coefficient. In industrial reactors where high superficial velocities are employed, the term $(1-u_{mf}/u)$ can be neglected.

Behie et al. (1976) have reported experimental data on mass transfer from a vertical-grid jet into a fluidized bed of cracking catalyst. The test nozzle diameter was varied threefold (6.4–19.1 mm) and the nozzle velocity from 15.2 to 91.5 m/sec. The axial concentration data and mass transport coefficient have been related to a Froude group (Fr), a nozzle number (No), and the Reynolds number (Re) by the following correlation:

$$\ln\left(\frac{C_j-C_e}{C_{j0}-C_e}\right) = -1.92(Fr)^{-0.504}(No)^{0.905}(Re)^{0.068}$$
$$= -\frac{4k_{je}}{d_{or}G}(l_j-l_{j0}) \quad (14.36)$$

where
$$Fr = \frac{u_{or}^2}{gl_j}, \quad No = \frac{l_j}{d_{or}}, \quad Re = \frac{d_{or}u_{or}\rho_g}{\mu}$$

and l_j is the length along the grid jet; l_{j0} the length from the nozzle mouth within which the concentration in the jet remains unchanged; and C_{j0} the concentration in the jet at the nozzle mouth.

Behie and Kehoe (1973) and Behie et al. (1976) used Orcutt's model in the bubbling region. Mori and Wen (1976) have incorporated the grid region in the original Kato–Wen bubble-assemblage model (Section 14.5.2). They have simulated the catalytic conversions and carbonization of char in a number of fluidized-bed reactors and found that the predictions from the model compare satisfactorily with experimentally observed values. The method is illustrated in Section 14.8.3.

The key assumptions of the Behie–Kehoe model are that the gas in the emulsion phase is completely mixed and that there are no solids dispersed in the jet phase. Grace and de Lasa (1978) have examined the sensitivity of modeling with respect to these two assumptions, and have established their essential qualitative correctness. For sufficiently high transfer rates from jet to emulsion, as shown by de Lasa and Grace (1977), the predictions of the model approach those for a single-stage CSTR. The presence of dispersed solids in the dense phase can have a significant effect on model predictions, particularly when fast reactions with smaller values of P_1 are involved. Also, if the reaction carried out is sufficiently exothermic, then the grid region will commonly represent a region of marked temperature gradients. It is necessary in such instances to write an appropriate energy balance equation. Errazu et al. (1979) have analyzed such a situation in the simulation of the fluid catalytic cracking (FCC) regenerator unit.

14.8.2. Freeboard Region Models

The experimental investigations of Chavarie and Grace (1975), Furusaki et al. (1976), Ford et al. (1977), and Pereira and Beer (1978) clearly indicate the need to consider regions other than the bubbling bed in modeling fluid-bed reactors. Of these, the grid region has been discussed in the preceding section. In this section we shall consider the models that take into account the freeboard region. Among such models, that of Miyauchi (1974) and Miyauchi and Furusaki (1974), which was discussed in Section 14.6, specifically includes the freeboard region. It will be recalled here that the model requires experimental determination of the concentration profile of dilute-phase solids and that it makes no allowance for recycling of entrained particles. The models considered in this section are somewhat different from Miyauchi's model, which is why it was considered in an earlier section.

The freeboard region, where the particles thrown up from the fluid bed are in contact with the unreacted gases, represents a region of high gas–solids contact area. However, high temperature gradients may exist in this region in certain exothermic reactions such as in the fluid-bed phthalic anhydride plant or in FCC regenerator units.

Particles may be present in the freeboard region for various reasons. While Kehoe (1969), Basov et al. (1969), and Leva and Wen (1971) suggest that these particles are ejected from the wake region, Do et al. (1972) report the observation that the ejected particles originate from the nose of the bubble. Clearly, more experimental work is necessary to resolve this problem.

Yates and Rowe (1977) have proposed a model which assumes that the particles are ejected from the bubble–wake. It further assumes that the particles are equally spaced and each particle is surrounded by a cell of fluid volume V_c. If we assume the validity of the two-phase theory, the particle holdup (PH), voidage of the freeboard region f_{fb}, and volume of the cell V_c can be obtained as

$$PH = \frac{L_f}{u - u_t} f \frac{(u - u_{mf}) A_c (1 - f_{mf})}{3} \quad (14.37)$$

$$f_{fb} = 1 - \frac{f(1 - f_{mf})}{3} \frac{u - u_{mf}}{u - u_t} \quad (14.38)$$

$$V_c = \frac{3V_p}{f(1 - f_{mf})} \frac{u - u_t}{u - u_{mf}} \quad (14.39)$$

where f represents the fraction of particles ejected from bubble–wake and V_p the volume of a particle.

For a first-order chemical reaction occurring in the freeboard region, Yates and Rowe (1977) have obtained the following concentration profile for species A on the assumption of plug flow and completely mixed reactant conditions:

$$\frac{C_A}{C_{AL}} = \exp\left(-\frac{fl_{fb}}{3(u - u_t)\left(\frac{d_p}{6k_g(1 - f_{mf})} + \frac{1}{k_v}\right)}\right) \quad (14.40)$$

$$\frac{C_A}{C_{AL}} = \left(1 + \frac{fL_{fb}}{3(u - u_t)\left(\frac{d_p}{6k_g(1 - f_{mf})} + \frac{1}{k_v}\right)}\right)^{-1} \quad (14.41)$$

In order to use this model, it is necessary to know the terminal velocity of the particles. For known $Re_t = (d_p u_t \rho_g / \mu)$, the mass transfer coefficient k_g can be estimated from the relationship

$$Sh = 2 + 0.69 Sc^{1/3} Re_t^{1/2} \quad (14.42)$$

where

$$Sh = \frac{k_g d_p}{D_e} \quad \text{and} \quad Sc = \frac{\mu}{\rho_g D_e}$$

The model of Yates and Rowe is simple, assumes isothermal conditions in the freeboard region, and for the purposes of estimating u_t assumes that the particles reach the velocity $u - u_t$ immediately after ejection. Only particles with $u_t < u$ are therefore considered in this model. Some of the recent experimental data report the presence of particles with $u < u_t$ in the downstream.

The subsequent model of de Lasa and Grace (1979) treats particles with u_t greater or less than u. The model also takes into account the recirculation of particles through the cyclone. The particle concentration in this model is predicted based on the work of Do et al. (1972) and the mechanistic model of entrainment by George and Grace (1978). The model also takes into account the thermal gradients in the freeboard region and is specifically demonstrated for the FCC regenerator unit.

The model assumes the gas in the bubble and emulsion phases to be completely mixed with no reaction occurring within the bubble. The dense bed is isothermal in nature and the temperature of the gas entering the freeboard region is equal to the dense-bed temperature. Piston flow is assumed for both gas and solid in the freeboard region. No reaction occurs in the cyclone or dip-leg. With these assumptions the conservation equations for mass and heat can be written for the dense and freeboard regions. De Lasa and Grace (1979) have specifically considered the case of the FCC regeneration unit and solved this set of equations. The analysis of their results indicates that the freeboard region can become important, especially for shallow beds and for particles that are close to the critical size at which their terminal velocity is equal to the superficial velocity of the gas. The freeboard also influences the steady-state temperature level in the dense bed.

14.8.3. Example: Use of the Grid Model in the Simulation of the Carbonization Reactor

Calculate the conversion in the carbonization reaction carried out in a fluidized bed by using the following experimental data:

$d_{or} = 12.7$ mm, $u_{or} = 45.7$ m/sec, $u_{mf} = 1.53$ cm/sec

$u = 60$ cm/sec, $f_{mf} = 0.5$, $\delta = 0.25$

$L_f = 60$ cm, $d_p = 60\mu$, $D = D_e = 0.5$ cm²/sec

$d_t = 60$ cm, $\rho_g = 1.29$ kg/m³, $\alpha = 0.33$

effective rate constant $K'' = 9.63$ sec^{-1}

Solution

In view of the high reaction rate, this reaction is expected to be grid controlled. The conversion calculated by using the conventional bubbling-bed models will thus be considerably in error. In this example we shall calculate the conversion by both the grid model and the bubbling-bed model, and compare the results.

Grid Model. Let us first calculate the penetration of the jet inside the bed and the diameter of the bubbles breaking away from the jets by using the simple equations of Basov et al. (1969) in appropriate units:

$$J_p = \frac{d_p}{0.0007 + 0.566 d_p}(Q_{or})^{0.35}$$

$$d_b = 0.45(Q_{or})^{0.375}$$

Substituting the values of d_p and Q_{or} in the equations, we obtain

$$J_p = \frac{60 \times 10^{-4}}{0.0007 + (0.566 \times 60 \times 10^{-4})}$$

$$\times \left(\frac{\pi}{4} \times 1.27^2 \times 4570\right)^{0.35}$$

$$= 30.83 \text{ cm}$$

$$d_b = 0.45 \left(\frac{\pi}{4} \times 1.27^2 \times 4570\right)^{0.375}$$

$$= 11.59 \text{ cm}$$

The mass transport coefficient from the bubble to the emulsion phase may be obtained from Eqs. 13.21 and 13.23 by using the following value of u_b:

$$u_b = 60 - 1.53 + 0.71 (gd_b)^{1/2} = 134.24 \text{ cm/sec}$$

$$K_{ce} = 6.78 \left[\frac{0.5 \times 0.5 \times 134.24}{(11.59)^3}\right]^{1/2} = 0.995 \text{ sec}^{-1}$$

$$K_{bc} = 0.624 \text{ sec}^{-1}$$

The value of K_{je} can be obtained from the data of Behie (1972) as 5.64 sec^{-1}. Note that $K_{je} = a_j k_{je}$ of Section 14.8.1.

Next we calculate the value of γ [$= P_1 + P_2(L_f/J_p - 1)$] which can be obtained from a knowledge of P_1 and P_2:

$$P_1 = \frac{K_{je} A_c J_p}{Q}$$

$$= \frac{5.64 \times A_c \times 30.83}{A_c \times 60} = 2.898$$

$$P_2 = \frac{K_{be} A_c J_p}{Q_b}$$

$$= \frac{0.624 \times A_c \times 30.83}{0.25 \times A_c \times (60 - 1.53)} = 1.316$$

This value of γ obtained at a bed height of 30 cm, when used in Eq. 14.35, yields a conversion value of 0.83.

As calculated from the Kunii–Levenspiel model,

$$\delta = \frac{u - u_{mf}}{u_b}$$

$$= 0.435$$

Substituting in Eqs. 13.21 and 13.23 we have

$$K_{bc} = 1.676 \text{ sec}^{-1}, \qquad K_{ce} = 0.995 \text{ sec}^{-1}$$

$$s_{bb} = 0.001 - 0.01$$

$$s_{cb} = (1 - f_{mf}) \left(\frac{3 u_{mf}/f_{mf}}{u_{bs} - u_{mf}/f_{mf}} + s_{wb} \right)$$

$$= 0.228$$

$$s_{eb} = \frac{(1 - f_{mf})(1 - \delta)}{\delta} - (s_{cb} + s_{bb})$$

$$= 0.420$$

$$K_f = s_{bb} + \left(\cfrac{1}{\cfrac{K''}{K_{bc}} + \cfrac{1}{s_{cb} + (K''/K_{ce} + 1/s_{eb})^{-1}}} \right)$$

$$\times \frac{K'' u}{(1 - f_{mf}) u_{bs}}$$

$$= 1.99 \text{ sec}^{-1}$$

For the fluidized bed

$$\ln \left(\frac{C_{A0}}{C_A} \right) = K_f t_f = K_f \left(\frac{W}{Q \rho_s} \right) = K_f \left(\frac{L_f}{u} \right)$$

The values of conversion for various bed heights are therefore

l (cm)	30	60	90	120	150
x	0.394	0.632	0.777	0.865	0.918

The values so obtained are considerably lower than those predicted by the grid model in the region close to the grid plate of the bed ($= 0.83$ at a bed height of 30 cm).

It is apparent that the grid arrangement can have a profound influence on the conversion for fast reactions. It can be seen from the expressions (Eq. 14.35) for γ and x that the parameter P_1 has a strong influence on the conversion. For values of P_1 close to 3 the conversion is close to its maximum value. Thus little improvement in the conversion can be effected by changing the value of this parameter—that is, by modifying the grid. On the other hand, for values of P_1 much less than 3, a large improvement can be brought about by modifying the grid. Also, P_1 tends to increase as the grid size is decreased and the number of holes is increased. This example therefore demonstrates a means of evaluating the role of grid-plate design on the performance of a fluidized-bed reactor for a fast reaction.

14.9. REACTOR DESIGN FOR A DEACTIVATING CATALYST

The use of fluidized-bed reactors has grown largely because of the successful operation of catalytic crackers involving rapidly deactivating catalysts with half-lives of the order of a few seconds. Continuous regeneration of such a catalyst is accomplished, as already pointed out, by recirculating it through a regenerator connected in series. Thus two basic additional parameters are involved in the design of a fluidized-bed reactor under conditions of catalyst fouling; the *average activity* of the catalyst during the period it spends in the reactor; and the catalyst *recirculation rate*, which determines its average residence time in the reactor. We shall describe briefly the procedure for calculating these parameters, show how they can be used in the design of a fluid-bed reactor, and touch upon the design of a circulation system.

14.9.1. Catalyst Activity and Recirculation Rate

The rate of deactivation $d\Omega/dt$ of a catalyst can be expressed in several ways (see Section 8.1), the two most common being (Voorhies, 1945; and Szepe and Levenspiel, 1968)

$$-\frac{d\Omega}{dt} = k_{v,f} \Omega \qquad (14.43a)$$

$$-\frac{d\Omega}{dt} = k_{v,f} \Omega^3 \qquad (14.43b)$$

where $k_{v,f}$ is the rate constant for deactivation. Equation 14.43a is simpler to use and more generally applicable, whereas 14.43b is specific to catalytic cracking reactions (which are by far the best candidates for fluidized-bed operation).

From the design point of view the more important parameter is the *average activity* $\bar{\Omega}$ of the catalyst over the period of time it spends in the reactor. This is given by

$$\bar{\Omega} = \int_0^\infty \Omega [E_s(t)] \, dt \qquad (14.44)$$

The term within the integral represents the product of the activity (Ω) of all the particles of age between t and $t + dt$ and the fraction of particles in the exit stream in this age interval $[E_s(t)]$. If it is assumed that the solids in the reactor are fully mixed, then the fraction of exit-stream particles with age between t and $t + dt$ is given by

$$E_s(t) = \frac{t}{t} \int \exp(-t/\bar{t}) \, dt \qquad (14.45)$$

where the mean residence time

$$\bar{t} = \frac{W}{F_s} \tag{14.46}$$

It will be noted that the mean residence time of solids $\bar{t}$ can be calculated from the weight of the solids and the recirculation rate F_s.

The activity Ω in the age interval t to $t + dt$ can be found by integrating Eq. 14.43a to obtain

$$\Omega = \exp(-k_{v,f}t) \tag{14.47}$$

or Eq. 14.43b to obtain

$$\Omega = \left(\frac{1}{1 + 2k_{v,f}t}\right)^{1/2} \tag{14.48}$$

Incorporating Eq. 14.47 or 14.48 in 14.44 and integrating, we arrive at the following expressions for the average activity based on the two forms of catalyst decay represented by Eqs. 14.43a, b;

$$\overline{\Omega} = \frac{1}{1 + k_{v,f}\bar{t}} \text{ or } \frac{1}{1 + k_{v,f}(W/F_s)} \tag{14.49}$$

In the derivations just outlined, it has been assumed that the catalyst is not permanently damaged and that it can be restored to its initial activity of unity after regeneration. If the initial activity is not unity but some fraction represented by Ω_0, then Eq. 14.49 becomes

$$\overline{\Omega} = \exp\left(\frac{1}{2k_{v,f}\bar{t}\Omega_0^2}\right) \text{erfc}\left[\frac{1}{\Omega(2k_{v,f}\bar{t})^{1/2}}\right]\left(\frac{\pi}{2k_{v,f}\bar{t}}\right)^{1/2} \tag{14.50}$$

We have thus far obtained equations for the average activity. Now we need only calculate the rate corresponding to this activity from the rate equation

$$r = k_{v,f}f(C_A) \tag{14.51}$$

in which $f(C_A)$ may be C_A, C_A^2, or any other function of C_A. The average activity being related to the recirculation rate of the solids, the latter can be calculated for a given set of reaction conditions and output requirements.

14.9.2. Reactor Design

In the rigorous application of a bubbling-bed model to a system involving catalyst deactivation as well as continuous recirculation of solids, two essential additional aspects of design have to be considered:

1. The average activity $\overline{\Omega}$ must be incorporated in the rate equation.
2. Allowance must be made for the solids velocity within the reactor in calculating the absolute rising velocity of the bubbles in the fluidized bed.

The average activity can be calculated from Eq. 14.49 or 14.50. In calculating the absolute bubble rising velocity u_b, the following velocities were considered in Section 13.3.2: the natural rising velocity given by Eq. 13.4, and the upward velocity due to the continuous flow of gas into the bubble phase $(u - u_{mf})$. These velocities do not, however, account for solids recirculation. Thus, if the solids velocity is u_s, this term will also appear in the equation for the absolute rising velocity, being negative in the case of countercurrent flow and positive for cocurrent flow. It can be readily seen that the absolute rising velocity will now be

$$u_b = u - u_{mf} + u_{bs} \pm u_s \tag{14.52}$$

The solids velocity u_s may be expressed as

$$u_s = \frac{F_s}{(1 - f_B)\rho_s A_c} \tag{14.53}$$

where F_s is the mass rate of solids flow, ρ_s the density of the solid, and A_c the cross-sectional area of the reactor.

With these modifications, and assuming complete mixing in the emulsion phase, equations can be developed for the fraction of reactant unconverted at the top of the bed by following a procedure exactly similar to that for a nonfouling catalyst described in Section 14.3. For a second-order reaction exemplified by Eq. 14.43b for catalytic cracking, this is, $r_{vA} = k_v C_A^2$, Tigrel and Pyle (1971) have derived the following equation:

$$\frac{C_A}{C_{A0}} = 1 - x_A = \frac{Ue^{-Y}}{1 \pm Z} + \frac{(1 - Ue^{-Y} \pm Z)^2}{[2(1 \pm Z)K_{fd}\overline{\Omega}]}$$

$$\left[-1 + \left(1 + \frac{4K_{fd}\Omega}{1 - Ue^{-Y} \pm Z}\right)^{1/2}\right] \tag{14.54}$$

where

$$U = 1 - \frac{u_{mf}}{u}, \qquad Z = \frac{u_s}{u}$$

$$\hspace{12cm} (14.55)$$

$$K_{0f} = \frac{L_{mf} k_v C_{A0}}{u}, \qquad Y = \frac{Q_{be} L_f}{u_b V_b}$$

and k_v has the units $cm^3/mol\ sec$. The sign of Z is negative for countercurrent flow and positive for cocurrent flow of solids. It will be noted that Eq. 14.54 is similar in form to the Davidson equation. The parameters of the equation are also indentical to those of Eq. 14.1, but an additional parameter Z, representing solids circulation, is involved.

It can be shown that an optimum circulation rate exists and methods have been proposed for estimating it (Tigrel and Pyle, 1971). It is, however, a relatively simple matter to calculate the conversion from Eq. 14.54 for several combinations of operating conditions and then to arrive at a reasonable optimum value of the solids circulation rate. Such a procedure should be adequate since the optimization of the conversion in the reactor may not always be the cost-determining step.

14.9.3. Example: Calculation of Catalyst Recirculation Rate for the Vinyl Acetate Reactor

A catalyst with a half-life (of activity) of 87.171 hr is to be used in the design of a reactor for monovinyl acetate with a production capacity of 100 tonnes/day. If the catalyst gives satisfactory conversion until the mean activity is 5% of the fresh catalyst, calculate the necessary rate of circulation of catalyst if the reactor contains 282.7 tonnes of regenerated catalyst.

Solution

The rate of deactivation is calculated from Eq. 14.47:

$$\Omega = \exp(-k_{v,f} t)$$
$$0.5 = \exp(-87.171 \times 3600 k_{v,f})$$

or

$$k_{v,f} = 0.1908\ day^{-1} = 2.208 \times 10^{-6}\ sec^{-1}$$

The mean activity of the catalyst is given to be

$$\overline{\Omega} = 0.05$$

The required circulation rate is then obtained by rearranging Eq. 14.49:

$$F_s = \frac{k_{v,f} W \overline{\Omega}}{1 - \overline{\Omega}}$$

$$= \frac{2.208 \times 10^{-6} \times 282.7 \times 0.05}{1 - 0.05}$$

$$= 32.85 \times 10^{-6}\ tonnes/sec$$

$$= 2.84\ tonnes/day$$

In view of the small rate of solids circulation involved (2.84 tonnes/day in a 282.7-tonne bed of regenerated catalyst) the Davidson model for calculating the conversion can be used. For a larger circulation rate, the circulation rate also becomes a parameter of the model, and Tigrel and Pyle's modified procedure, as outlined in Section 14.9.2, should be used.

14.9.4. Design of Circulation System

In the design of circulation systems several circuits are possible; for example (1) the reactor and regenerator can be operated as dense phases, with circulation between the two occurring in the lean phase through transport pipes; (2) the reaction can occur in the dense phase (i.e., within the reactor), with regeneration occurring in the lean-phase transport line; and (3) the reaction can occur in the lean-phase transport line, with regeneration occurring in the dense phase. The first system is perhaps the most common and we shall briefly consider this below. Reference may be made to Kunii and Levenspiel (1969), Leung and Wilson (1973), and Leung (1977) for a detailed analysis of this and other systems.

Let us consider Figure 14.4, which represents a reactor and a regenerator connected by transport lines that enable the solids to flow from one to the other. The operation of the circulation loop is clearly brought out in part b of the figure. In the regenerator, the carbon deposit on the catalyst is burned off by air which also keeps the solids in a state of fluidization. The solids in the reactor are fluidized by the feed. The chief parameter of the circulation system is of course the rate at which the solids should be circulated, which is directly a function of the residence time of the solids in the reactor for obtaining the desired average conversion per pass. This rate is calculated by the procedure outlined in Section 14.9.1. The next most important design parameter is the level difference that must be maintained between the vessels to cause the desired circulation rate; it can be calculated as follows.

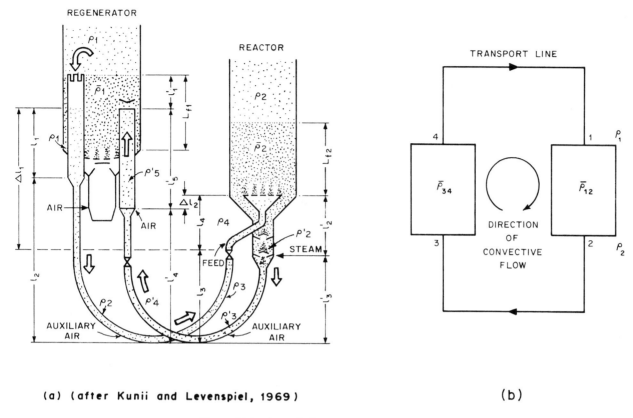

Figure 14.4. A two-bed reactor–regenerator system.

Circulation Loop

The circulation loop consists of several sections, and the flow in each section is made up of a static term and a friction term. The following final expression can be written in terms of the two heights H_1 and H_2 on the assumption that $\rho_1 = \rho_2 = \rho_3 = \rho'_3 = \rho$:

$$\rho(H_1 + H_2) = \rho_4 h_4 + \rho'_5 h'_5 + \bar{\rho}_1 h'_1 + \frac{g_c}{g} \sum p_{f,1} = 0 \tag{14.56}$$

where $\sum p_{f,1}$ is the total frictional loss. Equation 14.56 can be solved for H_1 and H_2 provided the densities and heights of sections 1, 4, and 5 of Figure 14.4a and the total frictional loss are known. The individual values of H_1 and H_2 can then be adjusted according to convenience.

Circulation of fluidized solids through standpipes and risers (the parts connecting the main reactor bodies in Figure 14.4a) is an essential part of such fluid-bed processes as catalytic cracking and fluid coking. The particulate solids are often transferred downward out of a fluidized bed in a standpipe to a regenerator. Most catalytic cracker standpipe circulation rates are in the range of 600–1500 kg/m² sec. The regenerated catalyst flows through another standpipe to be transported through a riser to the reactor. The diameter of the riser is usually the same as or somewhat larger than that of the associated standpipe, with a corresponding effect on mass flow rate.

Two types of flow pattern are possible for the downflowing solids in the standpipe (Leung and Wilson, 1973): fluidized-bed flow and packed-bed flow. In fluid-bed flow the particles are in suspension, whereas in packed-bed flow the particles move *en bloc* at the voidage of the packed bed. Under certain conditions both types of flow can exist simultaneously in the standpipe, or the perturbation in operating conditions might lead to transition from one flow pattern to another without change in mass flow rate. In the design of a circulation system it is necessary to identify the flow pattern, since such a transition can cause abnormal operations.

Among other parameters, besides the flow pattern, a knowledge of the aeration velocity, frictional losses, and problems associated with the formation of large bubbles is necessary for the rational design of the circulation system.

Aeration

The net gas flow in the standpipe is almost always down and is compressed as it is carried downward. It is therefore necessary to add aeration gas in order to counteract gas compression and to maintain proper fluidization. The basic equation relating the standpipe density to gas and solids flow rates suggested by Matsen (1973) can be used to calculate the required aeration:

$$\begin{pmatrix} \text{aeration} \\ \text{rate} \end{pmatrix} = A_{st} \frac{T_0}{P_0 T} \left[u_b(\rho_{fs,\,mf} - \rho_{fs}) + W_s\left(1 - \frac{\rho_{fs}}{\rho_s}\right) \right.$$
$$\left. + u_{mf}\rho_g \right] \tag{14.57}$$

where A_{st} is the standpipe cross-sectional area; T_0 and P_0 are respectively, the standard temperature and pressure; ρ_{fs} is the density of the fluidized solids; $\rho_{fs,\,mf}$ the density of the fluidized solids at minimum fluidization; and ρ_s the true particle density.

When the net gas flow is upward in the standpipe, the gas expands as it rises. The top of the standpipe will thus be highly aerated, leading to a low pressure buildup. As for the riser, the gas flow rates in it are usually adjusted to give pressure gradients of 0.5 $\rho_{fs,\,mf}$ or less.

Friction

Friction between particles and walls can sometimes be a significant part of transfer-line pressure drop and is mainly a function of catalyst density and hence of the aeration rate. Relatively little pertinent experimental data and theoretical predictions are, however, available. Matsen (1976) has compiled the normal and unusual operating experiments of some 65 cat crackers and fluid-cokers and is perhaps the only guide available. The data cover standard pipe diameters up to 1.5 m, heights up to 45 m, circulation rates to 120 tonnes/min, and a pressure buildup of almost 3 atm.

The data in general show that friction can approach zero below the minimum fluidization density but becomes quite significant as the density exceeds the minimum fluidization density. Pressure gradients under packed-bed flow are very sensitive to aeration, and a large scatter in the friction measurement is normally observed.

In fluidized standpipes, as evidenced in several commercial cat crackers and cokers, pressure gradients as high as 95% of the minimum fluidization density have been measured. The friction losses in fluidized zones are quite low (5% of $\rho_{fs,\,mf}$) in large standpipes, probably as a result of defluidization and/or aeration to a density less than the minimum fluidization density.

The friction losses can therefore be neglected in most design work, and pressure gradients of 70–80% of the minimum fluidization density can be selected as a safe base to allow for operating flexibility.

Measurements on catalyst density in U-bends show a higher density in the upflowing portion of the bend than in the downflowing limb. The density difference is mainly due to the slow movement of bubbles as compared to the catalyst in the down leg and faster movement in the up leg of the bend. This density difference gives rise to much of the pressure drop (1–2 psi) that is observed and sometimes (wrongly) attributed to friction.

Bubbles

The formation of large bubbles in the standpipe causes noticeable vibrations and oscillations in the circulation rates. The U-bend provides a suitable environment for the formation of large bubbles, which may dissipate the pressure buildup. The bubbles may also be held stationary in the constricted section in the standpipe. The flow rates of the solids could be such that a bubble can rise in the larger-diameter section of the standpipe but be held stationary in the smaller-diameter section as a result of the increased mass flow rate there. The bubbles trapped in the standpipe can grow, reduce the pressure buildup, and upset the entire circulation pattern.

Unstable circulation has also been reported in the case of standpipes with sloped sections. Both Type-A and Type-B slugs described by Stewart and Davidson (1967; see Section 14.12.1) have been noticed a short distance below the sloped section of the standpipe. The phenomenon called "bridging" or "trapping" can lead to unwanted situations. This difficulty may be overcome by using a high aeration rate to the sloped section; this stabilizes the circulation and gives high pressure buildups. Too high aeration rates can, however, cause upward gas flow into the regenerator. Thus there exists a definite upper limit for circulation.

The transition in flow pattern from fluidized-bed flow to packed-bed flow can also upset the circulation control system. The use of slide valves at the bottom part of the standpipe (a common practice) provides the constriction necessary to provoke this changeover of the flow pattern. The typical pressure profiles observed by Leung (1976) for some practical systems are sketched qualitatively in Figure 14.5. Ideally one would desire to maintain fluidized-bed flow in the pipe. As can be seen from Figure 14.5, however, a sudden loss of pressure above the slide valve can be triggered by a change in

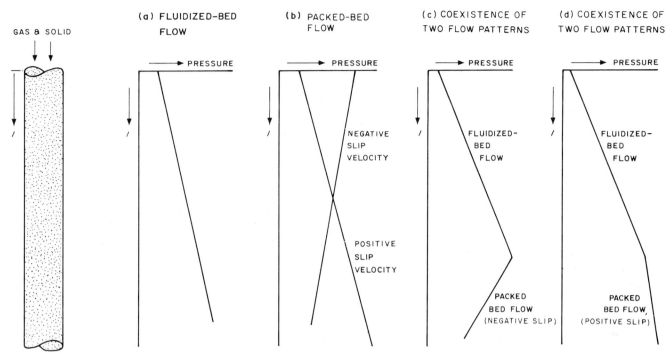

Figure 14.5. Pressure profiles for cocurrent gas–solid downflow.

flow pattern, leading to pressure reversal in the system with a consequent instability in flow.

14.10. DESIGN FOR COMPLEX REACTIONS

In the treatment presented earlier, attention was restricted to simple reactions. In this section we shall consider the effect of change in volume due to reaction, and cases where more than one reaction step is involved. For multistep reactions carried out in fluid beds the selectivity of the intermediate product is greatly reduced. We shall consider the use of catalyst dilution as an effective method of bringing about an improvement in selectivity.

Case Studies

Despite the use of these models, the prediction of conversion and selectivity in a complex reaction scheme remains uncertain. Further, the more rigorous models based on bubble behavior may not be usable for the simulation of complex reactions. The conventional approach in such cases is to formulate a simple model for the particular scheme at hand. In this connection the following case studies on the simulation of industrially important complex reactions are noteworthy:

1. gasification reaction based on the Kunii–Levenspiel model (Kunii and Yoshida, 1974);

2. cracking of methyl cyclohexane based on the Kunii–Levenspiel model (Tone et al., 1974);

3. catalytic conversions and carbonization of char in a number of fluid-bed reactors based on the modified bubble-assemblage model of Mori and Wen (1976);

4. hydrogenolysis of *n*-butane over Ni–silica catalyst (Shaw et al., 1974);

5. FCC regeneration unit using the grid model (Errazu et al., 1979; de Lasa and Grace, 1979).

14.10.1. Analysis of Reactions with Volume Change

Many industrially important systems involve a change in volume of the gas phase due to reaction, as typified by such systems as the dehydration of alcohol, many hydrogenation and dehydrogenation reactions, the manufacture of high-density polyethylene, and the absorption of sulfur dioxide in lime. The conversion-predicting equations available for a simple first-order system are inapplicable for these cases. For reactions involving a change in the number of moles, there can occur a substantial change in the volume of the gas phase, which alters the superficial velocity in the bed. This effect has been accounted for by Irani et al. (1980a) by considering a reaction

$$A \rightarrow v_B B$$

The constitutive equation that describes the concentration of species A can be written, based on the Kunii–Levenspiel model, as

$$-\frac{d}{dl}(u_b C_{Ab}) = kE_c C_{Ab} \quad (14.58)$$

The term E_c represents the efficiency of contact and depends on the superficial velocity. However, the variation of E_c with u_0 is very insignificant and Eq. 14.58 can be written as

$$-\frac{d}{dl}(u_b C_{Ab}) = K_f C_{Ab} \quad (14.59)$$

without any serious loss of accuracy. The concentration variable in this equation can be expressed in terms of mole fraction as

$$C_{Ab} = P y_{Ab} \quad (14.60)$$

where the total pressure P at any cross section in the bed is given by

$$P = P_t + (1-\delta)\rho_B g(L_f - l) \quad (14.61)$$

where P_t is the pressure at the top of the fluid-bed.

$k_v = 10\,\mathrm{sec}^{-1}$,	$u_{mf} = 3\,\mathrm{cm/sec}$,		$\rho_B = 2.0\,\mathrm{g/cm^3}$,		
$d_b = 8\,\mathrm{cm}$,	$f_{mf} \simeq f_{packed} = 0.5$,		$D_e = 0.2\,\mathrm{cm^2/sec}$,		$x_{A0} = 0.98, 0.5$
$L_f = 100\,\mathrm{cm}$,	$D_t = 100\,\mathrm{cm}$,		$P = 1.0\,\mathrm{atm}$,		$\alpha_w = 0.33$
$f_e = 0.5$,	$v_B = 0,1,2,3$,	$u_0\vert_{z=0} = 30\,\mathrm{cm/sec}$		reaction temperature $= 250°C$	

Incorporating Eqs. 14.60 and 14.61 in 14.58 and integrating yields

$$\frac{l}{L_f} = \frac{1+\alpha_1}{\alpha_1} + \left[\frac{(1+\alpha_1)^2 + 2\alpha_1\beta_1}{\alpha_1^2}\right]^{1/2} \quad (14.62)$$

where α_1 and β_1 are defined as

$$\alpha_1 = \frac{(1-\delta)\rho_B g L_f}{P_t} \quad (14.63a)$$

$$\beta_1 = \alpha_2 \left(\frac{(1-v_B)(y_A - y_{A0})}{[1-(1-v_B)y_A][1-(1-v_B)y_{A0}]} \right.$$
$$\left. + \ln\left\{\frac{y_A[1-(1-v_B)y_{A0}]}{y_{A0}[1-(1-v_B)y_A]}\right\}\right) \quad (14.63b)$$

and

$$\alpha_2 = \frac{u_b}{K_f L_f} \frac{v_B N_{A0} + N_j}{N_{A0} + N_j}(1+\alpha_1) \quad (14.63c)$$

In Eq. 14.63c N_{A0} and N_j refer to inlet molar flows of species A and inerts, respectively, per unit time, and y_A refers to the mole fraction of A.

Equation 14.62 is of the correct form and can be solved to obtain the mole fraction of A remaining unconverted at any distance l in the reactor. For $v_B = 1$ it reduces to 14.15 for the simple K–L model.

14.10.2. Example: Illustration of the Effect of Volume Change in the Modeling of a Fluid-Bed Reactor by the Irani-Kulkarni-Doraiswamy Method (1980a)

For the first-order reaction

$$A \xrightarrow{k} v_B B$$

establish the effects of the stoichiometric coefficient v_B, molar percentage of inerts in feed, and pressure variation along the bed height by employing the volume-change model developed above and using the following data for fluid-bed operation.

Solution

The conversion profile along the bed is given in terms of the volume-change model by Eq. 14.62, in which α_1 and β_1 are given by Eq. 14.63.

The calculation of the parameters proceeds along the lines of the K–L model. Substitution of the relevant parameters in the basic Equation 14.62 yields results as presented in Figures 14.6 and 14.7.

1. Effect of stoichiometric coefficient v_B:

From Figure 14.6 it can be seen that for a reaction with a change in the number of moles $(v_B > 1)$ the conversion at the bed exit is always less than that for a reaction with no volume change. This is in consonance with the physical situation, since the presence of extra moles dilutes the reactant species A in the reactor, thereby decreasing the conversion.

2. Effect of pressure variation along the bed height:

The broken lines in Figure 14.6 represent the inclusion of the effect of pressure along the bed height. The effect on the conversion profile is seen to be marginal. Thus α_1 can conveniently be taken to equal zero, with negligible change in the predictions of the performance equations and with considerable simplification in the volume-change model equation.

3. Effect of inerts in feed stream:

The effect of inerts is to reduce the effect of change in moles in the reactor. Figure 14.7 shows the effect of 2 mole % inerts and 50 mole % inerts on the exit conversion in the bed for different values of the stoichiometric coefficient v_B. The presence of inerts is seen to have no effect on the conversion level for the case in which $v_B = 1$. Here the inerts alter the inlet and exit-stream concentrations but the overall conversion level is unchanged. However, this statement is valid only for the case where $v_B = 1$ (representing no change in moles with reaction), as can be seen from Figure 14.7.

14.10.3. Analysis Based on the Kunii–Levenspiel Model

The treatment for a simple first-order reaction can be extended to a variety of first-order complex reaction

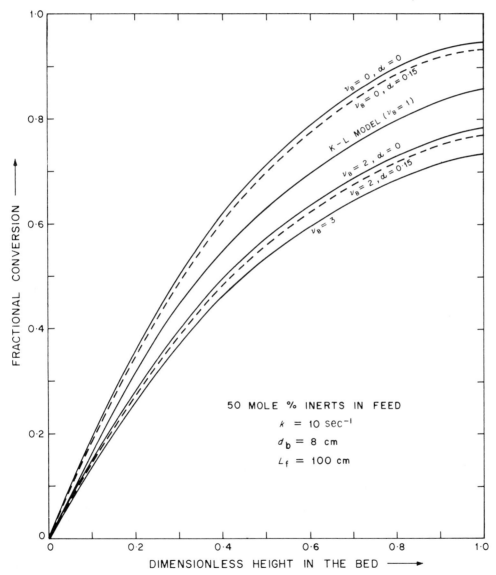

Figure 14.6. Influence of stoichiometric coefficient on the conversion profile along the bed height (Irani et al., 1980b).

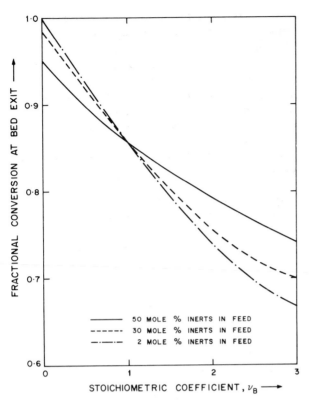

Figure 14.7. Influence of stoichiometric coefficient and presence of inerts in the feed stream on the exit conversion in the bed (Irani et al., 1980b).

Levenspiel et al. (1978) have formulated the basic equations for the Denbigh reaction scheme represented by scheme I above. Analytical solutions to these equations for a feed containing all the components assume negligible reaction in the bubble phase (i.e., s_{bb} is negligibly small). Irani et al. (1980b) have extended the analysis to the reversible consecutive reaction scheme II and to scheme III.

In any complex reaction scheme we are interested in more than just the disappearance of reactant. The concentration changes of the other reacting species, the product distribution, the amount of catalyst needed to maximize the intermediate R, and in general the performance of a fluid-bed reactor in relation to that of a plug-flow reactor are equally important considerations. In view of the general nature of the Irani–Kulkarni–Doraiswamy (1980b) expressions, solutions for simplified cases corresponding to a variety of first-order reaction schemes can be readily deduced from them.

Elnashaie and Yates (1972) have carried out the analysis for a parallel reaction scheme by using the Partridge and Rowe model. The analysis shows that delayed addition of reactant improves the yield of desired product. Also, the yield is improved when the order of the undesired side reaction is more than that of the desired reaction.

14.10.4. Influence of the Dilute Phase on the Selectivity of Complex Reactions (Miyauchi Model)

The dilute phase containing the suspended catalyst particles has been included in the analysis of Miyauchi

networks. In this section we shall present the performance equations for both conversion and product distribution in fluidized-bed reactors according to the Kunii–Levenspiel·model for the rather general first-order reaction schemes:

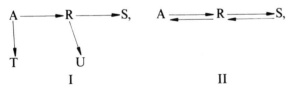

I   II

By setting the appropriate rate constants to zero in the solution for these networks, we can obtain the performance equations for a variety of simpler reaction schemes, such as

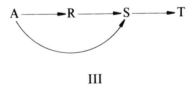

III

(1974) in order to account for the experimentally observed increase in the apparent mass transfer rates between the bubble and the emulsion phases for the case of a fast reaction. This model was discussed in

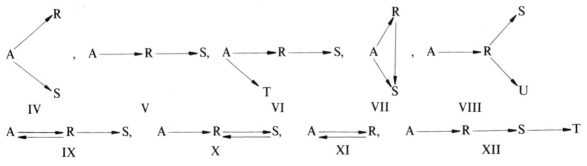

IV   V   VI   VII   VIII

IX   X   XI   XII

Section 14.6.2. The dilute phase has a lower heat capacity and the energy transport in this phase is considerably smaller than in the dense phase. As a result, a temperature distribution in the dilute phase exists and becomes an important factor in controlling the selectivity of the reaction.

For a consecutive reaction $A \rightarrow B \rightarrow C$ the results of Miyauchi and Furusaki (1974) show that in the case of isothermal reactions the selectivity of B will be greater for the case of larger K_m. It will also be greater if the integral $\int_1^{z_{ft}}(1-\delta)dz_f$ in Eq. 14.27 is chosen larger because of the direct contact in the dilute phase. For nonisothermal reactions, the presence of the dilute phase adversely affects the selectivity if the reaction is exothermic; hence it should be kept to a minimum. This can be readily achieved by operating at moderate to low superficial velocities. The use of internals (which are always used in a highly exothermic reaction) favors the situation by reducing the circulation in the dense phase and subsequent carryover of the particles in the dilute phase. The yield of product B does not vary significantly from the isothermal case when the reaction is endothermic; the role of the dilute phase is to enhance the selectivity for reactions with $E_2 > 30$. Since in many practical cases the activation energy for the second stage is high, the use of the dilute phase would seem to be advantageous.

Let us now reconsider reaction scheme I proposed by Denbigh (1958). The numerical computations of Miyauchi and Furusaki (1974) show that for an exothermic reaction ($\beta_m = 0.1$) with $E_1 < E_2$, $E_3 > E_4$, a considerable increase in the yield of R can be observed. For an endothermic reaction ($\beta_m = -0.1$) there is no significant effect of the dilute phase on selectivity. For the case in which $E_1 > E_2$ and $E_3 < E_4$, the selectivity is again not improved significantly for either endothermic or exothermic reactions. In general the effect of the dilute phase on this type of reaction is not unfavorable. Even if the selectivity is not improved significantly, the larger throughput can be an advantage.

14.10.5. Selectivity Considerations in Complex Reactions

Frequently, in reactions such as oxidation and chlorination the intermediate happens to be the desired product. The analysis presented in the previous sections indicates that the selectivity of the intermediate product is severely reduced in the fluid bed. In the present section we therefore focus our attention on methods that could be used to bring about an improvement in selectivity.

Catalyst Dilution

The concept of catalyst dilution has been applied conventionally to exothermic reactions carried out in tubular reactors packed with catalyst (see Section 12.5.1). Here the rate of heat generation varies markedly along the length of the reactor, giving rise to undesirable temperature peaks. The dilution of the catalyst can be effectively used to establish a specific temperature profile, and in certain cases this strategy can be used to improve the reactor performance. Temperature control, however, poses no problem in a fluid bed, since the establishment of near-isothermal conditions is one of the main features of this type of reactor. The chief drawback of the fluid-bed reactor is the reduced conversion for the same weight of catalyst as compared to a fixed bed; it is also characterized by low selectivity in the case of complex reactions. The concept of catalyst dilution can be applied to the fluid bed in order to control these two effects. For the purpose of illustrating the influence of catalyst dilution we shall consider the applicability of the simple K–L model to describe the behavioral features of the bed.

We define a catalyst dilution ratio as

$$R' = \frac{\text{total weight of solids in the bed}}{\text{weight of active catalyst in the bed}}$$

(14.64)

Thus if k represents the rate constant for the reaction in units of cubic centimeters per gram of active catalyst per second, the effective rate constant based on total solids becomes

$$k_{eff} = \frac{k}{R'} \quad \frac{cm^3}{g \text{ solid sec}}$$

(14.65)

It is clear from this definition that $R' \geqslant 1$, and therefore the effective rate constant k'_{eff} falls with dilution, while the efficiency of contact given by Eq. 14.13 increases. The modified rate constant for the fluid bed represents the product of the effective rate constant k_{eff} and the efficiency of contact E_c:

$$K_f = k_{eff} E_c$$

(14.66)

and for a given dilution ratio the decrease in k_{eff} is greater than the increase in E_c. The overall effect of catalyst dilution is thus to lower the value of the modified rate constant and for a first-order reaction the conversion in the fluid bed would fall with increasing dilution ratios.

We next consider a complex reaction such as the consecutive reaction $A \xrightarrow{1} B \xrightarrow{2} C$. For a plug-flow reactor the selectivity for this scheme is defined by k_1/k_2, while for a fluid-bed reactor it becomes $k_1 E_{c1}/k_2 E_{c2}$ where E_{c1} and E_{c2} represent the efficiencies of contact for the two reaction steps. For the plug-flow case catalyst dilution will have no influence on selectivity. In the case of a fluid-bed reactor, however, the efficiencies of contact E_{c1} and E_{c2} will be affected to different extents and the selectivity would change. Irani et al. (1979) have

considered a consecutive reaction scheme and studied the influence of catalyst dilution on the conversion of A and selectivity to B for the cases where $k_1/k_2 < 1$ and $k_1/k_2 > 1$. For $k_1/k_2 < 1$, as in the case of fixed-bed reactors, higher selectivities are realizable only at low conversion levels. For the case in which $k_1/k_2 > 1$, Figure 14.8 shows the results obtained. For the sake of comparison the selectivity in a plug-flow reactor is also presented. It is seen that the maximum selectivity is obtained in the plug-flow reactor and the lowest in the

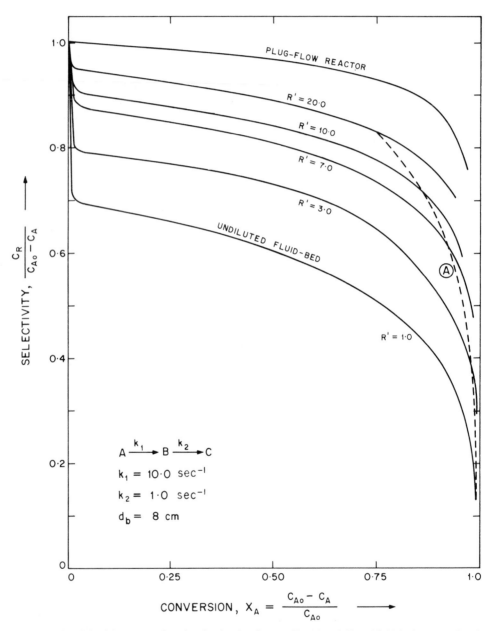

Figure 14.8. Selectivity–conversion plots for the plug-flow, undiluted, and diluted fluid-bed reactors (Irani et al., 1979).

undiluted fluid-bed reactor. With increasing dilution the selectivity improves for the same conversion, until for an infinitely diluted bed (corresponding to near-zero conversion) it approaches that for plug flow. Curve A in the figure shows the path along which the undiluted fluid-bed operation moves when it is diluted to different extents. For the parameter values employed in this figure, for a dilution ratio of 20 the reduction in conversion is seen to be small (from 99 to 75%), whereas the improvement in selectivity of intermediate is from 14 to 83%. Thus, when the intermediate is the desired product, it seems preferable to use high dilution rates with high levels of conversion in the bed.

Effect of Temperature on Selectivity

It is well known that the performance of a fixed-bed reactor for complex reactions can be improved by imposing an optimum temperature profile (see Section 12.4.2). In the case of a fluid-bed reactor, the vigorous mixing of the solids precludes the possibility of a temperature profile along the reactor and ensures isothermal operation. However, there is still the choice of using two (or more) fluidized-bed reactors in series, each operating isothermally at a different temperature. It would be of interest to know the extent of improvement in performance that can be obtained in this manner. An upper limit to the possible improvement can be obtained by considering the limiting case of a large number of fluidized-bed reactors operating in series. Mathematically this is equivalent to a fluidized-bed reactor with a temperature profile.

Kulkarni and Patwardhan (1980) have analyzed this problem for consecutive and parallel reaction schemes and derived expressions for optimum temperature profiles based on the K–L model description of the bed. Their analysis shows an interesting feature: the temperature derivative (dT/dl) can be either positive or negative depending on the parameter values chosen, indicating that the operating temperature profile could be a rising or a decreasing one for the same reaction. This is in contrast to what is observed in fixed-bed reactors and is clearly due to the interaction of mass transfer and chemical reaction.

A comparison of the fluid-bed reactor operated at a single bed temperature with that in which an optimal temperature progression is used confirms the superiority of the reactor system with temperature progression. However, the improvement in performance is so marginal for most parameter values of practical interest that its use is not warranted.

14.10.6. Example: Illustration of the Effect of Catalyst Dilution for a Hypothetical Consecutive Reaction using the Irani-Kulkarni-Doraiswamy analysis (1979)

We wish to carry out the irreversible successive reaction

$$A \xrightarrow{\ k_1\ } B \xrightarrow{\ k_1\ } C$$

in a fluidized-bed reactor with the following parameter values:

$$k_{v1} = 10 \sec^{-1}, \qquad k_{v2} = 1.0 \sec^{-1},$$

$$\rho_c = 2.0 \, \mathrm{g/cm^3}, \qquad \frac{W}{Q_0} = 8.0 \, \mathrm{g \, sec/cm^3},$$

$$u_{mf} = 3 \, \mathrm{cm/sec}, \qquad u = 30 \, \mathrm{cm/sec}, \qquad f_{mf} = 0.4,$$

$$D_e = 0.2 \, \mathrm{cm^2/sec}; \qquad d_b = 8 \, \mathrm{cm}$$

Catalyst dilution with unimpregnated support is proposed to be employed. The effect of catalyst dilution on the exiting stream from the bed, keeping the total weight of solids in the bed unchanged, is proposed to be studied.

Solution

The values of the various parameters of the K–L model are obtained as follows:

$$u_{bs} = 0.711 \, (g d_b)^{1/2} = 62.95 \quad \mathrm{cm}$$

$$u_b = u_0 - u_{mf} + u_{bs} = 89.95 \quad \mathrm{cm/sec}$$

$$\delta = \frac{u_0 - u_{mf}}{u_b} = 0.300$$

$$K_{bc} = 4.5 \left(\frac{u_{mf}}{u_b} \right) + 5.85 \left(\frac{D_e^{1/2} g^{1/4}}{d_b^{5/4}} \right) = 2.77 \quad \sec^{-1}$$

$$K_{ce} = 6.78 \left(\frac{f_{mf} D_e u_b}{d_b^3} \right)^{1/2} = 0.803 \quad \sec^{-1}$$

$$s_{bb} \simeq 0.01\text{--}0.001$$

$$s_{cb} = (1 - f_{mf}) \left(\frac{3u_{mf} - f_{mf}}{u_{bs} - u_{mf}/f_{mf}} + \frac{V_w}{V_b} \right) = 0.423$$

$$s_{eb} = (1 - f_{mf}) \left(\frac{1 - \delta}{\delta} \right) - s_{cb} = 0.974$$

The performance equations for the reaction are obtained from (Irani et al., 1979, 1980b):

$$K_{11} = \left[\frac{k_{v1}}{K_{bc}} + \frac{1}{s_{cb} + (k_{v1}/K_{ce} + 1/s_{eb})^{-1}} \right]^{-1} \frac{k_V u}{(1 - f_{mf})u_{bs}}$$

$$= 1.413 \quad \text{sec}^{-1}$$

and similarly

$$K_{22} = 0.523 \text{ sec}^{-1}$$

with

$$K'_{11} = K_1 \left(1 - \frac{K_{22}\psi_1}{K_{11}} \right) = 0.989 \text{ sec}^{-1}$$

$$\tau = \frac{W}{Q\rho_c} z$$

and hence the concentration profiles in the bed are obtained as

$$\hat{C}_{Ab} = \exp(-1.413\tau)$$

$$\hat{C}_{Bb} = 1.111 \left[\exp(-0.523\tau) - \exp(-1.413\tau) \right]$$

$$\hat{C}_{Cb} = 1.0 - (\hat{C}_{Ab} + \hat{C}_{Bb})$$

These profiles have been employed to plot selectivity vs. conversion for the undiluted fluid bed in Figure 14.8.

The effect of catalyst dilution is to alter the rate constant in terms of a modified rate constant k_v/R'. Thus, the rate constants K_{11}, K_{22}, K'_{11} are modified in the analysis above to the extent that k is replaced by k/R' in the mathematical computations, and the analysis then proceeds along similar lines.

Dilution of catalyst results in a drastic improvement in selectivity with only a marginal drop in conversion, as can be observed in Figure 14.8 from path A, which represents the exit conditions for undiluted and diluted fluid-bed operations. For the values of k_1 and k_2 employed, with $R' = 20$ and throughput unchanged, the conversion is reduced from 99 to 75 %, whereas the corresponding increase in the selectivity of intermediate is from 14 to 83 %. At the hypothetical limit of infinite dilution the fluid-bed reactor performance approaches that of the corresponding plug-flow reactor.

14.10.7. Example: Experimental Observations on the Effect of Catalyst Dilution in the Dehydration of Ethyl Alcohol (Irani et al., 1980c)

Commercial Flykalumina catalyst was employed and kinetic runs free from pore diffusion and external mass transfer effects were initially carried out in an integral reactor. Runs were subsequently conducted at the same temperature (338°C) in a fluidized-bed reactor (10 cm in diameter). Initially, undiluted catalyst was employed in the bed and further runs were conducted employing glass powder as the inert catalyst diluent. The dilution ratio was thus varied while keeping the operating parameters and overall residence time in the bed unchanged. The effect of catalyst dilution on overall conversion and formation of intermediate ether was then experimentally obtained for the undiluted and diluted fluid beds. The results are presented as a selectivity–conversion plot in Figure 14.9.

The integral reactor corresponds to plug–flow conditions, and hence at any given level of conversion the corresponding selectivity of intermediate (ether) will always be higher than for the fluid bed. Dilution of catalyst results in substantial improvement in the selectivity of intermediate in the fluid bed, a more than twofold rise (7 % to 16 %), in consonance with the qualitative predictions of the model. For the operating parameters employed, an optimum catalyst dilution ratio R' is shown to exist such that intermediate production at the bed exit is maximized. In the present study $(R')_{opt} = 3.67$.

14.11. BUBBLE DIAMETER AND REACTOR INTERNALS

In all the models based on the assumption of a uniform bubble size, a knowledge of this size is of fundamental importance in designing the fluid-bed reactor. One easy way of estimating an average effective diameter is to carry out a simple fluidization experiment and observe the size of the bubble passing through the bed. In order to get meaningful results, it is necessary to use a column 10–15 cm in diameter, since smaller-diameter columns promote slugging, particularly when the height of the column is more than about 50–60 cm. In such a column, although bubbles may be small at the bottom, they can increase almost to the size of the column as they travel through the bed. A 10-cm column with a height of less than 20 cm should normally give reproducible results.

The large-scale reactor would, however, operate under entirely different hydrodynamic conditions. Thus, in order to simulate industrial conditions, it would be more desirable to set up a reactor that is 0.5–1 m in diameter and determine the bubble size under anticipated conditions of feed rate. It is common practice to use internal baffles of various designs and orientations to break up the bubbles and restrict them

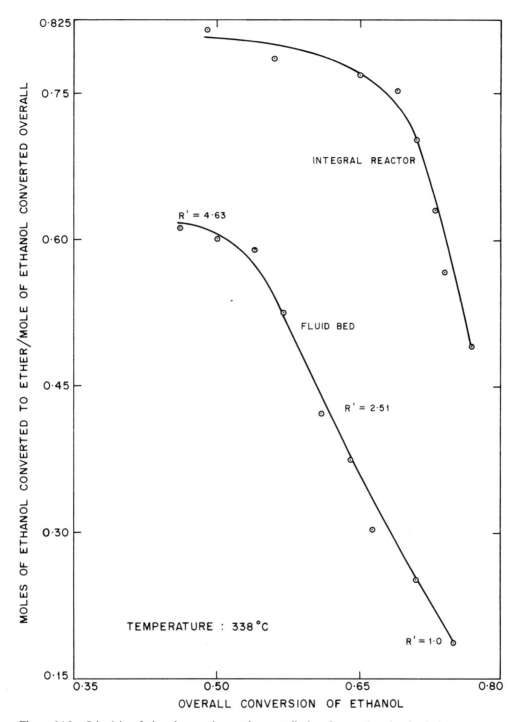

Figure 14.9. Selectivity of ethanol conversion to ether: overall ethanol conversion plots for the integral reactor and for undiluted and diluted fluid beds (Irani et al., 1982).

to a reasonable size (4–10 cm). When this is not done a free bubbling bed results, in which case the design of the distributor plate becomes the primary consideration (it has already been discussed in Chapter 13). With internals the problem is, to what extent can the bubble size be uniformly restricted and the reactor operation made to approach that of a bench-scale reactor? It is therefore clear that bubble size and the use of internals are interrelated and will thus be considered together in this section.

14.11.1. Bubble Diameter

In the absence of an experimental facility to determine the bubble size, it does not appear possible at present to undertake the *a priori* design of a fluid-bed reactor based on any of the bubbling-bed models. The only reliable method of obtaining an estimate of the bubble size is through experiment. For this purpose large-diameter cold models (with internals if necessary) should be used. A relatively easy method of determining the bubble size is by measuring the bubble frequency by a capacitance probe inserted in the bed. Another method is to photograph the bed from the top with a motion picture camera. The bubble size can then be calculated on the following assumptions: (1) the size of the eruption at the surface is 50% larger than the bubble causing it (Botterill et al., 1966); (2) the estimated bubble size at the surface truly represents the size at the corresponding depth in a deeper bed (Geldart, 1967); and (3) the bubble size is independent of temperature. This method can be advantageously used where facilities for motion picture photography are available.

Based on the state of experimental work as well as its reliability, the frequency measurement method is recommended. As a second alternative, the equation of Mori and Wen given in Table 13.1 may be used.

14.11.2. Reactor Internals

We shall now consider the use of internals in promoting smooth operation. Several types of internals have been used: corrugated baffles, horizontal screens and perforated plates, horizontal tubes or rods, vertical tubes or rods, inclined surfaces, fixed packings, floating packings, and perforated tubes.

The most successful arrangement of internals is a plurality of vertical tubes immersed in the bed. A horizontal arrangement of tubes is also possible, but the vertical orientation is to be preferred, since it promotes smooth circulation of solids; the horizontal arrangement may lead to nonuniform heat transfer coefficients on the tube surface and excessive temperature gradients in the bed.

Depending on the size of the internals, two types of bubble behavior are possible: (1) the bubbles pass through the vertical region between adjacent internals, or (2) they enclose the internals and move upward along their sides. In practice, the first type of behavior, characterized by a less congested arrangement of internals, is more commonly encountered.

Volk et al. (1962) have reported extensive studies on the use of vertical internals for the first type of behavior.

They found that any large-scale reactor can be made to behave like a laboratory or small pilot plant reactor by adjusting the size and arrangement of internals in such a manner that the equivalent diameter of the reactor would correspond to the actual diameter of the smaller reactor. This equivalent diameter is defined as

$$d_e = \frac{\text{free cross-sectional area of the bed}}{\begin{array}{c}\text{wetted perimeter of all vertical}\\\text{surfaces exposed to the bed}\end{array}} \quad (14.67)$$

Examples

Results of studies on four reactions carried out in laboratory- and large-scale reactors are summarized in Table 14.2. It may be seen that as long as the equivalent diameter of the large-scale reactor calculated from Eq. 14.67 is nearly the same as the actual diameter of the smaller reactor, the performances of the two reactors are comparable. In the case of the chloromethanes reactor (Doraiswamy et al., 1974), results are available for three sizes, namely, a small pilot-scale reactor (20 cm), a semicommercial reactor (60 cm), and a full-size commercial reactor (180 cm). It can be seen from the table that in all three cases the same degree of conversion of methane to the two principal products (carbon tetrachloride and chloroform) is obtained, even though the equivalent diameters of the semicommercial and commercial reactors are not equal. This is because of some novel features incorporated in the larger reactor, such as provision for dissipating the jets. In the absence of such a provision, reactant gas streaks through the bed (often as pencils of flame), resulting in lower conversion and damage to internals.

Based on the observations presented here as well as some unreported results on the design of fluidized-bed reactors for a few other reactions, it is believed that a reactor of equivalent diameter between 15 and 40 cm would give good stable operation.

It may be noted that the equivalent diameter really corresponds to the space between the tubes. In the absence of experimental data, the bubble size can be taken as equal to this diameter. In other words, each vertical enclosure between internals can be regarded as a slugging laboratory reactor of diameter equal to d_e with a minimum slugging velocity given by Eq. 14.69, in which the tube diameter is replaced by d_e. Under these conditions there is an increase in bed expansion (Grace and Harrison, 1968a) and a decrease in the heat transfer coefficient between the fluidized bed and reactor wall (Botterill, 1966). The latter deficiency is usually made up by using the internals themselves for heat transfer.

TABLE 14.2. Performance Data on the Use of Internals for Typical Industrial Reactions

Reaction	Laboratory or Pilot Plant Reactor Diameter (cm)	Large Reactor Diameter (cm)		Conversion	Reference
		Actual	Equivalent		
Hydrocarbons and oxygenated hydrocarbons from H_2 and CO on iron (Hydrocol reaction)	18 30[a]	30 500	16 31	94% in both reactors 83–85% in both reactors	Volk et al. (1962)
Iron by reduction of iron oxide by H_2 (H–iron process)	15	90	14	80% in all three reactors	—
Hexachloroethane (HCE) from Cl_2 and C_2H_4 on activated carbon	10	40	12	88–90% (C_2H_4 to HCE) in both reactors	Doraiswamy et al. (1966)
Chloromethanes from Cl_2 and CH_4 on activated carbon	20	60 180	45[b] 90[b]	85–88% (CH_4 to carbon tetrachloride and chloroform in the ratio 3:1) in all three reactors	Doraiswamy et al. (1974)

[a]Extrapolated.
[b]The equivalent diameter mentioned refers only to the vertical inserts. The conversions obtained in a pilot plant reactor have been reproduced in the larger-scale reactors even though the equivalent diameter was increased from 20 cm to 90 cm. This is probably because of other unique features like multilevel gas distributors, which acted as horizontal inserts, and tapering of the reactor.

While the use of vertical internals as described above is generally recommended for many types of industrial reactors, it is interesting to note an entirely contrary finding reported by Agarwal and Davis (1966). These investigators propose that small-scale laboratory fluidized beds should indeed be baffled to simulate conditions in a large reactor. In other words, the smaller reactor should be made to behave like the contemplated larger one, instead of considering the larger reactor as an assemblage of vertical, parallel laboratory reactors. There is, however, no evidence to show that Agarwal and Davis's postulation has any practical significance.

Before concluding this brief treatment of internal baffles, a few pertinent observations are outlined with respect to vertical rods enclosed by bubbles (behavior of the second type) and the possibility of channeling between internals.

1. Bubble-enclosed rods lead to decreased bubble coalescence and increased bubble size uniformity in a given cross section (Grace and Harrison, 1968; Botton, 1970).

2. Too congested an arrangement of baffles can lead to channeling. According to Grace and Harrison (1969), for reasons that are not very clear, the gap between vertical surfaces becomes devoid of particles, with the result that gas channeling occurs through these gaps. In order

to avoid this channeling, a distance of at least 30 particle diameters should be maintained between any pair of adjacent verticals. If this distance is too large, then a large proportion of the bubbles will tend to pass between the gaps so that the reactor would shift to a configuration in which behavior of the first type prevailed.

In spite of the vast volume of literature on the subject of internals (Harrison and Grace, 1971), no entirely rational procedure is yet available for the sizing and arrangement of internals. The observations made above can only serve as guidelines.

14.12. SOME PRACTICAL CONSIDERATIONS

Having outlined the various models for calculating the conversion in a fluidized bed and the methods for determining the bubble diameter, we are now in a position to consider the practical aspects of fluidized-bed operation. In this section we shall cover slugging, defluidization, "gulf streaming," gas bypassing, premixing of gases, and reactor start-up.

14.12.1. Slugging

The rising velocity of a gas bubble increases with its size d_b, with the result that the bubble residence time is proportional to $1/d_b$. Thus in a large commercial reactor bubbles grow to a very large size, whereas in a small laboratory reactor they approach the diameter of the tube. The latter situation is referred to as slugging. This type of slugging, referred to as Type A, in which particles rain down between the slug and the sides of the vessel, is the more common type. In another type of slugging, Type B, alternate zones of dense and void regions are formed, which is an indication of poor bed performance. Here we shall be concerned with only the first kind of slugging.

From the results of Davidson and Harrison (1963) and Collins (1967), Hovmand and Davidson (1971) conclude that slugging behavior can be assumed when the effective bubble diameter exceeds one third of the bed diameter. According to Stewart and Davidson (1967), a bubble is recognizable as a slug only when d_b/d_t is greater than about 0.9. Under slugging conditions the rising velocity of the bubble is governed by the bed diameter rather than its own.

Generally, gas–solid contact, and therefore the conversion that can be obtained, is higher in a narrow slugging bed than in a freely bubbling bed of the same depth (Hovmand and Davidson, 1968). No fully acceptable equation is yet available for estimating the slugging limit. Stewart and Davidson (1967) suggest that slugging is likely to occur if

$$\frac{u - u_{mf}}{0.35 (d_t g)^{1/2}} > 0.2 \tag{14.68}$$

and the velocity for the onset of slugging (u_{ms}) is given by

$$u_{ms} = u_{mf} + 0.07 (d_{ts} g)^{1/2} \tag{14.69}$$

It will be seen from this condition that for coarse particles, say greater than $100 \, m\mu$ (for which u_{mf} would be relatively high), slugging is less likely to occur than for small particles. It also depends inversely on the square root of diameter of the reactor, that is, $u_{ms} \propto \sqrt{1/d_t}$.

Baeyens and Geldart (1974) have shown that Stewart and Davidson's criterion is valid for bed heights exceeding a critical value. For lower bed heights they proposed the following criterion, which satisfies the literature data reasonably well for bed depths of about 30 cm:

$$(u - u_{mf}) - 0.07 (g d_t)^{1/2} = (1.6 \times 10^{-3}) (L_c - L_{mf})^2 \tag{14.70}$$

where L_c is the critical bed height defined as the limiting height at which coalescence is complete and a stable slug spacing is achieved.

A slugging reactor gives higher conversions than a normal bubbling-bed reactor, and it is essential that this difference be recognized in designing a freely bubbling commercial reactor. The reduced gas–solid contact in the latter, and the consequent lowering in conversion, can be accepted for certain reactions for which high conversion is not an economic necessity— for example, the cracking of hydrocarbons. On the other hand, for reactions like the oxidation of o-xylene to phthalic anhydride or ammoxidation of propylene to acrylonitrile, where high conversions are essential for economic operation, slug flow in the commercial reactor may indeed be advantageous. The high conversions obtained by the use of vertical internals (see Section 14.11.2) may actually be due to the promotion of slug flow in the reactor. The use of slug-flow reactors as model reactors in fluidized-bed reactor development is described in Section 14.15.

14.12.2. Defluidization of Bed

During the operation of a fluidized-bed reactor care should be taken to avoid any sudden rise in pressure caused by malfunctioning of the valves downstream. Reactor pressure can also increase as a result of choking of downcomers from equipment by solid particles. If the pressure is permitted to rise unchecked, the mass velocity for minimum fluidization G_{mf} will increase at the same total flow rate G, and a stage may be reached when G_{mf} at the disturbed condition may actually become higher than G. When this happens defluidization of the bed occurs. For exothermic reactions whose selectivity is a strong function of temperature, this can have a detrimental effect on selectivity.

More important is the permanent effect this may have on the fluidization behavior of the particular batch of solids. In the interval when defluidization occurs, hot spots in certain regions of the bed cause permanent deactivation or caking of catalyst into sizes that cannot be fluidized.

Defluidization can also occur if the temperature of the bed is permitted to rise suddenly. The Froude number falls sharply with increase in temperature at the same pressure, and a stage may be reached when it would correspond to a velocity lower than that for minimum fluidization. In view of the rise in temperature necessarily associated with this type of defluidization, the danger of catalyst caking and permanent damage is even greater than in the case of defluidization caused by a rise in pressure.

The danger of defluidization, not commonly realized in plant operation, must be guarded against by proper control of temperature and pressure. Control of flow rate, though important, is not adequate, since, as pointed out earlier, we are really concerned with G_{mf}; it is this parameter that is affected, not the operating flow rate G.

Progressive decrease in fluidization efficiency leading ultimately to defluidization can also occur in batch fluidization when the catalyst is not regenerated continuously. The particle density increases with time as a result of deposition of pyrolytic carbon and tarry matter; this increases the velocity required for minimum fluidization, and may eventually lead to defluidization of the bed. For such a situation, the design should be based on the ultimate density of the used particles (and not fresh particles). In fact, the temperature–velocity relation could be followed to mark this region of operation. The pressure drop at this point falls sharply, with gases taking preferential channels. On increasing the velocity, the bed can be restored to the normal state of operation.

The phenomenon is also noticed sometimes during reactor start-up. In the beginning the reactor is normally fluidized with an inert gas until steady conditions corresponding to temperature, pressure, and so on are attained. The reactants are then introduced gradually to replace the inert gas flow. During this transition, there might sometimes occur an instability that would cause the bed to drop dead. The phenomenon, explained as being due to changes in the molecular weights and viscosities of the two gases, is akin to defluidization and is dangerous, especially when an exothermic reaction is involved. During the unstable state of the bed, the reaction might occur locally and help to aggravate the conditions that favor runaway, the return from which (to the normal state) seems improbable.

14.12.3. Minimizing Entrainment Losses

Most commercial fluid-bed reactors operate at velocities far in excess of that required for minimum fluidization. When no provision is made for bubble breakers or internals, large bubbles are formed in the bed. These bubbles rapidly rise to the surface and burst there, entraining all sizes of particles. Particles that have terminal velocities greater than the fluidization velocity gradually settle back into the bed, while smaller particles are carried away with the gas to the cyclones. Methods of minimizing entrainment losses are outlined in Section 13.9.

14.12.4. Premixing of Gases

In most practical situations, more than one reactant gas is involved. These gases may be either premixed outside and then fed to the reactor, or introduced separately. The mode of introducing the gases into the bed is dependent on reaction conditions. Thus, if the gases do not react in the absence of catalyst, they can be premixed outside the bed (see Section 12.8.2). In the case where the gases react in the homogeneous phase, it is desirable to introduce them separately in the bed. This becomes particularly important for gases like methane and chlorine, which form explosive mixtures.

These reactions normally involve successive steps, occur in the homogeneous phase, and are highly exothermic. A bed of inert solid particles fluidized by one of the reactant species introduced at the bottom grid plate can be advantageously used, with the second reactant species introduced at a certain height from the grid plate. The points pertinent to this situation are:

1. the choice of gas to be introduced at the bottom plate;
2. location of the inlet positions (one or more) for the second gas; and
3. the type of mixing that results and its influence on product distribution and reactor performance.

The fluid bed tends to favor the formation of end products. Thus chlorination of methane would produce more of carbon tetrachloride, and chlorination of cyclopentadiene more of the octachlorinated product. In many cases it is also desirable to produce the intermediates.

If it is assumed that the gas introduced at the bottom plate is well in excess of G_{mf}, then the second gas introduced directly into the bed will go preferentially into the bubble phase, and the extent of this gas in the emulsion phase will be restricted to the quantity passing from bubbles by diffusion. The emulsion gas being completely mixed (probably) and the reaction being very fast, the concentration would approach zero in that phase. As far as the bubble phase is concerned, the mixing would depend on the rate of coalescence of bubbles of the two reactant gases. The rate of coalescence will depend on the diameter of the bubbles, the nature and size of the particles used, and the properties of the gases. In view of the complex influence of these factors on the performance of the fluid bed, it has not yet been possible to quantify the level of mixing, and resort to experiments on the actual scale of operation seems the only way to do so.

14.12.5. Caking

The problem of catalyst caking is sometimes encountered in the operation of industrial fluid-bed reactors, and is intimately connected with the type of distribution and mode of introduction of the gases. This is especially important when highly exothermic reactions are to be carried out. For instance, in the operation of a pilot plant for the chlorination of ethylene to hexachloroethane (Doraiswamy et al., 1966) it was observed that malfunctioning of the reactor (in particular, poor fluidization just above the grid plate) resulted in catalyst cakes the size of the reactor diameter; understandably, they rendered the reactor inoperable.

14.12.6. Gulf Streaming

Another major difference between the operation of large and small fluidized beds is the phenomenon of gulf streaming. The term "gulf streaming" implies the formation of violent circulating currents induced by bubbles. These currents are not usually observed in small fluidized-bed reactors but can be a significant feature of commercial beds. In the case of large shallow beds, because of the high viscosity of the fluidized bed (10 poise or more), gulf streaming is not significant.

Davidson (1973) has described the results of some experiments carried out in a draft tube by Bromley and Burgess (1970), in which the circulation patterns were observed through the movement of a small sphere containing a transmitter. The results transmitted by the radio pill showed that the circulation rate would be of the order of 200 mm/sec. According to Davidson, circulation rates of at least this magnitude can be expected in larger beds. As a result, the bubble residence time would be smaller, with corresponding loss in conversion, It is therefore not possible to predict the performance of a larger fluidized bed from data obtained in a laboratory reactor, which highlights once again the need for large-scale pilot plant trials before designing a commercial reactor.

14.12.7. Gas Bypassing and Use of High Fluidizing Velocities

In Chapter 13 we pointed out the detrimental effect of bypassing in larger fluidized beds. Laboratory-scale experiments on the ozone reaction conducted by Hovmand and Davidson (1968) and Hovmand et al. (1971) show that the reactor approaches CSTR performance. This happens because of the violent mixing caused by small bubbles and the rapid mass exchange between the bubbles and the surrounding emulsion. On the other hand, with larger beds (460 mm in diameter) the bubble diameter sometimes approaches the diameter of the bed itself. The bubble residence time is thus drastically reduced (to a few seconds), leading to conversions much lower than in a CSTR. The assumption of CSTR performance on the basis of laboratory-scale data can therefore be completely misleading.

Kehoe and Davidson (1971) have also reported the complete breakdown of the bubbling regime in small (30–60 microns) catalyst particles at velocities exceeding $5u_{mf}$.

14.12.8. Attrition of Particles in the Bed

Attrition of solids in a fluid bed might be the result of a variety of processes, such as decrepitation, thermal shock, abrasion among solid particles and in circulation ducts, distributor jet impingement, and fragmentation during reaction, and represents one of the major sources of particulate loss. Comparatively little attention has been paid to this problem, and there is hardly any correlation available for design purposes. Among the few studies reported, the graphical correlation for the approximate rate of attrition in a cyclone presented by Zenz (1971) appears useful. Gwyn and Colthart (1969) also developed a relationship for the attrition rate of a few catalysts (> 75 microns), and suggested the following equation for silica–alumina catalysts:

$$(\text{mass fraction of solids attrition}) = md_p^{-2/3}t^n$$

$$(14.71)$$

where t represents the duration of fluidization and m and n are constants dependent on the operating conditions and the catalyst particle, respectively. This study covers only the influence of particle size on the rate of attrition and ignores the influence of operating conditions and gas distributor design.

A more detailed analysis covering the influence of gas velocity, size and shape of particles, mechanical strength of solids, grid design, and duration of fluidization has been reported by Blinichev et al (1967). Based on the experimental data on solids like silica gel, limestone, anthracite, and chalk, they derived the following empirical correlation:

$$\begin{pmatrix} \text{mass fraction of} \\ \text{solids attrition} \end{pmatrix} = 1750(CF)(SF)\,\text{Fr}^{1.3}$$

$$\times \left(\frac{u_t}{d_p}\right)^{0.6} \left(\frac{\sigma \rho_g}{u^2}\right)^{-1.3} \qquad (14.72)$$

$$\times \left(\frac{\rho_g}{\rho_s}\right)^{1.3} 10^{\left(\frac{0.21 d_{or}}{d_p - 18f}\right)}$$

where (SF) is the shape factor, σ the limiting compression strength of the solid, and (CF) a correction factor equal to 1 for a batch-operated fluidized bed and to 0.849, 0.947, and 0.967, respectively, for single-stage, two-stage, and three-stage continuously operated fluidized beds. The correlation, tested for the following ranges of parameter values,

$$40 < \text{Fr} < 3000, \qquad 0 < \frac{u_t}{d_p} < 6 \times 10^6,$$

$$2 < \frac{d_{or}}{d_p} < 5, \qquad 0.006 < f < 0.05$$

$$1.4 \times 10^5 < \frac{\sigma}{u^2 \rho_g} < 1.5 \times 10^7$$

is the best available and is recommended for use in design.

14.12.9. Start-Up of Fluidized-Bed Reactors

Three main factors have to be considered: (1) preheating of the bed to the reaction temperature, (2) conditioning of catalyst (dehydration, activation, de-airing, etc.) and (3) mode of introduction of reactant gases. The procedure to be followed for each of these depends greatly on the specific process under consideration, but some general comments can be made.

1. Preheating of the reactor bed can be achieved by one or more of the following methods:
   a. fluidizing the bed with hot flue gas;
   b. burning a suitable fuel in the bed under fluidizing conditions;
   c. circulating a preheated inert (with respect to the system) through the bed;
   d. passing steam, Dowtherm, or any other heat transfer fluid through the heat transfer rig (tubular inserts, coils, jacket, etc.) of the reactor which is normally used for heat extraction; and
   e. electrical induction or resistance heating.

2. In the case of some batch-operated fluidized-bed reactors, the catalyst is activated inside the reactor just before the reaction is started. Some dry gas, inert to the reaction, is passed through the reactor until the concentration of an undesirable substance in the bed (e.g., water vapor) is brought down to some predetermined level.

3. The preheated bed is maintained under fluidized conditions with inert gas. The flow rates are adjusted so as to get the desired hydrodynamic conditions in the reactor, as noted by differential pressure and temperature readings. Downstream equipment is started or kept in a state of readiness to receive the product vapors. When the plant as a whole is tuned to the desired conditions, the reactant gases are started at low rates in the preferred order and gradually increased to the normal levels. During the start-up, care is taken to keep the linear velocity of the gas at various points at the normal levels. When the reactant flow rate is gradually increased, the flow rate of the inert is correspondingly reduced, so that the hydrodynamic conditions are not disturbed.

Initially all the controls are taken to manual, and when the preliminary adjustments have been completed they are switched to automatic.

14.13. FLUID-BED MODELING AND DESIGN: AN APPRAISAL

14.13.1. General Comments

Several models of the fluid-bed reactor have been presented in this chapter. Their main features are summarized in Table 14.3. The choice of the right model is often difficult, since it is by no means clear which model would correspond best to the widely differing physical situations encountered. Most of these models are in surprisingly good accord with one another and with the experimental findings on certain major points such as conversion and selectivity. However, the more important consideration is the predicted concentration profile, and not just the exit concentration (or selectivity); and here we are confronted with sharp disagreement between certain models and the experimental findings. The experimen-

TABLE 14.3. Chief Features of Some Bubbling-Bed Models for Evaluating the Conversion

Model	Bubble Size	Bubble–Cloud Phase	Emulsion Phase	Mass Transfer From Bubble to Cloud–Emulsion Phase	Mass Transfer From Cloud to Emulsion Phase	Phase in Which Reaction Occurs	Remarks
Davidson–Harrison (1963)	Single effective diameter	No cloud; complete mixing in bubble	No mixing, or complete mixing	Yes, to emulsion (through a film inside the bubble)	—	Emulsion; contributions from both bubble- and emulsion-phase gases considered	Simple model with minimal computations
Partridge–Rowe (1966)	Average size estimated in each section from X-ray data on bubble size distribution	Complete mixing in the bubble–cloud phase	No mixing	Yes (according to the rigid-sphere model)	—	Cloud and emulsion; contributions from both bubble- and emulsion-phase gases considered	Experimental X-ray data essential; cumbersome calculations
Toor–Calderbank (1967)	Increase in size due to coalescence accounted for by an empirical equation	No cloud; complete mixing in the bubble	No mixing, or complete mixing	Yes, to emulsion (through a film inside the bubble)	—	Emulsion; contributions from both bubble- and emulsion-phase gases considered	Experimental data on bubble size distribution are required.
Kunii–Levenspiel (1968a,b)	Single effective diameter	Plug flow in bubble phase	No flow through cloud–emulsion phases	Yes, to cloud (through a film inside the bubble)	Yes (no circulation)	Contributions from bubble-phase gas alone considered	Elegant model requiring a single experimental value
Kato–Wen (1969)	Increase in bubble size accounted for by dividing the bed into several compartments and estimating the size in each compartment by an empirical equation	Complete mixing in the bubble–cloud phase	Complete mixing	Yes, from bubble–cloud to emulsion	—	Emulsion; contribution from bubble-phase gas alone considered	No backmixing between compartments; elaborate computation with no adjustable parameters

Hovmand–Davidson (1971)	Single effective diameter	No cloud; complete mixing in bubble	No mixing or complete mixing	Yes, to emulsion (through films inside and outside the bubble)	—	Emulsion; contributions from both bubble- and emulsion-phase gases considered	Simple model with minimal computation
Orcutt–Carpenter (1971)	Increase in size due to coalescence accounted for by a hydrodynamic simulation method	No mixing	No mixing	Yes, to cloud (through a film inside the bubble)	Yes (through a film inside the cloud) with no circulation	Cloud and emulsion; contributions from both bubble- and emulsion-phase gases considered	Reactor divided into several sections; solution requires elaborate computation
Fryer–Potter (1972a)	Single effective diameter	Plug flow in bubble and cloud–wake phases	No mixing	Yes, to cloud (through a film inside the bubble)	Yes (circulation allowed for)	Contributions from both bubble- and emulsion-phase gases considered	Elegant model requiring a single experimental value
Fryer–Potter (1972b)	Increase in size of bubble accounted for by an empirical relationship	—	—	Fryer–Potter (1972a) model	—	—	—
Chiba–Kobayashi (1973); also Miyauchi–Morooka (1969)	Increase in bubble size with bed height accounted for by an empirical equation	Complete mixing in the bubble-cloud phase	No mixing	Yes (through films on the bubble and emulsion sides)	—	Bubble and emulsion; contributions from both bubble- and emulsion-phase gases considered	Adsorption on catalyst is also accounted for through a partition function (m), which appears in the equation for the overall mass transfer coefficient
Behie–Kehoe (1973)	Orcutt's model (1960) in the bubbling bed, with jetting zone accounted for, based on the experiments of Behie et al. (1971)	Plug flow	Complete mixing	Yes	—	—	Suitable for fast reactions; the model accounts for reaction in the grid zone by accounting for mass transfer from jet to emulsion phase

(continued)

359

TABLE 14.3. (continued)

Model	Bubble Size	Bubble Cloud Phase	Emulsion Phase	Mass Transfer		Phase in Which Reaction Occurs	Remarks
				From Bubble to Cloud Emulsion Phase	From Cloud to Emulsion Phase		
Mori–Wen (1976)	Increase in bubble size with bed height accounted for by a modified empirical equation of Mori and Wen (1976)	Rest as per Kato–Wen (1969) bubble-assemblage model					This represents a modification of the Kato–Wen (1969) model. Thus jetting region is included in the model besides the improved methods in estimating bubble diameter and sizing of compartments
Werther (1977)	Increase in bubble size accounted for	Plug flow in bubble; mixing in cloud phase	No flow through cloud–emulsion phase	Yes, to cloud (related through flux and interchange coefficient)	Yes, to emulsion (related through flux and interchange coefficient)	Cloud and emulsion	Simple model with minimal computation
Kuhne–Wippern (1980)	Increase in bubble size implicitly accounted for	No cloud phase; dispersion in bubble phase	Dispersion in emulsion phase	Yes, to emulsion	Transfer from emulsion to solid surface	No reaction considered	Model considers adsorption of gas on solids, thereby affecting residence time behavior of system
Jayaraman–Kulkarni–Doraiswamy (1981)	Single effective bubble diameter	Plug flow in bubble and cloud-wake phase	complete mixing	Yes, to cloud through a film inside the bubble	Yes	—	Similar to Fryer–Potter model giving similar results but computationally simpler

360

tal concentration profiles of Chavarie and Grace (1975) and the minima in the profiles observed by Fryer and Potter (1976) lend strength to this ambiguity. Within the framework of the assumptions made, these models yield results that may or may not be in agreement with the experimental findings. When there exists a correspondence between the assumptions in a model and the actual experimental conditions, that model, among all others, gives the best predictions; but the experimental conditions differ over a wide range and satisfy different assumptions. Different models are thus expected to simulate the fluid-bed behavior under different sets of conditions. It is futile then to compare these models for a given set of conditions. Since the reactor models are ultimately to be used in the *a priori* design of large-scale industrial units, an inquiry into the nature of the assumptions and approximations associated with the models is clearly indicated.

14.13.2. Major Assumptions and Their Validity

1. As stated earlier, several assumptions have been made, many of which are common to most models. Thus the two-phase theory of fluidization discussed in Section 13.2.2 is assumed to be valid at all conditions. In addition to the limitations of this theory pointed out in that section, the experimental results of Werther (1974) indicate that this theory is valid only for tall beds. For beds of larger diameter, the experimental results of several investigators in two- and three-dimensional beds suggest that the visible bubble flow has the smallest value at the bottom of the bed, and increases with height from the grid to that predicted by the two-phase theory at the top of the bed. Proper accounting of bubble flow thus seems necessary for a rational design.

2. The second common assumption concerns the rise velocity of bubbles (Section 13.3.2). In analogy with gas–liquid systems, the rise velocity is calculated by using the Davies–Taylor relationship. This, however, ignores the inviscid nature of the bed. Also, even in gas–liquid systems the relation is known to be valid only for relatively large bubbles. The influence of size and shape of particles and the fluidization number on the rise velocity of bubbles and the disparity between the values calculated with Davidson's equation and those observed experimentally suggest that we should at least be aware of these limitations in applying this equation to reactor modeling.

3. Particle size has a profound influence on the state of fluidization. As already pointed out in Section 14.1, the fluidization characteristics of particles have been classified into four groups by Geldart (1973). A typical example of group A particles is the FCC catalyst, and most of the reactor models are applicable to this group of particles. For group B particles, which are somewhat larger in size and density and are also commonly encountered, the application of the two-phase theory has been questioned by some investigators (Geldart, 1972, 1973; Baumgarten and Pigford, 1960). The two-phase theory over-predicts the throughflow via bubbles, which actually decreases with the increase in average particle diameter in the bed (Baeyans and Geldart, 1973). The experimental data of Baeyans and Geldart show the bubble diameter to be independent of d_p, thus indicating that the bubble frequency also must decrease with increase in d_p.

    Bubble-growth studies for the two groups of particles (A and B) show quite different characteristics. The rate of growth decreases with height for group A particles and shows considerable influence of bubble splitting. Bubble growth for group B particles, however, is dominated by the coalescence phenomenon. The statistical model proposed for this group of particles shows good agreement with the X-ray measurements of Botterill et al. (1966) and yields a simple correlation for three-dimensional bubbling beds. The net inflow of gas from the dense phase into the bubble, and under certain conditions the net outflow of gas observed during the coalescence process, seem to be of major importance during the modeling of fluid-bed reactors.

4. The various interphase mass transfer mechanisms, discussed in Chapter 13 and employed in the formulation of models in the present chapter also need to be evaluated critically. Drinkenburg and Rietema (1972) have discussed some of these mechanisms at length. The magnitude of the coefficient is usually in the range of 1.0–3.0 cm/sec for bubble sizes varying from 5 to 8 cm. Over a dozen mass transfer correlations are available. From the experimental data of Kobayashi et al. (1967a), Calderbank et al. (1976), Rietema and Hoebink (1976), and Bohle and van Swaaij (1978), it appears that the correlation of Calderbank et al. (1976) rep-

resents the results most satisfactorily. The relationship between the mass transfer coefficient and the bubble diameter based on the experimental data just mentioned can be represented by a band, as shown in Figure 14.10. The Kunii–Levenspiel correlation falls (not shown) below the band, as do most others.

The assumption in all these mechanisms that the interphase mass transport in the bed is well represented by transport across a single isolated bubble seems to be highly restrictive. Also, transport as a result of the unsteady nature of the interphase and additional factors like transpiration (Brian and Hales, 1969) and oscillatory stretching of the interface (Toei et al., 1969) has been ignored.

In a freely bubbling bed, the velocities in the bubble and emulsion phases differ in magnitude and direction. The viscous dense phase thus offers resistance to the counterflowing bubble phase. Since the differently directed velocity vectors form a couple of forces at the interphase, the layers of fluid adjoining the interphase begin to rotate. At sufficient intensity these forces pierce the surface, bring about a disruption, and can lead to an abrupt increase in transport. Interphase transport thus depends on the flow field existing, and within the same bed different mechanisms could be operative.

Chavarie and Grace (1976) have conducted some experiments in two-dimensional beds to discern the mechanism of transport. The results are, however, inconclusive. The experimental results of Rietema and Hoebink (1976) show an increase in mass transfer with increased bubble diameter. The inconsistency between

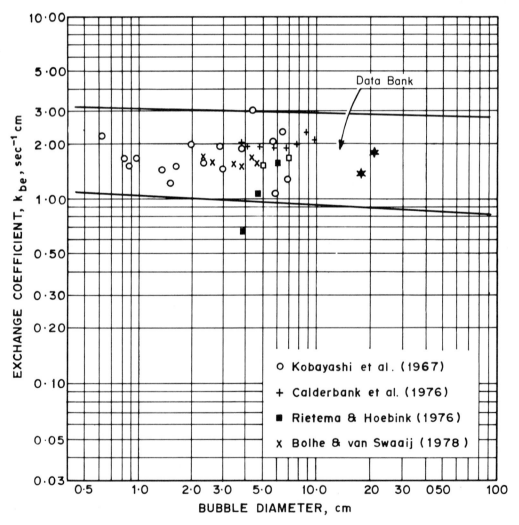

Figure 14.10. Ranges of the mass transfer coefficient between the bubble and emulsion phases for varying bubble diameters.

a crucial experimental observation like this and the theoretical prediction points to the need for more work in this area.

Examples of Fluid Bed Modeling

Based on the comments presented above it is clear that no single model is likely to represent the behavior of a fluid bed for the variety of physical situations encountered. Under a given set of conditions any one (or more) of these models may predict the gross results satisfactorily, but such a matching could well be fortuitous. A summary of the results on the modeling of a number of reactions by different models is presented in Table 14.4. The errors in the predictions of many phenomena (interphase transport, reaction in the phases, throughflow, etc.) occurring simultaneously in the fluid bed might well cancel each other, giving a gross overall matching.

14.13.3. Recommended Bubbling Bed Model

For purposes of preliminary design, however, it would be desirable to choose a model that is satisfactory in many ways and requires relatively little input data in spite of its basic limitations as outlined above. The uniform size models thus logically commend themselves, particularly since most fluid-bed reactors use internals, which are believed to lead to bubbles of practically uniform size. Even if there is a reasonable variation in size, an average size corresponding to that at the midpoint between the grid and bed surface may be used (Fryer and Potter, 1972b). The Kunii–Levenspiel model is recommended not only because it is conceptually simple and is based on an average bubble size, but also because the interphase mass exchange mechanism assumed is logical and the predicted concentration profiles match the experimental profiles over a fairly wide range of conditions (Chavarie and Grace, 1975). Modifications of the Kunii–Levenspiel model, such as those of Fryer and Potter (1972 a, b) and Jayaraman et al. (1980), may be used when reaction in the bubble phase cannot be ignored, or when there is definite evidence of significant variation in bubble size. Behie and Kehoe's model

(1973) is recommended for very fast reactions, which tend to be grid controlled. Nonisothermal and three-phase models, while in theory more correct than the simpler models, do not appear to be of any practical relevance at this stage.

14.13.4. A Model Based on Analogy with Gas–Liquid Reactors

Werther (1980) has proposed a model that, though not as rigorous as some of the countercurrent backmixing models, offers certain advantages in the scale-up of a fluid-bed reactor. The model visualizes the fluid bed as consisting of two regions (bubble and emulsion phases) and describes the catalytic reaction occurring in a fluid bed as an absorption from the bubble phase with subsequent reaction in the emulsion phase. Akin to gas–liquid absorption, the model assumes a fictitious film in the emulsion phase where all the resistance to transfer lies. The model involves a knowledge of the interphase transport coefficient, specific interfacial area, and volume fractions of the phases.

The conservation equations for the species can be readily written and for a first-order reaction the simple equation presented in Table 14.5 can be obtained. Several parameters involved in this model have been empirically correlated. The equations, which take into account the geometry of the reactor and are thus suitable for scale-up of reactors, are summarized in Table 14.5. Although this model lacks the rigor of the bubbling-bed models, it is conceptually simple, and can be used for preliminary reactor sizing. The model has subsequently been extended to include more complex reaction situations (Werther and Hegner, 1981).

14.13.5. Example: Design of Reactor for Aniline: Comparison of the Conversions Computed from the Two Recommended Models

Calculate the conversion in a fluid-bed reactor for the hydrogenation of nitrobenzene to aniline on a silica-based catalyst using (1) the Werther model and (2) the Kunii–Levenspiel model. The following data may be assumed:

$$
\begin{array}{lll}
u = 0.3 \text{ m/sec}, & u_{\text{mf}} = 0.02 \text{ m/sec}, & D = 9 \times 10^{-5} \text{ m}^2/\text{sec} \\
T = 270°\text{C}, & k_v = 8.77 \exp\left(\dfrac{-2631}{R_g T}\right), & W = 30.5 \text{ tonnes} \\
L_f = 2.9 \text{ m}, & \rho_s = 2.2 \text{ g/cm}^3, \quad d_t = 3.55 \text{ m}, & Q = 2.4643 \text{ m}^3/\text{sec (reaction condition)}
\end{array}
$$

TABLE 14.4. Examples of Fluid-Bed Modeling

Reaction	Models Tested[a]	Remarks	Reference
1. Decomposition of ozone	ODP; PR; KW; KL	KL model found best; however, purely on statistical grounds none of these models accurately represents the data; modifications and extensions of the conventional two-phase models are suggested.	Chavarie and Grace (1975)
2. Decomposition of ozone according to data of Fryer (1974)	ODP; ODP with Kunii–Levenspiel equation for gas exchange; Fryer–Potter	Minima in the axial concentration profiles as predicted by Fryer–Potter model verified in this study; similar observations are reported by Schugerl (1974)	Fryer and Potter (1976)
3. Decomposition of ozone according to data of Hovmand et al. (1971)	Grid model of Behie and Kehoe; simple bubble model	The grid model and the simple bubble model were compared to show the influence of the jetting region.	Behie and Kehoe (1973)
4. Isomerization of n-butenes over a silica–alumina catalyst	PR	Model predictions were compared with experimental results.	Yates et al. (1970)
5. Hydrogenolysis of n-butene over a nickel–silica catalyst	ODP; KW; PR; KL	All models show good match with experimental results.	Shaw et al. (1974)
6. (a) Hydrogenation of ethylene according to data of Lewis et al. (1959)	Miyauchi model	Incorporation of a dilute phase was shown to improve prediction.	Miyauchi and Furusaki (1974)
(b) Decomposition of ozone according to data of van Swaaij and Zuiderweg (1972, 1973)			
7. Hydrogenolysis of n-butene over a nickel–silica catalyst	ODP; PR; KW; modified KW	A model discrimination criterion was developed; the KW model appears to provide the best fit.	Shaw et al. (1974)
8. Ammoxidation of propylene to acrylonitrile in an industrial fluidized bed	Tasev–Georgieva–Doichev model	A mathematical model was proposed; calculated and experimental total conversions show good agreement.	Tasev et al. (1974)

364

Reaction	Model	Comment	Reference
9. (a) Decomposition of nitrous oxide according to data of Shen and Johnstone (1955) (b) Synthesis of acrylonitrile according to data of Ogasawara et al. (1959) (c) Hydrogenation of ethylene according to data of Lewis et al. (1959) (d) Oxidation of ammonia according to data of Massimilla and Johnstone (1961) (e) Decomposition of ozone according to data of Kobayashi et al. (1969) (f) Decomposition of ozone according to data of Fryer (1974)	ODP; KW; KL; Mori–Wen model	The KW model as modified by Mori and Wen was shown to agree with experimental results over a wide range for slow and fast reactions.	Mori and Wen (1976)
10. Catalytic cracking of methyl cycohexane over silica–alumina catalyst	Tone–Seko–Maruyama–Otake model	A simulation model was developed that can predict the conversion and selectivity in a fluid bed with catalyst fouling.	Tone et al. (1974)

[a]The short form ODP denotes the Orcutt–Davidson–Pigford model, PR the Partridge–Rowe model, KW the Kato–Wen model, KL the Kunii–Levenspiel model.

TABLE 14.5. Conversion Equation for First-order Reaction and Empirical Correlations for Parameters in Werther's Model

Conversion Equation

$$x_A = 1 - \exp\left(-\left|\frac{(\beta^{-1}-1)E + \tanh E}{(\beta^{-1}-1)E \tanh E + 1}\right| E\alpha\right) \tag{1}$$

where

$$E = \frac{\sqrt{k_v D_e}}{k_{be}}, \qquad \alpha = \frac{K_{be} a_b L_f}{(u - u_{mf})}, \qquad \beta = \frac{\text{volume of film}}{\text{volume of emulsion}} = \frac{a\delta_f}{(1-\delta)} \quad {}^b$$

Empirical Correlations for the Parameters of Eq. 1

$$\alpha = \frac{L_f}{A\psi_1(d_t)\psi_2(L_f, h^*)} \tag{2}$$

where

$$A = \begin{cases} 0.156 & \text{for sand} \\ 0.055 & \text{for porous catalyst carriers (silica–alumina)} \end{cases}$$

$$\psi_1(d_t) = \begin{cases} 0.64, & d_t \leqslant 0.1 \text{ m} \\ 1.60(d_t)^{0.4}, & 0.1 \leqslant d_t \leqslant 1 \text{ m} \\ 1.60, & d_t \geqslant 1 \text{ m} \end{cases}$$

$$\psi_2(L_f, h^*) = \begin{cases} \dfrac{L_f}{0.18[1 - (1 + 6.84L_f)^{-0.8}]}, & L_f \leqslant h^* \\[4mm] \dfrac{L_f}{0.18[1 - (1 + 6.84h^*)^{-0.8}] + (1 + 6.84h^*)^{-1.8}(L_f - h^*)}, & L_f > h^* \end{cases}$$

$$a_b = \frac{29.1(u - u_{mf})(\psi_1\psi_2)^{-1}}{[1 + 27.2(u - u_{mf})]^{1/2}} \tag{3}$$

$$K_{be} = \frac{3.44 \times 10^{-4}}{A}[1 + 27.2(u - u_{mf})]^{1/2} \tag{4}$$

$$\delta = \frac{2.47(u - u_{mf})}{[1 + 27.2(u - u_{mf})]^{1/6}} \frac{\psi_3(L_f, h^*)}{\psi_1(d_t)} \tag{5}$$

where

$$\psi_3(L_f, h^*) = \frac{0.37}{L_f}\left[(1 + 6.84L_f)^{0.4} - 1\right], \qquad L_f \leqslant h^*$$

$$= \frac{0.37}{L_f}\left[(1 + 6.84h^*)^{0.4} - 1\right] + (1 + 6.84h^*)^{-0.6}\left(1 - \frac{h^*}{L_f}\right), \qquad L_f \geqslant h^*$$

[a] In these equations h^* is the height above the distributor at which bubble growth stops.
[b] In this table and in Section 14.13.5 δ_f represents the film thickness as against the notation δ generally employed throughout the book.

Solution

1. *Conversion using Werther's model:*

The conversion in a fluid-bed reactor may be computed from the equation presented in Table 14.4:

$$x_A = 1 - \exp\left\{-\left[\frac{(\beta^{-1}-1)E + \tanh E}{(\beta^{-1}-1)E\tanh E + 1}\right]E\alpha\right\}$$

The various quantities appearing in this equation are defined in Table 14.5 and can be obtained as follows.

Calculation of α. The parameter α can be calculated by using the equation $\alpha = L_f/A\psi_1\psi_2$. The parameters A, ψ_1, and ψ_2 are first obtained as

$$A = 0.055 \qquad \text{(for silica-based catalyst)}$$

$$\psi_1 = 1.6 \qquad \text{(since } d_t = 1 \text{ m)}$$

$$\psi_2 = \frac{2.9}{0.18\left[1 - (1 + 6.84 \times 2.9)^{-0.8}\right]} = 17.667$$

Substituting these values in the equation, we obtain

$$\alpha = \frac{2.9}{0.055 \times 1.6 \times 17.667} = 1.865$$

Calculation of E. The parameter E can be calculated by using the equation $E = \sqrt{k_v D/K_{be}}$. The parameter K_{be} is obtained from the empirical correlation presented in Table 14.5 as

$$K_{be} = \frac{3.44 \times 10^{-4}}{0.055}\left[1 + 27.2(0.3 - 0.02)\right]^{1/2} = 0.018$$

The parameter E is then obtained as

$$E = \frac{(0.7655 \times 9 \times 10^{-5})^{1/2}}{0.018} = 0.452$$

Calculation of β. The parameter β can be calculated by using the equation $\beta = a_b(\text{film thickness})/(1 - \delta)$. This requires a knowledge of the interfacial area a_b, film thickness, and volume fraction of bubbles in the bed, which can be obtained from the correlations

$$a_b = \frac{29.1(0.3 - 0.02)(1.6 \times 17.667)^{-1}}{\left[1 + 27.2(0.3 - 0.02)\right]^{0.5}} = 0.09819$$

$$\text{film thickness} = \frac{D}{K_{be}} = \frac{9 \times 10^{-5}}{0.0183} = 4.9022 \times 10^{-3}$$

The calculation of the volume fraction of bubbles requires a knowledge of ψ_3, which can be obtained by using the empirical correlation

$$\psi_3 = \frac{0.37}{2.9}\left[(1 + 6.84 \times 2.9)^{0.4} - 1\right] = 0.3022$$

The volume fraction can now be obtained as

$$\delta = \frac{2.47(0.3 - 0.02)(0.3022)}{\left[1 + 27.2(0.3 - 0.02)\right]^{1/6} \times 1.6} = 0.09125$$

Knowing the interfacial area, film thickness, and volume fraction of bubbles, we can calculate β as

$$\beta = \frac{(0.09819)(4.9022 \times 10^{-3})}{1 - 0.09125} = 5.2973 \times 10^{-4}$$

The parameters α, β, and E are thus evaluated. We can now proceed to calculate the conversion:

$$x_A = 1 - \exp$$

$$\left[-\frac{\left(\dfrac{1}{5.2973 \times 10^4} - 1\right) \times 0.452 + \tanh(0.452)}{\left(\dfrac{1}{5.2973 \times 10^4} - 1\right) \times 0.452 \times \tanh(0.452) + 1}\right]$$

$$\times (0.452 \times 1.865)$$

$$= 0.863$$

2. *Conversion using the K–L model:*

In the calculation of conversion using Werther's model the parameter d_b did not appear explicitly. As a result, this parameter was not specified in the data. The K–L model, however, requires a knowledge of this parameter for the calculation of conversion. In order that the two models be compared on the same basis, we assume that the volume fraction of the bubbles in the bed is the same as obtained in Werther's model. Thus $\delta = 0.0912$. Using this value of δ we obtain d_b as follows:

$$\delta = \frac{u - u_{mf}}{u_b} = \frac{0.3 - 0.02}{u_b} = 0.0912$$

or

$$u_b = 3.0685$$

The rise velocity of a single bubble is then obtained as

$$u_b = u - u_{mf} + u_{bs}$$

$$3.0685 = 0.3 - 0.02 + u_{bs} \text{ or } u_{bs} = 2.7884.$$

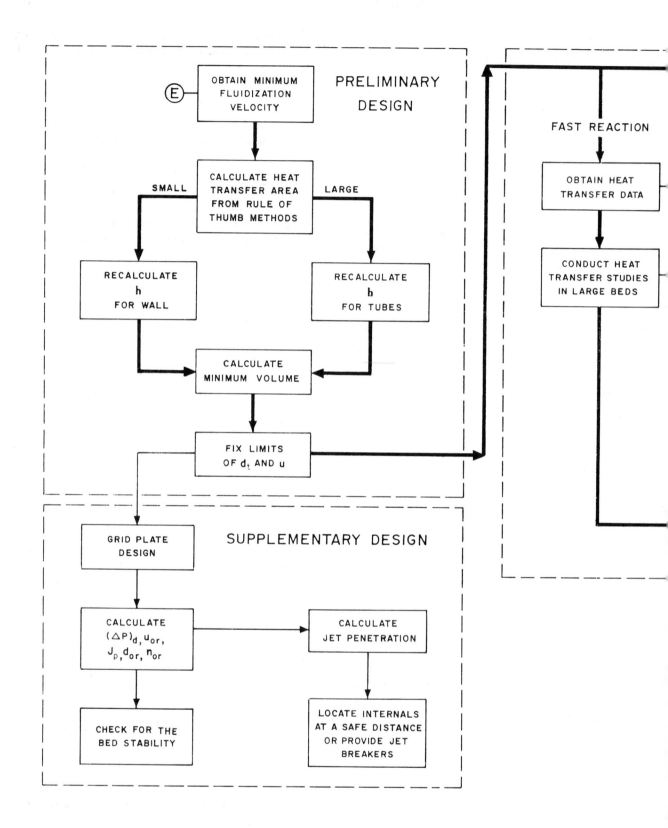

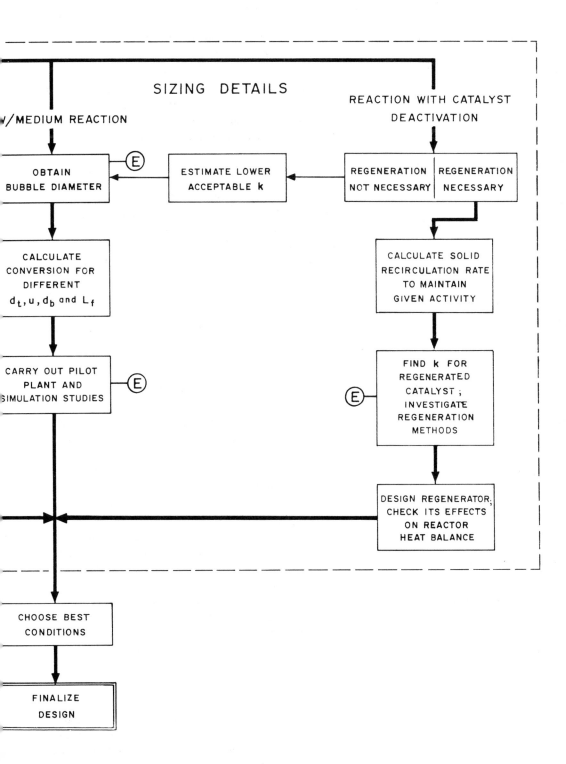

Figure 14.11. Morphology of fluidized-bed reactor design.

The diameter of the bubble is now calculated as

$$u_{bs} = (0.711)(9.8 \times d_b)^{1/2} = 2.7884$$

or

$$d_b = 1.5679$$

Once the parameter d_b is obtained, we can proceed to calculate the following parameters (see Section 14.3):

$$K_{bc} = 4.50\left(\frac{0.02}{1.5679}\right) + 5.85\frac{(9 \times 10^{-5})^{0.5}(9.8)^{0.25}}{(1.5679)^{1.25}}$$

$$= 0.11338 \ \text{sec}^{-1}$$

$$K_{ce} = 6.78\left[\frac{0.4 \times 9 \times 10^{-5} \times 3.0685}{(1.5679)^3}\right]^{0.5}$$

$$= 0.03629 \ \text{sec}^{-1}$$

$$s_{bb} = 1 \times 10^{-3}$$

$$s_{cb} = (1 - 0.4)\left(\frac{3 \times 0.02/0.4}{2.7884 - 0.02/0.4} + 0.33\right)$$

$$= 0.2308$$

$$s_{eb} = \frac{(1 - 0.4)(1 - 0.912)}{0.0912} - s_{cb} - s_{bb} = 5.7434$$

The modified rate constant can be obtained by using Eq. 14.13:

$$K_f = 1.4649 \times 10^{-4}$$

The conversion can now be obtained as

$$x_A = 1 - \exp\left(-\frac{K_f W}{Q}\right)$$

$$= 1 - \exp\left[-\frac{(1.4649 \times 10^{-4}) \times (30.5 \times 10^3)}{2.4643}\right]$$

$$= 0.837$$

Comments

It is interesting to note that both models predict conversion in almost the same range. However, the assumption of the same volume fraction of bubbles in the bed as the basis of comparison of the two models leads to a value of d_b that seems unrealistic.

14.14. A STRATEGY FOR REACTOR DESIGN AND DEVELOPMENT

Modeling of a fluid-bed reactor for predicting conversion and selectivity is just one of the steps in the complete design and development of the reactor. In fact, in some cases, such as the chlorination of methane, modeling to predict conversion is not involved in the overall strategy of design, since heat transfer and mechanical considerations assume overriding importance. As pointed out in Section 14.13.3, for the purpose of pure modeling, the Kunii–Levenspiel model (or the grid model of Behie and Kehoe, 1973) can be conveniently used.

The overall strategy for the design of a fluidized-bed reactor involves a combination of theoretical predictions and experimental determinations on both bench and pilot plant scales. The complete procedure, outlined in Figure 14.11, consists of the following steps.

1. Select a suitable particle size range, and estimate (or determine experimentally) the minimum fluidization velocity.

2. Make a preliminary study of the heat transfer problem involved and decide whether cooling tubes are necessary or jacket cooling is adequate. If cooling tubes are necessary, decide on their orientation. Vertical placement is usually to be preferred, since the tubes can then also be regarded as internals (see step 4).

3. Depending on decision in step 2, calculate the minimum bed volume to contain the exchange surface and estimate the limits of bed diameter and gas velocity.

4. If vertical tubes are used for heat transfer, calculate the number of tubes of the diameter chosen in step 2 that will give the desired equivalent diameter, and compare with the number required for heat transfer alone.

5. a. Measure the average bubble diameter experimentally, using a cold model.

   b. If reaction is very fast (i.e., if heat transfer is the determining factor) make experimental heat transfer studies in a prototype fluidization unit (without reaction).

6. Make a tentative selection of bed dimensions by using any of the bubbling-bed models or from heat transfer results.

7. Estimate the freeboard height and the size of the cyclone dip-leg.

8. Design the grid plate and check for bed instability (i.e., channeling).

9. Finalize design based on practical and economic considerations—by carrying out pilot plant trials if necessary.

10. In the case of a rapidly deactivating catalyst:

  a. Calculate the solids recirculation rate from a knowledge of the mean allowable activity of the existing catalyst.

  b. Estimate the reactor size from a knowledge of the deactivation and regeneration kinetics (in addition to that of the main reaction).

  c. Calculate the size of the regenerator necessary to reactivate the catalyst back to the stipulated level.

  d. Decide on the circulation system to be used and estimate the height necessary to provide the driving force for the circulation rate calculated in step 10a.

14.15. USE OF SLUGGING-BED REACTORS AS MODEL REACTORS

Bubbles being an inevitable feature of fluidized-bed reactors, a laboratory reactor characterized by stable and predictable bubble behavior should provide the best means of obtaining scale-up data. Hovmand and Davidson (1968) suggest that model studies might be based on laboratory reactors operated intentionally and fully in the slugging regime since these reactors ensure stable bubble behavior. Another advantage of slug-flow reactors is that the size of the slug can be predicted from theory. Thus, if a single-effective-diameter model (such as the Davidson or Kunii–Levenspiel model) is to be used for design, experimental determination of bubble size would no longer be necessary, provided it can be ensured that the larger reactor would also operate in the slugging regime. There is some evidence to show that reactors with vertical baffles conform to this requirement.

14.15.1 Hovmand–Davidson and Raghuraman–Potter Models

We shall be concerned with slug flow where there are alternate zones of dense and void regions (see Section 14.12.1) in model studies for scale-up, the criterion for which is given by 14.68, namely, that $(u - u_{mf})/0.35(d_t g)^{1/2}$ should be greater than 0.2.

The various steps involved in calculating the conver-

sion in a first-order reaction occurring in a slug-flow reactor have been discussed by Hovmand and Davidson (1971) and may be summarized as follows.

1. Having established that slug flow occurs, we estimate the average slug length l_s from

$$\frac{l_s}{d_t} - 0.495 \left(\frac{l_s}{d_t} \right)^{1/2} \left[1 - \frac{u - u_{mf}}{0.35(gd_t)^{1/2}} \right] + 0.061$$
$$- \frac{1.939(u - u_{mf})}{0.35(gd_t)^{1/2}} = 0 \qquad (14.73)$$

It may be noted that in slug flow the tube diameter d_t replaces the bubble diameter d_b. This is because the rising velocity of a bubble (in this case a slug) is governed by the bed diameter rather than the bubble diameter when $d_b/d_t > \frac{1}{3}$.

2. The transfer factor Y defined by Eq. 14.4 is rewritten in terms of the slug volume V_s and calculated from the reaction

$$Y = \frac{QL_f}{uV_s} = \frac{L_{mf}}{0.35(gd_t)^{1/2} d_t \text{SF}}$$
$$\left\{ u_{mf} + 16 I \left(\frac{f_{mf}}{1 + f_{mf}} \right) \left(\frac{D_g}{\pi} \right)^{1/2} \left(\frac{g}{d_t} \right)^{1/4} \right\}$$
$$(14.74)$$

where the shape factor SF for the slug is given by

$$\text{SF} = \frac{l_s}{d_t} - 0.495 \left(\frac{l_s}{d_t} \right)^{1/2} + 0.061 \quad (14.75)$$

and I is an integral along the slug surface and can be obtained from Figure 14.12, in which it is plotted as a function of l_s/d_t. This equation is valid only for relatively fast reactions, that is, where $k_v > 2 \text{ sec}^{-1}$.

3. The maximum height of the slugging bed is then calculated from

$$\frac{L_f - L_{mf}}{L_{mf}} = \frac{u - u_{mf}}{0.35(gd_t)^{1/2}} \qquad (14.76)$$

4. On the assumption of plug flow in both the gas and emulsion phases, the conversion can be calculated from Eq. 14.5 by using the values of Y obtained from Eq. 14.74.

In another model Raghuraman and Potter (1978) assume that two thirds of the interslug region is occupied by a well-mixed emulsion traveling as a wake

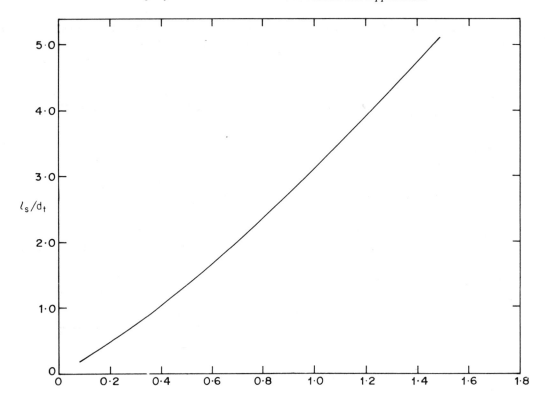

Figure 14.12.   The integral *I* along the slug surface (Hovmand and Davidson, 1971).

behind the rising slugs. Gas exchange occurs between slug and wake, and between wake and particulate phase, which is in plug flow. The transfer factor *Y* given by Eq. 14.74 is used to describe the slug–wake exchange, whereas exchange between the wake and the emulsion is given by

$$K_{we} = \left(\frac{1-\delta}{\check{Z}d_t\delta}\right)(u_s f_{mf} - u_{mf}) \qquad (14.77)$$

where $\check{Z}$ is the ratio of the interslug distance to the bed diameter d_t. Once in the particulate phase, the reactant is assumed to react according to the model of Fryer and Potter (1972a, b) presented in Section 14.4.2.

As a first step in design based on the prior model, the equivalent diameter of a proposed commercial reactor is calculated. Then, on the assumption that the larger reactor with this equivalent diameter operates in the slugging regime much like a laboratory model reactor with the same open diameter, the conversion may be calculated from Eq. 14.5. More experimental evidence is required, however, to justify the assumption that a reactor provided with vertical internal baffles to give equivalent diameters in the normal range of laboratory reactor diameters does indeed exhibit the charac-

teristics of the latter—and operates in the slugging regime.

14.15.2.   Example: Comparison of Models: Catalytic Oxidation of *o*-Xylene

In this example, the predictions of conversion based on theoretical reactor models are compared with the experimental results obtained for *o*-xylene oxidation in a slugging fluidized bed (Yates and Gregoire, 1980). The criterion employed for slugging is that given by Eq. 14.68 and the velocity for the onset of slugging is given by Eq. 14.69.

The Raghuraman–Potter model takes no account of the reaction in the bubbling region of the bed, the assumption being that slugs are formed at the distributor. The Hovmand–Davidson model, however, assumes the gas to be in plug flow above the distributor and to react on the basis of the model of Orcutt et al. (1962).

A 5 wt. % V_2O_5 catalyst on silica gel support was employed for the kinetic runs in a 0.10-m-i.d. stainless steel reactor 1.33 m long that was operated at a bed temperature of 372°C. The gas velocity (0.10–0.30 m/sec) and initial bed height (0.21 m, 0.48 m, and 0.62 m) were varied. Slug velocities u_s, and l_s, were

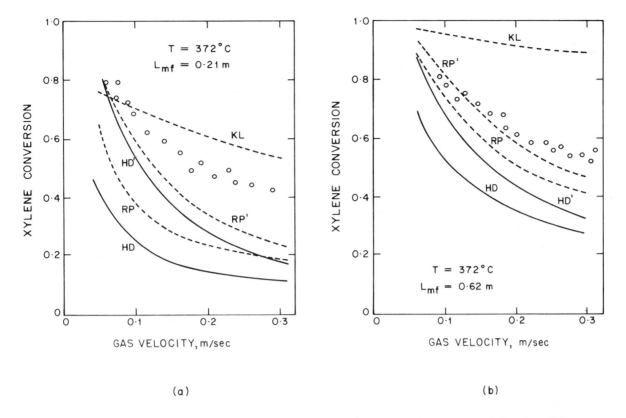

Figure 14.13. Comparison of slugging-bed model predictions with experimental data (Yates and Gregoire, 1980).

measured using a Perspex bed of the same dimensions and with the same distributor as employed in the main reactor. The passage of gas slugs was observed by means of two light probes fitted to an amplifier and ultraviolet recorder.

Figure 14.13 shows a comparison of the Hovmand–Davidson (HD); Raghuraman–Potter (RP), and basic Kunii–Levenspiel (K–L) model predictions with the experimentally observed results. The HD and RP models have also been modified to include the preslugging region of the bed by applying the K–L model to this region with an effective bubble diameter equal to 0.025 m. This has the effect of increasing the predicted conversion, and the modified models are represented as HD′ and RP′, respectively. First-order kinetics was considered, with

$$k = (1.28 \times 10^6) \exp\left(-\frac{80.67}{R_g T}\right) \quad \text{kJ/mol}$$

An empirical value of $\check{Z} = 2.8$ was used throughout. For the K–L model predictions the effective bubble diameter for the entire bed was taken to be 0.025 m. Thus, for the deeper beds where the system would

deviate considerably from the freely bubbling regime, the K–L model would overpredict the conversion because the previous assumption becomes increasingly untenable.

Figure 14.13a shows a comparison of the various models for $L_{mf} = 0.21$ m and Figure 14.13b for a deeper bed having $L_{mf} = 0.62$ m. Thus it is seen that for a deep slugging bed the modified RP model incorporating reaction in the preslugging zone gives the closest agreement with experimental data. At higher gas velocities the deviation is greater, possibly because both the RP and HD models neglect reaction in the slugs due to solid particles raining through them. At higher slug velocities this phenomenon is more pronounced, which conceivably explains the underprediction of conversion by these models.

14.16. INDUSTRIAL EXAMPLES

14.16.1. Phthalic Anhydride

The fluid-bed has been used for the production of phthalic anhydride, although the fixed-bed reactor appears to be the preferred one. It is claimed that the development of large capacity fluid-beds with naphthalene as feed has substantially improved the economics of phthalic anhydride production with attendant reduction in unexplained explosions experienced in fixed-bed operation, which may have been due to build-up of naphthaquinone concentration as a result of nonuniformity of temperature in the fixed-bed.

Badger Co. Inc. now offers a process based on *o*-xylene. Convertible (naphthalene or *o*-xylene) fluid-bed plants have also been operated successfully. Thus the initial Sherwin Williams–Badger process based on a feed stock of naphthalene has been "converted" to include *o*-xylene as feed.

14.16.2. Acrylonitrile

The Sohio process for the production of acrylonitrile by ammoxidation of propylene is considered one of the most successful applications of the fluid-bed to a synthesis reaction. The feed consists of propylene, ammonia, and air. Hydrocyanic acid and acetonitrile are obtained as salable byproducts. Montecatini–Edison and Badger Co. Inc. have also developed fluid-bed processes for this reaction. A feature of the former process is the high conversion obtained on a once-through basis. Troublesome operations such as separation and recycling of unreacted raw materials are thus avoided.

14.16.3. Isophthalonitrile

Isophthalonitrile is a raw material in the manufacture of agricultural chemicals and aromatic diamines. *m*-Xylene, ammonia, and air are introduced into a fluid-bed catalytic reactor where the *m*-xylene is primarily converted to aromatic nitriles. Hydrogen cyanide and small quantities of benzonitrile and *m*-tolunitrile are obtained as byproducts which may be recovered and are salable products.

This technology, developed by Mitusubishi Gas Chemical Badger, has been in commercial operation in Japan since 1970, and the first U.S. installation went onstream in 1973.

14.16.4. Vinyl Acetate

The production of vinyl acetate, an important monomer, from ethylene and acetic acid has been carried out in a fluid-bed reactor, although vapor phase fixed-bed operation appears to be the preferred mode of production. Three fiber companies with independently developed fluid-bed processes account for a considerable proportion of the vinyl acetate produced in Japan.

14.16.5. Fluid Catalytic Cracking (FCC)

Fluid catalytic cracking (FCC) represents one of the earliest uses of the fluid-bed, and the feed stocks used range from naphthas to vacuum gas oils and coker distillates. The cracking reactions are endothermic and are accompanied by catalyst fouling through carbon deposition. The FCC operation consists of two interconnected fluid-beds in the form of a reactor and a regenerator (where the carbon is burnt in air), with continuous circulation of solids between the two. Different designs have been employed, such as the model developed by Universal Oil Products Co. It features a higher pressure in the regenerator than in the reactor, a single riser, and the use of microspherical catalyst. M. W. Kellogg and Co. has developed the Orthoflow model A in which a single vessel construction is used to incorporate both reactor and regenerator. Many other designs have also been put into practice.

14.16.6. Fluid Catalytic Reforming

Fluid catalytic reforming, in which isomerization of a light naphtha feed is carried out to improve the octane rating, is also widely in use. The SOD Hydroformer was first introduced in 1953 and is similar in design to the Model IV FCC unit except that high pressures of 14–15 atm are used. A reactor/regenerator system is employed, and the endothermic reforming reaction is carried out on lines similar to the FCC operation.

14.16.7. Ethanol Dehydration

C. E. Lummus has recently developed a fluid-bed process for the dehydration of ethanol which is claimed to reduce plant investment and operating costs in relation to the conventional fixed-bed process. Multiple fixed-bed reactors (five in the case of a 60,000 tonnes/year ethylene plant, for example) and their auxiliary equipment are replaced by a single fluidized-bed reactor and regeneration system. The higher cost for catalyst and chemicals consumption of the fluid-bed process and the slight increase in power consumption necessary for fluidizing the catalyst are small compared to savings in investment and ethanol consumption.

14.16.8. Chloromethanes

Chlorination of methane to chloromethanes has been conventionally carried out in an empty tube reactor. The reaction involved is extremely complex and is associated with evolution of a large amount of heat. A primary requirement is close temperature control within a narrow range to avoid hot spots which lead to pyrolysis and formation of unwanted side products. National Chemical Laboratory, India, has developed a fluid-bed process which minimizes pyrolysis and other side reactions, and a commercial plant has been established at Standard Alkali and Chemicals Ltd., Bombay.

The principal features of the process are: mixing of methane and chlorine within the reactor at several levels by fluidizing with one gas and introducing the second at different levels, control of reaction temperature by use of recycle HCl, and use of jet breakers to prevent damage to reactor internals by the second gas introduced within the reactor.

14.16.9. Chlorofluoromethanes [Freons]

Montecatini–Edison has patented a new process for the production of chlorofluoromethanes, commercially known as Freons, in a fluidized-bed via direct methane halogenation. The earlier conventional process involved catalytic fluorination of carbon tetrachloride in liquid or vapor phase. The new process is operated by reacting methane, chlorine, and hydrogen fluoride over a catalyst in a fluidized-bed reactor in the presence of a mixture of recycled halogenated hydrocarbons. The use of such recycle allows close temperature control, high reaction selectivity, and excellent conversion efficiency. A 5000 tonnes/year plant employing this catalytic fluid-bed process went into production at Porto Marghera, Italy, in 1969, and expansion to 10,000 tonnes/year is reported to be underway.

14.16.10. Per- and Trichlorethylene

Diamond Shamrock has developed a fluid-bed process for perchlorethylene from ethylene dichloride (EDC) or other chlorinated C_2 organics. In this process involving simultaneous chlorination and dehydrochlorination of EDC, liquid EDC and chlorine form the reactants to a fluidized-bed catalytic reactor. Organics from the product distillation section are also recycled to the reactor. With different operating conditions tri-chlorethylene and perchlorethylene can be produced in the desired proportions.

Pittsburg Plate Glass (PPG) Industries also has an oxychlorination/oxyhydrochlorination process for the production of per- and trichlorethylene from ethylene or chlorinated C_2 organics. The catalyst is contained in a vertical bundle of tubes, each being a separate fluid-bed unit (the so-called longitudinal fluid-bed reactor). The heat of reaction is removed by boiling liquid in the jacket. World-wide production capacity of per- and trichlorethylene by PPG's single-step oxychlorination/oxyhydrochlorination process is rapidly approaching one billion pounds per year since being introduced in 1964.

A spin-off from this technology has been the development of an oxyhydrochlorination process for ethylene dichloride which can be integrated with vinyl chloride monomer plants. Rhone-Poulenc S. A. and Goodrich Chemicals Co. operate a two-step process in which ethylene dichloride is first produced from ethylene, air, and hydrogen chloride (from the second step) in a catalytic fluidized-bed reactor. The ethylene dichloride is then purified and thermally cracked (in the conventional manner) in a pyrolysis furnace to yield vinyl chloride monomer.

14.16.11. Terephthalic Acid

Lummus Co. has developed a fluid-bed process for the production of high-purity terephthalic acid from p-xylene. In the first of the two steps involved, p-xylene and recycled p-tolunitrile are reacted with ammonia in a fluidized bed to form terephthalonitrile. A novel metal oxide catalyst is used which provides the oxygen for the reaction in addition to catalyzing it at very high selectivity (90%). In the second step, purified terephthalonitrile is hydrolyzed and steam stripped to form solid mono-ammonium terephthalate, which is then thermally decomposed to terephthalic acid. Mother liquor and ammonia are recycled.

14.16.12. Aniline

American Cyanamid Co. has developed a vapor-phase fluid-bed catalytic process for the reduction of nitrobenzene to aniline. A mixture of nitrobenzene and excess hydrogen is passed into a fluidized-bed reactor containing a copper catalyst. The heat of reaction is removed by circulating a heat transfer liquid through tube bundles suspended in the catalyst bed.

14.16.13. Monoethylaniline

The conventional method of producing diethyl-aniline by reaction between ethanol and aniline in an autoclave is not applicable when the desired product is monoethylaniline (MEA). National Chemical Laboratory, India, has developed a process involving fluid-bed ethylation of aniline over a bauxite/alumina catalyst which produces MEA in a conversion of 40% based on aniline (Doraiswamy et al, 1970). A plant of semicommercial size (100 tonnes per annum) is in operation in India.

14.16.14. High Density Polyethylene

High-density polyethylene (HDPE) is produced in a fluidized-bed using a new series of supported chromium based catalyst. This novel process differs from conventional processes in that no solvent or diluent is required in the polymerization step. Gaseous ethylene and catalyst in the form of a dry powder are fed continuously to a fluidized-bed reactor. As the polymer is formed, the circulating gas keeps the growing polymer in a state of fluidization, supplies monomer for the reaction, and provides a medium for heat removal. The heat of reaction is removed from the stream in an air cooler. Granular polymer product is periodically discharged from the reactor through an appropriate arrangement.

14.16.15. Chlorosilanes

A Japanese process employs a stirred fluidized-bed reactor operating at about 6 atm pressure and 250–450°C for the production of chlorosilanes, important raw materials for a variety of silicone products, from methyl chloride. The fluid-bed contains silicon powder mixed with copper catalyst, having comparable size distributions, so that the two types of particles are thoroughly mixed and no segregation occurs. To accommodate the requirement of a high residence time in the reactor, low gas velocities are used in beds of very fine solids (−270 mesh). An agitator is employed to ensure satisfactory fluidization under these conditions.

National Chemical Laboratory, India, has also developed a fluid-bed process for chlorosilanes. In this process, a contact mass of Si–Cu is prepared by heating a mixture of enriched ferrosilicon particles and cuprous chloride along with promoters. The contact mass is then reacted with methyl chloride in a stirred fluidized-bed reactor at atmospheric pressure.

14.16.16. Use of Fluidized Beds for Immobilized Enzyme Systems

Although there are at present no commercial fluidized-bed installations for immobilized enzyme systems, several studies have been reported which suggest that fluidized-bed reactors could be commercially attractive for immobilized enzyme catalysts. Some of these studies are mentioned in Section 18.3 concerned specifically with immobilized enzyme catalysts.

REFERENCES

Agarwal, J. C., and Davis, W. L. (1966), *Chem. Eng. Prog. Symp. Ser.*, **62**, 101.

Baeyens, J., and Geldart, D. (1973), *Proc. Int. Symp. Fluidization*, p. 182.

Baeyens, J., and Geldart, D. (1974), *Chem. Eng. Sci.*, **29**, 255.

Basov, V. A., Markhevka, V. I., Melik-Akhnazarov, T. K., and Orochko, D. I. (1969), *Int. Chem. Eng.*, **9**, 263.

Baumgarten, P. K., and Pigford, R. L. (1960), *AIChE J.*, **6**, 115.

Behie, L. A. (1972), Ph.D. dissertation, University of Western Ontario, Canada.

Behie, L. A., and Kehoe, P. (1973), *AIChE J.*, **19**, 1070.

Behie, L. A., Bergougnou, M. A., Baker, C. G. J., and Base, T. E. (1971), *Can. J. Chem. Eng.*, **49**, 557.

Behie, L. A., Bergougnou, M. A., and Baker, C. G. J. (1976), *Fluidization Technology*, Vol. 1, ed. Keairns, D. L., Hemisphere, Washington, D.C., p. 261.

Blinichev, V. N., Streltsov, V. V., and Lebedeva, E. S. (1967), *Int. Chem. Eng.*, **8**, 615 (1968).

Bohle, W., and van Swaaij, W. P. M. (1978), *Fluidization*, eds. Davidson, J. F., and Keairns, D. L., Cambridge Univ. Press, Cambridge, p. 167.

Botterill, J. S. M. (1966), *Brit. Chem. Eng.*, **11**, 122.

Botterill, J. S. M., George, J. S., and Besford, H. (1966), *Chem. Eng. Prog. Symp. Ser.*, **62**, 7.

Botton, R. J. (1970), *Chem. Eng. Prog. Symp. Ser.* **66**, 8.

Brammer, K. R., Schugerl, K., and Schiemann, G. (1967), *Chem. Eng. Sci.*, **22**, 573.

Brian, P. L. T., and Hales, H. B. (1969), *AIChE J.*, **15**, 419.

Bromley, P. G., and Burgess, L. B. (1970); Jerry, I. G., and Scott, A. M. (1971), Research Project Reports for Chemical Engineering Tripos, Cambridge, England.

Bukur, D. B. (1978), *Chem. Eng. Sci.*, **33**, 1055.

Bukur, D. B., and Amundson, N. R. (1975a), *Chem. Eng. Sci.*, **30**, 847.

Bukur, D. B., and Amundson, N. R. (1975b), *Chem. Eng. Sci.*, **30**, 1159.

Bukur, D. B., Wittman, C. V., and Amundson, N. R. (1974), *Chem. Eng. Sci.*, **29**, 1173.

Bukur, D. B., Caram, H. S., and Amundson, N. R. (1977), *Chemical Reactor Theory—A Review*, eds. Lapidus, L., and Amundson, N. R., Prentice-Hall, Englewood Cliffs, New Jersey, p. 686.

Calderbank, P. H., Pereira, J., and Burgess, J. M. (1976), *Fluidization Technology*, Vol. 1, ed., Keairns, D. L., Hemisphere, Washington, p. 115.

Chavarie, C., and Grace, J. R. (1975), *Ind. Eng. Chem. Fundam.*, **14**, 75, 79, 86.

Chavarie, C., and Grace, J. R. (1976), *Chem. Eng. Sci.*, **31**, 741.

Chiba, T., and Kobayashi, H. (1973), *Proc. Int. Symp. Fluidization*, p. 468.

Collins, R. C. (1967), *J. Fluid. Mech.*, **28**, 97.

Cooke, M. J., Harris, W., Highley, J., and Williams, D. F. (1968), *Proc. Symp. Fluidization*, **1**, 14.

Danckwerts, P. V. (1952), *Trans. Faraday Soc.*, **46**, 300.

Davidson, J. F. (1973), *AIChE Symp. Ser.* (No. 128), **69**, 16.

Davidson, J. F., and Harrison, D. (1963), *Fluidized Particles*, Cambridge Univ. Press, Cambridge.

Davidson, J. F., and Harrison, D. (1971), *Fluidization*, Academic Press, London.

Davidson, J. F., and Keairns, D. L. (1978), *Fluidization*, Cambridge Univ. Press, Cambridge.

de Lasa, H. I., and Grace, J. R. (1977), Paper presented at 70th annual AIChE meeting, New York.

de Lasa, H. I., and Grace, J. R. (1979), *AIChE J.*, **25**, 984.

Denbigh, K. G. (1958), *Chem. Eng. Sci.*, **8**, 125.

Do, H. T., Grace, J. R., and Clift, R. (1972), *Powder Technol.*, **6**, 195.

Doraiswamy, L. K., Mukherjee, S. P., and Sadasivan, N. (1966), unpublished work, NCL.

Doraiswamy, L. K., Krishnan, V., and Krishnan, G. R. V. (1970), unpublished work, NCL.

Doraiswamy, L. K., Krishnan, V., and Krishnan, G. R. V. (1972), unpublished work, NCL.

Doraiswamy, L. K., Sadasivan, N., and Krishnan, G. R. V. (1974), unpublished work, NCL.

Drinkenburg, A. A. H., and Rietema, K. (1972), *Chem. Eng. Sci.*, **27**, 1765.

Elnashaie, S., and Yates, J. G. (1972), *Chem. Eng. Sci.*, **27**, 1757.

Elnashaie, S., and Yates, J. G. (1973), *Chem. Eng. Sci.*, **28**, 515.

Errazu, A. F., de Lasa, H. I., and Sarti, F. (1979), *Can. J. Chem. Eng.*, **57**, 191.

Fan, L. T., Lee, C. J., and Bailie, R. C. (1962), *AIChE J.*, **8**, 239.

Ford, W., Reineman, R. C., Vasalos, I. A., and Fahrig, R. J. (1977), *Chem. Eng. Prog.*, **73**, (No. 4), 92.

Fryer, C. (1974), Fluidized Bed Reactors, Ph.D. thesis, Monash University, Australia.

Fryer, C., and Potter, O. E. (1972a) *Ind. Eng. Chem. Fundam.*, **11**, 338.

Fryer, C., and Potter, O. E. (1972b), *Powder Technol.*, **6**, 317.

Fryer, C., and Potter, O. E. (1973), *Proc. Int. Symp. Fluidization*, and its Applications, Toulouse, p. 440.

Fryer, C., and Potter, O. E. (1976), *Fluidization Technology*, Vol. 1, ed. Keairns, D. L., Hemisphere, Washington, D.C., p. 171.

Furusaki, S., Kikuchi, T., and Miyauchi, T. (1976), *AIChE J.*, **22**, 354.

Geldart, D. (1967), *Chem. Ind.*, 1474.

Geldart, D. (1972), *Powder Technol.*, **6**, 201.

Geldart, D. (1973), *Powder Technol.*, **7**, 285.

George, S. E., and Grace, J. R. (1978), *AIChE Symp. Ser.* (No. 176), **74**, 67.

Gilliland, E. A., and Knudsen, C. W. (1971), *AIChE Symp. Ser.* (No. 116), **67**, 169.

Grace, J. R., and de Lasa, H. I. (1978), *AIChE J.*, **24**, 364.

Grace, J. R., and Harrison, D. (1968), *Proc. Inst. Chem. Eng.—VTG/VGI* (Brighton), p. 93.

Grace, J. R., and Harrison, D. (1969), *Chem. Eng. Sci.*, **24**, 499.

Grace, J. R., and Matsen, J. M., eds. (1981), *Fluidization*, Plenum Press, New York.

Gwyn, J. E., and Colthart, J. D. (1969), *AIChE J.*, **15**, 932.

Harrison, D., and Grace, J. R. (1971), *Fluidization*, eds. Davidson, J. R., and Harrison, D., Academic Press, London, p. 599.

Hatfield, W. B., and Amundson, N. R. (1971), *AIChE Symp. Ser.* (No. 116), **67**, 54.

Horio, M., and Wen, C. Y. (1977), *AIChE Symp. Ser*, **73**, (No. 161), 9.

Hovmand, S., and Davidson, J. F. (1968), *Trans. Inst. Chem. Eng. (London)*, **46**, 190.

Hovmand, S., and Davidson, J. F. (1971), *Fluidization*, eds. Davidson, J. F., and Harrison, D., Academic Press, London, p. 193.

Hovmand, S., Freedman, W., and Davidson, J. F. (1971), *Trans. Inst. Chem. Eng. (London)*, **49**, 149.

Irani, R. K., Kulkarni, B. D., and Doraiswamy, L. K. (1979), *Ind. Eng. Chem. Process Design Dev.*, **18**, 648.

Irani, R. K., Kulkarni, B. D., and Doraiswamy, L. K. (1980a) *Ind. Eng. Chem. Fundam.*, **19**, 424.

Irani, R. K., Kulkarni, B. D., and Doraiswamy, L. K. (1980b), *Ind. Eng. Chem. Process Des. Dev.*, **19**, 24.

Irani, R. K., Jayaraman, V. K., Kulkarni, B. D., and Doraiswamy, L. K. (1980c), *Chem. Eng. Sci.*, **36**, 29.

Irani, R. K., Kulkarni, B. D., Doraiswamy, L. K., and Hussain, S. Z. (1982), *Ind. Eng. Chem. Process Des. Dev.*, **21**, 188.

Ishii, T., and Osberg, G. L. (1965), *AIChE J.*, **11**, 279.

Jayaraman, V. K., Kulkarni, B. D., and Doraiswamy, L. K. (1981), *ACS Symp. Ser.* (No. 168), p. 19.

Kato, K., and Wen, C. Y. (1969), *Chem. Eng. Sci.*, **24**, 1351.

Keairns, D. L., ed. (1975), *Fluidization Technology*, Vols. 1 and 2, Hemisphere, Washington, D.C.

Kehoe, P. W. K. (1969), Ph.D. thesis, Cambridge University.

Kehoe, P. W. K., and Davidson, J. F. (1971), *Inst. Chem. Eng. Symp. Ser.*, **33**, 97.

Kobayashi, H., Arai, F., and Chiba, T. (1966), *Kagaku Kogaku*, **4**, 147.

Kobayashi, H., Arai, F., Izawa, N., and Miya, T. (1967a), *Kagaku Kogaku*, **30**, 656.

Kobayashi, H., Arai, F., and Sunakawa, T. (1967b), *Kagaku Kogaku*, **31**, 239.

Kobayashi, H., Arai, F., Chiba, T., and Tanaka, Y. (1969), *Chem. Eng. (Tokyo)*, **33**, 274.

Kuhne, J., and Wippern, D. (1980), *Can. J. Chem. Eng.*, **58**, 527.

Kulkarni, B. D., and Patwardhan, V. S. (1980), *Chem. Eng. J.*, **21**, 195.

Kulkarni, B. D., Ramachandran, P. A., and Doraiswamy, L. K. (1981), *Fluidization*, eds. Grace, J. R., and Matsen, J. M., Plenum Press, New York, p. 589.

Kunii, D. (1973), *AIChE Symp. Ser.* (No. 128), **69**, 24.

Kunii, D., and Levenspiel, O. (1968a), *Ind. Eng. Chem. Fundam.*, **7**, 446.

Kunii, D., and Levenspiel, O. (1968b), *Ind. Eng. Chem. Process Design Dev.*, **7**, 481.

Kunii, D., and Levenspiel, O. (1969), *Fluidization Engineering*, Wiley, New York.

Kunii, D., and Yoshida, K. (1974), *J. Chem. Eng. Japan*, **7**, 34.

Lee, W., and Kugelman, A. M. (1973), *Ind. Eng. Chem. Process Design Dev.*, **12**, 197.

Leung, L. S., and Wilson, L. A. (1973), *Powder Technol.*, **7**, 343.

Leung, L. S. (1976), *Fluidization Technology*, Vol. 2, ed. Keairns, D. L., Hemisphere, Washington, D.C., p. 125.

Leung, L. S. (1977), *Powder Technol.*, **16**, 1.

Leva, M., and Wen. C. Y. (1971), *Fluidization*, eds. Davidson, J. F., and Harrison, D., Academic Press, London, p. 627.

Levenspiel, O., Baden, N., and Kulkarni, B. D. (1978), *Ind. Eng. Chem. Process Design Dev.*, **17**, 478.

Lewis, W. K., Gilliland, E. R., and Glass, W. (1959), *AIChE J.*, **5**, 419.

Lewis, W. K., Gilliland, E. R., and Girouard, H. (1962), *Chem. Eng. Prog. Symp. Ser.* (No. 38), **58**, 87.

Luss, D., and Amundson, N. R. (1968), *AIChE J.*, **14**, 211.

Luyben, W. L., and Lamb, D. E. (1963), *Chem. Eng. Prog. Symp. Ser.* (No. 46), **59**, 165.

Massimilla, L. H., and Johnstone, H. F. (1961), *Chem. Eng. Sci.*, **16**, 105.

Matsen, J. M. (1973), *Powder Technol.*, **7**, 93.

Matsen, J. M. (1976), *Fluidization Technology*, Vol. 2, ed. Keairns, D. L., Hemisphere, Washington, D.C., p. 135.

Matsuura, T., and Kato, M. (1967), *Chem. Eng. Sci.*, **22**, 171.

Mireur, J. P., and Bischoff, K. B. (1967), *AIChE J.*, **13**, 839.

Miyauchi, T. (1974), *J. Chem. Eng. Japan*, **7**, 201.

Miyauchi, T., and Furusaki, S. (1974), *AIChE J.*, **20**, 1087.

Miyauchi, T., and Morooka, S. (1969), *Int. Chem. Eng.*, **9**, 713.

Mori, S., and Wen, C. Y. (1975), *AIChE J.*, **21**, 109.

Mori, S., and Wen, C. Y. (1976), *Fluidization Technology*, Vol. 1, ed. Keairns, D. L., Hemisphere, Washington, D.C., p. 179.

Nguyen, H. V., Whitehead, A. B., and Potter, O. E. (1977), *AIChE J.*, **23**, 913.

Ogasawara, S., Saraki, A., Hojo, K., Sirai, T., and Morikawa, K. (1959), *Kagaku Kogaku*, **23**, 299.

Orcutt, J. C. (1960), Ph.D. dissertation, University of Delaware, Newark.

Orcutt, J. C., and Carpenter, B. H. (1971), *Chem. Eng. Sci.*, **26**, 1049.

Orcutt, J. C., Davidson, J. F., and Pigford, R. L. (1962), *Chem. Eng. Prog. Symp. Ser.* (No. 38), **58**, 1.

Partridge, B. A., and Rowe, P. N. (1966), *Trans. Inst. Chem. Eng.* (*London*), **44**, T 335.

Pereira, F. J., and Beer, J. M. (1978), *Fluidization*, eds. Davidson, J. F., and Keairns, D. L., Cambridge Univ. Press, Cambridge.

Potter, O. E. (1978), *Catal. Rev. Sci. Eng.*, **17**, 155.

Raghuraman, J. A., and Potter, O. E. (1978), *AIChE J.*, **24**, 698.

Rietema, K., and Hoebink, J. (1976), *Fluidization Technology*, Vol. 1, ed. Keairns, D. L., Hemisphere, Washington, D.C., p. 279.

Schugerl, K. (1974), *Discussion in Proc. Int. Symp. Fluidization*, p. 712.

Shaw, I. D., Hoffman, T. W., and Reilly, P. M. (1974), *AIChE Symp. Ser.* (No. 141), **70**, 41.

Shen, C. Y., and Johnstone, H. F. (1955), *AIChE J.*, **1**, 349.

Sheplev, V. S., and Luss, D. (1979), *Chem. Eng. Sci.*, **34**, 515.

Stewart, P. S. B., and Davidson, J. F. (1967), *Powder Technol.*, **1**, 61.

Szepe, S., and Levenspiel, O. (1968), Paper presented at 4th European Symp. Chem. React. Eng. European Fed. Chem. Eng.

Tasev, Zh., Georgieva, S., and Doechev, I. (1974), *God. Vysshikh Khim. Tekhnol. Inst. Burgas Bulg.*, **10**, 555.

Tigrel, A. Z., and Pyle, D. L. (1971), *Chem. Eng. Sci.*, **26**, 133.

Toei, R., Matsuno, R., Miyagawa, H., Nishitani, K., and Kamagawa, Y. (1969), *Int. Chem. Eng.*, **9**, 358.

Tone, S., Seko, H., Matsuyama, A., and Otake, T. (1974), *J. Chem. Eng. Japan*, **7**, 44.

Toor, F. D., and Calderbank, P. H. (1967), *Proc. Int. Symp. Fluidization*, Netherlands Univ. Press, Amsterdam, p. 373.

van Deemter, J. J. (1961), *Chem. Eng. Sci.*, **13**, 143.

van Swaaij, W. P. M., and Zuiderweg, F. J. (1972), *Proc. 5th European Symp. Chem. React. Eng.*, Elsevier Publ. Co., Amsterdam, p. B9–25.

van Swaaij, W. P. M., and Zuiderweg, F. J. (1973), *Proc. Int. Symp. Fluidization*, Toulouse, p. 454.

Volk, W., Johnson, C. A., and Stotler, H. H. (1962), *Chem. Eng. Prog. Symp. Ser.* (No. 38), **58**, 44.

Voorhies, A. (1945), *Ind. Eng. Chem.*, **37**, 318.

Werther, J. (1974). *AIChE Symp. Ser.* (No. 141), **70**, 53.

Werther, J. (1977), *Chem. Ing.-Tech.*, **49**, 777.

Werther, J. (1980), *Chem. Eng. Sci.*, **35**(1), 372; *Int. Chem. Eng.*, **20**, 529.

Werther, J., and Hegner, B. (1981), *Int. Chem. Eng.*, **21**, 585.

Westerterp, K. R. (1962), *Chem. Eng. Sci.*, **17**, 423.

Yates, J. G. (1975), *Chem. Eng.* (*London*), **303**, 671.

Yates, J. G., and Rowe, P. N. (1977), *Trans. Inst. Chem. Eng.* (*London*), **55**, 137.

Yates, J. G., Rowe, P. N., and Whang, S. T. (1970), *Chem. Eng. Sci.*, **25**, 1387.

Yates, J. G., and Gregoire, J. Y. (1980), *Chem. Eng. Sci.*, **35**, 380.

Zenz, F. A. (1971), *Fluidization*, eds. Davidson, J. F., and Harrison, D., Academic Press, London, p. 1.

CHAPTER FIFTEEN

Fluidized-Bed Reactors: Variations in Design

In Chapter 14 a strategy for the design of fluidized-bed reactors based on the bubbling-bed concept was outlined. There are, however, several other types of fluidized-bed reactors for which no rational design procedures are yet available. These reactors, some of quite recent origin, offer advantages over the more conventional bubbling-bed systems. In this chapter about half a dozen variations in the design of fluidized-bed reactors are considered, including the moving-bed reactor, which is not really a fluidized-bed reactor but bears superficial resemblance to it in the sense that it also involves movement of both the solid and fluid phases.

The various categories of fluid-bed reactors that will be considered in this chapter are staged fluidized-bed reactors; packed-fluidized-bed reactors; fast fluidized-bed reactors; transport reactors; and moving-bed reactors. Each of these reactor types has its own specific advantages. In particular, the transport reactors are being increasingly used in the petroleum industry. In fact, during the last ten years, a so-called operation revamp has been launched for converting existing catalytic crackers to transport systems, in which the major portion of the reaction occurs in the transport lines. The packed-fluidized-bed reactor is another important variation in which the fluid bed contains fixed packings (in other words, fluidization is allowed to occur in the interstices of fixed packings). The design of such a reactor is therefore related directly to the packing size and not to the reactor dimensions, which leads to considerable simplification in scale-up.

15.1. MULTISTAGE FLUIDIZED BED

The need for multistage operation of a fluidized bed arises mainly out of one basic shortcoming of the conventional single-stage reactor: low efficiency. No matter in what manner the fluidized bed is operated (i.e.,

as a heat exchanger, calciner, adsorber, dryer, etc.) the efficiency of a continuous single-stage reactor is always lower than that of a continuous multistage reactor. Two practical ways of achieving multistage contacting are crosscurrent and countercurrent operation. In both the cases the single bed is divided into several stages with downcomers for solids flow from one stage to the next. The conclusions from operating as a heat exchanger are, however, valid for other cases (calciner, adsorber, etc.) as well.

An energy balance for the two modes of operation can be readily written, and the following simple expression for the efficiencies obtained (Kunii and Levenspiel, 1969):

$$\eta_s = \eta_g = \frac{N}{N+1} \qquad \text{for countercurrent flow}$$

$$\eta_s = N\phi_F\eta_g \qquad \text{for crosscurrent flow}$$

(15.1)

where

$$\phi_F = \frac{A_c u \rho_g C_p}{F C_{ps}}$$

and η_s and η_g are the efficiencies for the solid and fluid phases, respectively. It is apparent from the definition of the efficiencies that they are practically independent of the inlet temperature of the gas and the solid streams. More recent experiments of Peyman and Laguerie (1980), however, indicate that the gas-to-solid heat capacity ratio and inlet temperature of the streams have substantial influence on the efficiencies. The dependence of the efficiencies on the inlet temperature arises as a result of the heat losses and the radiative heat exchange between consecutive stages.

Each type of contacting has its own advantages, but for a fixed number of stages countercurrent contacting leads to higher efficiency. This, however, needs careful design of the downcomer. In a single-stage fluidized bed the residence time of solids is very low, as a result of which a large volume of the reactor is necessary to achieve high conversions. Multistaging the solids gives a distribution of residence time approaching plug flow and thus improving the efficiency of operation. In practical cases the solids feed (or catalyst) used contains a wide size distribution, and during fluidization different stages will have different size distributions, with the fines accumulating in the upper stages. Also, the reaction conditions such as temperature and gas composition may vary from stage to stage, so that a stagewise calculation of conversions becomes necessary. A procedure for making these calculations has been presented by Tone et al. (1967).

The measurement and accurate analysis of residence time distribution (RTD) and contact time distribution are essential in any theoretical study of multistage fluidized beds. The RTD of the phases in a single-stage fluidized bed has been described by Kunii and Levenspiel (1969). A few theoretical and experimental studies on RTD in multistage reactors are also available in the literature. These studies have largely been restricted to multistage reactors without downcomers. Thus Winterstein and Rose (1961), Ketteridge (1962), Bowing and Watts (1963), and Valchar (1965) assumed the tanks-in-series model with ideal mixing in each stage, while Raghuraman and Verma (1973a) included the effect of backmixing and dead space in the analysis. Wolf and Resnik (1963) defined the mixing efficiency and the phase shift for a system consisting of plug flow, perfect mixing, dead space, and short-circuiting. Brauner et al. (1970) proposed a dispersion model. The conclusions from all these studies point to the fact that axial mixing decreases with increase in the number of stages. Real systems, however, seldom correspond to either the tank-in-series model or the dispersion model. Thus nonidealities are being increasingly incorporated in the newer models proposed (Buffham and Gibilaro, 1968; van Swaaij et al. 1969; Raghuraman and Verma, 1973b, 1974).

From the practical point of view the crossflow arrangement appears simple and is recommended for use. This avoids the uncertainty associated with the design of the downcomer, which is often critical, since stable downflow of solids is difficult to maintain. An alternative to multistage fluidized beds (Kunii and Levenspiel, 1969) is a system involving a fluidized bed followed by a moving-bed reactor. The higher conversions (say, 80–90%) are achieved in the fluidized bed, where heat effects can be properly taken care of; the remaining conversion is then achieved in the lean phase (moving-bed reactor) with minor problems of heat

removal. An interesting variation in the design of multistage beds, namely, pneumatically controlled downcomers, has been suggested by Liu et al. (1980). This arrangement avoids the frequent problem of bridging or spouting of solids in the downcomer.

15.2. PACKED-FLUIDIZED-BED REACTORS

An important variation in the design of a fluidized-bed reactor is realized when fixed packings are introduced into a conventional fluid-bed reactor. The addition of packings to the fluid-bed reactor results in significant changes in the quality and performance of the fluidized bed. For instance, the packings decrease the vertical mixing of particles and inhibit the growth of bubbles and their coalescence, thus reducing the slugging tendency of the fluidized bed. On the other hand, certain features of the conventional fluidized bed, such as the minimum fluidization velocity and eddy diffusivity of the gas, remain unaltered.

From the design point of view, the most attractive feature of the packed fluidized bed is that the bed performance is related to the packing size, which is independent of the scale of operation. Thus the scale-up of this reactor would be considerably simpler than that of the more common bubbling-bed reactor.

15.2.1. Properties of the bed

Pressure Drop

The pressure drop in a packed fluidized bed is calculated on the assumption that the Blake–Kozeny equation for laminar flow through a fixed bed is valid for the packed section. The resulting expressions are

$$\frac{\Delta P}{L_f} = \frac{150}{\rho_s d_p^2 (\mathrm{SF})^2} \frac{\left\{ 1 - \left[1 - \dfrac{(1 - f_B)}{1 + (R_B - 1)/f_{pa}} \right] \right\}^2 G_{mf}}{\left[1 - \dfrac{(1 - f_B)}{1 + (R_B - 1)/f_{pa}} \right]^3}$$

$$+ \frac{(1 - f_B)(R_B - 1)\rho_s f_{pa}}{f_{pa} + R_B - 1} \tag{15.2a}$$

for a packed fluidized bed with solid packings, and

$$\frac{\Delta P}{L_f} = \frac{f_{pa}(1 - f_B)\rho_s}{f_{pa} + R_B - 1} + \frac{(1 - f_B)(R_B - 1)f_{pa}\rho_s}{f_{pa} + R_B - 1}(1 - \mathrm{CF}) \tag{15.2b}$$

for a packed fluidized bed with open-ended screen cylinders. In these equations f_{pa} represents the voidage of the empty packed bed, f_B the porosity of the bed without packing, SF the shape factor, R_B the bed expansion ratio, and CF the correction factor for channeling.

The pressure drop in a bed with flooded packing (neglecting the resistance of the packing itself to the flow) can be calculated from the equation (Udilov et al., 1972)

$$\frac{\Delta P}{L_f} = \frac{36\mu u}{d_{pa}^2}\left[\frac{1.5 K_1 (1-f_{fp})^2}{f_{pa} f_{fp}} + \frac{(1-f_B)^2}{f_B^3} K_2 \frac{L_h}{L_f}\right]$$

(15.3)

where L_h is the initial height of the fixed layer over the packing, f_{fp} is the voidage of the bed with unflooded packing, and K_1 and K_2 are constants.

In the case of flooded packing, the physical picture changes. The velocity of the gas phase in the layer over the packing is considerably lower than in its pores. Thus the material in this layer remains unfluidized, while that in the pores might have reached incipient fluidization. Any increase in the gas throughput therefore increases the resistance sharply, and this continues until the material in the layer over the packing also starts fluidizing. The maximum resistance of the bed is observed at a gas velocity corresponding to the start of fluidization in the layer over the packing and is given by

$$\frac{(\Delta P)_{max}}{L_f} = g(\rho_s - \rho_g)\left[1.5\frac{K_1}{K_2}\frac{(1-f_{fp})^2 f_B^3}{f_{pa} f_{fp}^3 (1-f_B)} \right. $$
$$\left. + (1-f_B)\frac{L_h}{L_f}\right]$$

(15.4)

The ratio K_1/K_2 varies within the limits 0.8–1.6.

Minimum Fluidization Velocity and Bed Expansion

The packing in a packed fluidized bed hinders the motion of the particles and thus alters the bed characteristics and minimum fluid-bed voidage. The minimum fluid-bed voidage thus becomes an important parameter in affecting the behavior of the packed fluidized-bed. The minimum fluid-bed voidage obviously depends on the type of packing used and greatly influences the value of the minimum fluidization velocity. The experimental data of Pillai and Raja Rao (1976) show an increase of from 5 to 45% in the minimum fluidization velocity, depending on the nature of the packing used. The increase in the minimum fluidization velocity has been expressed in terms of a

bed voidage group by the following relationship:

$$\frac{G'_{mf}/f_{pa}}{G_{mf}} = \frac{(1-f_{mf}^2)/f_{mf}^3}{(1-f_B)^2/f_B^3}$$

(15.5)

where G'_{mf} denotes the mass flow rate at minimum fluidization velocity in the packed fluidized bed. The minimum fluidization velocity in the case of flooded packing has been correlated by Udilov et al. (1972) as

$$u'_{mf} = \frac{g d_p^2 (\rho_s - \rho_g) f_B^3}{36 K \mu (1-f_B)}$$

(15.6)

where K is a constant.

The packing restricts the motion of particles and inhibits the growth of bubbles, thus reducing the bed expansion. Evidently the reduction in bed expansion would be less severe in the case of open-ended screens than in the case of solid packing.

Gas Mixing

Gas-mixing studies in a packed fluidized bed have been carried out by Gabor and Mecham (1964) and Chen and Osberg (1967) using the tracer injection technique. The experimental data of Chen and Osberg with hollow open-wall packings (less than 3 in) indicate that the reciprocal mixing length (u/D_e) first decreases and then increases with the expansion ratio, passing through a minimum, in contrast to the behavior of an unpacked fluidized bed. The minimum length $(u/D_e)_{min}$ always occurs at bed expansion $R_B = 0.30$, independent of the column diameter and of the shape and size of the packing and of the particle. On the other hand, $(u/D_e)_{min}$ for a fluidized bed packed with open-screen cylinders depends on the size of the particle and the packing used. Chen and Osberg (1967) correlated the experimental data by means of the equation

$$\left(\frac{u}{D_e}\right)_{min} = 11858\left(\frac{d_{pa}}{d_p}\right) - 18.4$$

(15.7)

There appears to be a basic difference between the mixing properties of open-mesh screens and solid packing. The data of Gabor and Mecham (1964) with $\frac{1}{8}$-in. spheres show that the gas eddy diffusivities in the packed fluidized bed are in fact nearly identical with those in beds with the same type of packing but without the fluidizing material.

Solids Mixing

The literature on solids mixing in packed fluidized beds is relatively scarce. Gabor (1964) investigated the lateral mixing of solids and Kang and Osberg (1966) studied the role of longitudinal mixing. In general, packed fluidized beds would be expeced to show features that are combinations of those of packed and fluidized beds. The primary motion of the particle is determined by its size and gas velocity, whereas the dimensions of the packed material determine the void structure within which the fluidization occurs. The experimental data of Gabor (1964) suggest that lateral particle mixing is related to the void structure of the packed bed; the following generalized correlation for the diffusivity (of solids) has been obtained:

$$(D_e)_s = d_{pa} \left(\frac{u - u_{mf}}{d_p f_{pa}} \right)^{1.15} (1.22 \times 10^{-6}) \quad \text{ft}^2/\text{sec}$$

$$(15.8)$$

where all the dimensions are in feet and seconds. The rate of solids mixing is independent of the height of the packed fluidized bed, in contrast to the variation observed for open-tube fluidization (without packing).

Longitudinal particle mixing in a screen-packed gas–solid fluidized bed has been studied by Kang and Osberg (1966). A simple cell model was formulated that correlates the eddy diffusivity $(D_e)_s$ with a void term regardless of the mesh size of the packing used:

$$(D_e)_s = 257 \left(\frac{v_b^2}{v_c} \right) \quad (15.9)$$

where v_b and v_c are the volumetric fractions of the bubble and continuous phases, respectively. The continuous phase can be calculated from

$$v_c = \left(\frac{l_{mf}}{L} \right)(1 - v_{pa}) \quad (15.10)$$

where v_{pa} is the volumetric fraction of the packing.

Heat Transfer in Packed Fluidized Beds

As with other studies in packed fluidized beds, the literature on heat transfer is also scarce. Gabor et al. (1965) have reported thermal conductivity measurements in a fluidized bed packed with solid spheres and cylinders, and have correlated the data according to the equation

$$\frac{k'_e}{\rho_s c_{ps}} = (0.0075)\ 0.0558\ d_{pa} \left(\frac{u_p - u_{mf}}{u_{mf}} \right) \quad (15.11a)$$

where u_p denotes the fluidizing-gas velocity corrected

for fixed packing voidage and all units are in feet and seconds.

Experimental data on heat transfer between spherical packings and the fluidized bed (Baskakov and Vershinina, 1964) reveal that the bed voidage has no significant effect on the heat transfer coefficient. The heat transfer coefficient ordinarily increases as the mean dimension of the particle decreases in a fluidized bed. However, no such observation is reported for a packed fluidized bed. This is obviously due to a deterioration in the flow properties in the packing. The heat transfer coefficient increases as the packing diameter is increased and also depends on the intensity of mixing.

Kato et al. (1979) report data on the drying of wet particles of activated alumina by hot air in the constant drying rate period in a packed fluidized bed. Their experimental data are correlated in the form of an empirical equation for the gas–particle heat transfer coefficient as

$$\text{Nu} = 0.59\ (\text{Re}_p)^{1.1} \left(\frac{d_p}{L_f} \right)^{0.9}, \qquad 3 < \text{Re}_p < 50$$

$$(15.11b)$$

where Re_p is the Reynolds number based on u_p.

Radial heat transfer measurements in a packed fluidized bed (Ziegler and Brazelton, 1963) indicate that the bed conductivity and surface heat transfer coefficients are improved with increase in packing size and decrease in particle size. A value of 30 as compared with 0.4 Btu/hr ft^2 °F/ft in a packed bed for the thermal conductivity is reported for a packed fluidized bed.

Hirama et al (1979), using random walk theory, have developed an equation for predicting the lateral thermal conductivity in a packed fluidized bed with horizontal flow:

$$(k'_{er})_{pf} = 0.112 f_{ps} (d_p) \left(\frac{L_f}{L_{mf}} \right) (u - u_{mf}) \quad (15.11c)$$

where f_{ps} represents the ratio of the volume of particles moving in conjunction with the bubbles to the volume of bubbles. In general this is an unknown quantity. By actual measurement of lateral temperature profiles the authors suggest a relation between f_{ps} and f_w, the wake fraction of solids (as distinct from s_{wb}) as

$$f_{ps} = \frac{f_w}{3(1 - f_w)}$$

Once f_{ps} is known it is possible to use Eq. 15.11c.

15.2.2. Reactor Model for Packed Fluidized Beds

As in the case of the conventional bubbling bed, the packed fluidized bed also consists of the bubble and emulsion phases, and the bubbles carry along with them solid particles in the wake. The size of the bubbles in the packed fluidized bed is, however, restricted by the packing, and any increase in the gas flow rate helps in increasing the number of gas bubbles per unit volume of the bed and the bubble rise velocity. A fluid flow model based on the size of the bubble in the bed thus seems unsuitable for the packed fluidized bed. Kato et al. (1974) have employed a fluid flow model similar to that of Mathis and Watson (1956) and Lewis et al. (1959) in calculating the conversion in packed-fluidized-bed reactors.

In view of the large effective thermal conductivity realized in packed fluidized beds, these reactors are more nearly isothermal than the conventional fluidized beds. On the assumption of plug-flow of the gas in the bubble and emulsion phases, the following conservation equations can be written:

Bubble phase:

$$\frac{dC_b}{dl} + \frac{(R_B - 1)k_{gv}}{R_B(u - u_{mf})}(C_b - C_e) + \frac{s_{bb}r_b}{R_B(u - u_{mf})} = 0$$

$$(15.12)$$

Emulsion phase:

$$\frac{dC_e}{dl} - \frac{(R_B - 1)k_{gv}}{R_B u_{mf}}(C_b - C_e) + \frac{(1 - s_{bb})r_e}{R_B u_{mf}} = 0$$

$$(15.13)$$

Boundary condition: $l = 0$, $C_b = C_e = C_0$ (15.14)

where r_b and r_e denote the reaction rates in the bubble and emulsion phases, respectively. The concentration at the exit of the packed fluidized bed can be calculated as

$$C = \left(\frac{u_{mf}}{u}C_e + \frac{u - u_{mf}}{u}C_b\right)_{z=1}$$

$$(15.15)$$

The mass balance equations 15.12 and 15.13 require a knowledge of two parameters: the fraction of solids contained in the bubble s_{bb}, and the gas interchange coefficient k_{gv} between the two phases. The gas interchange coefficient can be obtained from the relation-

ship (Kato et al., 1967)

$$k_{gv} = (4.2 \times 10^{-3})\left(\frac{u - u_{mf}}{u_{mf}}\right)^{-0.6} \text{cm/sec}$$

$$(15.16)$$

Equations 15.12 and 15.13 can be integrated analytically for a first-order reaction. Numerical solutions have been presented by Kato et al. (1974) for a general-order reaction with the fraction of solids in the bubble s_{bb} as the parameter. Their results indicate that the reactant concentrations in the bubble and emulsion phases are quite sensitive to s_{bb} when the reaction rate constant is high. Increasing the fraction of solids in the bubble increases the conversion in the bubble phase, and the larger the rate constant, the larger the concentration difference between the bubble and the emulsion phases. Reactant conversion calculated from this model for a first-order reaction with reaction rate constant as a parameter is reproduced in Figure 15.1. The figure shows the conversion in a packed fluidized bed to be almost the same as that in the plug-flow reactor when the rate constant is small ($\sim 0.1 \text{ sec}^{-1}$). For intermediate values of the rate constant ($\sim 1.0 \text{ sec}^{-1}$) the conversion lies between those in the plug-flow and perfectly mixed reactors. When the reaction rate constant is large ($\sim 10.0 \text{ sec}^{-1}$), the conversion calculated from this model is smaller than that in the perfectly mixed flow reactor.

The fluid flow model is perhaps the most important single consideration in the design of a packed-fluidized-bed reactor. The model as discussed above requires a knowledge of two parameters, s_{bb} and k_{gv}. For a reaction with known kinetics (determined separately in a fixed bed) the model parameter s_{bb} can be calculated with the equations presented above. Since the bed diameter and bed height have no influence on the fluid flow pattern, this parameter, calculated from a simple laboratory experiment, can be directly used in the design of a commercial packed-fluidized-bed reactor. Thus the scaling-up problems are considerably simplified in these reactors.

Examples

This model has been successfully used by Kato et al. (1979) to interpret the results for the reactant conversion in the case of the catalytic cracking of cumene in a packed fluidized bed under conditions of no deactivation. When catalyst deactivation exists the authors have shown that for reactions with high initial rate constants ($> 1.0 \text{ liter/sec}$) the reactant conversion can be calculated by assuming perfect mixing of particles in

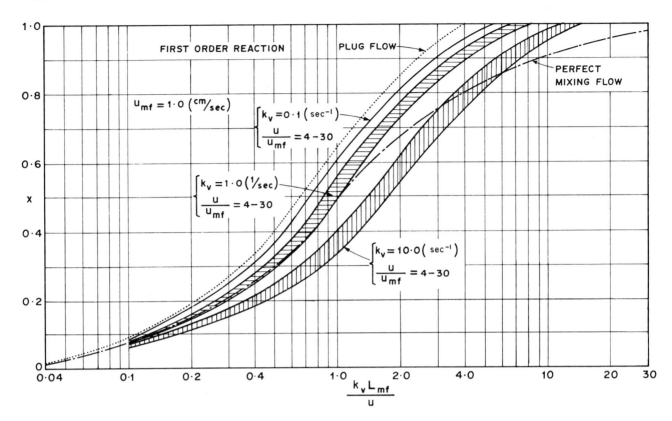

Figure 15.1. Performance of a packed-fluidized-bed reactor in relation to that of plug-flow and fully mixed reactors (Kato et al., 1974).

the bed. For reaction with lower initial rate constants (< 0.3 liter/sec) the assumption of plug flow of particles within the bed seems adequate.

Another potential application of the packed fluidized bed is in the regeneration of spent activated carbon after its use in municipal waste water treatment and in the removal of odors from waste gases (Kato et al., 1980). The adsorption capacity of the regenerated catalyst as measured by the Iodine number was found to agree well with the values predicted by the model described above.

15.3. FAST-FLUIDIZED-BED REACTORS

The operation of a fluid bed in the bubbling regime has several disadvantages: (1) considerable backmixing of the gas occurs, which lowers the conversion and may promote side reactions; (2) a major portion of the gas bypasses the solids in the bed in the form of bubbles; (3) insufficient solids mixing restricts its use to free-flowing solids; (4) in view of the large number of parameters involved, the scaling-up is usually a difficult task; and (5) since the fluidization velocities employed in the

bubbling regime are usually in the range $u/u_{mf} = 3$–10, the processing capacities of such reactors are low. A solution to most of these problems is offered by the so-called fast-fluidized-bed reactors.

Fast fluidization is a technique in which a gas at very high velocity is brought into contact with an entrained dense phase of solids. The operation is characterized by heavy turbulent conditions in the bed. The use of very high gas velocities coupled with backmixing of the solids makes these reactors attractive for use in industry. However, this process of fast fluidization has not been fully explored, and information available is restricted to the use of fine particles.

The operation of a fast-fluidized bed may be more clearly understood by referring to Figure 15.2, which shows the usual pressure gradient as a function of the superficial velocity. The gradient first increases sharply as a result of the rise of pressure across the fixed bed of solids. At minimum fluidization velocity the bed begins to expand, and further increase in velocity leads to the formation of bubbles, giving the appearance of a bubbling bed. As the velocity is increased further, a point is reached where there is a sharp drop in the bed density over a narrow velocity range due to excessive

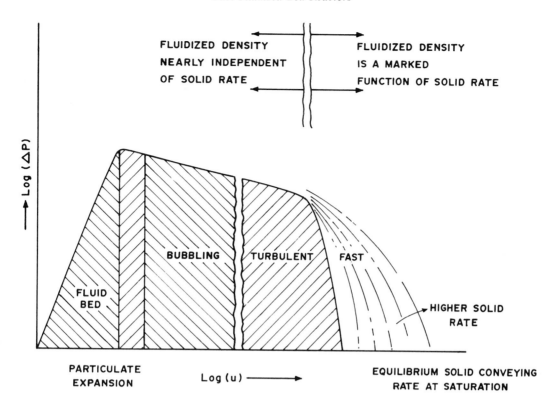

Figure 15.2. Different regimes of fluidization (Yerushalmi et al., 1976).

carryover. This transition from the bubbling to the turbulent regime has been demonstrated by Kehoe and Davidson (1971), while Massimilla's results (1973) demonstrate the higher contacting efficiencies of the turbulent beds. A bubbling bed is characterized by a distinct upper surface level, which is considerably diffused in the case of a turbulent bed owing to the higher gas velocities employed. In either mode of operation, the carryover of solids is constant above the TDH.

In the turbulent regime, in order to maintain the upper level of solids at a constant position, it is necessary to feed the solids into the bed at the same rate at which they are removed at the top. The addition of solids to the fluidized bed operating in the bubbling or turbulent regime does not influence the fluid-bed density significantly. For velocities greater than u_{tr} (the velocity at which a sharp decrease in pressure gradient is recorded), the fluid-bed density is, however, a strong function of the rate of solids fed in at the bottom of the bed. It is this property of a fluidized bed which is made use of in fast-bed reactors. Fluid-bed densities as high as those encountered in the bubbling regime of operation can be maintained in fast reactors by adjusting the solids flow rate to the reactor.

The fast fluidization regime is transformed to that of pneumatic transport as the gas velocity is increased to that for pneumatic transport. As this velocity is greatly influenced by solids circulation rate, it can be readily seen that for low solids circulation rate the fast fluidization region would be practically nonexistent, and with increasing solids rate there would be an expanding region of fast fluidization (Li and Kwauk, 1980).

Experimental investigations on fast reactors (Yerushalmi et al., 1976) point to solids distribution between the two phases accompanied by rapid interchange. The higher slip velocities reported are a measure of the high degree of backmixing of solids (see also Yerushalmi et al., 1978; Yerushalmi and Cankurt, 1978; Cankurt and Yerushalmi, 1978). Li and Kwauk (1980) have considered the dynamics of fast-fluidized beds and obtained the vertical voidage distribution of solids in terms of model parameters. The backmixing can be understood on the physical reasoning that the dense packets of particles (a characteristic of fast reactors) would naturally have a terminal velocity much higher than the individual particles, and hence cannot be sustained by the rising gas. They would therefore fall back, with subsequent disintegration,

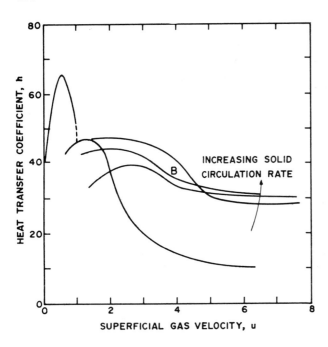

Figure 15.3. Heat transfer coefficient in fast-fluidized-bed reactors (Yerushalmi et al., 1978).

giving rise to a higher slip velocity and hence backmixing. Available experimental data (pressure gradient at different heights in the bed) also confirm the high fluidized-bed densities.

Figure 15.3 shows the heat transfer coefficient for a fast-fluidized bed as a function of superficial gas velocity for various solids recirculation rates. For the sake of comparison, the h–u curve for the dense bed is also shown. The heat transfer coefficient decreases in general as the gas velocity is increased for all solids circulation rates. The relative decrease in the coefficient, however, is less marked at higher solids circulation rates. For a sufficiently high circulation rate, as marked by curve B in Figure 15.3, a more or less uniform coefficient, independent of the circulation rate and comparable with that in the dense bed, can be realized.

In fast-bed reactors a close control of temperature all through the bed is easy to obtain because of the high degree of solids mixing. The heat transfer rates to walls and immersed surfaces have been shown (Kiang et al., 1975) to be comparable to those for a bubbling bed. Since high gas velocities are employed in fast reactors, the diameter of the vessel, which represents a significant cost factor, can be cut down significantly. The use of high gas velocities precludes appreciable gas backmixing, thus making it possible to approach plug-flow conditions.

In conclusion, it can be said that the fast-fluidized bed appears to be a natural extension of the bubbling and turbulent regimes of fluidization. In the bubbling regime, gas is brought into contact with solids at relatively high concentrations in the bed. In the turbulent regime, the solids concentration in the bed is low, and excessive carryover is an essential feature. In fast beds, solid concentrations as high as that in the bubbling bed can be achieved by regulating the solids inflow. The higher gas velocities employed make it possible to achieve better gas–solids contact and thus increase the reactor capacity. The scaling-up of fast reactors might indeed prove to be easier than that of bubbling-bed reactors. However, data on large-scale equipment with coarser particles is still lacking and should form the subject of further research in this promising field.

15.4. TRANSPORT REACTORS

15.4.1. General Features

In transport reactors the catalyst or solid reactant is carried through a pipeline by the reacting gas. The solids proportion is usually less than 2% by volume. As in the case of moving-bed reactors, transport reactors owe their development to the petroleum industry. Since the fluidized-bed reactor offered several advantages, however, the development of this reactor was taken up at the expense of the transport and other possible reactor types.

The transport reactor is ideally suited for consecutive reactions where an intermediate product is desired and a close control of residence time becomes necessary. Another advantage is that the catalyst loading can be varied independently of the gas flow. In the catalytic cracking of petroleum feedstocks, where the conventional fluidized-bed system would normally be unable to accommodate any sizable fluctuation in the rate and composition of the feedstocks, practically 90% of the cracking can be achieved in a so-called riser (which corresponds to a transport reactor) before the gas–solid mixture enters the conventional dense-phase fluidized-bed system (Bryson et al., 1972; Strother et al., 1972; Pierce et al., 1972).

A particularly advantageous feature of the transport reactor is the increased heat transfer rate characteristic of these systems (Sadek, 1972). Thus a predetermined temperature gradient can be imposed in the pipeline to establish an optimum profile (see Section 12.4).

Another obvious advantage is that the probability of hot spots developing in the reactor is minimized.

The availability of high-activity zeolites to replace the conventional silica–alumina catalyst in catalytic cracking has revived interest in transport reactors. The riser crackers (an accepted terminology for transport reactors) are claimed to perform better than the conventional reactor–regenerator system operated as dense-phase fluid beds.

The ease of fluidization and pneumatic conveyance of solid particles varies with the size, bulk density, and nature of the solids handled, as well as with the material of construction of the pipe. Bourquet et al. (1961) have described how the critical catalyst velocity was determined in the case of the Thermofor catalytic cracking (TCC) airlift.

Notwithstanding the several advantages of transport reactors, virtually no theoretical studies have been undertaken on these reactors. Only recently has some interest been shown, which has led to the development of a model for predicting the influence of the major operating variables on the performance of a transport reactor. A brief description of this model follows.

15.4.2. An Unsteady-State Model

Pratt (1974) has proposed a model of the transport reactor based on the following assumptions: (1) the catalyst particles are spherical and of uniform size, (2) both the solid and gas phases are in plug flow, axial diffusion being negligible (Jepson et al., 1965), (3) isothermal conditions prevail, and (4) a simple first-order irreversible reaction with no volume change is involved.

The continuity equation for the reacting component A in the gas phase can be written as

$$-u_i \frac{dC_{Ab}}{dl} = \frac{1-f_B}{f_B} r_{vA} \qquad (15.17)$$

which is an equivalent form of Eq. 11.4 with $u_i f_B$ replacing u and $r_{vA}(1-f_B)$ replacing the global rate r_{vA}. As for the reaction term r_{vA}, it can be expressed in terms of the concentration of A in the pellet as (see Section 4.3.1)

$$r_{vA} = \frac{3D_{eA}}{R} \left(\frac{dC_A}{dr}\right)_{r=R} \qquad (15.18)$$

where C_A represents the concentration within the

pellet. Combining 15.17 and 15.18, we have

$$-u_i \frac{dC_{Ab}}{dl} = \left(\frac{1-f_B}{f_B}\right) \frac{3D_{eA}}{R} \left(\frac{dC_A}{dr}\right)_{r=R} \qquad (15.19)$$

The mass balance equation for the particle is given by

$$\frac{f_c}{D_{eA}} \frac{\partial C_A}{\partial t} = \frac{\partial^2 C_A}{\partial r^2} + \frac{2}{r} \frac{\partial C_A}{\partial r} - \frac{k_v C_A}{D_{eA}} \qquad (15.20)$$

Equations 15.19 and 15.20 may be solved simultaneously subject to the following boundary conditions:

1. At the reactor entrance, catalyst particles do not contain A:

$$C_A(0, r) = 0, \qquad 0 \leqslant r \leqslant R \qquad (15.21)$$

2. The reacting component has a concentration C_{Ab0} at the entrance:

$$C_A(0) = C_{A0} = C_A(0, R) \qquad (15.22)$$

3. Spherical symmetry in the particles:

$$\frac{\partial C_A}{\partial r}(t, 0) = 0 \qquad (15.23)$$

4. External mass transfer resistance is absent:

$$C_A(t, R) = C_{Ab}(l) \qquad (15.24)$$

Solution to this set of equations provide the concentration profiles in a catalyst particle as well as in the gas phase in the reactor. Pratt (1974) has obtained numerical solutions in terms of the following dimensionless groups:

$$P_X = \frac{D_{eA}L}{f_c R^2 u_s}, \qquad P_Y = \frac{f_c u_s}{u_i}\left(\frac{1-f_B}{f_B}\right) \qquad (15.25)$$

and the usual Thiele modulus $\phi_{s1} = R(k_v/D_{eA})^{1/2}$.

The parameter P_X may be regarded as a measure of the number of diffusion time constants that a particle spends in the reactor (single R^2/D_{eA} represents the time constant for diffusion). Thus increasing P_X will always lead to increasing conversion. The parameter P_Y is a direct measure of the catalyst circulation rate W_s (i.e., weight of catalyst circulated per unit weight of gas).

Typical axial concentration profiles in the reactor at different values of P_Y (and fixed values of P_X and ϕ_{s1}) are shown in Figure 15.4a, while the concentration profiles within the pellet at different axial positions (and fixed values of P_X and P_Y) appear in Figure 15.4b and c at a high ($\phi_{s1}^2 = 100$) and a low ($\phi_{s1}^2 = 1$) value, respectively, of the Thiele modulus. The latter is consistent with the usteady-state behavior of the solids under transport.

It can be seen that the axial profile in the reactor is almost linear at low catalyst circulations (corresponding to $P = 0.005$). The profiles within the pellet are steep at $\phi_{s1}^2 = 100$, whereas they are flat at $\phi_{s1}^2 = 1.0$.

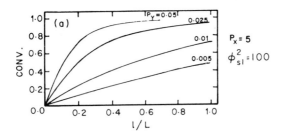

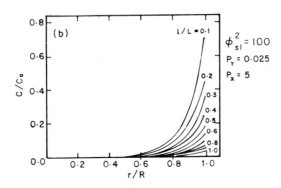

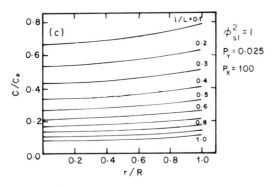

Figure 15.4. Axial conversion profiles (a) and concentration profiles within the catalyst particles (b,c) for various values of the parameters P_X, P_Y, and ϕ_{s1} (Pratt, 1974).

Thus, even though an unsteady-state situation is involved in transport reactors, the relationship between the Thiele modulus and catalyst effectiveness is similar to that obtained from a steady-state analysis (see Chapter 4).

In fact, a subsequent study (Robertson and Pratt, 1975) has shown that steady-state effectiveness factors represent a reasonable approximation to the actual transient values. The effectiveness factor calculated from the methods outlined in Chapter 4 can thus be advantageously used in the case of transport reactors also.

Varghese and Varma (1979) have also analyzed the transport-line reactor by considering a simple first-order reaction and assuming the gas and solid phases to move in plug flow. They specifically investigated the influence of particle diameter on reactor conversion and showed that if the velocity behavior of solids in a high-voidage reactor (2–3% solid) can be approximated to the behavior of a single particle, then a significant advantage can be gained by using larger particles.

More recently Robertson and Pratt (1981) have considered a consecutive reaction scheme and obtained the optimal axial temperature distribution in a riser reactor to maximize the yield of the intermediate product. The general results indicate that imposition of an axial temperature profile does not improve the yield over that obtainable by operating at the optimal isothermal temperature. The result appears sufficiently general to suggest operation at the optimal isothermal temperature rather than the imposition of a specified temperature distribution.

15.5. MOVING-BED REACTORS

A moving-bed reactor is essentially one in which solids are conveyed mechanically or under the force of gravity and the gas is allowed to flow either cocurrently or countercurrently (or even crosscurrently). Although the solids move downward, they do so *en masse*, with each particle remaining in the same position relative to its neighbors during its entire transit through the reactor. Both the gas and solid phases are assumed to be in plug flow, which constitutes perhaps the chief distinguishing feature of the moving-bed reactor. The unsteady-state behavior of the individual solid particles coupled with the steady-state behavior of the reactor as a whole makes the moving bed an attractive reactor system.

After a promising start in the forties, the moving-bed reactor has made practically no progress. There has, however, been a revival of interest in recent years, but this appears to be largely restricted to theoretical modeling and development. It seems distinctly likely that with its established viability in the petroleum industry and the increasing use of more active zeolite catalysts for cracking, the moving-bed reactor will regain some of its importance.

The two original processes based on the moving-bed system are the Thermofor and Houdry flow processes. During catalytic cracking of petroleum crude, the catalyst is rapidly covered by a layer of coke, which leads to substantial deactivation, but the original activity can be restored by burning off the coke (see Chapters 8 and 16). In contrast to the fluid catalytic cracking (FCC) catalyst, which cannot be burned entirely free of carbon because of mixing, the Thermofor catalytic cracking (TCC) catalyst can be burned to give a catalyst of low and uniform carbon content. The first commercial installation of a TCC system was based on the airlift principle and was put on-stream at the Mobil Oil Company's Beaumont (Texas) refinery in 1950, while the first Houdry flow process commercial unit, based on the gas-lift principle, was put on-stream at the Sun Oil Company's Toledo (Ohio) refinery, also in 1950 (Faragher et al., 1951; Bland and Davidson, 1967).

Details of the constructional, operational, and other features of the moving-bed reactor have been described by Newton et al. (1945), Sittig (1950), Ardern et al. (1951), and Vener (1955). The main advantages of this reactor may be summarized as follows:

Ease of true cocurrent or countercurrent operation

Ability to handle a wide variety of solids

Little or no problem of dust enrichment in the effluent gas stream

Reasonable pressure drop across the bed

Provision for establishing a temperature gradient along the reactor to permit optimum conversion

Ease of start-up and shutdown

Suitability for handling available and solids material

Provision for a wide range of solids retention time

Uniformity of catalyst residence time

Negligible erosion

Ready adaptability to high-pressure operation

Suitable temperature control by various heat transfer techniques

Several designs of moving-bed reactors have been used.

The blast furnace is perhaps the best-known example. Multiple-hearth furnaces, lime kilns, moving trays, sintering machines, and cat crackers constitute other examples of moving-bed systems. From the point of view of catalytic reactions, a moving-bed reactor (cat cracker) consists of a lift pipe with a hopper at both the lower and upper ends and a reactor system in which reaction and regeneration take place consecutively.

Since moving-bed reactors are largely used for gas–solid reactions, the modeling of these reactors will be considered in Chapter 20. In the case of catalytic reactions, their use is restricted to catalysts subject to rapid deactivation. The design of such reactors for isothermal diffusion-free conditions is considered in Chapter 17.

REFERENCES

Ardern, D. B., Dart, J. C., and Lassiat, R. C. (1951), *Adv. Chem. Ser.*, **5**, 13

Baskakov, A. P., and Vershinina, V. S. (1964), *Int. Chem. Eng.*, **4**, 119.

Bland, W. F., and Davidson, R. L. (1967), *Petroleum Processing Handbook*, McGraw-Hill, New York.

Bourquet, J. M., Drew, R. D., and Valentine, S., III, (1961), *Chem. Eng. Prog. Symp. Ser.*, **57**, (No. 34) 29.

Bowing, K. M., and Watts, A. (1963), *Aust. J. Appl. Sci.*, **14**, 57.

Brauner, H., Muehle, J., and Schmidt, M. (1970), *Chem. Ing.—Tech.*, **42**, 494.

Bryson, M. C., Huling, G. P., and Glausser, W. E. (1972), *Hydrocarbon Process., Petrol. Refiner.* **51**(5), 85.

Buffham, B. A., and Gibilaro, L. G. (1968), *Chem. Eng. Sci.*, **23**, 1399.

Cankurt, N. T., and Yerushalmi, J. (1978), *Fluidization*, eds. Davidson, J. F., and Keairns, D., Cambridge Univ. Press, New York and London, p. 387.

Chen, B. H., and Osberg, G. L. (1967), *Can. J. Chem. Eng.*, **45**, 91.

Faragher, W. F., Noll, H. D., and Bland, R. E. (1951), *Proc. 3d World Petrol. Cong.*, Section IV, p. 138.

Gabor, J. D. (1964), *AIChE J.*, **10**, 345.

Gabor, J. D., and Mecham, W. J. (1964), *Ind. Eng. Chem. Fundam.*, **3**, 60.

Gabor, J. D., Strangeland, B. E., and Mecham, W. J. (1965), *AIChE J.*, **11**, 130.

Hirama, T., Yumiyama, M., Tomita, M., and Yamaguchi, H. (1979), *Int. Chem. Eng.*, **19**, 102.

Jepson, G., Poll, A., and Smith, W. (1965), paper presented at *AIChE Inst. Chem. Eng.* Joint Meeting, London.

Kang, W. K., and Osberg, G. L. (1966), *Can. J. Chem. Eng.*, **44**, 142.

Kato, K., Imafuku, K., and Kubota, H. (1967), *Chem. Eng. Japan*, **31**, 967.

Kato, K., Arai, H., and Ito, U. (1974), *Adv. Chem. Ser.*, **133**, 270.

Kato, K., Ito, H., and Omura, S. (1979), *J. Chem. Eng. Japan*, **12**, 403.

Kato, K., Matsuura, K., and Hanzawa, T. (1980), *Fluidization*, eds. Grace, J. R., and Matsen, J. M., Plenum Press, New York, p. 555.

Kehoe, P. W. K., and Davidson, J. F. (1971), *Inst. Chem. Eng. Symp. Ser.*, **33**, 97.

Ketteridge, I. B. (1962), *Brit. Chem. Eng.*, **7**, 326.

Kiang, K. D., Liu, K. T., Nack, H., and Oxley, J. H. (1975), Paper presented at Int. Conf. Fluidization, Asilomar, California.

Kunii, D., and Levenspiel, O. (1969), *Fluidization Engineering*, Wiley, New York.

Lewis, W. K., Gilliland, E. R., and Glass, W. (1959), *AIChE J.*, **5**, 419.

Li, Y., and Kwauk, M. (1980), *Fluidization*, eds. Grace, J. R., and Matsen, J. M., Plenum Press, New York, p. 537.

Liu, D., Li, X., and Kwauk, M. (1980), *Fluidization*, eds. Grace, J. R., and Matsen, J. M., Plenum Press, New York, p. 485.

Massimilla, L. (1973), *AIChE Symp. Ser.* (No. 128), **69**, 11.

Mathis, J. F., and Watson, C. C. (1956), *AIChE J.*, **2**, 518.

Newton, R. H., Dunham, G. S., and Simpson, T. P. (1945), *Trans. AIChE*, **41**, 215.

Peyman, M., and Laguerie, C. (1980), *Fluidization* eds. Grace, J. R., and Matsen, J. M., Plenum Press, New York, p. 243.

Pierce, W. L., Souther, R. P., and Kaufman, T. G. (1972), *Hydrocarbon Process. Petrol. Refiner.*, **51** (5), 92.

Pillai, B. C., and Raja Rao, M. (1976), *Ind. Eng. Chem. Process Design Dev.*, **15**, 250.

Pratt, K. C. (1974), *Chem. Eng. Sci.*, **29**, 747.

Raghuraman, J., and Verma, Y. B. G. (1973a), *Chem. Eng. Sci.*, **28**, 305.

Raghuraman, J., and Verma, Y. B. G. (1973b), *Chem. Eng. Sci.*, **28**, 585.

Raghuraman, J., and Verma, Y. B. G. (1974), *Chem. Eng. Sci.*, **29**, 697.

Robertson, A. D., and Pratt, K. C. (1975), *Chem. Eng. Sci.*, **30**, 1185.

Robertson, A. D., and Pratt, K. C. (1981), *Chem. Eng. Sci.*, **36**, 471.

Sadek, S. E. (1972), *Ind. Eng. Chem. Process Design Dev.*, **11**, 133.

Sittig, M. (1950), *Chem. Eng.*, **57**, 106.

Strother, C. W., Vermillion, W. L., and Conner, A. J. (1972), *Hydrocarbon Process. Petrol. Refiner*, **51** (5), 89.

Tone, S., Kawamura, K., and Otake, T. (1967), *Kakagu Kogaku*, **31**, 77.

Udilov, V. M., Baskakov, A. P., and Rubtsov, G. K. (1972), *Theoret. Found. Chem. Eng.*, **6**, 443.

Valchar, J. (1965), *Brit. Chem. Eng.*, **10**, 532.

Van Swaaij, W. P. M., Charpentier, J. C., and Villermaux, J. (1969), *Chem. Eng. Sci.*, **24**, 1083, 1097.

Varghese, P., and Varma, A. (1979), *Chem. Eng. Sci.*, **34**, 337.

Vener, R. E. (1955), *Chem. Eng.*, **62**, 175.

Winterstein, G., and Rose, K. (1961), *Chem. Tech. (Berlin)*, **13**, 590.

Wolf, D., and Resnik, W. (1963), *Ind. Eng. Chem. Fundam.*, **2**, 287.

Yerushalmi, J., and Cankurt, N. T. (1978), *Chemtech.*, **8**, 564.

Yerushalmi, J., and Cankurt, N. T. (1980), "Further Studies on Regimes of Fast Fluidization," *Personal communication*.

Yerushalmi, J., Turner, D. H., and Squires, A. M. (1976), *Ind. Eng. Chem. Process Des. Dev.*, **15**, 47.

Yerushalmi, J., Cankurt, N. T., Geldart, D., and Liss, B. (1978), *AIChE Symp. Ser.*, **176**, 1.

Zeigler, E. N., and Brazelton, W. T. (1963), *Ind. Eng. Chem. Process Design Dev.*, **2**, 276.

CHAPTER SIXTEEN

Design and Regeneration of a Fixed-Bed Reactor Subject to Catalyst Deactivation

In Chapter 11 methods of designing a fixed-bed reactor were presented, and in Chapter 8 the mechanism of catalyst deactivation and regeneration with respect to a single pellet was explained. For the purpose of design, the influence of catalyst deactivation on the performance of a catalyst *bed* as a whole is of prime importance. Also, a knowledge of the profiles of the deactivating impurity (mostly carbon) and temperature in the bed during regeneration is necessary for the calculation of heat transfer loads (during regeneration). From the practical point of view, other important considerations are the determination of optimum on-stream and regeneration cycles and of optimum temperature policies. This chapter is concerned with these aspects of fixed-bed reactor design. A few illustrative examples are also solved.

16.1. DESIGN AND PERFORMANCE OF A DEACTIVATING REACTOR

16.1.1. Isothermal Reactor with No Diffusion

In the absence of all diffusional effects, the continuity equation for reactant A based on the dispersion model of the packed-bed reactor developed in Chapter 11 reduces to the simple form

$$\frac{\partial C_A}{\partial t} + u \frac{\partial C_A}{\partial l} = -r_{vA} \tag{16.1}$$

On the assumption of constant density

$$\left(\frac{f_B \rho_g}{M_A}\right) \frac{\partial y_A}{\partial t} + \left(u \frac{\rho_g}{M_A}\right) \frac{\partial y_A}{\partial l} = -r_{vA} \tag{16.2}$$

or

$$\frac{\partial y_A}{\partial t} + \left(\frac{G}{\rho_g f_B}\right) \frac{\partial y_A}{\partial l} = -\frac{\rho_B}{\rho_g}\left(\frac{1}{f_B}\right) r_{wA} \tag{16.3}$$

where ρ_B is the bulk density of the catalyst bed and r'_{wA} the reaction rate in mass per unit mass of catalyst per unit of time ($\sec^{-1}$).

Equation 16.3 can be rendered dimensionless by introducing the following nondimensional time:

$$t_C = \frac{Gt}{f_B \rho_g d_p} \tag{16.4}$$

and the usual reduced length $z' = l/d_p$:

$$\frac{\partial y_A}{\partial t_C} + \frac{\partial y_A}{\partial z'} = -\frac{\rho_B d_p}{G} r'_{wA} \tag{16.5}$$

In a similar manner we can write the continuity equation for carbon deposited in the bed, remembering that the term containing the feed rate will now drop out:

$$\frac{\partial W_c}{\partial t_C} = \frac{f_B \rho_g d_p}{G} r'_{wc} \tag{16.6}$$

where W_c is the weight fraction of carbon on the catalyst and r'_{wc} is the rate of carbon deposition ($\sec^{-1}$).

The solution of the partial differential equations 16.5 and 16.6 is considerably facilitated if we define a time variable such that it has a value of zero at the displacement front of the flowing fluid, rather than at the commencement (Froment and Bischoff, 1961, 1962). Such a time variable is given by

$$t_D = \frac{G}{\rho_g f_B d_p} t' \tag{16.7}$$

where $t' = t - l/u_i$. With this definition of time, the continuity equations for A and carbon become

$$\frac{\partial y_A}{\partial z'} = -\left(\frac{\rho_B d_p}{G}\right) r'_{wA} \tag{16.8}$$

$$\frac{\partial W_c}{\partial t_D} = \left(\frac{\rho_g f_B d_p}{G}\right) r'_{wc} \tag{16.9}$$

To solve Eqs. 16.8 and 16.9, expressions for the rates r'_{wA} and r'_{wc} must be available. Assuming first-order kinetics, we have[†]

Parallel:

$$r'_{wA} = ky_A + k_f y_A \qquad (16.10)$$

or, if $k_f \ll k$,

$$r'_{wA} = ky_A \qquad (16.11)$$

$$r'_{wc} = k_f y_A \qquad (16.12)$$

Consecutive:

$$r'_{wA} = ky_A \qquad (16.13)$$

$$r'_{wc} = k_f(1 - y_A) \qquad (16.14)$$

Solutions for Specific Forms

In its most general form, the rate constant for a decaying catalyst can be expressed as

$$k = \Omega k^0 \qquad (16.15)$$

where Ω has different functional forms (see Table 8.2), and k^0 is the value of the rate constant at zero carbon content. Restricting our attention now to parallel fouling and assuming inverse proportionality ($\Omega = 1/W_c$), we have

$$\frac{\partial y_A}{\partial z'} = -K_1 \frac{y_A}{W_c} \qquad (16.16)$$

$$\frac{\partial W_c}{\partial t_D} = K_2 \frac{y_A}{W_c} \qquad (16.17)$$

where

$$K_1 = \frac{\rho_B d_p}{G} k^0 \qquad (16.18)$$

$$K_2 = \frac{\rho_g f_B d_p}{G} k_f^0 \qquad (16.19)$$

In writing the forgoing equations it was assumed that both reactions are affected to the same extent by fouling. It is possible that the second reaction, which is the actual fouling reaction, remains unaffected by

[†] The general notation k is used for the rate constants (which, in this case, have the dimensions of rate); k_f is the rate constant for the fouling reaction.

carbon deposition. In that case the factor $1/W_c$, or any other functional representation of Ω, will not appear in Eq. 16.17.

In a similar manner, solutions can be obtained for other forms such as linear, exponential, and hyperbolic. These are all presented in Table 16.1. Froment and Bischoff (1961) have given typical plots of the reactant mole fraction as a function of the time group ($K_1 K_2 t_D$) for different positions ($K_1 z'$), and of the carbon content as a function of position at different values of the time group.

Experimental coke profiles have been reported by Van Zoonen (1965) on the hydroisomerization of olefins on silica–alumina catalyst, and by Eberly et al. (1966), Campbell and Wojciechowski (1969), and Pachovsky et al. (1973) on the cracking of gas oil. The results of Lambrecht et al. (1972) and de Pauw and Froment (1975) on the isomerization of *n*-pentane on Pt–Al$_2$O$_3$ show an ascending profile with a certain coke content at the inlet of the reactor, suggesting a series–parallel mechanism. The profile would be descending for parallel coking and ascending (starting from zero) for series coking.

General Solution

The continuity equations can be expressed in their most general form as

$$\frac{\partial y_A}{\partial z'} = -f(p)y_A \qquad (16.20)$$

$$\frac{\partial p}{\partial t_D} = f(p)y_A \qquad (16.21)$$

where $f(p)$ is a function of the poison concentration variable p [which in the case of coke formation is $f(W_c)$]. The boundary conditions are

$$y_A(0, t_D) = y_{A0}(t_D), \qquad p(z', 0) = p_0 \quad (16.22)$$

Using the method of du Domaine et al. (1943) as the basis, Bischoff (1969) has given the following intuitive solution:

$$y_A(z', t_D) = \frac{p(z', t_D) - p_0}{p(0, t_D) - p_0} y_A(0, t_D) \qquad (16.23)$$

The reactant and poison profiles can then be obtained by substituting Eq. 16.23 in 16.20 and 16.21.

$$\int_0^{t_D} y_{A0}(t_D) dg = \int_{p(0,0)}^{p(0,t_D)} \frac{1}{f(p)} dh \qquad (16.24)$$

TABLE 16.1. Solutions to the Continuity Equations for a Catalytic Bed Subject to Parallel Fouling and Operating under Isothermal Diffusion-Free Conditions

Form of the Decay function Ω	Continuity Equations	Solution	Assumptions and/or Remarks
Reciprocal: $\Omega = \dfrac{1}{W_c}$	$\dfrac{\partial y_A}{\partial z'} = -K_1\left(\dfrac{y_A}{W_c}\right)$ $\dfrac{\partial W_c}{\partial t_D} = K_2\left(\dfrac{y_A}{W_c}\right)$	$y_A = 1 - \dfrac{K_1 z'}{\sqrt{2K_2 t_D}}$ $W_c = \sqrt{2K_2 t_D} - K_1 z'$	1. Both reactions are affected to the same extent. 2. $k_f^0 \ll k^0$
Linear: $\Omega = 1 - \hat{W}_c$	$\dfrac{\partial y_A}{\partial z'} = -K_1 y_A(1 - \hat{W}_c)$ $\dfrac{\partial \hat{W}_c}{\partial t_D} = K_2(1 - \hat{W}_c)$	$y_A = \dfrac{\exp(K_2 t_D - K_1 z') - \exp(-K_1 z')}{[1 - \exp(-K_2 t_D)][1 + \exp(K_2 t_D - K_1 z') - \exp(-K_1 z')]}$ $\hat{W}_c = \dfrac{\exp(K_2 t_D - K_1 z') - \exp(-K_1 z')}{1 + \exp(K_2 t_D - K_1 z') - \exp(-K_1 z')}$	1. Both reactions are affected to the same extent. 2. $k_f^0 \ll k^0$
Exponential: $\Omega = e^{-\alpha W_c}$	$\dfrac{\partial y_A}{\partial z'} = -K_1[\exp(-\alpha W_c)]y_A$ $\dfrac{\partial W_c}{\partial t_D} = K_2 y_A$	$y_A = \{1 + \exp(-\alpha K_2 t_D)[\exp(K_1 z') - 1]\}^{-1}$ $\exp(-\alpha W_c) = \{1 + \exp(-K_1 z')[\exp(\alpha K_2 t_D) - 1]\}^{-1}$	1. Main reaction alone is affected. 2. $k_f^0 \ll k^0$
Hyperbolic $\Omega = \dfrac{1}{1 + \beta W_c}$	$\dfrac{\partial y_A}{\partial z'} = -\dfrac{K_1 y_A}{1 + \beta W_c}$ $\dfrac{\partial W_c}{\partial t_D} = \dfrac{K_2 y_A}{1 + \beta W_c}$	$y_A = \exp\left[-K_1 z' + (1 + 2K_2 \beta t_D)^{1/2} - 1 - \beta W_c\right]$ $\beta W_c \exp(\beta W_c) = \left[(1 + 2K_2 \beta t_D)^{1/2} - 1\right]\exp\left[-K_1 z' + (1 + 2K_2 \beta t_D)^{1/2} - 1\right]$	1. Both reactions are affected to the same extent. 2. $k_f^0 \ll k^0$, so that $K_1 = \dfrac{\rho_c d_p(1 - f_B)}{G}\left(k^0 + k_f^0\right)$
General: $\Omega = f(W_c)$	$\dfrac{\partial y_A}{\partial z'} = -f(W_c)y_A$ $\dfrac{\partial W_c}{\partial t_D} = f(W_c)y_A$	$\displaystyle\int_0^{t_D} y_A(0, t_D)\,dy' = \int_0^{W_c(0, t_D)} \dfrac{dW_c}{f(W)}$ (a) $\displaystyle -z' = \int_{W_c(0, t_D)}^{W_c(z', t_D)} \dfrac{dW_c}{(W_c - W_{c0})f(W_c)}$ (b) $y_A(z', t_D) = \dfrac{W_c'(z', t_D) - W_{c0}}{W_c(0, t_D) - W_{c0}}\, y_A(0, t_D)$ (c)	In Eqs. (a), (b), and (c) $f(W_c)$ has been used in place of general functional form $f(p)$, used in Section 16.1.1.

Use (a), (b), and (c) to obtain $W_c(z', t_D)$ and $y_A(z', t_D)$.

$$-z' = \int_{p(0,t_D)}^{p(z',t_D)} \frac{dw}{(w-p_0)f(w)} \qquad (16.25)$$

where g, h, and w are dummy variables.

Using Eqs. 16.23–16.25, the desired expression for $y_A(z',t_D)$ and $p(z',t_D)$ can be obtained. In view of the usefulness of this general solution its application to a specific decay form will be illustrated next. Ozawa (1970) has extended the treatment to reactions of general order.

16.1.2. Example: Use of the General Solution for Obtaining the Solution for a Particular Decay Form

Using the general solution presented above (Eqs. 16.23–16.25), derive the solution for the specific case of reciprocal decay.

Solution

The function $f(p)$ in the general equations developed will now become

$$f(p) = \frac{1}{W_c} \qquad (16.26)$$

The initial conditions for the problem can be written as

$$y_A(0,t_D) = 1, \qquad W_c(z',0) = W_{c0} = 0 \qquad (16.27)$$

Substitution of these values in Eqs. 16.24 and 16.25 leads to

$$\int_0^{t_D} dy' = \int_0^{W_c(0,t_D)} \frac{W_c}{K_2} dW_c \qquad (16.28)$$

and

$$-z' = \int_{W_c(0,t_D)}^{W_c(z',t_D)} \frac{W_c}{K_1(W_c - W_{c0})} dW_c \qquad (16.29)$$

Equations 16.28 and 16.29 can be simplified to

$$W_c(0,t_D) = (2K_2 t_D)^{1/2} \qquad (16.30)$$

$$W_c(z',t_D) - W_c(0,t_D) = -K_1 z' \qquad (16.31)$$

Equations 16.30 and 16.31 when combined yield

$$W_c(z',t_D) = (2K_2 t_D)^{1/2} - K_1 z' \qquad (16.32)$$

The concentrations of reactant and carbon are

related through the equation

$$y_A(z',t_D) = \frac{W_c(z',t_D) - W_{c0}}{W_c(0,t_D) - W_{c0}} y(0,t_D) \qquad (16.33)$$

which for the initial conditions stipulated in Eq. 16.27 reduces to

$$y_A(z',t_D) = \frac{W_c(z',t_D)}{W_c(0,t_D)} \qquad (16.34)$$

Combining Eqs. 16.30, 16.32, and 16.34, we obtain the reactant concentration profile as

$$y_A(z',t_D) = 1 - \frac{K_1 z'}{(2K_2 t_D)^{1/2}} \qquad (16.35)$$

Equations 16.32 and 16.35 represent the carbon and reactant profiles for the case of reciprocal decay.

16.1.3. Isothermal Reactor with Radial Dispersion

The approach to this case consists in combining the continuity equation for reactant A with the rate equation corrected for diffusion through the effectiveness factor for a fouling catalyst. The continuity equation for A is given by

$$-u\frac{\partial C_A}{\partial l} + \frac{D_{eA}}{r}\frac{\partial}{\partial r}\left(r\frac{\partial C_A}{\partial r}\right) - r_{vA} = \frac{\partial C_A}{\partial t} \qquad (16.36)$$

Boundary conditions:

$$l = 0, \qquad C_A = C_{A0}, \qquad R < r < 0, \quad t > 0 \qquad (16.37)$$

$$r = 0 \text{ and } r = R, \qquad \frac{\partial C_A}{\partial r} = 0, \qquad L < l < 0, \quad t > 0 \qquad (16.38)$$

Initial condition:

$$t = 0, \quad L < l < 0, \quad R < r < 0, \quad C_A = 0 \qquad (16.39)$$

Using the transformations

$$z' = \frac{l}{d_p}, \qquad \omega' = \frac{r}{d_p}, \qquad \text{Da} = \frac{k_v d_p}{u_i}$$

$$c = \frac{C_A}{C_{A0}}, \qquad \text{Pe}'_m = \frac{u d_p}{D_{eA}} \qquad (16.40)$$

we can recast Eq. 16.36 along with its boundary conditions as

$$-\frac{\partial c}{\partial z'} + \frac{1}{Pe'_m}\frac{\partial^2 c}{\partial(\omega')^2} - \varepsilon_f \, Da \, c^n = 0 \quad (16.41)$$

$$z' = 0, \qquad c = 1, \qquad \frac{R}{d_p} < \omega' < 0 \quad (16.42)$$

$$\omega' = 0 \text{ and } \omega' = \frac{R}{d_p}, \quad \frac{\partial c}{\partial \omega'} = 0, \qquad \frac{L}{d_p} < z' < 0$$
$$(16.43)$$

where ε_f is the effectiveness factor for fouling.

Equation 16.41 can be solved when the order of the main reaction n is known. For a first-order reaction

$$c(z', \omega') = \sum_{j=1}^{\infty} \exp\left(\frac{Z}{Pe'_m} + Da\varepsilon_f\right)z' \quad (16.44)$$

where

$$Z = \frac{j\pi\omega'd_p}{R} \quad (16.45)$$

Equation 16.44 represents the concentration profile in a fixed-bed reactor for a first-order reaction. It can be reduced to certain well-known expressions. For a high value of Peclet number the reactor behaves as a plug-flow reactor and Eq. 16.44 becomes

$$c(z') = \exp(-Da\varepsilon_f z') \quad (16.46)$$

which can be written in the conventional form as

$$c(z') = \exp\left(-\frac{kl}{u}\varepsilon_f\right) \quad (16.47)$$

This equation reduces to that for the concentration profile for a plug-flow reactor (Chapter 11) when ε_f equals unity. Equation 16.44 is plotted in Figure 16.1, where the variation of concentration in radial and axial directions is shown as a function of system parameters ε_f and Da. This equation is highly convergent, so a value of $j = 1, 2$ can be used without causing much error in computation.

Masamune and Smith (1966) have obtained numerical solutions to the one-dimensional model which gives the normalized concentration of reactant A as a function of reduced position, time, and Thiele modulus. They also assumed that the diffusivity in the poisoned shell is the same as that in the unpoisoned core (which

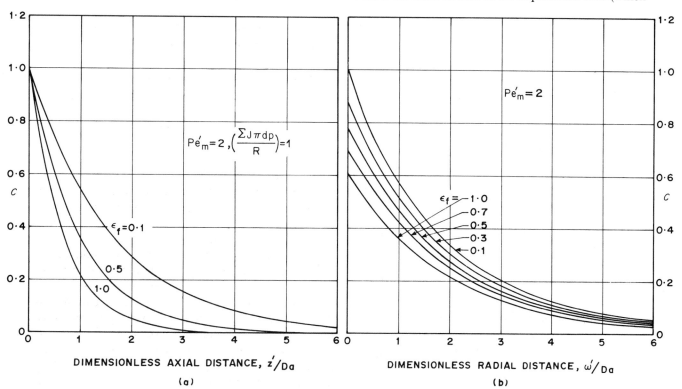

Figure 16.1. Concentration profiles in a deactivating reactor for a case of parallel fouling. (a) Axial profiles in a fixed-bed reactor at various values of ε_f; (b) radial profiles at various values of ε_f.

has a sharp boundary with the shell). The results of Masamune and Smith (1966) show that catalysts with relatively large intraparticle resistance lose their activity at a slow rate even though the initial activity is low. Clearly, then, the effect of intraparticle diffusion should not be viewed as something detrimental to the performance of a fixed-bed reactor. Rather, the optimum design of a reactor might require a catalyst with significant diffusion resistance.

It is also possible to apply the shell-progressive model described earlier (Section 8.2.3) to the fixed bed as a whole. Thus, in keeping with the concept of pore-mouth poisoning, it can be postulated that the inlet region of the reactor has developed pore-mouth poisoning which decreases progressively toward the reactor exit. A quantitative description of this model by Olson (1968) enables us to determine the allowable time of reaction (i.e., on-stream time) before regeneration is undertaken.

By analogy with the behavior of an ion-exchange column, Olson (1968) has shown that the performance of an isothermal deactivating bed is determined by the following groups:

1. Number of solid diffusion transfer units

$$N_s = \frac{3D_{ep}l}{R^2 u_i} \qquad (16.48)$$

2. Damköhler number for the poisoning reaction

$$Da = \frac{k_{Sp}R}{D_{ep}} \qquad (16.49)$$

3. Thiele modulus, defined in the usual manner,

$$\phi_{s1} = R\left(\frac{k_v}{D_{eA}}\right)^{1/2} \qquad (16.50)$$

4. Biot number for mass transfer (based on pellet radius)

$$(Bi)_m = \frac{k_g R}{D_{eA}} \qquad (16.51)$$

5. A dimensionless time

$$t_E = \left(\frac{u_i t}{L} - \frac{l}{L}\right)\frac{f_B v_s C_{A0}}{(1-f_B)(C_A^*)_s} \qquad (16.52)$$

where $(C_A^*)_s$ is the equilibrium surface concentration of A.

The new concept introduced is that of the diffusion transfer unit N_s, which is a measure of the penetration of poison within the pellet at a given position in the bed. The Damköhler number is also based on the reaction rate of the poison, and is a measure of the relative rates of reaction and diffusion in the pellet. On the other hand, the Thiele modulus is a measure of the main reaction in relation to intraphase diffusion.

The effectiveness factor for a pellet with an unpoisoned core radius of r_i is given by

$$\varepsilon = 3\phi_{s1}^{-2}\left[(Bi)_m^{-1} + \frac{1-r_i}{r_i} + \frac{1}{r_i[r_i\phi_{s1}\coth(r_i\phi_{s1})-1]}\right]^{-1} \qquad (16.53)$$

Olson defined an activity ratio for the bed by averaging the effectiveness factor calculated from this equation and dividing by the effectiveness factor in the absence of deactivation:

$$\text{activity ratio} = \frac{\bar{\varepsilon}_f}{\varepsilon} \qquad (16.54)$$

Plots of the activity ratio as a function of time for different values of the reaction and poison parameters ϕ_{s1} and N_s were prepared, based on which the following qualitative rule may be formulated for increasing the on-stream time of the fixed bed:

Thiele Modulus	Effective Diffusivity of Poison (N_s)
Low (< 10)	Decrease
High (~ 100)	Increase

In the case of gels this can be accomplished by varying the gelation procedure, while in the case of pellets the adjustable parameter is pressure. Thus, in theory, a catalyst can be more or less tailored to give the most economical on-stream performance.

It must be clear from the forgoing discussion that an important consideration in determining the optimum value of N_s is the fractional penetration of poison into a pellet as a function of pellet distance (from entrance) in the bed. Olson has prepared plots for low (0.1), intermediate (0.33), and high (1.0) values of N_s for two values of the dimensionless time t_E: 0.4 and 0.6. The conclusion from these plots is that the magnitude of N_s has a profound influence on the extent of poisoning at a given position in the bed. For large N_s (i.e., $N_s = 1$) and for $t_E = 0.4$ there is practically no poisoned shell

$(\hat{R} \simeq 1)$ beyond a bed distance of $z = 0.7$. On the other hand, for $t_E = 0.6$ and for the same value of N_s, the initial portion of the bed is saturated with the poison.

16.1.4. Simulation of Sulfur Poisoning in Steam Reforming: A Case Study

The steam-reforming process is used to produce hydrogen by converting hydrocarbons with steam in heated tubes filled with nickel catalyst. The catalyst is subject to poisoning by the reversible adsorption of sulfur in the bed. The reaction may be described by a Langmuir isotherm where the concentration term is the ratio of the mole fractions of hydrogen sulfide (A) and hydrogen (B). Thus

$$\theta = \frac{K y_A / y_B}{1 + (K y_A / y_B)} \quad (16.55)$$

where A is hydrogen sulfide, B hydrogen, and K the equilibrium constant.

The operating conditions and the calculated transport properties may be assumed to vary with axial distance but are time invariant. The transient profiles may thus be calculated with fixed conversion and axial temperature profile. Also, the temperature and concentration of hydrogen are assumed to be invariant in the catalyst particle.

Mathematical Model[†]

The mass balance for hydrogen sulfide in the gas phase is

$$\frac{-\partial(G_M y_{Ab})}{\partial l} = \frac{3k_g}{R}(1 - f_B)(y_{Ab} - y_{As}) \quad (16.56)$$

where G_M is the molar flow rate, R the pellet radius, and f_B the void fraction; y_{Ab} and y_{As} are, respectively, the bulk and surface mole fractions in the gas phase; and l the axial distance.

The mass balance for hydrogen sulfide in the particle is

$$\frac{\partial C_s}{\partial t} = D_e\left(\frac{\partial^2 C_A}{\partial r^2} + \frac{2}{r}\frac{\partial C_A}{\partial r}\right) \quad (16.57)$$

where C_s is the concentration of adsorbed sulfur in

[†] Note that in this section the gas-phase concentration c_A is expressed as the mole fraction y_A; consequently the phenomenological mass transfer coefficient, k_g has the units of mol/sec cm^2.

moles per unit volume of catalyst, and C_A is the concentration in the gas phase inside the particle.

The boundary conditions in the particle center and at the particle surface are

$$\frac{\partial C_A}{\partial r}\bigg|_{r=0} = 0; \quad -D_e\frac{\partial C_A}{\partial r}\bigg|_{r=R} = k_g(y_{As} - y_{Ab}) \quad (16.58)$$

The initial conditions for the two partial differential equations are

$$C_A = 0 \quad \text{for} \quad t = 0, \quad \text{all } l, r \quad (16.59a)$$

$$y_{Ab} = y_{A0} \quad \text{for} \quad l = 0, \quad \text{all } t \quad (16.59b)$$

The dependent variables may be transformed by using the relationship between the solid concentration C_s and the sulfur coverage θ. Thus Eqs. 16.56 and 16.57 reduce to

$$-\frac{\partial x_b}{\partial l} = K_3 K_4(x_b - x_s) + x_b\frac{\partial \ln G_M}{\partial z} \quad (16.60)$$

and

$$\frac{\partial \theta}{\partial t} = K_1 K_2\left(4\hat{R}\frac{\partial^2 x}{\partial \hat{R}^2} + 6\frac{\partial x}{\partial \hat{R}}\right) \quad (16.61)$$

with the boundary conditions

$$-\frac{\partial x}{\partial \hat{R}}\bigg|_{\hat{R}=1} = \frac{(\text{Bi})_m}{2}(x_b - x_s) \quad (16.62)$$

and the initial conditions

$$x = 0 \quad \text{for} \quad t = 0, \quad \text{all } z, \hat{R} \quad (16.63a)$$

$$x_b = 1 \quad \text{for} \quad z = 0, \quad \text{all } t \quad (16.63b)$$

Here $K_1 = y_{A0}D_eC_t$; $K_2 = M_sC_s/(s_c\rho_gR^2)$; $K_3 = k_g/G_M$; $K_4 = 3(1 - f_B)/R$; $(\text{Bi})_m = k_gR/C_tD_e$; x is the dimensionless gas concentration y_A/y_{A0}; $C_s = (\theta\rho_g s_c)/M_s$; s_c is the capacity factor for sulfur in kg S/kg particle; and M_s is the molecular weight of sulfur.

Solution

The number of variables may be reduced by the discretization of the pellet equation (16.63) in the radial direction by means of the orthogonal collocation technique (for details see Villadsen and Michelsen, 1978). The equation then reduces to N coupled first-order differential equations with time as the independent variable; N is the number of interior collocation points and $NT = N + 1$ is the total number of collocation points. Thus Eqs. 16.62 and 16.63 after rearrangement reduce to

$$-\frac{\partial x_b}{\partial z} = K_3 K_4 \left(\sum_{j=1}^{N} E_j x_j + E_{NT} x_b \right) \qquad (16.64)$$

and

$$\frac{\partial \theta}{\partial t} = K_1 K_2 \sum_{j=1}^{NT} C_{ij} x_j + C_{i,NT} x_b) + x_b \frac{\partial \ln G_M}{\partial z} \qquad (16.65)$$

where the following auxiliary variables have been introduced:

$$C_{ij} = D_{ij} - \frac{D_{i,NT} A_{NT,j}}{K_5}; \qquad C_{i,NT} = D_{i,NT}\left(1 - \frac{(Bi)_m}{2K_5}\right)$$

$$E_j = \frac{A_{NT,j}}{K_5}, \qquad D_{ij} = 4R_i B_{ij} + 6A_{ij}$$

and

$$K_5 = \frac{(Bi)_m}{2} + A_{NT,NT} \qquad (16.66)$$

The resulting coupled first-order differential equations 16.64 and 16.65 are solved for the dependent variables, θ and x, by division of the t–z plane into a rectangular grid. The integration is then performed through the grid for a given time using a numerically stable modified Eüler method (for details see Christiansen and Andersen, 1980).

The average sulfur coverage is determined by integration of θ in spherical geometry

$$\bar{\theta} = \frac{3}{2} \int_0^1 \theta(R) R^{1/2} \, dR \qquad (16.67)$$

The integral is evaluated by Gaussian quadrature.

The authors present sulfur poisoning profiles in a typical naphtha-based 1000 tonnes/day ammonia plant. Three different cases with 0.03, 0.3, 3.0 weight ppm sulfur on naptha basis are considered. Figure 16.2a shows the breakthrough curves for the three cases. It may be noted that breakthrough occurs immediately, indicating the serious effect of a sulfur peak in the reformer feed on the downstream catalyst. Figure 16.2b shows the catalyst poisoning profiles in the tubes. This is a plot of the average sulfur coverage as a function of axial distance with time as a parameter. Note that there is no sulfur front moving through the bed. Figure 16.2c shows pellet profiles for different times at the axial distance 2.5 m. For low concentrations the poisoning is almost homogeneous, whereas at higher concentrations it is a core–shell model.

Christiansen and Andersen admit, however, that the practical verification of the model is difficult because of the very limited experimental information. They do note that calculated and experimental results from data available from a plant with 0.03 ppm sulfur show fair agreement. They note that besides sulfur poisoning of the catalyst, sintering also takes place, resulting in a decrease in the nickel surface area, and hence in the sulfur capacity of the tubes. Also, the agreement between the calculated and experimental profiles is not good, the experimental profile being steeper than the calculated one.

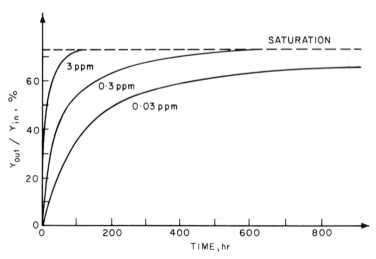

Figure 16.2a. Calculated breakthrough curves (Christiansen and Andersen, 1980).

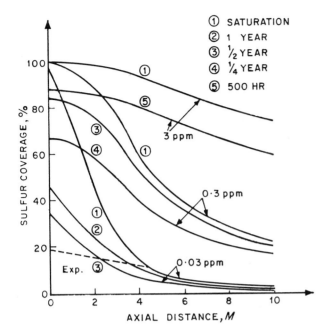

Figure 16.2b. Catalyst poisoning profiles in the reactor tube (Christiansen and Andersen, 1980).

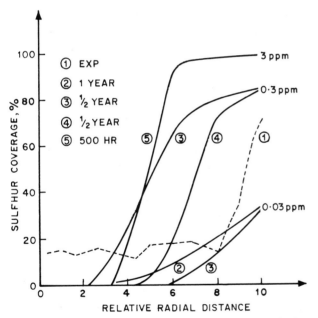

Figure 16.2c. Catalyst pellet profiles at a distance of 2.5 m in the tube (Christiansen and Andersen, 1980).

In conclusion, the model is in fair quantitative agreement with scarce measurements, but in good qualitative agreement with industrial experience.

16.1.5. Guard-Bed Reactor

In the case of poison that can be removed before the reactant is introduced into the reactor, the so-called

guard-bed reactor can be used, which will permit only a predetermined leakage of poison in the main reactor. The primary consideration then in the design of the guard-bed reactor is the time required for the poison to breakthrough the bed. Either the catalyst can be replaced immediately thereafter, or the same catalyst can be continued until the poison in the effluent exceeds the permitted level.

Here again the breakthrough concentration of poison is greatly influenced by the magnitude of the diffusion transfer unit N_s. If it is required to reduce the poison concentration to 1 % of its initial value (say from 500 to 5 ppm) any value of $N_s < 1.0$ is clearly ineffective. A threefold increase in N_s can increase the guard-bed life twelvefold. Thus, for efficient operation of the guard bed, D_e should be increased and particle diameter decreased, since both these changes lead to an increased value of N_s.

It is also possible to determine the overall average activity of a reactor on the assumption of a specific distribution of poison (Anderson and Whitehouse, 1961; Wheeler and Robell, 1969), but this method is of limited practical use.

16.1.6. Nonisothermal Reactors

In the case of a reactor operating nonisothermally it becomes necessary to take into account the heat balance equation in addition to the species conservation equations. Here again we may consider a reactor with no dispersion operated in an adiabatic or nonadiabatic mode. Alternatively, if dispersion is important, we can use any of the pseudohomogeneous dispersion models.

The nonisothermal (adiabatic) fixed-bed reactor with no dispersion gives rise to the following set of equations:

$$\frac{\partial y_A}{\partial z'} = -M_1 r'_{wA}, \quad \frac{\partial w_c}{\partial t_D} = M_1 f_B r'_{wc}, \quad \frac{\partial T}{\partial t_D} = M_2 r_{wA}$$

$$(16.68)$$

where

$$M_1 = \frac{\rho_B f_B d_p}{G}, \qquad M_2 = \frac{-\Delta H}{C_{ps} T_0} \frac{\rho_g f_B d_p}{G} \qquad (16.69)$$

In writing these equations the assumption has been made that the accumulation of energy over a period t in the void space of the bed is negligible in comparison with that in the solids. Solutions for parallel fouling and inverse proportionality ($\Omega = 1/w_c$) using the rate equations described by 16.16 and 16.17 have been given by Bischoff (1969). A conclusion from these solutions is that the maximum temperature, and the distance at which it occurs, are functions of time. This is consistent with the experimental finding of Sreeramamurty and Menon (1968) on the oxidation (by air) of hydrogen sulfide on a deactivating carbon catalyst.

The case of a pseudohomogeneous model with axial dispersion of heat and mass has been analyzed by Price and Butt (1977), Weng et al. (1975), and Billimoria and Butt (1981) with a view to explaining the observed reactor behavior in the hydrogenation of benzene on kieselguhr. The computer simulation of Weng et al. (1975) was based on nonadiabatic operation, and the results were extended to adiabatic operation by Price and Butt (1977). The case of start-up and quasi-steady-state profiles in fresh and deactivated adiabatic beds has been analyzed by Billimoria and Butt (1981). The mathematical models assumed linear poisoning kinetics, that is, kinetics of the separable type (see 16.3), and interphase and intraparticle gradients were ignored. While these analyses were restricted to specific systems, some general conclusions from the numerical results can be stated.

1. Much like the case of nondeactivating beds, the dynamics of deactivating reactors is dominated by the thermal properties of the bed. The magnitude of the exotherms and the speed of the response are both severely affected as a result of variations in the thermal properties of the bed.

2. The transient analysis of the deactivating reactor indicates the presence of three distinct periods: a fast concentration response period, where the concentration rapidly attains a steady-state value while the temperature keeps changing slowly; a slow response period, where both concentration and temperature evolve slowly toward a steady state; and a third period, where the developed profiles move down the bed with constant velocity.

A more rigorous analysis using a heterogeneous model of the fixed bed is possible but tedious, and is not warranted unless a specific reaction system demands it.

16.2. REGENERATION OF A FOULED FIXED-BED REACTOR

In developing the differential equations for regeneration, two cases have to be considered: (1) where the carbon burning rate is independent of temperature; and (2) where the rate is temperature sensitive. The first case corresponds to diffusional limitation of the regeneration process and the second to intrinsic reaction control. The transition between the two regions occurs around 500°C (Weisz and Goodwin, 1966). Whereas in many cases of practical importance regeneration is carried out in the diffusional regime (e.g., regeneration of silica–alumina cracking catalysts), in a few instances (regeneration of zeolite catalysts) regeneration is done in the kinetic regime in view of the deleterious effect of high temperature on the catalyst. In the present treatment we shall first consider the mathematically simpler diffusion regime and then proceed to the kinetic regime. While in both the regimes the burning rate is proportional to the oxygen partial pressure and fraction of carbon remaining and follows an Arrhenius-type equation (Zhorov et al., 1967), in the diffusion-controlled region the activation energies are of the order of 6–16 kcal/mol (Gonzales and Spencer, 1963; Johnson et al., 1962; Schulman, 1963) and the assumption of temperature independence is largely justified. According to Weisz and Goodwin (1966), activation energies of the order of 40 kcal/mol are involved in the kinetic regime. It may be noted that van Deemter's assumption (1953, 1954) of zero-order burning is an oversimplification and is therefore not considered.

16.2.1. Regeneration in the Diffusion Regime

The following continuity equation can be written for oxygen directly from the general plug-flow model (Eq. 11.3) developed in Chapter 11 with all the radial and axial dispersion terms omitted:

$$\left(\frac{f_B \rho_g}{M_g}\right)\frac{\partial y}{\partial t} + \left(\frac{G}{M_g}\right)\frac{\partial y}{\partial l} = -\frac{r'_{cb}\rho_B}{M_c} \quad (16.70)$$

with

$$y = y_0 \quad \text{at} \quad l = 0$$

where y is the mole fraction of oxygen and r'_{cb} (sec^{-1}) is the carbon burning rate. The carbon balance is given by

$$\frac{\partial W_c}{\partial t} = r'_{cb} \quad (16.71)$$

with

$$W_c = W_{c0} \quad \text{at} \quad t < 0 \quad (16.72)$$

The rate of carbon burning is

$$r'_{cb} = k_{cb} y \hat{W}_c \quad (16.73)$$

where k_{cb} has the units sec^{-1} and will be assumed to be independent of temperature.

Equations 16.70 and 16.71 can now be solved by substituting Eq. 16.73 for the burning rate and normalizing through the following new dimensionless groups (Johnson et al., 1962):

$$z_F = l\left(\frac{p_B}{G}\frac{M_g}{M_c}k_{cb}\right) \quad \text{(a new normalized length)}$$

$$(16.74)$$

$$t_F = \left(\frac{y_0}{W_{c0}}k_{cb}\right)\left[t - l\left(\frac{f_B\rho_g}{G}\right)\right] = \left(\frac{y_0}{W_{c0}}k_{cb}\right)t'$$

$$(16.75)$$

(a new normalized time)

The final solution is

$$\hat{W}_c = \frac{W_c}{W_{c0}} = [1 + e^{-c'z_F}(e^{t_F}-1)]^{-1} \quad (16.76)$$

$$\hat{y} = \frac{y}{y_0} = [1 + e^{-t_F}(e^{c'z_F}-1)]^{-1} \quad (16.77)$$

where c' is a stoichiometric correction factor.

To account for temperature variation in the bed as a function of time and distance, the following energy balance equation can be written:

$$\left[(1-f_B)\rho_c C_{ps} + f_B\rho_g C_p\right]\frac{\partial T}{\partial t} + (C_p G)\frac{\partial T}{\partial l}$$

$$+ \frac{UA_s\Delta T}{A_c} = r'_{cb}\rho_B(-\Delta H') \quad (16.78)$$

where C_p and C_{ps} are the heat capacities of the fluid and catalyst packing, respectively; A_s is the surface area per unit length, and A_c the cross-sectional area of the reactor. This equation can be simplified by making use of the fact that the term $(1-f_B)\rho_c C_{ps}$ is much larger than the term of $f_B\rho_g C_p$. Thus the energy balance becomes

$$\rho_B C_{ps}\frac{\partial T}{\partial t} + C_p G\frac{\partial T}{\partial l} + \frac{UA_s\Delta T}{A_c} = \rho_B k_{cb}y\hat{W}_c(-\Delta H')$$

$$(16.79)$$

With this equation as the basis, we shall now consider two types of regeneration, adiabatic and isothermal.

Adiabatic Regeneration

If the regeneration is adiabatic, the term $UA_s\Delta T/A_c$ vanishes, and we obtain

$$A\frac{\partial T_F}{\partial t_F} + B\frac{\partial T_F}{\partial z_F} = \hat{y}\hat{W}_c \quad (16.80)$$

where T_F is a dimensionless temperature:

$$T_F = T\frac{C_{ps}}{(-\Delta H')W_{c0}} \quad (16.81)$$

$$A = 1 - \frac{C_p}{C_{ps}}\frac{f_B\rho_g}{p_B} \quad (16.82)$$

$$B = \frac{M_g}{M_c}\frac{C_p}{C_{ps}}\frac{W_{c0}}{y_0} \quad (16.83)$$

On the assumption that there is no initial profile in the bed (which is valid for most engineering purposes) Johnson et al. (1962) give the following solution to this equation:

$$T_F = \frac{1}{2(c'B-A)}\left[\tanh\frac{c'z_F-t_F}{2} - \tanh\frac{z_F-(B/A)t_F}{2\,B/A}\right]$$

$$(16.84)$$

Figure 16.3 shows a plot of the temperature rise T_F as a function of the time t_F for various positions z_F, while 16.4 is a plot of T_F as a function of position for various times. An inspection of Eq. 16.82 shows that A is approximately unity; also, B usually has a value between 1 and 2. Thus, in preparing these figures it has been assumed that $A = 1$ and $B = 1.5$. It can be noticed that each position in the reactor experiences an instantaneous temperature maximum, the actual value depending on the magnitude of the parameters c', A, and B. Similarly, there is a maximum temperature experienced by the bed as a whole (at any one position) at a given time.

The maxima represented by the curves of Figures 16.3 and 16.4 may be defined, respectively, as

$$(T_F)_{max,\,z_F} \quad \text{given by} \quad \left.\frac{\partial T_F}{\partial t_F}\right|_{z_F} = 0 \quad (16.85)$$

and

$$(T_F)_{max,\,t_F} \quad \text{given by} \quad \left.\frac{\partial T_F}{\partial z_F}\right|_{t_F} = 0 \quad (16.86)$$

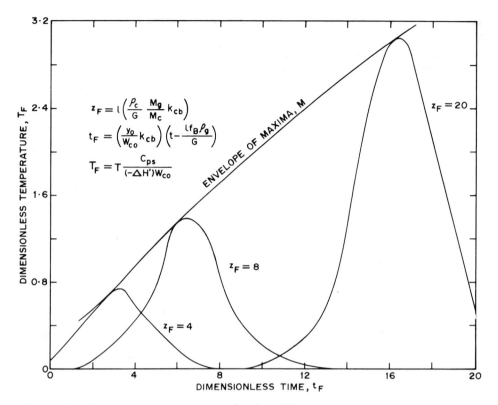

Figure 16.3. Dimensionless temperature as a function of dimensionless time at various positions (z_F denote dimensionless distance) (Johnson et al., 1962).

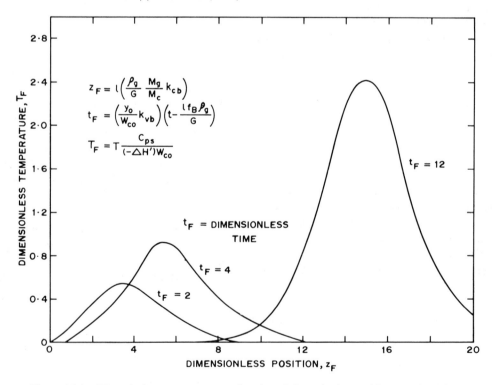

Figure 16.4. Dimensionless temperature as a function of dimensionless position at various times (Johnson et al., 1962)

From the design point of view $(T_F)_{max, z_F}$, which represents the maximum temperature rise experienced at any position during the entire regeneration cycle, is the more important parameter, and is obtained simply by differentiating Eq. 16.84 with respect to t_F at constant z_F:

$$(T_F)_{max, z_F} = \frac{1}{c'B - A}\left[\tanh\frac{z_F A}{4B}(c'B - 1)\right] \quad (16.87)$$

Curve M of Figure 16.3 represents the envelope of the instantaneous maxima as given by Eq. 16.87. Not only does the maximum shift toward the outlet, but its value also increases progressively.

Johnson et al. (1962) suggest the following equation for calculating the total regeneration time t_{Ft}:

$$t_{Ft} = c'z_F + \ln\left[\frac{1 - (W_c/W_{c0})_f}{(W_c/W_{c0})_f}\right] \quad (16.88)$$

If the final carbon content $(W_c/W_{c0})_f$ is to be zero, then evidently the total regeneration time t_{Ft} would be infinity.

A plot of $(T_F)_{max, z_F}$ as a function of z_F for various combinations of the parameters A and B can be prepared. The value of B can be altered (mainly by manipulating the proportion of oxygen y_0, since the other terms in B cannot be easily changed) such that the temperature does not rise beyond a stipulated maximum anywhere within the reactor.

Regeneration under Isothermal Cooling

In this case the conservation equation is given by Eq. 16.79, which assumes no radial profile. The solution is, for $A \simeq 1$, $c'B \simeq 1$:

$$T_F = \frac{1 - \exp(Dt_F)}{2 + \exp(c'z_F - t_F) + \exp(-c'z_F - t_F)} \quad (16.89)$$

where

$$D = \frac{UA_s W_0}{A_c C_{ps} k_{cb} \rho_B y_0} \quad (16.90)$$

The assumption that both A and $c'B$ are equal to unity is justified in the majority of cases. The parameter D defines the cooling rate. A plot of this equation shows that the profile at any position travels in a fixed pattern with time, and that beyond a certain value of D the cooling rate does not have any influence on the profile. The maximum value of the reduced temperature is 0.25.

The problem of regeneration has also been consi-

dered by van Deemter (1953), Schulman (1963), and Olson et al. (1968). Johnson et al.'s procedure outlined in this chapter is perhaps the most relevant from the design point of view. It correctly predicts the moving temperature maxima in a regenerator and emphasizes the importance of the partial pressure of oxygen in the regenerating gas; the latter is borne out by another mathematical model proposed by Panchenkov et al. (1969).

16.2.2. Regeneration in the Kinetic Regime

With the basic equations given in Section 16.2.1 it is also possible to analyze the regeneration process in the kinetic region provided the rate constant for carbon burning (Eq. 16.73) can be expressed in the usual Arrhenius form,

$$k_{cb} = A_{f,cb}\exp\left(-\frac{E}{R_g T}\right) \quad (16.91)$$

With this modification, Ozawa (1969) has solved the equations in terms of an oxygen transit time t_O and a heat transit time t_H:

Oxygen transit time: $\quad t_{O_2} = \frac{f_B \rho_g}{G}L = \frac{V_r f_B}{F_w} \quad (16.92)$

Heat transit time: $\quad t_H = \frac{L}{u_H} = 1 + \frac{(1 - f_B)}{f_B}\frac{\rho_B C_{ps}}{\rho_g C_p}\frac{V_r f_B}{F_w}$ $\quad (16.93)$

These represent the total time required for the oxygen or heat front to travel one full length of the reactor.

The gas transit time is normally very much smaller than the heat transit time because of the far greater density of catalyst particles compared with that of a gas. Ozawa has shown that $t_{O_2} < 0.01\, t_H$. It is therefore adequate to express regeneration time as a function of the heat transit time t_H. Profiles of temperature, oxygen, and coke for a typical set of values have been calculated by Ozawa for $t = t_H$, $t = 0.75t_H$, and $t = 0.5t_H$. His calculations show that the coke concentration exhibits a minimum in the axial direction; obviously this corresponds to a maximum in the carbon burning rate (caused by the effect of increasing temperature along the reactor). Ozawa has further shown that this minimum moves toward the end of the reactor and that the combustion zone does not travel with a constant velocity, as it does during regeneration in the kinetic regime.

16.3. SEPARABLE AND NONSEPARABLE KINETICS

In all the cases considered so far in Chapters 8 and 16, the catalyst activity was assumed to be a monotonic function of time and the equations describing the activity were obtained under constant operating conditions. It is advantageous sometimes to operate a commercial reactor under changing conditions (Butt, 1972), in which case the reaction rate at any instant would be a function not merely of time, but also of the entire past history of the catalyst. Thus

$$r = r \text{ (present conditions, past history)} \quad (16.94)$$

Equation 16.94 has been called a "complete rate equation" by Szepe and Levenspiel (1968). In studying the kinetics of catalytic reactions under conditions of catalyst deactivation, the general objective is to determine this functional relationship. The attendant phenomenon of deactivation makes it more difficult to arrive at the exact form of this equation, which is already difficult even for a simple catalytic reaction without deactivation. We are forced therefore to search for a suitable functional form that is simple and theoretically acceptable.

Thus we define the simplest possible form of Eq. 16.94 as

$$r = r_0 \text{ (present conditions)} \times \Omega \text{ (past history)} \quad (16.95)$$

This rate equation is called a complete separable rate equation, and under these conditions the kinetics of the reaction remains unchanged by deactivation. The activity, which is a function of both the present condition and past history of the catalyst, becomes (under this restriction of separable kinetics) a function only of the past history. Hence its variation can be studied independently of the reaction time. This was the underlying assumption in all the developments presented in the previous sections.

Szepe and Levenspiel (1968) proposed a deactivation rate expression

$$\frac{d\Omega}{dt} = (T, C, \Omega) = \psi_1(T)\psi_2(C)\psi_3(\Omega) \quad (16.96)$$

similar to the rate equation

$$r_0 = \psi'(T,C) = \psi'_1(T)\psi'_2(C) \quad (16.97)$$

where $\psi'_1(T) = A_f \exp(-E/R_g T)$ and $\psi'_2(C)$ is some function of concentration. If ρ_a represents the density of active sites (the unoccupied sites plus the sites on which reversible adsorption of reactant, product, or intermediate species occurs), Eq. 16.96 can be written as

$$\frac{d\Omega}{dt} = \psi''(T, C, \rho_a) \quad (16.98)$$

Under the assumption that the surface is uniform and the adsorbed species on adjacent sites are noninteracting, this equation becomes

$$r = \rho_a \psi''_1(T, C) \quad \text{(constant activity distribution)} \quad (16.99)$$

or

$$r = \psi''_2(\rho_a)\psi''_1(T, C) \quad \text{(varying activity distribution)} \quad (16.100)$$

These equations (Gavalas, 1971) include the H–W type of kinetics also.

The plug-flow equation under conditions of no axial or radial dispersion and at steady state (deactivation being assumed to be very slow in comparison with the main reaction) can be described by the following nonseparable form of kinetics:

$$\frac{d(uC)}{dl} = G_M, \quad \frac{dy}{dl} = f_1(\rho_a, T, C) \quad (16.101)$$

Equation 16.101 has been solved by Gavalas (1971) to give

$$y = f_2\left(\frac{S}{G_M, y_0, T_0}\right) \quad (16.102)$$

where

$$S = \int_0^L \rho_a(l)\, dl$$

represents the cumulative activity, and f_1 and f_2 represent the general functional forms 16.101 and 16.102, respectively. Equation 16.102, based on separable kinetics, states that the trajectory in the concentration space is independent of the molar velocity G_M and the activity profile $\rho_a(l)$. The reactor output depends on SL/G_M, which remains unchanged upon flow reversal, and upon mixing and repacking of catalyst.

Thus separable kinetics fails to describe the dependence of product distribution upon the direction of flow through a reactor with a poison gradient. It has been found, however, that in the case of dehydration of alcohol over poisoned silica–alumina catalyst, the rates of formation of ether and ethanol change by 10–30% each upon flow reversal in the reactor with a poison gradient (Bakshi and Gavalas, 1975). The selectivity is reported to vary by 17%. Such flow-directional effects can only be explained on the basis of nonseparable kinetics and can in theory be exploited to optimize the product distribution in a commercial reactor subject to catalyst deactivation.

16.4. OPTIMUM TEMPERATURE POLICIES IN REACTORS SUBJECT TO CATALYST DEACTIVATION

One of the frequently used methods of counteracting the effect of deactivation is to progressively increase the temperature. An example of this is the catalytic hydrogenation of nitrobenzene where the temperature is raised from 180°C to 260°C over a period of 200 days. Another is the catalytic reforming of naphtha, but in this case there is an upper limit of temperature dictated by selectivity considerations.

Several authors (see Chapter 12) have determined the optimum temperature policies for nondeactivating reactors; more recently the anlaysis has been extended to include catalyst deactivation for a first-order irreversible reaction (Szepe, 1966; Chou et al., 1967; Butt and Rohan, 1968; Butt, 1970; Szepe and Levenspiel, 1968; Ogunye and Ray, 1968; Crowe, 1970; Lee and Crowe, 1970a, b; Crowe and Lee, 1971; Park and Levenspiel, 1976; Levenspiel and Sadana, 1978; Sadana, 1980) as well as for a reversible reaction (Jackson, 1965; Chou et al., 1967; Drouin, 1969; Haas et al., 1974).

In what follows we shall consider the performance of a plug-flow reactor as well as of a fully mixed batch reactor. Both reactors yield analytical solutions that can be directly used in the optimization of flow reactors.

16.4.1. Irreversible Reaction

For a solid-catalyzed reaction in an isothermal plug-flow fixed-bed reactor and for first-order reaction ($A \rightarrow B$) and deactivation kinetics with Arrhenius temperature dependencies, that is,

$$-r_A = kC_A\Omega \quad \text{with} \quad k = A_f \exp\left(\frac{-E}{R_g T}\right)$$
(16.103)

$$\frac{-d\Omega}{dt_p} = k_f\Omega \quad \text{with} \quad k_f = A_{f,f} \exp\left(\frac{-E}{R_g T}\right)$$
(16.104)

the optimum temperature operations criterion may be obtained from the Pontryagin maximum principle (Pontryagin et al., 1962) as

$$k\Omega = \text{constant}$$
(16.105)

This optimum operations criterion was initially proposed by Szepe and Levenspiel (1968) for a batch reactor and later by others (Chou et al., 1967; Ogunye and Ray, 1968; Crowe and Lee, 1971; Crowe, 1970; Crowe and Gruyaert, 1976; Lee and Crowe, 1970a, b; Levenspiel and Sadana, 1978; Park and Levenspiel, 1976) for a fixed-bed reactor. The net effect of this criterion is to keep the conversion of the reactant unchanged during the run. Recently, a modified optimum temperature operations criterion

$$\frac{k\Omega}{T^n} = \text{constant}$$
(16.106)

has been proposed (Sadana, 1980) for constant catalyst loading and feed rate in deactivating fixed-bed reactors. By optimum operations we mean that way of running the reactor which gives

highest $\bar{x}_A$ for given Ω_f and t_{p1},
highest Ω_f for given $\bar{x}_A$ and t_{p1},
longest t_{p1} for given $\bar{x}_A$ and Ω_f.

More likely than not an upper temperature constraint T^* will be specified. In such a case it has been widely stated (Szepe, 1966; Szepe and Levenspiel, 1968; Ogunye and Ray, 1968; Park and Levenspiel, 1976; Lee and Crowe, 1970a, b; Crowe, 1970; Gruyaert and Crowe, 1976) that the optimum operating policy requires following a rising curve of T vs. constant x_A to T^*, then following T^* with falling x_A as shown by path CDB of Figure 16.5.

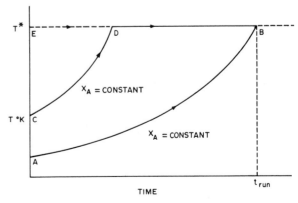

Figure 16.5. Three operating policies for a deactivating fixed-bed reactor with upper temperature constraint T.

However, as pointed out by Levenspiel and Sadana (1978), this policy is not in accord with intuition, as is illustrated in Figure 16.5. The path *AB* represents intuitively the best that can be done if we start at *A*. But according to the earlier policy, path *CDB* is the best if we start at *C*. By extension then, path *EDB* should be the best if we start at *E*.

It is most unlikely that all three paths give the same optimum. If not, then only one of them is the real optimum. By taking the second derivative of the Hamiltonian and examining its sign, we can show that this conclusion holds in general whenever $E_f > E$ (Levenspiel and Sadana, 1978).

The optimum temperature–time path for constant catalyst loading and feed rate is obtained from Eq. 16.106 by setting

$$\frac{d}{dt_p}\left(\frac{k\Omega}{T^n}\right) = 0 \qquad (16.107)$$

or

$$\frac{dT}{dt_p} = \frac{A_{f,f}\exp(-E_f/R_g T)}{E/R_g T^2 - n/T} \qquad (16.108)$$

On integrating Eq. 16.108 with boundary condition $T = T_{start}$ at $t = 0$ the optimum temperature–time path is obtained as

$$t = \frac{n}{A_{f,f}}\left[\mathrm{Ei}\left(\frac{E_f}{R_g T}\right) - \mathrm{Ei}\left(\frac{E_f}{R_g T_{start}}\right)\right]$$

$$- \left(\frac{E}{E_f A_{f,f}}\right)\left[\exp\left(\frac{E_f}{R_g T}\right) - \exp\left(\frac{E_f}{R_g T_{start}}\right)\right] \qquad (16.109)$$

where

$$\mathrm{Ei}(y) = \int_{-\infty}^{y}\frac{e^{-t}}{t}dt$$

is the exponential integral.

The average conversion along an isothermal path, for example *EDB*, may be obtained from Eqs. 16.103 and 16.104, since

$$\frac{C_A}{C_{A0}} \equiv \exp(-k\tau\Omega) \qquad (16.110)$$

or

$$1 - x_A = \exp\left[-k\tau\exp(-k_f t_p)\right] \qquad (16.111)$$

where τ is the space-time in the reactor. Taking the

average over time of run (t_{p1}) gives

$$\overline{1 - x_A} = 1 - \bar{x}_A = \frac{1}{t_{p1}}\int_0^{t_{p1}}\exp\left[-k\tau\exp(-k_f t_p)\right]dt_p \qquad (16.112)$$

Equation 16.112 integrates to

$$\bar{x}_A = 1 - \frac{1}{k_f t_{p1}}\left\{\mathrm{Ei}^*\left[ki\tau\exp(-k_f t_{p1})\right] - \mathrm{Ei}^*(k\tau)\right\} \qquad (16.113)$$

where

$$\mathrm{Ei}^*(h) = -\mathrm{Ei}(-h) = \int_h^t \frac{e^{-t}}{t}dt$$

is the exponential integral. We can obtain T_{start} from Eq. 16.111 by specifying x_A and τ at $t = 0$.

Equation 16.105 is the optimum temperature operations criterion when deactivation is independent of the reactant concentration. When deactivation is dependent on reactant concentration, the optimum temperature operations criterion for a batch reactor using the Hamiltonian method is (Lee and Crowe, 1970a, b):

$$k\Omega g_2(x_A)^p = \text{constant} \qquad (16.114)$$

where the deactivation is given by

$$-\frac{d\Omega}{dt_p} = k_f \Omega g_2(x_A)^p \qquad (16.115)$$

No such simple criterion is available for a fixed-bed reactor

16.4.2. Reversible Reaction

The optimal temperature policy for a reversible reaction in a batch-operated reactor was determined by Haas et al. (1974) by the formulation of a calculus of variations problem following the technique of Szepe and Levenspiel (1968). The two-point boundary value variational problem was reformulated in terms of an initial value problem with a parameter that includes the initial value of temperature. This initial value problem was solved by a regression technique.

The industrially important enzymatic reaction involving the isomerization of D-glucose to D-fructose catalyzed by glucose isomerase in solution was chosen as the reaction system. The optimal temperature operational policy gave 10% less denaturation of glucose isomerase when compared with the final isomerase activity obtained with the reactor operated at the optimal isothermal temperature for the same conversion and reaction time.

Recently, the optimum temperature operations criterion for Michaelis–Menten kinetics for substrate-independent and substrate-dependent enzyme deactivation in a batch reactor was obtained (Sadana, 1979).

In the forgoing treatment there is the distinct stipulation that the catalyst activity should be uniform throughout the reactor, which is referred to as the *constant activity* policy. This policy, very difficult to achieve practically except in a fully mixed reactor, would be desirable where the economics of the process demand the elimination of bed activity profiles. In contrast, the so-called *constant reactivity* policy, which is more realistic, demands that the bed operate under conditions of nonuniform activity. In other words, whereas in the former the activity is a function only of time, in the latter it is a function of both time and position in the bed.

16.5. OPTIMUM TEMPERATURE POLICIES FOR PRODUCTION AND REGENERATION FOR REACTORS SUBJECT TO CATALYST DEACTIVATION

Besides determining the optimum temperature policy of the reactor (production problem) it is also important to know when to stop the run and regenerate or replace the catalyst (regeneration problem).

16.5.1 Relay Variable: Park–Levenspiel (1976) Method (1976)

In the overall optimization strategy, where one needs to look at both production and regeneration, the selection of a relay variable is important to interconnect the two problems. Initially Szepe (1966) formulated a three-variable relay problem (initial activity Ω_0, final activity Ω_f, and run time t_p). Miertschin and Jackson (1970) showed that only two relay variables were needed (Ω_0 and t_p). Park and Levenspiel (1976) have simplified the approach even further and proposed a one-variable formulation. Since the search for the optimum production cycle for a single-variable formulation proceeds directly without the excessive computation needed in multivariable formulations, it is worthwhile discussing this procedure briefly.

Caution must be exercised, however, for Park and Levenspiel used a temperature policy in which the run is continued even at the maximum allowable operating temperature, T^*. This policy has since been shown to be incorrect by Levenspiel and Sadana (1978), who proposed that the run should be stopped as soon as T^* is reached (see Section 16.4). In all other respects Park and Levenspiel's procedure is correct.

Park and Levenspiel set up a profit function **P**, defined as

$$\mathbf{P} = \frac{\left(\begin{array}{c}\text{money inflow from sale of}\\ \text{product in production phase}\end{array}\right) - \left(\begin{array}{c}\text{cost of}\\ \text{regeneration}\end{array}\right)}{\text{production time} + \text{regeneration time}}$$

The cost and time of regeneration are each divided into two parts—a variable and a fixed part. The optimization problem is to determine the best interval of catalyst activity (regeneration problem) and the best temperature policy (production problem).

The single relay variable is the profit function $\tilde{\mathbf{P}}$ redefined to account for the optimal temperature policy $\mathbf{T}(\Omega)$. Thus the regeneration problem reduces to one of finding the largest **P** and the corresponding interval of catalyst activity.

Park and Levenspiel (1976) also set up an optimum operation index **I** and a regeneration index **J**. The optimal temperature policy to use during the production cycle is that where the optimum operation index stays as high as possible.

Very briefly the iterative procedure suggested is as follows. (The various equations used, are summarized in Table 16.2.)

1. Guess maximum profit $\mathbf{P}_{max} = \mathbf{P}_{guess}$.
2. Obtain the operation index **I** from Eq. 5a.
3. Maximize the operation index by changing temperature $T(\Omega)$. That $T(\Omega)$ which maximizes the operation index is $\mathbf{T}(\Omega)$, the optimal temperature progression.
4. Obtain the optimum operation index from Eq. 5b.
5. Obtain the regeneration index from Eq. 6.
6. Repeat steps 1–5 until an equation containing the regeneration index, maximum profit, and fixed cost for regeneration (Eq. 7) is satisfied.

16.5.2. Plant Optimization Method of Douglas-Reef-Kittrell (1980)

Douglas et al. (1980) have proposed a procedure for a preliminary evaluation of process designs with catalyst deactivation. Their design considers not only the reactor but the rest of the plant also. Thus, we have economic trade-offs between the reaction and the rest

TABLE 16.2. Summary of Equations for Optimum Temperature Policies during Production and Regeneration

	Equation
1. Profit function	$$\mathbf{P} = \frac{\displaystyle\int_0^{t_p} M_p\,dt - \mathbf{C}_r(\Omega_0,\Omega_f)}{t_p + t_r(\Omega_0,\Omega_f)}$$
2. Variable and fixed regeneration costs and times	$\mathbf{C}_r(\Omega_0,\Omega_f) = \mathbf{C}_{r,v}(\Omega_0,\Omega_f) + \mathbf{C}_{r,f}$ $t_r = t_{r,v} + t_{r,f}$
3. Deactivation rate and profit rate	$\dfrac{-d\Omega}{dt} = \alpha(\Omega, T); \quad M_p = M_p(\Omega, T)$
4. Evaluating $\mathbf{P}$ along the decaying catalyst activity instead of time[a]	$$\mathbf{P} = \frac{-\displaystyle\int_{\Omega_0}^{\Omega_f} \frac{M_p[\Omega, \mathbf{T}(\Omega)]}{\alpha[\Omega, \mathbf{T}(\Omega)]}\,d\Omega - \mathbf{C}_r(\Omega_0,\Omega_f)}{-\displaystyle\int_{\Omega_0}^{\Omega_f} \frac{d\Omega}{\alpha[\Omega, \mathbf{T}(\Omega)]} + t_r(\Omega_0,\Omega_f)}$$
5a. Operation index	$\mathbf{I}(\Omega, T, \mathbf{P}) = \dfrac{M_p(\Omega, T) - \mathbf{P}}{\alpha(\Omega, T)}$
5b. Optimum operation index	$\mathbf{I}(\Omega, \mathbf{P}) = \max\left[\mathbf{I}(\Omega, T, \mathbf{P})\right]$ $\qquad\quad = \dfrac{M_p(\Omega, \mathbf{T}) - \mathbf{P}}{\alpha(\Omega, \mathbf{T})}$
6. Regeneration index	$\mathbf{J}(\Omega, \mathbf{P}) = \mathbf{I}(\Omega, \mathbf{P}) - \dfrac{d\mathbf{C}_{r,v}}{d\Omega}(\Omega, 0) - \mathbf{P}\dfrac{dt_{r,v}}{d\Omega}(\Omega, 0)$
7.	$\displaystyle\int_{\Omega_f}^{\Omega_0} \mathbf{J}(\Omega, \mathbf{P})\,d\Omega = \mathbf{C}_{r,f} + \mathbf{P}t_{r,f}$

[a]The term $\mathbf{T}(\Omega)$ represents the optimal temperature policy and $\mathbf{P}$ the single relay variable which fixes the temperature progression.

of the plant. This appears rational since, for example, if the reactor is operated at a constant total feed rate and temperature, the conversion will decline with time in a deactivating reaction. This will result in a continually increasing recycle rate. Thus, the plant design must be such as to accommodate the varying flow and heat duties over the entire catalyst life cycle. The specific example considered next highlights the salient features of the method.

16.5.3. Example: Process Design of a Deactivating Reactor for the Dehydrogenation of Isopropanol to Acetone

Consider a 25-million-kg per year acetone production facility. Acetone is produced by the dehydro-genation of isopropanol.

$$C_3H_7OH \rightarrow C_3H_6O + H_2 \qquad (16.116)$$

At the outset we define an objective function, the total annualized cost, which we want to minimize:

$$\mathbf{C}_{tot} = \alpha_c I + \mathbf{C}_{op} \qquad (16.117)$$

where I is the total process investment, α_c is the annual capital charge factor, and $\mathbf{C}_{op}$ is the annual operating cost.

The first step in the design is to present the capital and operating-cost components in terms of variables relevant to the deactivation process. Thus

$$\mathbf{C}_{tot} = f_1(F_T, x) \tag{16.118}$$

and

$$\mathbf{C}_{tot} = f_2(F_T, x_1, x_2) \tag{16.119}$$

for nondeactivating and deactivating reactors, respectively. Here F_T is the total feed rate through the reactor, x is the conversion, and x_1 and x_2 are the start-of-run and end-of-run conversions, respectively.

In designing the deactivating reactor, the cost elements subject to deactivation must be further specified. For example, consider the case wherein the reactor is operated at constant temperature and feed rate. Conversion decreases with time. Thus, the acetone and isopropanol distillation columns must be designed at the start-of-run and end-of-run conversion levels. Catalyst costs are fixed by initial conversion and utilities costs by the average conversion level during the run.

Assume that the regeneration cost is $1000 and the time for regeneration is 48 hr, and that the catalyst can be regenerated to regain its <u>initial</u> activity. The average production rate of acetone (PR) during one operating interval is then

$$(\overline{PR})_A = \frac{1}{t_p} \int_0^{t_p} 0.685 F_T \times dt = 0.685 F_T \bar{x}_A \tag{16.120}$$

where F_T is a fixed constant, t_p is the operating period, and $\bar{x}_A$ is the average conversion over the operating interval.

Let n be the annual number of operating intervals; then

$$51.23 \times 8400 = P_A(8400 - 48n)$$

$$= 0.685 F_T x_A (8400 - 48n) \tag{16.121}$$

where 51.23 is the moles per hour of acetone required in 8400 hr (per year) to attain the 25 million kg basis of acetone per year. Also $(t + 48)n = 8400$ and $x_A = (x_1 + x_2)/2$. Thus,

$$F_T = \frac{149.50}{(x_1 + x_2)} \frac{(t + 48)}{t} \tag{16.122}$$

Finally, we need a catalyst deactivation expression. Use

$$\Omega = \exp\left[(-2.854 \times 10^{-3})\frac{F_T t}{V_r}\right] \tag{16.123}$$

Equation 16.123, after incorporation into the reactor model, yields

$$t = 115[-3.46 \ln(1 - x_1) - x_1] \ln\left[\frac{-3.46 \ln(1 - x_1) - x_1}{-3.46 \ln(1 - x_2) - x_2}\right] \tag{16.124}$$

On substituting this equation into the cost model, we get

$$\mathbf{C}_{tot} = f_1'(x_1, x_2) \tag{16.125}$$

where the total annualized cost is a function of the start-of-run and end-of-run conversion levels, x_1 and x_2, respectively.

The dependence of the total annualized cost on x_2 for three values of x_1 is shown in Figure 16.6a (Douglas et al., 1980). The optimal conversion levels are $x_1 = 0.982$ and $x_2 = 0.93$. The conversion time trajectories are shown in Figure 16.6b. Note that (a) as $x_1 \to 1$, the catalyst and reactor costs increase significantly; (b) as $x_2 \to x_1$, then the total flow rate through the process has to be increased significantly to compensate for downtime, and therefore all of the equipment must be oversized; (c) as the final conversion decreases, the recycle flow of unconverted material increases significantly so that all of the equipment in the recycle loop must be oversized.

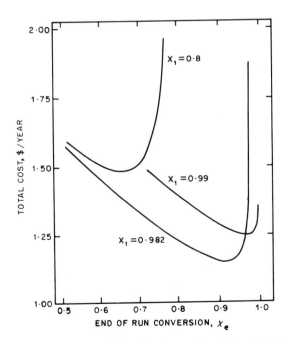

Figure 16.6a. Optimization of total annual cost (Douglas et al., 1980).

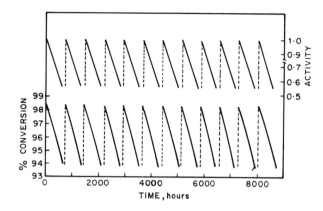

Figure 16.6b. Reactor–conversion and catalyst–activity profiles (Douglas et al., 1980).

The advantage of this methodology is that it is readily extended to include more exact cost representations, reaction selectivity effects, and other operating strategies.

16.6. OPTIMUM PRODUCTION–REGENERATION CYCLES

In this section we shall be concerned with optimum combinations of production and regeneration times for different situations without imposing optimum tem-perature policies during the production and regener-ation cycles. We shall consider three cases: deactivation with no regeneration, negligible regeneration period, and regeneration period $\simeq$ production period, (see Figure 16.7).

16.6.1. Deactivation with No Regeneration

There are catalysts that last for several weeks (or months) and that are discarded after their activity has fallen to a certain level either because they cannot be regenerated or because the cost of regeneration does not warrant it. For catalysts falling in this category, the type of aging is shown in Figure 16.7a.

The chief consideration in the economics of catalyst replacement is that the contribution to the total operat-ing cost due to catalyst must be minimum, the catalyst being discarded after a production cycle corresponding to this cost. This total contribution (C_T) is made up of two parts: (1) cost of periodic replacement of catalyst necessitated by attrition, loss, etc. (C_r); and (2) cost of total replacement of the catalyst after it has deteriorated to the optimum limit (C_a).

Let the rate at which the catalyst deterioration increases the cost of production, namely,

$$r_a = \frac{\text{cost}/t_p}{t_p} \tag{16.126}$$

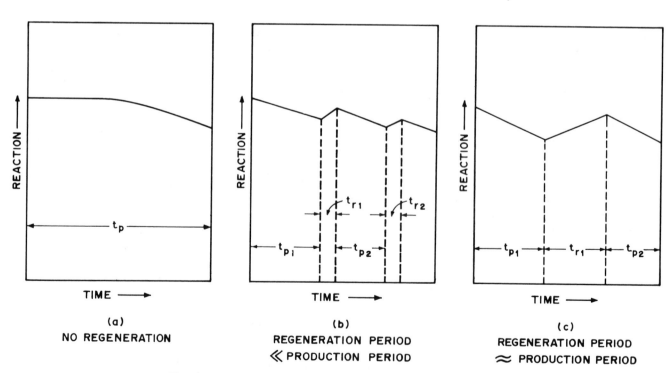

Figure 16.7. Patterns of reactor operation with a deteriorating catalyst.

be represented by line A in Figure 16.9 (corresponding to case a of Figure 16.8). The total cost per unit time is given by

$$\overline{C}_T = \frac{C_T}{t_p} = \frac{C_r}{t_p} + \frac{C_a}{t_p} \qquad (16.127)$$

where $\overline{C}_T$ is the average total cost per unit time (usually a day). Since the cost of production rises at a steady rate r_a as a result of catalyst deterioration, the average aging cost over a period t_p would be $r_a t_p/2$. Thus

$$\overline{C}_T = \frac{C_r}{t_p} + \frac{r_a t_p}{2} \qquad (16.128)$$

Differentiation of Eq. 16.128 and setting the result to zero now lead to the optimum value of t_p (corresponding to minimum $\overline{C}_T$). Thus

$$t_{pm} = \frac{2C_r}{r_a} \qquad (16.129)$$

and

$$\overline{C}_{Tm} = \frac{C_r}{t_{pm}} + \frac{t_{pm} r_a}{2} \qquad (16.130)$$

The forgoing analysis is for deactivation corresponding to type (a) in Figure 16.8. The following equations can be easily written for the most general type of deactivation represented by Figure 16.8b, in which the aging rate changes with time:

$$t_{pm} = \left[\frac{2C_r}{r_a} - \frac{r_a' t_f^2}{r_a} + (t_f + t_c) \right] \qquad (16.131)$$

and

$$\overline{C}_{Tm} = \left(\frac{C_r}{t_{pm}} + \frac{t_{pm} r_a}{2} + t_f r_a' \, 1 - \frac{t_f}{2t_{pm}} \right)$$
$$- r_a(t_f + t_c)\left(1 - \frac{t_f + t_c}{2t_{pm}} \right) \qquad (16.132)$$

where r_a' = rate of increase in cost during the initial aging period

t_f and t_c = initial aging and constant activity loss periods, respectively

16.6.2. Production Period ≫ Regeneration Period

Let us now consider the case where the catalyst activity can be restored to a point close to its original activity by a brief period of regeneration (which is negligible compared to the on-stream time). The total rise in operating cost for catalysts of this type can be attributed to (1) a lowering of the reaction rate due to aging during any on-stream period, and (2) the cost associated with regeneration. Walton (1961) has presented analytical solutions for the case of linear cost rise due to catalyst decay.

A typical replacement policy is shown in Figure 16.9. In general, it may be assumed that the original activity of the catalyst can never be fully restored. There are exceptions to this, however, as noted by George (1978) in the regeneration of the Claus alumina catalyst used to convert H_2S to sulfur. The cost increase associated with this shortfall has a constant value (in terms of cost per day) independent of the number of regenerations, although (and this is true in many practical cases) the time between regenerations decreases continuously.

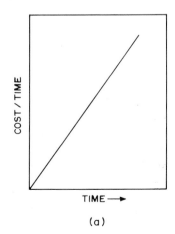
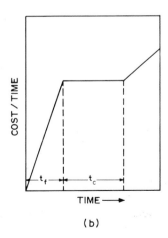

(a)   (b)

Figure 16.8. Patterns of constant-rate fouling of catalysts.

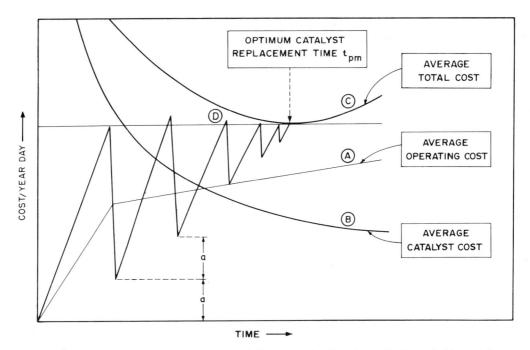

Figure 16.9. Determination of optimum catalyst life for the case in which the production period is much longer than the regeneration period (Walton, 1961).

Stated alternatively, the end of every production cycle brings the instantaneous operating cost up to D. A curve drawn through these operation–regeneration cycles represents the average operating cost. As in the case of nonregenerative deactivation, curve B represents the average catalyst replacement cost. The total time after which the catalyst should be replaced is then given by the minimum in the total cost curve **C**, and can be estimated from the equation

$$t_{\text{pm}} = \frac{an_c}{2r_a}(n_c - 1) \qquad (16.133)$$

where a = loss in product value due to regeneration,
n_c = number of on-stream cycles.

The value of n_c to be used in this equation is calculated by a trial-and-error solution of the equation for minimum cost.

$$\mathbf{C}_{\text{Tm}} = \frac{2\mathbf{C}_r r_a}{an_c(n_c - T)} + \frac{a(4n_c - 5)}{6} \qquad (16.134)$$

A plot of $\mathbf{C}_{\text{Tm}}$ vs. n_c shows a minimum, and the value of n corresponding to this minimum gives the optimum number of on-stream cycles. By incorporating this value of n in Eq. 16.133, the optimum life t_{pm} is calculated.

16.6.3.   Production Period $\simeq$ Regeneration Period

The basic feature of the development for the case in which the production period is about same as the regeneration period is the definition and the use of an efficiency criterion. A simple and straightforward criterion for a reactor subject to alternate cycles of operation and regeneration is the following definition of reactor efficiency (E_r):

$$E_r = \frac{\substack{\text{total product obtained over any number of}\\ \text{operation–regeneration cycles}}}{\substack{\text{total product obtained over the same period}\\ \text{if the reactor were to operate without}\\ \text{regeneration and at } 100\% \text{ conversion}}}$$

$$= \frac{n_c t_p x_{\text{av}}}{F t_t} = \frac{n_c t_p x_{\text{av}}}{F[n_c(t_p + t_r)]} \qquad (16.135)$$

Equation 16.135 contains the term for average conversion x_{av} at the end of an operation cycle. An expression for this will be derived in Chapter 17, in which the performances of different types of reactors are compared. For the present we shall anticipate the result, which is

$$x_{\text{av}} = \frac{1}{k_{v,f} t_p} \log\left[\frac{1 + k_{v,f}/S_v}{1 + (k_{v,f}/S_v)\exp(-k_{v,f} t_p)}\right]$$

$$(16.136)$$

The term $k_{v,f}t_p$ gives the extent of catalyst decay for the production period t_p, and $k_{v,f}/S_v$ is a measure of the extent of reaction. Although Eq. 16.136 is limited to isothermal reactors, it can also be used with reasonable accuracy for nonisothermal reactors involving reactions of relatively low activation energy (say < 10,000 cal/mol). Substituting Eq. 16.136 in 16.135, we obtain the following relation for reactor efficiency:

$$E_r = \frac{1}{k_{v,f}(t_p + t_r)} \log\left[\frac{1 + k_{v,f}/S_v}{1 + (k_{v,f}/S_v)\exp(-k_{v,f}t_p)}\right]$$

(16.137)

This is the basic equation for establishing optimum combinations of t_p and t_r for different values of space velocity and of the rate constants for the main and deactivation reactions. The operation time can be regarded as independent of regeneration time for some reactions, say certain types of rapid feed-stream poisoning, or in reactor schemes where the time for mechanical changeover is very large in relation to the regeneration time. For other reactions the two are interdependent. Thus for reactions subject to catalyst fouling by carbon deposition, the time required to burn off the carbon (regeneration cycle) would be strongly dependent on the extent of carbon deposition, that is, on the operation time. Therefore, in applying Eq. 16.137, two cases must be considered: that in which the regeneration time is independent of operation time, and that in which the regeneration time depends on operation time.

Regeneration Time Independent of Operation Time

It is useful to examine Eq. 16.137 by plotting the efficiency E_r as a function of two of the variables, holding the others constant. Based on such plots the following main conclusions can be drawn:

1. Shorter regeneration times lead to higher efficiencies, with the optimum efficiency occurring at reduced operation times.

2. Lower space velocities lead to higher efficiencies, with the optimum efficiency occurring at higher operation times. Since lower space velocities would mean higher equipment (capital) cost, an optimum space velocity exists.

3. More slowly decaying catalysts give higher efficiencies, with the optimum efficiency occurring at longer operation times.

Regeneration Time Dependent on Operation Time

In the case where regeneration time depends on operation time—t_r is dependent on t_p—it is necessary to formulate a relationship between the two parameters before Eq. 16.137 can be used to predict the behavior of reactor efficiency. For this purpose, the following kinetic expressions for deactivation (carbon deposition in the case of most catalysts) and of regeneration (carbon burn-off) may be employed:

$$\frac{dW_c}{dt_p} = k_{v,f}W_c^{-m}$$

(16.138)

The regeneration rate under conditions of chemical control is given by

$$\frac{dW_c}{dt_r} = -k_{cb}C_{O_2}W_c^n$$

(16.139)

If at the start of the production cycle the carbon content is assumed to be negligible, then Eq. 16.138 becomes

$$W_c = at_p^b$$

(16.140)

where

$$a = (1 + mk_{v,f})^b; \qquad b = (1 + m)^{-1}$$

(16.141)

Equation 16.139 can now be solved at constant oxygen concentration and under the boundary condition $W_c = 0$, $t_p = 0$ at $t_r = 0$ with the result

$$t_r = \frac{(a^{1-n})\left(t_p\dfrac{1-n}{m-1}\right)}{(k_{cb}C_{O_2})(1-n)}, \qquad n \neq 1$$

(16.142)

According to Weekman (1968), the reactor efficiency E_r will show a maximum as t_p is increased provided that

$$m > 0, \qquad n \geqslant 0$$

(16.143)

REFERENCES

Anderson, R. B., and Whitehouse, A. M. (1961), *Ind. Eng. Sci.*, **53**, 1011.

Bakshi, K. R., and Gavalas, G. R. (1975), *AIChE J.*, **21**, 494.

Billimoria, R. M., and Butt, J. B. (1981), *Chem. Eng. J.*, **22**, 71.

Bischoff, K. B. (1969), *Ind. Eng. Chem. Fundam.*, **8**, 665..

Butt, J. B. (1970), *Chem. Eng. Sci.*, **25**, 801.

Butt, J. B. (1972), *Adv. Chem. Ser.*, **109**, 259.

Butt, J. B., and Rohan, D. M. (1968), *Chem. Eng. Sci.*, **23**, 489.

Campbell, D. R., and Wojciechowski, B. W. (1969), *Can. J. Chem. Eng.*, **47**, 413.

Chou, A., Ray, W. H., and Aris, R. (1967), *Trans. Inst. Chem. Eng. (London)*, **45**, T 153.

Christiansen, L. J., and Andersen, S. L. (1980), *Chem. Eng. Sci.*, **35**, 314.

Crowe, C. M. (1970), *Can. J. Chem. Eng.*, **48**, 576.

Crowe, C. M., and Lee, S. I. (1971), *Can. J. Chem. Eng.*, **49**, 385.

De Pauw, R. P., and Froment, G. F. (1975), *Chem. Eng. Sci.*, **30**, 789.

Douglas, J. M., Reeff, E. K., Jr., and Kittrell, J. R. (1980), *Chem. Eng. Sci.*, **35**, 322.

Drouin, J. G. (1969), Computational Studies of the Optimization of a Catalytic Reactor for a Reversible Reaction with a Catalyst Decay, M.S. thesis, McMaster University.

du Domaine, J., Swain, R. L., and Hougen, O. A. (1943), *Ind. Eng. Chem.*, **35**, 546.

Eberly, P. E., Kimberlin, C. N., Miller, W. H., and Drushel, H. V. (1966), *Ind. Eng. Chem. Process Design Dev.*, **5**, 193.

Froment, G. F., and Bischoff, K. B. (1961), *Chem. Eng. Sci.*, **16**, 189.

Froment, G. F., and Bischoff, K. B. (1962), *Chem. Eng. Sci.*, **17**, 105.

Gavalas, G. R. (1971), *Ind. Eng. Chem. Fundam.*, **10**, 621.

George, Z. M. (1978), *Can. J. Chem. Eng.*, **56**, 711.

Gonzalez, L. O., and Spencer, E. H. (1963), *Chem. Eng. Sci.*, **18**, 753.

Gruyaert, F., and Crowe, C. M. (1976), *Can. J. Chem. Eng.*, **54**, 617.

Haas, W. R., Tavlarides, L. L., and Wnek, W. J. (1974), *AIChE J.*, **20**, 707.

Jackson, R. (1965), *Optimum Temperature Gradients in Tubular Reactors with Decaying Catalysts*, ed. Pirie, J. M., Institution of Chemical Engineers, London, p. 33.

Johnson, B. M., Froment, G. F., and Watson, C. C. (1962), *Chem. Eng. Sci.*, **17**, 835.

Kulkarni, B. D., and Doraiswamy, L. K. (1977), unpublished work.

Kulkarni, B. D., and Doraiswamy, L. K. (1980), unpublished work.

Lambrecht, G. C., Nussey, C., and Froment, G. F. (1972) *Proc. 5th Int. Symp. Chem. React. Eng.*, B-2-19.

Lee, S. I., and Crowe, C. M. (1970a), *Chem. Eng. Sci.*, **25**, 743.

Lee, S. I., and Crowe, C. M. (1970b), *Can. J. Chem. Eng.*, **48**, 192.

Levenspiel, O., and Sadana, A. (1978), *Chem. Eng. Sci.*, **33**, 1393.

Masamune, S., and Smith, J. M. (1966), *AIChE J.*, **12**, 384.

Miertschin, G. N., and Jackson, R. (1970), *Can. J. Chem. Eng.*, **48**, 702.

Ogunye, A. F., and Ray, W. H. (1968), *Trans. Inst. Chem. Eng. (London)*, **46**, T 225.

Olson, J. H. (1968), *Ind. Eng. Chem. Fundam.*, **7**, 185.

Olson, K. E., Luss, D., and Amundson, N. R. (1968), *Ind. Eng. Chem. Process Design Dev.*, **7**, 96.

Ozawa, Y. (1969), *Ind. Eng. Chem. Process Design Dev.*, **8**, 378.

Ozawa, Y. (1970), *Chem. Eng. Sci.*, **25**, 529.

Pachovsky, R. A., John, T. M., and Wojciechowski, B. W. (1973), *AIChE J.*, **19**, 802.

Panchenkov, G. M., Laz'yan, Yu. I., Kozlov, M. U., and Zorov, Yu. M. (1969), *Khim. Tech. Topl. Masel*, **14** (7), 4.

Park, J. Y., and Levenspiel, O. (1976), *Ind. Eng. Chem. Process Design Dev.*, **15**, 538.

Pontryagin, L. S., Boltyanskii, V. G., Gamkrelidze, R. V., and Mischenko, E. F. (1962), *The Mathematical Theory of Optimal Processes*, (trans.) K. N. Trircgoff, Wiley (Interscience), New York.

Price, T. H., and Butt, J. B. (1977), *Chem. Eng. Sci.*, **32**, 393.

Sadana, A. (1979), *AIChE J.*, **25**, 535.

Sadana, A. (1980), *Chem. Eng. Commun.*, **4**, 51.

Schulman, B. L. (1963), *Ind. Eng. Chem.*, **55** (12), 44.

Sreeramamurty, R., and Menon, P. G. (1968), *J. Catal.*, **8**, 95.

Szepe S. (1966), Ph.D. thesis, Illinois Institute of Technonogy.

Szepe, S. and Levenspiel, O. (1968), *Chem. Eng. Sci.*, **23**, 881.

van Deemter, J. J. (1953), *Ind. Eng. Chem.*, **45**, 1227.

van Deemter, J. J. (1954), *Ind. Eng. Chem.*, **46**, 2300.

van Zoonen, D. D. (1965), *Proc. 3d Int. Conq. Catal.*, p. 1319.

Villadsen, J., and Michelsen, M. L. (1978), *The Solution of Differential Equation Models by Polynomial Approximation*, Prentice-Hall, Englewood Cliffs, New Jersey.

Walton, P. R. (1961), *Chem. Eng. Prog.*, **57** (8), 42.

Weekman, V. W., Jr. (1968), *Ind. Eng. Chem. Process Design Dev.*, **7**, 90.

Weisz, P. B., and Goodwin, R. B. (1966), *J. Catal.*, **6**, 227.

Weng, H. S., Eigenberger, G., and Butt, J. B. (1975), *Chem. Eng. Sci.*, **30**, 1281.

Wheeler, A., and Robell, A. J. (1969), *J. Catal.*, **13**, 299.

Zhorov, Yu. M., Panchenkov, G. M., and Laz'yan, Yu. I. (1967), *Zh. Fiz. Khim.*, **41**, 1574.

CHAPTER SEVENTEEN

Performance of Catalytic Reactors Under Conditions of Catalyst Deactivation

It is often necessary to decide on the type of reactor to be used for a deactivating catalyst. While the final choice of reactor would involve such considerations as reaction type, thermal behavior of the reaction, and necessity for isothermal operation, an important consideration—one that might easily outweigh others—is the performance of the reactor with a deactivating catalyst. For this purpose a quantitative means of assessing the performance of a reactor subject to catalyst deactivation and regeneration is desirable.

The continuity equation for a plug-flow reactor containing a time-decaying catalyst through which a reactant (say A) is passing and reacting under diffusion-free conditions is given by

$$\left(\frac{\partial C_A}{\partial t_p}\right) + u\left(\frac{\partial C_A}{\partial l}\right) = -r_{VA}(C_A, t_p) \quad (17.1)$$

or

$$\frac{f_B \rho_g}{M}\left(\frac{\partial y_A}{\partial t_p}\right) + \frac{G}{M}\left(\frac{\partial y_A}{\partial l}\right) = -r_{VA}(y_A, t_p) \quad (17.2)$$

where t_p is the production (or on-stream) time. This equation is identical to Eq. 16.1, used for catalyst deactivation with the assumption that the reaction rate (i.e., the instantaneous rate of disappearance of reactant A to products) is a function not only of the mole fraction of A but also of the reaction time t_p.

Making use of this basic equation, we shall derive the governing equations for the three reactor types discussed earlier (fixed bed, moving bed, and fluid bed), and compare their performances under conditions of catalyst deactivation.

17.1. GOVERNING EQUATIONS

All three reactor types can be employed for a deactivating catalyst. In the case of a fixed-bed (or batch-operated fluidized-bed) reactor, two identical reactors would have to be used alternatively for reaction and regeneration. On the other hand, in the case of a moving-bed or fluid-bed reactor, the catalyst itself moves through the reactor at a predetermined rate and a steady-state operation is involved.

17.1.1. Fixed-Bed Reactor

The reaction time in the case of a fixed-bed reactor is evidently the same as the total decay (or reaction) time t_{p1} when viewed from the standpoint of catalyst decay. Thus Eq. 17.2 can be rewritten in terms of the normalized time

$$\hat{t} = \frac{t_p}{t_{p1}} \quad (17.3)$$

and the usual normalized length $z = l/L$ to give

$$\frac{f_B \rho_g L}{G t_{p1}}\frac{\partial y_A}{\partial \hat{t}} + \frac{\partial y_A}{\partial z} = -r_{VA}(y_A, \hat{t})\frac{ML}{G} \quad (17.4)$$

The mass velocity G can be expressed in terms of the feed space velocity $S_F \cdot (\sec^{-1})$ as

$$G = \begin{cases} \rho_F S_F L & \text{for any feed} \\ \rho_l S_l L & \text{for liquid feed} \end{cases} \quad (17.5)$$

and Eq. 17.4 can then be recast into the form

$$A'\frac{\partial y_A}{\partial \hat{t}} + \frac{\partial y_A}{\partial z} = -B r_{VA}(y_A, \hat{t}) \quad (17.6)$$

where

$$A' = \frac{f_B \rho_g}{\rho_F S_F t_{p1}} = \frac{f_B \rho_g}{\rho_l S_l t_{p1}} \quad (17.7)$$

and

$$B = \frac{M}{\rho_F S_F} = \frac{M}{\rho_l S_l} \quad \frac{cm^3 \, \sec}{mol} \quad (17.8)$$

415

Equation 17.6 can be greatly simplified if it is assumed that the first term can be neglected. Since the constant A' of this term represents the ratio of the feed transit time through the reactor to the catalyst decay time, and this ratio is usually negligibly small, such an assumption would be a reasonable one. Accordingly, Eq. 17.6 reduces to

$$\frac{dy_A}{dz} = -B r_{VA}(y_A, \hat{t})$$ (17.9)

If now an expression for the reaction rate (which is a function of the decay time t_p and the fraction of reactant unconverted y_A) can be found, Eq. 17.9 can be solved. For this purpose the rate of disappearance of A can be written as

$$r_{VA}(y_A, t_p) = k_v(t_p)(1 - f_B) y_A^m$$ (17.10)

where $k_v(t_p)$ is a time-dependent rate constant (mol/cm^3 sec) that accounts for catalyst decay and can be expressed in exponential form as

$$k_v(t_p) = k_{v0} \exp(-a t_p)$$ (17.11)

Several other decay forms are possible, as summarized in Table 8.2, but the exponential form, as given by Eq. 17.11, and the linear form (to be considered later) are usually adequate. Thus, for exponential decay, combining Eqs. 17.9, 17.10, and 17.11 leads to the following result:

$$\frac{dy_A}{dz} = -B' \exp(-\lambda \hat{t}) y_A^m$$ (17.12)

where

$$\lambda = a t_{p1}$$ (17.13)

$$B' = k_{v0}\left(\frac{M}{\rho_F S_F}\right) = k_{v0} B(1 - f_B)$$ (17.14)

Equation 17.12 is the basic nondimensional differential equation describing the mole fraction of A in a fixed-bed reactor containing a time-decaying catalyst as a function of position and time in terms of two dimensionless constants, λ and B'. It can be seen that λ is a decay parameter, while B' is a reaction parameter. The solutions discussed below are for the reactor exit, that is, $z = 1$. The solution to Eq. 17.12 is given by expression 1a of Table 17.1 (Sadana and Doraiswamy, 1971), which reduces to expression 1b for a first-order case (Weekman, 1968).

The average conversion obtained from the reactor (at its outlet—i.e., at $z = 1$) can be computed from the instantaneous conversion x_A by the relation

$$\bar{x}_A = 1 - \bar{y}_A = 1 - \int_0^1 y_A \, d\hat{t}$$ (17.15)

The average conversion thus found represents the time-averaged value obtained by collecting the reactor effluent over a given interval of time, determining the concentration of A in this entire quantity, and then computing the conversion. Solutions to this equation (for $z = 1$) are also presented in Table 17.1 as expressions 1c and 1d. The table also includes solutions, expressions 4 and 7, for two other decay forms.

17.1.2. Moving-Bed Reactor

In a moving-bed reactor, the residence time of the decaying catalyst in the chamber corresponds to the total decay time t_{p1} of the fixed-bed reactor. At a reduced position z in this steady-state reactor, the catalyst would have been exposed simultaneously to reaction and decay for a time $z t_{p1}$, which corresponds to t_p of the fixed-bed reactor. Therefore the reduced decay time in a moving-bed reactor is given by the position variable z. The following expression for the reaction rate may now be readily written:

$$r_{VA}(y_A, \hat{t}) = k_{v0}(1 - f_B)\exp(-\lambda z) y_A^m$$ (17.16)

In order to obtain an equation for y_A for the moving-bed reactor, we can substitute Eq. 17.16 in 17.9, with the result

$$-\frac{dy_A}{dz} = B' \exp(-\lambda z) y_A^m$$ (17.17)

Solutions to Eq. 17.17 for $m \neq 1$ and $m = 1$ are included in Table 17.1 as expressions 2a and 2b, respectively.

As in the case of the fixed-bed reactor, conversion can be determined from Eq. 17.15, with this difference: there is no average value to be considered now, since steady-state operation is involved. Thus,

$$x_A = 1 - y_A$$ (17.18)

The final equations are also included in the table as expressions 2c and 2d.

Expressions for y_A and x_A can also be derived by using the two other decay forms considered, and the resulting equations are included in Table 17.1 as expressions 5 and 8.

TABLE 17.1. Expressions for Mole Fraction and Conversion of Reactant A for Various Decay Forms (Sadana and Doraiswamy, 1971)

Decay Form	Expression for y_A — mth Order ($m \neq 1$)	Expression for y_A — First Order	Expression for Conversion — mth Order ($m \neq 1$)	Expression for Conversion — First Order
$k_v = k_{v0}e^{-\lambda\hat{t}}$				
Fixed	$\left[\dfrac{1}{(m-1)B'e^{-\lambda\hat{t}}+1}\right]^{1/(m-1)}$ (1a)	$\exp\{-[B'\exp(-\lambda\hat{t})]\}$ (1b)	$1-\int_0^1\left[\dfrac{1}{(m-1)B'e^{-\lambda\hat{t}}+1}\right]^{1/(m-1)}d\hat{t}$ (1c)	$1+\dfrac{1}{\lambda}\mathrm{E_i^*}(B')-\mathrm{Ei^*}(B'-\lambda)$ (1d)
Moving	$\left[\dfrac{\lambda}{(m-1)B'(1-e^{-\lambda})+\lambda}\right]^{1/(m-1)}$ (2a)	$\exp\left\{\dfrac{B'}{\lambda}\left[\exp(-\lambda)-1\right]\right\}$ (2b)	$1-\left[\dfrac{\lambda}{(m-1)B'(1-e^{-\lambda})+\lambda}\right]^{1/(m-1)}$ (2c)	$1-\exp\left\{\dfrac{B'}{\lambda}\left[\exp(-\lambda)-1\right]\right\}$ (2d)
Fluid	$\left[\dfrac{(\lambda+1)}{(m-1)B'+(\lambda+1)}\right]^{1/(m-1)}$ (3a)	$\exp\left(-\dfrac{B'}{\lambda+1}\right)$ (3b)	$1-\left[\dfrac{(\lambda+1)}{(m-1)B'+(\lambda+1)}\right]^{1/(m-1)}$ (3c)	$1-\exp\left(-\dfrac{B'}{\lambda+1}\right)$ (3d)
$k_v = k_{v0}-\lambda\hat{t}$				
Fixed	$\left[\dfrac{k_{v0}}{B'(m-1)(k_{v0}-\lambda\hat{t})+k_{v0}}\right]^{1/(m-1)}$ (4a)	$\exp\left[-\left(B'-\dfrac{B'\lambda\hat{t}}{k_{v0}}\right)\right]$ (4b)	$1-\int_0^1\left[\dfrac{k_{v0}}{B'(m-1)(k_{v0}-\lambda\hat{t})+k_{v0}}\right]^{1/(m-1)}d\hat{t}$ (4c)	$1-\int_0^1\exp\left[-\left(B'-\dfrac{B'\lambda\hat{t}}{k_{v0}}\right)\right]d\hat{t}$ (4d)
Moving	$\left[\dfrac{2k_{v0}}{B'(m-1)(2k_{v0}-\lambda)+2k_{v0}}\right]^{1/(m-1)}$ (5a)	$\exp\left[B'\left(\dfrac{\lambda}{2k_{v0}}-1\right)\right]$ (5b)	$1-\left[\dfrac{2k_{v0}}{B'(m-1)(2k_{v0}-\lambda)+2k_{v0}}\right]^{1/(m-1)}$ (5c)	$1-\exp\left[B'\left(\dfrac{\lambda}{2k_{v0}}-1\right)\right]$ (5d)
Fluid	$\left[\dfrac{k_{v0}}{B'(k_{v0}-\lambda)(m-1)+k_{v0}}\right]^{1/(m-1)}$ (6a)	$\exp\left[\left(\dfrac{B'}{k_{v0}}\right)(k_{v0}-\lambda)\right]$ (6b)	$1-\left[\dfrac{k_{v0}}{B'(k_{v0}-\lambda)(m-1)+k_{v0}}\right]^{1/(m-1)}$ (6c)	$1-\exp\left[\left(-\dfrac{B'}{k_{v0}}\right)(k_{v0}-\lambda)\right]$ (6d)
$k_v = k_{v0}-\lambda\hat{t}^d$				
Fixed	$\left[\dfrac{k_{v0}}{(B'k_{v0}-B'\lambda\hat{t}^d)(m-1)+k_{v0}}\right]^{1/(m-1)}$ (7a)	$\exp\left(\dfrac{B'}{k_{v0}}\lambda\hat{t}^d-B'\right)$ (7b)	$1-\int_0^1\left[\dfrac{k_{v0}}{(B'k_{v0}-B'\lambda\hat{t}^d)(m-1)+k_{v0}}\right]^{1/(m-1)}d\hat{t}$ (7c)	$1-\int_0^1\exp\left(\dfrac{B'}{k_{v0}}\lambda\hat{t}^d-B'\right)d\hat{t}$ (7d)
Moving	$\left[\dfrac{(d+1)k_{v0}}{(m-1)B'[(d+1)k_{v0}-\lambda]+(d+1)k_{v0}}\right]^{1/(m-1)}$ (8a)	$\exp\left\{B'\left[\dfrac{\lambda}{k_{v0}(d+1)}-1\right]\right\}$ (8b)	$1-\left[\dfrac{(d+1)k_{v0}}{(m-1)B'[(d+1)k_{v0}-\lambda]+(d+1)k_{v0}}\right]^{1/(m-1)}$ (8c)	$1-\exp\left\{B'\left[\dfrac{\lambda}{k_{v0}(d+1)}-1\right]\right\}$ (8d)
Fluid	$\left[\dfrac{k_{v0}}{k_{v0}+(m-1)B'(k_{v0}-d!\lambda)}\right]^{1/(m-1)}$ (9a)	$\exp\left[B'\left(\dfrac{d!\lambda}{k_{v0}}-1\right)\right]$ (9b)	$1-\left[\dfrac{k_{v0}}{k_{v0}+(m-1)B'(k_{v0}-d!\lambda)}\right]^{1/(m-1)}$ (9c)	$1-\exp\left[B'\left(\dfrac{d!\lambda}{k_{v0}}-1\right)\right]$ (9d)

17.1.3. Fluid-Bed Reactor

It was pointed out in Chapter 13 that in a fluid-bed reactor neither the fluid nor the solids are truly in piston flow. For the purpose of the present development, however, it will be assumed that, as in the case of fixed- and moving-bed reactors, the fluid is in piston flow, but contrary to the assumption made for moving-bed reactors, the solid phase is assumed to be fully mixed.

For a perfectly mixed catalyst system, the distribution is given by e^{-t}. Considering the existence of such an age distribution and assuming exponential decay, we can postulate an average value for the rate constant and can use this value in solving Eq. 17.12. The average rate constant may be expressed as

$$[k_{v}(\hat{t})]_{av} = k_{v0} \int_{0}^{\infty} \exp(-\hat{t}) \exp(-\lambda\hat{t}) \, d\hat{t} \qquad (17.19)$$

This equation results directly from modification of Eq. 17.11 for exponential decay to account for age distribution in a perfectly mixed system. The solution of Eq. 17.19, as given by Weekman (1968), is

$$[k_{v}(\hat{t})]_{av} = \frac{k_{v0}}{1+\lambda} \qquad (17.20)$$

Substituting Eq. 17.20 for the rate constant in 17.10, we obtain

$$r_{VA}(y_{A}, \hat{t}) = \frac{k_{v0}(1-f_{B})}{1+\lambda} y_{A}^{m} \qquad (17.21)$$

for a reaction of order m taking place in a steady-state fluid-bed reactor. Then, upon incorporation of this equation for $r_{VA}(y_{A}, t)$ in 17.9, there results

$$\frac{dy_{A}}{dz} = -\frac{B'}{\lambda+1} y_{A}^{m} \qquad (17.22)$$

Solution to this equation (as well as for the other decay forms) for $m \neq 1$ and $m = 1$, as given by Sadana and Doraiswamy (1971), are included in Table 17.1, along with the corresponding expressions for conversion (expressions 3, 6, and 9).

17.2. COMPARISON OF PERFORMANCES

The developments presented above were restricted to plug flow of the reactant in the reactor. We shall now compare the performances of the three reactor types

under plug-flow conditions, and then examine the effect of axial dispersion.

17.2.1. Plug Flow of Fluid

Equations for the fractional conversion x_{A} given in Table 17.1 for the three reactor types undergoing exponential decay are graphically displayed in Figure 17.1, where the conversion x_{A} is plotted as a function of the reaction parameter B' for two different values of the decay parameter λ.

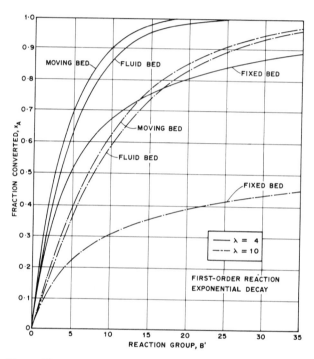

Figure 17.1. Conversions obtained in fixed-bed, moving-bed, and fluid-bed reactors for different values of the decay parameters.

When the decay parameter is zero, the equations for all three reactor types reduce to

$$\text{conversion} = \begin{cases} 1 - \left[\dfrac{1}{B'(m-1)z + 1} \right]^{1/(m-1)}, & m \neq 1 \\[2mm] 1 - \exp(-B'z), & m = 1 \end{cases} \qquad (17.23)$$

regardless of the decay form employed. These equations can be easily derived by setting $\lambda = 0$ in all the conversion expressions listed in Table 17.1 and then performing simple algebraic manipulations.

It is clear from Eqs. 17.23 that when a nondeactivating catalyst is used, all three reactor models considered above (operating under plug-flow conditions) would give the same conversion. A significant corollary of this fact is that the nature of catalyst presence, whether in fixed or fluidized form, mixed or unmixed, is of no consequence as long as it is ensured that the fluid is in plug flow and the catalyst exhibits a constant time-independent activity. It is only when the catalyst is subject to decay that the performances of the three reactor types are different, and the nature of the decay equation then plays a significant role in determining the conversions achievable.

An inspection of Figure 17.1 shows that the three reactor types can be characterized in terms of the parameters B' and λ. The reactor variables that usually determine performance are space velocity, residence time, reactor volume, and the ratio of catalyst to feed. Equations 17.13 and 17.14, which define the parameters λ and B', respectively, show that all these variables can be characterized by these groups.

Attention so far has been restricted to first-order reactions. Sadana and Doraiswamy (1971) have studied the effect of reaction order on conversion for a deactivating catalyst bed. Since the conversion is uninfluenced by reactor type in a nondeactivating bed under conditions of plug flow, a single line is obtained for all reactor types at $\lambda = 0$. On the other hand, for different finite values of the decay group λ, different curves are obtained for fixed-, moving- and fluidized-bed reactors. The effect of deactivation is far less severe for reactions of higher order in a moving-bed or fluidized-bed reactor than in a fixed-bed reactor. The conclusion follows, therefore, that the moving- and fluid-bed reactors, which are in general (but not always) superior to the fixed-bed reactor for a deactivating catalyst, are even more preferable for reactions of higher order under conditions of severe deactivation.

17.2.2. Example: Cracking of Gas Oil: Comparison of Fixed- and Moving-Bed Reactors for Slow and Fast Fouling

Weekman (1968) has given the following parameter values for the cracking of midcontinent gas oil on commercial TCC catalyst: $a = 2.96 \, \text{hr}^{-1}$, $t_{p1} =$ catalyst residence time $= 20 \, \text{min}$; B' (reaction parameter) $= 1.03$, 1.30, and 1.61 at $455°$, $482°$, and $510°C$, respectively. Compare the conversions in moving- and fixed-bed reactors at $455°$, $482°$, and $510°C$ at a low value of the decay parameter ($\lambda \simeq 1.0$) and at a high value ($\lambda = 15$).

Solution

Gas oil cracking generally follows second-order dependence on reactant concentration. The expressions for conversion for exponential decay are given in Table 17.1. For $m = 2$, these become

$$\text{Moving bed:} \quad x_A = \frac{B'(1 - e^{-\lambda})}{\lambda + B'(1 - e^{-\lambda})}$$

$$\text{Fixed bed:} \quad x_A = \frac{1}{\lambda} \ln\left(\frac{1 + B'}{1 + B'e^{-\lambda}}\right)$$

At $482°C$ the decay parameter

$$\lambda = at_{p1} = 2.96 \times \frac{20}{60} = 0.987$$

Conversions in the two reactors are then

$$\text{Fixed bed:} \quad x_A = \frac{1}{0.987} \ln\left(\frac{1 + 1.30}{1 + 1.30e^{-0.987}}\right) = 0.444$$

$$\text{Moving bed:} \quad x_A = \frac{1.30(1 - e^{-0.987})}{0.987 + 1.30(1 - e^{-0.987})} = 0.453$$

Because of the low value of $\lambda(0.987)$, there is no significant difference in the performances of the two reactors, and choice between the two is dictated by considerations other than catalyst deactivation.

By repeating the forgoing calculations for the other two temperatures, $455°C$ and $510°C$, and for the higher (assumed) value of the decay parameter ($\lambda = 15$), a table of conversions (x_A) can be prepared (see Table 17.2). It will be noticed that the effect of deactivation is most marked at the highest values of fouling and temperature, that is, at $\lambda = 15$ and $510°C$, with the moving bed being distinctly superior to the fixed bed. The difference in the performances of the reactors is less pronounced at a lower λ and temperature.

17.2.3. Effect of Axial Diffusion

So far we have considered the plug-flow model. It would be instructive to examine the role of axial diffusion in the performance of fixed-, moving-, and fluid-bed reactors subject to catalyst deactivation. For this purpose the axial dispersion model developed in Chapter 11 provides a ready basis. Written in terms of the mole fraction of reactant A, this model can be

TABLE 17.2.   Conversions at Three Temperatures for Two Decay-Parameter Values

Temperature (°C)	Reaction Parameter B'	Conversion x_A			
		$\lambda = 0.987$		$\lambda = 15$	
		Moving bed	Fixed bed	Moving bed	Fixed bed
455	1.03	0.396	0.389	0.0642	0.0474
482	1.30	0.453	0.444	0.080	0.0558
510	1.61	0.506	0.496	0.097	0.0642

expressed as

$$\frac{dy_A}{dz} - \frac{1}{Pe}\left(\frac{d^2 y_A}{dz^2}\right) = -B r_{VA}(y_A, \hat{t}) \qquad (17.24)$$

where, as before, the group A' (for a fixed bed) has been neglected.

For a reaction of mth order in a fluidized-bed reactor under conditions of plug flow of the fluid and complete solids mixing, the reaction rate is given by Eq. 17.21. Combining this equation (for $m = 1$) with Eq. 17.24, we obtain the following expression, which includes the effect of axial diffusion:

$$\frac{dy_A}{dz} - \frac{1}{Pe}\left(\frac{d^2 y_A}{dz^2}\right) = -\left(\frac{B'}{\lambda+1}\right)y_A \qquad (17.25)$$

With the usual Danckwerts boundary conditions (see Chapter 11), the solution to Eq. 17.25 at $z = 1$ may be obtained as

$$y_A = 4pq \exp\left(\frac{Pe_m}{2}\right) \qquad (17.26)$$

where

$$q = \left[(1+p)^2 \exp\left(\frac{p\,Pe}{2}\right) - (1-p)^2 \exp\left(-\frac{p\,Pe}{2}\right)\right]^{-1}$$

$$p = \left(1 + \frac{4B'}{(\lambda+1)Pe}\right)^{1/2}$$

The conversion x_A at the reactor exit is then obtained by substituting Eq. 17.26 in 17.18:

$$x_A = 1 - 4pq \exp\left(\frac{Pe}{2}\right) \qquad (17.27)$$

At $Pe = 0$ (corresponding to backmix flow), Eq. 17.27

reduces to

$$x_A = \frac{B'/(\lambda+1)}{1 + B'/(\lambda+1)} \qquad (17.28)$$

By plotting conversion as a function of the Peclet number for various values of the decay parameter λ at fixed B', it may be concluded that the increased conversion obtained at relatively low values of the reaction group due to plug flow practically vanishes as the decay parameter is increased.

The effect of axial dispersion can be more clearly brought out by expanding Eq. 17.26 in series. For large values of Pe (i.e., small deviations from plug flow) the following result is obtained for a fluid-bed reactor:

$$\frac{y_A}{y_{A,p}} = 1 + \left(\frac{B'}{\lambda+1}\right)^2 \frac{1}{Pe} \qquad (17.29)$$

where $y_{A,p}$ is the fraction unconverted under plug-flow conditions. Similarly, for a fixed-bed reactor

$$\frac{y_A}{y_{A,p}} = 1 + [B' \exp(-\lambda\hat{t})]^2 \frac{1}{Pe} \qquad (17.30)$$

The effect of axial dispersion in a moving-bed reactor for first-order kinetics is given by

$$\frac{dy_A}{dz} - \frac{1}{Pe}\frac{d^2 y_A}{dz^2} = B' \exp(-\lambda z)y_A \qquad (17.31)$$

Lin (1978) has obtained an approximate solution to Eq. 17.31 subject to the usual Danckwerts boundary conditions (see Chapter 11) at high Peclet numbers by using the singular perturbation method of matched asymptotic expansions (van Dyke, 1964).

Lin (1978) formulates an inner and outer expansion for $y_A(z)$, utilizes both of the Danckwerts boundary conditions, and provides the matching condition, which stipulates that the inner expansion be in agreement with the outer expansion at the edge of the "boundary layer." Then, the fraction of reactant unconverted at the reactor exit may be approximated by

$$
y_A(1) \simeq \exp\left[-\frac{B'}{\lambda}(e^\lambda - 1)\right]\left(1 + \frac{1}{Pe}\frac{(B')^2}{2\lambda}(e^{2\lambda} - 1)\right.
$$
$$
+ \left(\frac{1}{Pe}\right)^2 \left\{ \frac{(B')^4}{8\lambda^2}e^{4\lambda} - \frac{7}{6}\frac{(B')^3}{\lambda}e^{3\lambda}\right.
$$
$$
+ \left[\frac{(B')^2}{2\lambda} - \frac{(B')^4}{4\lambda^2} + \frac{(B')^2}{2} + \frac{(B')^3}{2\lambda} - B'\lambda\right]e^{2\lambda}
$$
$$
\left.\left. + B'\lambda e^\lambda - 1 - \frac{(B')^2}{2\lambda} + \frac{2(B')^3}{3\lambda} + \frac{(B')^4}{16\lambda^2}\frac{(B')^2}{2}\right\}\right)
$$

$$(17.32)$$

This treatment may be repeated for the appropriate equation for a fixed-bed reactor.

17.2.4. Example: Effect of Axial Mixing on Gasoline Cracking in a Fixed-Bed Reactor

Weekman (1969) has given the following parameter values for gasoline cracking on a commercial catalyst at 482°C in a fixed-bed reactor: B' (reaction parameter) = 0.85, and λ (decay parameter) = 5.34. Since gasoline is a narrow boiling fraction, a first-order assumption for gasoline cracking is reasonable. Evaluate the effect of axial diffusion for large values of the Peclet number on gasoline cracking for a low value of the reaction parameter ($B' = 0.85$) and a high value of the decay parameter ($\lambda = 5.34$), and for a high value of the reaction parameter ($B' = 8.5$) and a low value of the decay parameter ($\lambda = 0.5$) in the fixed-bed reactor.

Solution

The effect of axial diffusion for large values of the Peclet number in a fixed-bed reactor is given by Eq. 17.30:

$$
\frac{y_A}{y_{A,p}} = 1 + (B'e^{-\lambda\hat{t}})^2\frac{1}{Pe}
$$

where $y_{A,p}$ represents the fraction converted under plug-flow conditions.

For turbulent flow of gasoline in fixed beds (Bischoff and Himmelblau, 1968)

$$
\frac{1}{Pe} = \frac{D}{uL} = \frac{D}{ud_p}\frac{d_p}{L} = \frac{d_p}{L}
$$

Let $L/d_p = 100$, that is, a short bed. Then, for $B' = 0.85$ and $\lambda = 5.34$,

$$
\frac{y_A}{y_{A,p}} = 1 + \frac{(0.85e^{-5.34\hat{t}})^2}{100}
$$
$$
= 1 + 0.00722e^{-10.68\hat{t}}
$$

Since the fixed-bed reaction is an unsteady-state operation, on time-averaging from $\hat{t} = 0$ to $\hat{t} = 1.0$ we obtain

$$
\left(\frac{y_A}{y_{A,p}}\right)_{ta} = 1 + \int_0^1 0.00722e^{-10.68\hat{t}}\,d\hat{t}
$$
$$
= 1.00067
$$

Similarly, for $B' = 8.5$ and $\lambda = 0.5$,

$$
\frac{y_A}{y_{A,p}} = 1 + \frac{(8.5e^{-0.5\hat{t}})^2}{100}
$$

and on time-averaging from $\hat{t} = 0$ to $\hat{t} = 1.0$ we obtain

$$
\left(\frac{y_A}{y_{A,p}}\right)_{ta} = 1 + \int_0^1 0.722e^{-\hat{t}}\,d\hat{t}
$$
$$
= 1.4567
$$

The following table summarizes the changes in gasoline conversion due to axial diffusion under the two conditions considered in a fixed-bed reactor.

Reaction Parameter B'	Decay Parameter λ	Change in Fraction of Unconverted Gasoline %
0.85	5.34	0.067
8.5	0.5	45.67

Thus, axial diffusion effects are predominant at low values of the decay parameter and high values of the reaction parameter.

17.3. EXTENSION TO A CONSECUTIVE REACTION SYSTEM

So far attention has been restricted to the effect of different decay forms on a simple reaction in an isothermal reactor. We shall now examine the effect of the decay parameter—and decay model—on the yield of the intermediate product B in a consecutive reaction $A \rightarrow B \rightarrow C$ for all the three reactor types (Sadana and Doraiswamy, 1971; Prasad and Doraiswamy, 1974).

17.3.1. Effect of Decay Parameter on Yield

For a first-order consecutive reaction the rate equations are

$$r_{VA} = -k_{v1}(1 - f_B)C_A$$

and

$$r_{VB} = (k_{v1}C_A - k_{v2}C_B)(1 - f_B) \quad (17.33)$$

The reaction is assumed to be carried out in an isothermal fluid-bed reactor with axial diffusion in the feed stream. At constant density, and on the assumption that the same value of λ holds for both steps, the following set of equations can be written:

$$\frac{d^2 y_A}{dz^2} - \text{Pe}\frac{dy_A}{dz} - \text{Pe}\left(\frac{B'}{\lambda+1}\right)y_A = 0$$

$$\frac{d^2 y_B}{dz^2} - \text{Pe}\frac{dy_B}{dz} + \text{Pe}\left(\frac{B'}{\lambda+1}\right)y_A$$

$$-\frac{\text{Pe}}{s}\left(\frac{B'}{\lambda+1}\right)y_B = 0 \quad (17.34)$$

with the boundary conditions

$$z = 0, \quad \text{Pe}(y_A - 1) - \frac{dy_A}{dz} = 0, \quad \text{Pe}\, y_B - \frac{dy_B}{dz} = 0$$

$$(17.35a)$$

$$z = 1.0, \quad \frac{dy_A}{dz} = \frac{dy_B}{dz} = 0 \quad (17.35b)$$

Assuming ideal plug flow (i.e., $\text{Pe} = \infty$) and solving the set of Eqs. 17.34 simultaneously, we get at the reactor exit

$$y_B = \frac{s}{1-s}\left\{\exp\left[-\left(\frac{B'}{\lambda+1}\right)\right] - \exp\left[-\frac{1}{s}\left(\frac{B'}{\lambda+1}\right)\right]\right\}$$

$$(17.36)$$

For $s \rightarrow \infty$, this reduces to expression 3b of Table 17.1 for a simple reaction.

Similarly, the following equation can be written for the moving-bed reactor:

$$y_B = \frac{s}{1-s}\left\{\exp\left[-\frac{B'}{\lambda}(1 - e^{-\lambda})\right]\right.$$

$$\left. - \exp\left[-\frac{B'}{s\lambda}(1 - e^{-\lambda})\right]\right\} \quad (17.37)$$

It is interesting to note that combination of Eq. 17.36 with expression 3b for y_A (at $z = 1$) listed in Table 17.1 leads to

$$y_B = \frac{s}{1-s}(y_A - y_A^{1/s}) \quad (17.38)$$

which is identical with Eq. 5.12 for a nondeactivating catalyst under diffusion-free conditions. The same equation is obtained by combining Eq. 17.37 with expression 2b in Table 17.1 for y_A for the moving-bed reactor. Further, the form of the deactivation equation makes no difference to the form of Eq. 17.38.

The observation cited above is significant, since a single equation is evidently adequate to completely define the selectivity behavior of both fluid-bed and moving-bed reactors for a first-order reaction irrespective of the decay model. The dependence of y_A on λ and B' is, however, different for the two reactor types, being determined by the corresponding expressions for y_A listed in Table 17.1.

For the fixed-bed reactor the instantaneous conversion of the intermediate y_B is also given by Eq. 17.40 for identical boundary and fluid flow conditions. But in the case of the fixed-bed reactor, as pointed out by Weekman and Nace (1970), time-averaged (ta) values of the mole fraction should be used. Thus Eq. 17.38 now becomes

$$(y_B)_{ta} = \frac{s}{1-s}\int_0^1 (y_A - y_A^{1/s})\, d\hat{t} \quad (17.39)$$

This equation integrates to give

$$(y_B)_{ta} = \left(\frac{s}{1-s}\right)\frac{1}{\lambda}\left\{\left[\text{Ei}^*(B'e^{-\lambda}) - \text{Ei}^*(B')\right]\right.$$

$$\left. - \left[\text{Ei}^*\left(\frac{B'}{s}e^{-\lambda}\right) - \text{Ei}^*\left(\frac{B'}{s}\right)\right]\right\} \quad (17.40)$$

By combining Eq. 17.38 with the appropriate equations for y_A listed in Table 17.1, we can draw plots for y_B vs. B' for different values of λ and s. The major conclusion provided by such plots is that y_B has the same maximum value for all degrees of fouling at a given selectivity for fluid- and moving-bed reactors. For the fixed-bed reactor, on the other hand, in view of the time averaging involved, y_B falls sharply with increasing λ for a given s. It follows therefore that intrinsically the fixed-bed reactor should be regarded as unsuited for a consecutive reaction system with a high value of the fouling parameter.

In this analysis the same value of λ was assumed for each step. Rickert and Wei (1968) and Campbell and

Wojciechowski (1970) have extended the analysis to the case where λ has different values for the two steps. This is referred to as selective aging as against nonselective aging for constant λ. Campbell and Wojciechowski (1970) employed essentially the same procedure as outlined above but used the time-on-stream theory for aging presented in Chapter 8. The chief conclusion from their study is that fluidized- and moving-bed reactors give better selectivity than the fixed-bed reactor.

17.3.2. Effect of Decay Model

The developments presented above were based on an exponential decay model. It is interesting to examine the behavior of the fluid-bed reactor for a catalyst that exhibits linear decay. In this case, in addition to fixing the values of the parameters B' and s, it is also necessary to know the value of the intrinsic rate constant $k_{v\theta}$. Figure 17.2 shows a plot of the intermediate mole fraction y_B as a function of the fouling parameter λ for $B' = 6.0$ and $k_{v0} = 20.0$ for various values of s. This plot

is meant particularly to show that the yield of the intermediate exhibits a maximum at a certain value of λ, and that irrespective of the mathematical model used for deactivation the same maximum value results for a given selectivity.

This observation is also true for a moving-bed reactor. On the other hand, it does not hold for the fixed-bed reactor, although with increasing selectivity the values of the maxima for the exponential and linear decay models move closer to each other.

17.3.3. Effect of Time Averaging

A quantitative study of the effect of time-averaging on the performance of a fixed-bed reactor has been made by Weekman and Nace (1970) and Weekman (1969) for the catalytic cracking of petroleum according to the model

$$C_1 \xrightarrow{k_{v1}} a_1 C_2 + a_2 C_3, \qquad C_2 \xrightarrow{k_{v2}} C_3$$

where C_1 is the gas oil charged, C_2 the C_5 fraction, and C_3 the butanes, dry gas, and coke. They developed

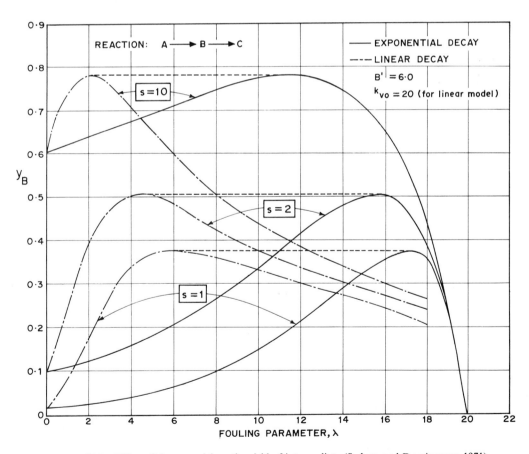

Figure 17.2. Effect of decay model on the yield of intermediate (Sadana and Doraiswamy, 1971).

expressions for determining time-averaged as well as instantaneous yields of C_2. The chief conclusion from their studies appears to be that the effect of time averaging is more pronounced at higher values of λ, particularly in a certain intermediate conversion range.

Another conclusion is that laboratory data taken in fixed-bed reactors should not be used as indicative of the performance of fluid- or moving-bed reactors. This is particularly important because fixed-bed reactors are easy to set up in the laboratory and one is tempted to extrapolate the fixed-bed performance of a fouling catalyst to reactors operating under steady-state conditions. The correct procedure would be to evaluate the decay parameter λ and the reaction group B' by organizing suitable kinetic studies and then examine the behavior of conversion or selectivity as a function of these parameters as described in the previous sections.

REFERENCES

Bischoff, K. B., and Himmelblau, D. M. (1968), *Process Analysis and Simulation: Deterministic System*, Wiley, New York.

Campbell, D. R., and Wojciechowski, B. W. (1970), *Can. J. Chem. Eng.*, **48**, 224.

Lin, C. H. (1978), *Chem. Eng. Sci.*, **33**, 614.

Prasad, K. B. S., and Doraiswamy, L. K. (1974), *J. Catal.*, **32**, 384.

Rickert, L., and Wei, J. (1968), *Ind. Eng. Chem.Fundam.*, **7**, 125.

Sadana, A., and Doraiswamy, L. K. (1971), *J. Catal.*, **23**, 147.

Van Dyke, M. V. (1964), *Perturbation Methods in Fluid Mechanics*, Academic Press, New York.

Weekman, V. W., Jr. (1968), *Ind. Eng. Chem. Process Design Dev.*, **7**, 90.

Weekman, V. W., Jr. (1969), *Ind. Eng. Chem. Process Design Dev.*, **8**, 385.

Weekman, V. W., Jr., and Nace, D. M. (1970), *AIChE J.*, **16**, 397.

CHAPTER EIGHTEEN

Heterogenizing (or Immobilization) of Homogeneous Catalysts (or Enzymes)

The field of catalysis can be divided into three broad categories: homogeneous, heterogeneous, and enzymatic. The preceding chapters dealt with the analysis of heterogeneous reactions (gas–solid catalytic) and design procedures for a variety of reactors.

In recent years it has been found that homogeneous catalysts consisting of transition metal complexes catalyze a variety of reactions with greater selectivity and activity than the conventional solid catalysts. The commercial importance of such catalysts has been firmly established by the success of the Wacker process for acetaldehyde and of the oxo process as well as the Monsanto process for acetic acid. Among the disadvantages of homogeneous catalysts are corrosion problems, product contamination, and difficult catalyst recovery. All these disadvantages can be eliminated, while the special advantageous features of homogeneous catalysts are retained, through a process of "heterogenizing" the catalysts by dispersing them in the pores of a solid support or by binding them to a polymer substrate. These two methods of heterogenizing have led to the so-called supported-liquid-phase catalysts (SPLCs) and polymer-bound catalysts, respectively.

Another important class of catalysts is provided by enzymes, which have the ability to catalyze certain reactions with a high degree of selectivity. However, the main disadvantage of enzymatic catalysts is that the enzymes have to be recovered for subsequent re-use, which often presents serious problems. One method of overcoming these problems is to immobilize the enzyme on a solid support. This is also a form of heterogenizing the catalyst system, but in this case the term "immobilization" has come to be used in the biochemical literature.

In this chapter we shall consider these three classes of heterogeneous catalysis, which have already begun to revolutionize the approach to catalysis in general. Since in all these cases solid catalysts are used, the general philosophy of design remains essentially unaltered, but the methods of catalyst preparation and the role of diffusion are significantly different.

18.1. SUPPORTED-LIQUID-PHASE CATALYSTS

18.1.1. The Concept of Heterogenization of Homogeneous Catalysts

It was found during the last decade that transition metal complexes in solution catalyze a variety of reactions (hydrogenation, isomerization, oxidation, hydroformylation, carbonylation, polymerization, etc.) with remarkable activity. Some of the notable examples of commercially important homogeneous catalysts are the Wacker process for acetaldehyde or vinyl acetate (Szonyi, 1968; Henry, 1968; Stern, 1967), the oxo process (Falbe, 1970) for hydroformylation of α-olefins, and the Monsanto process (Roth, 1975) for acetic acid manufacture. An interesting feature of homogeneous catalysis is higher activity and selectivity under mild conditions.

There have been attempts to "heterogenize" the homogeneous catalysts so that they would combine the advantages of both homogeneous and heterogeneous catalysis. Considerable progress has already been made in this direction (Delmon and Jannes, 1975; Bailar, 1974; Manassen and Whitehurst, 1973).

The supported-liquid-phase catalyst (SLPC) represents one type of heterogenized homogeneous catalyst (Jannes, 1975); it consists of a homogeneous solution of the catalyst dispersed in the pore volume of the solid support. In this unique method of gas–liquid contact, the state of the liquid-phase catalyst does not change, but the mode of its use does. The SLPC is a free-flowing solid and can be used like any conventional heterogeneous catalyst in a fixed- or fluidized-bed reactor (Acres et al., 1966).

Molten-salt catalysis is another area in which the SLPC system can be applicable. It has been shown by Topsøe and Nielsen (1948), and recently by Livbjerg et al. (1974, 1976), that the SO_2 oxidation catalyst consists of a molten electrolyte of V_2O_5–$K_2S_2O_7$ supported on silica under reaction conditions. Kenney (1975) has listed several examples of this type in which an SLPC could have potential application. Gorin et al. (1948) have indicated the possibility of using a melt of $CuCl_2$–Cu_2Cl_2–KCl dispersed on solid supports (such as pumice) in the oxidative chlorination of hydrocarbons.

There is enough evidence now to suggest that almost any kind of liquid-phase catalyst can be heterogenized and that the SLPC system is of potential importance in commercial processes based on homogeneous and molten-salt catalysis.

18.1.2. Methods of Preparation

Normally, an SLPC is prepared by dissolving the catalyst in a nonvolatile solvent and then supporting the solution in the porous solid. For example, Acres et al. (1966) prepared a catalyst by supporting an $RhCl_3$ solution in ethylene glycol on a silocel support. In this method a stock solution of the liquid-phase catalyst is prepared by first dissolving the catalyst in a nonvolatile solvent. The required amount of this solution (usually less than the total pore volume of the support material) is then diluted by a volatile solvent (acetone, chloroform, or benzene). The porous solid support is added to this solution and the solid–liquid composite is refluxed to remove the volatile solvent. The resulting dried particles are free flowing and can be used as solid catalysts. This method has also been used by Rony (1969; Rony and Roth, 1975) for preparing SLPCs for hydrogenation, isomerization, and hydroformylation reactions.

This method is not convenient for supporting molten-salt catalysts in porous solids, since most of the molten-salt catalysts exist in the liquid state at high temperatures (the reaction temperature is usually around 400°C). Livbjerg et al. (1976) used a method conventionally employed in preparing heterogeneous catalysts; in this method the support material is impregnated with catalyst solution and is then dried. When activated at elevated temperatures (480–490°C) this dry catalyst forms a supported molten-salt catalyst. It is by no means certain, however, that the catalyst will

redistribute itself in the same manner when the reaction is stopped and restarted.

18.1.3. Dispersion of Liquid in SLPC Pores

The dispersion of the liquid phase within the solid pores is clearly an important factor affecting the SLPC activity. It has been reported (Livbjerg et al., 1976; Rony, 1969) that at lower values of liquid loading the liquid is likely to be dispersed as a thin film, whereas at higher loadings liquid plugs and clusters may form. It is also possible that some pores are filled with liquid plugs while others are coated with thin film.

Rony (1968) was the first to propose a model for the SLPC system. The main assumption of the model is that the liquid is dispersed as a thin layer at the pore wall in one region or as a plug at the bottom of a single pore. Since several empirical constants are involved in the model (which can only be determined experimentally), *a priori* prediction of catalyst activity for a given liquid loading and pore structure is not possible.

Livbjerg et al. (1974, 1976; Villadsen and Livbjerg, 1978) have proposed empirical models for various states of liquid dispersion. They assumed that the degree of dispersion can be characterized by a single length parameter δ_{sl}, which denotes the average liquid-layer thickness. The empirical models for two states of dispersion—uniform liquid film and dispersed liquid plugs—follow:

Uniform liquid film:

$$\delta_{sl} = \frac{\frac{1}{2} r_p f_l}{(1 - f_l)^{1/2}} \qquad (18.1a)$$

Dispersed liquid plugs:

$$\delta_{sl} = \frac{\frac{1}{2} r_p \left[(1/V_v' \rho_s) + f_1 \right]}{1 - f_l} \qquad (18.1b)$$

These models have no general applicability, but can certainly give a rough estimate of δ_{sl}. No reliable theory for predicting δ_{sl} has been proposed so far.

18.1.4. Kinetic Studies

Supported-liquid-phase catalyst systems have the unique advantage that the kinetics of the reactions can be separately studied in the liquid-phase catalysts and then compared with the results of the SLPC systems to enable us to understand the effect of transport resistances. However, Rony (1969) and Rony and Roth

(1975) have shown, by studying the hydrogenation, isomerization, and hydroformylation reactions, that the intrinsic kinetics can also be studied directly in SLPC systems.

An interesting observation in the theoretical and experimental studies reported so far is that the rate of reaction in an SLPC as a function of liquid loading goes through a maximum, which suggests that for a set of conditions an optimum catalyst loading exists. Rony (1968, 1969) has demonstrated this in the case of hydroformylation, while Livbjerg et al. (1974, 1976) and Villadsen and Livbjerg (1978) found a similar trend in SO₂ oxidation.

Hjortkjaer et al. (1981) have investigated the kinetics of hydroformylation of propylene by using a supported liquid-phase rhodium complex catalyst. They found that excess triphenylphosphine has an influence on the kinetics as well as the selectivity of *n*-butanol formation. The following equation has been proposed for the rate of conversion of propylene:

$$r_1 = k p_{C_3H_6}^a p_{CO}^b p_{H_2}^c \qquad (18.2)$$

These authors have also proposed rate equations for the formation of *n*-butanol and isobutanol.

18.1.5. Mass Transfer Effects

There are three types of mass transfer processes that can influence SLPC activity: intraparticle gas phase, intraparticle liquid phase, and external gas film. Of these, the intraparticle gas-phase and external mass transfer effects are similar to those of solid-catalyzed systems but the liquid-phase mass transfer is an additional resistance in SLPC systems. A qualitative discussion of these processes follows.

Intraparticle Mass Transfer (Gas Phase)

The main difference in the intraparticle diffusion in an SLPC and in solid catalysts is that the porosity in SLPC systems is strongly dependent on the liquid loading. Liquid loading is therefore a new variable in SLPC systems. Rony (1968) analyzed the case of diffusion with first-order reaction in an SLPC by extending the single-pore concept of porous catalysts, and obtained solutions for the effectiveness factor. Later Abed and Rinker (1973) obtained solutions, again for a first-order reaction, assuming that the liquid is distributed uniformly in the solid support. Livbjerg et al. (1974, 1976) and Villadsen and Livbjerg (1978) have obtained theoretical as well as experimental effectiveness factors for SO₂ oxidation on an SLPC.

Chen and Rinker (1978) proposed a model for describing the mass transfer of gaseous reactants and products in an SLPC. This model incorporates the pore size distribution of the dry support and nonuniform distribution of the liquid phase. The model was tested experimentally by using the 1-chloronaphthalene–dibutylphthalate–α-alumina system.

It has been observed from these studies that the variation of the effectiveness factor with the Thiele modulus is similar to that in solid-catalyzed systems, but the effect of liquid loading is interesting. As the liquid loading increases, the pore diffusion (gas-phase) resistance increases, since the gas-phase porosity is reduced. Rony (1969) has experimentally observed that at higher liquid loadings the conversions decrease drastically as a result of pore diffusion resistance.

Intraparticle Mass Transfer (Liquid Phase)

In most cases of diffusion and reaction in SLPCs considered so far, it has been assumed that the liquid-phase transport resistance is negligible. Since the reaction occurs in the liquid phase, in may cases the diffusional resistance would be significant (Livbjerg et al., 1976). Depending on the catalyst concentration, distribution of liquids, diffusivity, solubility, and other system properties, the magnitude of the reaction rate constant and the mass transfer coefficient would vary drastically. It has been found, for instance, that even at negligible pore diffusion resistance, the effectiveness factor can be much less than unity. This is due to liquid-phase mass transfer resistance.

External Mass Transfer

When the overall reaction rate in an SLPC is much faster than the rate of transport through the external film of the catalyst pellet, the controlling step would be diffusion through the external film. Rony (1969) has observed that under certain conditions the hydroformylation of propylene becomes external mass transfer controlled.

18.1.6. SLPC Reactors

Supported-liquid-phase catalysts may be used in fixed- or fluidized-bed reactors. Slurry- or trickle-bed reactors cannot be used for these systems, since the presence of the liquid phase in these reactors would tend to leach out the active species from the porous solid by dissolution or diffusion.

Almost all the investigations using supported-liquid-phase-catalyzed reactions have been carried out in fixed-bed reactors (Acres et al., 1966; Rony, 1969; Rony and Roth, 1975; Moravec et al., 1941; Ciapetta, 1947; Komiyama and Inoue, 1975, 1977; Livbjerg et al., 1974, 1976; Villadsen and Livbjerg, 1978). The effectiveness factor relations would be different for the SLPC system (see Section 4.10 for a discussion of the latter).

Rony (1968, 1969) has derived equations for the conversion in a fixed-bed reactor of SLPCs and has used these equations in interpreting propylene hydroformylation data. The theoretical analysis given by Rony is summarized by the following equations:

$$x = 1 - \exp\left[-\left(\frac{KkV_l}{Q}\right)R_{ov}\right] \quad (18.3)$$

$$\frac{1}{R_{ov}} = \frac{1}{R_{liq}} + \frac{1}{R_{gas}} \quad (18.4)$$

$$R_{liq}f_l = \frac{f_{l,1}[\tanh A(1-f_{l,2})/A] + [\tanh(Bf_{l,2})/2]}{1 + \left(\frac{A}{Bf_{l,1}}\right)\tanh Bf_{l,2}[\tanh A(1-f_{l,2})]} \quad (18.5)$$

where R_{ov}, R_{liq}, and R_{gas} are the overall, liquid and gas-film resistances, f_l is the fractional liquid loading, and V_l is the volume of the catalyst solution;

$$B^2 = \frac{kl_p^2}{D_L} \quad (18.6a)$$

$$A^2 = B^2\left(\frac{f_{l,1}KD_L}{f_{l,1}KD_L + f_g D_g}\right) \quad (18.6b)$$

$$f_l = f_{l,2} + (1-f_{l,2})f_{l,1} \quad (18.6c)$$

and

$$f_{l,2} \propto (f_l) = f_l[1 - \exp(-af_l^n) + \exp(-a)]. \quad (18.6d)$$

Data on the isomerization, hydrogenation, and hydroformylation of olefins in a fixed-bed continuous-flow reactor have been correlated by using these equations (Rony, 1969, Rony and Roth, 1975). It has also been pointed out (Jannes, 1975) that the flushing of volatile products, which could lower the activity of an SLPC, in a continuous-flow reactor is an important factor in the operation of an SLPC reactor. The main weakness of Eqs. 18.3–18.6 is that they contain constants that cannot be estimated from physical considerations but must be determined experimentally.

Acres et al. (1966) also used a fixed-bed reactor in studying the isomerization of pentene, and observed

that the conversion decreased rapidly with contact time. This study represents an interesting example of a deactivating SLPC reactor.

For reactions where the heat of reaction is very high a fluidized-bed reactor would be more appropriate, whereas for kinetic studies a basket-type reactor would be more suitable. When the activity of an SLPC is very poor, a continuous-flow fixed-bed reactor would lead to low conversions, and in such cases either a batch or continuous-recycle reactor (Section 12.6.5) could be used. However, there are no reports of the use of these reactors in SLPC studies. Since the SLPC system is identical to a gas–liquid chromatographic column, a chromatographic reactor can also be used (Section 9.5.2).

18.1.7. Example: Oxidation of Ethylene on Supported Palladium Complex Catalyst: Calculation of Conversion

It is required to oxidize ethylene by using a supported-liquid-phase catalyst. Komiyama and Inoue (1975) studied this reaction using a Pd complex catalyst supported in a porous carrier. Predict for the following conditions the conversion in a fixed-bed reactor if plug-flow behavior is assumed.

Reaction rate constant	k	$= 2.5 \times 10^{-2} \sec^{-1}$
Gas–liquid partition coefficient	K_{par}	$= 4.46 \dfrac{cm^3 \text{ gas}}{cm^3 \text{ liquid}}$
Length of the pore	l_p	$= 0.5\,cm$
Effective diffusivity in liquid	D_L	$= 3.18 \times 10^{-5}\,cm^2/\sec$
Gas velocity	Q	$= 0.06\,cm^3/\sec$
Effective diffusivity in gas phase	D_g	$= 2.5 \times 10^{-2}\,cm^2/\sec$
Liquid loading	f_l	$= 0.4$
Volume of the catalyst solution	V_l	$= 0.22\,cm^3$

Assume that gas-film diffusional resistance is negligible.

Solution

The parameters required for Eq. 18.3 can be calculated as follows. To calculate the parameter $f_{l,2}$, the correlation proposed by Rony (1969) is assumed to be valid ($a = 5.9$, $n = 3.3$). Then

$$f_{l,2} = 0.4\left[1 - \exp\left(-5.9 \times (0.4)^{3.3}\right) + \exp\left(-5.9\right)\right]$$

$$= 0.1$$

$$f_{l,1} = \frac{0.4 - 0.1}{1 - 0.1}$$

$$= 0.3327$$

$$B^2 = \frac{2.5 \times 10^{-2} \times 0.5 \times 0.5}{3.18 \times 10^{-5}} = 196.54$$

$$A^2 = \frac{196.54 \times 0.3327 \times 4.46 \times 3.18 \times 10^{-5}}{0.3327 \times 4.46 \times 3.18 \times 10^{-5} + (1 - 0.3327) \times 2.5 \times 10^{-2}} = 0.555$$

Hence

$$R_{\text{liq}} = \frac{0.3327\left\{\dfrac{\left[\tanh\sqrt{0.555}(1-0.1)\right]}{\sqrt{0.555}}\right\} + \left[\dfrac{\tanh\left(\sqrt{196.54}\times 0.1\right)}{\sqrt{196.54}}\right]}{0.4\left\{1 + \dfrac{\sqrt{0.555}}{\sqrt{196.54 \times 0.3327}}\left[\tanh\left(\sqrt{196.54}\times 0.1\right)\right]\left[\tanh\sqrt{0.555}(1-0.1)\right]\right\}}$$

$$= 0.749$$

The conversion x by Eq. 18.3 is given as

$$x = 1 - \exp\left(-\frac{4.46 \times 2.5 \times 10^{-2} \times 0.22 \times 0.749}{0.06}\right)$$

$$= 0.2638 \quad \text{or} \quad 26.38\%$$

18.1.8. Examples of SLPC Systems

A wide range of hydrocarbon reactions (hydrogenation, isomerization, hydroformulation, oxidation, and polymerization) are known to be catalyzed by supported-liquid-phase catalysts. Some homogeneously catalyzed reactions need a liquid environment for reaction; in such cases an SLPC can be very useful. Several examples of reactions using SLPC systems are listed in Table 18.1.

18.2. POLYMER-BOUND CATALYSTS

Polymer-bound catalysts represent another class of heterogenized catalysts; they combine the potential versatility and selectivity of homogeneous catalysts with the practical advantages associated with the use of solid catalysts. These catalysts, which are prepared by chemically binding metal complexes to polymers, can (unlike the homogeneous catalysts) be used in continuous systems. The subject of polymer-bound catalysis in general has been reviewed by Burwell (1974), Heinemann (1971), Manassen (1969), Michalska and Webster (1974), Pittman and Evans (1973), Bailar (1974), Delmon and Jannes (1975), Grubbs (1977), Basset and Smith (1977), Whitehurst (1980), and Bailey and Langer (1981).

18.2.1. Methods of Preparation

In order for chemical linkage between the catalyst and the support to be effected, the latter must have functionality. This may be an intrinsic property of the support or it can be introduced by chemical modification. In most cases the catalyst has to be built up from its components, namely, the support, the cross-linking groups, and the catalytic complex. Jannes (1975), Pittman and Evans (1973), and Michalska and Webster (1974) have reviewed various methods of preparing polymer-bound catalysts.

18.2.2. Reactors for Polymer-Bound Catalysts

Polymer-bound catalysts are identical to the conventional heterogeneous catalysts insofar as handling is concerned, and therefore many types of reactors suggested for heterogeneous catalysts would also be suitable for these catalysts.

Some polymer-bound catalysts are more effective in the presence of solvents or require a solvent environ-

TABLE 18.1. Examples of Reactions on Solid Supported–Liquid–Phase Catalysts

System	Catalyst	Reactor	Reaction Conditions	Reference
Hydrogenation of propylene	$(PPh_3)_3RhCl$ in trichlorobenzene on silica gel	Fixed bed	24°C, 2.4 atm	Rony and Roth (1975)
Isomerization of 1-pentene	$RhCl_3 \cdot 3H_2O$ in ethylene glycol on 44 to 60-mesh silocel	Fixed bed, continuous and pulsed flow	25°C, 1.0 atm	Acres et al. (1966)
Isomerization of 1-butene	$RhCl_3 \cdot 3H_2O$ in *n*-heptanol on silica gel	Fixed bed	25°C, 1.0 atm	Rony and Roth (1975)
Isomerization of quadricyclene	Cobalt–tetraphenylporphyrin	IR cell, fixed bed	20–50°C	Wilson and Rinker (1976)
Hydroformulation of propylene	$(PPh_3)_2 RhCOCl$ in butyl benzyl phthalate on silica gel, carbon, and alumina	Fixed bed	136°C, 50 atm	Rony (1969); Rony and Roth (1975)
	$HRh(CO)(PPh_3)_3$ on silica	Fixed bed	100°C, 12 atm	Hjortkjaer et al. (1981)
Oxidation of ethylene	$PdCl_2$–$CuCl_2$ solution on porous glass	Fixed bed	70°C, 1.0 atm	Komiyama and Inoue (1975, 1977)
Oxidation of SO_2	Molten V_2O_5–$K_2S_2O_7$ on silica glass support	Fixed bed	435–530°C, 1.0 atm	Livbjerg et al. (1976)
Polymerization of isobutene	H_2SO_4 on silica gel	Fixed bed	22°C, 1 atm	Ciapetta (1947)
Polymerization of olefin–paraffin mixture	H_2SO_4 on majolica chips, carbon, or silica gel	Fixed bed	100°C, 4 atm	Moravec et al. (1939)

ment, in which case vapor-phase catalytic reactors cannot be used. The reactors for gas–liquid–solid contacting, such as agitated slurry reactors, bubble column slurry reactors, trickle-bed reactors, or packed bubble column reactors (see Volume 2) would be most suitable for such systems. It is also interesting to note that in most of the experimental studies (Pittman et al., 1975; Grubbs and Kroll, 1971; Delmon and Jannes, 1975; Bailar, 1974; Collman et al., 1972) on polymer-bound catalysts, slurry reactors at atmospheric or higher pressures have been used. No attempt has yet been made, however, to investigate the role of different reactor types in these catalytic systems. Mass transfer effects and kinetics have also not been studied in detail, since the current emphasis is on developing methods for binding homogeneous catalysts to polymers. Pittman et al. (1975) have shown that diffusion plays an important role in the activity and selectivity of polymer-bound catalysts, and in general the effect of mass transfer would be similar to that in fixed-bed or slurry reactors.

Only a few studies on the kinetics of reactions using polymer-catalysts have been reported, and the rate models proposed are summarized in Table 18.2. Since the mechanism of homogeneously catalyzed reactions is highly complex, further work on the kinetics of reactions and the role of mass transfer in polymer-bound catalysts is necessary.

18.2.3. Examples of Polymer-Bound Catalysts

Polymer-bound catalysts have been shown to catalyze a variety of hydrocarbon reactions, some of which are the hydrogenation of cyclohexene, 1,5-cyclooctadiene, cyclohexene (Grubbs and Kroll, 1971; Pittman et al., 1975; Jarrell and Gates, 1978), and ethylene (Otero-Schipper et al., 1980); the hydroformylation of 1-hexene (Allum et al., 1972; Hancock et al., 1975), allyl alcohol, and methyl methacrylate (Pittman and Honnick, 1980; Pittman et al., 1980) and of ethylene, propylene, and 1-

TABLE 18.2. Summary of Kinetic Studies Using Polymer-Bound Catalysts

Reaction	Catalyst	Rate Equation	References
Hydrogenation of cyclohexene	$Rh_6(CO)_{16}$ bound to a polystrene DVB membrane	$kp_{CH}^2 p_{H_2}^{0.5}$	Jarrell and Gates (1978)
Hydrogenation of ethylene	$Rh_6(CO)_{16}$ bound to a polystyrene DVB membrane	$kp_E^{0.8} p_{H_2}^{0.2}$	Jarrell and Gates (1978)
Hydrogenation of styrene and cyclohexene	Polymer–Rh(II)Cl$_2$ (bivalent Rh–polystyrene)	$\dfrac{k_3 K_2 (C_{cat})(p_{H_2})(p_{olefin})}{1 + K_2 p_{olefin}}$	Imanaka et al. (1976)
Hydrogenation of olefins and acetylenes	Polymer–CH$_2$–PPh$_2$–PdCl$_2$	$k_2 p_{H_2} p_{olefin}$	Terasawa et al (1978, 1979)

butene (Arai, 1978; Carlock, 1980); the carbonylation of methanol (Jarrell and Gates, 1975); the oxidation of ethylene (Linarte-Lazcano et al., 1975; Arai and Yashiro, 1978); the dimerization of ethylene and propylene (Mizoroki et al., 1975); the oligomerization of butadiene (Pittman et al., 1975, 1976); and the cyclo-oligomerization and hydrogenation (or hydroformylation) of butadiene (Pittman and Smith, 1975).

18.2.4. Membrane and Hollow-Fiber Reactors

The principles of homogeneous catalysis can also be applied to biological systems. Continuous operation with biocatalysts is not only possible by using the (more popular) carrier-fixed enzymes (which will be dealt with subsequently), but also by using ultrafiltration membranes (Wandrey, 1979). The ultrafiltration membrane in a continuously operated enzyme reactor retains the biocatalyst. Continuous homogeneous catalysis with enzymes in membrane reactors was suggested by Michaels (1968). The specific advantages of the membrane technique are the following.

1. Continuous homogeneous catalysis without mass transport limitations can be established.

2. It is applicable to multienzyme systems, since it is possible to retain even coenzymes by means of ultrafiltration membranes when these compounds are bound to soluble polymers of sufficient molecular weight.

Symmetric and asymmetric membranes are available in flat-sheet or hollow-fiber design. The trend is to use asymmetric membranes, since the pressure drop across an asymmetric membrane is less than that across a symmetric membrane. The average pore diameter for enzyme membrane reactor is 1.5–15 nm. Most enzymes

are retained by membranes with pores of such size. Since low molecular weight substrates are normally less than 0.5 nm in diameter, effective separation is possible.

Typically, the enzyme membrane reactor consists of a recycle loop with an effective recirculating pump. Product leaves the reactor as a result of enforced flow across the ultrafiltration membrane. Plug flow may be approached by using two or three such systems in series. To minimize enzyme polarization it is essential to establish turbulent conditions as near to the membrane as possible (Strathmann, 1978).

Another specific advantage of membrane reactors is that they permit measurement of catalyst deactivation at operational conditions, since it is possible to supply continuously as much free enzyme to the reactor as is deactivated in the reactor. Wandrey (1979) has suggested a possible control system whereby constant productivity is possible at constant substrate flow by supplying the required fresh catalyst to "replace" the deactivated catalyst.

18.3. IMMOBILIZED-ENZYME SYSTEMS

In the last decade, immobilized-enzyme (IME) reactors have come to occupy a prominent place in the field of catalytic engineering. Earlier, the large-scale commercial application of enzymes as catalysts for chemical reactions was hampered by the inability to recover the enzyme from the reaction system. This situation necessitated the use of enzymes in high dilutions, with consequent large reactor volumes. Further, continuous reactor techniques were not practicable. With the rapid growth of immobilization techniques in the last few years, these drawbacks have largely been removed and IME reactors are finding increasing industrial applications.

The reason for including IME reactors in this book is that in many instances they are similar to fixed-bed catalytic reactors where the interplay between diffusional effects and reaction kinetics is important. Some examples are the DEAE–Sephadex–L-aminoacylase system for the resolution of amino acid mixtures, the use of immobilized glucose isomerase for the conversion of glucose to fructose, the use of immobilized penicillin acylase for the manufacture of 6-aminopenicillanic acid (6-APA) and the potential use of microencapsulated urease in artificial kidney machines.

Some special features that distinguish IME reactors

from the conventional chemical reactors are the high specificity of enzyme reactions and their high sensitivity to parameters like temperature and pH. The most important distinguishing feature of IME reactors, however, is the nature of the system itself. As against the gas–solid systems considered in heterogeneous catalytic reactions, here we are concerned with liquid–solid catalytic systems, but the methods of analysis are substantially the same.

Before we proceed with the analysis of immobilized enzymes, a brief account of the nature of enzymes seems desirable. Enzymes are macromolecules, with molecular weights of the order of 10^6, which are synthesized by living cells using 20 different monomer species. An

Figure 18.1.   Three-dimensional structure of an enzyme: carboxypentidase A (Lipscomb, 1972).

enzyme is characterized by the number and types of monomers joined together and the sequence of joining. Interactions between the functional groups on the monomers, among themselves and with the micro-environment of the enzyme, lead to a three-dimensional enzyme configuration, as shown in Figure 18.1. It will be noticed that the enzyme molecule is not straight, but has a folded structure. As in the case of a catalyst, only a small region of the molecule is catalytically active. The catalytic effectiveness of the molecule depends on its configuration, the charges of the functional group, and the local environment. Sometimes the activity can be increased by the association of small ions or molecules with the enzyme; in biochemical terminology these are called cofactors.

Two distinguishing features of enzyme catalysis are noteworthy. There is usually an optimum pH value at which the enzyme activity is maximal. This is sub-stantiated by many systems—for example, the enzyme-catalyzed oxidation of glucose, as shown in Figure 18.2. This behavior is characteristic of both immobilized and soluble forms of glucose oxidase, with the immobilized form showing less sensitivity, but cannot be taken as general. The second feature is the temperature dependence of the activity. For instance, in the de-composition of hydrogen peroxide catalyzed by the enzyme catalase, there is a maximum activity at a certain temperature. Indeed, Johnson et al. (1954) have pro-posed a model for this type of rate–temperature behavior that uses the absolute rate theory.

While dealing with enzymes the following situations must be recognized: They are produced in the cell and retained in the cell liquid (cytosol), attached to the cell membrane, or secreted into the microenvironment surrounding the cell. Enzymes that are retained in the cell are referred to as intracellular enzymes, and those that are secreted are called extracellular enzymes. In the case of intracellular enzymes, it is sometimes profitable to employ the cell as a whole for catalysis. These cells can either be used in solution or immobilized on a solid, as in the case of enzymes. Use of the whole cells obviates the need to break up the cells in order to isolate the enzyme. The analysis and design of live immobilized cell systems is considerably more difficult. Hardly any information is available beyond a preliminary analysis (Venkatsubramanian et al., 1983). This is not con-sidered in the present treatment.

Examples of Immobilization Techniques

Immobilization of enzymes on solid surfaces can be achieved in a variety of ways, such as adsorption,

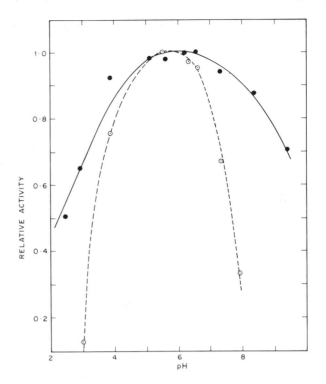

Figure 18.2. The pH dependence of glucose oxidase activity: soluble ($\bigcirc$) and immobilized ($\bullet$) forms (Cho and Bailey, 1978).

adsorption followed by cross-linking, and covalent attachment and encapsulation in pores, spun fibers, and gels. Tables 18.3 and 18.4 list the techniques used for immobilizing certain selected enzymes and micro-organisms, respectively.

A typical industrial enzyme immobilization, carried out at the National Chemical Laboratory in collabor-ation with Hindustan Antibiotics Ltd., Poona, on penicillin acylase (penicillin amidohydrolase, EC 3.5.1.11) for the production of 6-aminopenicillanic acid (a key intermediate in the manufacture of semisynthetic penicillins), is described next. The partially purified enzyme from *Escherichia coli* NCIM 2400 was immobilized by covalent attachment to cyanogen-bromide(CNBr)-activated cellulose. Enhanced binding was obtained after preliminary alkali treatment of the support prior to activation (Subramanian et al., 1978). Although CNBr-activated polysaccharides have been used extensively for enzyme immobilization, relatively little is known about the nature of the covalent attachment (Porath, 1974). The reaction proceeds pre-sumably through an isocyanate derivative of the carrier as an unstable intermediate that in the coupling step can yield at least three products: imidocarbonate, sub-stituted carbonate, and isourea derivatives. The enzyme immobilized on the CNBr-activated cellulose has been

TABLE 18.3. Illustrative Examples of Methods Used for Enzyme Immobilization

Enzyme	Carrier	Method of Immobilization	Application	Reference
Glucose isomerase	DEAE cellulose	Adsorption	Production of high-fructose corn syrup from glucose	Davis (1970)
Glucoamylase	ZrO_2-coated porous glass	Covalent attachment after activation by silanization	Hydrolysis of starch to glucose	Weetall and Havewala (1972)
Papain	—	Intermolecular cross-linking with glutar-aldehyde	Hydrolysis of proteins	Jansen and Olson (1969)
Pencillin acylase	Cellulose triacetate fibers	Droplet encapsulation	Hydrolysis of biosynthetic penicillins to 6-aminopenicillanic acid	Dinelli (1972)
Amino acid acylase	k-Carrageenan	Gel entrapment	Resolution of α-amino acids	Tosa et al. (1979)

TABLE 18.4. Illustrative Examples of Methods used for Immobilization of Microbial Cells

Microorganism	Method of Immobilization	Application	Reference
Streptomyces griseus	Containment within dialysis tubing in fermentor	Cycloheximide production	Kominek (1975)
Aspergillus foetoidus	Adhesion on stainless steel wire crushed into spheres	Citric acid production	Atkinson et al. (1979)
Streptomyces sp.	Ion-exchange adsorption on DEAE–Sephadex	High-fructose syrup production from glucose	Shigesada et al. (1975)
Micrococcus luteus	Cells covalently linked to carbodiimide-activated CM cellulose	Urocanic acid production from L-histidine	Jack and Zajic (1977)
Bacillus coagulans	Cells cross linked with glutaraldehyde	High-fructose syrup production from glucose	Novo Industri (1976)
Escherichia coli	Entrapment in k-carrageenan	Production of L-aspartic acid from ammonium fumarate	Tosa et al. (1979)
Escherichia coli	Droplet encapsulation in cellulose triacetate fibers	Hydrolysis of benzylpenicillin to 6-aminopenicillanic acid	Marconi et al. (1975)

used in batch-type stirred-tank reactors in pilot plant trials over 20 cycles of use with negligible loss in activity (Borkar et al., 1978).

18.3.1. Modeling of IME Reactions

Because of the similarity between IME reaction systems and the more common heterogeneous catalytic systems, the existing information in the heterogeneous catalytic literature has been fully utilized in developing not only reaction models but also design equations for IME systems. However, due consideration should be given to the peculiarities of enzyme systems, such as the effect of the method and conditions of immobilization on the true kinetics of the reaction.

The simplest and the most common rate equation

used for IME reactions is the Briggs–Haldane model:

$$r = -\frac{dC_S}{dt} = \frac{V_m C_S}{K_m + C_S} \qquad (18.7)$$

where C_S is the concentration of the reactant, generally referred to in biochemical terminology as the substrate concentration; V_m is the product of the rate constant and enzyme concentration ($k\breve{E}_0$) and represents the maximum rate of substrate utilization; and K_m is the Michaelis–Menten constant. Rate is denoted by v in the biochemical literature, but for consistency the notation r is retained here. For a detailed discussion of the kinetics of immobilized-enzyme reactions, reference may be made to a review by Carbonell and Kostin (1972). Although true values of V_m and K_m can be obtained when the enzyme is in free solution, immobilization generally leads to lower values of the rate.

The changes caused by immobilization can be attributed to two major factors. The first of these arises directly from the attachment of the enzyme molecule to the support and can lead to structural changes that result in restrictions on the accessibility of the active sites. The second is the direct result of the heterogeneous nature of the system, in which the concentration of the substrate in the enzyme environment may be different from that in the bulk solution. These are again broadly divided into two effects, namely, partition phenomena and diffusional effects.

18.3.2. Structural Changes on Immobilization; Partition Effects

Where an enzyme is adsorbed on or covalently bound to a solid support, the interaction with the support may result in a modification of the enzyme conformation (Gabel and Hofsten, 1971; Cho and Swaisgood, 1974). The covalent bonds can stretch the whole molecule and thus alter the three-dimensional structure of the active sites (which is known to be responsible for catalytic action). On the other hand, the attachment of the enzyme to the carrier matrix may reduce the accessibility of the enzyme molecule to the substrate, thus lowering the overall rate of reaction (Porath et al. 1967; Axen and Ernbach, 1971). This is referred to as steric hindrance. Clearly, the alteration in the kinetic behavior due to these factors is strongly dependent on the nature of the enzyme and support and the technique of immobilization.

The interaction between the macroenvironment (i.e., the bulk solution) and the microenvironment (i.e., the immediate vicinity of the bound enzyme) can cause sharp concentration differences between the two areas, leading to the so-called partition effect. Such an effect is not present in the nonenzymatic catalytic reactions considered in the earlier chapters. It is caused mainly by the hydrophobic or hydrophillic and electrostatic interactions between the carrier and the substrate or products. For example, a relatively nonpolar substance can be more soluble in a hydrophobic membrane than in the bulk solution, which causes its concentration on the membrane to be higher than in the surrounding solution. An example of the partition effect is the observed change in the degree of inhibition of β-fructofuranosidase by aniline and Tris when the enzyme is attached to a polystyrene surface (Filippusson and Hornby, 1970). This leads to a lower value of the Michaelis–Menten constant. In fact, the use of a modified Michaelis–Menten parameter K_m' appears to be a popular practice in enzyme engineering.

The partition effect is also caused by the interaction of the electrostatic charges on the support and the mobile charged species. For example, a polyanionic support increases the hydrogen ion concentration in the enzyme microenvironment and thus causes the optimum value in the pH activity curve to shift toward more alkaline values. This situation was observed in the case of chymotrypsin by Goldstein and Katchalski (1968), who also noted that the opposite effect is obtained when the same enzyme is immobilized on a polycationic derivative.

18.3.3. Diffusional Resistances (External and Internal

The situation corresponding to simultaneous diffusion and chemical reaction in porous catalysts is also encountered in IME systems. However, the effect of diffusion is far more significant than in conventional catalytic reactions, since molecular diffusivities in aqueous solutions and gels are lower than in gas–solid systems and the catalytic activity of enzymes is also usually much higher than that of a solid catalyst. Another important factor is the size of the species being transported; large molecules with relatively small diffusivity in the porous medium encounter significant diffusional resistance (Mosbach and Mosbach, 1966).

The data obtained from laboratory IME reactors reveal the presence of significant resistance to the transport of species in the film surrounding the IME particles. The apparent reaction kinetics has been observed to be dependent on the flow rate of the substrate over the IME bed. The apparent Michaelis constant K_m' varies with the flow rate, approaching at very high flow rates the value of the free enzyme (Wilson et al., 1968; Hornby et al., 1968). These and other studies (e.g., Rovito and Kittrell, 1973; Brams and McLaven, 1974) bring out the importance of external or bulk diffusion in IME reactions.

In the case of a single IME particle, the theoretical treatment of the effect of external mass transfer limitations is similar to that in heterogeneous catalysis. For an enzymatic reaction of simple Michaelis–Menten kinetics, the material balance equation is written as

$$k_L(C_{Sb} - C_{Ss}) = \frac{V_m C_{Ss}}{K'_m + C_{Ss}} \qquad (18.8)$$

where k_L is the external liquid-film mass transfer coefficient, and C_{Sb} and C_{Ss} are the substrate concentrations in the bulk and at the surface, respectively. Defining the dimensionless substrate concentration β_S as C_S/K'_m, we obtain the following expression for β_S:

$$\beta_S = \frac{(\beta_{Sb} - 1 - \mu) + [(\beta_{Sb} - 1 - \mu)^2 + 4\beta_{Sb}]^{1/2}}{2} \qquad (18.9)$$

where the substrate modulus (Horvath and Engasser, 1974) $\mu = V'_m/k_L K'_m$. The external effectiveness factor is then given by

$$\eta = \frac{\beta_S(1 + \beta_{Sb})}{\beta_{Sb}(1 + \beta_{Ss})} \qquad (18.10)$$

where $\beta_{Ss} = C_{Ss}/K'_m$ and $\beta_{Sb} = C_{Sb}/K'_m$.

The variation of η is conveniently expressed by the dimensionless parameters β_{Sb} and μ and has been presented as a plot by Horvath and Engasser (1974). Kinetic control giving η of unity at low values of μ and diffusion control at high values of μ can be clearly observed in such a plot. The low and high values of β_{Sb} represent the approach to first- and zero-order kinetics, respectively.

Greenfield et al. (1975) have studied the effect of film diffusion on Michaelis–Menten kinetics in a packed-bed reactor. They present a figure from which we can determine η, and that tells us if film diffusion effects are significant or not. In general, if $\eta > 0.95$, film diffusion effects are insignificant; and if $\eta < 0.6$, film diffusion effects are dominant.

Toda (1975) and Lee et al. (1979) carried out a theoretical investigation of the packed-bed IME reactor and predicted an approximate quantitative relationship between the apparent kinetic constants and

external film mass transfer. However, the basis of the approximations indicates stringent limitations. Patwardhan and Karanth (1982) have given a simple approximation that predicts the often observed linear behavior of the kinetic plots namely $C_{S0} x$ vs. $\ln(1 - x)$ of packed-bed IME reactors. They give a new approach for calculating the intrinsic kinetic parameters of the IME reactor based on the intercepts of the kinetic plots which are film diffusion influenced. The advantage of this method over the earlier ones is that it is applicable even when the plots are nonlinear. The linearity or otherwise of such plots can be determined with a simple analytical criterion.

Diffusional limitations often affect the kinetics of single-substrate reactions by decreasing the apparent affinity of the enzyme for its substrate. With enzymatic reactions involving two substrates or one substrate and a cofactor, apparent affinities of the bound enzyme may increase or decrease as compared to the affinity of the soluble enzyme.

Subramanian (1978) has used a modified form of the Frank–Kamentskii solution to study the rate of internal substrate diffusion and simultaneous biochemical reaction in the biological-film reactor.

While the external-film diffusional effects are significant in some systems, the diffusional influence within the porous solid matrix is more important. The influence of internal diffusion has been demonstrated in several cases—for instance, polyacrylamide film (Bunting and Laidler, 1972), collodion membranes (Goldman et al., 1968), porous glass particles (Marsh et al., 1973), kieselguhr (Krishnaswamy and Kittrell, 1978), ion-exchange resin beads (Kobayashi and Moo-Young, 1973), and controlled pore glass (Lee et al., 1980), all of which have been used as supports for the enzymes. In all these experiments, in accordance with theory (see Chapter 4), an increase in the characteristic length of the particle resulted in a corresponding decrease in the effectiveness factor. It is noteworthy that in an immobilized porous-glass–glucose-oxidase system, particles as small as 30 μm and an average pore size of less than 2000 Å must be used in order to avoid pore diffusional resistance (Rovito and Kittrell, 1973).

The recently developed technique of immobilizing microbial intact cells has drawn considerable interest. Kobayashi and Suzuki (1976) have determined the intraparticle effectiveness factors for α-galactosidase containing spherical pellets formed naturally under given conditions in a submerged culture of *Mortierella vinacea*. The experimental effectiveness factors were found to be represented as a single function of the modified Thiele modulus, including such parameters as pellet size, enzyme concentration in the pellet, and substrate concentration.

Mathematically the problem of simultaneous mass transfer and reaction of the substrate in the matrix of a single catalyst element for a simple Michaelis–Menten kinetics can be analyzed by a second-order differential equation:

$$D_e \nabla^2 C_S - \frac{V'_m C_S}{K'_m + C_S} = 0 \qquad (18.11)$$

with boundary conditions defined appropriately, according to whether external mass transfer resistance is negligible or not (see Section 6.3). Defining the Thiele modulus ϕ as

$$\phi = L \left(\frac{V'_m}{K'_m D_e} \right)^{1/2} \qquad (18.12)$$

Eq. 18.11 can be numerically solved for the effectiveness factor. In the limiting case of a pseudo-first-order reaction ($K'_m \gg C_S$) the expression for the effectiveness factor ε for a spherical pellet is given by Eq. 4.18. For such a case, the experimental data on effectiveness factors and conversions in a packed-bed reactor seem to agree reasonably well with the theoretical predictions (Rovito and Kittrell, 1973; Bunting and Laidler, 1974).

Numerical solutions for Michaelis–Menten kinetics and effectiveness factor charts have been developed by several authors (Marsh et al., 1973; Engasser and Horvath, 1973; Blanch and Dunn, 1974). These are in general similar to the solutions presented in Chapter 4 for reactions following H–W kinetics under isothermal conditions.

Dahodwala et al. (1976) have developed a pore diffusion model involving a two-substrate enzymatic reaction. The theoretically calculated effectiveness factors compared reasonably well with the experimentally determined effectiveness factors for the galactose-oxidase-catalyzed oxidation of galactose. Swanson et al. (1978) measured the effectiveness factors for glucoamylase immobilized on two agarose carriers and designed several reactors to show the impact of pore diffusion. Lin (1979) analyzed the effectiveness factor for an encapsulated enzyme particle by considering internal as well as external mass transfer resistances. The interaction of diffusion and reaction in complex consecutive reactions involving a two-substrate system including deactivation has been studied by Reuss and Buchholz (1979).

Lineweaver–Burk plots r vs. C_{Sb} are typically drawn for immobilized-enzyme reactions to determine the apparent values of the Michaelis–Menten constant K'_m and the maximum rate V''_m. Gondo et al. (1975) and Ngian et al. (1977) have shown that both the slope and the intercept are influenced by internal diffusion.

Marrazzo et al. (1975) have developed theoretical models to describe axial dispersion, particle–film mass transfer, intraparticle diffusion, and chemical reaction of the substrate for enzymes immobilized on porous particles in packed columns for first-order and zero-order limits of Michaelis–Menten kinetics. Many practical reactor systems are likely to exhibit both intra- and interparticle mass transfer effects, and a theoretical analysis of this situation has been made by Horvath et al. (1973), Waterlands et al. (1974), and Fink et al. (1973). In a complex case where Michaelis–Menten kinetics is followed along with inhibition by substrate

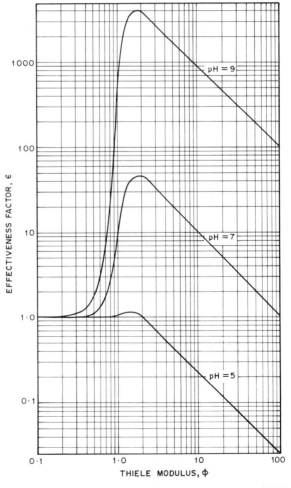

Figure 18.3. Influence of pH on the effectiveness factor–Thiele modulus relationship (Bailey and Chow, 1974).

or products, the analysis of the combined effects of internal and external diffusion involves numerical calculations and a large number of dimensionless parameters.

It is interesting to note that for reactions such as glucose oxidation and hydrolysis of proteins, esters, and urea, which involve a change in the pH of the reaction mixture, the effectiveness factor can be increased over a 1000-fold by immobilization, as shown in Figure 18.3 (Bailey and Chow, 1974). This can be explained by reference to Figure 18.2. Diffusional factors create local pH values that would be closer to the optimum than the bulk values, leading to increased activity.

18.3.4. Diffusional Effects and Chemical Inhibition

Many enzymatic reactions are characterized by inhibition by the substrate, products, or other inhibitors, and the reaction rate expression depends on the type of inhibition. Diffusion has the effect of moderating the decrease in enzymatic activity due to chemical inhibition. This is because an enzyme inhibitor that decreases the inherent enzymatic activity tends to enhance the rate of substrate transport (Engasser and Horvath, 1974). Thus two opposing influences are simultaneously operative. The net result is a decrease in the degree of sensitivity of the immobilized-enzyme activity to changes in the macroenvironmental concentration of the inhibitor.

18.3.5. pH–Activity and Temperature–Activity Behavior as Influenced by Diffusion

One important characteristic of enzymes is their extreme sensitivity to certain parameters, the chief among which are pH and temperature. On immobilization, the apparent thermal and pH stability can be increased, diminished, or preserved relative to the native enzyme; further, the stability can be increased on one side of the optimum and decreased on the other side. Ollis (1972) theoretically examined the effect of internal diffusion on the apparent thermal stability of a reversibly denaturable IME catalyst and showed that the apparent thermal stability is increased as a result of diffusion. The analysis of Karanth and Bailey (1978) proves analytically Ollis's assertions based on numerical calculations in a restricted case and places them in a much broader context. With very few assumptions they showed that the effect of diffusion in general is to increase the apparent thermal and pH stability. In fact, the sensitivity of the IME catalyst to any parameter invariant

within the pellet is decreased as a result of diffusional influences (without, however, changing the optimal values of that parameter). An analysis along similar lines with the same conclusions has been reported by Reilly and Lee (1981).

18.3.6. Example: A Model for pH-Dependent Deactivation and Its Application to Various Immobilized (and Soluble) Enzymes

Protein denaturation is not a function of temperature only, since pH may also play an important role. For the profitable and large-scale commercial utilization of immobilized enzymes it is clearly desirable to evaluate the role of unavoidable and deleterious pH effects on enzyme stability and hence on substrate conversion.

We develop a pH-dependent deactivation model (Sadana, 1980) applicable to both immobilized and soluble enzymes. The model is then applied to the data of Flynn and Johnson (1977a, b, 1978), Johnson (1978), Johnson and Coughlan (1978), and Higgins and Johnson (1977), and its usefulness demonstrated.

As a first approximation we may ignore both diffusional effects and the effects of the electrical characteristics of the support upon the microenvironment of the catalyst, which in the case of immobilized enzymes includes the pH of the environment. Diffusional resistances, if present, may be attenuated by buffers. Furthermore, the electrical properties may be neglected if the ionic strength of the medium is maintained sufficiently high by means of a neutral electrolyte like KCl.

Assuming that the enzyme exhibits three alternative ionization states and that the rate of enzyme decay is proportional to the concentration of the active enzyme, we can write the following deactivation scheme:

$$\breve{E}H_2^+ \underset{}{\overset{K_a}{\rightleftharpoons}} \breve{E}H \underset{}{\overset{K_b}{\rightleftharpoons}} \breve{E}^-$$

$$\downarrow k_d \qquad\qquad (18.13)$$

$$\text{inactive enzyme, } \breve{E}_d$$

From the conservation equation for the enzyme species in the system we obtain

$$\breve{E}_t = \breve{E}H + \breve{E}H_2^+ + \breve{E} + \breve{E}_d \qquad (18.14)$$

On rearranging, we have

$$\breve{E}H = \frac{\breve{E}_t - \breve{E}_d}{1 + C_{H^+}/K_a + K_b/C_{H^+}} \qquad (18.15)$$

where

$$K_b = \frac{\breve{E}C_{H^+}}{\breve{E}H}, \qquad K_a = \frac{\breve{E}H C_{H^+}}{\breve{E}H_2^+}$$

and $\breve{E}_t - \breve{E}_d$ is the concentration of the residual enzyme.

If it is assumed that the rate of enzyme decay is proportional to the concentration of the active enzyme $\breve{E}H$, then

$$\frac{d\breve{E}_d}{dt} = \frac{k_d(\breve{E}_t - \breve{E}_d)}{1 + C_{H^+}/K_a + K_b/C_{H^+}} \qquad (18.16)$$

On integration of Eq. 18.16 we get

$$\ln\left(\frac{\breve{E}_t - \breve{E}_d}{\breve{E}_t}\right) = \frac{k_d t}{1 + C_{H^+}/K_a + K_b/C_{H^+}} = -k_d' t \qquad (18.17)$$

Figure 18.4 shows that the decay of both immobilized (Johnson, 1978; Johnson and Coughlan,

1978) and soluble (Flynn and Johnson, 1977a, b; 1978) enzymes fits the proposed first-order decay model well. Table 18.5 gives k_d' and $t_{1/2}$ (half-life) values for the various enzymes along with their working conditions of temperature and pH.

The proposed model facilitates judicious comparison of deactivation rate constants for different enzymes and provides a rational approach to deactivation whenever applicable. It should prove helpful in the design and analysis of deactivating fixed-bed and batch enzyme reactors. If sufficient data were available at different values of pH for a particular temperature, then K_a and K_b would also be obtainable.

18.3.7. Deactivation of IME Catalysts

The loss of activity in IME reactors can be due to the physical loss of the enzyme from the support or the denaturation (i.e., deactivation) of the enzyme molecule due to irreversible structural rearrangements. Enzyme

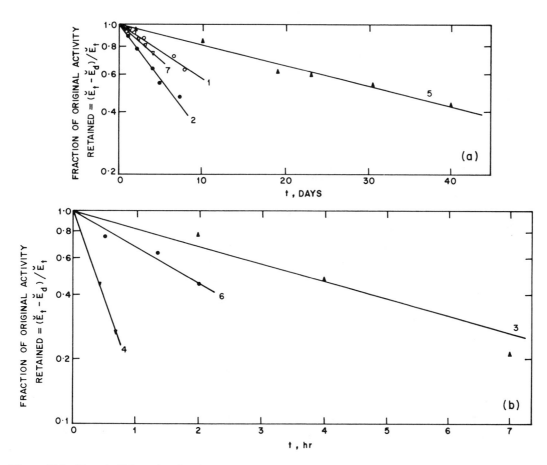

Figure 18.4. Test of pH-dependent deactivation model for the enzyme deactivation data of Johnson and co-workers. Numbers correspond to the enzymes in Table 18.5.

TABLE 18.5. Values of k_d' and $t_{1/2}$ for Immobilized and Soluble Enzymes

Enzyme	Enzyme Conc. (mg/ml)	pH	Temp. (°C)	k_d' (day^{-1})	$t_{1/2}$ (day)	Condition[a]	Reference
1. Urate oxidase on hornblende	3	8.5	30	0.057	12.1	i	Johnson and Coughlan (1978)
2. Glucoamylase	2700[b]	4.3	50	0.115	6	s	Flynn and Johnson (1978)
3. Glucoamylase	0.33	4.9	50	4.75	0.146	s	Flynn and Johnson (1977)
4. HL–ADH	0.0067	7.5	30	46.2	0.015	s	Higgins and Johnson (1977)
5. Invertase	50[b]	4.5	45	0.022	32	s	Flynn and Johnson (1977b)
6. HL–ADH on enazcryl-TIO	1	7.5	30	0.075	9.17	i	Johnson (1978)
7. HL–ADH on PG-1000-400	2.5	8.5	50	9.9	0.07	i	Johnson (1978)

[a]Immobilized enzymes are denoted by i, soluble enzymes by s.
[b]This quantity is in units per gram of protein.

activity usually exhibits an exponential decay, which for some systems is not true at lower temperatures (Havewala and Pitcher, 1974).

Under internal diffusion-limited conditions, IME reactions exhibit slower apparent loss of activity than if no diffusion effects were present. As the enzymatic activity decreases, the effectiveness factor increases, thus slowing the apparent activity-loss rate. This phenomenon has been discussed in detail by Ollis (1972). The effect of diffusion thus is to extend the half-life of IME and under extreme conditions of internal diffusion control the half-life is doubled. Ollis's calculations are restricted to $\beta > 9$ at substrate concentrations $C_S \gg K_m'$. Korus and Driscoll (1975) have extended these results to a wide range of values of β and C_S. The example in Section 18.3.9 demonstrates the role of diffusion in prolonging the period of enzyme activity by immobilization. This period is further extended when external diffusion effects are also present (Naik and Karanth, 1978).

Reilly and Lee (1981) have analyzed numerically the stabilities of IME reactors at varying pH and under different levels of intraparticle diffusional limitation. They found that when diffusional limitations are either absent or very severe, pH–stability profiles are identical in shape. On the other hand, those under severe limitation have half-lives exactly double those under diffusion-free conditions at all pHs.

For the case of substrate-independent thermal denaturation of the enzyme, on the assumption of first-order decay the following expression has been derived for a fixed-bed reactor (Vieth et al., 1976):

$$k_d \frac{V_r}{Q} = \ln \left\{ \frac{C_{Sb} x(0) - K_m' \ln[1 - x(0)]}{C_{Sb} x(t) - K_m' \ln[1 - x(t)]} \right\} \qquad (18.18)$$

where $x(0)$ and $x(t)$ are the conversions at $t = 0$ (no deactivation) and at $t = t$, respectively. Expressions can also be readily derived for several other models of deactivation, such as denaturation dependent on both enzyme and substrate concentrations, substrate inhibition, and product inhibition.

Do and Weiland (1980) have advocated that the biochemical reaction and the deactivation rate expressions have certain related forms if they are to be logically consistent. In another paper (1981) the same authors used rate expressions for enzyme poisoning that are consistent with a Michaelis–Menten kinetic scheme to analyze the performance of an IME fixed-bed reactor. They showed that a minimum in the enzyme activity can occur in the interior of the bed, well away from the ends.

A comparison of fixed-bed and fluid-bed reactor performances for an IME reaction with enzyme denaturation can be made by the methods outlined in Chapter 17. The main feature of these equations is that

Michaelis–Menten kinetics has been employed in place of power law equations.

18.3.8. Optimization of Deactivating Enzyme Reactor Systems

The performance of deactivating-enzyme-catalyzed reactor systems may be optimized by standard mathematical techniques. We outline common methods (Reilly, 1979) to optimize one- and two-enzyme systems, both when the enzymes are in soluble form and when they are immobilized.

One-Enzyme Systems

Reactions involving single soluble enzymes are generally conducted batchwise and at high temperatures. This is done for two main reasons:

1. The possibility of microbial contamination is reduced, leading to an increase in reaction rate.

2. Since the separation of the enzyme from the mixture is difficult, a one-time-only run is preferred irrespective of the high decay rate.

Since soluble-enzyme costs are high, low enzyme concentrations are preferable even though this implies higher capital and labor costs. Immobilization may reduce enzyme cost, since it can be used for long periods to produce large amounts of product. However, the savings in enzyme are counteracted by the cost of the carrier and whatever activating procedure is used; thus low-cost enzymes are not suitable candidates for immobilization.

Reilly (1979) suggests three methods to be used in industrial practice for a deactivating immobilized-enzyme fixed-bed reactor for obtaining conversion near a specified level at all times:

1. Maintain a constant flow rate and constant temperature. If the conversion is initially above the desired level, it may be allowed to fall below that level before reaction is stopped and the enzyme is replaced, with the total production being blended to achieve what is required.

2. Decrease the flow rate in order to hold the conversion at the set value for isothermal operation. This is in keeping with the reactor operation policy for a deactivating catalyst suggested by Prasad and Doraiswamy (1974).

3. Maintain the flow rate constant but increase the temperature continually to hold the conversion constant.

In the first two cases, there is an optimum residual enzyme activity at which the enzyme is replaced, usually one quarter or one eighth of the original activity. In the third case, the constantly increasing temperature causes more rapid enzyme decay, and therefore more rapid increase in temperature. At some point the enzyme activity crashes or there is an upper limit for temperature operation.

Two-Enzyme Systems

Although dual-enzyme systems open up a number of areas amenable to optimization, one area has attracted study. This is the problem of placement of two immobilized enzymes catalyzing sequential reactions in a tubular reactor (see Section 5.3.1).

Two extreme cases may be formulated. In the first, the two catalysts are packed uniformly throughout the reactor. This has the disadvantage that toward the reactor entrance the second enzyme is not fully utilized, since its substrate is not present in high amounts.

In the other extreme case, the two enzymes are packed sequentially. If we want product C in the reaction sequence

$$A \xrightleftharpoons{\check{E}_A} B \xrightleftharpoons{\check{E}_B} C$$

then enzyme $\check{E}_A$ is followed by enzyme $\check{E}_B$, and the relative residence times are optimized (Chang and Reilly, 1976; 1978). This system is known as the *bang-bang* case. If we want to minimize component A, then we have the *bang-bang-bang* case, where $\check{E}_A$ is followed by $\check{E}_B$ and then by more $\check{E}_A$. Again, the relative length of each section is to be optimized.

Chang and Reilly (1978) used glucoamylase and glucose isomerase immobilized on different alkylamine porous silica beads to convert maltose to glucose and then to fructose. A bang-bang profile was optimal for the maximum production of fructose. A bang-bang-bed system was constructed and found to give results at two different residence times very close to those predicted.

18.3.9. Example: Problem to Illustrate the Effect of Diffusion on Immobilized-Enzyme Activity

The resistance to diffusion of substrate may cause a decrease in the rate at which the activity of an immobilized-enzyme decays. This example quantifies the role of diffusion in determining the life of immobilized-enzyme activity.

The two parameters to consider are the Thiele modulus ϕ and the Damköhler number Da:

$$\phi = \left(\frac{\mathrm{Da}\, K'_m}{K'_m + C_S}\right)^{1/2} \qquad (18.19)$$

where K'_m is the apparent Michaelis–Menten constant and C_S is the substrate concentration. Korus and Driscoll (1975) have provided plots of the effectiveness factor as a function of the dimensionless substrate concentration C_S/K'_m at various values of Da for enzymes immobilized on spherical gel particles (see Figure 18.5).

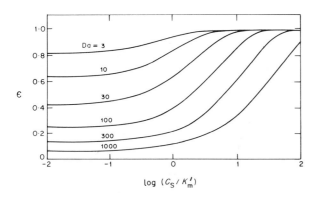

Figure 18.5. Effectiveness factor ε as a function of dimensionless substrate concentration C_S/K'_m at various values of Da for enzymes immobilized on spherical gel particles (Sizer, 1944).

Consider first-order deactivation of the free enzyme:

$$-\frac{d\breve{E}}{dt} = 0.1\,\breve{E}, \quad \text{days} \qquad (18.20)$$

and two values of Da: (a) Da = 300, and (b) Da = 100. Assume that $K'_m = C_S$.

1. Da = 300:

From Eq. 18.20 the half-life $t_{1/2}$ is obtained as

$$t_{1/2} = \frac{\ln 2}{0.1} = 6.93 \quad \text{days}$$

After 11 days (an assumed period) the enzyme activity remaining equals 33.3 % of the initial activity, E_0 (from eq. 18.20). Therefore, Da decreases to $300 \times 33.3/100 \simeq 100$. At zero time $\varepsilon = 0.22$ (from Figure 18.5). Thus

$$r_{\text{immob}} = 0.22\, r_{\text{free}}$$

At $t = 11$ days for Da = 100, $\varepsilon = 0.4$, and

$$r_{\text{immob}} = 0.4\, r_{\text{free}}$$

Since we have first-order deactivation of the free enzyme,

$$r_{\text{free}}(t = 11 \text{ days}) = 0.333\, r_{\text{free}}(t = 0)$$

and

$$\frac{r_{\text{immob}}(t = 11 \text{ days})}{r_{\text{immob}}(t = 0)} = \frac{0.333 \times 0.4}{0.2} = 0.6$$

Thus, the immobilized enzyme would lose 40 % of its activity in 11 days as compared to 50 % loss in activity of the free enzyme in 6.93 days.

2. Da = 100:

On repeating the calculations above for Da = 100 we obtain the results summarized in Table 18.6 along with those for Da = 300. Clearly, increasing the diffusional limitation reduces the loss of enzyme activity.

TABLE 18.6. Effect of Diffusion on Enzyme Activity

Enzyme	Damköhler Number	Loss in Activity %	Time for Loss in Activity (days)
Free	0	50	6.9
Immobilized	100	46.4	11
Immobilized	300	40	11

18.3.10. Design of IME Reactors

The basic principles of design for IME reactors are the same as for gas–solid (catalytic) reactions outlined in Chapter 11. A major simplification is possible, however, in that enzymatic reactions do not involve significant heat effects, so that isothermal conditions can be assumed. With the further assumption of plug flow, the following simple reactor equation can be written using the notation of IME reactors:

$$\frac{V_r}{Q} = C_{S0} \int_0^x \frac{dx}{r} \qquad (18.21)$$

Substituting Eq. 18.7 for r, we have (Bar-Eli and Katchalski, 1963)

$$C_{S0}x - K'_m \ln(1-x) = k\breve{E}_0 f_B \frac{V_r}{Q} \qquad (18.22)$$

For a CSTR the equation takes the form

$$C_{So}x + K'_m\left(\frac{x}{1-x}\right) = k\breve{E}_0\frac{V_r}{Q} \qquad (18.23)$$

For systems in which external mass transfer resistance is also an important consideration along with chemical reaction, it is expedient to assume first-order kinetics. The following simple form of reactor equation then results:

$$\frac{V_r f_B}{Q} = K''_m[-\ln(1-x)] \qquad (18.24)$$

where

$$K''_m = \frac{(K'_m/V'_m)k_L a}{K'_m/V'_m + k_L a} \qquad (18.25)$$

and represents a combined coefficient incorporating the effects of reaction and external mass transfer. The liquid–solid mass transfer coefficient k_L can be estimated from any of the available correlations, such as the following.

1. Spherical microencapsulated enzymes (Mogensen and Vieth, 1973):

$$k_L = \frac{0.000464\,G^{1/3}}{f_B d_p^{2/3}} \qquad (18.26)$$

2. Porous glass beads (Rovito and Kittrell, 1973):

$$j_d = 1.625(Re')^{-0.507} \qquad (18.27)$$

3. Collagen–enzyme chips (Davidson et al., 1974; Fernandes, 1974):

$$j_d = 5.7(Re')^{-0.78} \qquad (18.28)$$

Unlike gas–solid catalytic reactions, IME reactions are often accompanied by electrostatic effects along with the usual external diffusion effect considered above. These effects have been reviewed by Goldman et al. (1971), Melrose (1971), and Katchalski et al. (1971). The combined electrostatic and boundary-layer diffusion effects are usually expressed in terms of an external effectiveness factor, in analogy with that for catalytic reactions (see Chapter 6).

Let us consider a first-order isothermal reaction in a fixed-bed reactor with internal diffusion. The following equation can be written for spherical particles:

$$\frac{V_r}{Q} = -\frac{\ln(1-x)}{K'_m/V'_m}\left[\frac{1}{\phi}\left(\frac{1}{\tanh 3\phi} - \frac{1}{3\phi}\right)\right] \qquad (18.29)$$

where ϕ is the usual Thiele modulus, given by $R[(K'_m/V'_m)/D_e]^{0.5}$. Expressions for other shapes can be found in Table 4.1. A more rigorous expression based on Michaelis–Menten kinetics has been developed by Davidson et al. (1974) by considering the internal and external field problems simultaneously. The internal field or membrane-phase equation for a flat plate is given by the nondimensional second-order diffusional equation

$$\frac{\partial^2 \hat{C}_S}{\partial \hat{L}^2} = \phi^2 \beta'_0\left(\frac{\hat{C}_S}{\beta'_0 + \hat{C}_S}\right) = 0 \qquad (18.30)$$

where

$$\hat{C}_S = \frac{C_S}{C_{So}}, \qquad \hat{L} = \frac{l}{L}, \qquad \beta'_0 = \frac{K'_m}{C_{So}} \qquad (18.31)$$

and

$$\phi = L\left(\frac{k\breve{E}_0}{K'_m D_e}\right)^{1/2}$$

a Thiele modulus. The boundary conditions are

$$\begin{aligned}\hat{L} &= 0, & \frac{\partial \hat{C}_S}{\partial \hat{L}} &= 0\\ \hat{L} &= 1, & C_S &= C_{Sb}\end{aligned} \qquad (18.32)$$

(no external film resistance) where $\hat{C}_S = C_S/C_{Sb}$.

Then, by writing a material balance equation for the external field and expressing the effectiveness factor as a polynomial in C_{Sb}, namely, $\hat{C}_{Sb} = a_0 + a_1\hat{C}_{Sb} + a_2\hat{C}_{Sb}^2$, the following reactor equation can be written:

$$\frac{d\hat{C}_{Sb}}{d\hat{z}} = (1-f_B)(a_0 + a_1\hat{C}_{sb} + a_2\hat{C}_{Sb}^2)\left(\frac{\phi^2\beta'_0\hat{C}_{Sb}}{\beta'_0 + \hat{C}_{Sb}}\right) \qquad (18.33)$$

Once the constants are known, this equation can be used in obtaining the concentration profile in the reactor.

The forgoing treatment can be readily extended to the case where both internal and external diffusion resistances are operative by suitably redefining the second boundary condition to include a Biot number for external mass transfer, that is, at

$$\hat{L} = 1, \qquad \frac{\partial \hat{C}_S}{\partial \hat{L}} = (Bi)_m(C_{Sb} - C_{Ss})$$

where $(Bi)_m = k_L L/D_e$.

18.3.11. Multiple Steady States in IME Reactions

When diffusion is coupled with reaction in non-enzymatic heterogeneous reactors, the occurrence of multiple steady states is usually due to the coupled heat and mass transport effects within the catalyst particles (see Chapter 12); but in enzyme reactions, the heat effects are negligible, mainly because of low exothermicities and the aqueous medium of reaction. A reaction can, however, be expected to exhibit multiple steady states if the reaction rate exhibits a maximum with reactant concentration. In this case the concentration gradient of substrate within the particle may actually mean an increase in the rate of reaction toward the center, giving rise to effectiveness factor values greater than unity. At certain values of kinetic and transport parameters, this situation may lead to three different steady states. However, multiplicity of steady states in IME systems seems to be of only academic interest at present.

18.3.12. IME Reactor Types

Because of the wide variety of immobilization techniques, several types of reactors can be used. Some of these are discussed below.

Packed-Bed Reactor

When the IME is in the form of spheres, chips, disks, sheets, beads, or pellets, a packed-bed reactor seems most appropriate. Some examples of packed-bed IME reactor systems are enzymes attached to porous glass beads (Weetall and Detar, 1974; Marsh et al., 1973), beads of ion-exchange resins such as DEAE-cellulose and DEAE–Sephadex (Chibata et al., 1972; Tosa et al., 1973), and chips of collagen membranes (Saini and Vieth, 1975).

Fixed-bed reactors packed with immobilized whole microbial cells have assumed industrial importance. L-Aspartic acid is being produced industrially in Japan with immobilized *Escherichia coli* cells having high aspartase activity (Takamatsu et al., 1980). A great deal of work is in progress on packed-bed reactors with immobilized yeast cells (Sitton et al., 1979; Ghose and Bandyopadhyay, 1980; Margaritis et al., 1981), which have the advantage of high productivity and consequently reduced reactor volume.

Several variations of the conventional packed-bed reactor have been proposed, including tubular reactors packed with filter paper (Kay et al., 1968), porous sheets (Reynolds, 1972), and porous blocks such as open-pored polyether or polyurethane foams (Lilly and Dunhill, 1972).

A novel type of enzyme packing is the microencapsulated enzyme. This consists of a soluble enzyme entrapped in semipermeable membrane capsules that allow for the transport of only the substrate and the products but not the enzyme molecules. These types of reactors are being tried for artificial kidney devices that are expected to replace the expensive dialysis machines used at present.

Continuous Stirred-Tank Reactors

Although for a given reactor volume the plug-flow reactor gives higher conversion than a CSTR, peculiarities of IME catalysis may necessitate use of the latter. These characteristics may be the large residence times and consequent difficulties involved in maintaining proper flow conditions, or the excessive pressure drop and clogging problems encountered in the case of fine-size IME catalysts. A basic operating requirement of CSTR systems is that the IME particles be retained within the reactor. This is readily achieved by providing a filter at the outlet. Retention can also be achieved magnetically when the enzymes are coupled to magnetically active supports (Heden 1973).

Tubular Reactors with Enzymatically Active Walls

Enzymes can be attached to the inner wall of a tubular reactor, or to a suitable matrix which forms an annulus at the tube wall. Such reactors are very useful in biomedical applications (as "extracorporeal shunts") for carrying out specific enzyme-catalyzed biological changes in body fluids. The relatively low pressure drop and unhindered flow associated with these reactors minimizes the possibilities of body fluid coagulation.

Fluidized-Bed Enzyme Reactors

Liquid fluidized-bed reactors operate with complex flow patterns that do not conform to perfect plug-flow or backmixing patterns. They may deviate from plug flow depending on the size, shape, and density of the catalyst particles, and on the viscosity, density, and flow rate of the liquid. The influence of particle size, flow rate, and viscosity on the deviation from plug flow has been examined by Emery and Revel-Chion (1978). Plug flow may be approached by using lower flow rates, larger particle size, and lower viscosity. Allen et al. (1979) indicate that the design of the liquid distributor is especially important in achieving a close approach to plug flow. Emery and Cardoso (1978) have shown that, for starch hydrolysis using low concentrations of substrate, fluidized beds give consistently better performance than fixed beds.

Fluidized-bed reactors have been shown to be particularly effective in the hydrolysis of lactose in whole whey (Coughlin and Charles, 1974a, b; Coughlin et al., 1974, 1976, 1978), waste treatment (Scott and Hancher, 1976; Jeris et al., 1974, 1977; Beer, 1970), and starch saccharification (Allen et al., 1979). The major advantage of the fluidized-bed systems for waste treatment is the enormous surface area provided by the fluidized media. For cornstarch hydrolysis the major advantage of fluidized-bed reactors is their ability to handle small catalyst particles (as small as $50\,\mu m$) without incurring sizable drops in pressure and plugging problems.

18.3.13. Examples of IME Reactions

Here we shall briefly describe two important examples of the application of immobilized enzymes. One is in industrial use, and the other represents a rather novel use of the immobilization technique. Other current applications are resolution of D- or L-amino acids, production of 6-APA from penicillin, and manufacture of L-aspartic acid from fumaric acid. Some of the potential applications are in artificial kidney machines, the conversion of starch to glucose, and hydrolysis of lactose in milk or whey.

Production of Fructose by Immobilized Whole Cells

It was pointed out earlier that where cellular or intracellular enzymes are involved, whole cells may be immobilized. Such a procedure would be economically more attractive than breaking open the cells, separating the enzyme, and then immobilizing it. A particularly successful application of immobilized whole cells is in the isomerization of glucose to fructose by the enzyme glucose isomerase:

$$\text{glucose} \underset{\text{isomerase}}{\overset{\text{glucose}}{\rightleftharpoons}} \text{fructose}$$

The conversion obtained at equilibrium is of the order of 50%. Since fructose is sweeter than glucose, even partially isomerized glucose would be far sweeter than glucose. The immobilized glucose isomerase offered by Novo Industri A/S is produced by spray drying a centrifuged concentrate of the organism *Bacillus coagulans*, cross-linking the cells with glutaraldehyde, extruding the resulting material, and drying the pellets (Poulsen and Zittan, 1976).

This is one of the few industrially important systems for which the effectiveness factor has been experimentally determined and compared with the predicted values. The data of Boersma et al. (1979) with pellets of immobilized glucose isomerase *Arthrobacter* show the usual trend—the effectiveness factor decreases with increase in Thiele modulus.

Oxidation of Glucose to Gluconolactone

Glucose can be oxidized in the presence of the enzyme glucose oxidase to give gluconolactone and hydrogen peroxide:

$$\text{glucose} + O_2 \xrightarrow[\text{oxidase}]{\text{glucose}} \text{gluconolactone} + H_2O_2$$

Bailey (1980) reports that this reaction has important analytical applications, and may be used to produce gluconic acid. The difficulty, however, is that H_2O_2, which is one of the products of reaction, causes deactivation of the enzyme (Kleppe, 1966; Greenfield et al., 1975; Cho and Bailey, 1977). This deleterious effect can be neutralized if the enzyme is immobilized on a solid that catalyzes the decomposition of hydrogen peroxide.

$$H_2O_2 \xrightarrow[\text{decomposition by support}]{\text{catalytic}} H_2O + \tfrac{1}{2}O_2$$

Several solids, such as activated carbon and manganese dioxide, have been used for this purpose.

One method of immobilization is to adsorb the catalyst on the solid and cross-link the adsorbed enzyme molecule with a bifunctional reagent such as glutaraldehyde. The second method (applicable to activated carbon and other solids with carboxyl groups on the surface) is to couple the enzyme molecule directly with the surface through an intermediate complex activated by a water-soluble carbodiimide (Bailey, 1980). The results obtained by these two methods of immobilization show that the second method leads to considerable improvement in the deactivation behavior of the system. This is probably because this method of immobilization leaves more of the solid surface available for catalyzing the decomposition of H_2O_2 than the first method.

REFERENCES

Abed, R., and Rinker, R. G. (1973), *J. Catal.*, **31**, 119.

Acres, G. J. K., Bond, G. C., Cooper, B. J., and Dawson, J. A. (1966), *J. Catal.*, **6**, 139.

Allen, B. R., Coughlin, R. W., and Charles, M. (1979), *Ann. N.Y. Acad. Sci.*, **326**, 105.

Allum, K. G., Hancock, R. D., McKenzie, S., and Pitkethly, R. C. (1972), *Proc. 5th Int. Cong. Catal.*, Palm Beach.

Arai, H. (1978), *J. Catal.*, **51**, 135.

Arai, H., and Yashiro, M. (1978), *J. Mol. Catal.*, **3**, 427.

Atkinson, B., Black, G. M., Lewis, P. J. S., and Pinches, A. (1979), *Biotechnol. Bioeng.*, **21**, 193.

Axen, R., and Ernbach, S. (1971), *European J. Biochem.*, **18**, 351.

Bailar, J. C., Jr. (1974), *Catal. Rev. Sci. Eng.*, **10**, 17.

Bailey, J. E. (1980), *Chem. Eng. Sci.*, **35**, 1854.

Bailey, J. E., and Chow, M. T. C. (1974), *Biotechnol. Bioeng.*, **16**, 1345.

Bailey, D. C., and Langer, S. H. (1981), *Chem. Rev.*, **81**, 109.

Bar-Eli, A., and Katchalski, E. (1963), *J. Biol. Chem.*, **238**, 1690.

Bassett, J. M., and Smith, A. K. (1977), *Fundamental Research in Homogeneous Catalysis*, eds. Tsutsui, M., and Vgo, R., Plenum Press, New York and London, p. 69.

Beer, C. (1970), *J. San. Eng. Div. Am. Soc. Civil Eng.*, **96** (SA6), 1452.

Blanch, H. W., and Dunn, I. J. (1974), *Adv. Biochem. Eng.*, **3**, 127.

Boersma, J. G., Vellanga, K., de Wilt, H. G. J., and Joosten, G. E. H. (1979), *Biotechenol. Bioeng.*, **21**, 1711.

Borkar, P. S., Thadani, S. B., and Ramachandran, S. (1978), *Hindustan Antibiotics Bull.*, **20** (3/4), 81.

Brams, W. H., and McLaven, A. D. (1974), *Soil Biol. Biochem.*, **6**, 183.

Bunting, P. S., and Laidler, K. J. (1972), *Biochemistry*, **11**, 4477.

Bunting, P. S., and Laidler, K. J. (1974), *Biotechnol. Bioeng.*, **16**, 119.

Burwell, R. L., Jr. (1974), *Chem. Technol.*, **4**, 370.

Carbonell, R. G., and Kostin, M. D. (1972), *AIChE J.*, **18**, 1.

Carlock, J. T., (1980), U.S. Patent 4, 183, 825.

Ciapetta, F. G. (1947), U.S. Patent 2, 430, 803.

Chang, H. N., and Reilly, P. J. (1976), *Chem. Eng. Sci.*, **31**, 413.

Chang, H. N., and Reilly, P. J. (1978), *Biotechnol. Bioeng.*, **20**, 243.

Chen, O. T., and Rinker, R. G. (1978), *Chem. Eng. Sci.*, **33**, 1201.

Chibata, I., Tosa, T., Sato, T., Mori, T., and Matu, Y. (1972), *Proc. 4th Int. Ferment. Symp. (Fermentation Technology Today)*, p. 383.

Cho, Y. K., and Bailey, J. E. (1977), *Biotechnol. Bioeng.*, **19**, 769.

Cho, Y. K., and Bailey, J. E. (1978), *Biotechnol. Bioeng.*, **20**, 1651.

Cho, I. C., and Swaisgood, H. (1974), *Biochim. Biophys. Acta*, **334**, 243.

Collman, J. P., Hegedus, L. S., Cooke, M. P., Norton, J. R., Docetti, G., and Marquardnt, D. N. (1972), *J. Am. Chem. Soc.*, **94**, 1789.

Coughlin, R. W., and Charles, M. (1974a), *Enzyme Engineering*, Vol. 2, ed. Pye, E. K., Plenum Press, New York, p. 339.

Coughlin, R. W., and Charles, M. (1974b), *Enzyme Technol. Digest* (No. 2), **3**, 69.

Coughlin, R. W., Charles, M., Allen, B. R., Paruchiri, E. K., and Hasselberger, F. X. (1974), *AIChE Symp. Ser.* (No. 144), **70**, 199.

Coughlin, R. W., Charles, M., and Julkowski, K. (1976), 69th Ann. Meeting. AIChE, Chicago.

Coughlin, R. W., Charles, M., and Julkowski, K. (1978), *AIChE Symp. Ser.*, **172**, 40.

Dahodwala, S. K., Humphrey, A. E., and Weibel, M. K. (1976), *Biotechnol. Bioeng.*, **18**, 987.

Davidson, B., Vieth, W. R., Wang, S. S., Zwiebel, S., and Gilmore, R. (1974), *AIChE Symp. Ser.*, **144**, 182.

Davis, J. C. (1970), *Chem. Eng.*, **19**, 52.

Delmon, B., and Jannes, G., eds. (1975), *Catalysis—Heterogeneous and Homogeneous*, Elsevier, Amsterdam.

Dinelli, D. (1972), *Process Biochem.*, **7**, 9.

Do, D. D., and Weiland, R. H. (1980), *Biotechnol. Bioeng.*, **22**, 1087.

Do, D. D., and Weiland, R. H. (1981), *Biotechnol. Bioeng.*, **23**, 691.

Emery, A. N., and Cardoso, J. P. (1978), *Biotechnol. Bioeng.*, **20**, 1903.

Emery, A. N., and Revel-Chion, L. (1978), Paper presented at AIChE 77th national meeting, Pittsburgh.

Engasser, J. M., and Horvath, C. (1973), *J. Theoret. Biol.*, **42**, 137.

Engasser, J. M., and Horvath, C. (1974), *Biochemistry*, **13**, 3845.

Falbe, J. (1970), *Carbon Monoxide in Organic Synthesis*, Springer-Verlag, Berlin.

Fernandes, P. M. (1974), Ph.D. thesis, Rutgers University, New Brunswick, New Jersey.

Filippusson, H., and Hornby, W. E. (1970), *Biochem. J.*, **120**, 215.

Fink, D. J., Na, T. Y., and Schultz, J. S. (1973), *Biotechnol. Bioeng.*, **15**, 879.

Flynn, A., and Johnson, D. B. (1977a), *Int. J. Biochem.*, **8**, 501.

Flynn, A., and Johnson, D. B. (1977b), *Int. J. Biochem.*, **8**, 243.

Flynn, A., and Johnson, D. B. (1978), *Biotechnol. Bioeng.*, **20**, 1445.

Gabel, D., and Hofsten, B. V. (1971), *European J. Biochem.*, **15**, 410.

Ghose, T. K., and, Bandyopadhyay, K. K. (1980), *Biotechnol. Bioeng.*, **22**, 1489.

Goldman, R., Kedem, O., and Katchalski, E. (1968), *Biochemistry*, **7**, 4518.

Goldman, R., Goldstein, L., and Katchalski, E. (1971), *Biochemical Aspects of Reactions on Solid Supports*, ed. Stark, G. R., Academic Press, New York, p. 36.

Goldstein, L., and Katchalski, E. (1968), *Z. Anal. Chem.*, **243**, 375.

Gondo, S., Isayama, S., and Kusunoki, K. (1975), *Biotechnol. Bioeng.*, **17**, 423.

Gorin, E., Fontana, C. M., and Kidder, G. A. (1948), *Ind. Eng. Chem.*, **40**, 11.

Greenfield, P. F., Kittrell, J. R., and Laurence, R. L. (1975), *Anal. Biochem.*, **65**, 109.

Greenfield, P. F., Kinzler, D. D., and Laurence, R. L. (1978), *Biotechnol. Bioeng.*, **17**, 1555.

Grubbs, R. H. (1977), *Chemtech*, **7**, 512.

Grubbs, R. H., and Kroll, L. C. (1971), *J. Am. Chem. Soc.*, **93**, 3062.

Hancock, R. D., Howell, I. V., Pithkethly, R. C., and Robinson, P. J. (1975), *Catalysis—Heterogeneous and Homogeneous*, eds. Delmon, B., and Jannes, G., Elsevier, Amsterdam, p. 361.

Havewala, N. B., and Pitcher, W. H., Jr. (1974), *Enzyme Engineering*, Vol. 2, eds. Pye, E. K., and Wingard, L. B., Jr. Plenum Press, New York, p. 315.

Heden, C. G. (1973), *Biotechnol. Bioeng. Symp.*, No. 3.

Heinemann, H. (1971), *Chem. Tech.*, p. 286.

Henry, P. M. (1968), *Adv. Chem. Ser.*, **70**, 126.

Higgins, A. C., and Johnson, D. B. (1977), *Int. J. Biochem.*, **8**, 807.

Hjortkjaer, J., Scurrell, M. S., Simonsen, P., and Svendsen, H. (1981), *J. Mol. Catal.*, **12**, 179.

Hornby, W. E., Lilly, M. D., and Crook, E. M. (1968), *Biochem. J.*, **107**, 669.

Horvath, C., and Engasser, J. M. (1974), *Biotechnol. Bioeng.*, **16**, 909.

Horvath, C., Solomon, B. A., and Engasser, J. M. (1973), *Ind. Eng. Chem. Fundam.*, **12**, 431.

Imanaka, T., Kaneda, K., Teranishi, S., and Terasawa, M. (1976), *Proc. 6th Int. Cong. Catal.*, **A-41**, 509.

Jack, T. R., and Zajic, J. R. (1977), *Biotechnol. Bioeng.*, **19**, 631.

Jannes, G. (1975), *Catalysis—Heterogeneous and Homogeneous*, eds., Delmon, B., and Jannes, G., Elsevier, Amsterdam, p. 83.

Jansen, E. F., and Olson, A. C. (1969), *Arch. Biochem. Biophys.*, **129**, 221.

Jarrell, M. S., and Gates, B. C. (1975), *J. Catal.*, **40**, 255.

Jarrell, M. S., and Gates, B. C. (1978), *J. Catal.*, **54**, 81.

Jeris, J. S., Beer, C., and Mueller, J. A. (1974), *J. Water Pollution Control Fed.*, **46**, 2118.

Jeris, J. S., Owens, R. W., and Hickey, R. (1977), *J. Water Pollution Control Fed.*, **49**, 816.

Johnson, D. B. (1978), *Biotechnol. Bioeng.*, **20**, 1117.

Johnson, D. B., and Coughlan, M. P. (1978), *Biotechnol. Bioeng.*, **20**, 1085.

Johnson, F. H., Eyring, H., and Polissar, M. J. (1954), *The Kinetic Basis of Molecular Biology*, Wiley, New York.

Karanth, N. G., and Bailey, J. E. (1978), *Biotechnol. Bioeng.*, **20**, 1817.

Katchalski, E., Silman, I., and Goldman, R. (1971), *Adv. Enzymol. Related Areas Mol. Biol.*, **34**, 445.

Kay, G., Lilly, M. D., Sharp, A. K., and Wilson, R. J. H. (1968), *Nature*, **217**, 641.

Kenney, C. N. (1975), *Catal. Rev. Sci. Eng.*, **11**, 197.

Kleppe, K. (1966), *Biochem.*, **5**, 139.

Kobayashi, T., and Moo-Young, M. (1973), *Biotechnol. Bioeng.*, **15**, 47.

Kobayashi, H., and Suzuki, H. (1976), *Biotechnol. Bioeng.*, **18**, 37.

Kominek, L. (1975), U.S. Patent 3, 915, 802.

Komiyama, H., and Inoue, H. (1975), *J. Chem. Eng. Japan*, **8**, 310.

Komiyama, H., and Inoue, H. (1977), *J. Chem. Eng. Japan*, **10**, 125.

Korus, R. A., and Driscoll, K. F. (1975), *Biotechnol. Bioeng.*, **17**, 441.

Krishnaswamy, S., and Kittrell, J. R. (1978), *Biotechnol. Bioeng.*, **20**, 821.

Lee, D. D., Lee, G. K., Reilly, P. J., and Lee, Y. Y. (1980), *Biotechnol. Bioeng.*, **22**, 1.

Lee, S. B., Kim, S. M., and Ryu, D. D. Y. (1979), *Biotechnol. Bioeng.*, **21**, 2023.

Lilly, M. D., and Dunhill, P. (1972), *Biotechnml. Bioeng. Symp.*, No. 3, 97.

Lin, S. H. (1979), *Chem. Eng. J.*, **17**, 55.

Linarte-Lazcano, R., Valle-Machorro, J., and Cuatecontzi-Santa Cruz, D. H. (1975), *Catalysis—Heterogeneous and Homogeneous*, Elsevier, Amsterdam, p. 467.

Lipscomb, W. N. (1972), *Bio-organic Chemistry and Mechanisms*, ed. Milligan, W. O., Academic Press, New York.

Livbjerg, H., Sorensen, B., and Villadsen, J. (1974), *Adv. Chem. Ser.*, **133**, 242.

Livbjerg, H., Jensen K. F., and Villadsen, J. (1976), *J. Catal.*, **45**, 216.

Manassen, J. (1969), *Chim. Ind. (Milan)*, **51**, 1058.

Manassen, J., and Whitehurst, D. D. (1973), *Catalysis: Progress in Research*, eds., Basolo, F., and Burwell, R. L. Plenum Press, New York, p. 177.

Marconi, W., Bartoli, F., Cecere, F., Galli, G., and Morisi, F. (1975), *Agr. Biol. Chem.*, **39**, 277.

Margaritis, A., Bajpai, P. L., and Wallace, J. B. (1981), *Biotechnol. Letters* **3**, 613.

Marrazzo, W. N., Merson, R. L., and McCoy, B. J. (1975), *Biotechnol. Bioeng.*, **17**, 1515.

Marsh, D. R., Lee, Y. Y., and Tsao, G. T. (1973), *Biotechnol. Bioeng.*, **15**, 483.

Melrose, G. J. H. (1971), *Rev. Pure Appl. Chem.*, **21**, 83.

Michaels, A. (1968), *Separation and Purification*, Vol. 1, Wiley, New York, p. 297.

Michalska, Z. M., and Webster, D. E. (1974), *Platinum Metals Rev.*, **18**, 65.

Mizoroki, T., Kawata, N., Hinata, S., Maruya, K., and Ozaki, A. (1975), *Catalysis—Heterogeneous and Homogeneous*, eds. Delmon, B., and Jannes, G., Elsevier, Amsterdam, p. 319.

Mogensen, A. O., and Vieth, W. R. (1973), *Biotechnol. Bioeng.*, **15**, 467.

Moravec, R. Z., Schelling, Wm. T., and Oldershaw, C. F. (1939), British Patent 511556.

Moravec, W. T., Schelling, C. F., and Oldershaw, C. F. (1941) Can., Patent **396**, 994.

Mosbach, K., and Mosbach, R. (1966), *Acta Chem. Scand.*, **20**, 2807.

Naik, S. S., and Karanth, N. G. (1978), *J. Appl. Chem. Biotechnol.*, **28**, 569.

Ngian, K. F., Lin, S. H., and Martin, W. R. B. (1977), *Biotechnol. Bioeng.*, **19**, 1773.

Novo Industri (1976), Japanese Patent Kokai, 76–51580.

Otero-Schipper, Z., Lieto, J., and Gates, B. C. (1980), *J. Catal.*, **63**, 175.

Ollis, D. F. (1972), *Biotechnol. Bioeng.*, **14**, 871.

Patwardhan, V. S., and Karanth, N. G., (1982), *Biotechnol. Bioeng.*, **24**, 763.

Pittman, C. U., Jr. (1968), *J. Polymer Sci.*, **B6**, 19.

Pittman, C. U., Jr. (1971), *Chem. Technol.*, p. 416.

Pittman, C. U., Jr., and Evans, G. O. (1973), *Chem. Tech.*, p. 560.

Pittman, C. U., Jr., and Honnick, W. D. (1980), *J. Org. Chem.*, **45**, 2132.

Pittman, C. U., Jr., Honnick, W. D., and Yang, J. J. (1980), *J. Org. Chem.*, **45**, 684.

Pittman, C. U., Jr., and Smith L. R. (1975), *J. Am. Chem. Soc.*, **97**, 1749.

Pittman, C. U., Jr., Smith, L. R., and Hanes, R. M. (1975), *J. Am. Chem. Soc.*, **97**, 1742.

Pittman, C. U., Jr., Wuu, S. K., and Jacobson, S. E. (1976), *J. Catal.*, **44**, 87.

Prasad, K. B. S., and Doraiswamy, L. K. (1974), *J. Catal.*, **32**, 384.

Porath, J. (1974), *J. Meth. Enzymol.*, **34**, 13.

Porath, J., Axen, R., and Ernback, S. (1967), *Nature*, **215**, 1491.

Poulsen, P. B., (1981), *Enzyme Microbiol. Technol.*, **3**, 271.

Poulsen, P. B., and Zittan, L. (1976), *Immobilized Enzymes*, Academic Press, New York, p. 809.

Reilly, P. J. (1979), *Annals. N.Y. Acad. Sci.*, **326**, 97.

Reilly, P. J., and Lee, G. K. (1981), *Chem. Eng. Commun.*, **12**, 195.

Reuss, M., and Buchholz, K. (1979), *Biotechnol. Bioeng.* **21**, 1061.

Reynolds, J. H. (1972), U.S. Patent 3, 705, 084.

Rony, P. R. (1968), *Chem. Eng. Sci.*, **23**, 1021.

Rony, P. R. (1969), *J. Catal.*, **14**, 142.

Rony, P. R., and Roth, J. F. (1975), *Catalysis—Heterogeneous and Homogeneous*, eds. Delmon, B., and Jannes, G., Elsevier, Amsterdam, p. 373.

Roth, J. F. (1975), *Platinum Metals Rev.*, **19**, 12.

Rovito, B. J., and Kittrell, J. R. (1973), *Biotechnol. Bioeng.*, **15**, 143.

Sadana, A. (1980), *Biotechnol. Letters*, **3**, 279.

Saini, R., and Vieth, W. R. (1975), *J. Appl. Chem. Biotechnol.*, **25**, 115.

Scott, C. D., and Hancher, C. W. (1976), *Biotechnol. Bioeng.*, **18**, 1393.

Shigesada, S., Ishmiatsu, Y., and Kimura, S. (1975), *Japanese Patent Kokai* 75-160475.

Sitton, O. C., Foutch, G. L., Book, N. L., and Gaddy, J. L. (1979), *Chem. Eng. Prog.*, Dec., **75**, 52.

Stern, E. W. (1967), *Catal. Rev.*, **1**, 73.

Strathmann, H. (1978), *Chem. Tech. Berlin.*, **7**, 333.

Subramanian, T. V. (1978), *Biotechnol. Bioeng.*, **20**, 601.

Subramanian, S. S., Sivaraman, H., Rao, B. S., Ratnaparkhi, R. R., and Sivaraman, C. (1978), *Hindustan Antibiotics Bull.*, **20** (3/4), 74.

Swanson, S. J., Emery, A., and Lim, H. C. (1978), *AIChE J.*, **24**, 30.

Swanson, S. J., Lim, H. C., and Emery, A. (1978), *Ind. Eng. Chem. Process Design Dev.*, **17**, 401.

Szonyi, G. (1968), *Adv. Chem. Ser.*, **70**, 53.

Takamatsu, S., Yamashita, K., and Sumi, A., (1980), *J. Ferment. Technol.*, **58**, 129.

Terasawa, M., Kaneda, K., Imanaka, T., and Teranishi, S. (1978), *J. Catal.*, **51**, 406.

Terasawa, M., Yamamoto, H., Kaneda, K., Imanaka, T., and Teranishi, S. (1979), *J. Catal.*, **57**, 315.

Thornton, D., Flynn, A., Johnson, D. B., and Ryan, P. D. (1975), *Biotechnol. Bioeng.*, **17**, 1679.

Toda, K. (1975), *Biotechnol. Bioeng.*, **17**, 1729.

Topsøe, H., and Nielsen, A. (1948), *Trans. Danish Acad. Tech. Sci.*, **1**, 18.

Tosa, T., Sato, T., Mori, T., Matuo, Y., and Chibata, I. (1973), *Biotechnol. Bioeng.*, **15**, 69.

Tosa, T., Sato, T., Mori, T., Yamamoto, K., Takata, I., Nishida, Y., and Chibata, I. (1979), *Biotechnol. Bioeng.*, **21**, 1697.

Venkatsubramanian, K., Karkare, S. B., and Vieth, W. R. (1983), Private communication.

Vieth, W. R., Venkatsubramanian, K., Constantinides, A., and Davidson, B. (1976), *Appl. Biochem. Bioeng.*, **1**, 221.

Villadsen, J., and Livbjerg, H. (1978), *Catal. Rev. Sci. Eng.*, **17**, 203.

Wandrey, C. (1979), *Annals. N.Y. Acad. Sci.*, **326**, 87.

Waterlands, L. R., Michaels, A. S., and Robertson, C. R. (1974), *AIChE J.*, **20**, 50.

Weetall, H. H., and Detar, C. C. (1974), *Biotechnol. Bioeng.*, **16**, 1095.

Weetall, H. H., and Havewala, N. B. (1972), *Enzyme Engineering*, ed. Wingward, L. B., Wiley Interscience, New York, p. 241.

Whitehurst, D. D. (1980), *Chem. Tech. Berlin.*, **10**, 44.

Wilson, H. D., and Rinker, R. G. (1976), *J. Catal.*, **42**, 268.

Wilson, R. J. H., Kay, G., and Lilly, M. D. (1968), *Biochem. J.*, **108**, 845.

CHAPTER NINETEEN

Gas–Solid Noncatalytic Reactions: Analysis and Modeling

Noncatalytic gas–solid reactions constitute an important class of heterogeneous reactions. Examples of these can be abundantly found in metallurgical processes, solid decompositions, and gasification reactions. A tabulated summary of a variety of gas–solid reactions encountered in industry has already been presented in Chapter 1. Like their catalytic counterparts, these reactions tend to be influenced to a great extent by heat and mass transport processes. Thus any mathematical analysis of these processes must take into account the simultaneous influence of reaction and of heat and mass transfer.

The most general type of heterogeneous reaction can be represented by

$$A + B = R + S$$

in which A, B, R, and S can each be either a solid or a gas; further, one of the reactants or products may not be present at all. The examples listed in Table 1.2 are based on this classification. The important types of reactions are the following.

Type A (Reduction and roasting of ores):

Fluid and solid reactants → fluid and solid products

Type B:

Fluid and solid reactants → solid products

Type C (Decomposition reactions):

Solid reactants → fluid and solid products

Type D (Oxidation, chlorination of ores; carbonyl formation):

Fluid and solid reactants → fluid products

Type E (Gasification reactions):

Solid reactants → fluid products

Type A is the most common and is exemplified by the reduction, oxidation, and roasting of ores. Decomposition reactions (e.g., decomposition of calcium carbonate to calcium oxide and carbon dioxide) belong to Type C. It is characteristic of Types D and E that only fluid products are involved and the solid disappears completely at the end of the reaction.

In view of the generality of Type A reactions and their importance in industrial practice, the major emphasis will be on this class of reactions. Although in theory this class represents the most general type of gas–solid reactions and can therefore be applied to other classes also by appropriate simplifications, there are certain features associated with other classes that need to be considered separately. Thus, we shall present the analysis and modeling of gas–solid reactions under the following major subheads: models for Types A and B, models for Type C, and models for Types D and E. Certain basic assumptions made in all these models will then be eliminated by describing models that incorporate the effect of structural changes, and also by removing the constraint of isothermicity. Modeling of complex reactions will also be considered, followed by a brief analysis of the stability of gas–solid reactions. Examples (both quantitative and qualitative) will be presented during the description of the more important models. An exhaustive review on modeling has been presented by Ramachandran and Doraiswamy (1982a) while the book by Szekely et al. (1976) covers various aspects of gas–solid reactions.

19.1. MODELS FOR REACTION TYPES A AND B

In this section we shall consider the models that have been developed for predicting the conversion as a function of time for solids undergoing reaction according to the following general scheme:

$$v_A A(g) + v_B B(s) \rightarrow v_R R(g) + v_S S(s)$$

This reaction scheme belongs to Type A, but can be easily applied to Type B, in which no gaseous product is involved ($v_R = 0$). Basically two types of models have been considered, one in which the solid is assumed to be nonporous and one in which it is assumed to be porous. Under these two basic categories several models have been proposed, particularly for porous solids. The more important ones are described below, with examples of their applications.

19.1.1. Sharp-Interface Model

In the sharp-interface model (SIM) which is restricted to a nonporous solid, the reaction is assumed to occur at a sharp interface between the exhausted outer shell and the unreacted core of the solid. The unreacted core shrinks in size as the reaction proceeds. A schematic representation of the model* is shown in Figure 19.1.

In deriving an equation for this model, we shall make

---

* This model is also referred to as the shrinking-core model or, as in Chapter 8, the shell-progressive model.

the following assumptions: isothermal conditions, constant pellet size, equimolal counterdiffusion of gaseous reactants and products, pseudo-steady-state approximation, and first-order irreversible reaction. The implications and the conditions of validity of these assumptions are discussed later. The various steps involved in the reaction, and the corresponding rates, are as follows.

1. Diffusion of A through the gas film:

$$\mathbf{R}_A \ (\text{mol/sec}) = 4\pi R^2 k_g (C_{Ab} - C_{As}) \tag{19.1}$$

2. Diffusion of A through the product layer (based on the assumption of equimolal counterdiffusion):[†]

$$\mathbf{R}_A = \frac{4\pi R R_i D_{eS}}{R - R_i}(C_{As} - C_{Ai}) \tag{19.2}$$

3. Chemical reaction at the interface:

$$\mathbf{R}_A = 4\pi R_i^2 k_S C_{Ai} \tag{19.3}$$

[†] In this chapter D_{ei} stands for the effective diffusivity of gas in solid i, unlike in catalytic reactions, where it represents the diffusivity of gas i in the catalyst.

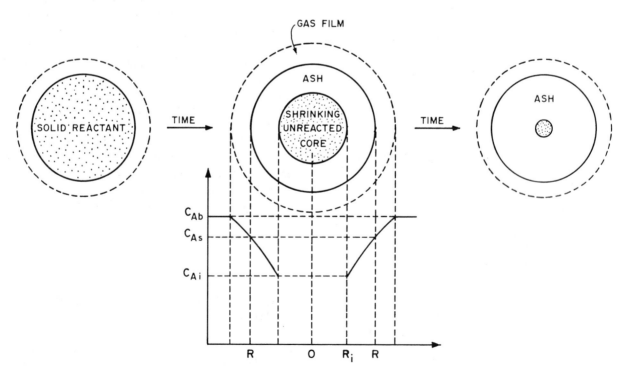

Figure 19.1. Schematic representation of the sharp-interface model.

The rate of reaction at a given time t (when the position of the reaction interface is at R_i) is given by combining Eqs. 19.1–19.3:

$$\mathbf{R}_A = \left(\frac{1}{4\pi R^2 k_g} + \frac{R - R_i}{4\pi R R_i D_{eS}} + \frac{1}{4\pi R_i^2 k_S} \right)^{-1} C_{Ab}$$

(19.4)

The rate of movement of the sharp interface can be related to the rate of reaction through a stoichiometric balance on B:

$$-\frac{d}{dt}\left(\frac{4}{3}\pi R_i^3 \frac{\rho_B}{M_B} \right) = \frac{v_B}{v_A}\mathbf{R}_A = v\mathbf{R}_A$$

(19.5)

where v is defined as v_B/v_A.* Substituting for $\mathbf{R}_A$ and integrating the resulting equations, we have

$$\frac{R}{3k_g}\left[1 - \left(\frac{R_i}{R}\right)^3 \right] + \frac{R^2}{6D_{eS}}\left[1 - 3\left(\frac{R_i}{R}\right)^2 + 2\left(\frac{R_i}{R}\right)^3 \right]$$

$$+ \frac{R}{k_S}\left(1 - \frac{R_i}{R} \right) = \frac{v C_{Ab} M_B}{\rho_B}t$$

(19.6)

In gas–solid reactions the quantity of interest is the fractional conversion (x) of the solid B. For a spherical solid this is related to R_i by

$$x = 1 - \left(\frac{R_i}{R}\right)^3$$

(19.7)

leading to the following relationship between conversion and time:

$$\frac{R}{3k_g}x + \frac{R^2}{6D_{eS}}\left[1 - 3(1-x)^{2/3} + 2(1-x) \right]$$

$$+ \frac{R}{k_S}\left[1 - (1-x)^{1/3} \right] = \frac{v C_{Ab} M_B}{\rho_B}t$$

(19.8)

Equation 19.8 can also be expressed as

$$t = t_M + t_{DP} + t_R$$

(19.9)

where t_M, t_{DP}, and $t_R = \rho_B/v M_B C_{Ab}$ times the first, second, and third terms, respectively, of Eq. 19.8. The physical meanings of these quantities are as follows: t_M is the time required to achieve a given conversion if the

---

* ρ_B represents the density of solid B, and not the bulk density of the bed as in previous chapters.

process were controlled only by external mass transfer; t_{DP} is the time required for the same conversion if the process is controlled entirely by ash-layer diffusion; and t_R is the time that would be needed if chemical reaction dominated the whole process. Equation 19.9 is based on the well-known concept of addition of resistances in series for a first-order process and is known as the law of addition of reaction times [see Sohn (1978) and Rao et al. (1979) for a more detailed discussion of this subject].

It is also seen, on comparing the various terms of Eq. 19.8, that the following $x - t$ relationships hold for different controlling regimes:

1. gas-film diffusion controls, $t \propto x$;
2. ash diffusion controls, $t \propto [1 - 3(1-x)^{2/3} + 2(1-x)]$;
3. chemical reactions control, $t \propto (1-x)$.

The utility of Eq. 19.8 is thus twofold. It can predict the conversion as a function of time for a system for which all the parameters are known. Alternatively, it can be used to interpret laboratory experimental data on gas–solid reactions with a view to identifying the controlling regime to obtain the model parameters.

The time required for complete conversion can be obtained from Eq. 19.8 by setting $x = 1$:

$$t_{x=1} = \left(\frac{R}{3k_g} + \frac{R^2}{6D_{es}} + \frac{R}{k_S} \right) \frac{\rho_B}{v M_B C_{Ab}}$$

(19.10)

Equations 19.8 and 19.10 are applicable to a spherical pellet. Equations for pellets of other shapes can be obtained in a similar manner, as shown by Gokarn and Doraiswamy (1973).

The shrinking-core model has been applied to a number of systems, such as the oxidation of nickel (Carter, 1961), reduction of iron oxides (Kawasaki et al., 1962), combustion of coke deposited on catalysts (Weisz and Goodwin, 1963), and oxidation of zinc sulfide (Gokarn and Doraiswamy, 1971). The validity of the various assumptions made in developing the model is discussed below.

1. The assumption of isothermicity is not valid for systems that involve a large heat of reaction. The models for nonisothermal systems are discussed in Section 19.1.5 and the situation where the pellet size changes during the course of reaction is discussed in Section 19.3.

2. The assumption of equimolal counterdiffusion results in some error when the moles of gaseous

product (R) formed is different from the moles of gaseous reactant (A) reacted. The error is greatest when the diffusion in the ash layer is the chief controlling mechanism and when the transport is dominated by bulk diffusion. An analysis of these effects has been made by Beveridge and Goldie (1968).

3. The pseudo-steady-state assumption has been critically examined by Bischoff (1963), Luss (1968), and Wen (1968). The physical significance of this assumption is that the interface can be assumed to remain stationary at any time, while a steady-state diffusion flux is calculated to find the concentration profile. The assumption is valid when the concentration of the reacting gas-phase species in the bulk is much less than the molal density of the solid or when

$$\frac{C_{Ab} M_B}{\rho_B} < 10^{-3} \qquad (19.11)$$

This generally holds for gas–solid reactions but may not hold for liquid–solid reactions when C_{Ab} has a value comparable to ρ_B / M_B.

The effect of reaction order has been analyzed by Sohn and Szekely (1972). The rate of interfacial reaction can now be expressed, for the power law model, as

$$\mathbf{R}_A = 4 \pi R_i^2 k_S C_{Ai}^m \qquad (19.12)$$

or for the Hougen–Watson model as

$$\mathbf{R}_A = \frac{4 \pi R_i^2 k_S C_{Ai}}{1 + K_A C_{Ai}} \qquad (19.13)$$

The equations for diffusion through the gas film and the product layer are the same as Eqs. 19.1 and 19.2 for a nonlinear case. A simplified equation for the overall rate similar to Eqs. 19.4 cannot be obtained for the nonlinear system. The expression for $\mathbf{R}_A$ would now be complex and the rate of movement of interface would have to be solved numerically. For convenience the equations are presented in dimensionless form in terms of an effectiveness factor defined as

$$\varepsilon = \frac{\mathbf{R}_A}{4 \pi R_i^2 k_S C_{Ab}^m} \qquad (19.14)$$

It should be noted here that the effectiveness factor for a noncatalytic reaction is a function of time, in contrast to a catalytic reaction. This concept of ε for noncatalytic

reactions was introduced by Ishida and Wen (1968). The equation for the rate of movement of the reaction plane can be expressed in terms of the effectiveness factor as

$$-\frac{d\hat{R}_i}{d\hat{t}} = \varepsilon \qquad (19.15)$$

where $\hat{t}$ is a dimensionless time, defined as

$$\hat{t} = \frac{v C_{Ab}^m M_B k_S t}{\rho_B R} \qquad (19.16)$$

and

$$\hat{R}_i = \frac{R_i}{R} \qquad (19.17)$$

The dimensionless time $\hat{t}$ required for a given conversion x can then be obtained by a numerical integration of Eq. 19.15 as

$$\hat{t} = \int_{\hat{R}_i = (1-x)^{1/3}}^{1} \frac{1}{\varepsilon} d\hat{R}_i \qquad (19.18)$$

In order to solve Eq. 19.18 the effectiveness factor ε has to be obtained from various values of $\hat{R}_i$. For a general mth-order reaction the following implicit relation can be derived:

$$\varepsilon = \left[1 - \frac{\varepsilon \, \mathrm{Da}}{\dfrac{\mathrm{Da}}{\mathrm{Sh}} + \dfrac{\mathrm{Da}(1 - \hat{R}_i)}{\hat{R}_i}} \right]^m \qquad (19.19)$$

where

$$\mathrm{Sh} = \frac{k_g R}{D_{eS}} \qquad (19.20)$$

and

$$\mathrm{Da} = \frac{k_S R}{D_{eS}} C_{Ab}^{m-1} \qquad (19.21)$$

An explicit solution for ε can be obtained for a first-order reaction:

$$\varepsilon = \left[1 + \hat{R}_i^2 \, \mathrm{Da} \left(\frac{1}{\mathrm{Sh}} + \frac{1 - \hat{R}_i}{\hat{R}_i} \right) \right]^{-1} \qquad (19.22)$$

19.1.2. Homogeneous Model*

When the solid is porous and the rate of diffusion of the reactant gas is rapid, the gas will penetrate everywhere

* This is also referred to as the volume reaction model.

into the solid and the reaction will take place throughout the pellet. In some cases diffusional gradients may exist inside the pellet, resulting in varying degrees of reaction within it. The homogeneous model takes these effects into account. The model equations can be represented as

$$D_{eB}\left(\frac{d^2C_A}{dr} + \frac{2}{r}\frac{dC_A}{dr}\right) = k_v C_A^m C_B^n \quad (19.23)$$

$$-\frac{dC_B}{dt} = v k_v C_A^m C_B^n \quad (19.24)$$

The order of reaction m with respect to the gaseous reactant (A) is generally observed to be unity in many gas–solid reactions. If the reaction proceeds through adsorption of A, then the reaction order may vary anywhere between 0 and 1, depending on the extent of adsorption. In such cases the kinetics may also be represented by a H–W model. The order of reaction with respect to the solid (B) may also vary: depending on the reaction mechanism it may be 1, $\frac{2}{3}$ (if the reaction is controlled by a surface process), or 0. In the oxidation of ZnS the reaction order with respect to ZnS was observed by Gokarn and Doraiswamy (1971) to be zero.

The boundary conditions required for the solution of the homogeneous model are, at

$$r = R, \qquad D_{eB}\frac{dC_A}{dr} = k_g(C_{AB} - C_{As}) \quad (19.25)$$

and at

$$r = 0, \qquad \frac{dC_A}{dr} = 0 \quad (19.26)$$

The initial condition is $t = 0, \quad C_B = C_{BO}$.

Equations 19.23 and 19.24 may be expressed in dimensionless form as

$$\frac{d^2\hat{C}_A}{d\hat{R}^2} + \frac{2}{\hat{R}}\frac{d\hat{C}_A}{d\hat{R}} = \phi^2 \hat{C}_A^m \hat{C}_B^n \quad (19.27)$$

$$-\frac{d\hat{C}_B}{d\hat{t}} = \hat{C}_A^m \hat{C}_B^n \quad (19.28)$$

where

$$\phi = \left(\frac{k_v C_{Ab}^{m-1} C_{B0}^n R^2}{D_{eB}}\right)^{1/2} \quad (19.29)$$

and

$$\hat{t} = v k_v C_{Ab}^m C_{B0}^{n-1} t \quad (19.30)$$

The concentration profile of A within the pellet depends on the value of ϕ. Three regimes may be identified:

1. $\phi < 0.2$: Here the concentration of A is uniform throughout the pellet (see Figure 19.2). The conversion vs. time relationship can be easily obtained for this situation. For the specific case in which $n = 1$

$$x = 1 - \exp(-\hat{t}) \quad (19.31)$$

and for $n = 0$

$$x = \hat{t} \quad (19.32)$$

2. $0.2 < \phi < 3$: Here the concentration distribution is as shown in Figure 19.2. It is necessary to solve the model equations numerically in order to predict the x–t relationship. In some cases analytical solutions can be obtained. The solutions for the cases in which (1) $m = 1, n = 0$; (2) $m = 1, n = 1$; and (3) $m = 0, n = 1$ are given later.

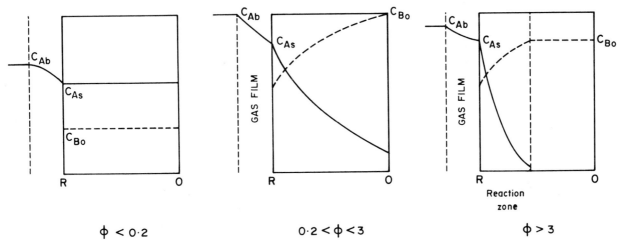

Figure 19.2. Concentration profiles for the homogeneous model for various values of the Thiele modulus.

3. $\phi > 3$: Here the reaction occurs in a zone of finite thickness in the pellet as the concentration of A falls to zero toward the center of the pellet. The model now reduces to the finite-thickness model proposed by Bowen and Cheng (1969) or the conceptually similar diffused-interface model (Tudose, 1970, Mantri et al., 1976). As ϕ increases the thickness of the zone decreases, and for large ϕ the model approaches the SIM. The rate of reaction in the absence of external diffusion (Sh $\to \infty$) and at $t = 0$ is given by

$$\mathbf{R}_A = 4\pi R^2 C_{Ab}\left(\frac{2}{m+1}D_{eB}k_v C_{Ab}^{m-1}C_{B0}^n\right)^{1/2}$$

(19.33)

Comparison with the SIM gives a relation between k_S and k_v that is valid at large values of ϕ:

$$k_S = \left(\frac{2}{m+1}D_{eB}k_v C_{Ab}^{m-1}C_{B0}^n\right)^{1/2}$$ (19.34)

Solutions to the Homogeneous Model

We shall now consider the solution of the homogeneous model for various cases.

Case (1): $m = 1$, $n = 0$. The problem has been analyzed by Ishida and Wen (1968), who divided the total reaction time into two periods: (a) the constant-rate period, and (b) the falling-rate period [see also the analysis of Ausman and Watson (1962) outlincd in Section 8.5 for the burning of carbon deposited in a

pellet]. The necessity for considering two such separate periods arises for $\phi > 3$ because, as the reaction proceeds, a zone of completely reacted product forms near the surface. Further reaction in the core (which occurs simultaneously with diffusion) has to be preceded by pure diffusion through the product layer, where no reaction occurs. The concentration distributions for the two periods are shown in Figure 19.3, which illustrates the situation quite clearly. The model is also known as the two-zone model, in contrast to the SIM, where also two zones exist but the reaction is restricted to the interface between the zones.

The constant-rate period extends up to a dimensionless time given by

$$\hat{t}_{crp} = 1 + \frac{\phi^2}{6}\left(1+\frac{2}{Sh}\right) - \left(1-\frac{D_{eB}}{D_{eS}}\right)\ln\left(\frac{\sinh\phi}{\phi}\right)$$

(19.35)

The conversion at the end of the constant rate period is given by

$$x(\hat{t}_{crp}) = \frac{3(\phi\coth\phi - 1)\hat{t}_{crp}}{\phi^2\left[1 + \dfrac{1}{Sh}(\phi\coth\phi - 1)\right]}$$ (19.36)

The conversion in the falling-rate period depends on the extent of the product layer. If at time $\hat{t}$ the product layer is present in the region $R < r < R_i$, then the conversion is given as

$$x = 1 - \hat{R}_i^3 + \frac{3\hat{R}_i}{\phi^2}[\phi\hat{R}_i\coth(\phi\hat{R}_i)-1]$$ (19.37)

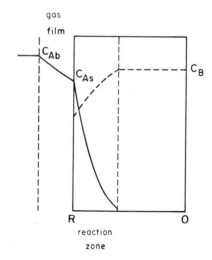 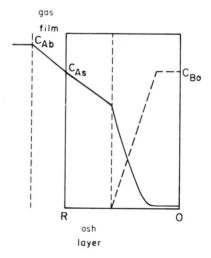

Figure 19.3. Concentration profile for the two-zone model.

The relation between $\hat{R}_i$ and $\hat{t}$ is given implicitly as

$$\hat{t} = 1 - \left[1 - \frac{D_{eB}}{D_{es}}\ln\left(\frac{\hat{R}_i \sinh\phi}{\sinh\phi\hat{R}_i}\right)\right]$$
$$+ \frac{\phi^2}{6}(1 - \hat{R}_i)^2(1 + 2\hat{R}_i) + \frac{\phi^3}{3\text{Sh}}(1 - \hat{R}_i^3)$$
$$+ \left[\frac{D_{eB}}{D_{es}}(1 - \hat{R}_i) + \frac{\hat{R}_i}{\text{Sh}}\right][\phi\hat{R}_i(\coth\phi\hat{R}_i) - 1]$$

$$(19.38)$$

Case 2: m = 1, n = 1. For this case there is no need to consider the presence of a completely reacted product layer, since theoretically for a first-order reaction ($n = 1$) the concentration C_B becomes zero only at infinity. The problem requires numerical solution and such solutions have been obtained by a number of workers.

A new technique of solving these problems, introduced by Del Borghi et al. (1976) and by Dudukovic and Lamba (1978a), reduces the computational effort considerably. The technique is based on the use of a new variable, defined as

$$\psi = \int_0^{\hat{t}} \hat{C}_A d\hat{t} \qquad (19.39)$$

The variable ψ represents a cumulative gas concentration. With this transformation Eqs. 19.23 and 19.24 can be combined into a single equation:

$$\frac{d^2\psi}{d\hat{R}^2} + \frac{2}{\hat{R}}\frac{d\psi}{d\hat{R}} = \phi^2[1 - \exp(-\psi)] \qquad (19.40)$$

The solution of this equation is simpler than that of the original set of equations. Dudukovic and Lamba (1978a) solved Eq. 19.40 by collocation methods. The final expression for conversion for a spherical pellet is

$$x = 1 - 3\int_0^1 \hat{R}^2 \exp(-\psi)d\hat{R} \qquad (19.41)$$

Ramachandran and Kulkarni (1980) used a single-point collocation on Eq. 19.40 and obtained the following implicit approximate analytical solution for spherical geometry:

$$\ln\left[\frac{(1-x)}{0.699} - \frac{\exp(-\hat{t})}{2.33}\right] + \hat{t} = \frac{\phi^2}{10.5}\left[1 - \frac{1-x}{0.699} + \frac{\exp(-\hat{t})}{2.33}\right]$$

$$(19.42)$$

The solution is accurate up to $\phi = 5$. Even for $\phi = 10$ the maximum error between the approximtion and the numerical solution is only 17%. Equation 19.42 can be used for predicting the x–$\hat{t}$ relationship and interpreting laboratory experimental data. The solution is based on the assumption that $D_{eB} = D_{es}$. If the assumption is not valid, an average value of D_e can be used as an approximation.

Case 3: m = 0, n = 1. When the gaseous species A is strongly adsorbed on the solid surface, the reaction may exhibit a zero-order rate. Dudukovic and Lamba (1978b) have made a detailed analysis of this problem, and have defined a critical Thiele modulus below which the concentration of A would be finite at all points:

$$(\phi)_{\text{crit}} = \frac{6}{2/\text{sh} + 1} \qquad (19.43)$$

The solid conversion is then given by Eq. 19.31. If $\phi > (\phi)_{\text{crit}}$ the concentration of A drops to zero at some point in the pellet. The reaction then occurs only in the region where $1 < \hat{R} < \hat{R}_i$. The position of $\hat{R}_i$ itself changes with time. Equations have been derived by Dudukovic and Lamba (1978a) to predict x–$\hat{t}$ relations for this situation.

The definitions of ϕ and $\hat{t}$ for the three models are given in Table 19.1. The differences between the models are not very significant when x is plotted as a function of dimensionless time. The corresponding definition of $\hat{t}$ is, however, different for each model as can be noted from Table 19.1. The differences in the models become more pronounced at high conversions because for $n = 0$ complete conversion is achieved in a finite time, whereas for $n = 1$ this is not so.

Case 4: m = 0, n = 0. (The Jumping Zone Model). Where the reaction is zero order with respect to both A and B, the analysis is quite straightforward if there is no diffusional resistance. However, for systems with significant diffusional gradients, the reaction zone gets frozen at a position near the interface until the solid here is completely converted. The reaction zone then jumps to an adjacent position and remains there, again until the solid of the new zone is completely exhausted. Thus, we obtain a new model for the situation, which may be called the jumping zone model (Ramachandran and Doraiswamy, 1983). The thickness of the zone corresponds to the depth of penetration of the reactant gas into the pellet, and the number of zones is obviously a function of the Thiele modulus. An illustrative plot of conversion as a function of dimensionless time is shown in Figure 19.4 for a Biot

TABLE 19.1. Definition of Parameters for Various Homogeneous Model

| Model | Order with respect to | | ϕ | $\hat{t}$ |
	A	B		
Two-zone model	1	0	$R\sqrt{\dfrac{k_v}{D_{eB}}}$	$\dfrac{vk_vC_{Ab}t}{C_{Bs}}$
Zone model for zero-order reaction	0	1	$R\sqrt{\dfrac{k_vC_{Bs}}{D_{eB}D_{As}}}$	vk_vt
Second-order reaction model	1	1	$R\sqrt{\dfrac{k_vC_{Bs}}{D_{eB}}}$	$vk_vC_{Ab}t$

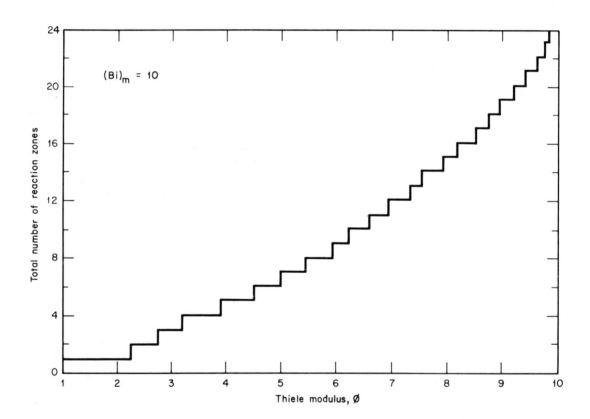

Figure 19.4. Number of reaction zones that develop in the pellet for various values of ϕ (Ramachandran and Doraiswamy, 1983)

number of 10 (Ramachandran and Doraiswamy, 1983). It will be seen that the number of zones increases significantly as the Thiele modulus increases. For $(Bi)_m = 10$, as many as 24 zones can develop. Clearly, the jumping zone model would give an almost smooth profile as the number of zones increases.

19.1.3. Finite-Reaction-Zone Model

In the finite-reaction-zone model the reaction interface is not a sharp boundary but is assumed to be a diffused boundary of finite thickness. Bowen and Cheng (1969) have presented an analysis that results in an equation similar to the sharp-interface model for $m = 1$, $n = 1$ with the difference that the term Da is now modified as

$$Da' = R \left(\frac{k_v C_{B0} \sigma'}{6 v D_{eS}} \right)^{1/2} \qquad (19.44)$$

where σ' is the ratio of the fluid concentration gradient at the diffusion–reaction interface to the linear gradient across the reaction zone. Although in general the numerical value of σ' is less than 1, for practical purposes a value of 1 may be assumed. The equation for the effectiveness factor for this model is the same as Eq. 19.22 with Da replaced by Da'. The conversion–time relationship can then be obtained by using Eq. 19.18.

Mantri et al. (1976) postulated for gas–solid reactions a three-zone model that has the merit of being general and reducible to simpler situations. The three zones that develop in a pellet are (a) a zone of the product or exhausted solid (ash layer) near the outer surface of the pellet, (b) a zone of finite thickness where reaction occurs, and (c) a zone of unreacted solid toward the center of the pellet. The three zones are shown schematically in Figure 19.5. The theoretical plot of the thickness of the reaction zone as a function of the Thiele modulus is shown in Figure 19.6. For high values of ϕ, the zone thickness approaches zero and the model reduces to the SIM. For low values of ϕ (i.e., for a porous pellet), the zone thickness assumes the dimension of the pellet and the model reduces to the homogeneous model. At a certain point during the reaction when the core completely disappears the model reduces to that of Ishida and Wen (1968).

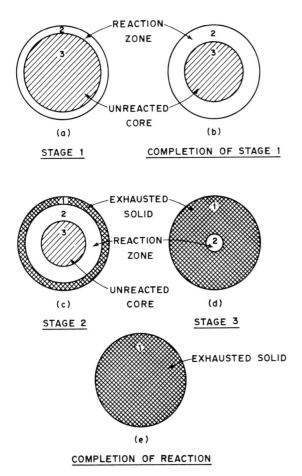

Figure 19.5. The three stages of the zone model (Mantri et al., 1976).

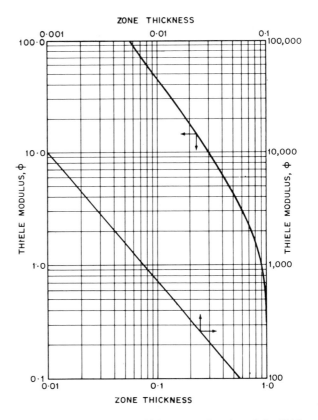

Figure 19.6. Plot of zone thickness as a function of the Thiele parameter (Mantri et al., 1976).

19.1.4. Particle–Pellet or Grain Model

The particle–pellet model postulates that the solid pellet consists of a number of small particles or grains, which are surrounded by macropores of the gas phase through which the gas has to diffuse to reach the various grains. The reaction occurs in each grain according to the sharp-interface model. A product layer will form with time around each grain and this will in turn offer some resistance to mass transfer. The mathematical analysis of such systems can be built up by first considering the rate of reaction of individual grains and then incorporating it in the mass balance of A in the macropores of the solid. A schematic diagram of the model is shown in Figure 19.7.

Rate of Reaction within an Individual Grain

The rate of diffusion through the ash layer of the grain is

$$\mathbf{R}_{GA} = \frac{4\pi D_{eG} r_{Gi} r_{G0}}{r_{G0} - r_{Gi}} (C_A - C_{Ai}) \tag{19.45}$$

where the subscript G represents the grain. The rate of reaction within the grain is

$$\mathbf{R}_{GA} = 4\pi r_{Gi}^2 k_S C_{Ai} \tag{19.46}$$

Eliminating the unknown concentration C_{Ai}, we obtain

$$\mathbf{R}_{GA} = \text{rate (in mol/sec) per unit grain}$$

$$= \frac{4\pi r_{Gi}^2 k_S C_A}{1 + \dfrac{r_{Gi} k_S}{D_{eG}}\left(1 - \dfrac{r_{Gi}}{r_{G0}}\right)} \tag{19.47}$$

The mass balance of A within the macropores can be written as

$$D_{e,ma}\left(\frac{d^2 C_A}{dr^2} + \frac{2}{r}\frac{dC_A}{dr}\right)$$

$$= \mathbf{R}_{GA} \times \text{number of grains per unit pellet volume} \tag{19.48}$$

The number of grains per unit volume is given by

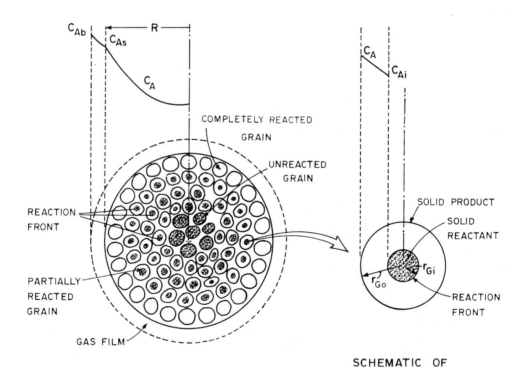

SCHEMATIC OF
A PARTIALLY REACTED PELLET

SCHEMATIC OF
A PARTIALLY REACTED GRAIN

Figure 19.7. Schematic of the grain model for the reaction of a porous solid with a gas (Ramachandran and Doraiswamy, 1982).

$3(1 - f_s)/4\pi r_{G0}^3$. Hence Eq. 19.48 becomes

$$D_{e,\,ma}\left(\frac{d^2 C_A}{dr^2} + \frac{2}{r}\frac{dC_A}{dr}\right)$$
$$= \frac{\dfrac{3(1 - f_S)}{r_{G0}}\dfrac{k_S}{}\left(\dfrac{r_{Gi}}{r_{G0}}\right)^2 C_A}{1 + \dfrac{r_{Gi}k_S}{D_{eG}}\left(1 - \dfrac{r_{Gi}}{r_{G0}}\right)} \qquad (19.49)$$

The quantity r_{Gi} would be a function both of time and position in the pellet, because grains at different radial positions will be exposed to different gas-phase concentrations. The change in r_{Gi} as a function of time can be obtained by a stoichiometric balance on B:

$$-\frac{dr_{Gi}}{dt} = \frac{vk_S C_A}{\dfrac{\rho_B}{M_B}\left[1 + \dfrac{r_{Gi}k_S}{D_{eG}}\left(1 - \dfrac{r_{Gi}}{r_{G0}}\right)\right]} \qquad (19.50)$$

The boundary conditions are, at

$$r = R, \qquad D_{e,\,ma}\frac{dC_A}{dr} = k_g(C_{Ab} - C_A) \qquad (19.51)$$

and at

$$r = 0, \qquad \frac{dC_A}{dr} = 0 \qquad (19.52)$$

If there is no resistance to mass transfer in the film surrounding the pellet, the boundary condition given by 19.51 may be modified to

$$C_A = C_{Ab} \qquad \text{at} \quad r = R \qquad (19.53)$$

The initial condition is, at

$$t = 0, \qquad \hat{r}_{Gi} = 1 \quad \text{for all } r \qquad (19.54)$$

These equations can be put in dimensionless form as

$$\frac{d^2\hat{C}_A}{d\hat{R}^2} + \frac{2}{\hat{R}}\frac{d\hat{C}_A}{d\hat{R}} = \frac{\phi^2 \hat{C}_A}{1 + \text{Da}_G\,\hat{r}_{Gi}(1 - \hat{r}_{Gi})} \qquad (19.55)$$

and

$$-\frac{d\hat{r}_{Gi}}{dt^*} = \frac{\hat{C}_A}{1 + \text{Da}_G\,\hat{r}_{Gi}(1 - \hat{r}_{Gi})} \qquad (19.56)$$

where

$$\hat{r}_{Gi} = \frac{r_{Gi}}{r_{G0}} \qquad (19.57)$$

$$\phi = R\left[\frac{3(1 - f_s)k_S}{D_{e,\,ma}r_{G0}}\right]^{1/2} \qquad (19.58)$$

$$\text{Da}_G = \frac{k_S r_{G0}}{D_{eG}} \qquad (19.59)$$

$$t^* = \frac{vM_B C_{Ab}k_S t}{\rho_B r_{G0}} \qquad (19.60)$$

It is seen that the conversion of the solid is given by

$$x = 1 - 3\int_0^1 (\hat{r}_{Gi})^3 (\hat{R})^2 \, d\hat{R} \qquad (19.61)$$

where (with $\text{Sh} = k_g R/D_{e,\,ma}$)

$$x = f(\phi, \text{Da}_G, \text{Sh}, t^*)$$

In order to obtain the conversion as a function of these parameters a numerical solution of these equations is necessary. This has been provided by Calvelo and Smith (1970).

A general dimensionless representation of the particle–pellet (or grain) model has been proposed by Sohn and Szekely (1972a) which allows for spherical and flat-platelike pellets made up of spherical or flat-plate like grains. The model is based on negligible diffusional gradients within the grains ($D_{eG} \rightarrow \infty$). The authors have noted two patterns of asymptotic behavior for the system.

1. $\phi \rightarrow 0$ (kinetic control)

Here the concentration within the pellets is uniform and all the grains are exposed to the same gas concentration. The conversion–time relationship for this case for $\text{Sh} \rightarrow \infty$ is

$$t^* = g(x) = 1 - (1 - x)^{1/3} \qquad (19.62)$$

2. $\phi \rightarrow \infty$ (diffusion control)

Here a sharp demarcation can be observed between the reacted and unreacted portions of the pellet and the behavior is similar to that of the sharp-interface model. The x vs. t^* relationship now is

$$\frac{18t^*}{\phi^2} = p(x) = 1 - 3(1 - x)^{2/3} + 2(1 - x) \qquad (19.63)$$

For intermediate values of ϕ, Sohn and Szekely (1972a) proposed the following approximate solution for conversions for a spherical pellet comprising spherical particles:

$$t^* = g(x) + \frac{\phi^2}{18}\left[p(x) + \frac{2x}{\text{Sh}}\right] \qquad (19.64)$$

In a subsequent study Sohn and Szekely (1973) extended this model to a system following H–W kinetics. The effect of the form of the rate equation was found to be largest in the intermediate region where both kinetics and diffusion contribute equally to the overall rate.

19.1.5. Nonisothermal Effects

When the chemical reaction is accompanied by a large heat effect, the assumption of isothermal conditions is not valid. As in the case of catalytic reactions, additional equations for heat balance and the dependence of the rate constant on temperature have to be incorporated into the model. In this section, we shall consider certain aspects of the modeling of nonisothermal gas–solid reactions as applied to some of the models described above.

Sharp-Interface Model

A simplified model can be built on the assumption of pseudo-steady state for heat transfer and a constant temperature in the solid reactant core. The temperature distribution in the pellet at an instant of time when the reactant core is at a position R_i is shown in Figure 19.8 for an exothermic reaction.

Heat is transferred from the reaction interface to the external surface of the solid mainly by conduction.

From the external surface it is transferred to the bulk gas by a process of convection and radiation. The rates of each of these steps are as follows:

1. Heat transferred through product layer:

$$Q_h = \frac{4\pi k'_e R_i R}{R - R_i}(T_i - T_s) \quad (19.65)$$

2. Heat transferred by convection and radiation from external surface to bulk gas:

$$Q_h = 4R^2(h_a + h_b)(T_s - T_b) \quad (19.66)$$

where h_a is the coefficient for convective heat transfer and h_b the coefficient for radiative heat transfer, which can be approximated as

$$h_b = 4\pi\sigma\varepsilon_M\left(\frac{T_b + T_s}{2}\right)^3 \quad (19.67)$$

where σ is the Stefan–Boltzmann constant and ε_M is the emissivity.

This approximation is valid if the temperature difference between the bulk gas and external surface of the solid is not very large. If these differences are significant, then the rate of heat transfer is given by

$$Q_h = 4\pi R^2[h_a(T_s - T_b) + \sigma\varepsilon_M(T_s^4 - T_b^4)] \quad (19.68)$$

The net rate of heat transfer can be obtained by combining Eqs. 19.65 and 19.66. This can be equated to the net rate of heat generation, which is given as

$$Q_h = 4\pi R_i^2(-\Delta H)k_S(T_b)\exp\left[\frac{E}{R_g T_b}\left(1 - \frac{T_b}{T_i}\right)\right]C_{Ai} \quad (19.69)$$

Thus the interfacial temperature can be calculated at each value of R_i by solving Eqs. 19.68 and 19.69 simultaneously.

For convenience of calculation the equations can be put in dimensionless form as follows:

$$\left(\frac{1 - \hat{R}_i}{\hat{R}_i} + \frac{1}{Nu}\right)^{-1}(\hat{T}_i - 1)$$
$$= \frac{(-\Delta H)k_S(T_b)C_{Ab}R}{T_b k'_e}\varepsilon\hat{R}^2 \quad (19.70)$$

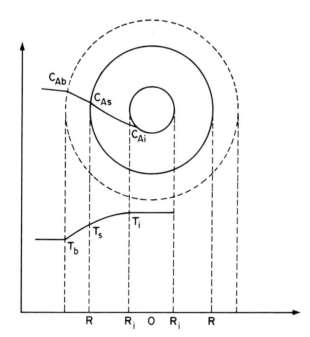

Figure 19.8. Temperature distribution in a solid particle (sharp-interface model).

where ε is the effectiveness factor under nonisothermal conditions and is given by

$$\varepsilon = \frac{\exp\left[\alpha\left(1 - \frac{1}{\hat{T}_i}\right)\right]}{1 + \text{Da}\left[\left(\frac{1}{\text{Sh}} - 1\right)\hat{R}_i^2 + \hat{R}_i\right]\exp\left[\alpha\left(1 - \frac{1}{\hat{T}_i}\right)\right]}$$

(19.71)

and

$$\text{Nu} = (h_a + h_b)\frac{R}{k_e'}$$

(19.72)

The rate of movement of the interface is related to ε by Eq. 19.15.

The calculation procedure can be conveniently summarized as follows:

1. At $\hat{t} = 0$, $\hat{R}_i = 1$; thus Eqs. 19.70 and 19.71 are solved simultaneously to obtain $\hat{T}_i$ and ε.
2. Knowing ε, we solve Eq. 19.15 for a chosen time increment $\Delta\hat{t}$ to obtain the value of $\hat{R}_i$ at time $\hat{t} + \Delta\hat{t}$.
3. Steps 1 and 2 are then repeated to obtain $\hat{R}_i$ at successive increments of time. The conversion at any given time $\hat{t}$ is then given as $1 - \hat{R}_i^3$.

The assumption of pseudo-steady state for heat transfer leads to a simple method of estimating the nonisothermal effects, as shown above. The assumption, however, is not very accurate. Although the accumulation terms in the mass balance are negligible, the corresponding assumption for the heat balance is not entirely justified. An improved model incorporating transient terms in the heat balance has been proposed by Luss and Amundson (1969). They have assumed the chemical reaction at the interface to be very rapid. Shettigar and Hughes (1972) have modified the equation of Luss and Amundson (1969) to incorporate the effect of chemical reaction. The solution of Luss and Amundson (1969) can be represented as

$$\hat{T}_i - 1 = \frac{2}{3}L_1 \sum_{n=1}^{\infty} \frac{\sin(\beta^*\hat{R}_i)}{\hat{R}_i}f(\beta^*)\int_{R_i}^1 Z^* \sin(\beta^*Z^*)$$

$$\hat{R}_i \exp(-\beta^{*2}\check{L}_2\check{F})dZ^*$$

(19.73)

where β^* are the positive roots of the equation

$$\beta^* \cot \beta^* + \text{Nu} - 1 = 0$$

(19.74)

and $\check{F}$ is a function given by

$$\check{F} = \frac{Z^* - \hat{R}_i}{\text{Da}\exp[\alpha(1 - 1/\hat{T}_i)]} + \frac{1}{2}(Z^{*2} - \hat{R}_i^2)$$

$$+ \frac{1}{3}\left(\frac{1}{\text{Sh}} - 1\right)(Z^{*3} - \hat{R}_i^3)$$

(19.75)

$\check{L}_1$ and $\check{L}_2$ are constants defined by

$$\check{L}_1 = \frac{3(-\Delta H)v\rho_B}{M_B\rho_s C_{pS}T}$$

(19.76)

$$\check{L}_2 = \frac{\rho_B k_e'}{v M_B\rho_s C_{pS}D_{eS}C_{Ab}}$$

(19.77)

and

$$f(\beta^*) = \frac{(\text{Nu} - 1)^2 + \beta^{*2}}{\text{Nu}(\text{Nu} - 1) + \beta^{*2}}$$

(19.78)

In Eq. 19.73, Z^* is a dummy variable for the purpose of integration.

The effectiveness factor is given again by Eq. 19.71, and the relation between $\hat{R}_i$ and $\hat{t}$ can be obtained in an analogous manner.

Homogeneous Model

The dimensionless equations describing the homogeneous model for a nonisothermal system are as follows:

$$N_1\frac{\partial\hat{C}_A}{\partial\hat{t}} = \frac{\partial^2\hat{C}_A}{\partial\hat{R}^2} + \frac{2}{\hat{R}}\frac{\partial\hat{C}_A}{\partial\hat{R}}$$

$$- \phi^2\hat{C}_A^m\hat{C}_B^n\exp\left[\alpha\left(1 - \frac{1}{\hat{T}}\right)\right]$$

(19.79)

$$N_2\frac{\partial\hat{T}}{\partial\hat{t}} = \frac{\partial^2\hat{T}}{\partial\hat{R}^2} + \frac{2}{\hat{R}}\frac{\partial\hat{T}}{\partial\hat{R}}$$

$$+ \beta_{mb}\phi^2\hat{C}_A^m\hat{C}_B^n\exp\left[\alpha\left(1 - \frac{1}{\hat{T}}\right)\right]$$

(19.80)

$$\frac{\partial\hat{C}_B}{\partial\hat{t}} = -\hat{C}_A\hat{C}_B\exp\left[\alpha\left(1 - \frac{1}{\hat{T}}\right)\right]$$

(19.81)

with these boundary conditions: at

$$\hat{R} = 1, \quad \frac{d\hat{C}_A}{d\hat{R}} = \text{Sh}(1 - \hat{C}_A)$$

(19.82)

$$\frac{d\hat{T}}{d\hat{R}} = \text{Nu}(1 - \hat{T})$$

(19.83)

and at

$$\hat{R} = 0, \qquad \frac{d\hat{C}_A}{d\hat{R}} = \frac{d\hat{T}}{d\hat{R}} = 0 \qquad (19.84)$$

The initial condition is that at $\hat{t} = 0$, $\hat{C}_A = 0$, $\hat{T} = 1$, and $\hat{C}_B = 1$.

The parameters N_1, N_2, β_{mb} are defined as follows:

$$N_1 = \frac{f_s}{D_{eB}} k_v C_{Ab}^m C_{Bs}^{n-1} t R^2 \qquad (19.85)$$

$$N_2 = \frac{\rho_s C_{ps} k_v C_{Ab}^m C_{Bs}^{n-1} t R^2}{k_e'} \qquad (19.86)$$

$$\beta_{mb} = (-\Delta H) \frac{D_e C_{Ab}}{k_e' T_b} \qquad (19.87)$$

(Equation 19.87 is the same as Eq. 4.44.) The model equations require numerical solution.

Finite-Reaction-Zone Model

The theory of Luss and Amundson (1969) was extended by Shettigar and Hughes (1972) to the finite-thickness model. The equations are similar to the SIM except that the definition of Da is now given by Eq. 19.44.

Particle-Pellet Model

A complete model including transient heat and mass balance equations for the particle–pellet model has been formulated and solved by Sampath et al. (1975). The equations are as follows:

$$\check{N}_1' \frac{\partial \hat{C}_A}{\partial \hat{t}} = \frac{\partial \hat{C}_A}{\partial \hat{R}^2} + \frac{2}{\hat{R}} \frac{\partial \hat{C}_A}{\partial \hat{R}} - \phi^2 \hat{r}_{Gi}^2 \hat{C}_A \check{F}_2 \quad (19.88)$$

$$\check{N}_2' \frac{\partial \hat{T}}{\partial \hat{t}} = \frac{\partial \hat{T}}{\partial \hat{R}^2} + \frac{2}{\hat{R}} \frac{\partial \hat{T}}{\partial \hat{R}} + \beta_{mb} \phi^2 \hat{r}_{Gi}^2 \hat{C}_A \check{F}_2$$

$$(19.89)$$

$$\frac{d\hat{r}_{Gi}^2}{\partial \hat{t}} = -\hat{C}_A \check{F}_2 \qquad (19.90)$$

where

$$\check{F}_2 = \frac{\exp\left[\alpha\left(1 - \frac{1}{\hat{T}}\right)\right]}{1 + \mathrm{Da}_G r_{Gi}(1 - \hat{r}_{Gi}) \exp\left[\alpha\left(1 - \frac{1}{\hat{T}}\right)\right]}$$

$$(19.91)$$

$$\check{N}_1' = \frac{v f_s R^2 M_B C_{Ab} k_S}{D_{e,ma} \rho_B r_{G0}} \qquad (19.92)$$

$$\check{N}_2' = \frac{v R^2 \rho_s C_{ps} M_B C_{Ab} k_S}{k_e' \rho_s r_{G0}} \qquad (19.93)$$

The boundary conditions are the same as given by the set of Eqs. 19.82–19.84.

A simpler model (pseudo-steady-state model) may be obtained by setting $N_1' = N_2' = 0$, and the resulting equations solved (Calvelo and Smith, 1970). The solutions show an interesting effect: for a certain range of parameter values complete conversion can first occur in some intermediate region of the pellet and not at the external surface. This occurs because the concentration of the reactant is a minimum at the center whereas the temperature is a maximum at this point. The combined effect can be such that the rate is a maximum neither at the surface nor at the center but at some intermediate point.

The complete transient model as formulated above was solved by Sampath et al. (1975) using orthogonal collocation. A comparison of the two approaches (transient and pseudo-steady-state) was also made. Typical results are shown in Figure 19.9. In the pseudo-steady-state model the maximum temperature is attained at $t = 0$, while in the transient model the temperature reaches a maximum at some intermediate time depending on the value of $\check{N}_2'/\check{N}_1'$. Another interesting result is that in the pseudo-steady-state analysis the maximum temperature is attained at the center of the pellet at any time, while in the transient analysis the maximum temperature occurs initially in a region close to the surface of the pellet that then moves toward the center as the time progresses. The conversions predicted by the two models were not compared in the work of Sampath et al. (1975).

19.2. MODELS INCORPORATING THE EFFECT OF STRUCTURAL CHANGES

Structural changes are known to take place in the solid during the course of reaction, and the effect of these changes should be included in any analysis if a more realistic model is to be formulated. The main structural changes that take place are due to (a) chemical reaction and (b) sintering. The molal volumes of the product are generally different from those of the reactants, and hence the porosity changes with reaction. Some models proposed to account for this change will be reviewed

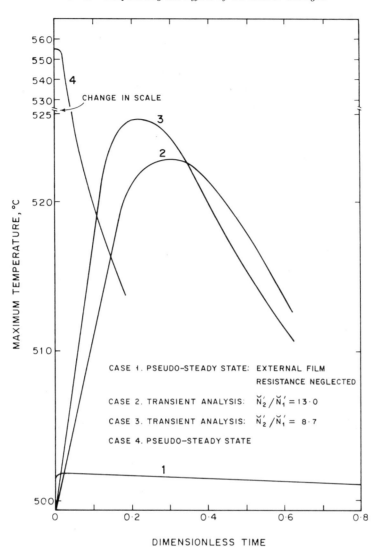

Figure 19.9. Comparison of the maximum temperature profile generated by the transient and pseudo-steady-state models (Sampath et al., 1975).

here. These are of a general type meant primarily for reaction types A and B; models for other reaction types will be discussed in the appropriate sections.

19.2.1. Modified SIM

Shen and Smith (1963) analyzed systems with changing porosity on the basis of the SIM by allowing for variation of pellet size with reaction. To characterize this change, let us consider a solid that has an initial radius R_0 in the reaction

$$v_A A(g) + v_B B(s) \rightarrow v_R R(g) + v_S S(s)$$

Let the radius be R at any given time when the interface is at R_i. The change in the radius is due to the different volumes occupied by the product and reactant for the same number of moles. The volume of reactant that has reacted is $\frac{4}{3}\pi(R_0^3 - R_i^3)$, which corresponds to $\frac{4}{3}\pi(R_0^3 - R_i^3)\,(\rho_B/M_B)$ moles of B. The corresponding amount of the product is $\frac{4}{3}\pi(R_0^3 - R_i^3)$ $(\rho_B/M_B)\,(M_S/v_S)\,(v_B/v_R)$, which occupies a volume of $\frac{4}{3}\pi(R^3 - R_i^3)$. If the ratio of molal volume of product to reactant is denoted by a parameter Z_v, the resulting radius R of the particle is

$$R = \left[\frac{R_0^3 + (Z_v - 1)R_i^3}{Z_v} \right]^{1/3} \quad (19.94)$$

where

$$Z_v = \frac{v_B}{v_R} \frac{M_S}{\rho_S} \frac{\rho_B}{M_B} \qquad (19.95)$$

It is seen that if Z_v is less than one, the radius of the particle shrinks with time; whereas if Z_v is greater than one, there is an increase in particle size during the course of reaction.

The time required to achieve a given conversion is the same when the reaction is controlled by chemical reaction at the sharp interface. When the reaction is controlled by external film diffusion, the time required to achieve a given conversion can be obtained as follows.

The mass transfer coefficient at any given time can be approximated for small particles as

$$k_g = \frac{k_{g0} R_0}{R} \qquad (19.96)$$

where k_{g0} is the mass transfer coefficient based on the initial radius R_0. Here k_g is assumed to be inversely proportional to R. Equating the rate of mass transfer to the rate of movement of the interface and substituting for R results in

$$-R_i^2 \frac{\rho_B}{M_B} \frac{dR_i}{dt} = v\left[\frac{R^3 + (Z_v - 1)R_i^3}{Z_v}\right]^{1/3} k_{g0} C_{Ab} R \qquad (19.97)$$

Integration of this equation gives

$$\frac{3v(Z_v - 1)k_{g0}C_{Ab}RM_Bt}{\rho_B(Z_v)^{1/3}} = \tfrac{2}{3}(Z_v)^{2/3} - \left[R^3 + (Z_v - 1)R_i^3\right]^{2/3} \qquad (19.98)$$

When diffusion through the ash layer is controlling, the corresponding equation for X vs. t is

$$\frac{6vD_{eS}C_{Ab}M_Bt}{\rho_B R^2} = 3\left[\frac{Z_v - (Z_v + (1 - Z_v)(1 - x)^{2/3}}{Z_v - 1} - (1 - x)^{2/3}\right] \qquad (19.99)$$

Rehmat and Saxena (1977) have extended this model to the case of a nonisothermal system.

19.2.2. Modified Homogeneous Model

In the homogeneous model the porosity is assumed to change with local solid conversion according to some empirical equation, such as

$$f_s = f_{s0} + \alpha\left(1 - \frac{C_B}{C_{B0}}\right) \qquad (19.100)$$

where α is a constant. The corresponding diffusivity change can be empirically related to the porosity change (Wakao and Smith, 1962; Wen, 1968; Calvelo and Cunningham, 1970; Fan et al., 1977). For instance, the relationship used by Fan et al. is

$$D_e = D_{e0}\left[\frac{\alpha_1}{\exp\left(1 - \frac{1}{\alpha_2 C_B/C_{B0}}\right) + \alpha_2 - 1}\right] \qquad (19.101)$$

where D_e and D_{e0} are diffusivities at any time t and at zero time respectively, and α_1 and α_2 are empirical constants. The sigmoid conversion–time behavior predicted by this model can also be attributed to nucleation effects discussed in Section 19.5.2.

Gidaspow et al. (1976) have considered the effect of structural changes based on the homogeneous model. For $Z_v > 1$ the possibility of pore closure is predicted. The following equation for the time for pore closure t_{pc} has been derived:

$$t_{pc} = \ln\left[\frac{f_{s0}(Z_v - 1)}{2f_{s0} - 1}\right] \qquad (19.102)$$

19.2.3. Modified Particle–Pellet Model

The effect of structural changes on the basis of the particle–pellet model has been analyzed by Ramachandran and Smith (1977b). The model attempts to account for structural changes due to both chemical reaction and sintering. The pellet size is assumed to be constant in the model, while the particle size is treated as a variable during the course of the reaction. The change in size of the particle is assumed to be in accordance with the value of the parameter Z_v and is represented as

$$r_G = \left[\frac{r_{G0}^3 + (Z_v - 1)r_{Gi}^3}{Z_v}\right]^{1/3} \qquad (19.103)$$

The porosity of the pellet is assumed to change with the extent of reaction because of two effects: (1) change in particle size caused by chemical reaction, and (2) sintering of the solid. To account for the former, the theory proposed by Kim and Smith (1974) has been used. The change in porosity as a result of chemical reaction (due to changes in r_G) can be expressed as

$$\frac{1 - f_s}{1 - f_{s0}} = \left(\frac{r_G}{r_{G0}}\right)^3 \quad (19.104)$$

Additional changes in porosity occur as a result of sintering and this has also been accounted for in the model. The application of the model to the hydrofluorination of uranium dioxide has been discussed by Ramachandran and Smith (1977b). This system has a value of $Z_v = 1.7$ and exhibits the phenomenon of incomplete asymptotic conversion.

A similar model has been proposed by Georgakis et al. (1979); it predicts the time for pore closure analytically. The maximum conversion of the solid is given by the following equation for $Z_v > 1$:

$$x_{max} = \frac{f_{s0}}{(Z_v - 1)(1 - f_{s0})} \quad (19.105)$$

19.2.4. Single-Pore Model

In order to account for the effect of structural changes in a porous solid, a model has been proposed by Ramachandran and Smith (1977a) and Chrostowski and Georgakis (1978) that focuses attention on the changes taking palce in a single pore during the course of reaction, and thus accounts for changes in pore geometry. If the parameter $Z_v = 1$, the pore geometry does not change with time. For $Z_v < 1$, the reaction leads to a more open structure. In this case the single-pore model predicts that the time required to achieve a given conversion is reduced. If $Z_v > 1$, a decrease in average pore radius is predicted, and large values of Z_v can lead to pore-mouth closure, which means that the conversion can approach an asymptotic value less than 100%. The maximum possible conversion of the solid under such conditions is given by

$$x_{max} = 1 - \frac{1 - f_{s0}[Z_v/(Z_v - 1)]}{1 - f_{s0}} \quad (19.106)$$

In actual practice the observed conversion is less than x_{max} because of diffusional effects. Such asymptotic conversions (less than 100%) have been observed in the

hydrofluorination of uranium dioxide and sulfation of calcium oxide. The single-pore model is able to predict this complex behavior. A disadvantage of the model is that it does not take into account the intersections of the reaction surfaces as the reaction progresses, an effect that cannot be ignored after an appreciable conversion has occurred.

An extension of this model by Ulrichson and Mahoney (1980) accounts for the effect of bulk flow and reversibility of the reaction. This is applicable to reactions like the chlorination of magnesium oxide:

$$MgO + Cl_2 \rightleftharpoons MgCl_2 + \tfrac{1}{2}O_2, \qquad K = 0.33$$

Lee (1980) has proposed a parallel-pore model, similar in concept to the single-pore model, and has obtained analytical relations that are valid for low and intermediate values of the Thiele modulus.

19.2.5. Distributed-Pore Model

The distributed-pore model is similar in concept to the single-pore model but incorporates the effect of pore size distribution. Three main approaches have been used.

1. Christman and Edgar (1980) considered a local position r in the pellet and a pore of radius r_{p1}. This pore is assumed to be covered with an inner concentric product layer of thickness $r_{p2} - r_{p1}$. The reaction interface is thus a cylindrical surface at r_{p2}. At time $t = 0$, $r_{p1} = r_{p1,0}$ (the initial pore radius) and $r_{p2} = r_{p1}$. r_{p1} and r_{p2} at any time are related by an equation that is the same as that for the single-pore model.

$$r_{p1}^2 = Z_v r_{p1,0}^2 + (1 - Z_v)r_{p2}^2 \quad (19.107)$$

When the average rate of reaction at position r, obtained by integrating the equation for the single pore over the entire pore size distribution, and an average value for effective diffusivity are incorporated into the mass balance, we obtain

$$\frac{1}{r^2}\frac{d}{dr}(r^2 \overline{D}_e)\frac{dC_A}{dr} = \overline{k}C_A \quad (19.108)$$

where $\overline{D}_e$ and $\overline{k}$ are the average values at position r, defined as

$$\overline{D}_e = \frac{1}{\tau^2}\int_0^\infty \pi r_{p1}^2 \left(\frac{1}{D_M} + \frac{1}{D_K(r_{p1})}\right)^{-1} f(r_{p1})\, dr_{p1} \quad (19.109)$$

and

$$\bar{k} = 2k_S \int_0^\infty \frac{\pi r_{p2} f(r_{p1}) dr_{p1}}{1 + \dfrac{r_{p2} k_S}{D_{eG}} \ln \dfrac{r_{p2}}{r_{p1}}} \quad (19.110)$$

Here the pore size distribution is represented by the function $f(r_{p1})$, which is defined such that $f(r_{p1}, r, t) dr_{p1}$ represents the number of pores of sizes between r_{p1} and $r_{p1} + dr_{p1}$ intersecting a unit surface at position r at time t. The change of pore size distribution with time can be accounted for by a population balance equation assuming that there is no net fractional increase in the number of pores by "birth" or "death" mechanisms; that is, the total number of pores is conserved. An important aspect of this approach is that the evolution of the pore structure with time and position can be predicted.

2. In a somewhat different approach, Simons and Rawlins (1980) postulated that each pore in the pellet reaches the surface as the trunk of a tree. Thus the external surface of the pellet may be assumed to have a number of holes (or pores), and a model can then be developed by considering each pore and averaging over the surface pore size distribution (or pore tree) to obtain the average flux of the gaseous reactant at the surface. This flux can be related to the average conversion of the solid B. With each pore of

radius r_p, Simons and Finson (1979) associated an internal surface area S_i defined by the equation

$$S_i = \frac{2\pi K_0 r_p^3 (1 - f_s)}{f_s^{1/3} r_{p\,min}} \quad (19.111)$$

where K_0 is the aspect ratio, normally assigned a value of about 5. The pore size distribution function $f(r_p)$ can be described by a suitable mathematical function, such as

$$f(r_p) = \frac{f_s}{2 \ln (r_{p\,max}/r_{p\,min}) \pi r_p^3} \quad (19.112)$$

The model has been successfully used in the interpretation of the initial rate data for the reaction of SO_2 and H_2S with calcined limestone (Simons and Rawlins, 1980) and for char gasification (Lewis and Simons, 1979).

3. Bhatia and Perlmutter (1980) have proposed a particularly attractive model that takes into account the intersection of the pores as reaction proceeds. This random-pore model assumes that the actual reaction surface of solid B is formed by a set of cylindrical surfaces (per unit volume of space), with size distribution $g(r_p)$ where $g(r_p) dr_p$ is the total length of the cylindrical surfaces. A schematic diagram of the model is shown in Figure 19.10, which depicts the pro-

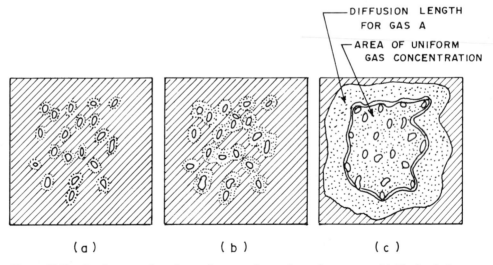

Figure 19.10. Development of reaction surface according to the random-pore model. The hatched area represents unreacted solid B, the dotted area the product layer (a) at an early stage with the product layer around each pore visible; (b) at an intermediate stage, showing some overlapping reaction surfaces; and (c) at a later stage, showing full development of the product layer and reaction surface for the view chosen (Bhatia and Perlmutter, 1981a).

duct layer surrounding the pores at different stages of the reaction. Using this model, the following equation for the conversion–time behavior for the case of kinetic control can be derived:

$$\frac{dx}{d\hat{t}} = (1-x)\left[1 - \psi \ln(1-x)\right]^{1/2}$$

$$(19.113)$$

where $\hat{t}$ is a dimensionless time given by

$$\frac{k_s C_{Ab}^m S_{v0} t}{(1-f_{s0})},$$

and ψ is a structural parameter defined as

$$\psi = \frac{4\pi L_{p0}(1-f_{s0})}{S_{v0}^2} \qquad (19.114)$$

where S_{v0} and L_{p0} are, respectively, the reaction surface and total length of the pore per unit of pellet volume at time zero.

For $\psi = 0$, Eq. 19.113 reduces to the homogeneous model with $n = 1$, and for $\psi = 1$ to the grain model. An important conclusion of the model is that the reaction order with respect to the solid is related to the pore structure of the solid (i.e., to the parameter ψ). Thus the random-pore model is able to rationalize the concept of reaction order with respect to the solid. Bhatia and Perlmutter (1981a, b) have also extended the model to the case of ash diffusion.

The model introduces a single parameter, ψ, to account for the pore size distribution of the solid. Equations have been derived for ψ for various types of pore size distribution, such as bimodal, square, triangular, lognormal, and normal. For pores of uniform size the relation for ψ is

$$\psi = -\frac{1}{\ln(1-f_{s0})} \qquad (19.115)$$

whereas for the general case it is

$$\psi = \frac{4\pi \displaystyle\int_0^\infty g(r_p)\, dr_p}{2\pi \displaystyle\int_0^\infty r_p g(r_p)\, dr_p} \qquad (19.116)$$

An analysis of different pore size distributions indicates that a uniform pore size leads to the lowest reactivity. For a bimodal distribution, an optimum structure exists for which the reactivity is maximum.

19.2.6. Effects of Sintering

An important structural change involved in a noncatalytic reaction accompanied by a large temperature rise in the pellet is the sintering of the pellet. Sintering causes a decrease in the effective diffusivity of the pellet, and any realistic modeling of gas–solid reactions must account for this effect. Evans et al. (1973) did this by assuming an exponential decay in the effective diffusivity with reaction time.

Kim and Smith (1974) measured the effective diffusivity in nickel oxide pellets during various stages of reduction, and correlated the fractional increase in tortuosity as a function of the fractions f_p of the pores removed.

$$F(f_p) = \frac{\tau(f_p)}{\tau(0)} \qquad (19.117)$$

where $\tau(f_p)$ is the tortuosity factor when a fraction f_p of the pores is removed and $\tau(0)$ is the factor at $f_p = 0$. A more complicated model for predicting the effect of sintering on the tortuosity factor has been developed by Chan and Smith (1976).

The following correlation has been proposed for incorporating the change in effective diffusivity in the pellet due to the combined effect of chemical reaction and sintering (Ramachandran and Smith, 1977b):

$$D_e = \left(\frac{1}{D_M} + \frac{1}{D_K}\right)^{-1} \frac{1}{F(f_p)}\left[1 - (1-f_{s0})\left(\frac{r_{Gi}}{r_{G0}}\right)(1-f_p)\right]$$

$$(19.118)$$

The only additional information required for the modeling of the effect of sintering is the variation of f_p with time. Ramachandran and Smith (1977b) modeled this as a first-order process thus:

$$\frac{df_p}{dt} = (1-f_p)A_{fp}\exp\left[\frac{E_s}{R_g(T-T_M)}\right] \qquad (19.119)$$

where E_s and A_{fp} are the Arrhenius parameters for sintering and T_M is a characteristic temperature corresponding to the onset of sintering. Generally T_M will correspond to the Tamman temperature, which is approximately half the melting point of the solid. The model parameters introduced to account for the effect of sintering are A_{fp}, E_S, and T_M, and these can probably be estimated by independent sintering measurements on the solid, that is, by measuring the porosity and the effective diffusivity for various sintering times.

In another approach (Ranade and Harrison, 1979, 1981), the sintering process is visualized as resulting in the combination of adjacent grains. The consequent change in grain size is related to the specific surface area of the grains, which is described as a function of time (Nicholson, 1965). The model has been applied to experimental results on the reaction of hydrogen sulfide with zinc oxide.

19.3. DECOMPOSITION REACTIONS (TYPE C)

19.3.1. General Isothermal Case

A decomposition reaction can be represented as

$$B(s) \rightleftharpoons R(g) + S(s)$$

The decomposition reaction is generally reversible and the equilibrium constant K may be represented as $K = p_{R,i}$ where $p_{R,i}$ is the partial pressure of R at the interface. The rate of decomposition can be assumed to be controlled by the rate at which R is desorbing from the interface:

$$\mathbf{R}_R = \left(\frac{1}{4\pi R^2 k_g} + \frac{R - R_i}{4\pi R R_i D_{eS}} \right)^{-1} (C_{Ri} - C_{Rb})$$

$$(19.120)$$

The rate of movement of the reaction interface is given by the same equation as for the sharp-interface model. Substituting for $\mathbf{R}_R$, integrating, and substituting for C_{Ri}, we have

$$\frac{R}{3k_g}\left[1 - \left(\frac{R_i}{R}\right)^3\right] + \frac{R^2}{6D_{eS}}\left[1 - 3\left(\frac{R_i}{R}\right)^2 + 2\left(\frac{R_i}{R}\right)^3\right]$$

$$= \left(\frac{K}{R_g T_b} - C_{Rb}\right)\frac{M_B t}{\rho_B} \qquad (19.121)$$

Equation 19.121 can be used to predict the rate of decomposition, but the following limitations should be noted: (1) It is restricted to isothermal systems. (2) The convective flow terms are neglected in the derivation of

the equation for diffusion through the ash layer; the effect of this is not significant if ash-layer diffusion is controlled by Knudsen transport, but the error becomes quite large when bulk diffusion predominates. A modification to take both these effects into account has been proposed by Wang (1971).

It is also possible that the decomposition reactions are controlled by the rate of heat transfer to the decomposition surface. A novel extension (Narsimhan, 1961) considers a two-phase solid system ($MgCO_3$–$CaCO_3$) characterized by finite but different equilibrium decomposition temperatures; a numerical solution has been suggested to solve the equations.

Mu and Perlmutter (1980) have proposed a model to account for the effect of changes in pellet size arising from possible density differences between reactant and product. Their results show that the effect is more significant than that of the variation of D_{eG} with temperature. In another recent study Prasannan et al. (1982) incorporated the effect of nonequimolal diffusion of the product gas on the rate of decomposition. This effect is not important if the diffusion is in the Knudsen regime, whereas if the process is controlled by bulk diffusion, then the analysis of the problem should include this effect.

An increasingly important decomposition reaction is the devolatilization or pyrolysis of coal, which may be represented by the simplified scheme

$$\text{coal} \rightarrow \text{residues} + \text{volatiles}$$

Although the detailed chemistry of the process is not well understood (Anthony and Howard, 1976), a model has been proposed (Gavalas and Wilks, 1980) based on this simple scheme in which the diffusing molecules are lumped into three species: tar, gases, and inerts. The concentration profiles of these species are then evaluated by describing the process by ternary diffusion and viscous flow in conjunction with a simple pore model for coal. The model has been compared with some limited experimental data on subbituminous coal.

19.3.2. Example: Decomposition of Calcium Carbonate—A Case of Heat Transfer Control

Predict the time required for complete decomposition of $CaCO_3$ assuming the process to be controlled by heat transfer. The following data may be used:

Radius of the pellet $R = 10^{-2}$ m

Density $= 1840 \, \text{kg/m}^3$

Molecular weight $= 10^{-1}$ kg/mol

$(-\Delta H) = 1.7 \times 10^5$ J/mol

$\Delta T = 102°$K

$h = 57$ W/m^2 $°$K

Solution

The heat transfer to the decomposing surface is given by

$$Q_h = \left(\frac{1}{4\pi R^2 (h_a + h_b)} + \frac{R - R_i}{4\pi k'_e R_i R} \right)^{-1} (T_b - T_i)$$

which can be rearranged as

$$Q_h = 4\pi R^2 h \left[1 + \frac{\text{Nu}(1 - \hat{R}_i)}{\hat{R}_i} \right] \Delta T$$

where

$$h = h_a + h_b, \qquad \text{Nu} = \frac{hR}{k'_e}$$

This is also equal to the endothermic heat of decomposition

$$-(\Delta H) \frac{d}{dt} \left(\frac{4}{3} \pi R_i^3 \frac{\rho_B}{M_B} \right)$$

or

$$\frac{-4\pi \rho_B (\Delta H)}{M_B} \left(\hat{R}_i^2 \frac{d\hat{R}_i}{dt} \right)$$

Equating the rate of heat transfer to the endothermic heat of decomposition and rearranging, we obtain

$$-\alpha_H R_i^2 \frac{dR_i}{dt} = \frac{R_i}{R_i + \text{Nu}(1 - R_i)}$$

where α_H is a quantity defined as $(\Delta H) R \rho_B / M_B h \Delta T$. Integrating from 1 to 0 for $\hat{R}_i$, and from 0 to t_d for t (where t_d is the time for complete decomposition), we obtain

$$t_d = \alpha_H \left(\frac{1}{3} + \frac{\text{Nu}}{6} \right)$$

For the given problem

$$\text{Nu} = 0.8233, \qquad \alpha_H = 5380.1168$$

Hence the time of decomposition t_d is 0.702 hr.

19.4. COMPLETE GASIFICATION REACTIONS (TYPES D AND E)

Many gas–solid reactions yield only gaseous products and the reaction scheme for such cases may be represented as

$$A(g) + v_{BB}(s) \to \text{gaseous products}$$

Examples of such reactions are found in the combustion of carbon, water–gas reaction of carbon, formation of metal carbonyls, and the chlorination of certain ores (e.g., ilmenite). The analysis of the reaction scheme is best done separately for nonporous and porous solids.

19.4.1. Nonporous Solid

A schematic diagram of a nonporous solid undergoing a gasification-type reaction is shown in Figure 19.11. A detailed mathematical analysis has been presented by Szekely et al. (1976) and by Levenspiel (1974). Two main features of this system may be noted.

1. Since no product layer exists, there is no ash diffusion resistance.
2. Since the radius of the pellet is now changing with time, the external film transfer coefficient (k_g) would also change with time. For example, if the equation of Ranz and Marshall (1952) is used, then we have the following relationship between k_g with R.

$$k_g = \frac{D_b}{R} + 0.6 \left(\frac{\mu}{\rho_g D_b} \right)^{1/3} \left(\frac{u \rho_g}{\mu} \right)^{1/2} D_b d_p^{-1/2} \tag{19.122}$$

The equation shows the following behavior: $k_g \simeq 1/R$ for small R and u, and $k_g \simeq u^{1/2}/R^{1/2}$ for large R and u. The x vs. t relations are thus different for small and large particles and are given in Table 19.2.

The overall reaction can also be kinetically controlled. In such a situation the kinetic parameters can be determined by carrying out the reaction in a fluidized-bed reactor with the gas rate very close to that for minimum fluidization. Most of the gas passes through the dense phase and the assumption of plug flow for the gas phase and complete mixing for the solid phase is quite valid. Here the conversion of the gas will vary with time and position in the reactor, whereas the solid conversion varies with time alone. Moreover, when the

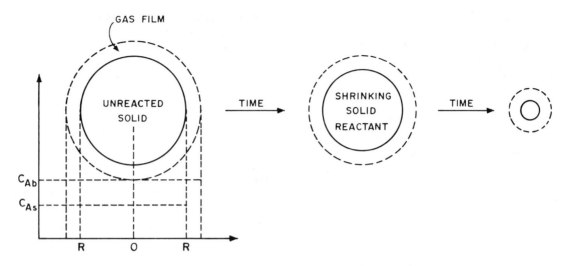

Figure 19.11. Representation of nonporous solid undergoing a gasification reaction in which no solid product is formed.

TABLE 19.2. Conversion–Time Relations for Shrinking Spheres

	Film Diffusion Control	Reaction Control
Small particle	$\dfrac{t}{\breve{t}} = 1 - (1-x)^{2/3}$	$\dfrac{t}{\breve{t}} = 1 - (1-x)^{1/3}$
	$\breve{t} = \dfrac{\rho_B R^2}{2 v D_{eB} C_{Ab} M_B}$	$\breve{t} = \dfrac{\rho_B R}{v k_S C_{Ab} M_B}$
Large particle	$\dfrac{t}{\breve{t}} = 1 - (1-x)^{1/2}$	$\dfrac{t}{\breve{t}} = 1 - (1-x)^{1/3}$
	$\breve{t} = (\text{constant})\dfrac{R^{3/2}}{C_{Ab} M_B}$	$\breve{t} = \dfrac{\rho_B R}{b k_S C_{Ab} M_B}$

products of reaction are gaseous, the progress of the reaction can be easily followed by the analysis of the product gas. Examples of such reactions are selective chlorination of ilmenite and fluorination of uranium tetrafluoride. Such reactions are controlled by either the rate of chemical reaction or the rate of supply of reactant gas. By properly controlling the feed rate of gas, mass transfer effects can be minimized. The kinetic parameters can then be readily evaluated by the methods of Doraiswamy et al. (1959), used in the chlorination of ilmenite, and of Corella (1980) outlined below.

Fluorination of UF₄.   Corella (1980) extended the treatment of Doraiswamy et al. (1959) to the determination of the kinetics of gas–solid reactions in a semicontinuous integral fluidized-bed reactor. He as-

sumed that the gas–solid reaction is chemically controlled and that the reaction follows first-order kinetics. Thus, if A represents fluorine, we have

$$r = k C_{A0} (1 - x) \qquad (19.123a)$$

or

$$\ln \frac{1}{1-x} = k C_{A0} \frac{W}{F} \qquad (19.123b)$$

where x is the conversion of the key component in the exit gas stream. Thus a plot of $\ln[1/(1-x)]$ vs. $C_{A0} W/F$ can be drawn easily at various temperatures. The slope of the line gives the rate constant k. This method was used by Corella (1980) in his study of the kinetics of fluorination of uranium tetrafluoride.

19.4.2. Porous Particles

For porous particles the change in surface area with reaction must be considered. Petersen (1957) assumed a random distribution of uniform-sized pores and proposed the following equation for the gasification of a porous pellet in the absence of intraparticle gradients (Szekely et al., 1976):

$$\frac{f_{s0}}{1-f_{s0}}\left[\left(1+\frac{k_S C_{Ab}^m t}{r_{p0}}\right)^2 \frac{\left(G-1-\dfrac{k_S C_{Ab}^m t}{r_{p0}}\right)}{G-1} - 1\right]$$

(19.124a)

where G is the solution of the cubic equation

$$\frac{4}{27}f_{s0}G^3 - G + 1 = 0 \qquad (19.124b)$$

and r_{p0} and f_{s0} are the initial pore radiqs and porosity, respectively.

Hashimoto and Silveston (1973a, b) analyzed the gasification reaction by allowing for pore growth, initiation of new pores, and coalescence of adjoining pores. Each of these leads to a change in the number of pores of a particular size and in the pore size density distribution function. These changes were incorporated using a population balance model along with appropriate equations for the rate of formation of new pores and rate of pore coalescence.

The effect of intraparticle diffusion was also incorporated through a Thiele modulus $R_0(k_S \rho_B S_{g0} f_{s0}/v_B D_{e0})^{1/2}$. If the modulus is greater than 10, the shrinking-core model can be used, whereas if it is less than 0.1, diffusional gradients may be ignored in the model formulation. The model predicts a maximum in the relative surface area vs. solid conversion behavior; it needs a large number of parameters to characterize the system, many of which can only be obtained by matching experimental data with theory.

Simons and Finson (1979) and Simons (1979) have proposed a similar model, which considers such important quantities as density of pore intersections and length of pore segments. The gasification rate is calculated by integrating the rate of gasification per "tree of pores" over the postulated pore size distribution (see Section 19.2.5).

Gavalas (1980) has proposed a random-capillary model in which the porous solid is described by a single probability density function $p(r_p)$ that is related to the commonly used pore size distribution. For low porosity

this function $p(r_p)$ is defined as

$$p(r_p) = \frac{f(r_p)}{2\pi r_p^2} \qquad (19.125)$$

Based on the probability distribution, two structural parameters have been defined:

$$\psi_I = \int_{r_{p\min}}^{r_{p\max}} p(r_{p0})\,dr_{p0} \qquad (19.126a)$$

$$\psi_{II} = \int_{r_{p\min}}^{r_{p\max}} r_{p0}\,p(r_{p0})\,dr_{p0} \qquad (19.126b)$$

where $p(r_{p0})$ is the density distribution based on the initial pore size distribution. This model gives the following expression for conversion in terms of the two structural parameters:

$$x(t) = 1 - \exp\left[-2\pi(\psi_I v^2 t^2 + 2\psi_{II} vt)\right] \qquad (19.127)$$

where v is the reaction velocity defined as the rate of change of pore radius with time.

A rearranged form of Eq. 19.127 indicates that a plot of $(1/t)\ln[1/(1-x)]$ vs. t is a straight line with a slope equal to $2\pi\psi_I v^2$ and an intercept equal to $4\pi\psi_{II} v$. Similarly, differentiation of Eq. (19.127) and elimination of t yields the following linear equation:

$$\frac{1}{4\pi(1-x)}\frac{dx}{dt} = \frac{\psi_I v^2}{2\pi}\ln\left(\frac{1}{1-x}\right) + (\psi_{II} v)^2$$

(19.128)

This equation is very convenient in correlating x–t or (dx/dt)–x data and has been used by Gavalas (1980) to correlate the data of Mahajan et al. (1976) and by Dutta and Wen (1977) for correlating the data on the gasification of char by oxygen.

The random-pore model of Bhatia and Perlmutter (1980) described in Section 19.2.5 can also be used to model gasification processes. (For such systems the parameter Z_v approaches zero.) The model predicts a semi log relationship between the groups $[S_v/(1-x)]^2$ and $1-x$, and this has been confirmed by the data of Hashimoto et al. (1979) on the steam activation of char.

The reaction of coal char with CO_2 was found by Dutta and Wen (1977) to follow H–W kinetics. Inclusion of this behavior is necessary for a correct modeling of the process, especially when significant intraparticle gradients exist in the pellet.

Srinivas and Amundson (1980a) have proposed a model for intraparticle diffusion effects in char combustion that incorporates the effects of three independent reactions:

$$C + \tfrac{1}{2}O_2 \rightarrow CO \qquad (I)$$

$$CO + \tfrac{1}{2}O_2 \rightarrow CO_2 \qquad (II)$$

$$CO_2 + C \rightarrow 2CO \qquad (III)$$

In a related study of char gasification, the same authors (Srinivas and Amundson, 1980b) assumed the process to consist of four independent reactions, namely, steam gasification, carbon dioxide gasification, hydrogasification, and the water–gas shift reaction, and made a detailed parametric study of the gasification process.

19.5. MISCELLANEOUS MODELS

19.5.1. Crackling-Core Model

Park and Levenspiel (1975) proposed a so-called crackling-core model to account for the sigmoidal behavior of the x vs. t plots in some observed systems. In this model the reaction is assumed to occur in two steps

$$A \text{ (nonporous)} \xrightarrow{\text{physical change}} A \text{ (porous)}$$

$$A \text{ (porous)} \longrightarrow \text{product}$$

The first stage is a physical transformation of A from a nonporous structure to a more reactive porous structure, which then undergoes reaction according to the second step. The model can also be used to represent a situation where the reaction of solid takes place through two or more consecutive steps. For example, in the reduction of hematite the steps are

$$Fe_3O_4 \xrightarrow{+H_2} Fe_2O_3 \xrightarrow{+H_2} FeO \xrightarrow{+H_2} Fe$$

The first step results in the formation of a porous structure that undergoes further reaction. A similar sequence occurs in the reduction of manganese dioxide by hydrogen (de Bruijn et al., 1980):

$$MnO_2 \xrightarrow{+H_2} Mn_2O_3 \xrightarrow{+H_2} Mn_3O_4 \xrightarrow{+H_2} MnO$$

Although the crackling-core model can predict the

qualitative features of a system reasonably well, it is only an approximation and does not consider the basic mechanisms responsible for the sigmoidal behavior. One such mechanism could be nucleation, which is discussed next.

19.5.2. Nucleation Model

Nucleation effects, more common in solid-solid reactions, can also be significant in some gas–solid reactions such as the reduction of metallic oxides. It is only recently that this problem is being analyzed. Typical x–t behavior at low temperatures shows three periods: (a) induction period, (b) acceleratory period, and (c) decaying period. These stages are the result, respectively, of (1) formation of nuclei of the metallic phase at localized sites on the oxide surface, (2) growth of these nuclei, and (3) overlap of the growing nuclei and a decrease in the metal–oxide interface. The length of the induction period is determined by the rate of formation of nuclei.

A rate equation has been developed by Avrami (1940) for the x–t relationship in the absence of pore diffusion:

$$\ln\left(\frac{1}{1-x}\right) = c't^N \qquad (19.129)$$

where c' and N are constants and N depends on the rate of formation of nuclei.

Equation 19.129 is a simplified version of a more detailed model proposed by Avrami. This equation has been found to correlate the data on the reduction of wustite (El-Rahaiby and Rao, 1979) with a value of 3 for the exponent N.

Since nucleation is essentially a solid-phase transformation that may or may not involve a gas phase it is considered in greater detail in Chapter 21.

19.5.3. Solid-Catalyzed Gas–Solid Reactions

Some important gas–solid reactions are catalyzed by solids, such as:

1. Hydrogasification of coal and coal char catalyzed by alkali metal or calcium compounds.
2. Formation of chlorosilanes (used in the manufacture of silicones) by reaction of methyl chloride and silicon catalyzed by metallic copper. Even though systematic modeling of this complex solid-catalyzed gas–solid system has not so far been attempted, Voorhoeve (1967) has suggested a plausible physical picture of the reac-

tion. The reaction is catalyzed by η-Cu_3Si, which is formed during the initial stages of the reaction. This initial solid–solid reaction was found to be autocatalytic by Tamhankar et al. (1981). Silicon from the η phase then reacts with methyl chloride and leaves the lattice as a gaseous product. The vacant lattice site is occupied by silicon diffusing from the silicon phase to the η phase. The silicon conversion is limited by the blocking of the silicon surface by the η phase, whereby the rate of solid-state diffusion of silicon to the copper-rich surface becomes the rate-limiting step.

Detailed mathematical analysis is not available for these systems. Lee (1979) proposed a model for such a system on the idealized assumption that the entire catalyst is present as an external layer surrounding the solid reactant. The reactant diffuses through the catalyst, undergoing a reaction, and then reacts with the solid reactant. The model is a simplified picture and more elaborate modeling along with systematic experiments is necessary.

Guzman and Wolf (1979) have also analyzed this problem on the more realistic assumption of a nonuniform distribution of catalyst within the pellet. Their computational results show that nonuniform catalyst distribution is detrimental to the solid conversion.

19.5.4. Reaction between Solids Proceeding through Gaseous Intermediates

Reactions between solids can be classified into two major groups: (a) true solid–solid reactions that take place in the solid state between two species in contact with each other, and (b) reactions between solid reactants that take place through gaseous intermediates. Both these aspects of solid–solid reactions are considered in Chapter 21, but a brief reference is made in what follows to reactions involving a gaseous intermediate.

A number of reactions, such as the reduction of metal oxides (MO) with carbon, fall under this category. The reaction can be represented as

$$MO + nC \rightleftharpoons M + nCO$$

Although this representation indicates that it is a solid–solid reaction, the actual mechanism involves a gaseous intermediate, namely, CO_2, and would be represented as

$$MO + nCO \rightarrow M + nCO_2$$
$$CO_2 + C \rightleftharpoons 2CO$$

A typical example is the reduction of iron ore (hematite) by the solid reductant coke (Rao, 1971, 1974; Sohn and Szekely, 1973):

$$Fe_gO_h(s) + CO(g) \rightarrow Fe_gO_{h-1}(s) + CO_2(g)$$
$$C(s) + CO_2(g) \rightarrow 2CO(g)$$

19.6. MODELING OF COMPLEX REACTIONS

Many reactions follow complex reaction schemes. No generalized approach seems possible at this stage, but the results of studies on a number of important reported systems are outlined below.

19.6.1. Consecutive Reactions

A consecutive gas–solid reaction is encountered in the reduction of hematite. Hence complex models are necessary to represent this problem, such as those of Spitzer et al. (1968) and Tsay et al. (1976), which account for three interfaces, namely, hematite–magnetite, magnetite–wustite, and wustite–iron.

19.6.2. Reaction of Two Gases

Reaction of two gases with the same solid is encountered in a number of situations.

General Analysis

A general formulation for reaction of two gases with a solid has been proposed by Wen and Wei (1971). The reaction scheme can be represented as

$$A_1(g) + v_1B(s) \rightarrow products$$
$$A_2(g) + v_2B(s) \rightarrow products$$

The system has been modeled using the SIM. Sohn and Braun (1980) have also developed a general model for such systems.

Examples

1. Reduction of iron ore with a mixture of CO and H_2:

This reduction has been modeled by Szekely et al. (1976). The reaction scheme is

$$CO(g) + FeO(s) \rightarrow CO_2(g) + Fe(s)$$
$$H_2(g) + FeO(s) \rightarrow H_2O(g) + Fe(s)$$

The multicomponent diffusional effects have been incorporated in the model. Croft (1979) has indicated that the effect of the water–gas shift reaction,

$$CO + H_2O \rightleftharpoons CO_2 + H_2$$

should also be included in the model.

2. Reaction of UO_2F_2 with a mixture of F_2 and BrF_5:

The reaction scheme for this case is

$$2F_2(g) + UO_2F_2(s) \rightarrow UF_6(g) + O_2(g)$$

$$4BrF_5(g) + 5UO_2F_2(s) \rightarrow 5UF_6(g) + 2Br_2(g) + 5O_2(g)$$

3. Reaction of CaO with a mixture of H_2S and CO:

This reaction is of importance in pollution control.

$$H_2S(g) + CaO(s) \rightarrow H_2O(g) + CaS(s)$$

$$CO(g) + H_2O(g) + CaO(s) \rightarrow H_2(g) + CaCO_3(s)$$

19.6.3. Reaction of Two Solids

Reaction of two components of the solid phase with the same gas occurs in a number of situations, for example, the chlorination of ilmenite and the sulfation of dolomite. Another interesting example is encountered in the burning of coke from catalyst particles. Here coke oxidation may be considered to be composed of two steps: oxidation of hydrogen, which occurs relatively rapidly in the initial time, leaving behind a solid of hydrogen-depleted coke, which gets oxidized subsequently. A model has been proposed for this process by Ramachandran et al. (1975). A schematic representation of the model is shown in Figure 19.12. This model predicts the maximum temperature rise in the pellet which is in good agreement with experimental data (see also Section 8.5.2).

19.6.4. Delayed Diffusion in Gas–Solid Reactions

Let us consider a simple gas–solid reaction represented schematically by

$$A(s) \rightarrow R(s) + S(g)$$

Normally in a gas–solid reaction of this type, either the reaction occurs homogeneously throughout the volume of the solid, or the shrinking-core model is followed. The intermediate case of a reaction-zone model is also possible. In certain cases, however, a high resistance to

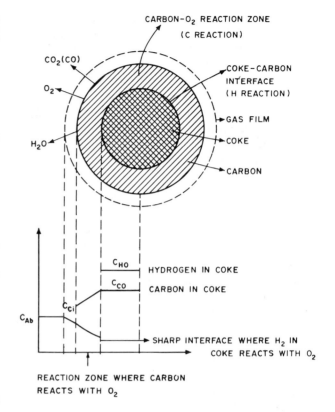

Figure 19.12. Model for simultaneous carbon and hydrogen reaction for coke oxidation (Ramachandran et al., 1975).

the diffusion of the evolved gas exists, and there is a buildup of the gas in the pellet before diffusion occurs. This is illustrated below for a reaction of industrial importance.

Example: Disproportionation of Potassium Benzoate to Terephthalate

The detailed mechanism of the formation of terephthalate has been shown by Ratusky and Sorm (1959) to involve the following steps:

STEP 1:

The negative ion formed in step 1 undergoes instantaneous reaction with 1 mole of potassium benzoate to give phenyl anion(II) and benzene.

Step 1a:

$$(-) \quad \text{COOK} \quad \text{COOK}$$

(reaction scheme: phenyl anion $(-)$ + benzoate $\rightarrow$ product $+$ benzene)

$(-)$

The anion further reacts with the CO_2 generated in step 1 to give the desired product of the reaction.

Step 2:

$$\text{COOK} + CO_2 + K \longrightarrow \text{COOK, COOK}$$

(reaction scheme: potassium benzoate + CO_2 + K $\rightarrow$ dicarboxylate product)

The decarboxylation reaction represented by step 1 is known to be reversible; hence, continuous removal of CO_2 from the reaction site would favor the forward step. Step 2 requires a definite concentration of CO_2 and phenyl anion(II).

The concept of delayed diffusion has been explained by Kulkarni and Doraiswamy (1980) by considering the decomposition of a spherical pellet of component A to give the gaseous product S. The concentration profile within the pellet has been obtained as

$$C_A = -\frac{2k_1 R^2}{\pi r D_{eA}} \sum_{n=1}^{\infty} (-1)^n D_{eA} \sin(n\pi r)$$

$$\left[\frac{\exp(-\alpha_1 t) - \exp(-D_{eA} n^2 \pi^2 t / R^2)}{D_{eA} n^2 \pi^2 - R^2 \alpha_1} \right] \quad (19.130)$$

where α_1 is a constant. It is seen from this equation that for higher resistance to the diffusion of the gas the concentration of the entrapped gas within the particle would be higher. In fact, in several of the decomposition reactions the diffusivity of the evolved gas can be of the order of $10^{-8} - 10^{-11}$ cm^2/sec. The resistance to this diffusion of the gas results in a buildup of pressure within the particle. The pressure, however, cannot build up indefinitely and finds relief either by increasing the porosity of the particle or even by breaking it. In the event, however, that the generated gas takes part in further reaction, the gas pressure builds up to a certain extent depending on its rate of generation and consumption due to reaction. An important point to note is that in a pressurized pellet the maximum pressure will be felt around the point of symmetry,

namely, the center of the pellet. The second reaction would therefore start preferentially at the center and move away toward the surface as the reaction proceeds. This simple argument leads us to an important conclusion: the delayed diffusion of the gas can result in a situation where an expanding reaction zone, with reaction starting at the center and moving outward, can be observed. Kulkarni and Doraiswamy (1980) have obtained the rate of advance of this zone as

$$\frac{dR_i}{dt} = \frac{1}{\rho_s} \left(\frac{k_1'}{k_2 C_g} - C_S^* \right) \left(1 + \phi_{R_i} \right) \frac{D_{eS}}{R}$$

$$\left\{ \frac{1 - \dfrac{\phi_{R_i} - 1}{\phi_{R_i} + 1} \dfrac{\phi_R + 1}{\phi_R - 1} \exp\left[2(\phi_{R_i} - \phi_R) \right]}{1 + \dfrac{\phi_R + 1}{\phi_R - 1} \exp[2(\phi_{R_i} - \phi_R)]} \right\} \quad (19.131)$$

where $k_1' = k_1 \exp(-\alpha_1 t)$ and k_1 and k_2 are the rate constants for steps 1 and 2, respectively. R_i represents the radius of the product phase and ϕ_{R_i} and ϕ_R the Thiele modulus based on R_i and R, respectively; C_S^* and ρ_S represent, respectively, the initial concentration and density of the newly formed product.

The results of Kulkarni and Doraiswamy (1980) may be tested against the experimental results of Gokhale et al. (1975). They carried out experiments with pellets of potassium benzoate prepared from (1) physically mixed charge, and (2) solution-mixed charge. With pellets from (1) a central zone of product that spread outward radially was clearly observed. In the pellets prepared from the solution-mixed charge, the reaction was found to occur over the entire cross section of the pellet. The outer crust of the pellet, however, remained unreacted, indicating that here again the same behavior is followed. To explain these observations Gokhale et al. (1975) postulated three regimes of operation based on a comparison of the rate of diffusion of CO_2 and the rate of its reaction. In the case of mixed pellets the reaction occurs in the regime where the rate of diffusion of CO_2 is much greater than the rate of its consumption, whereas in the case of pellets from the physically mixed charge the reaction occurs in the regime where the rate of diffusion of CO_2 is much less than the rate of its consumption.

This expanding-core behavior can also be observed in simple reactions with a H–W-type rate equation (Erk and Dudukovic, 1981), and has been termed the rotten apple model (since the apple rots from the center).

19.7. STABILITY OF GAS–SOLID REACTIONS

The stability of gas–solid reactions has been analyzed by a number of workers. The instability occurring in these systems may be classified into four categories:

1. Geometric instability;
2. Thermal instability due to metastable temperature;
3. Discontinuous transition in the rate-controlling step;
4. Gradual transition of the rate-controlling steps.

19.7.1. Geometric Instability

When the rate of reaction decreases as the interfacial area of reaction reduces, then the system is stable. In this case any irregularity on the surface disappears because the rate is less at the points of deeper penetration. If the reverse is true, then greater unevenness will result. This phenomenon is termed geometric instability and was first pointed out by Cannon and Denbigh (1957).

The condition for the existence of geometric instability is

$$\left(1 - \frac{1}{Sh}\right)\frac{R_i}{R} = 0.5 \tag{19.132}$$

For nonisothermal systems the criteria are, for exothermic reactions,

$$\left(1 - \frac{1}{Nu}\right)\frac{R_i}{R} > 0.5 \tag{19.133}$$

and for endothermic reactions

$$\left(1 - \frac{1}{Nu}\right)\frac{R_i}{R} < 0.5 \tag{19.134}$$

19.7.2. Thermal Instabilities

Instabilities 2, 3, and 4 are thermal in nature and occur only when the reaction is exothermic. In general, the interfacial temperature is given by the point(s) of intersection of the heat generation vs. heat removal curves, as already explained in Section 12.2. An illustrative plot of heat loss lines is shown as *OA* and *OB* in Figure 19.13, corresponding, respectively, to the kinetic and diffusion-controlled situations. The heat generation line is a sigmoid curve, and the intersection of

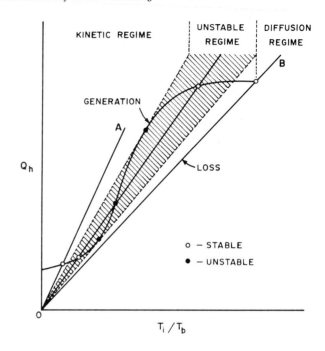

Figure 19.13. Thermal balance for exothermic irreversible gas–solid reactions.

these curves gives the reaction surface temperature. The dotted lines represent the situations within which multiple steady states exist.

Because of the dependence of the rate of reaction on time, the shape of the heat generation curve and slope of the heat loss curve will change for a given system with time, even when the surrounding conditions are kept constant (contrast this with gas–solid catalytic systems, which attain a true steady state). This leads to sudden transitions of the rate-controlling step, as recognized by Shen and Smith (1963) and Aris (1967).

Beveridge and Goldie (1968) derived the following criterion for the absence of thermal instabilities, based on the assumption of a psuedo-steady state and using the sharp-interface model:

$$\frac{E}{R_g T_b} < 4 \tag{19.135}$$

Some aspects of the stability of gas–solid reactions, again based on the pseudo-steady-state assumption, but using the particle–pellet model, have been considered by Calvelo and Smith (1970). A transient analysis of the problem has been attempted by Wen and Wang (1970).

19.7.3. Oscillatory Behavior

Oscillatory instability is often the result of some complex reactions occurring in the solid, and can be identified by oscillations in the weight loss vs. time data. An example is the reaction of ZnO with H_2S at high temperatures (Gibson and Harrison, 1980), in which the oscillations are due to the decomposition of ZnO at high temperatures. The Zn vapor produced in this decomposition diffuses outward and blocks the pores of the catalyst. Another example of oscillatory behavior is the oxidation of CaS (Lynch and Elliott, 1980), probably due to the formation of $CaSO_4$ in the pores and its subsequent decomposition.

19.8. MODEL EVALUATION

It is clear from the analysis presented in the previous sections that the structure of the solid, effective intraparticle diffusivity, and sintering are important considerations in the modeling of a gas–solid reaction. A complete characterization of the solid during the various stages of reaction is necessary to provide the required data for using some of the models just described. The pore size and porosity distribution of the solid are among the parameters that must be measured. Effective intraparticle diffusivity is an equally important parameter. This can often be estimated or measured under nonreacting conditions by a number of techniques (see Chapters 3 and 9). In systems characterized by sintering, diffusivity measurements of partly and fully sintered pellets are necessary.

The most important aspect of gas–solid reactions is prediction of the conversion–time behavior. For this purpose a knowledge of the order of reaction, both with respect to the gas and solid phases, is necessary. The latter is difficult to obtain, and may have to be arrived at by curve-fitting the experimental data.

Electron probe microanalysis (EPMA) is being increasingly used for the purpose of model discrimination and parameter estimation in gas–solid systems. EPMA provides a detailed concentration profile in the solid for various degrees of conversion. This information, together with independent measurements of porosity and effective diffusivity, can be effectively used in

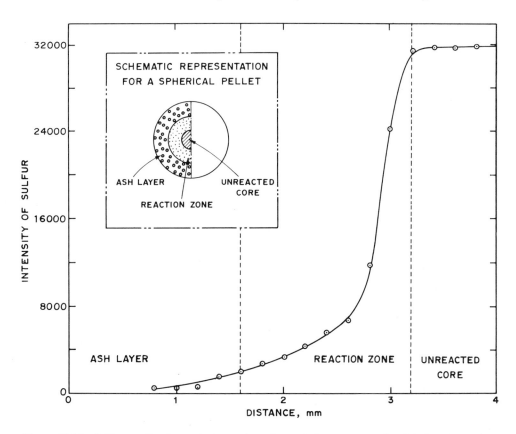

Figure 19.14. Sulfur profile from electronprobe microanalysis measurements: sintering temperature = 1200°C; $x_B = 0.72$ (Prasannan and Doraiswamy, 1982).

the precise modeling of gas–solid systems. Prasannan and Doraiswamy (1982) have discussed this technique and used it to obtain solid profiles in pellets of zinc sulfide at various stages of oxidation. Gibson and Harrison (1980) and Hartman and Coughlin (1976) have also used this technique.

Of the various models described in this chapter, the sharp interface model is often found to fit the experimental data for nonporous pellets over a limited range of operating variables, but is usually not satisfactory when extrapolated over a wide range of parameter values. The homogeneous model, normally applicable to porous solids, suffers from the disadvantage that it accounts for structural changes in a purely empirical way. The three-stage model of Mantri et al. (1976) appears to be quite attractive, particularly since the evolution of the three stages has been experimentally verified (Prasannan and Doraiswamy, 1982). A typical stage in the model for the oxidation of zinc sulfide is shown in Figure 19.14. The particle–pellet models are mathematically elegant but fail to carry conviction when the pellet is not formed by compaction; in such a case a fictitious grain size (and shape) will have to be assumed. The single-pore models are mathematically simple, but the assumption of uniform pore size renders them suspect. The distributed-pore models and the random-pore models incorporate the structural features of the solid in greater detail, but the data inputs for using these models can only be obtained by sophisticated experimentation.

In conclusion it would be seen that no single model can be recommended as the most useful or probable one. The adequacy of a model determined from a limited range of parameter values can be misleading particularly in the absence of data on the solid structure. Indeed the structure of the solid itself should provide the basic guidelines for eliminating a number of models and restricting the choice to a few.

REFERENCES

Anthony, D. B., and Howard, J. B. (1976). *AIChE J.*, **22**, 625.

Aris, R. (1967), *Ind. Eng. Chem. Fundam.*, **6**, 319.

Ausman, J. M., and Watson, C. C. (1962), *Chem. Eng. Sci.*, **17**, 323.

Avrami, M. (1940), *J. Chem. Phys.*, **8**, 212.

Beveridge, G. S. G., and Goldie, P. J. (1968), *Chem. Eng. Sci.*, **23**, 913.

Bhatia, S. K., and Perlmutter, D. D. (1980), *AIChE J.*, **26**, 379.

Bhatia, S. K., and Perlmutter, D. D. (1981a), *AIChE J.*, **27**, 247.

Bhatia, S. K., and Perlmutter, D. D., (1981b), *AIChE J.*, **27**, 226.

Bischoff, K. B. (1963), *Chem. Eng. Sci.*, **18**, 711.

Bowen, J. H., and Cheng, C. K. (1969), *Chem. Eng. Sci.*, **24**, 1829.

Calvelo, A., and Cunningham, R. E. (1970), *J. Catal*, **17**, 1.

Calvelo, A., and Smith, J. M. (1970), *CHEMECA '70 (Proc.) Conf.*, No. 3, 1.

Cannon, K. J., and Denbigh, K. G. (1957), *Chem. Eng. Sci.*, **6**, 145.

Carter, R. E. (1961), *J. Chem. Phys.*, **34**, 2010.

Chan, S. F., and Smith J. M. (1976), *Ind. Chem. Eng.*, **18**, 42.

Christman, P. G., and Edgar, T. F. (1980), Paper presented at 78th annual AIChE meeting, Chicago.

Chrostowski, J. W., and Georgakis, C. (1978), *ACS Symp. Ser.*, **65**, 225.

Corella, J. (1980), *Chem. Eng. Sci.*, **35**, 25.

Croft, V. (1979), *Met. Trans.*, **10B**, 121.

de Bruijn, T. J. W., Soeravidjaya, T. H., Jong, W. A., and Vandenberg, P. J. (1980), *Chem. Eng. Sci.*, **35**, 1591.

del Borghi, M., Dunn, J. C., and Bischoff, K. B. (1976), *Chem. Eng. Sci.*, **31**, 1065.

Doraiswamy, L. K., Bijawat, H. C., and Kunte, M. V. (1959), *Chem. Eng. Prog.*, **55**, 80.

Dudukovic, M. P., and Lamba, H. S. (1978a), *Chem. Eng. Sci.*, **33**, 303.
    Dudukovic, M. P., and Lamba, H. S. (1978b), *Chem. Eng. Sci.*, **33**, 471.

Dutta, S., and Wen, C. Y. (1977), *Ind. Eng. Chem. Pracess Design Dev.*, **16**, 31.

El-Rahaiby, S. K., and Rao, Y. K. (1979), *Met. Trans.*, **10B**, 257

Evans, J. W., Szekely, J., Ray, W. H., and Chaung, Y. K. (1973), *Chem. Eng. Sci.*, **28**, 683.

Erk, H. F., and Dudukovic, M. P. (1981), Paper presented at the 79th annual AIChE meeting, Chicago.

Fan, L. S., Miyanami, K., and Fan L. T. (1977), *Chem. Eng. J.*, **13**, 13.

Gavalas, G. R. (1980), *AIChE J.*, **26**, 577.

Gavalas, G. R., and Wilks, K. A. (1980), *AIChE J.*, **26**, 201.

Georgakis, C., Chang, C. W., and Szekely, J. (1979), *Chem. Eng. Sci.* **34**, 1072.

Gibson, J. B., and Harrison, D. P. (1980), *Ind. Eng. Chem. Process Design Dev.*, **19**, 231.

Gidaspow, D., Dharia, D., and Leung, L. (1976), *Chem. Eng. Sci.*, **31**, 337.

Gokarn, A. N., and Doraiswamy, L. K. (1971), *Chem. Eng. Sci.*, **26**, 1521.

Gokarn, A. N., and Doraiswamy, L. K. (1973), *Chem. Eng. Sci.*, **28**, 401.

Gokhale, M. V., Naik, A. T., and Doraiswamy, L. K. (1975), *Chem. Eng. Sci.*, **30**, 1409.

Guzman, G. L., and Wolf, E. E. (1979), *Ind. Eng. Chem. Fundam.*, **18**, 7

Hartman, M., and Coughlin, R. W. (1976), *AIChE J.*, **22**, 490.

Hashimoto, K., and Silveston, P. L. (1973a), *AIChE J.*, **19**, 259.

Hashimoto, K., and Silveston, P. L. (1973b), *AIChE J.*, **19**, 368.

Hashimoto, K., Miura, K., Yoshikawa, F., and Imai, I. (1979), *Ind. Eng. Chem. Process Des. Dev.*, **18**, 73.

Ishida, M., and Wen, C. Y. (1968), *AIChE J.*, **14**, 311.

Kawasaki, E., Sanscranite, J., and Walsh, T. J. (1962), *AIChE J.*, **8**, 48.

Kim, K. K., and Smith, J. M. (1974), *AIChE J.*, **20**, 670.

Klinger, N., Strauss, E. L., and Komarek, K. L. (1966), *J. Am. Ceram. Soc.*, **49**, 369.

Kulkarni, B. D., and Doraiswamy, L. K. (1980), *Chem. Eng. Sci.*, **35**, 817.

Lee, H. H. (1979), *Chem. Eng. Sci.*, **34**, 5.

Lee, H. H. (1980), *Ind. Eng. Chem. Process Design Dev.*, **19**, 237.

Levenspiel, O. (1974), *Chemical Reaction Engineering*, 2nd ed., Wiley, New York.

Lewis, P. F., and Simons, G. A. (1979), *Combust. Sci. Technol*, **20**, 117.

Luss, D. (1968), *Can. J. Chem. Eng.*, **46**, 154.

Luss, D., and Amundson, N. R. (1969), *AIChE J.*, **15**, 194.

Lynch, D. C., and Elliott, J. F. (1980), *Met. Trans.*, **11B**, 415.

Mantri, V. B., Gokarn, A. N., and Doraiswamy, L. K. (1976), *Chem. Eng. Sci.*, **31**, 779.

Mahajan, O. P., Yarzal, R., and Walker, P. L., Jr. (1976), *Chem. Eng. Sci.*, **31**, 779.

Mu, J., and Perlmutter, D. D. (1980), *Chem. Eng. Sci.*, **35**, 1645.

Narsimhan, G. (1961), *Chem. Eng. Sci.*, **16**, 7.

Nicholson, D. (1965), *Trans. Faraday Soc.*, **61**, 990.

Park, J. Y., and Levenspiel, O. (1975), *Chem. Eng. Sci.*, **32**, 233.

Petersen, E. E. (1957), *AIChE J.*, **3**, 443.

Prasannan, P. C., and Doraiswamy, L. K., (1982), *Chem. Eng. Sci.*, **36**, 925.

Prasannan, P. C., Ramachandran, P. A., and Doraiswamy, L. K. (1982), to be published.

Ramachandran, P. A., and Doraiswamy, L. K. (1982a), *AIChE J.*, **28**, 881.

Ramachandran, P. A., and Doraiswamy, L. K. (1982b), *AIChE J.* (in press).

Ramachandran, P. A., and Kulkarni, B. D. (1980), *Ind. Eng. Chem. Process Design Dev.*, **19**, 717.

Ramachandran, P. A., and Smith, J. M. (1977a), *AIChE J.*, **23**, 353.

Ramachandran, P. A., and Smith, J. M. (1977b), *Chem. Eng. J.*, **14**, 137.

Ramachandran, P. A., Rashid, M. H., and Hughes, R. (1975), *Chem. Eng. Sci.*, **30**, 1391.

Ranade, P. V., and Harrison, D. P. (1979), *Chem. Eng. Sci.*, **34**, 427.

Ranade, P. V., and Harrison, D. P. (1981), *Chem. Eng. Sci.*, **36**, 1079.

Ranz, W. E., and Marshall, N. R. (1952), *Chem. Eng. Prog.*, **48**, 141.

Rao, Y. K. (1971), *Met. Trans.*, **2**, 1439.

Rao, Y. K. (1974), *Chem. Eng. Sci.*, **29**, 1435.

Rao, Y. K., El-Rahaiby, S. K., and Al-Kahatany, M. M. (1979), *Met. Trans.*, **10B**, 295.

Ratusky, J., and Sorm, F. (1959), *Collect. Czech. Chem. Commun.*, **24**, 2553.

Rehmat, A., and Saxena, S. C. (1977), *Ind. Eng. Chem. Process Design Dev.*, **16**, 502.

Sampath, B. S., Ramachandran, P. A., and Hughes, R. (1975), *Chem. Eng. Sci.*, **30**, 125.

Shen, J., and Smith, J. M. (1963), *Ind. Eng. Chem. Fundam.*, **4**, 293.

Shettigar, U. R., and Hughes, R. (1972), *Chem. Eng. J.*, **3**, 93.

Simons, G. A. (1979), *Combust. Sci. Technol.*, **19**, 227.

Simons, G. A., and Finson, M. L. (1979), *Combust Sci. Technol.*, **19**, 217.

Simons, G. A., and Rawlins, W. T. (1980), *Ind. Eng. Chem. Process Design Dev.*, **19**, 565.

Sohn, H. Y. (1978), *Met. Trans.*, **9B**, 89.

Sohn, H. Y., and Braun, R. L. (1980), *Chem. Eng. Sci.*, **35**, 1625.

Sohn, H. Y., and Szekely, J. (1972a), *Chem. Eng. Sci.*, **27**, 763.

Sohn, H. Y., and Szekely, J. (1972b), *Can. J. Chem. Eng.*, **50**, 674.

Sohn, H. Y., and Szekely, J. (1973), *Chem. Eng. Sci.*, **28**, 1169.

Spitzer, R. H., Manning, F. S., and Philbrook, W. O. (1966), *Trans. Met. Soc. AIME*, **236**, 726.

Spitzer, R. H., Manning, F. S., and Philbrook, W. O. (1968), *Trans. TMS AIME.*, **242**, 618.

Srinivas, B., and Amundson, N. R. (1980a), *Can. J. Chem. Eng.*, **58**, 476.

Srinivas, B., and Amundson, N. R. (1980b), *AIChE J.*, **26**, 487.

Szekely, J., Evans, J. W., and Sohn, H. Y. (1976), *Gas–Solid Reactions*, Academic Press, New York.

Tamhankar, S. S., Gokarn, A. N. and Doraiswamy, L. K. (1981), *Chem. Eng. Sci.*, **36**, 1365.

Tsay, Q. T., Ray, W. H., and Szekely, J., (1976), *AIChE J.*, **22**, 1064.

Ulrichson, D. L., and Mahoney, D. J. (1980), *Chem. Eng. Sci.*, **35**, 567

Tudose, R. Z. (1970), *Bul. Inst. Politeh. Iasi*, **16**, (20), 241.

Voorhoeve, R. J. H. (1967), *Organohalosilanes, Precursors to Silicones*, Elsevier, Amsterdam.

Wakao, N., and Smith, J. M. (1962), *Chem. Eng. Sci.*, **17**, 825.

Wang, S. (1971), Ph.D. thesis, West Virginia University.

Weisz, P. B., and Goodwin, R. D. (1963), *J. Catal.*, **2**, 397.

Wen, C. Y. (1968), *Ind. Eng. Chem.*, **60**, 34.

Wen, C. Y., and Wang, S. C. (1970), *Ind. Eng. Chem.*, **62**(8), 30.

Wen, C. Y., and Wei, L. Y. (1971), *AIChE J.*, **17**, 272.

CHAPTER TWENTY

Design of Noncatalytic Gas–Solid Reactors

20.1. MODES OF CONTACT IN GAS–SOLID SYSTEMS

Modeling of gas–solid reactions was considered in Chapter 19. Equally important from the design point of view is the selection of the mode of contact between solids and gas, as well as the method of supplying the required thermal energy to the reactor. The design of industrial reactors for gas–solid systems has been based on a number of contacting modes. This may be exemplified by the variety of designs available for the calcination of limestone. The types of reactors used include the rotary kiln, flat hearth, fluidized bed, pneumatic conveyor, and moving bed. Another noteworthy example is the reduction of iron ore. The blast furnance is usually associated with iron ore reduction, but it represents only one of the modes of contact, namely, a vertical moving bed. At least three other modes of contact have been used: the fixed bed, fluidized bed, and rotary cylinder.

This chapter is concerned with the principles of reactor design for three major modes of contact, namely, the fixed bed, moving bed, and fluidized bed, using any of the models in Chapter 19 for describing the gas–solid reaction within the particle. Although these three modes are perhaps the most important, and have received much attention from both the theoretical and practical points of view, several other modes of contact have also been employed, including the horizontal or inclined moving bed, the pneumatic conveyor, the rotating cylinder, and the flat hearth. Kunii (1980) has summarized the use of various modes of contact with respect to six major endothermic gas–solid reaction systems, some of which are still in the developmental stage on a pilot plant scale.

20.2. FIXED-BED REACTORS

In modeling fixed-bed noncatalytic reactors the continuity equation for the gas phase (external field equation) has to be coupled with the equation for the reaction of single particles. The problem is basically of a transient nature, since the rate of reaction decreases with time owing to the consumption of the solid reactant B. The major contributions in the modeling of fixed-bed noncatalytic systems have been made by Moriyama (1971), Evans and Song (1974), Sampath et al. (1975), and Ranade and Evans (1980). A number of related experimental and theoretical studies are summarized in Table 20.1.

$$v_A A + v_B B \rightarrow v_R R + v_S S$$

For an isothermal system the model equation for the reaction can be formulated as follows:

$$f_{\text{bed}} \frac{\partial C_{\text{Ab}}}{\partial t} - u \frac{\partial C_{\text{Ab}}}{\partial l} = r_V \qquad (20.1)$$

where r_V is the rate of reaction of A per unit volume of the reactor.* The rate r_V can be related to the conversion of B by the equation

$$r_V = v(1 - f_{\text{bed}}) C_{\text{BO}} \frac{dx_B}{dt} \qquad (20.2)$$

where dx_B/dt is the rate of change of conversion of B for a pellet exposed to a bulk gas-phase concentration of C_{Ab}. The transient term $f_{\text{bed}} \partial C_{\text{Ab}}/\partial t$ is usually negligible compared to other terms in Eq. 20.1; hence this equation can be expressed as

$$-u \frac{dC_{\text{Ab}}}{dl} = v(1 - f_{\text{bed}}) C_{\text{BO}} \frac{dx_B}{dt} \qquad (20.3)$$

For a reaction that is first order with respect to A, the

---

* In this chapter the notation f_{bed} denotes bed voidage and f_B denotes the voidage of solid B.

TABLE 20.1. Some Examples of Studies of Gas–Solid Reactions in Fixed-Bed Reactors

System	Remarks	Reference
Iron oxide reduction by hydrogen	Isothermal fixed-bed model for kinetic control developed and verified experimentally	Barner et al. (1963)
Iron oxide reduction with CO	Isothermal mixed control model involving one-step reduction developed and experimentally verified	Osman et al. (1966)
Decomposition of calcium carbonate	Heat and mass transfer control model data for single particles applied to predict performance of fixed-bed reactor	Hill (1968)
Sorption of moisture by activated carbon	Solution obtained for predicting packed-bed behavior based on single-pellet data	Bowen and Lacey (1969)
Oxidation of pyrite	Thermal stability criterion for highly exothermic reaction in a packed bed discussed	Agnew and Narsimhan (1970)
Sulfidation of ilmenite	Preferential sulfidization of ilmenite in a packed bed by H_2S studied at various temperatures; the merits of the process for commercial utilization examined	Jain et al. (1970)
Oxidation of zinc sulfide	Simulation of packed-bed roasting of ZnS carried out by using a pellet model based on a particle effectiveness factor representing the transport rates within the pellet; the validity of the fixed-bed model in moving-bed systems and general industrial reactor applications considered and the ZnS sinter strand roasting application extensively examined	Beveridge and Kawamura (1974)
Regeneration of coked catalyst	Particle–pellet model used to simulate a fixed-bed reactor for coke regeneration	Sampath et al. (1975) (see also Mariovet and Wajc, 1977)
Sorption of SO_2 by copper oxide in presence of O_2 (air pollution control)	A homogeneous dispersion model applied to the fixed-bed reactor to obtain kinetic parameters	Yates and Best (1976) (see also Bourgeois et al. 1974)
Oxidation of pyrites	Shrinking-core model with inter- and intraparticle mass transfer control developed and used to design large plants with large heat effects	Ciliberti and Lancaster (1977)
Reduction of hematite with hydrogen	Kinetic studies on the reduction of hematite pellets in a fixed bed carried out; a modified form of SIM for single-particle reduction combined with the equations of fluid flow and continuity to describe the fixed-bed reduction	Ohmi et al. (1978)
Fluorination of $CaCO_3$ (liquid–solid reaction)	A process investigated in which fluoride in waste water can be reduced by limestone in a fixed-bed reactor; equilibrium concentrations of flouride and the conversion rate investigated experimentally; the macrokinetics described by the SIM	Simonsson (1979)
Reduction of molybdenite	The kinetics of reduction by H_2 in static-bed conversion of MoO_3 powder to MoO_2 investigated; several simple models including the SIM employed to characterize the reduction kinetics but none found to account for all the observed parameter dependence	Orehotsky and Kaczenski (1979)

term dx_B/dt can be expressed as

$$\frac{dx_B}{dt} = C_{Ab}\breve{\beta}(x_B) \qquad (20.4)$$

where $\breve{\beta}(x_B)$ is a function of x_B that depends on the type of model used to describe the reaction and the model parameters and has the units $cm^3/mol\ sec$. To illustrate this point let us assume that the sharp-interface model is applicable. Then from Eq. 19.8, $\breve{\beta}(x_B)$ takes on the form

$$\breve{\beta}(x_B) = \frac{\nu M_B}{\rho_B}\left\{\frac{R}{3k_g} + \frac{R^2}{3D_{eS}}\left[(1-x_B)^{-1/3}-1\right]\right.$$
$$\left. + \frac{R}{3k_S}(1-x_B)^{-2/3}\right\}^{-1} \qquad (20.5)$$

Substituting the relation for dx_B/dt given by Eq. 20.4 in Eq. 20.3 and integrating, we obtain

$$\int_{C_{Ab}}^{C_{A0}}\frac{dC_{Ab}}{C_{Ab}} = \frac{\nu(1-f_{bed})C_{B0}}{u}\int_0^l \breve{\beta}(x_B)dl$$

or

$$C_{Ab} = \frac{\nu(1-f_{bed})C_{B0}}{u}C_{A0}\exp\left[-\int_0^l\breve{\beta}(x_B)dl\right] \qquad (20.6)$$

Substituting in Eq. 20.4 we have

$$\frac{dx_B}{dt} = \frac{\nu(1-f_{bed})C_{B0}}{u}C_{A0}\breve{\beta}(x_B)\exp\left[-\int_0^l\breve{\beta}(x_B)dl\right] \qquad (20.7)$$

The conversion–time profiles at various positions in the reactor can then be obtained by a combination of Simpson's rule and Runge–Kutta integrations. Evans and Song (1974) have applied this procedure to simulate packed-bed systems on the basis of the particle–pellet model. The governing equations can be put in dimensionless form:

$$-\frac{d\hat{C}_{Ab}}{dl^*} = \frac{dx_B}{dt^*} = \hat{C}_{Ab}\hat{\beta} \qquad (20.8)$$

where

$$l^* = \frac{k_S C_{A0}}{r_{G0}u\rho_{MA}}(1-f_{bed})l \qquad (20.9)$$

$$t^* = \frac{\nu k_S C_{A0}M_B t}{\rho_B r_{G0}} \qquad (20.10)$$

and $\hat{\beta}$ is the dimensionless version of the parameter $\breve{\beta}$ and ρ_{MA} is the molal density of the gas.

For simplifying the computation, the approximate relation for conversion vs. time (Eq. 19.63) can be used. Then $\hat{\beta}$ can be obtained by neglecting gas–solid mass transfer:

$$\hat{\beta} = \frac{1}{3}(1-x_B)^{-2/3} + \frac{\phi^2}{18}\left[2(1-x_B)^{-1/3}-2\right]^{-1} \qquad (20.11)$$

Further, tables of $\hat{\beta}$ vs. x_B have been prepared (Evans and Song, 1974) for various values of ϕ to simplify the numerical calculations. These tables may be used for evaluating the exponential term in Eq. 20.6. For low values of ϕ the conversion vs. distance plots exhibit a sigmoid shape. These profiles travel down the bed with increase in dimensionless time. The reaction zone is confined to this sigmoidal portion. For l^* lying outside this portion of the curve, the solids are either completely converted or unconverted. For $\phi > 3$ the reaction zone extends throughout the bed.

This model has been successfully applied to the reduction of Fe_2O_3 in a packed bed by Ranade and Evans (1980). For this simulation the various parameters were obtained independently by single-particle studies. The packed-bed simulation was then compared with experiments and was found to be satisfactory.

Moriyama (1971) has analyzed a general formulation of isothermal reactions and has proposed integral equations for the concentration of reactant and conversion of solid. The mathematical analysis was based on an earlier treatment by Bischoff (1969) for the problem of catalyst fouling (see Chapter 16). It may be noted here that the problems of catalyst deactivation and gas–solid reaction are both transient in nature and there is considerable similarity in the model equations. The work of Moriyama (1971) is mainly useful in the absence of intraparticle gradients.

A detailed model for nonisothermal systems has been proposed by Sampath et al. (1975). The particle–pellet or grain model was used to describe the solid behavior, while the bed behavior was simulated by a unidirectional axial dispersion model. The model equations in dimensionless variables are as follows:

Bulk fluid equations:

$$\frac{1}{Pe_m}\frac{L}{d_t}\frac{d^2\hat{C}_{Ab}}{dz^2} - \frac{d\hat{C}_{Ab}}{dz} - \frac{k_g aL}{f_{bed}u}(\hat{C}_{Ab}-\hat{C}_{As}) = \frac{d\hat{C}_{Ab}}{dt^*} \qquad (20.12)$$

$$\frac{1}{\text{Pe}_h} \frac{L}{d_t} \frac{d^2\hat{T}}{dz^2} - \frac{d\hat{T}}{dz} - \frac{h_{fp}aL}{f_{bed}u\rho_G C_p}(\hat{T} - \hat{T}_s)$$

$$+ W_F(\hat{T}_w - \hat{T}) = \frac{d\hat{T}}{dt^*} \qquad (20.13)$$

where z is the dimensionless axial distance l/L; Pe_m and Pe_h are the Peclet numbers for mass and heat transfer, respectively; W_F is a dimensionless wall heat transfer coefficient, defined as $h_w L/d_t f_{bed} u\rho_G C_p$; and $\hat{T}_w$ is the dimensionless wall temperature.

The bulk fluid equations are coupled with the pellet equations through the terms $\hat{C}_{As}$ and $\hat{T}_s$. These terms, as determined by the differential equations for concentration and temperature profiles within the pellet, are discussed in Chapter 19 for various models.

The simultaneous solution of the fluid and particle equations requires considerable computational effort for a nonisothermal system. The difficulty can be overcome to some extent by use of orthogonal collocation. The problem then reduces to one of solving a set of first-order differential equations of the initial value type, which can be more easily solved than the original set. Standard methods are available for this purpose.

Sampath et al. (1975) applied this model to the regeneration of coked catalyst in a fixed bed. The maximum temperature that could be attained in the reactor was predicted and a number of operating strategies to keep this temperature within permissible limits were suggested as a consequence of the model simulations.

20.3. MOVING-BED REACTORS

A number of contacting arrangements can be envisaged for a moving-bed system:

1. Countercurrent flow of gas and solids, as in a shaft furnace.
2. Sinter-bed operation with crossflow of gas and solids.
3. Continuous fluid-bed operation.

In operations 1 and 2 the flow pattern of the solids is closer to plug flow, whereas in a continuous fluid bed the solids may be assumed to be completely mixed. In each case, the conversion of the gas may be small in the reactor, so that all solids see the same concentration, which approximately corresponds to the inlet gas concentration. Alternatively there may be significant conversion of the gas in the reactor, and solids will see

different gas concentrations as they traverse the bed. These two cases can be treated separately for the purpose of modeling. A further complication may be that there would be a size distribution of solids in the feed to the reactor (see Section 20.4.1). This will result in a conversion distribution in the effluent solids as small particles get converted to a greater extent than large particles for the same residence time.

The modeling of moving-bed systems for the case of constant gas-phase concentration is straightforward. The conversion of the solids leaving the reactor is the same as that for a single particle exposed to gas for a time equal to the residence time of the solids in the reactor.

For the case of varying gas-phase concentration, the model equations for a moving bed can be formulated as follows. For an isothermal system the material balance equation can be formulated as

$$-u\frac{\partial C_{Ab}}{\partial l} = \frac{F_{Ms}}{v} \frac{dx_B}{dl} \qquad (20.14)$$

where F_{Ms} is the molal flow rate of solids per unit area of the reactor. The boundary conditions are

$$l = 0, \qquad C_A = C_{A0}$$
$$l = 0, \qquad x_B = 0 \qquad \text{for cocurrent flow} \qquad (20.15a)$$

or

$$l = L, \qquad x_B = 0 \quad \text{for countercurrent flow} \qquad (20.15b)$$

The integrated form of Eq. 20.14 can be written as

$$C_{Ab} - C_{A0} = -\frac{F_{Ms}}{vu}x_B \qquad \text{for cocurrent flow} \qquad (20.16)$$

and

$$C_{Ab} - C_{A0} = -\frac{F_{Ms}}{vu}(x_B - x_{BL}) \quad \text{for countercurrent flow} \qquad (20.17)$$

where x_{BL} is the solids conversion at the reactor exit.

The value dx_B/dl can also be related to the extent of conversion in the particle:

$$\frac{dx_B}{dl} = \frac{C_{Ab}}{F_{Ms}}\breve{\beta}(x_B) \qquad (20.18)$$

The quantity C_{Ab} is now a function of x_B because, as the particle travels down the bed, it experiences different

TABLE 20.2. Some Examples of Studies of Gas–Solid Reactions in Moving-Bed Reactors

System	Remarks	Reference
1. Iron oxide reduction by hydrogen	Countercurrent isothermal system; theoretical model verified experimentally	Spitzer et al. (1968)
	Effect of gas and solids maldistribution on reactor performance evaluated for a countercurrent system	Yagi and Szekely (1979)
2. Gas–solid reactions in general	Steady-state behavior of countercurrent reactor discussed	Schaefer et al. (1974) (see also Ishida and Wen, 1971)
3. Reduction of hematite with H_2 and CO	Model to determine optimum operation conditions for countercurrent reactor developed and verified experimentally	Tsay et al (1976)
4. Coal gasification by CO_2 and steam	Steady-state reaction model developed and compared with performance of pilot plant and commercial units like Lurgi reactors	Yoon et al. (1978)
5. Coal gasification	Transient response in dry-ash Lurgi reactor computed for both oxygen and air operation with low-activity Illinois and high-activity Wyoming coals	Yoon et al. (1979)

concentrations of gas A. The functional relationship of C_{Ab} vs. x_B is given by Eqs. 20.16 and 20.17. Substituting this relation in 20.18 and integrating the resulting equation, we can obtain the conversion in the moving bed at position $l = L$. For the case of countercurrent flow the integration has to be coupled with a trial-and-error procedure because the quantity x_{BL} in Eq. 20.17 is not known at the start of the calculations. Illustrative results for the conversion as a function of various operating parameters in a moving bed have been presented by Evans and Song (1974).

A mathematical model of the moving-bed reactor for direct reduction of iron ore has been proposed by Tsay et al. (1976) based on countercurrent flow. The reduction of hematite with a mixture of CO and H_2 was used in this work. The results show that an optimum inlet-gas composition exists for a certain range of operating variables. A number of related studies on moving-bed reactors have been reported and are summarized in Table 20.2.

20.4. FLUIDIZED-BED REACTORS

20.4.1. General Considerations

The factors that have to be considered in the overall strategy for the simulation of a fluidized-bed reactor for a noncatalytic gas–solid reaction are as follows:

1. For catalytic reactions, gas mixing and mass transfer between the bubble and emulsion phases play an important part in determining the conversion of reactant gas. For noncatalytic reactions, however, solids mixing has also to be considered in order to obtain a model for the fluid-bed reactor.

2. The particles entering the reactor may have a wide range of sizes in the general case. Here solids are involved in the reaction, and the size of particles has a bearing on the residence time of solids in the reactor. In catalytic reactions, size distribution of the catalyst has an effect on the fluidizing characteristics but not directly on the extent of the reaction occurring, as in the case of noncatalytic particles.

3. During reaction the particles may grow, shrink, or remain unchanged in size. The thermal cracking of crude oil in a bed of carbon is one of the few examples of particle growth during reaction. The combustion or gasification of carbonaceous materials is an example of shrinkage in particle size during reaction. The roasting of sulfide ores and reduction of iron ore are examples of reactions in which particle size remains unchanged. In describing fluidized non-catalytic systems two important phenomena should be taken into consideration: solids attrition, which leads to particle shrinkage,

and solids self-agglomeration, which may be considered a special case of particle growth. Unfortunately the information on such phenomena is so limited that it cannot be incorporated with confidence in the mathematical description of the bed.

4. The right model has to be chosen to represent the kinetics for a single particle. A variety of models have been developed for a single particle and these have been summarized in Chapter 19. The entering solids may have a size distribution, and as a consequence gaseous diffusion may have an effect in the case of larger particles and no effect in the case of smaller particles.

Once a kinetic model for the conversion of a particle has been obtained, the strategy for fluid-bed reactor design revolves around applying a suitable model for a non-catalytic fluid bed, which would incorporate the phenomena of elutriation and varying size distribution. In the simpler models, gas-phase concentration is assumed constant and the solid phase alone is considered. In the more complex models variation in gas-phase concentration is also included. As in the case of catalytic reactions these models are based essentially on the two-phase concept of the fluidized-bed. Thus the Kunii–Levenspiel model (1968a, 1968b) has been extended to noncatalytic reactions by the same authors (1969), and the bubble-assemblage model of Kato and Wen (1969) has been extended by Yoshida and Wen (1970, 1971). Horio and Wen (1975) developed a mathematical model of a fluidized-bed combustor based on the modified bubble-assemblage model of Mori and Wen (1975). The two-phase models for catalytic fluid beds are applicable to catalytic reactions in fluid beds in general, whereas for noncatalytic reactions the models proposed usually simulate only specific reactions in a fluid bed. Table 20.3 summarizes some of the more recent noncatalytic fluid-bed reactor models that have been proposed and the systems to which they have been applied.

20.4.2. Models Based on Constant Gas-Phase Concentration

In this section we analyze the case where the gas-phase concentration is constant throughout the reactor. This will be valid for situations where a large excess of reactant gas is available in the system. All the particles are then exposed to the same concentration C_{Ab} and the analysis will have to be confined mainly to the effect of solids mixing on conversion. Also, it is assumed that all

the fines are returned to the reactor so that there is no carryover and all particles leave along with the exit steam.

The mixing pattern of the solids can be assumed to be close to backmixing. Thus there is a spread in the residence time distribution of the solids that can be expressed as:

$$E(t) = \frac{1}{\bar{t}} \exp\left(-\frac{t}{\bar{t}} \right) \qquad (20.19)$$

where $\bar{t}$ is the average residence time and $E(t)\,dt$ represents the fraction of the particles that have a residence time between t and $t + dt$. The extent of conversion of a particle residing in a reactor for time t can again be predicted by using the equations given in Chapter 19 provided that a suitable model is chosen. Let the conversion be denoted by $x_B(t)$; then the average conversion of solids in the reactor is

$$\bar{x}_B = \int_0^\infty x_B(t) E(t)\,dt \qquad (20.20)$$

Equation 20.20 is restricted to solids feed of only one size. If a size distribution exists then an appropriate model has to be developed. Let $x_B(R_j)$ represent the conversion of a particle of size R_j. Also, it will be assumed that all sizes have the same mean residence time, given by $\bar{t}(R_j)$ (no carryover). This assumption is not valid for a fluidized bed at high gas velocities. Small particles are carried away from the bed rapidly. Thus these particles spend much less time than larger particles. The consequences of this situation are considered subsequently. The conversion equations for the case of a feed with nonuniform size distribution can then be derived as follows. For a particle of size R_j, the residence time distribution in the bed is given by

$$\frac{1}{\bar{t}(R_j)} \exp\left(-\frac{t}{\bar{t}(R_j)} \right) \qquad (20.21)$$

The average conversion of such particles is

$$\bar{x}_B(R_j) = \int_0^\infty x_B(R_j, t) \frac{1}{\bar{t}(R_j)} \exp\left(-\frac{t}{\bar{t}(R_j)} \right) dt \qquad (20.22)$$

Hence the average conversion in the reactor is

$$\bar{x}_B = \sum_0^{R_M} \frac{F_s(R_j)}{F_s} \int_0^\infty x_B(R_j, t) \frac{1}{\bar{t}(R_j)} \exp\left(-\frac{t}{\bar{t}(R_j)} \right) dt \qquad (20.23)$$

TABLE 20.3. Some Examples of Studies of Gas–Solid Reactions in Fluid-Bed Reactors

System	Remarks	Reference
1. Noncatalytic gas–solid reactions in general	The Kunii–Levenspiel model is extended to include simple noncatalytic reaction with a size distribution of particle feed.	Kunii and Levenspiel (1969)
	The bubble-assemblage model of Kato and Wen·is extended to noncatalytic fluid beds with a uniform solid feed. Selectivities of parallel and successive reactions based on this model are presented	Yoshida and Wen (1970, 1971)
2. Selective chlorination of ilmenite in the presence of CO	Hougen–Watson model for catalytic reaction is applied to a noncatalytic system and kinetic parameters are evaluated.	Doraiswamy et al. (1959)
3. Selective chlorination of roasted ilmenite in the presence of carbon	The bubble-assemblage model is used and verified for roasted ilmenite following the SIM under kinetic control.	Fuwa et al. (1978)
4. Chlorination of rutile in the presence of CO	The shrinking-core model is applied and its usefulness for commercial reactor design discussed.	Morris and Jensen (1976)
5. Reduction of iron oxide (ore)	The SIM under chemical reaction control for single particle is applied to a fluidized-bed reactor.	Ahner and Feinman (1964)
	A continuous multistage fluidized system for iron ore reduction is mathematically simulated; size distribution of solids and elutriation are considered.	Doheim (1973)
6. Roasting (oxidation) of zinc sulfide	A mathematical model for reaction in a fluid bed is presented.	Yoshida and Wen (1972) (also Natesan and Philbrook, 1970)
	Kinetics of oxidation of ZnS is studied in a batch-type fluidized bed; a two-phase model is employed for the fluid bed using a rate equation for a single particle based on the SIM.	Fukunaka et al., 1976)
7. Thermal decomposition of limestone	Heat transfer and chemical reaction control model is applied to the experimental data.	Asaki et al. (1974)
8. Combustion of char	A mathematical model for the combustion of char particles in a fluidized bed is described; the model gives predictions for the burnout time of a batch of carbon and the size distribution of carbon particles in a bed continuously fed with uniform-size particles.	Avedesian and Davidson (1973)
9. Gasification of coal	The Kunii–Levenspiel model is extended to the reactions involved in this operation.	Yoshida and Kunii (1974)
	An analytical model is developed that incorporates the effect of the mixing of coal and carrier-gas stream with hot entraining gases; heat transport to coal particles, gas-phase reaction, and thermal cracking of volatiles are considered.	Ubhayakar et al. (1977)
10. Combustion of coal	The model proposed by Avedesian and Davidson (1973) is modified to give predictions in good agreement with the experimental result.	Campbell and Davidson (1975)

TABLE 20.3. (continued)

System	Remarks	Reference
	A model is developed based on varying bubble size, solids population balance relating the feed, overflow, and elutriation; physicochemical changes of particles in the bed incorporated for coal and limestone; the model gives good agreement with reported pilot plant data.	Chen and Saxena (1977)
11. Combustion of coal with limestone injection	The fluidized-bed combustion is analyzed on the basis of the modified bubble-assemblage model of Mori and Wen; the distributions of coal particle size and of limestone conversion are also considered.	Horio and Wen (1975)
	A model capable of calculating the combustion efficiency, axial temperature, carbon holdup in the bed, concentrations in the bubble and emulsion phases, and particulate carryover of elutriation is developed.	Horio et al. (1978)
12. Combustion of carbon particles	A nonisothermal, continuous fluidized bed is considered with particle size distribution of entering feed; sufficient conditions for the existence of a unique steady state are derived.	Gordon and Amundson (1976)
13. Sulfation of limestone	The reaction is simulated for fluid-bed operation.	Bethell et al. (1973)
14. Reduction of nickel oxide	Grain model for a single particle is applied to the fluidized-bed data.	Evans et al. (1976)
15. Roasting of molybdenite	The SIM under chemical reaction control is applied to the experimental data.	Doheim et al. (1976)
16. Flourination of uranium and plutonium compounds	A reaction model is fitted to the reaction of UO_2 pellets with fluorine, to the reactions of U_3O_8 with fluorine and BrF_5, and to the reaction of PuF_4 with fluorine in fluid-bed reactors; a diminishing sphere model is used to fit the kinetic data.	Anastasia et al. (1971)
17. Fluorination of uranium tetrafluoride	A simple model is developed for fluidized-bed operation near minimum fluidization velocity; kinetic parameters can easily be evaluated by using this homogeneous model.	Corella (1980)
18. SO_2 sorption by cupric oxide in the presence of O_2	Details of design of large-scale fluid-bed desulfurization reactors are discussed.	Best and Yates (1977)
19. SO_2 sorption by dolomite	A simple analytical model is derived based on experimental data; the model is used to study the economic feasibility of using different types of stones.	Lee et al. (1980)
20. Hydrogasification by the Hydrane process	The bubble-assemblage concept is utilized in fitting existing data for producing substitute natural gas in a fluidized-bed reactor.	Wen et al (1977)
21. Granulation with selective product discharge	A model is developed based on arbitrary particle size and a material balance of the granulatable material; relations are obtained for the particle size distribution in the bed and the product being discharged, together with the recirculation rate in the separator.	Rotkin et al. (1977)

where R_M is the maximum size in the feed and $F_s(R_j)$ is the quantity of particles of size R_j entering with the feed.

The procedure just outlined can be used to calculate the solids conversion in a gas–solid fluidized bed uncomplicated by gas-phase behavior.

Example: Reduction of Iron Ore in a Fluidized Bed

The reduction of iron ore of density 4600 kg/m³ and 5-mm diameter by hydrogen can be approximated by the sharp-interface model. The stoichiometry of the reaction is

$$4H_2 + Fe_3O_4 \rightarrow 4H_2O + 3Fe$$

The first-order rate constant has been measured by Otake et al. (1967) to be

$$k_S = 1930 e^{-22400/R_g T} \text{ m/sec}$$

Find the conversion that can be achieved in a fluid-bed reactor at a temperature of 913°K for the following conditions: Pure H_2 feed, total pressure = 1 atm = 101 kPa; $D_e = 3 \times 10^{-6}$ m²/sec; solids feed rate = 0.1 kg/sec; total weight of solids = 7000 kg.

Solution

For various conversions x_B, the time required can be calculated from Eq. 19.8. The corresponding age distribution $E(t)$ is then given by

$$E(t) = \frac{1}{\bar{t}} \exp\left(-\frac{t}{\bar{t}}\right)$$

where

$$\bar{t} \text{ (average residence time in the reactor)} = \frac{7000}{0.1}$$

$$= 70000 \text{ sec}$$

The quantity $x_B(t)E(t)$ can then be calculated for each value of time.

The average conversion in the reactor can be calculated by using Eq. 20.20, and the integral can be evaluated graphically by plotting x_B vs. $tE(t)$ and measuring the area under the curve.

This gives $\bar{x}_B = 0.645$.

Effect of Elutriation

Elutriation rates from fluidized beds depend on many factors that are often complex and interrelated; they include

1. Width of the particle size distribution;
2. Occurrence of slugging in small-diameter columns, which tends to increase elutriation;
3. Particle shape, surface characteristics, particle attrition, and fluid viscosity;
4. Gas dispersion and the quality of fluidization;
5. Freeboard height above the bed;
6. Internals and stirrers.

The rate of elutriation is defined on the basis of an elutriation constant E^* (see also Section 13.9):

$$\left(\begin{array}{c}\text{rate of removal of solids}\\\text{of size } R_j\end{array}\right) = E^* \left(\begin{array}{c}\text{weight of that size of}\\\text{solid in the bed}\end{array}\right)$$

where E^* is the elutriation constant.
A more specific definition is

$$\left(\begin{array}{c}\text{rate of removal of solids}\\\text{of size } R_j \text{ per unit}\\\text{bed area}\end{array}\right) = E_s^* \left(\begin{array}{c}\text{fraction of bed weight}\\\text{consisting of size } R_j\end{array}\right)$$

or

$$-\frac{1}{A_c} \frac{dW(R_j)}{dt} = E_s^* \frac{W(R_j)}{W} \tag{20.24}$$

where E_s^* may now be viewed as a specific elutriation constant. The two quantities E^* and E_s^* are related simply as follows:

$$E^* = E_s^* \frac{A_c}{W} \tag{20.25}$$

Based on the experimental data for batch operations with binary mixtures, Wen and Hashinger (1960) proposed an empirical correlation for the specific elutriation rate constant E_s^*:

$$\frac{E_s^*}{g(u-u_t)} = (1.7 \times 10^{-5})\left(\frac{u-u_t}{gd_p}\right)^{0.5}\left(\frac{d_p u_t \rho_g}{\mu}\right)^{0.725}$$

$$\times \left(\frac{\rho_s - \rho_g}{\rho_g}\right)^{1.15}\left(\frac{u-u_t}{u_t}\right)^{0.10} \quad (20.26)$$

The average residence time of solids of size R_j in a fluidized bed can be predicted in terms of the elutriation constant (Levenspiel, 1979):

$$\bar{t}(R_j) = \left[\frac{F_1}{W} + E^*(R_j)\right]^{-1} \quad (20.27)$$

Knowing $\bar{t}(R_j)$, we can use Eq. 20.23 to predict the solids conversion in the presence of elutriation.

Other studies on elutriation include those of Sycheva and Donat (1974), Tanaka and Shinohara (1972), Gontanev and Paulin ((1975), Lange et al. (1977), and Horio et al. (1980).

Models Based on Particle Growth or Shrinkage

As has been pointed out earlier, change in particle size is one of the basic features of gas–solid reactions. In the previous sections, the effect of particle size distribution was included by defining an average residence time for each particle size, leading finally to Eq. 20.23. An alternative approach is to account for change in particle size by assuming a simple topochemical model for particle growth or shrinkage as a first approximation. Equations can then be developed for the quantity of material leaving the reactor as well as for particle size distribution in the exiting stream.

The rate of change of particle radius can be represented by a simple linear model:

$$\frac{dR}{dt} = -\check{k} = \text{constant} \quad (20.28)$$

The parameter $\check{k}$ is related to the rate constant k_S for the SIM by

$$\check{k} = \frac{v k_S C_{Ai} M_B}{\rho_B} \quad (20.29)$$

Another simple topochemical model to account for

size change is the inverse shrinkage model, which can be represented as

$$\frac{dR}{dt} = -\frac{\check{k}}{R} \quad (20.30)$$

This model arises for a gas-film-controlled reaction (e.g., oxidation of carbon). Here the rate is proportional to the gas-film mass transfer coefficient k_g, which is inversely proportional to R (see Section 19.4).

Equations 20.28 and 20.30 represent the changing particle size for most practical situations. In cases where these equations are inadequate, other empirical rate forms, such as first-order shrinkage and/or growth, can be used.

The equations to predict changing particle size can be utilized to calculate the quantity of material of a given size ($F_{s,1}$) leaving the reactor. Thus for a feed ($F_{s,0}$) of uniform size R_0 and for a system with no elutriation, the quantity of material leaving the reactor can be calculated, if the linear model is assumed, as (Levenspiel, 1979)

$$\frac{F_{s,1}}{F_{s,0}} = 1 - 3\check{j} + 6\check{j}^2 - 6\check{j}^3\left[1 - \exp\left(-\frac{1}{\check{j}}\right)\right] \quad (20.31)$$

where

$$\check{j} = \frac{W\check{k}}{F_{s,1}R_0} \quad (20.32)$$

The size distribution function in the exit stream is given by

$$\check{\phi}_1 = \frac{F_{s,0}}{W\check{k}}\frac{R^3}{R_0^3}\exp\left[-\frac{F_{s,1}(R_0 - R)}{W\check{k}}\right] \quad (20.33)$$

Various other rate forms for the changing particle size can be employed to calculate the corresponding size distribution functions.

The development of the equations as illustrated above suffers from the assumptions of uniform feed (spherical particles), specific form of rate equation depicting changing particle size, and no elutriation. These assumptions can be relaxed and a more general set of equations formulated. Thus for a system with feed $F_{s,0}$ having a size distribution $\check{\phi}_0(R)$, an exit stream $F_{s,1}$ with a size distribution $\check{\phi}_1(R)$, an elutriation stream $F_{s,2}$, and an arbitrary rate law for the changing particle size, a material balance on solids of size between R and $R + dR$ yields

(solids in feed, kg/sec)	−	(solids in outflow, kg/sec)	−	(solids leaving in elutriation stream, kg/sec)

	(growth of solids into − and out of the interval, kg/sec)	+	(mass increase of solids within the interval, kg/sec)	= 0 (20.34)

On the assumption of complete backmixing of the solid, so that the size distribution of the outflow stream also represents the size distribution of solids within the bed, this equation can be written as

$$F_{s,0}\check\phi_0(R)\,dR - F_{s,1}\check\phi_1(R)\,dR - WE^*(R)\check\phi_1(R)\,dR$$

$$-W\frac{d}{dR}[r(R)\check\phi_1(R)\,dR]$$

$$+\frac{(SF)W}{R}r(R)\check\phi_1(R)\,dR = 0 \quad (20.35)$$

The terms $E^*(R)$, SF, and $r(R)$ represent, respectively, the elutriation constant defined by Eq. 20.25, the shape factor of the particle, and the rate of change of particle size. Equation 20.35, valid for a particular size range, can be supplemented by an overall balance equation over all sizes to give

$$F_{s,2} + F_{s,1} - F_{s,0} = (SF)W\int_{\text{all }R}\frac{\check\phi_1(R)r(R)}{R}\,dR \quad (20.36)$$

Equation 20.36 assumes positive or negative values depending on whether the particle grows or shrinks. Equations 20.35 and 20.36 can be rearranged to obtain the outflow stream in terms of the input stream as

$$\frac{W}{F_{s,0}} = \int_{R_M}^{R}\frac{R^3}{r(R)}I\left[\int_{R_M}^{R}\frac{\check\phi_0(R)\,dR}{R^3 I}\right]dR \quad (20.37)$$

The corresponding outflow size distribution is given by

$$\check\phi_1(R) = \frac{F_{s,0}R^3}{Wr(R)}I\left[\int_{R_M}^{R}\frac{\check\phi_0(R)\,dR}{R^3 I}\right] \quad (20.38)$$

In these equations R_M represents the smallest feed size for a growing particle or largest size for a shrinking

particle, and I is an integral defined as

$$I = \exp\left[\int_{R_M}^{R}\frac{F_{s,1}/W + E^*(R)}{r(R)}\,dR\right] \quad (20.39)$$

The set of eqs. 20.37–20.39 cannot in general be solved analytically. However, they represent a total generalization with respect to feed size distribution, reaction kinetics, and presence of an elutriation stream. For specific situations, such as constant-size feed, linear kinetics, or no elutriation, these equations can be simplified to obtain analytical solutions. Levenspiel (1979) has summarized some of these analytical solutions.

Example: Analysis of a Fluid-Bed Reactor for Chlorosilanes

The production of chlorosilanes from ferrosilicon (or silicon) and methyl chloride proceeds in two steps. In the first step the solids CuCl and FeSi (or silicon) are thoroughly mixed in the bed, which is continuously fluidized at a velocity close to u_{mf} by an inert gas (nitrogen). The following solid–solid reaction takes place:

$$\text{CuCl}(s) + \text{FeSi}(s) \to \eta\text{-phase (Cu}_3\text{Si)} \quad \text{(I)}$$

$$\text{(15 parts)} \quad \text{(100 parts)}$$

The η phase is a necessary intermediate in the reaction. Assuming the particle to be spherical and representing CuCl by C, we can write the rate of reaction for step I as

$$r_C = -\frac{dC_C}{dt} = k_1 C_C(1 + k''C_\eta) \quad (20.40)$$

In the second step of the operation, the inert gas is replaced by methyl chloride (CH_3Cl) and the fluidizing conditions are so maintained that the gas-phase concentration is uniform (Doraiswamy et al., 1979). The following reaction takes place.

$$CH_3Cl(g) + b\,Cu_3Si(s) \to CH_3SiCl_3 \quad \text{(about 70\%)} \quad \text{(II)}$$

$$+ (CH_2)_2SiCl_2, (CH)_3SiCl, \dots$$

with the kinetics described by the SIM. Find the exit size distribution of particles for an initial feed distribution $\check\phi_0(R)$.

Solution

By a material balance over the fluidized bed for step I we have

$$\frac{d(N\check\phi)}{dt} = \eta_0\check\phi_0 - \eta_1\check\phi_1 - \eta_2\check\phi_2 + \overline{N\frac{d\check\phi}{dt}} \quad (20.41)$$

where the bar over the last term denotes an average over the bed; η_0 is the number of particles (CuCl + FeSi) per unit time; and η_1, η_2 are the number of particles in the outflow and the number elutriated per unit time, respectively; N is the total number of particles in the feed (since a batch process is being considered); and $\breve{\phi}_0$, $\breve{\phi}_1, \breve{\phi}_2$ are the size distributions in the feed, outflow, and elutriation streams, respectively.

If we consider the elutriated fines to be reintroduced into the bed, and no size distribution of FeSi particles in the feed, then Eq. 20.41 becomes

$$\frac{dN\breve{\phi}(d_p)}{dt} = \eta_0 \breve{\phi}_0 - \eta_1 \breve{\phi}_1 + \overline{\frac{Nd\breve{\phi}}{dd_p} \frac{dd_p}{dt}} \quad (20.42)$$

The particle size in the reactor changes with time and this brings about a variation in the distribution of particles, and the term

$$\frac{Nd\breve{\phi}}{dd_p} \frac{dd_p}{dt} = \frac{N\,d\breve{\phi}}{dt}$$

expresses this variation in size distribution for a given number of particles in the bed.

The term dd_p/dt appearing in Eq. 20.42 can be evaluated if the velocity constant for the reactor is known. Thus

$$-\frac{dd_p}{dt} = \frac{M_C r_C}{\pi d_p^2 \rho_C} \quad (20.43)$$

$$\overline{\frac{Nd\breve{\phi}}{dd_p} \frac{dd_p}{dt}} = -\frac{NM_C}{\pi d_p^2 \rho_C} k_1 C_C (1 + k'' C_\eta) \frac{d\breve{\phi}}{dd_p} \quad (20.44)$$

For the case where there is no feed, or exit stream to and from the bed, Eq. 20.45 reduces to

$$\eta_2 \breve{\phi}_1 = -\frac{NM_C}{\pi d_p^2 \rho_C} k_1 C_C (1 + k'' C_\eta) \frac{d\breve{\phi}}{dd_p} \quad (20.45)$$

Rearranging Eq. 20.45 we obtain

$$\frac{d\breve{\phi}}{dd_p} + \frac{\eta_2 \pi d_p^2 \rho_C}{NM_C k_1 C_C (1 + k'' C_\eta)} \breve{\phi} = 0 \quad (20.46)$$

The solution of this differential equation is

$$\breve{\phi} = A \exp\left[-\frac{\eta_2 \pi d_p^3 \rho_C}{3NM_C k_1 C_C (1 + k'' C_\eta)}\right] \quad (20.47)$$

We consider the boundary condition to be given from the initial size distribution condition,

$$\breve{\phi} = \breve{\phi}_0, \qquad d_p = d_{pm} \quad (20.48)$$

or

$$\breve{\phi}(d_p) = \breve{\phi}_0 \exp\left\{\frac{\eta_2 \pi \rho_C}{3NM_C r_C} d_{pm}^3 \left[1 - \left(\frac{d_p}{d_{pm}}\right)^3\right]\right\} \quad (20.49)$$

Now

$$\bar{r}_C = \frac{1}{t_e} \int_0^{t_1} r_C \, dt \quad (20.50)$$

where $\bar{r}_C$ is the average rate of reaction over the time interval 0 to t_1. From Eq. 20.43 we have

$$-\int d_p^2 \, dd_p = \int \frac{M_C}{\rho_C \pi} r_C \, dt \quad (20.51)$$

or

$$\frac{d_p^3}{3} = -\frac{M_C}{\rho_C \pi} r_C t + \text{constant}$$

with the boundary condition

$$d_p = 0 \quad \text{at} \quad t = t_1 \quad (20.52)$$

Thus

$$d_p = \left[\frac{3M_C}{\pi \rho_C} r_C (t_1 - t)\right]^{1/3} \quad (20.53)$$

Substituting this expression for d_p in Eq. 20.49 we obtain

$$\breve{\phi} = \breve{\phi}_0 \exp\left\{\frac{\eta_2 \pi \rho_C}{3NM_C r_C} d_{pm}^3 \left[1 - \frac{3M_C r_C (t_1 - t)}{\pi d_{pm}^3 \rho_C}\right]\right\} \quad (20.54)$$

Thus an expression for the size distribution $\breve{\phi}$ as a function of initial size distribution $\breve{\phi}_0$, time t, and rate of reaction r_C has been obtained.

As regards step II of the operation, the inlet particle size distribution is $\breve{\phi}_1$ given by

$$\breve{\phi}_1 = \breve{\phi}(t = t_1) \qquad \text{of step I} \quad (20.55)$$

The size distribution of the exit stream can then be obtained by using Eq. 20.35 as

$$\breve{\phi}_1(R) = \frac{F_{s,0} R^3}{W\breve{k}} \exp\left[\frac{F_1}{W\breve{k}} (R - R_M)\right]\left[\int_{R_M}^R \frac{\breve{\phi}(t = t_1)}{R^3 \breve{k}} \, dR\right] \quad (20.56)$$

where $\check{k}$ refers to the topochemical constant for the second step and R_M the largest feed size for the shrinking particle. The constant $\check{k}$ is related to the surface rate constant by Eq. 20.29. The size distribution function (20.56) can therefore be written as

implies that each particle has the same *a priori* probability of removal from the reactor, and the same position in the reactor. At any instant, however, the reactor contains particles that have spent different lengths of time inside the bed and thus have a wide

$$\check{\phi}_1(R) = \frac{F_{s,0}R^3\rho_\eta}{Wvk_sC_{CH_3Cl}M_\eta}\exp\left[\frac{F_1\rho_\eta(R-R_M)}{Wvk_sC_{CH_3Cl}M_\eta}\right]\left[\int_{R_M}^R \frac{\check{\phi}(t=t_e)\rho_\eta}{R^3vk_sC_{CH_3Cl}M_\eta}dR\right] \qquad (20.57)$$

Equation 20.57 gives the size distribution of the exit stream.

20.4.3. Models with Varying Gas-Phase Concentration

The chief assumption made in the development presented in Section 20.4.2 was that the gas environment is known and close to constant everywhere in the reactor. This assumption simplified the treatment considerably, since an analysis based on the solid phase alone could be used to describe the bed behavior. Many practical systems using large-particle beds fluidized with a large excess of gas conform to this situation. If vigorously fluidized beds of fine particles are involved, however, the composition of the gas seen by the solids would vary. A simple analysis like that in Section 20.4.2 would then be inapplicable. In these instances, the conservation equations for the gas-phase species have to be incorporated into the model and separate equations for the bubble and cloud–wake phases are necessary.

A number of models to depict the behavior of the bed are available and are described in Chapter 14. Most of these models assume batchwise operation with no change in the solids mass. Their application is therefore primarily restricted to catalytic systems. For a gas–solid noncatalytic reaction the solid reactant is constantly consumed and a solids makeup feed is required for steady-state operation. The nature of these reactors is thus, of necessity, continuous, and the bubbling-bed models with modification to account for the solids feed can be used to describe their behavior. While some of the bubbling-bed models (such as those of Davidson and Harrison, 1963; Kunii and Levenspiel, 1968a,b; Kato and Wen, 1969) have been extended to gas–solid noncatalytic reactors, no general model that takes account of the realities of the situation in such beds is avaialble.

In gas–solid noncatalytic reactors the degree of backmixing and the mean residence time of the particles are important factors. It is generally assumed that the particles are completely and uniformly mixed. This

distribution of particle age. During the course of reaction the solid reactant in the particles is gradually displaced to different extents by the solid product formed. The volume and specific density of the product formed may not be the same as those of the reactant. For a realistic representation of the bed, it is necessary therefore to account for variation in both size and density of the particles.

Thus the modeling of the gas–solid noncatalytic reactor is far more complex than that of its counterpart, the catalytic reactor. Even so, simplified models have been used to get a qualitative (and to some extent, quantitative) feel for the performance of the reactor. Thus the two-phase model of Davidson and Harrison (1963) has been used by Campbell and Davidson (1975) to analyze the data on the combustion of carbon particles for short periods of combustion in a batch reactor. The model has also been used and considerably extended by Amundson (see Bukur et al., 1977). Tigrel and Pyle (1971) have used this model for the not-too-different problem of catalyst deactivation. Kunii and Levenspiel (1969) and Kato and Wen (1969) have extended their models to gas–solid noncatalytic systems. As pointed out in the introductory section of this chapter, most of these models were developed for specific systems (see Table 20.3). A particularly useful model that takes account of some of the complexities in practical systems has been suggested by Chen and Saxena (1978).

It has been shown by Chen and Saxena that the solids in the emulsion phase may have a net downward, stationary, or net upward movement, depending on the rate of solids feed. As a result, other parameters, such as bubble volume fraction and rise velocity, become dependent on the solids feed rate. A detailed solids population balance has been formulated and a computer simulation presented by these authors. The results show that the extent of gas bypassing through bubbles is reduced when the solids feed rate is increased.

These authors have also proposed a criterion to predict the direction of solids movement in the emulsion phase. An illustrative plot of this criterion is reproduced in Figure 20.1. The quantities F_{c1} and F_{c2} in this figure are defined as follows:

$$F_{c1} = \left[\frac{\rho_s A_c (1 - f_{mf}) s_{wb}}{1 + s_{wb} f_{mf}} \right] u - \left[\frac{\bar{\rho}_s A_c s_{wb} (1 - f_{mf})(1 - \delta - s_{wb}\delta)}{1 + s_{wb} f_{mf}} \right] u_{mf} - \frac{s_{wb}\delta}{1 - \delta} \frac{dW_r}{dt} \qquad (20.58)$$

and

$$F_{c2} = F_{c1} + \frac{s_{wb}\delta - 1 + \delta}{1 - \delta} \frac{dW_r}{dt} \qquad (20.59)$$

where W_r is the weight of the solids in the reactor, and $\bar{\rho}_s$ is the average particle density over the entire bed.

If $F_0 < F_{c1}$, solids move down the emulsion, whereas for $F_0 > F_{c2}$ they move upward.

20.4.4. Solids Circulation Systems

Many industrial systems involve two units between which the solids continuously circulate. The reactor–regenerator system for a deactivating catalyst is a well-known example (see Section 14.9). The removal of H_2S from a coal gasification stream by circulating iron-oxide–iron-sulfide particles is another example of such a process. The reactions in this case are

In reactor:

$$Fe_2O_3 \xrightarrow{\quad H_2S \quad} FeS$$

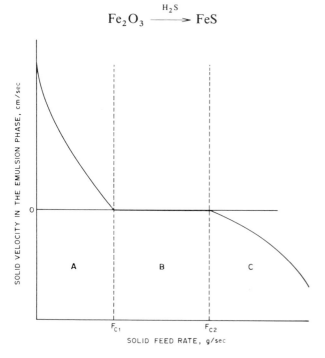

Figure 20.1. Classification of fluidization according to the solids feed rate. Areas A, B, and C refer to downflow, no net flow, and upflow of solids, respectively (Chen and Saxena, 1978).

In regenerator:

$$FeS \xrightarrow{\quad O_2 \quad} Fe_2O_3 + SO_2$$

The iron oxide utilizes the H_2S to form iron sulfide, which is oxidized back to Fe_2O_3 with the release of sulfur as SO_2. Depending on the residence time in each unit, the particles can undergo various degrees of reaction. This type of operation, using a reactive set of circulating solids, is also possible for the cleaning (by removal of SO_2) of product streams from combustion processes.

We shall assume that the changes occurring in a particle can be reasonably represented by the SIM. It can easily be visualized that the particles will be blackened in the reactor (as a result of reaction) and whitened in the regenerator (as a result of regeneration); these units can therefore be termed the blackener and the whitener, respectively. Additionally, it may be assumed that (1) a single size of particles circulates between the blackener and the whitener, (2) the solids are in mixed flow, and (3) whenever an advancing black front reaches a black region in a particle, it jumps across and continues advancing through the next white region.

Upon definition of a new dimensionless radial distance R' measured from the outer surface of a particle

$$\hat{R}' = 1 - \frac{r}{R} = 1 - \hat{R} \qquad (20.60)$$

the advance of a front located at any radial position within a spherical particle can be expressed as

$$\frac{t}{t_t} = \hat{R}' \qquad (20.61)$$

where t_t is the time required for the complete conversion of an all-black particle (t_{tB}) or of an all-white particle (t_{tw}).

Now the average volume fraction of black in particles leaving the blackener and that in particles leaving the whitener, respectively, are given by the following expressions:

$$\overline{X}_{bB}^{v} = 3 \int_0^1 (1 - \hat{R}')^2 P_{bB}(\hat{R}') \, d\hat{R}' \qquad (20.62)$$

$$\overline{X}_{bW}^{v} = 3 \int_0^1 (1 - \hat{R}')^2 P_{bW}(\hat{R}') \, d\hat{R}' \qquad (20.63)$$

where $P_{bB}(\hat{R}')$ is the probability that a particle leaving the blackener is black at depth $\hat{R}'$, and $P_{bW}(\hat{R}')$ is the probability that a particle leaving the whitener is black at depth $\hat{R}'$.

Upon suitable definition of the probabilities in terms of expected values and density functions, they can be evaluated and expressed in terms of a time parameter $\hat{\theta}$ defined as

$$\hat{\theta} = \frac{\text{time for complete conversion of a fresh particle in a unit}}{\text{mean residence time in that unit}}$$

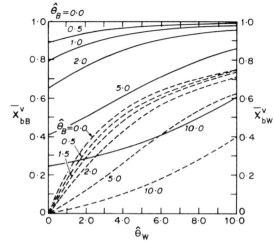

Figure 20.2. Chart for finding the fraction of black in the blackener and white in the whitener, given the system parameters θ_B and θ_W (Kimura et al., 1979).

The time parameter in the blackener is termed $\hat{\theta}_B$ and that in the whitener is termed $\hat{\theta}_W$.

The fraction of entirely black particles leaving the blackener (η_B) and the fraction of entirely white particles (η_W) can be obtained as

$$\eta_B = \frac{\hat{\theta}_W - \hat{\theta}_B}{\hat{\theta}_W - \hat{\theta}_B \exp\left[-(\hat{\theta}_W - \hat{\theta}_B)\right]} \qquad (20.64)$$

and

$$\eta_W = \frac{\hat{\theta}_B - \hat{\theta}_W}{\hat{\theta}_B - \hat{\theta}_W \exp\left[-(\hat{\theta}_B - \hat{\theta}_W)\right]} \qquad (20.65)$$

For the special case where $\hat{\theta}_B = \hat{\theta}_W = \hat{\theta}$ these reduce to

$$\eta_B = \eta_W = \frac{1}{1 + \hat{\theta}} \qquad (20.66)$$

For this case integration of Eqs. 20.62 and 20.63 leads to the following analytical solution:

$$\overline{X}_{bB}^{v} = \frac{1}{2} + \frac{3\left[\hat{\theta}\left(1 + \dfrac{\hat{\theta}}{2}\right) - (1 + \hat{\theta}) \ln(1 + \hat{\theta})\right]}{\hat{\theta}^3}$$

and $\qquad (20.67)$

$$\overline{X}_{bW}^{v} = 1 - \overline{X}_{bB}^{v}$$

For the case when $\hat{\theta}_B \neq \hat{\theta}_W$ the equations must be integrated numerically.

From the forgoing information pertaining to the properties of the circulating solid streams, design charts like the one in Figure 20.2 can be prepared relating the mean composition of the solid stream $\overline{X}_{bB}^{v}$ to the two

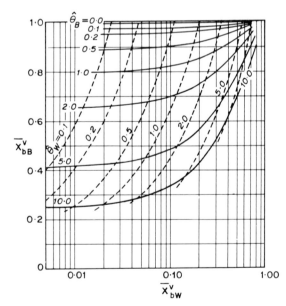

Figure 20.3. Chart for finding the kinetic rate constants given the composition of the circulating solid streams $\overline{X}_{bB}^{v}$ and $\overline{X}_{bW}^{v}$ (Kimura et al., 1979).

time parameters of the system, $\hat{\theta}_W$ and $\hat{\theta}_B$. Plots of $\overline{X}_{bB}^v$ vs. $\overline{X}_{bW}^v$ (Figure 20.3), from which the kinetic rate constants can be extracted, can also be prepared.

Kimura et al. (1979) note that the charts presented in Figures 20.2 and 20.3 can also be used, without serious loss of accuracy, for gas–solid systems controlled by ash diffusion or any other intermediate kinetics.

20.5. GENERAL CONSIDERATIONS IN THE DESIGN OF GAS–SOLID REACTORS

In the case of gas–solid catalytic reactions, the mode of contact is relatively simple to choose, because the alternatives are few. Mainly the fixed-bed multitubular, fixed-bed adiabatic, and fluidized-bed reactors have to be considered. In the case of gas–solid noncatalytic reactions, however, the choice is wider, since as pointed out at the beginning of the chapter, several alternative modes of contact are possible. Since each gas–solid system has its own characteristic features, even if a certain type of contact appears to be the logical choice, a number of bench-scale and pilot plant studies must be carried out with cold reactor models in order to ascertain the flow patterns of both the solids and gases. As in the case of catalytic reactors, no fixed strategy is possible for the design of gas–solid reactors, but the following steps in the overall design (Kunii, 1980) are clearly indicated:

1. Modeling of the gas–solid reaction concerned, with particular reference to transitions in the controlling regime;
2. Preliminary choice of the mode of contact between solids and gases, together with a strategy for the supply of heat to the reactor;
3. Conceptual design of the proposed reactor system;
4. Experimental studies with a bench-scale reactor;
5. Studies of flow patterns of solids and gases using cold reactors;
6. Simulation of the proposed reactor system;
7. Design of a pilot plant of suitable size to confirm the feasibility of the conceptual design;
8. Complete process design and cost estimation;
9. Construction and operation of pilot plant and identification of trouble spots;
10. Design of commercial reactor using the confirmed simulation models (which could be modified if needed).

REFERENCES

Agnew, J. B., and Narsimhan, G. (1970), *Chem. Eng. Sci.*, **25**, 685.

Ahner, W. D., and Feinman, J. (1964), *AIChE J.*, **10**, 652.

Anastasia, L. J., Alfredson, P. G., and Steindler, M. J. (1971), *Ind. Eng. Chem. Process Design Dev.*, **10**, 150.

Asaki, Z., Fukunaka, Y., Nagase, T., and Kondo, Y. (1974), *Met. Trans.*, **5B**, 381.

Avedesian, M. M., and Davidson, J. F. (1973), *Trans. Inst. Chem. Eng. (London)*, **51**, 121.

Barner, H. E., Manning, F. S., and Philbrook, W. O. (1963), *Trans. Met. Soc. AIME*, **227**, 897.

Best, R. J., and Yates, J. G. (1977), *Ind. Eng. Chem. Process Design Dev.*, **16**, 347.

Bethell, F. V., Denis, W., and Morgan, B. B. (1973), *Fuel*, **52**, 121.

Beveridge, G. S. G., and Kawamura, M. (1974), *Process in Pyrometallurgy*, ed. Jones, M. J., Institute of Mineral Metereology, London.

Bischoff, K. B. (1969), *Ind. Eng. Chem. Fundam.*, **8**, 665.

Bourgeois, S. V., Jr. Groves, F. R., and Wehe, A. H. (1974), *AIChE J.*, **20**, 94.

Bowen, J. H., and Lacey, D. T. (1969), *Chem. Eng. Sci.*, **24**, 965.

Bukur, D. V., Caram, H. S., and Amundson, N. R. (1977), *Chemical Reactor Theory—A Review*, eds. Lapidus, L., and Amundson, N. R., Prentice-Hall, Englewood Cliffs, New Jersey, p. 686.

Campbell, E. K., and Davidson, J. F. (1975), *Fluidization Technology*, Vol. 2, ed. Keairns, D. F., Hemisphere, Washington, D.C., p. 285.

Chen, T. P., and Saxena, S. C. (1977), *Fuel*, **56**, 401.

Chen, T. P., and Saxena, S. C. (1978), *AIChE Symp. Ser.* (No. 176), **74**, 149.

Ciliberti, D. F., and Lancaster, B. W. (1977), *Ind. Eng. Chem. Process Design Dev.*, **16**, 215.

Corella, J. (1980), *Chem. Eng. Sci.*, **35**, 25.

Davidson, J. F., and Harrison, D. (1963), *Fluidized Particles*, Cambridge Univ. Press, New York and London.

Doheim, M. A. (1973), *J. Appl. Chem. Biotechnol.*, **23**, 378.

Doheim, M. A., Abdel-Wahab, M. Z., and Rassoul, S. A. (1976), *Met. Trans.*, **7B**, 477.

Doraiswamy, L. K., Bijawat, H. C., and Kunte, M. V. (1959), *Chem. Eng. Prog.*, **55**, 80.

Doraiswamy, L. K., Mukherjee, S. P., Ramachandran, S., and Gokarn, A. N. (1979), NCL report.

Evans, J. W., and Song, S. (1974), *Ind. Eng. Chem. Process Design Dev.*, **13**, 146.

Evans, J. W., Song, S., and Leon-Sucre, C. E. (1976), *Met. Trans.*, **7B**, 55.

Fuwa, A., Kimura, E., and Fukushima, S. (1978), *Met.Trans.*, **9B**, 643.

Fukunaka, Y., Monta, T., Asaki, Z., and Kondo, Y. (1976), *Met. Trans.*, **7B**, 307.

Gontanev, V., and Paulin, A. (1975), *Rud-Metal Zb.*, **4**, 379.

Gordon, A. L., and Amundson, N. R. (1976), *Chem. Eng. Sci.*, **31**, 1163.

Hills, A. W. D. (1968), *Inst. Chem. Eng. Symp. Ser.*, **27**, 28.

Horio, M., and Wen, C. Y. (1975), *Fluidization Technology*, Vol. 2, ed. Keairns, D. F., Hemisphere, Washington, D. C., p. 289.

Horio, M., Rengarajan, P., Krishnan, R., and Wen, C. Y. (1978), *Sci. Tech. Aerasp. Rep.*, **16**, No. N 78-14119.

Horio, M., Taki, A., Hsieh, Y. S., and Muchi, I. (1980), *Fluidization*, eds. Grace, J. R., and Matsen, J. M. p. 509.

Ishida, M., and Wen, C. Y. (1971), *Ind. Eng. Chem. Process Design Dev.*, **10**, 164.

Jain, S. K., Prasad, P. M., and Jena, P. K. (1970), *Met. Trans.*, **1**, 1527.

Kato, K., and Wen, C. Y. (1969), *Chem. Eng. Sci.*, **24**, 1351.

Kimura, S., Fitzgerald, T. J., Levenspiel, O., and Kottman, C. (1979), *Chem. Eng. Sci.*, **34**, 1195.

Kunii, D. (1980), *Chem. Eng. Sci.*, **35**, 1887.

Kunii, D., and Levenspiel, O. (1968a), *Ind. Eng. Chem. Fundam.*, **7**, 446.

Kunii, Đ., and Levenspiel, O. (1968b), *Ind. Eng. Chem. Process Design Dev.*, **7**, 481.

Kunii, D., and Levenspiel, O. (1969), *Fluidization Engineering*, Wiley, New York.

Lange, J. F., Martinie, Y., and Bergougnou, M. A. (1977), *J. Powder Bulk Solids Technol.*, **1**, 15.

Lee, D. C., Hodges, J. L., and Georgakis, C. (1980), *Chem. Eng. Sci.*, **35**, 302.

Levenspiel, O. (1979), *The Chemical Reactor Omnibook*, Oregon State University, Corvallis.

Mariovet, J., and Wajc, S. J. (1977), *Chem. Eng. Sci.*, **32**, 779.

Mori, S., and Wen, C. Y. (1975), *AIChE J.*, **21**, 109.

Moriyama, A. (1971), *J. Iron and Steel Inst. Japan*, **11**, 176.

Morris, A. J., and Jensen, R. F. (1976), *Met. Trans.*, **7B**, 89.

Natesan, K., and Philbrook, W. O. (1970), *Met. Trans.*, **1**, 1353.

Ohmi, M., Usui, T., Minamide, Y., and Naito, M. (1978), *Proc. 3d Int. Iron and Steel Cong.*, p. 472.

Orehotsky, J., and Kaczenski, M. (1979), *Material Sci. Eng.*, **40**, 245.

Osman, M. A., Manning, F. S., and Philbrook, W. O. (1966), *AIChE J.*, **12**, 685.

Otake, T., Tone, S., and Oda, S. (1967), *Chem. Eng. (Japan)*, **31**, 71.

Ranade, M. G., and Evans, J. W. (1980), *Ind. Eng. Chem. Process Design Dev.*, **19**, 118.

Rotkin, V., Stepanouskii, V. M., and Scherbaker, A. Z. (1977), *Zh. Prikl. Khim.*, **50**, 1180.

Sampath, B. S., Ramachandran, P. A., and Hughes, R. (1975), *Chem. Eng. Sci.*, **30**, 135.

Schaefer, R. J., Vortmeyer, D., and Watson, C. C. (1974), *Chem. Eng. Sci.*, **29**, 119.

Simonsson, D. (1979), *Ind. Eng. Chem. Processes Design Dev.*, **18**, 288.

Spitzer, R. H., Manning, F. S., and Philbrook, W. O. (1966), *Trans. Met. Soc.-AIME*, **236**, 1715.

Sycheva, T. N., and Donat, E. V. (1974), *Khim. Prom.*, **6**, 459.

Tanaka, I., and Shinohara, H. (1972), *Mem. Fac. Eng. Kyushu Univ.*, **32**, 117.

Tigrel, A. Z., and Pyle, D. L. (1971), *Chem. Eng. Sci.*, **26**, 133.

Tsay, Q. T., Ray, W. H., and Szekely, J. (1976), *AIChE J.*, **22**, 1072.

Ubhayakar, S. K., Stickler, D. B., and Gannon, R. E. (1977), *Fuel*, **56**, 281.

Wen, C. Y., and Hashinger, R. F. (1960), *AIChE J.*, **6**, 220.

Wen, C. Y., Mori, S., Gray, J. A., and Yavorsky, P. M. (1977), *AIChE Symp. Ser.*, **73** (No. 161), 86.

Yagi, J., and Szekely, J. (1979), *AIChE J.*, **25**, 800.

Yates, J. G., and Best, R. S. (1976), *Ind. Eng. Chem. Process Design Dev.*, **15**, 243.

Yoon, H., Wei, J., and Denn, M. M. (1978), *AIChE J.*, **24**, 885.

Yoon, H., Wei, J., and Denn, M. M. (1979), *Ind. Eng. Chem. Process Design Dev.*, **18**, 306.

Yoshida, K., and Kunii, D. (1974), *J. Chem. Eng. Japan*, **7**, 34.

Yoshida, K., and Wen. C. Y. (1970), *Chem. Eng. Sci.*, **25**, 1395.

Yoshida, K., and Wen, C. Y. (1971), *AIChE Symp. Ser.* (No. 116), **67**, 151.

Yoshida, K., and Wen, C. Y. (1972), *Adv. Chem. Ser.*, **109**, 138.

CHAPTER TWENTY-ONE

Solid–Solid Reactions

In spite of the importance of solid–solid reactions, modeling studies on this system have been relatively few. When solid A reacts with solid B, an increasing layer of product AB is formed. One or both of the reactant solids must now diffuse through the product AB in order for further reaction to occur. Reactants A and B are usually in the form of well-mixed powder to enable faster completion of reaction. On the other hand, fundamental understanding of the simultaneous role of diffusion and reaction is possible only through studies with single pellets.

The various aspects of solid–solid reactions have been reviewed by Tamhankar and Doraiswamy (1979). In this chapter, which is based essentially on their review, we shall briefly discuss diffusion, analysis of pellet systems, analysis of mixed powder systems, the role of gas phase, and reactor development.

21.1. ROLE OF DIFFUSION

In a solid–solid reaction, it is essential that either or both of the reactants diffuse through an immobile product layer. This calls for a rigorous consideration of diffusion in solids. Two kinds of diffusion are involved in solid–solid reacting systems: self-diffusion and diffusion of a reactant through the product layer. Some good reviews have been published on solid–solid diffusion (e.g., that of Compaan and Haven, 1957, and of Frischat, 1974).

Solid–solid reactions generally tend to be diffusion controlled; thus the analysis of these reactions has largely been restricted to this aspect of the reactions. But surprisingly all these studies have been further confined to diffusion of reactants through the *product* layer. Studies on self-diffusion have been very few. A notable example may be mentioned: Schwab et al. (1961) in their studies on exchange reactions and the effect of dopents on these reactions concluded that the rate-determining step is self-diffusion in the starting substance.

It is worthwhile mentioning an important point concerning solid diffusivity measurements. Consider a product AB synthesized independently by a known method from solids A and B. This would have its own definite stoichiometry, lattice structure, defect characteristics, and impurity concentrations. The product formed during reaction, on the other hand, may have very different characteristics. Consequently the diffusivities in the two cases may be different, since diffusion in polycrystalline solids is known to be influenced by these factors. An example illustrating this fact is the difference in the diffusivity values reported by Tomas et al. (1969) and by Arrowsmith and Smith (1966) for the system phthalic anhydride + sulfathiazole. The values reported by Arrowsmith and Smith from independent diffusion experiments are of the order of 10^{-7}–10^{-8} cm^2/sec, whereas those reported by Tomas et al. from *in situ* evaluation during reaction and analyzed by the Serin–Ellickson expression (see Section 21.3.1) are of the order of 10^{-11} cm^2/sec.

Another interesting example is the reaction system (Schmalzried and Rogalla, 1963)

$$Al_2O_3 \text{ (single crystal)} + NiO \text{ (powder)} \rightarrow NiAl_2O_4$$

The product phase formed between the reactants is polycrystalline on the NiO side but single crystal on the Al_2O_3 side.

Studies on diffusion in solids are thus complicated by many factors. Even in independent diffusion studies, factors such as grain boundaries, electroneutrality conditions, Kirkendall effect of mass flow, and sintering make correct evaluation difficult. Moreover, in solid–solid diffusion, it is not really the concentration gradient but rather the chemical potential gradient that is responsible for diffusion. As such, the validity of Fick's laws for the analysis of these systems should occasionally be carefully checked. With these different points in mind, we shall describe below the principal methods of analysis used in the diffusion studies. Since extensive literature is available on this subject, we shall present only the salient features germane to our purpose.

21.1.1. Self-Diffusion

In almost all studies of self-diffusion, the radioactive tracer technique is used. Accordingly, single crystal or polycrystalline compacted samples are coated on one face with a thin film of the same compound containing a radioactive isotope. The samples are then annealed at required temperatures for definite time intervals, cooled, sectioned, and analyzed by counting the radioactivity. Concentration profiles are thus established.

Theoretical expressions for the concentration profiles can be readily obtained by Fick's law (although this may not be strictly applicable):

$$\frac{\partial C}{\partial t} = D \frac{\partial^2 C}{\partial l^2} \tag{21.1}$$

Solutions to this equation for different boundary conditions can be obtained from standard books (e.g. that of Crank, 1956). For the particular boundary conditions applicable to the experimental work at hand and the corresponding concentration profiles available, the appropriate solutions may be chosen and the diffusivities determined. A similar procedure may be adopted for studying diffusion of foreign atoms.

21.1.2. Interdiffusion

Interdiffusion is more important than self-diffusion from the practical viewpoint, since solid–solid reaction systems are often found to be controlled by this step. In a binary reaction system, when both the cations diffuse in opposite directions, it is impossible to find the individual diffusivities. Hence what is called an interdiffusion coefficient, generally denoted by $\tilde{D}$, is defined. Determination of $\tilde{D}$ is quite difficult because of the complexity introduced by its dependence on concentration. The differential equation involved is

$$\frac{\partial C}{\partial t} = \frac{\partial}{\partial l}\left(\tilde{D} \frac{\partial C}{\partial l} \right) \tag{21.2}$$

which may be solved by the Boltzmann–Matano method (see Shewmon, 1963). This method has been applied to NiO–MgO and NiO–CaO systems by Appel and Pask (1971).

Wagner (1969) developed a theory for the counterdiffusion of cations through the product layer. He calculated the individual fluxes of the two based on chemical potential gradient, the respective valencies, activity coefficients, and transport numbers. This method has been applied to the system MgO–MgAl$_2$O$_4$ by Whitney and Stubican (1971), who also correlated the $\tilde{D}$ values thus calculated with the individual self-diffusivities. Greskovich and Stubican (1970) applied the method to the system MgO–Cr$_2$O$_3$ and found it superior to that of Boltzmann and Matano.

In ionic solids, the condition of electroneutrality requires that for every cation diffusing there should be an adequate number of electrons or anions diffusing in the same direction, or an adequate number of cations diffusing in the opposite direction to balance the charges. An additional complexity is thus sometimes introduced in the analysis because of this simultaneous diffusion of cations and anions. In counterdiffusion, cations diffuse through the rigid lattice of anions, whereas when anions also contribute to the diffusion process, movement of the interface is observed. Moreover, the anions may diffuse as anions by the usual vacancy or interstitial mechanism or they may diffuse via the gas phase. In the latter case, the atmosphere will also influence the diffusion process. The effect of gas-phase diffusion on the reaction is discussed in Section 21.5.

21.2. ANALYSIS OF MIXED-POWDER REACTIONS

As in the case of other heterogeneous reactions, solid–solid reactions (with solids present as mixed powders) can also be controlled by one or more of several steps involved in the overall reaction. In essence, in any mixed-powder reaction, the solid particles should contact one another and at least one of them must then diffuse through an increasing product shell, after the initial surface reaction. This situation gives rise to several possibilities (Tamhankar and Doraiswamy, 1979):

1. Product growth controlled by diffusion of reactants through a continuous product layer.

2. Product growth controlled by nucleation and nuclei growth.

3. Product growth controlled by phase-boundary reactions.

4. Product growth controlled by kinetic equations based on the concept of an order of reaction.

We shall now outline the main features of the different models proposed based on these as controlling mechanisms, and of a few other models. The effect of particle size distribution and that of additives will then be considered.

21.2.1. Reaction Models

Product-Layer Diffusion Control

Models based on the product-layer diffusion mechanism involve three basic assumptions:

1. The reactant particles are spheres.
2. Surface diffusion rapidly covers reactant particles with a continuous product layer during the initial stages of the reaction.
3. Further reaction takes place by bulk diffusion of a mobile reactant species through this product layer, which is the rate-controlling step.

Jander (1927) was probably the first to treat a solid–solid reaction mathematically. In addition to the assumptions above he also made the rather drastic assumption that the cross-sectional area A_c is constant throughout and that changes in volume or density due to reaction are not significant. He derived the following expression for the fractional conversion x:

$$[1 - (1 - x)^{1/3}]^2 = kt \qquad (21.3)$$

Here k is the rate constant, given by

$$k = \frac{2k'' A_c}{R_i^2} \qquad (21.4)$$

where k'' is the local value of the rate constant. Thus, if an experiment is designed in such a way that all the assumptions are valid, k can be calculated for different particle sizes. One assumption that cannot be experimentally realized in a mixed-powder system is that of constant reaction cross section, since the surface area of the unreacted particle changes continuously as the reaction progresses.

Serin and Ellickson (1941) modified the rate expression of Jander by eliminating the assumption of constant reaction cross section. A number of other rate expressions have also been derived based on different approaches. These are summarized in Table 21.1,

which is from Tamhankar and Doraiswamy's review (1979).

In the Valensi–Carter model, change in volume due to reaction is accounted for by introducing a parameter Z_v in the expression. This and the first two (Jander and Serin–Ellickson) are the most widely used models in mixed-powder reaction rate studies.

Nucleation and Nuclei Growth

The general empirical form of the equation is

$$\ln (1 - x) = - (kt)^m \qquad (21.5)$$

The parameter m accounts for reaction mechanism, number of nuclei present, composition of the parent and product phases, and geometry of the nuclei. If a reaction is represented by this equation, a plot of $\ln [\ln(1 - x)]$ vs. $\ln t$ should give a straight line, with slope m and intercept $m \ln k$.

Another empirical form is that of Erofeev (1961):

$$\frac{dx}{dt} = kx^a (1 - x)^b \qquad (21.6)$$

This three-parameter equation has been successfully used by Neuburg (1970) in analyzing experimental data on the oxidation of cuprous iodide.

In a more rational approach to solid-phase transformations, germ nuclei are supposed to exist in the solid; these may consist of foreign particles, adsorbed layers of product, or embryos of the new solid phase. The transformation of a germ nucleus to a growth nucleus and the process of growth are both activated processes leading to product formation (Avrami, 1939–1941; Turnbull, 1956; Turnbull and Cohen, 1958; Young, 1966). This approach is generally applicable to phase-change phenomena of the type $A(s) \rightarrow S(s)$ exemplified by austenite–pearlite or austenite–bainite transitions in steel. It is also applicable to gas–solid systems of the type $A(s) \rightarrow R(g) + S(s)$, discussed in Section 19.2.1.

The Avrami model may be expressed in the form

$$V_{gn} = (SF)k_{growth} (t - t_{nf}) \qquad (21.7)$$

where V_{gn} is the volume of the growth nucleus and t_{nf} the time for nucleus formation. Note that the volume of germ nucleus is neglected, that is, although the nucleation is assumed to be heterogeneous, the nucleus is assumed to grow from zero volume. This leads to the

TABLE 21.1. Models for Mixed Powder Reactions (Tamhankar and Doraiswamy, 1979)

No.	Reaction Model	Mathematical Expression[a]	Reference
1.	Product-layer diffusion control:		
	a. Jander	$kt = [1-(1-x)^{1/3}]^2$	Jander (1927)
	b. Serin–Ellickson	$1-x = \dfrac{6}{\pi^2} \sum\limits_{n \text{ odd}} \left(\dfrac{1}{n^2}\right) \exp\left(\dfrac{-n^2\pi^2 Dt}{r^2}\right)$	Serin and Ellickson (1941)
	c. Kroger–Ziegler	$k \ln t = [1-(1-x)^{1/3}]^2$	Kroger and Ziegler (1954)
	d. Zuravlev–Lesokhin–Temple'man	$kt = \left[\dfrac{1}{(1-x)^{1/3}} - 1\right]^2$	Zuravlev et al. (1948)
	e. Ginstling–Brounshtein	$kt = 1 - \dfrac{2x}{3} - (1-x)^{2/3}$	Ginstling and Brounshtein (1950)
	f. Dunwald–Wagner	$kt = \dfrac{6}{\ln \pi^2 (1-x)}$	Dunwald and Wagner (1934)
	g. Valensi–Carter	$kt = \dfrac{Z_v - [1-(Z_v-1)x]^{2/3} - (Z_v-1)(1-x)^{2/3}}{(Z_v-1)}$	Valensi (1936); Carter (1961b)
2.	Nuclei-growth control:	$(kt)^m = -\ln(1-x)$	Avrami (1939, 1940, 1941)
3.	Phase-boundary reaction control:		
	a. Sphere reacting from the surface inward	$kt = 1 - (1-x)^{1/3}$	Jach (1963)
	b. Circular disk reacting from edge inward or a cylinder	$kt = 1 - (1-x)^{1/2}$	Jach (1963)
	c. Contracting cube	$x = 8k^3t^3 - 12k^2t^2 + 6kt$	Sharp et al. (1966)
4.	Based on the concept of an order of reaction	$kt = \dfrac{1}{n-1}\left[\dfrac{1}{(1-x)^{n-1}} - 1\right]$	
5.	Based on the concept of an index of reaction (Taplin)	$\dfrac{dx}{d(t^{z'})} = k(1-\beta x)^{m_i}$	Taplin (1974)
6.	Empirical:		
	a. Blum and Li	$\dfrac{dx}{dt} = \dfrac{a-x}{t}$	Blum and Li (1961)
	b. Patai et al.	$\dfrac{dx}{dt} = \dfrac{k(a-x)^m}{x^n}$	Patai et al. (1961)
		$\dfrac{dx}{dt} = k(a_1-x)(a_2-0.5x)$	Patai et al. (1962)
7.	Stochastic: Waite	$kt^{1/2} + I_0 = \left(\dfrac{x}{1-x}\right)\dfrac{1}{at^{1/2}}$	Wen et al. (1974)

[a]In these expressions a, a_1 and a_2 represent initial concentration of solid; k is the rate constant; and β, m_i and z' are constants.

following Volterra integral equation for fractional conversion:

$$x = (\text{SF})k_{\text{growth}}^3 N_{n,0} k_{\text{nf}} \int_0^t \exp(-k_{\text{nf}} t_{\text{nf}})$$

$$(t - t_{\text{nf}})^3 [1 - x(t_{\text{nf}})] dt_{\text{nf}} \quad (21.8)$$

where $N_{n,0}$ is the number of germ nuclei per unit volume of old phase (at $t = 0$), and k_{nf} is the rate constant for nucleus formation. This equation can only be solved numerically or by approximation. Some interesting features brought out by the solution of Ruckenstein and Vavanellos (1975) are outlined below.

Ruckenstein and Vavanellos showed that for parameter values of practical interest, a single equation is adequate to describe the kinetics of the reaction. This equation is based on a dimensionless time

$$\hat{t} = kt \quad (21.9)$$

and an activation-cum-growth parameter

$$\lambda = \frac{4\pi r_g N_{n,0}}{3k_{\text{act}}} \quad (21.10)$$

where r_g is the constant growth rate of the nucleus, and k_{act} is the rate constant for nuclei activation. The equation in its final form is

$$\log \hat{t}_{\text{max}} = 0.1525 - 0.25 \log(-\lambda) \quad (21.11)$$

where $\hat{t}_{\text{max}}$ is the time required for complete conversion (kt_{max}). The value of λ is usually higher than 10^9.

For a nonisothermal system, a heat balance equation should also be written. Several other parameters would then have to be considered, such as heat of reaction (ΔH), activation energy for nucleus formation (E), frequency factor (A_f), change in chemical potential per mole of new phase accompanying the transformation (ΔG), and free energy of activation per mole of nucleus growth ($\Delta G'$). On the assumption of adiabatic behavior and using the system properties of barium azide, Ruckenstein and Vavanellos (1975) have solved the mass and energy balance equations. At high values of ΔH (5 and 10 kcal/mol), catastrophic behavior is observed: a period of slow reaction rate followed by a sudden rise in rate. This behavior is characteristic of explosive transformations in solids, such as the well-known decomposition of mercury fulminate (Garner and Hailes, 1933). At lower values of ΔH, the rise in rate is less abrupt, with decreased explosion hazard.

Bhatia and Perlmutter (1980) have analyzed the same problem using the population balance approach. In addition to heterogeneous nucleation around germ nuclei, they also considered the possibility of homogeneous nucleation, in which nucleation is assumed to start from centers of zero volume (Göler and Sachs, 1932). Further, they modified the Avrami equation to include a term for the initial volume of a growth nucleus.

A factor often neglected is the interaction among nuclei during the growth process (Dehoff, 1972). Avrami's attempts (1940) to include this effect through the concept of an extended volume is not borne out by experimental results (Nakamori et al., 1974). An alternative approach (Bhatia and Perlmutter, 1980) is to treat impingement as a size-dependent process in which it is assumed that two nuclei of shape factor SF_1 and volume V_N each give rise to a single nucleus of shape factor SF_1 and volume $2V_N$.

The experimental results of Neuburg (1970) on the oxidation of cuprous iodide particles have been compared with predictions of the population balance model and the modified Avrami model in Figure 21.1. It will be noticed that the agreement is quite satisfactory, although the sigmoidal shape of the conversion–time curves is not predicted by either of the models.

Phase-Boundary Reaction Control

When the diffusion of the reactant species through the product layer is fast compared to reaction, the kinetics is controlled by phase-boundary reactions. Models have been developed for different geometries and corresponding boundary conditions. Thus, for a sphere reacting from the surface inward, the fractional reaction completed x and time t are related by the expression

$$kt = 1 - (1 - x)^{1/3} \quad (21.12)$$

which is identical with the expression derived for gas–solid reactions (Chapter 19). For a circular disk reacting from the edge inward or for a cylinder the relation is

$$kt = 1 - (1 - x)^{1/2} \quad (21.13)$$

and for a contracting cube

$$x = 8k^3 t^3 - 12k^2 t^2 + 6kt \quad (21.14)$$

In all these models, it is assumed that reaction is slow compared to diffusion, but fast enough to occur in a very shallow layer near the interface. If A is the diffusing

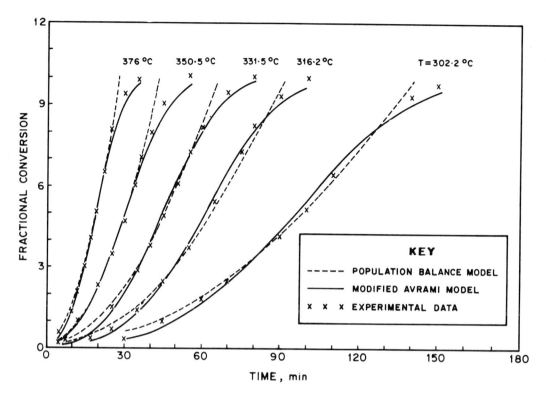

Figure 21.1. Conversion vs. time for Cu_2I_2 oxidation at various temperatures using different models (Bhatia and Perlmutter, 1979).

species, reaction in the bulk of the solid B is not given any consideration. A more rigorous treatment is thus required.

Kinetic Equations Based on the Concept of an Order of Reaction

The general rate equation for an nth-order reaction is represented by

$$\frac{1}{n-1}\left[\frac{1}{(1-x)^{n-1}} - 1\right] = kt \qquad (21.15)$$

The differential analysis of van't Hoff can be used to determine whether a reaction could be classified by a reaction order.

Thus, Kutty and Murthy (1974) have applied first-order kinetics to the reaction between urea nitrate and tricalcium phosphate in an ensemble of fine particles. They have attributed this to slow nucleation of one of the products and fast spreading of it further, because of negligible diffusion resistance. Branson (1965) observed that for the reaction between ZnO and Al_2O_3 the Valensi model is applicable only at high temperatures ($> 1000°C$), whereas below $1000°C$ the rate constant k

is time dependent. This was considered to be due to the possibility of phase-boundary-controlled kinetics in the early stages of the reaction. Hence, the data were correlated by the second-order kinetic equation

$$kt = \frac{1}{a}\left(\frac{x}{1-x}\right) \qquad (21.16)$$

where a is the initial concentration of the solid.

Phase-boundary-controlled kinetics suggest two possible mechanisms:

1. Diffusion through the product layer is so rapid that no reaction is possible at spinel–alumina interface;

2. Nucleation occurs from a supersaturated solid solution of ZnO in Al_2O_3.

The first mechanism suggests that reaction occurs not at the phase boundary but in the bulk of the solid, whereas the second suggests the possibility of nuclei growth control. If the second mechanism is operative, the nuclei growth model would probably correlate well with the data; but this has not been attempted. The first mechanism clearly suggests kinetic control, which has been shown to be the case with $n = 2$ (Eq. 21.15).

General Approach

The rate equations mentioned above and grouped under different controlling mechanisms can all be represented by a single expression:

$$F(x) = kt \qquad (21.17)$$

We can determine $F(x)$ for $x = 0.5$ and the ratio $t/t_{0.5}$, where $t_{0.5}$ is the time required for 50% conversion. Sharp et al. (1966) have prepared numerical tables that give the values of $F(x)$ and $t/t_{0.5}$ for different values of x between 0 and 1.

Taplin (1974) has suggested a unified treatment for most kinetic expressions used for the reaction of powdered materials by writing the rate equation in the generalized form

$$\frac{dx}{d(t^{z'})} = k(1 - \beta x)^{m_i} \qquad (21.18)$$

The exponent of time z' depends on the kinetic regime; the constant m_i has been termed the "index of reaction," which in many of the previous studies was misinterpreted as the order of reaction. By numerical values and different plots, the expression has been shown to be appropriate with different values of z', m_i, and β, depending on the kinetics. Thus, for linear kinetics where the reaction rate varies as the area of the reaction interface, the equation is shown to be valid for different particle shapes, with z' and β values of 1 and index of reaction (m_i) values of 2/3, 1/2, and 0, respectively, for sphere, elongated cylinder, and thin disk.

Of course, because of particle size distribution, values of m_i other than those mentioned for individual particles are obtained. Thus values as high as 1.5 have been reported. Moreover, the shape of a real particle is usually irregular. Generalization can be achieved by considering a right rectangular prism of variable dimensions. In such a configuration the kinetics of irregular-shaped particles can be approximated only by the index-of-reaction equation. In similar fashion, other types of kinetics, such as parabolic kinetics, nucleation-based kinetics, and limited-extent kinetics, can be generalized by Eq. 21.18 with different values assigned to z', β, and m_i.

Remarkably, coupling between linear and parabolic processes leading to joint kinetics has also been ap-proximated by the index-of-reaction equation with $z' = 2/3$ (Taplin, 1973a). Except for this, however, in almost all other studies it has been assumed that only one type of kinetics dominates for the entire reaction, which is obviously not the case. Often it is not realized that different regimes could be operative in a single reaction during its progress. For instance, as has been pointed out by Taplin (1973b), parabolic kinetics for a spherical particle cannot exist at either the beginning or the end of a reaction.

Empirical Equations

Apart from the different models presented above, a number of authors have proposed different empirical equations to describe the kinetics of a mixed-powder reaction. Thus, Blum and Li (1961) found that the rate equation

$$\frac{dx}{dt} = \frac{a - x}{t} \qquad (21.19)$$

describes the kinetics of nickel ferrite formation from the observation of a straight-line plot obtained for xt vs. t. The equation was found to hold for all ferrite formation reactions.

Patai et al. (1961, 1962), in their studies on the oxidation of p-divinyl benzene (p-DVB) and carbon black by $KClO_4$, have used the empirical rate expressions

$$\frac{dx}{dt} = \frac{k(a - x)^m}{x^n} \qquad (21.20)$$

and

$$\frac{dx}{dt} = k_1(a_1 - x)(a_2 - 0.5x) \qquad (21.21)$$

where the values of the empirical constants m and n are 1 and 2/3, respectively, and a_1 and a_2 are the initial concentrations of the reactants.

A Stochastic Model

From the forgoing discussion it is evident that the nature of solid–solid reactions is statistical, because of the inhomogeneity of mixing and other factors. Hence, a stochastic model would be more realistic in describing the course of a solid–solid reaction. Such a model has been developed by Waite (1957, 1958, 1960) in which a probabilistic distribution of pairs of A and B particles is considered, and different probability densities are calculated. The final form of the expression obtained is

$$\left(\frac{x}{1 - x} \right) \frac{1}{at^{1/2}} = kt^{1/2} + I_0 \qquad (21.22)$$

where I_0 is the distance of separation between the particles A and B within which they react. This reduces to the oridinary second-order rate equation when $I_0 = 0$. It would be particularly interesting if the data for different particle sizes were analyzed by using Eq. 21.22 in terms of an effectiveness factor (similar to that for other heterogeneous reactions), since the diffusional resistance is separated here in the factor I_0.

In spite of the drawbacks in some of the models, and the empiricism in the others, they have been used to analyze a number of solid–solid mixed-powder reactions.

21.2.2. Effect of Particle Size Distribution

Particle size distribution, invariably encountered in practice, can have important effects on kinetics. The smaller particles in the ensemble will be consumed in a short period of time, while the bigger particles are still reacting. Hence, the reaction rate per unit volume, which is based on the radius of an individual particle, will be affected; in the areas associated with smaller particles, reaction would be complete and the region may be considered dead with respect to the progress of reaction.

Particle size distribution will also have an effect on voidage and hence on the effective contact area, since smaller particles can go into the interstitial spaces formed by bigger particles. The effect of compaction pressure may also be influenced by size distribution.

Sasaki (1964) first attempted to account for the effect of particle size distribution on kinetics. The basic rate expression assumed was the one developed by Carter (1961b). Sasaki considered groups of different-sized particles designated by the mean radius $(_i r_0)$ for the ith group, with the fraction of the group by x_i. Carter's expression was then applied to obtain, for the ith group, an expression

$$\frac{Z_v - [1 + (Z_v - 1)x_i]^{2/3} - (Z_v - 1)(1 - x_i)^{2/3}}{2(Z_v - 1)}$$
$$= \frac{k}{a_0^{2(i-1)} \, _i r_0^2} t \qquad (21.23)$$

where a_0 is the ratio of the mean radii of adjacent groups. Hence, upon summing for all the groups, an expression for overall conversion was obtained. In this treatment, however, it was assumed *a priori* that the Valensi–Carter model fits with $Z_v = 2$, where Z_v is the volume of the product formed per unit volume of the reactants consumed. This assumption has been

eliminated in the more elaborate treatment of Kapur (1973).

21.2.3. Effect of Additives

Both catalytic and inhibitory effects are found to be exhibited by additives in a solid–solid mixed-powder system. These may be due to a number of different interactions of the additives with the reaction system. Such effects are well known in solid-state chemistry, particularly in connection with the semiconducting properties of solids. In fact, "doping" is an accepted and important technique in semiconductor technology. The conductivity of a sample can be increased or decreased, as desired, by addition of the dopents. This technique has an important application in catalysis, too, in view of similar dopent effects.

Certain additives are known to promote sintering on the surface, facilitating material transport, whereas some others may promote reaction by acting as oxygen transfer agents. They may inhibit the reaction physically by acting as a barrier between reactants, or in some cases chemically by trapping the active intermediates.

Schwab and Rau (1958) studied the exchange reaction $ZnO + CuSO_4$. The reaction rate was found to be accelerated by the addition of Li^+ to ZnO and retarded by the addition of Ga^{3+} to it. On the other hand, in the reaction $NiO + MoO_3 \rightarrow NiMoO_4$, addition of Cr^{3+} to NiO enhanced the rate, whereas addition of Li^+ retarded it.

An interesting observation has been reported by Szabo et al. (1961). In their experiments with pure NiO and Fe_2O_3, the activation energy was found to be 95 kcal/mol; addition of 1 % Cr_2O_3 to the NiO reduced it to 59 kcal/mol, whereas addition of 1 % TiO_2 to the Fe_2O_3 raised it to 132 kcal/mol.

21.2.4. The Concept of Effective Contact; Analogy with Gas–Solid Reactions

Solid–solid reactions differ from other types of heterogeneous reactions (gas–solid, gas–liquid, etc.) mainly in the nature of the contact between the reactant partners. Whereas in other cases at least one of the reactants is in a flow pattern, in solid–solid systems both reactants are stagnant, unless one of them (or part of it) is transported as a gas. Obviously, therefore, the progress of a solid–solid mixed-powder reaction depends to a significant extent on the initial contact between the reactants. The initial contact has to be achieved through proper mixing.

Apart from mixing, other factors, like the relative

particle sizes of the two, particle size distribution, particle shape, and sintering characteristics, also affect the contact between the reactants. Sintering can sometimes introduce complexity into the analysis, since it may change this contact area during the course of a reaction as a result of contraction or expansion and neck growth (for a detailed discussion of sintering, see Coble and Burke, 1961; Kuczynski and Stablein, 1961; see also Section 19.2.6).

Patai et al. (1961, 1962) considered the effect of contact surface area on th rate constants in their studies on the oxidation of p-divinyl benzene (p-DVB) and carbon by $KClO_4$. They observed that the rate constants (k) obtained by fitting the data with empirical rate equations were found to depend on the surface areas of the two. New rate constants were obtained by using the following expressions:

$$k_1 = k\left(\frac{A_{sA}}{A_{sB}}\right) \quad \text{for} \quad A_{sA} > A_{sB} \quad (21.24)$$

and

$$k_2 = kr^2\left(\frac{A_{sA}}{A_{sB}}\right) \quad \text{for} \quad A_{sA} < A_{sB} \quad (21.25)$$

where A_{sA} and A_{sB} are the surface areas of $KClO_4$ and p-DVB or carbon, respectively, and r is the radius of a $KClO_4$ particle.

Komatsu (1965) developed a theory based on the number of contact points. By simple geometric considerations, the number of contact points $N^0(A/B)$ between one central B particle and the surrounding A particles in an ideal system can be expressed as

$$N^0(A/B) = N_B^0\left(\frac{\alpha\hat{w}}{1 + \alpha\hat{w}}\right) \quad (21.26)$$

where N_B^0 is the total number of particles of A and B surrounding the central B particle, $\hat{w}$ is the weight ratio of the components (W_A/W_B), and α is given by

$$\alpha = \frac{r_B^3 \rho_B}{r_A^3 \rho_A} \quad (21.27)$$

where r_A, r_B, and ρ_A, ρ_B are the radii and densities of the components A and B, respectively.

Irregular shapes are encountered in real systems; also, the packing in real systems is not perfect. Hence, the actual number $N(A/B)$ would be less than $N^0(A/B)$ and is given by

$$N(A/B) = N_B^0\left(\frac{\alpha\hat{w}}{1 + \alpha\hat{w}}\right)^{m_2}, \quad m_2 > 1 \quad (21.28)$$

where the parameter m_2 is introduced to account for the difference in packing state. Based on this theory, Jander's equation can be modified and the following expression for the rate constant of the Jander equation developed:

$$k(T, \alpha, \hat{w}) = k^0(T)\left(\frac{\alpha\hat{w}}{1 + \alpha\hat{w}}\right)^{m_2} \quad (21.29)$$

This has been successfully applied to the systems $CaCO_3$–MoO_3, $BaCO_3$–SiO_2, CuCl–Si, and $PbCl_2$–Si (Komatsu, 1965). Komatsu and Uemura (1970) subsequently extended the theory to account for counterdiffusion.

Komatsu's theory, though remarkable, has been applied to the Jander model, which is known to have its own drawbacks. Also, since different shapes and sizes are encountered in practice, it is more logical to consider the total contact surface area rather than the number of contact points. This area of contact between the particles of A and B may be designated the effective contact area.

It is possible to develop a model based on the concept of effective contact area. If a geometry similar to that of the Komatsu model is considered, the effective contact area $A_{sB,e}$ of the B particle will be a fraction of its true surface area A_{sB}; this fraction (f_{eff}) is given by the ratio of the number of A particles surrounding to the total number of particles surrounding, and can be readily obtained from Eq. 21.28.

$$f_{eff} = \frac{N(A/B)}{N_B^0} = \left(\frac{\alpha\hat{w}}{1 + \alpha\hat{w}}\right)^{m_2} \quad (21.30)$$

A hypothetical particle B is now considered with surface area $A_{sB,e}$ and radius $r = \sqrt{A_{sB,e}/4\pi}$. The surface of this hypothetical particle will be completely in contact with the component A, with conditions closely approximating those for a gas–solid system. Such an analogy between solid–solid and gas–solid reactions will become clearer from the discussion presented in the next section.

The advantage of the concept introduced above is that the well-known equations developed for gas–solid reactions (see Chapter 19) can be readily applied and important system parameters then determined. Thus, equations for gas–solid reactions have been successfully applied by Tamhankar (1976) to the data of Ramachandran et al. (1974). An important conclusion is that, as for other heterogeneous systems, solid–solid mixed-powder reactions should also be analyzed according to the proper controlling regime. Surprisingly, in many of the studies reported, a single rate expression has been used to correlate the data throughout the range of conversions, which is not always adequate.

21.2.5. Multicomponent Systems

In some cases (e.g., the cement industry) multicomponent solid–solid reactions are encountered. Even when one goes from a binary to a ternary system, the phase diagram, if available, reveals the increased complexity in the system. An additional composition term is involved, and the behavior of the system depends on the region of the phase diagram in which one is operating. If the additional third component is present in a minute quantity ($< 1\%$), it is treated as a dopant that affects the system in a particular fashion. This is discussed separately. When all three components are present in considerable amounts a variety of new compounds may be formed. Systems consisting of more than three components will obviously be considerably more complex.

So far, only a few phase equilibrium studies have been reported in the literature. For example, Reijnen (1965) has studied the system MgO–FeO–Fe_2O_3; Winkler (1965) has reported structure studies in the systems BaO–MeO–Fe_2O_3 ($Me = Co, Ni, Cu, Zn, Mn, Mg$). Analysis of coupled diffusion and reaction in such systems has not yet been attempted.

21.3. ANALYSIS OF PELLET–PELLET REACTIONS

Studies in pellet–pellet systems are mainly aimed at elucidating the transport mechanism in detail and at gaining insight into the phenomenon of coupled diffusion and reaction. The second aspect, which is important from the engineering point of view, has received relatively little attention.

As mentioned earlier, the changes occurring in a mixed-powder system have a complex time dependence. Also, the geometries of individual particles are not well defined. Although, particles are often assumed to be spherical, in actual practice all kinds of shapes (flakes, plates, cylinders, needles, etc.) are encountered. Moreover, size distribution is also a factor. The results and conclusions based on these studies therefore tend to be dubious, and the rate constants and activation energies erratic.

The obvious method of overcoming these difficulties is to take a system with a fixed geometry and hence a well-defined contact surface area. This can most conveniently be achieved by taking cylindrical pellets of the reactants in contact.

21.3.1. The Role of Diffusion

In a pellet–pellet system, once the product layer has formed by the phase-boundary processes, further reaction is controlled by the diffusion of one or both of the reactants through this product layer. Depending on whether the diffusion is one sided or counterdiffusion is involved, the product growth will occur on only one side or on both sides of the original interface. Again, in the counterdiffusion case, diffusion rates of the two may be different. Accordingly, the ratio of the product-layer thicknesses on the two sides of the original interface will differ. Moreover, in systems in which ionic diffusion occurs, the condition of electroneutrality restricts the diffusion process, and consequently the product growth, on the two sides of the original interface.

A number of systems have been studied to determine the mode of diffusion. Most of them make use of the marker experiments to follow the interface movement. The markers used are normally inert platinum wires which are originally placed at the interface. But, as pointed out by Pettit et al. (1966), in high-temperature studies these markers may not really be inert. Also, slippage and breakage of markers may occur as a result of the uneven stresses developed at the interface. Hence, it has been suggested that texture composition, which is found to be different in the two parts in the case of counterdiffusion, can serve as a natural marker. Further, because of faster diffusion of one of the cations, vacancies are created that are visible under a microscope as pores. This layer of pores can be treated as a marker (Carter, 1961a; Hardel, 1972). Kooy (1965) has suggested a method of restricted contact, so that depending on diffusion rates, uneven product growth as well as swelling and contraction on two sides can be observed.

21.3.2. Sintering

Kuczynski (1965) has extensively studied the phenomenon of sintering and developed mathematical expressions of the form $(r')^n = kt$, where r' is the neck radius and n takes different values depending on the mechanism of sintering. The studies were then extended to the problem of sintering and reaction. Thus, the systems NiO–Fe_2O_3 and MgO–Fe_2O_3 were studied, with spheres of one in contact with plates of the other. Because of fast diffusion of the MgO, swelling on the

Fe_2O_3 side was observed; but some Fe_2O_3 was also found to have diffused in the MgO, which was neglected in previous studies. From the shape and large radii of the necks, it was concluded that neck growth is not controlled by vacancy gradient due to surface tension, as in a purely sintering process. Various aspects of sintering have been brought out in the monograph by Ristic (1979).

21.3.3. Reaction Models

A system in the pellet–pellet form is attractive from the point of view of mathematical analysis of coupled diffusion and reaction, particularly because it retains the structures and concentration profiles in a clearly distinguishable pattern after the reaction is quenched.

The simple parabolic law of Tammann (1920) is commonly used to follow the overall rate of reaction. According to this law, the square of the product-zone thickness is directly proportional to time:

$$(\Delta l)^2 = kt \qquad (21.31)$$

The expression is more or less an empirical one and is based on the assumption that the diffusion of the reacting species through the product layer is the controlling step, so that

$$\frac{d(\Delta l)}{dt} \propto \left(D, \frac{1}{\Delta l}, A_c \right)$$

or

$$\frac{d(\Delta l)}{dt} = k \frac{A_c D}{\Delta l} \qquad (21.32)$$

Integration of this expression yields Eq. 21.31, the parabolic rate law, where k is the overall rate constant.

Because of the simplicity of the expression and the easily obtainable product-layer thickness (PLT) values, the parabolic rate law has been in use for a long while. Parabolic rate constants, and hence the activation energies and other thermodynamic parameters, have been obtained in several studies; typical examples are presented in Table 21.2. More rigorous models, also based on diffusion in the product layer, have been proposed by Wagner (1936), Schmalzried (1962, 1965), Rastogi et al. (1962, 1963), Rastogi and Dube (1967), and Rastogi and Singh (1966; see also Rastogi, 1970).

In the models discussed above, the interactions taking place at the phase boundaries have been neglected. Schmalzried (1974) first recognized this fact, and introduced the concept of phase-boundary reactions as cation rearrangements. These rearrangements, restricted to a narrow zone, yield the product, and the rate of rearrangement determines the reaction rate.

If a relaxation time τ for the cation rearrangement is defined, then for a reaction of the type

$$AO + B_2O_3 \rightarrow AB_2O_4$$

TABLE 21.2. Examples of Systems for Which the Kinetics Have Been Studied by Pellet–Pellet Experiments (Tamhankar and Doraiswamy, 1979)

System	Rate Law[a]	Reference
$NiO-Al_2O_3$	$(\Delta l)^2 = kt$	Pettit et al. (1966)
$NiO-Al_2O_3$	$(\Delta l)^2 = kt$	Minford and Stubican (1974)
$MgO-Cr_2O_3$	$(\Delta l)^2 = kt$	Greskovich and Stubican (1970)
$Ag-TlI$	$(\Delta l)^2 = kt$	Flor et al. (1975)
$AgI-(K, Rb)I$	$\Delta l = kt$ (initial)	
	$(\Delta l)^2 = kt$ (later)	Brandley and Greene (1967)
$Hg_2Cl_2-I_2$	$(\Delta l)^2 = 2kt \exp(-c\,\Delta l)$	
$Hg_2Br_2-I_2$	$(\Delta l)^2 = kt$	Rastogi and Dube (1967)
$Hg_2I_2-I_2$		
Picric acid with hydrocarbons	$(\Delta l)^2 = 2kt \exp(-c\,\Delta l)$	Rastogi et al. (1963)
Picric acid with naphthols	$(\Delta l)^2 = 2kt \exp(-c\,\Delta l)$	Rastogi and Singh (1966)
p-Dimethyl aminobenzaldehyde–diphenylamine hydrochloride	$(\Delta l)^3 = kt$	Qureshi et al. (1975)

[a]The symbol c denotes constant.

the reaction zone is given by

$$\Delta l_r = (2D_B^{(AO)}\tau)^{1/2} \qquad (21.33)$$

where $D_B^{(AO)}$ is the diffusivity of B ions in AO.

The technique of electron probe microanalysis (EPMA), when used in studying solid–solid reactions, has clearly indicated the existence of such a reaction zone. Greskovich and Stubican (1969) studied the system $MgO–Cr_2O_3$ using the EPMA technique, and obtained concentration profiles of Cr^{3+} in MgO in a reaction zone.

Arrowsmith and Smith (1966) attempted an analysis in which they considered simultaneous diffusion and reaction in an organic pellet–pellet system. In their analysis, an irreversible second-order reaction was assumed ($A + B \rightarrow C$). Thus:

$$\frac{\partial C_A}{\partial t} = D\frac{\partial^2 C_A}{\partial l^2} - kC_A C_B \qquad (21.34)$$

and

$$\frac{\partial C_B}{\partial t} = D\frac{\partial^2 C_B}{\partial l^2} - kC_A C_B \qquad (21.35)$$

with the initial and boundary conditions

$$
\begin{aligned}
C_A = C_{A0}, \qquad C_B = C_C = 0, \qquad l < 0 \\
C_A = C_C = 0, \qquad C_B = C_{B0}, \qquad l > 0
\end{aligned}
\qquad (21.36)
$$

Assuming semi-infinite geometry we have

$$
\begin{aligned}
C_A = C_{A0}, \qquad C_B = C_C = 0, \qquad l = -\infty \\
C_A = C_C = 0, \quad C_B = C_{B0}, \qquad l = +\infty
\end{aligned}
\qquad (21.37)
$$

The nonlinear equations 21.34 and 21.35 can be converted into dimensionless form and solved numerically if constant density is assumed. Thus plots of the dimensionless concentration $\hat{C}_A$ vs. dimensionless distance in the solid $\hat{L}$ can be prepared, as in the case of a catalytic reaction. With such plots and independently determined diffusivity values, rate constants can be evaluated. However, the treatment assumes an infinite system, which is equivalent to assuming pseudo-steady state. Thus, no separate zones are considered. The assumption of second-order reaction has not been tested. Also, as pointed out in Section 21.1, it may not be correct to use the diffusivity value determined by independent methods.

21.3.4. Model Based on Analogy with Gas–Solid Systems

A more realistic analysis has been presented by Tamhankar and Doraiswamy (1978), in which two separate zones, namely, a product zone and a reaction zone, have been considered. This model is based on the concepts outlined in Chapter 19 for gas–solid reactions. The reaction considered is, again

$$AO + B_2O_3 \rightarrow AB_2O_4$$

The course of the reaction, following the concentration of one of the species, say B^{3+}, can be represented as shown in Figure 21.2.

Three possible mechanisms of diffusion may be considered: (1) counterdiffusion of cations A^{2+} and B^{3+}, (2) one-way diffusion of A^{2+} and O^{2-} or of B^{3+} and O^{2-}, and (3) one-way diffusion of, say, A^{2+} and $2e^-$ in the solid matrix and of oxygen as gas (O_2). Depending on which of these mechanisms is operative, one or both of the boundaries of the product layer will be moving. The reference axis is, therefore, positioned at one of the boundaries. Now we have only one moving boundary, followed by a reaction zone of finite width that moves along with the boundary.

As in the case of gas–solid (catalytic) reactions, where a number of modes of diffusion of the gas phase through the solid phase are possible (such as bulk diffusion, Knudsen diffusion, and surface diffusion), several modes of diffusion in solid–solid systems are also possible. The more important among these are (1) bulk diffusion via vacancy or interstitial mechanism; (2) surface diffusion; (3) grain-boundary diffusion; and (4) vapor-phase diffusion.

In practice, where polycrystalline materials are involved, different mechanisms may be operative simultaneously. Hence an effective diffusivity D_e has been defined. With this gross approach, we are justified in applying Fick's laws of diffusion.

The following mathematical analysis, based on two zones, the product zone and the reaction zone, has been presented by Tamhankar (1978).

Product Zone

Assuming diffusion to be independent of concentration, we can write the governing differential equation as

$$\frac{C_A}{\partial t} = D_{e,p}\frac{\partial^2 C_A}{\partial l^2} \qquad (21.38)$$

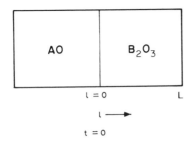

NO REACTION

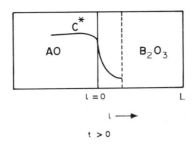

REACTION ZONE FORMATION
WITH NO PRODUCT LAYER

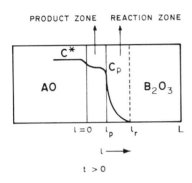

REACTION ZONE ALONG WITH
A PRODUCT LAYER WITH NO
DIFFUSION RESISTANCE

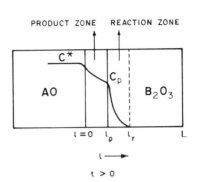

REACTION ZONE ALONG WITH
A PRODUCT LAYER WITH A
FINITE DIFFUSION RESISTANCE

Figure 21.2. Stages in the reaction between the solids AO and B_2O_3 in the reaction-zone model (Tamhankar and Doraiswamy 1978).

with the boundary conditions

$$C_A = C_{A0}, \quad l = 0$$
$$C_A = C_{Ap}, \quad l = l_p \tag{21.39}$$

where the subscript p denotes the product zone. The unsteady-state equation 21.38 is solved for these boundary conditions on the assumption that the zone extends to infinity, which introduces the additional boundary condition

$$C_A = 0, \quad l = \infty \tag{21.40}$$

The solution is given by Crank (1956) as

$$C_A = C_{A0} \, \text{erfc} \, (4D_p t)^{-1/2} \tag{21.41}$$

In dimensionless form this can be written as

$$\hat{C}_A = \text{erfc}\left[\frac{\hat{L}}{(4\theta)^{1/2}}\right] = 1 - \text{erf}\left[\frac{\hat{L}}{(4\theta)^{1/2}}\right] \tag{21.42}$$

where $\hat{C}_A = C_A/C_{A0}$, $\hat{L} = l/L$, and $\theta = D_{e,p}t/L^2$. At the end of the zone (i.e., at $l = l_p$) the concentration will be given by

$$\hat{C}_{Ap} = 1 - \text{erf}\left[\frac{\hat{L}_p}{(4\theta)^{1/2}}\right] \tag{21.43}$$

By virtue of the fact that in almost all the oxide reactions the growth of the product layer follows the parabolic rate law, this concentration $\hat{C}_{Ap}$ becomes constant, since by parabolic law $\hat{L}_p/\sqrt{\theta} = K_1$, a constant, and hence

$$\hat{C}_{Ap} = 1 - \text{erf}\left(\frac{K_1}{2}\right) = \text{a constant} \tag{21.44}$$

Reaction Zone

The reaction-zone thickness is assumed constant, so that the concentration profile in this zone is unaffected by time and we can write

$$\frac{C_A}{\partial t} = 0 = D_{er} \frac{\partial^2 C_A}{\partial l^2} - kC_A \qquad (21.45)$$

where k is the first-order reaction rate constant. The boundary conditions are

$$
\begin{aligned}
C_A &= C_{Ap}, & l &= l_p \\
C_A &= 0, & l &= l_r
\end{aligned} \qquad (21.46)
$$

In dimensionless form Eq. 21.45 becomes

$$\frac{\partial^2 C_A}{\partial \hat{L}^2} - \phi_{er}^2 C_A = 0 \qquad (21.47)$$

with the boundary conditions

$$
\begin{aligned}
\hat{C}_A &= \hat{C}_{Ap}, & \hat{L} &= \hat{L}_p \\
\hat{C}_A &= 0, & \hat{L} &= 0
\end{aligned} \qquad (21.48)
$$

where the Thiele modulus ϕ_r is defined as

$$\phi_r = L \left(\frac{k}{D_{er}} \right)^{1/2} \qquad (21.49)$$

The solution of Eq. 21.47 is given by

$$\hat{C}_A = \hat{C}_{Ap} \frac{\sinh \phi_r (\hat{L}_r - \hat{L})}{\sinh (\phi_r \Delta)} \qquad (21.50)$$

where

$$\Delta = \hat{L}_r - \hat{L}_p \qquad (21.51)$$

This model is similar to the one developed by Mantri et al. (1976) for gas–solid noncatalytic reactions. From Eq. 21.50 an expression for the reaction-zone thickness can be readily developed. Thus

$$\Delta = \frac{1}{\phi_r} \ln \left[\phi_r + (\phi_r^2 + 1)^{1/2} \right] \qquad (21.52)$$

The model has been verified by using some of the EPMA results reported in the literature. A typical comparison of theoretical and experimental concentration profiles is presented in Figure 21.3, from which good agreement between the two may be observed. The

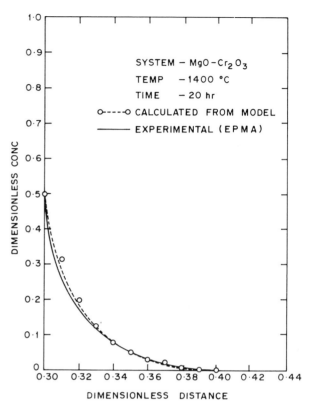

Figure 21.3. Comparison of the theoretical profile calculated from the zone model with the experimental profile reported (Tamhankar and Doraiswamy, 1979).

model has been further extended for a general nth-order reaction.

Equation 21.52 shows that the reaction-zone thickness depends only on the Thiele modulus ϕ_r. Normally, for a given system, ϕ_r is constant, but factors like sintering and exothermicity might change the value of ϕ_r during the reaction.

21.3.5. Heat Effects; Multiphase Product Layers

Cohn (1948) has derived a simple expression to account for the exothermicity of a solid–solid reaction; a more rigorous treatment that takes into account the effect of temperature has been reported by Gray and Harper (1960). From the solutions given, numerical criteria for the occurrence of explosion can be determined (see Section 21.2.1, "Nucleation and Nuclei Growth"). Apart from this special case of explosion, heat effects have not been studied in solid–solid reactions, particularly since at high temperatures these effects are negligible. At low temperatures, however, they may be significant, since most of the solids are known to have low heat conductivity.

Formation of multiphase product layers is another aspect of solid–solid reactions that has not received much attention. Multiphase product layers appear in the case of silicates and some aluminates (which are important components of cement) and of titanates. Hence, kinetic studies in these cases play a significant role in cement production. Titanates are important in semiconductors.

Kohatsu and Brindley (1968) and Brindley and Mayami (1965) studied the systems CaO–α-Al$_2$O$_3$ and MgO–SiO$_2$, respectively. The following sequence of phases has been observed in the product layer for the former:

$$C/C_3A/C_{12}A_7 \quad \text{or} \quad C_{12}A_7 \; H/CA/CA_2/CA_6/Al_2O_3$$

where
$$C = CaO \quad \text{and} \quad A = Al_2O_3.$$

The thicknesses of different phases have been measured and the corresponding rate constants calculated by applying the parabolic rate law to the growth of each phase. It has been observed that different rates are associated with different phases (in a given system). These rates could be correlated with the melting point and oxygen packing density of the corresponding phases. Thus, low melting point and low density will generally facilitate diffusion and thereby growth.

Sockel (1972) studied the system PbO–SiO$_2$ and observed the sequence to be

$$PbO/4PbO \cdot SiO_2/2PbO \cdot SiO_2/PbO \cdot SiO_2/SiO_2$$

Again the parabolic rate law could be applied for individual phase growths. Also, an expression has been derived relating the parabolic rate constant for a given phase in multiphase growth to that in a single-phase growth.

It is usually difficult to assign a precise mechanism of mass transport in such systems. Yamaguchi and Tokuda (1967) have reported EPMA profiles for a few silicate and titanate systems. The nature of these profiles reflects the complexity involved in the analysis. At present no rigorous analysis is available.

21.4. ROLE OF THE GAS PHASE

The role of the gas phase in solid–solid reactions has long been recognized. Many solids (particularly organic substances) are known to sublime or have considerable vapor pressure. The chemical reactivity of particles is often accounted for by Thomson's relation (Parravano, 1962):

$$\ln \frac{p}{p_0} = \frac{M}{R_g T} \frac{2r}{\rho_s^2} \tag{21.53}$$

which gives the vapor pressure p over the surface of a solid with a radius of curvature r, solid density ρ_s, and molecular weight M, as a function of the pressure in equilibrium over a flat surface p_0. Thus for ice, $M = 18$, $\rho_s = 1$, $T = 273°K$; for $r = 50$ Å

$$\frac{p}{p_0} = 10$$

and for $r = 5$ Å

$$\frac{p}{p_0} = 10^{10}$$

Clearly, with decreasing particle size, the reactivity of the system increases tremendously. This would suggest the possibility of gas-phase reaction for small particle sizes.

Konoyuk and Vashuk (1974) treated the kinetics of solid-phase reactions whose rate is limited by the sublimation of one of the reactants. The final expression obtained is the same as that of Ginstling and Brounshtein (see Table 21.1) under certain conditions. Whether it is solid-phase diffusion or transport via sublimation of one of the reactants, the barrier of the product layer has to be crossed for the reaction to occur.

Apart from vapor pressure and sublimation, the other process that contributes to gas-phase diffusion is dissociation. Borchardt (1959) examined the role of dissociation by deriving an expression for the lowest temperature of dissociation at which at least 1% conversion should occur. Applying this to the reactions between U$_3$O$_8$ and Fe, Cr, Nb, and Ni, he concluded that for at least some systems (e.g., U$_3$O$_8$–Fe and U$_3$O$_8$–Nb) gas-phase diffusion is absent, since reaction occurs even below the lowest dissociation temperature calculated.

Gluzman and Milner (1960) studied some organic solid–solid reactions with a view to investigating the role of the gas phase. In their experiments they separated the two reactants with a fine copper mesh so that minimum contact was allowed between particles of the two. They analyzed samples of the two for product formation and concluded that there is little contribution of gas-phase diffusion to the reaction, though it was not completely ruled out.

On the other hand, several studies using inorganic

oxides have been reported, with reactants kept apart, that clearly suggest gas-phase transport (e.g., Ginstling, 1951; Ginstling and Fradkina, 1952; Pozin et al., 1954). Elaborate studies by Rastogi and co-workers (1962; Rastogi and Singh, 1966; Rastogi and Dube, 1967) seem to confirm this finding for a variety of systems. The experimental technique used was one in which two reactants were filled in a capillary with a known distance of separation. The kinetics was followed using the parabolic rate law and an attempt was also made to correlate the rate constant with the distance of separation according to the relation $k = ae^{-bs}$, where a and b are constants and s the distance of separation.

The most common way by which the gas phase enters the reaction is through decomposition of one of the solids. There are also a number of situations where a gas (say CO_2) is introduced into the system to react with one of the solids (say carbon) to produce a gaseous intermediate (say CO), which then reacts with the other solid (say hematite). The methods of analysis presented in Chapter 19 would be applicable to such systems, and many solid–solid reactions belonging to this category have already been considered in Section 19.5.4.

Certain solid–solid reactions of industrial importance are known to proceed via gaseous intermediates. Thus in the reaction between silica and graphites, Klinger et al. (1966) have shown that the reaction proceeds through the dissociation of silica

$$SiO_2 \rightarrow SiO + \tfrac{1}{2}O_2$$

and

$$SiO + C \rightarrow SiC + \tfrac{1}{2}O_2$$

The reduction of iron ore (e.g., hematite) by solid reductant coke (carbon) proceeds via intermediate carbon monoxide formation except for the initial stage. Thus

$$Fe_gO_h(s) + CO(g) \rightarrow Fe_gO_{h-1} + CO_2(g)$$

and

$$C(s) + CO_2(g) \rightarrow 2CO(g)$$

This reaction has been treated mathematically in detail by Rao (1971), Rao and Chaung (1974) and Sohn and Szekely (1973). The second reaction is taken to be controlling, so that a treatment based on first-order gas–solid reaction is followed.

Gas-phase diffusion is faster than solid-phase diffusion, since it occurs through contact areas as well as through pores. This fact could be utilized in synthesizing certain materials by deliberately introducing a gas

phase into the system. Schäfer (1971) has suggested some interesting synthetic routes. The reaction

$$2CaO + SnO_2 \xrightarrow{900^\circ C} Ca_2SnO_4$$

can be accelerated by adding CO or H_2 to the system, since

$$SnO_2(s) + CO \rightarrow SnO(g) + CO_2$$

and $SnO(g)$ can further react more easily with CaO. Also, in the reaction

$$NiO + Cr_2O_3 \xrightarrow{1100^\circ C} NiCr_2O_4$$

if O_2 is introduced, we have

$$Cr_2O_3 + \tfrac{3}{2}O_2 \rightarrow 2CrO_3(g)$$

and $CrO_3(g)$ can further react to give the desired product $NiCr_2O_4$.

21.5. REACTOR DEVELOPMENT

21.5.1. Basic Design Considerations

Standard design procedures for solid–solid reactors have not yet been developed. In most cases, the conventional ceramic technique is used for manufacturing purposes. In this, the finely powdered reactants are first intimately mixed, usually in a ball mill; the powder mixture is pelletized if necessary and heated at the desired temperature for a predetermined time interval. Often the material thus obtained is reground, pelletized, and heated again. The procedure is repeated two or three times to ensure complete conversion.

The main difficulties in solid–solid reactions are the requirement of complete mixing (and hence initial ideal contact between reactant particles) and continuous removal of the product layer formed in order to promote better contact between yet unreacted phases. Sintering, as mentioned earlier, presents additional complexity. In the conventional ceramic technique, these difficulties are overcome by the repeat cycle procedure.

Unlike in other heterogeneous processes, here none of the reactants can be in a flow pattern; contact between unreacted phases can be achieved only through solid-state diffusion, which is known to be extremely slow. It can be remedied either by having a device in which the processes of mixing, grinding, and heating can be carried out simultaneously (e.g., a rotating furnace-cum-ball mill), or by deliberately introducing a gas phase in the reaction, as suggested by Schäfer (1964) (see Section 21.4). The work of Fiegl et al. (1944) is worth noting in this connection; they carried out solid–solid reactions even at high temperatures with constant stirring by a platinum stirrer.

The similarity between solid–solid and gas–solid (noncatalytic) reactions, as pointed out in Section 21.3.4, can be exploited in developing basic design procedures, since rational reactor design methods are already being practiced (although to a very limited extent) in the case of gas–solid (noncatalytic) reactions. Thus for a given system, conversion–time studies using proper expressions can be directly applied for design purposes. Yet the basic difference—that in a gas–solid system at least one of the reactants is in a flow pattern, whereas in solid–solid systems it is not so—cannot be overlooked. In fact, even in gas–solid systems most of the reactors today are designed on the basis of previous experience and by modifying the existing reactors.

In solid–solid systems the important step that hampers the overall process is the initial contact. Therefore, the concept of effective contact discussed in Section 21.2.4 may be usefully employed in predicting reactor performance. The effective contact area can then be chosen as the basis for any model. Since most of the reactions are initially kinetically controlled, rate studies during the initial time interval will provide the volumetric rate r_v, which may be expressed as

$$r_v = kA_{s,e} \qquad (21.54)$$

where k is the specific rate constant and $A_{s,e}$ is the effective contact area.

From model experiments with known contact areas, a plot of rate vs. contact area can be prepared. The maximum possible contact area can be calculated from the known system properties, such as densities and particle sizes, and on the assumption of an ideal arrangement.

Compaction pressure is also known to affect the rates. Hence a study of the rate per unit contact area as a function of pressure would also be helpful. Another factor that would greatly affect the initial rates is the temperature. Solids are known to have low heat conductivities and high heat capacities. Obviously, therefore, the reaction would initially be faster at the surface than in the interior. If the reaction is exothermic, additional complexity may be introduced.

Nothing concrete has yet been done in this direction; there certainly appears to be scope for future work.

21.5.2. Mechanical Design

Information on mechanical design is scanty. There are no standard design procedures; most of the commercial plants are designed from experience and on the basis of some of the empirical expressions. The industries in which this problem is encountered are ceramics, ferrites, semiconductors, metals, and cement. Reactors are either batch type or continuous. Of these industries, the ceramic, ferrites, and semiconductor industries mostly use batch-type reactors, following the procedure described above. The problem of reactor design, therefore, reduces to one of designing a furnace and a mixer for the initial mixing stage. This is done mostly from experience.

In the metal industry, the primary concern is the reduction of ores (e.g., hematite reduction by coke), wherein solid–solid reactions are encountered. In the cement industry, solid–solid reaction is a major step, since the important constituents are all present as solids and react in the solid state. In both the cases, a rotary kiln is often used as a reactor. Hence, the process is normally a continuous one; the reaction mixture is fed in at one end of the kiln and discharged at the other end. A gas is also sometimes introduced in cocurrent or countercurrent flow.

The rotating kilns used in the cement industry are very long (of the order of 70–150 m; diameters range up to 7–10 m); they are positioned at a slight incline from the horizontal (Pollitt, 1964). The entire length is divided into zones for drying, reaction, and cooling. The residence time of the cement raw mix in such kilns is typically $2\frac{1}{2}$ hours in a 70-m-long kiln and 6 hours in a 150 m-long kiln.

Equations have been developed for calculating the length (L_k) and diameter (d_k); for example, for a dry process the length required is given by (Martin, 1932)

$$L_k = 20(d_k - 1.5) + 0.2(d_k - 1.5)^2 \qquad (21.55)$$

For design purposes, much depends on the heat requirements in different zones, and as such on the heat capacities of the reactants and products. Calculations are thus made on the basis of available data, and what is known as a heat treatment schedule is prepared.

Vertical tube reactors are also in use in the cement industry; the heights are of the same order as the length of the rotary kiln. Air is introduced from the bottom and the raw mix is fed at the top. This somewhat resembles a fluidized-bed reactor; the airflow is used to adjust the resistance time of the raw mix in the reactor.

21.5.3. Recent Developments

From the forgoing discussion it may be seen that there is need to develop newer and better processes for the manufacture of mixed oxides. The conventional ceramic technique has obvious drawbacks; also, because the steps involved are slow, it cannot be adapted to continuous production.

Novel processes have recently been developed, particularly for the manufacture of ceramic nuclear fuels; these methods can be readily adapted to other industries. These processes are (1) the sol–gel process, (2) gel precipitation, and (3) spray processes (pyrogel process and flame reactor). All of these essentially start with an aqueous phase containing the metal ions in the desired proportions. The main advantages are (a) a lower temperature requirement, (b) the need for a shorter time of contact for conversion, and (c) control over particle size. A brief account of the processes is presented by Dell (1972)

Sol–Gel Process

In essence, the sol–gel process involves forming a concentrated colloidal sol of the metallic oxides or hydroxides and converting the sol to a semirigid gel by a convenient method. The gelation stage is important since it determines the particle shape and size of the final product. The gel is then dried and calcined to obtained the desired product. The process is best suited to the preparation of coarse products.

REFERENCES

Appel, M., and Pask, J. A. (1971), *J. Am. Ceram. Soc.*, **54**, 152.

Arrowsmith, R. J., and Smith, J. M. (1966), *Ind. Eng. Chem. Fundam.*, **5**, 327.

Avrami, M. (1939), *J. Chem. Phys.*, **7**, 1103.

Avrami, M. (1940), *J. Chem. Phys.*, **8**, 212.

Avrami, M. (1941), *J. Chem. Phys.*, **9**, 177.

Bhatia, S. K., and Perlmutter, D. D. (1979), *AIChE J.*, **25**, 298.

Bhatia, S. K., and Perlmutter, D. D. (1980), *AIChE J.*, **26**, 379.

Blum, S. L., and Li, P. C. (1961), *J. Am. Ceram. Soc.*, **44**, 611.

Borchardt, H. J. (1959), *J. Am. Chem. Soc.*, **81**, 1529.

Brandley, J. N., and Greene, P. D. (1967), *Trans Faraday Soc.*, **63**, 1023.

Branson, D. L. (1965), *J. Am. Ceram. Soc.*, **48**, 591.

Brindley, G. W., and Mayami, R. (1965), *Phil. Mag.*, **12**, 505.

Carter, R. E. (1961a), *J. Am. Ceram. Soc.*, **44**, 116.

Carter, R. E. (1961b), *J. Chem. Phys.*, **34**, 2010.

Coble, R. L., and Burke, J. E. (1961), *Proc. 4th Int. Symp. Reactivity of Solids*, p. 38.

Cohn, G. (1948), *Chem. Rev.*, **42**, 527.

Compaan, K., and Haven, Y. (1957), *Disc. Faraday Soc.*, **23**, 105.

Crank, J. (1956), *The Mathematics of diffusion*, Oxford (Clarendon Press), New York and London.

Dehoff, R. T. (1972), *Treatise on Materials Science and Technology*, Academic Press, New York.

Dell, R. M. (1972), *Proc. 7th Int. Symp. Reactivity of Solids*, p. 553.

Dunwald, H., and Wagner, C. (1934), *Z. Phys. Chem. (Leipzig)*, **B24**, 53.

Erofeev, B. V. (1961), *Proc. 4th Int. Symp. Reactivity of Solids*, p. 000

Fiegl, F., Miranda, L. I., and Suter, H. A. (1944), *J. Chem. Educ.*, **21**, 18.

Flor, G. V., Massarotti, V., and Riccardi, R. (1974), *Z. Naturforsch.* **29a**, 503.

Flor, G., Massarotti, V., and Riccardi, R. (1975), *Z. Naturforsch.* **30a**, 304.

Frischat, G. H. (1974), *Angew. Chem. Int. Ed. Engl.*, **13**, 384.

Garner, W. E., and Hailes, H. R. (1933), *Proc, Roy. Soc. (London)*, **AI39**, 576.

Ginstling, A. M. (1951), *J. Appl. Chem. USSR*, **24**, 629.

Ginstling, A. M., and Brounshtein, B. I. (1950), *J. Appl. Chem. USSR*, **23**, 1327.

Ginstling, A. M., and Fradkina, T. R. (1952), *J. Appl. Chem. USSR*, **25**, 1199.

Gluzman, M. Kh., and Milner, R. S. (1960), *Chem. Abstr.*, **54**, 4349e.

Göler, F. V., and Sachs, G. (1932), *Z. Physik*, **77**, 281.

Gray, P., and Harper, M. J. (1960), *Proc 4th Int. Symp. Reactivity of Solids*, p. 238.

Greskovich, C., and Stubican, V. S. (1969), *J. Phys. Chem. Solids*, **30**, 909.

Greskovich, C., and Stubican, V. S. (1970), *J. Am. Ceram. Soc.*, **53**, 251.

Hardel, K. (1972), *Angew. Chem. Int. Ed. Engl.*, **11**, 173.

Jach, J. (1963), *J. Phys. Chem. Solids*, **24**, 63.

Jander, W. (1927), *Z. Anorg. Allgem. Chem.*, **163**, 1.

Kapur, P. C. (1973), *J. Am. Ceram. Soc.*, **56**, 79.

Klinger, N., Strauss, E. L., and Komarek, K. L. (1966), *J. Am. Ceram. Soc.*, **49**, 369.

Kohatsu, I., and Brindley, G. W. (1968), *Z. Phys. Chem.*, **60**, 79.

Komatsu, W. (1965), *Proc. 5th Int. Symp. Reactivity of Solids*, p. 182.

Komatsu, W., and Uemura, T. (1970), *Z. Phys. Chem. (N.F.)*, **72**, 59.

Konoyuk, I. F., and Vashuk, V. V. (1974), *Russian J. Phys. Chem.*, **48**, 1072.

Kooy, C. (1965), *Proc. 5th Int. Symp. Reactivity of Solids*, p. 21.

Kroger, C., and Ziegler, G. (1954), *Glastech. Ber.*, **27**, 199.

Kuczynski, G. C., and Stablein, P. F., Jr. (1961), *Proc. 4th Int. Symp. Reactivity of Solids*, p. 91.

Kuczynski, G. C. (1965), *Proc. 5th Int. Symp. Reactivity of Solids*, p. 352.

Kutty, T. R. N., and Murthy, A. R. V. (1974), *Ind. J. Technol.*, **12**, 447.

Mantri, V. B., Gokarn, A. N., and Doraiswamy, L. K. (1976), *Chem. Eng. Sci.*, **31**, 779.

Martin, G. (1932), Chemical Engineering and Thermodynamics Applied to the Cement Rotary Kiln, The Technical Press, England, p. 62.

Minford, W. J., and Stubican, V. S. (1974), *J. Am. Ceram. Soc.*, **57**, 363.

Nakamori, I., Nakamura, H., Hayano, T., and Kagawa, S. (1974), *Bull. Chem. soc. Japan*, **47**, 1827.

Neuburg, H. J. (1970), *Ind. Eng. Chem. Process Design Dev.*, **9**, 285.

Parravano, G. (1962), *Chem. Eng. News*, **40**, 111.

Patai, S., Albeck, M., and Cross, H. (1961), *Proc. 4th Int. Symp. Reactivity of Solids*, p. 138.

Patai, S., Albeck, M., and Cross, H. (1962), *J. Appl. Chem.* **12**, 217.

Pettit, F. S., Randklev, E. H., and Felton, E. J. (1966), *J. Am. Ceram. Soc.*, **49**, 199.

Pollitt, H. W. (1964), *The Chemistry of Cements*, Vol. 1, ed. Taylor, H. F. W., Academic Press, New York., p. 27.

Pozin, M. E., Ginstling, A. M., and Pechkovsky, V. V. (1954), *J. Appl. Chem. USSR*, **27**, 261.

Qureshi, S. Z., Rothore, H. S., and Mohammad, A. (1975), *J. Phys. Chem.*, **79**, 116.

Ramachandran, S., Baradarajan, A., and Satyanarayana, M. (1974), *Can. J. Chem. Eng.*, **52**, 364.

Rao, Y. K. (1971), *Met. Trans.*, **2**, 1439.

Rao, Y. K., and Chuang, Y. K. (1974), *Chem. Eng. Sci.*, **29**, 1933.

Rastogi, R. P. (1970), *J. Sci. Ind. Res.*, **29**, 177.

Rastogi, R. P., and Dube, B. L. (1967), *J. Am. Ceram. Soc.*, **89**, 200.

Rastogi, R. P., and Singh, N. B. (1966), *J. Phys. Chem.*, **70**, 3315.

Rastogi, R. P., Bassi, P. S., and Chadha, S. L. (1962), *J. Phys. Chem.*, **66**, 2707.

Rastogi, R. P., Bassi, P. S., and Chadha, S. L. (1963), *J. Phys. Chem.*, **67**, 2569.

Reijnen, P. (1965), *Proc. 5th Int. Symp. Reactivity of Solids*, p. 562.

Ristic, M. M. (1979), Sintering—New Developments (Proc. 4th Int. Round Table Conf. Sintering, Dubrovnik, 1977), Elsevier, Amsterdam.

Ruckenstein, E., and Vavanellos, T. (1975), *AIChE J.*, **21**, 756.

Sasaki, H. (1964), *J. Am, Ceram. Soc.*, **47**, 512.

Schäfer, H. (1964), *Chemical Transport Reactions*, Academic Press, New York, p. 122.

Schäfer, H. (1971), *Agnew. Chem. Int. Ed. Engl.*, **10**, 43.

Schmalzried, H. (1962), *Z. Physik. Chem. (N.F.)*, **33**, 111.

Schmalzried, H. (1965), *Progress in Solid State Chemistry*, Vol. 2, ed. Reiss, H., Pergamon Press, Oxford, p. 265.

Schmalzried, H. (1974), *Battelle Institute Materials Science Colloquia*, Plenum Press, New York, p. 83.

Schmalzried, H., and Rogalla, W. (1963), *Naturwiss.*, **50**, 593.

Schwab, G. M., and Rau, M. (1958), *Z. Physik. Chem.*, **17**, 257.

Schwab, G. M., Rau, M. K., and Ehrenstorfer, S. (1961), *Proc. 4th Int. Symp. Reactivity of Solids*, p. 392.

Serin, B., and Ellickson, R. T. (1941), *J. Chem. Phys.*, **9**, 742.

Sharp, J. H., Brindley, G. W., and Achar, B. N. N. (1966), *J. Am. Ceram. Soc.*, **49**, 379.

Shewmon, P. G. (1963), *Diffusion in Solids*, McGraw-Hill, New York, p. 29.

Sockel, H. G. (1972), *J. Crystal Growth*, **12**, 106.

Sohn, H. Y., and Szekely, J. (1973), *Chem. Eng. Sci.*, **28**, 1789.

Szabo, Z. G., Batta, I., and Solymosi, F. (1961), *Proc. 4th Int. Symp. Reactivity of Solids*, p. 409.

Tamhankar, S. S. (1976), *Can. J. Chem. Eng.*, **54**, 655.

Tamhankar, S. S. (1978), Ph.D. thesis, Poona University, India.

Tamhankar, S. S., and Doraiswamy, L. K. *Ind. Eng. Chem. Fundam.*, **17**, 84 (1978).

Tamhankar, S. S., and Doraiswamy, L. K. (1979), *AIChE J.*, **25**, 561.

Tamhankar, S. S., and Doraiswamy, L. K. (1982), *Ind. Chem. Eng.* (to be published).

Tammann, G. (1920), *Z. Anorg. Allgem. Chem.*, **111**, 78.

Taplin, J. H. (1973a), *J. Chem. Phys.*, **59**, 194.

Taplin, J. H. (1973b), *J. Am. Ceram. Soc.*, **56**, 390.

Taplin, J. H. (1974), *J. Am. Ceram. Soc.*, **57**, 140.

Tomas, J., Pereira, E., and Ronco, J. (1969), *Ind. Eng. Chem. Process Design Dev.*, **8**, 120.

Turnbull, D. (1956), *Solid State Physics*, Vol. 3, eds. Seitz, F., and Turnbull, D., Academic Press, New York.

Turnbull, D., and Cohen, M. H. (1958), *J. Chem. Phys.*, **29**, 1049.

Valensi, G. (1936), *Compt. Rend.*, **202**, 309.

Wagner, C. (1936), *Z. Physik. Chem.*, **B34**, 309.

Wagner, C. (1969), *Acta. Met.*, **17**, 99.

Waite, T. R. (1957), *Phys. Rev.*, **107**, 463.

Waite, T. R. (1958), *J. Chem. Phys.*, **28**, 103.

Waite, T. R. (1960), *J. Chem. Phys.*, **32**, 21.

Wen, W. Y., Johnson, D. R., and Dole, M. (1974), *J. Phys. Chem.*, **78**, 1798.

Whitney, W. P., II, and Stubican, V. S. (1971), *J. Am. Ceram. Soc.*, **54**, 349.

Winkler, G. (1965), *Proc. 5th Int. Symp. Reactivity of Solids*, p. 572.

Yamaguchi, G., and Tokuda, T. (1967), *Bull. Chem. soc. Japan*, **40**, 843.

Young, D. A. (1966), *Decomposition of Solids*, Pergamon Press, Oxford.

Zuravlev, V. F., Lesokchin, I. G., and Templeman, R. G. (1948), *J. Appl. Chem. USSR*, **21**, 887.

Author Index

Subject Index